# 13th International Conference on Tribology (ROTRIB'16)



IOP Conference Series: Materials Science and Engineering
Volume 174

Galati, Romania
22 – 24 September 2016

ISBN: 978-1-5108-3714-0
ISSN: 1757-8981

**Printed from e-media with permission by:**

Curran Associates, Inc.
57 Morehouse Lane
Red Hook, NY 12571

Some format issues inherent in the e-media version may also appear in this print version.



Printed by Curran Associates, Inc. (2017)

For permission requests, please contact the Institute of Physics
at the address below.

Institute of Physics
Dirac House, Temple Back
Bristol BS1 6BE UK

Phone: 44 1 17 929 7481
Fax: 44 1 17 920 0979

techtracking@iop.org

**Additional copies of this publication are available from:**

Curran Associates, Inc.
57 Morehouse Lane
Red Hook, NY 12571 USA
Phone: 845-758-0400
Fax: 845-758-2633
Email: curran@proceedings.com
Web: www.proceedings.com

# Table of contents

## Volume 174

**13th International Conference on Tribology, ROTRIB'16**
**22–24 September 2016, Galati, Romania**

**Accepted papers received: 17 January 2017**
**Published online: 1 March 2017**

### Preface

011001
OPEN ACCESS
13th International Conference on Tribology

011002
OPEN ACCESS
Peer review statement

### Papers

012001
OPEN ACCESS
Cylindrical roller bearings with profiled contacting surfaces

S S Crețu.....1

012002
OPEN ACCESS
Triboelectrical charge generated by frictional sliding contact between polymeric materials

T Zeghloul, M B Neagoe, Y E Prawatya and L Dascalescu.....9

012003
OPEN ACCESS
Influence of surface roughness on the tribo-electric process for a sliding contact between polymeric plate materials

M B Neagoe, Y E Prawatya, T Zeghloul and L Dascalescu.....15

012004
OPEN ACCESS
Experimental analysis of axial and radial stress distribution in soft materials used for petrochemical valve stem sealing package

R G Ripeanu, A Ispas and D Ispas.....21

012005
OPEN ACCESS
Influence of tooth profile on the noncircular gear tooth contact

A Cristescu, L Andrei and B Cristescu.....28

012006
OPEN ACCESS
Dynamic analysis of six-bar mechanical press for deep drawing

S Mitsi, I Tsiafis and K D Bouzakis.....37

012007
OPEN ACCESS
The effects of mechanical properties and sintering duration on the wear behaviour of silicon nitride

A Brusilová, P Švec, Z Gábrišová and M Pokusová.....45

012008
OPEN ACCESS
The effect of transport velocity upon spin torque

S Alaci, F C Ciornei, D Amarandei, L Irimescu, I C Romanu and M A Rotar.....53

012009
OPEN ACCESS
Tribological behaviour of unidirectional carbon fibre-reinforced epoxy composites

Y Şahin and P De Baets.....60

012010
OPEN ACCESS
Abrasive wear of high velocity oxygen fuel (HVOF) superalloy coatings under vibration load

M Kandeva, B Ivanova, D Karastoyanov, T Grozdanova and E Assenova.....68

012011
OPEN ACCESS
Experimental research on energy losses through friction in order to increase lifetime of hydraulic cylinders

C Cristescu, R Radoi, C Dumitrescu and L Dumitrescu.....79

012012
OPEN ACCESS
Structural-energetic regularities of tribocontact evolution*

S Fedorov.....87

012013
OPEN ACCESS
The dynamic behavior of sliding tribosystems in unstable operating conditions

A Postaru, Ghe Postaru, V Ceban, C Spânu and P Stoicev.....96

012014
OPEN ACCESS
Experimental studies regarding wear processes through dry friction of the superficial layer for an unconventional treated steel

C P Papadatu.....106

012015
OPEN ACCESS
A comparative study on tribological behavior between metal and polymeric composites used to repair bronze made parts in dry reciprocating sliding tests

V Iliuță, M Rîpă, J Javorova and G Andrei.....116

012016
OPEN ACCESS
Study on friction behaviour of brake shoe materials for mining hoist

M Ungureanu, N S Ungureanu and I Crăciun.....126

012017
OPEN ACCESS
Micro-scratching tests of a rolled aluminium alloy AA2024-T351 thick plate using a diamond micro-blade

E Pirva, A Tudor, A Gavrus, G Chisiu, N Stoica and A Predescu.....134

012018
OPEN ACCESS
The adhesion and hysteresis effect in friction skin with artificial materials

K A Subhi, A Tudor, E K Hussein and H S Wahad.....142

012019
OPEN ACCESS
Tribological properties of the disc brake friction couple materials in the range of small and very small speeds

N A Stoica, A M Petrescu, A Tudor and A Predescu.....148

012020
OPEN ACCESS
Analysis of progressive cavity pumps specific wear processes using Finnie models

M Popescu, A I Popovici, F N Petrescu and N N Antonescu.....157

012021
OPEN ACCESS
The effect of friction in coulombian damper

H S Wahad, A Tudor, M Vlase, N Cerbu and K A Subhi.....164

012022
OPEN ACCESS
Analysis of progressive cavity pumps specific wear processes using Bitter and Hutchings models

A I Popovici, M Popescu, N N Antonescu and F N Petrescu.....173

012023
OPEN ACCESS
A study regarding friction behaviour of lysine and isoleucine modified epoxy matrix

I Bălan, R Bosoancă, A Căpățînă, I Graur, V Bria and C Ungureanu.....180

012024
OPEN ACCESS
Comparative study regarding friction coefficient for three epoxy resins

G Mihu, I Mihalache, I Graur, C Ungureanu and V Bria.....187

012025
OPEN ACCESS
A study of friction wear behaviour of nano-ferite modified epoxy resins

C Munteniță, C Eni, I Graur, C Ungureanu and M Bodor.....191

012026
OPEN ACCESS
The determination of the thickness of the layers deposited on the electronic circuit boards through tribological methods

A M Petrescu, A Tudor, G Chişiu, N A Stoica and U Cihak Bayr.....196

012027
OPEN ACCESS
Aspects of the tribological behaviour of powders recycled from rapid steel treated sub-zero*

S Radu and M Ciobanu.....202

012028
OPEN ACCESS
Investigation on the contact behaviour of MEMS micromembrane with serpentine hinges

M Pustan, C Birleanu, C Dudescu and F Rusu.....208

012029
OPEN ACCESS
Analysis of the bending stiffness and adhesion effect in RF-MEMS structures

C Birleanu, M Pustan, C Dudescu, V Merie and I Pintea.....216

012030
OPEN ACCESS
Two-way coupled Reynolds, Rayleigh-Plesset-Scriven and energy equations for fully transient cavitation and heat transfer modeling

M Braun, K Pierson and T Snyder.....228

012031
OPEN ACCESS
Saturated porous layers squeezed between parallel disks in enclosed cells

I C Melciu, T Cicone and M D Pascovici.....236

012032
OPEN ACCESS
Performance evolution of fully and partially textured hydrodynamic journal bearings lubricated with two lubricants

N Tala-ighil and M Fillon.....244

012033
OPEN ACCESS
The effect of lubricant supply and frequency upon the behaviour of EHD films subjected to vibrations

R Glovnea, X Zhang and J Sugimura.....250

012034
OPEN ACCESS
Tribological study of an aerodynamic thrust bearing in the supersonic regime

F. Dupuy, B. Bou-Saïd and M. Garcia.....258

012035
OPEN ACCESS
Down to the roughness scale assessment of piston-ring/liner contacts

H M Checo, A Jaramillo, R F Ausas, M Jai and G C Buscaglia.....264

012036
OPEN ACCESS
The behaviour of lubricated EHD contacts subjected to vibrations

X Zhang and R P Glovnea.....274

012037
OPEN ACCESS
A finite element solution of the Reynolds equation of lubrication with film discontinuities: application to helical groove seals

M Jarray, D Souchet, Y Henry and A Fatu.....282

012038
OPEN ACCESS
Analysis of squeeze film process between non-parallel circular surfaces

A V Radulescu and I Radulescu.....290

012039
OPEN ACCESS
Effects of fluid inertia and bearing flexibility on the performance of finite length journal bearing

Juliana Javorova and Vassil Alexandrov.....300

012040
OPEN ACCESS
Some considerations concerning four-ball machine testing of the polyacrylamide solutions

A C Drumeanu.....308

012041
OPEN ACCESS
Method and device for measurement of dynamic viscosity

F C Ciornei, S Alaci, D Amarandei, L Irimescu, I C Romanu and L I Acsinte.....315

012042
OPEN ACCESS
Bio and nanomaterials in tribocorrosion systems

Lidia Benea.....324

012043
OPEN ACCESS
Design and technology parameters influence on durability for heat exchangers tube to tubesheet joints

R G Ripeanu.....336

012044
OPEN ACCESS
Porous $TiO_2$-$ZrO_2$ thin film formed by electrochemical technique to improve the biocompatibility of titanium alloy in physiological environment

L Benea, E Dănăilă and P Ponthiaux.....345

012045
OPEN ACCESS
Improving tribocorroson behaviour by electro-codeposition of TiC nano-dispersed particles with nickel as hybrid layers for energy applications

Lidia Benea, Eliza Dănăilă, Pierre Ponthiaux and Jean-Pierre Celis.....351

012046
OPEN ACCESS
Numerical investigation of the effect of friction conditions to increase die life

M O Mutlu, C G Guleryuz and Z Parlar.....359

012047
OPEN ACCESS
Tribo-thermal fatigue of the steel used for the forging die construction

A C Drumeanu.....367

012048
OPEN ACCESS
Fault tree analysis of most common rolling bearing tribological failures

Aleksandar Vencl, Vlada Gašić and Blaža Stojanović.....375

012049
OPEN ACCESS
Geometric modelling of the contact point between the bushing and sprocket in chain drives

R Saulescu, R Velicu and M Lates.....385

012050
OPEN ACCESS
Penetration tests to study the mechanical tribological properties of chisel type knife

L Vlăduţoiu, G Chişiu, T Andrei, A Predescu, C Muraru and V Vlăduţ.....393

012051
OPEN ACCESS
Experimental study on guide friction contribution in global power loss of a tooth chain transmission

R Velicu, R Papuc, C C Gavrila and S Popa.....401

012052
OPEN ACCESS
Studies on Tribological Behavior of Aluminum Nitride-Coated Steel*

G C Ionescu, I Nae, R G Ripeanu, A Dinita and G Stan.....408

012053
OPEN ACCESS
Theoretical and experimental analysis of the lubricating system of a high speed multiplier

V G Marian, R F Mirică and T Prisecaru.....416

012054
OPEN ACCESS
Correlations of the instantaneous carrying force and pressure distribution in the case of the narrow sliding radial bearing under hard shocks*

I M Alexandrescu, R Cotețiu and D Darabă.....424

012055
OPEN ACCESS
Wear behavior of the plasma and thermal oxidized Ti-15Mo and Ti-6Al-4V alloys

I Hacisalioglu, F Yildiz, A Alsaran and G Purcek.....432

012056
OPEN ACCESS
Determination of residual stresses and natural frequencies of roll-tensioned disc by a dynamic simulation of the rolling process

G Skordaris, K-D Bouzakis and D Tasoulas.....441

012057
OPEN ACCESS
Evaluating lubricating capacity of vegetal oils using Abbott-Firestone curve

C Georgescu, G C Cristea, C Dima and L Deleanu.....449

012058
OPEN ACCESS
Improvement the wear behavior of low carbon steels by friction stir processing

D M Sekban, S M Aktarer, H Yanar, A Alsaran and G Purcek.....457

012059
OPEN ACCESS
Coating multilayer material with improved tribological properties obtained by magnetron sputtering

A O Mateescu, G Mateescu, M Balasoiu, G O Pompilian and M Lungu.....464

012060
OPEN ACCESS
Wear resistance of WC/Co HVOF-coatings and galvanic Cr coatings modified by diamond nanoparticles

M Kandeva, T Grozdanova, D Karastoyanov and E Assenova.....470

012061
OPEN ACCESS
Effect of friction stir processing on tribological properties of Al-Si alloys

S M Aktarer, D M Sekban, H Yanar and G Purçek.....478

012062
OPEN ACCESS
HVOF sprayed WC-CoCr coatings on aluminum: tensile and tribological properties

A Koutsomichalis, M Vardavoulias and N Vaxevanidis.....486

012063
OPEN ACCESS
Study on surface finish of AISI 2080 steel based on the Taguchi method

S Yalcinkaya and Y Şahin.....496

012064
OPEN ACCESS
Increasing in the wear resistance of injection molds made of 1.2343 steel using Ti/TiN/TiCN/nc-TiCN:a-C/nc-TiC:a-C/a-C nanocomposite coating

Velko Rupetsov, Georgi Mishev, Stefan Dishliev, Viktor Kopanov, Vassiliy Chitanov, Lilyana Kolaklieva and Chavdar Pashinski.....502

012065
OPEN ACCESS
A method for determining the thickness of tribological performing thin layers formed by selective transfer

Filip Ilie, Georgiana Chisiu and George Ipate.....510

012066
OPEN ACCESS
Some considerations on noise monitoring for air handling equipments

C Bujoreanu and M Benchea.....517

012067
OPEN ACCESS
Optimization of continuous triboelectrification process for polymeric materials in dry contact

Y E Prawatya, M B Neagoe, T Zeghloul and L Dascalescu.....524

012068
OPEN ACCESS
Functioning condition monitoring of industrial equipment

N S Ungureanu, A Petrovan, M Ungureanu and M Alexandrescu.....531

012069
OPEN ACCESS
Conformity assessment of the measurement accuracy in testing laboratories using a software application*

A Diniță.....537

IOP Conf. Series: Materials Science and Engineering **174** (2017) 011001 doi:10.1088/1757-899X/174/1/011001

# ROTRIB'16

The 13th International Conference on Tribology
September 22-24, 2016 – Galați, ROMANIA

## PROCEEDINGS EDITED BY

**Minodora RIPA**, Professor, PhD, "Dunarea de Jos" University of Galati, ROMANIA

**Razvan George RIPEANU**, Professor, PhD, Petroleum-Gas University of Ploiesti, ROMANIA

**Traian CICONE**, Professor, PhD, University POLITEHNICA of Bucharest, ROMANIA

Selected, peer reviewed papers from the 13th International Conference on Tribology ROTRIB'16

## FOREWORD

The Conference ROTRIB'16 is the 13th in a series of international scientific events on TRIBOLOGY organized in Romania. The general objective of this conference is to contribute and demonstrate fundamentals, applications and experiences in the field of Tribology. The conference provided an interdisciplinary forum for discussion, networking and initiating collaborations among specialists, about materials and Tribology and their application to product design and reliability. The exchange of ideas, theories, developments, innovations, improved methods and future trends was a key goal of this conference, by bringing together specialists from industry and academia, leading and young researchers, from various engineering and science fields.

Papers presented at ROTRIB 2016 are included in the following topics:

| | |
|---|---|
| *Friction and Wear* | *Tribochemistry and tribocorrosion* |
| *Contact Mechanics* | *Tribology of machine elements* |
| *Micro and nanotribology* | *Tribology in manufacturing processes* |
| *Lubrication* | *Surface metrology, surface engineering, coatings* |
| *Lubricants* | *Condition monitoring* |

These topics were discussed in sessions and two satellite symposia, VAREHD 2016 and DIPRE 2016. The official language of ROTRIB'16 for presenting and writing papers was English.

We have received an excellent number of submissions out of which, after a thorough peer reviewing process, a number of 75 scientific contributions has been accepted for presentation and publication. The authors of the accepted papers have been invited to present their work in the Conference and we are pleased to welcome 110 participants from 9 countries, presenting 69 papers.

The Conference program is enriched significantly by plenary presentations of renowned experts in their respective domains: Distinguished Professor. dr. Minel Braun (USA), Professor dr. Michel Fillon (France) and Emeritus Professor dr. Mircea D. Pascovici (Romania).

Conference ROTRIB'16 was organized by "Dunarea de Jos" University of Galati and the Romanian Tribology Association (ART), under the aegis of the Romanian Ministry of Education and Scientific Research. The Conference was held in Galati, ROMANIA at the Natural Sciences Museum Complex, September 22 – 24, 2016.

After 38 years since the first Romanian Conference on Tribology, "Tribotehnica", Galati, 1978, we organized ROTRIB'16 again in Galati, celebrating the $50^{th}$ anniversary of Tribology. Unfortunately, in this jubilee year of Tribology, Professor Peter Jost (the "father" of Tribology and President of International Tribology Council) and Professor Branko Ivkovic (former President of Serbian Tribology Association and Balkan Tribology Association) passed away.

The Conference is financially supported by "Dunarea de Jos" University of Galati, the Romanian Tribology Association, the Romanian Ministry of Education and Scientific Research and our sponsors, which participated in ROTRIB'16 with company-profile presentations and/or product exhibitions.

We would like to express our gratitude to the authors for submitting their work. We would like to thank to all those who have contributed, in one way or another, to the success of the Conference. We are grateful to the Scientific Committee members as well as to the external reviewers for their volunteer hard work of reading and discussing the submissions. Last but not least, we would like to thank to the members of the Organizing Committee for their technical and non-technical support.

It has been a great pleasure for us to organize the Conference ROTRIB'16 and we do hope that this scientific event will stimulate future developments and innovations in the field of tribology.

We also hope that all participants have had a pleasant and fruitful time in Galati, on the banks of the Danube River.

Organising Committee
Galati, September, 2016

## SCIENTIFIC COMMITTEES AND CHAIRS:

## NATIONAL ORGANISING COMMITTEE:



### Thank to our reviewers

Stelian Alaci, Gabriel Andrei, Laurentia Andrei, Emilia Assenova, Özgü Bayrak, Blaza Stojanovic, Jean Bouyer, Mihaela Buciumeanu, Sorin Cananau, Jean-Pierre Celis, Shriniwas Chippa, George Chirita, Traian Cicone, Florina Ciornei, Sorin Ciortan, Adrian Circiumaru, Liviu Ciupitu, Radu Cotetiu, Spiridon Cretu, Ana Cristescu, Lorena Deleanu, Adrian Catalin Drumeanu, Aurelian Fatu, Michel Fillon, Jean Geringer, Romeo Glovnea, Luminita Irimescu, Juliana Javorova, Baki Karamis, Magda Lakatos-Varsanyi, Victor Marian, Mathew T. Mathew, Ilie Musca, Nikolai Myshkin, Bogdan Nedic, Dumitru Olaru, Liviu Palaghian, Viorel Paleu, Chris Papadopoulos, Mircea Pascovici, Pierre Ponthiaux, Adrian Predescu, Pandora Psyllaki, Gencaga Purcek, Marius Pustan, Alexandru Radulescu, Irina Radulescu, Minodora Ripa, Razvan George Ripeanu, Yusuf Sahin, Adolfo Senatore, Samion Syahrullail, Nacer Tala-Ighil, Andrei Tudor, Radu Velicu, Mehmet Yazar, Thami Zeghloul, Peter Zentay.

**Thank to the Sponsors of Conference ROTRIB'16**

SC Etansari Grafex SRL
www.grafex.ro

S.C. TOTAL LUBRICANTS ROMANIA S.A.
http://www.total.com.ro/

SC NASDIS Consulting SRL
http://www.nasdis.ro/

Nanomechanics, Inc.
www.nanomechanicsinc.com

SC TECHNO VOLT SRL
www.technovolt.ro

SC RO-MEGA CONTROL SRL
www.romegacontrol.ro

SC MECRO SYSTEM SRL
www.mecrosystem.ro

Gamax Laboratory Solutions Ltd.
http://gamaxlabsol.com/

SC Sartorom Impex SRL
www.sartorom.ro

TTZH Tribologie & Hochtechnologie GmbH
www.ttzh.de www.ttzh.de

Nanovea Inc.
www.nanovea.com

SC HISTERESIS SRL
www.histeresis.ro

SC NavGAL PROFESSIONAL TEAM SRL
www.navgal.ro

Ducom Instruments Europe B.V.
www.ducom.com

SC BRD Groupe Societe Generale SA
www.brd.ro

**Participants at the 13th International Conference on Tribology ROTRIB'16**

IOP Conf. Series: Materials Science and Engineering **174** (2017) 011002 doi:10.1088/1757-899X/174/1/011002

# Peer review statement

All papers published in this volume of *Journal of Physics: Conference Series* have been peer reviewed through processes administered by the proceedings Editors. Reviews were conducted by expert referees to the professional and scientific standards expected of a proceedings journal published by IOP Publishing.

IOP Conf. Series: Materials Science and Engineering **174** (2017) 012001 doi:10.1088/1757-899X/174/1/012001

# Cylindrical roller bearings with profiled contacting surfaces

**S S Crețu**

Mechanical Engineering and Mechatronics Department, "Gheorghe Asachi" Technical University of Iași 61-63 D. Mangeron Bv., 700050, Iași, Romania

E-mail: spcretu@mail.tuiasi.ro

**Abstract**. An initial loading of cylindrical roller bearings in the elastic-plastic domain was performed to induce elastic shakedown phenomena able to change the basic profiles of both, rollers and raceways, which endorses a different value for the basic reference rating life. Fatigue life tests carried out on four lots of NJ206 cylindrical roller bearings revealed much higher values of $L_{10}$ and $L_m$ criteria for the bearings lots which experienced a suitable initial loading operation in the elastic-plastic domain. The reference rating lives, evaluated by using the lamina technique, confirmed the superiority of bearings lots which undergone an appropriate primary loading in the elastic-plastic domain.

## 1. Introduction

Reviewing papers by Hoeprich [1], de Mull *et al.* [2], and Reusner [3] pointed out that the contact pressures have a tendency to be significantly higher around the profile's discontinuities than those attained in the middle of the contact area. The edge loading effect is considerably amplified for one end and diminished at the other when the bearing operates with misalignment between the inner and outer raceways. Consequently, the primary target was to reduce the stresses in the roller-raceway contact by modifying the roller profile. The diversity of crowning profiles includes: the single straight line with chamfer ends, single circular arc or a combination of multiple circular arcs, cylindrical-crowned (ZB), and the logarithmic profile.

Kapoor and Johnson [4], and Williams [5] found that after few cycles of primary rolling contact loading in the elastic-plastic domain, the material elastically shakedowns to a slightly modified roller axial profile and a stable state of compressive residual stresses.

These plastic deformations, achieved in primary loading cycles or in transient overloads, alter the initial contact geometry and diminish the edge effect.

## 2. Cylindrical roller bearings primary loaded in elastic-plastic domain

### *2.1. Rolling bearing and equipment for the primary loading operation*

The NJ206 cylindrical roller bearing was selected for this study. Bearing's internal data were as follows: inner raceway diameter F=38.5 mm, roller diameter: $D_w$=7.5 mm, total roller length: $L_w$=7.5 mm, roller's end chamfer: $R_{ch}$=(0.3 - 0.5) mm, number of cylindrical rollers: $Z_w$=12, C2 internal clearance: S=(0 - 25) μm, P0 precision class. The bearings rings and cylindrical rollers were manufactured from 100Cr6 through hardened rolling bearing steel with the traction ultimate strength $R_m$ = 2200 MPa, and the traction yield limit $R_{p\,0.2}$ = 1700 MPa.

The target was to establish new cylindrical roller profiles after bearing's running a number of 200 inner ring rotations at high radial loads. The equipment used for the primary loading of NJ206 bearing assured small rotation speeds of both bearing's rings during the gradual increasing of radial load to the required level. During the primary loading operation the NJ206 bearing was symmetrical positioned to avoid shaft misalignment and unequal deformations.

### *2.2. Experimental procedure and results*

The radial loads were chosen to create, in the contact achieved between the most loaded roller and inner ring raceway, the following values for the Hertzian pressure $p_{H0} \in \{3.5$ GPa, 4 GPa, 4.5 GPa, 5 GPa$\}$. For each load level, the primary loading procedure was accomplished for two bearings. The results of measurements, performed along cylindrical roller generatrix, are encapsulated in figure 1.

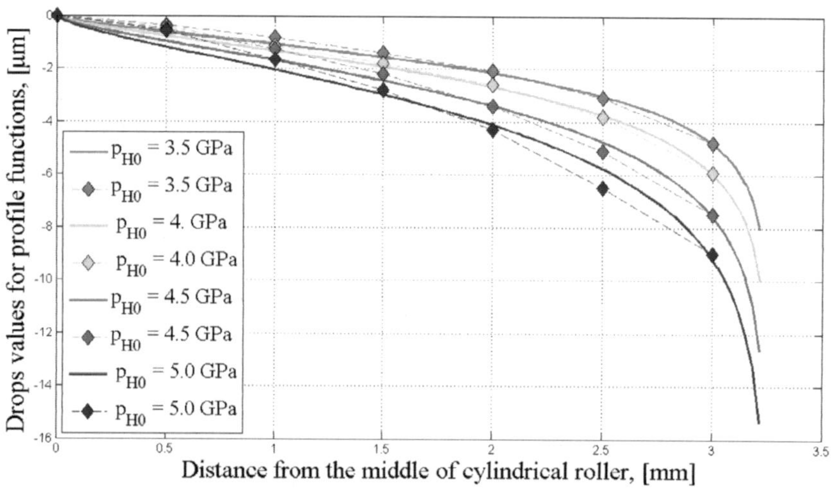


**Figure 1.** Plastic deformations caused by primary loadings and profile functions.

### *2.3. Profile functions for cylindrical rollers subjected to elastic-plastic primary loading*

Particular profile functions $P_k(x)$ were obtained for each level of the primary loading force (figure 1). The interpolations of the experimental data were accomplished using the model of logarithmic function revealed in standard ISO 16281 [6]:

$$P_k(x) = A_k \cdot D_w \cdot ln\left[\frac{1}{1-(2\cdot x/L_{we})^{0.6}}\right] \quad (1)$$

## 3. Fatigue life tests

### *3.1. Testing conditions*

*3.1.1. Testing machines.* The fatigue life tests were performed on a battery of 12 machines. In each machine (figure 2) two specimen bearings were positioned at the ends of the machine's shaft, so that the specimen bearings supported the same operating values regarding the load, misalignment angle, rotation speed, temperature of the lubrication oil, temperatures of inner rings, outer rings and rollers, respectively.

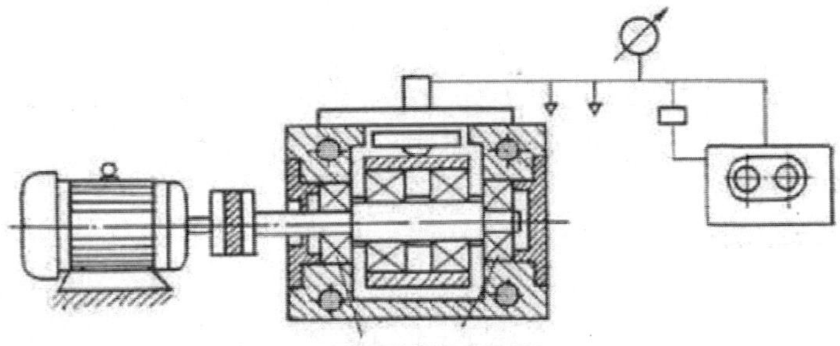


**Figure 2**. Scheme of machine for fatigue life testing of cylindrical roller bearings.

*3.1.2. Cylindrical roller bearings*. Four lots of 24 bearings each lot were durability tested. Two lots of bearings had cylindrical rollers with straight line profile (SLP), terminated with chamfers, and the other two lots, as result of a primary loading operation, had the cylindrical rollers and inner raceways with slightly modified profiles (PMP). The primary loading was performed a number of 200 inner ring rotations with a radial force able to generate, in the contact between the most loaded roller and inner raceway, a hertzian pressure $P_{H0}$ = 4.5 GPa. Before the primary loading, the roughnesses of the active surfaces were as follows: for inner and outer raceways $R_{a\_r}$ = 0.14 μm, and for cylindrical surface of rollers $R_{a\_w}$ = 0.1 μm. After the primary loading the roughness diminishes to $R_{a\_r}$ = 0.1 μm for the inner and outer raceways, and to $R_{a\_w}$ = 0.08 μm for cylindrical surface of rollers.

*3.1.3. Operating conditions*. Radial load on a bearing $F_r$ = 8500 N, misalignment due to shaft's elastic bending $\psi = 1.6\ minutes$, rotation speed of the inner ring $n_i$=2880 rev./min., oil bath lubrication (e_c = 0.6), lubrication oil- ISO VG 46 ($\nu_{40} = 46\frac{mm^2}{s}$, $\nu_{80} = 12\frac{mm^2}{s}$ ), operating temperatures: inner ring $t_i$ = 65 °C, outer ring $t_e$ = 45 °C, ambient $t_a$= 20 °C, shaft tolerance – k6, housing tolerance - H7.

*3.1.4. Testing method.* In order to shorten the testing time, the accelerated durability test method [7] was used. To maintain a high confidence level, the fatigue life test of each lot were performed until minimum 8 bearings of the lot failed by rolling contact fatigue.

*3.2. Results provided by the fatigue life testing program*

*3.2.1. Types of failures.* All bearings failed by rolling contact fatigue phenomena manifested practically complete as pitting on inner raceways (figure 3). The statistic of failures is revealed in table 1.

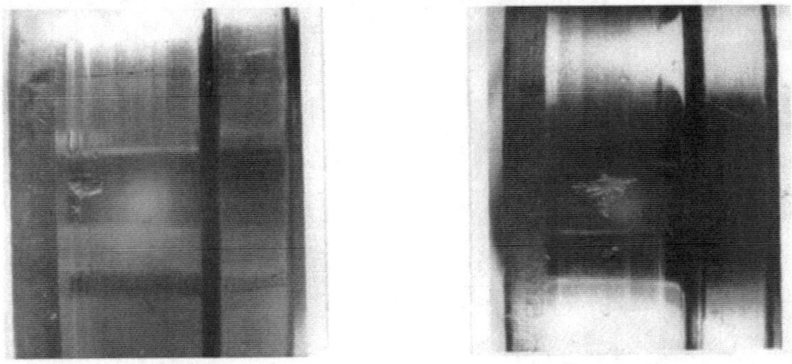

(a) (b)

**Figure 3.** Failures by rolling contact fatigue (pitting) on inner raceway
(a)- at the edge of raceway, (b)-at the middle of raceway.

**Table 1.** Statistics of failures by rolling contact fatigue.

| Bearing lots | Number of tested bearings | Number of failed bearings | Bearing's element failed | | | |
|---|---|---|---|---|---|---|
| | | | Inner raceway | | Outer raceway | Roller |
| | | | Edge | Middle | | |
| SLP 1 | 22 | 11 | 11 | 0 | 0 | 0 |
| SLP 2 | 26 | 14 | 13 | 1 | 0 | 0 |
| SLP 1+2 | 48 | 25 | 24 | 1 | 0 | 0 |
| PDP 1 | 24 | 9 | 2 | 7 | 0 | 0 |
| PDP 2 | 20 | 8 | 1 | 6 | 0 | 1 |
| PDP 1+2 | 44 | 17 | 3 | 13 | 0 | 1 |

*3.2.2. The distribution function and parameters estimation.* The Weibull distribution function has been hypothesized as describing the distribution of the fatigue lives. The estimation of parameters for each lot was accomplished by: (a)-the graphical method in which the values *P(i)* of the distribution function are plotted on Weibull probability papers (WPP), [7, 8]; (b)-the maximum likelihood method (ML), [8, 9]. The summary of fatigue lives is presented in table 2.

**Table 2.** Results of fatigue life tests.

| Bearing Group[a] | Hardness (roller/ ring) (HRC) | Primary loading pressure (MPa) | Method of estimation[b] | Shape factor e | $L_{10}$ rating lives (hours) | $L_{10}$ rating lives 90% confidence limits | $L_m$ median lives (hours) |
|---|---|---|---|---|---|---|---|
| SLP 1 | 61 / 61 | | WPP | 1.96 | 18 | 11.....30 | 50,3 |
| | | | ML | 1.76 | 18.3 | | 58.8 |
| SLP 2 | 63 / 61 | | WPP | 1.73 | 20.3 | 11.....38 | 66,5 |
| | | | ML | 2.47 | 23.5 | | 51,9 |
| PDP 1 | 61 / 61 | 4500 | WPP | 3.24 | 112 | 85.....168 | 202 |
| | | | ML | 3.84 | 118.5 | | 193 |
| PDP 2 | 63 / 61 | 4500 | WPP | 1.49 | 102 | 50.....209 | 416 |
| | | | ML | 0.92 | 59.6 | | 762 |

[a] SLP, straight line profiles; PDP, primary deformed profiles.
[b] WPP, Weibull probability paper; ML, maximum likelihood.

## 4. Rating Lives

### *4.1. Basic rating life*

The basic rating life is provided by equation:

$$L_{10} = \left(\frac{C_r}{P}\right)^p \tag{2}$$

where $P$ is the dynamic equivalent radial load and the exponent $p$ has the value $p=10/3$ for line contact. For the case of bearings with unmodified line profiles the value of the basic dynamic load rating $C_r$ was diminished by the correction factor $\lambda v=0.83$, ISO 281 [10].

IOP Conf. Series: Materials Science and Engineering **174** (2017) 012001 doi:10.1088/1757-899X/174/1/012001

*4.2. Modified rating life*

To evaluate the modified rating life of bearings, the Standard ISO 281 [10] uses the equation:

$$L_{10\,\mathrm{m}} = a_1 a_{\mathrm{ISO}} \cdot L_{10} = a_1 \cdot a_{\mathrm{ISO}} \cdot \left(\frac{C}{F_r}\right)^p \tag{3}$$

where $a_1 = 1$ when a 90% reliability is admitted.

The life modification factor $a_{\mathrm{ISO}}$ can be derived from equation:

$$a_{\mathrm{ISO}} = f\left(\frac{e_C \cdot C_u}{P}, \kappa\right) \tag{4}$$

where $C_u$ is the fatigue load limit whereas the parameters $e_C$ and κ take into account the contamination and lubrication conditions. Operating circumstances as: lubrication (the lubricant type, viscosity, bearing speed, bearing size, additives), environment (the contamination level, seals), contaminant particles (the hardness and particle size, lubrication method, filtration) and mounting, are considered by standard ISO 281[24] in selecting values for parameters $e_C$ and κ. For mineral oil lubrication and bearing raceway surfaces machined with *good manufacturing quality*, the condition of lubricant separation is described by the complex parameter $\kappa$ , defined as the ratio of the actual kinematic viscosity $\nu$ to the reference kinematic viscosity $\nu_1$ :

$$\kappa = \frac{\nu}{\nu_1} \tag{5}$$

For a more detailed estimation of the $\kappa$ value, e.g. for especially machined surfaces, $\lambda$ -ratio is first evaluated as the ratio between the minimum thickness of the EHD oil film, $h_{EHD\,\mathrm{min}}$ (formed between the most loaded roller and inner raceway) and the composed roughness of the involved surfaces:

$$\lambda = \frac{h_{EHD\,\mathrm{min}}}{1.25 \cdot \left(Ra^2_{raceway} + Ra^2_{roller}\right)^{1/2}} \tag{6}$$

Further, the complex parameter $\kappa$ value is estimated with the equation:

$$\kappa = \lambda^{1.3} \tag{7}$$

*4.3. Modified reference rating life*

Additional to the influencing parameters described by ISO 281 [10] the standard ISO 16281 [6] takes into account three more operating circumstances: the misalignment, operating internal clearance and internal load distribution on rolling elements.

For bearings which sustained a primary loading, the elastic-shakedown provided permanent deformations of inner raceways roughly one third of values experienced by cylindrical rollers, while the permanent deformations of the outer raceways attained irrelevant values and were neglected.

The effect of residual stresses is not considered in the mentioned ISO standards.

*4.3.1. The lamina technique.* The lamina technique, as presented in standard ISO-16281 [6] was used to evaluate the modified reference rating life. In this procedure the elastic pressures distributions developed in contact areas of each loaded roller with both, the inner raceway $(i)$ and outer raceway $(e)$, have to be considered. For computations presented in the paper, a number of 256 uniformly spaced rectangular laminae were built.

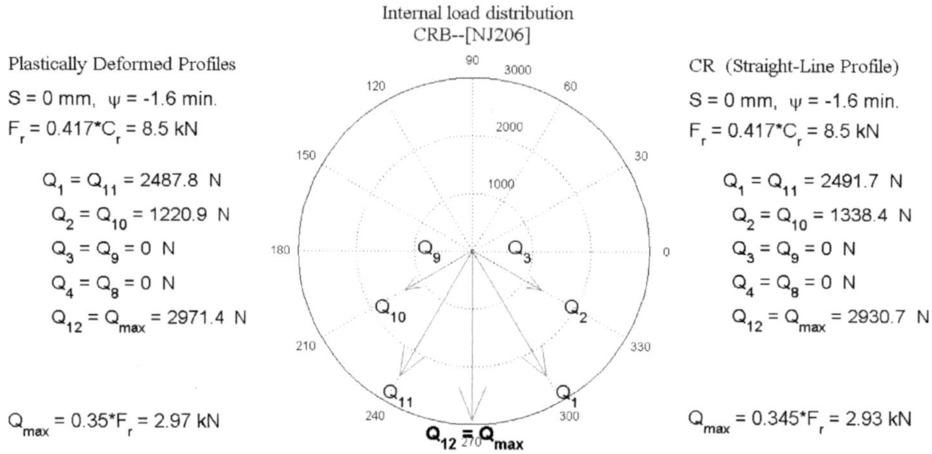


**Figure 4**. Internal load distribution of the external load.

*4.3.2. Internal load distribution.* The method used to determine the internal load distribution took into account operating radial clearance and rollers' tilt, figure 4. The operating clearance considers the different operating temperatures of inner and outer raceways as well as the interference fits.

*4.3.3. Pressures distributions.* A semi-analytical method (SAM) was used to solve non-hertzian contacts achieved between the cylindrical roller and corresponding raceways, [11, 12]. Using an elastic-perfect plastic constitutive law, the SAM provides 3D pressures distributions achieved between each loaded roller with inner and outer raceways. For each lamina, the maximum value of the pressure was needed only. The 2D pressures distributions, developed by the internal loads $Q_2$, $Q_1$ and $Q_{12}$ on contact areas situated on inner raceway are exemplified in figure 5, figure 6 and figure 7, respectively, for bearings with straight line profiles and bearings with profiles modified by the primary loading procedure.

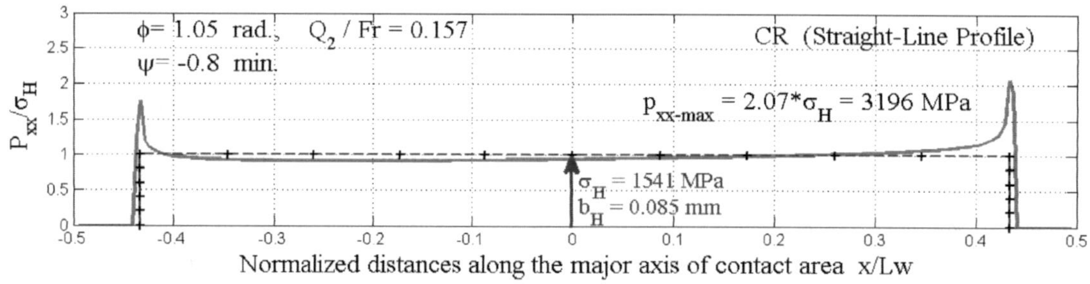


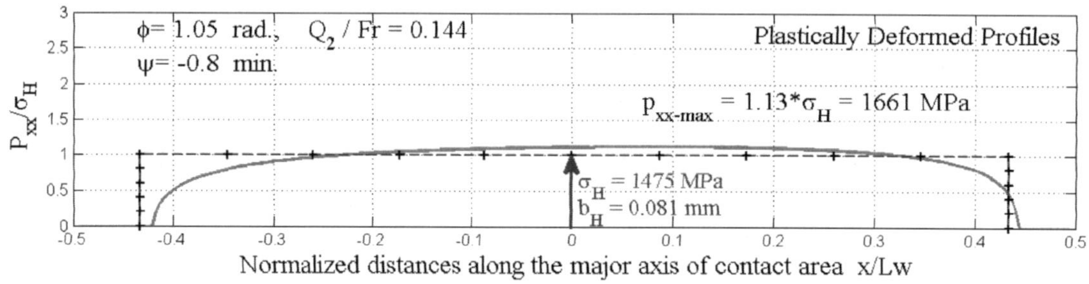


**Figure 5.** Pressures distributions along rollers 2 and 10.

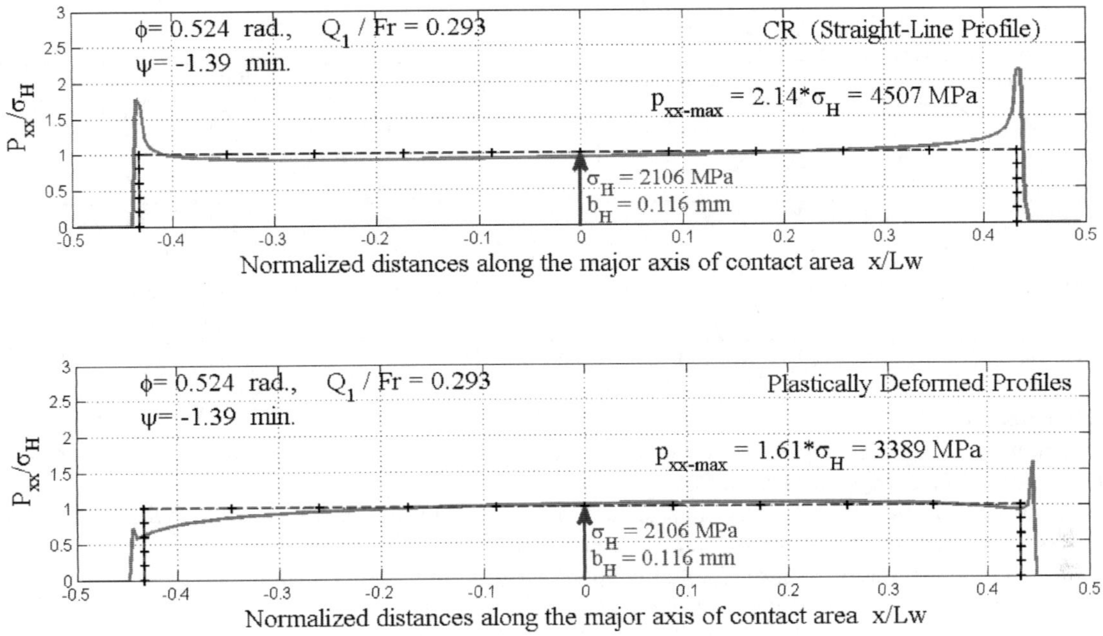


**Figure 6.** Pressures distributions along rollers 1 and 11.

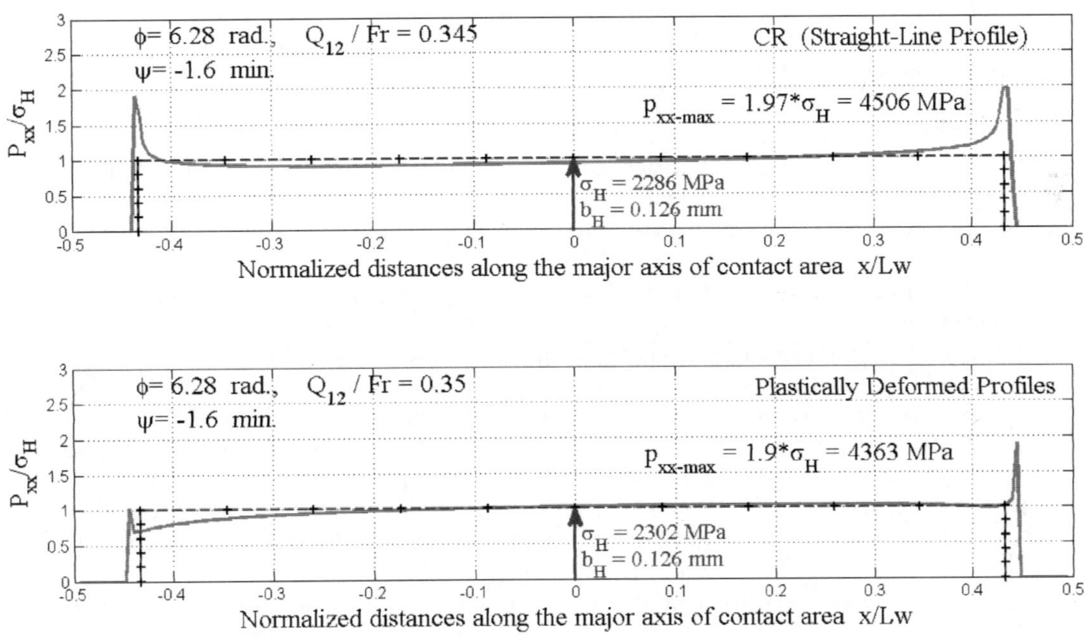


**Figure 7.** Pressures distributions along the most loaded roller.

*4.3.4 Rating lives comparison.* The same operating conditions as used in fatigue life tests were considered as the input data in the computations of rating lives. The results are summarized in table 3.

For bearings with straight line profiles, as well as for bearings with modified profiles by primary loading procedure, the corresponding value of the modified reference rating life is well situated inside 90% confidence band of the $L_{10}$ life provided by the fatigue life tests.

Table 3. Results of rating lives computations.

| Bearing Group[a] | Rating lives (hours) | | | Life tests (hours) | |
|---|---|---|---|---|---|
| | Basic rating life $L_{10}$ | Modified rating life $L_{10\mathrm{m}}$ | Modified reference rating life $L_{10\mathrm{m\,r}}$ | 90% life $L_{10}$ | 90% confidence limits |
| SLP | 57.2 | 40.1 | 19 | 18 | 11…..30 |
| | | | | 20.3 | 11…..38 |
| PDP | 107.1 | 90.3 | 135 | 112 | 85…..168 |
| | | | | 102 | 50…..209 |

[a] SLP, straight line profiles; PDP, primary deformed profiles.

## 6. Conclusions

A few number of loading cycles in elastic-plastic domain causes an elastic shakedown phenomenon able to create favorable changes for profiles of rolling elements.

Fatigue life tests performed on four lots of NJ206 cylindrical roller bearings revealed the superiority of bearings lots that primary supported a few cycles of rolling loading in the elastic-plastic domain.

The modified reference rating lives, evaluated using the lamina technique, exposed a good agreement with the $L_{10}$ lives accomplished by durability tests.

## References

[1] Hoeprich M 1985 Numerical procedure for designing rolling element contact geometry as a function of load cycle *SAE Technical Paper* Series 850764

[2] de Mul J M, Kalker J and Fredriksson B 1986 The contact between arbitrarily curved bodies of finite dimension *J TRIBOL T-ASME* **108** pp 140-148

[3] Reusner H 1987 The logarithmic roller profile the key to superior performance of cylindrical and taper roller bearings *Ball Bearing Journal* **230** pp 2-10

[4] Kapoor A and Johnson K L 1992 Effect of changes in contact geometry on shakedown of surfaces in rolling/sliding contact *INT J MECH SCI* **34** pp 223-239

[5] Williams J A 2005 The influence of repeated loading, residual stresses and shakedown on the behavior of tribological contacts *TRIBOL INT* **38** pp 786-797

[6] ISO 2008 Rolling bearings—methods for calculating the modified reference rating life for universally loaded bearings *ISO/TS 16281:2008(E)* (Geneva: ISO)

[7] Johnson G L 1964 *Statistical treatment of fatigue experiments* (New-York: Elsevier)

[8] Gafițanu M, Nastase D, Crețu S, Coman G, Racocea C, Nestor T and Olaru D 1985 *Rulmenți proiectare si tehnologie vol II* (Bucuresti: Editura Tehnica)

[9] McCool J 1971 Evaluating Weibull endurance data by the method of maximum likelihood *J LUBRC TECH-T ASME* **13** pp 189 – 202

[10] ISO 2007 Rolling bearings—dynamic load rating and rating life *ISO 281:2007* (Geneva: ISO)

[11] Crețu S and Antaluca E 2003 The study of non-hertzian concentrated contacts by a GC-DFFT technique *Annals of University of Galați* **VIII** pp 39-47

[12] Crețu S 2009 *Contactul concentrat elastic-plastic* (Iași: Politehnium)

# Triboelectrical charge generated by frictional sliding contact between polymeric materials

**T Zeghloul[1], M B Neagoe[1], Y E Prawatya[1, 2] and L Dascalescu[1]**

[1] Institut P', UPR 3346 CNRS - Université de Poitiers - ENSMA
IUT, 4 avenue de Varsovie, 16021 Angoulême, France
[2] Department of Industrial Engineering, Tanjungpura University
Jl. Ahmad Yani, 78124 Pontianak, Indonesia

E-mail : thami.zeghloul@univ-poitiers.fr

**Abstract**. The polymers used regularly in mechanical assemblies are brought up in relative sliding. The electrostatic charges generated in these functional conditions are merely known. Many factors are involved in the triboelectric charging process: normal load, the sliding velocity. The aim of this paper is to analyse the influence of these factors in the repartition and evolution of the electric potential at the surface in contact. The tribocharging experiments are carried out with samples cut from three polymers: sample A (5 mm x 15 mm x100 mm) from Acrylonitrile Butadiene Styrene (ABS) or Polypropylene (PP), and sample B (5 mm x 50 mm x 180 mm) from Polyvinyl Chloride (PVC). The normal load is set to four values in the range 2 to 14 N, and the sliding velocity is varied between 70 and 122 mm/s. The results point out that the variation of relative velocity between samples is not changing the average potential for the sample B. The surface potential has a linear increase with the normal load.

## 1. Introduction

The increasingly use of polymers and polymer composites in numerous industrial applications, such as aerospace, automotive, railway, etc., require their better characterization from different points of view. Compared with other materials, they can provide self-lubricant conditions, high strength/weight ratio, good mechanical and thermal properties, as well as a remarkable ease of manufacturing [1-5]. It is well known that rubbing together two insulators can provide an electrostatic charge generated at the surface of the bodies [6-8]. Studies are made to better control the electrostatic charge with the purposes to minimize it and avoid hazards (i.e. ignition in flammable environments) [9] or to maximize for industrial applications (i.e. electrostatic particle separation) [10].

The tribology theory is not simple, adding electrostatic phenomena increases its complexity. The study of electric charges through contact and friction between insulating materials has been an interest for many researchers. Because of multitude of factors involved, sliding charging is more complicated than simple contact charging [11-12].

The polymers used regularly in mechanical assemblies are brought up in relative sliding. The electrostatic charges generated in these functional conditions are merely knew. Many parameters are involved in this triboelectric charging mechanism: normal load, roughness, sliding velocity, rubbing time, etc. [13-23]. The aim of this study is to evaluate the effect of the normal load and the sliding velocity on the tribocharging of pairs of samples cut from Acrylonitrile butadiene styrene (ABS), Polyvinyl Chloride (PVC), and Polypropylene (PP).

## 2. Methods and materials

Two types of samples (A): 5 mm x 15 mm x 100 mm, and (B): 5 mm x 50 mm x 180 mm are employed for the electric tribo-charging experiments. They are cut from ABS, PVC, and PP. The samples (A) are cut from ABS or PP, while (B) is made of-PVC.The residual charge at the surface of these samples is neutralized using an ionizing system (electrode ECA 88 BS and high-voltage supply SC 04 B, 5 kV, 7 mA, manufacturer: ELCOWA, Mulhouse, France).

### *2.1. Charge generation*

Tribocharging process is carried out using a linear tribometer (figure 1) to generate the electric charge by friction between a pair of samples (A) and (B). The (A)-type sample is connected to a top holder (2) fixed in a vertical guiding system and using an adjustable arm (3) to vary the normal load. The (B)-type sample is placed in a holder attached to a back-and-forth sliding motion system (1), driven by an electric actuator (4). The tangential and normal loads, as well as the relative displacement of the samples are continuously measured. This laboratory bench enables the control of several factors that affect the electrostatic charge generation: relative sliding speed, normal contact load, and number of back-and-forth cycles. Room temperature and relative humidity are important factors in electric surface charging experiments and also for wear generation. The two are constantly read with a thermo-hygrometer and are kept in a certain interval of variation 23-25 C° respectively 35-45% RH, using a dehumidifier- heating system.

The experiments are conducted to show the effect of the relative velocity and normal load for the samples (A) and (B) brought in sliding contact in the tribocharging process. For the first experiment velocity is setup at 70 mm/s, 82 mm/s, 95 mm/s, 110 mm/s and 122 mm/s respectively. The normal load is $FN$ = 10 N and the number of sliding cycles is 10. The stroke used is 55 mm. The contact surface on sample B has the dimensions 50 mm (given by the bottom (B) sample width) and 70 mm (i.e., the stroke plus the width of the top (A) sample A). A second experiment is conducted at four values of the normal load: 2 N, 6 N, 10 N and 14 N. The relative velocity is setup at 82 mm/s, the number of cycles and the stroke are the same as in the previous experiment. Each experiment is repeated three times.

The surface of the sample (B) is investigated with a thermal infrared camera (FLIR E60) to monitor the increasing of temperature on the surface.

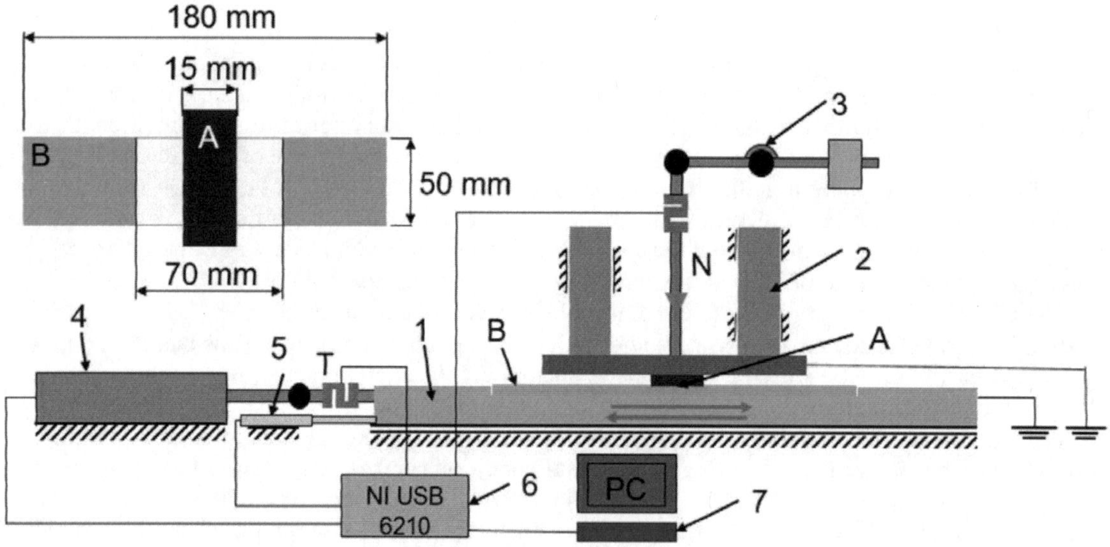


**Figure 1.** Schematic representation of the laboratory bench for triboelectric charging.

### 2.2. *Charge measurement*

The electric charge generation is monitored by measuring the surface electric potential distribution at the end of rubbing process on sample (B). This method has already been employed in another paper [24]. The measurements are performed with an induction probe (Trek, model 3450) of a 10 kV electrostatic voltmeter (Trek, model P0865). After charging, the sample is transferred in a 2-axis positioning device that facilitates the cartography of the surface potential. The measurement is performed in 5 x 19 grid beginning with line L1 (figure 2). The probe is measuring each 10 mm in a straight line. The distance between the sample and probe is 3 mm.

The probe "sees" the potential on a surface that has a diameter of approximately 10 mm. The surface potential measurements are conducted on the entire surface of the sample (B), including the area that comes not in contact with the sample (A). The data used for the analysis are the values of the average surface potential of the contact area.

Roughness measurements are performed using the roughness meter Mitutoyo SJ210 according to the standard ISO-1997. The measurements are repeated three times, for a sampling length of 15 mm. The roughness tests are made in the same direction as the sliding.

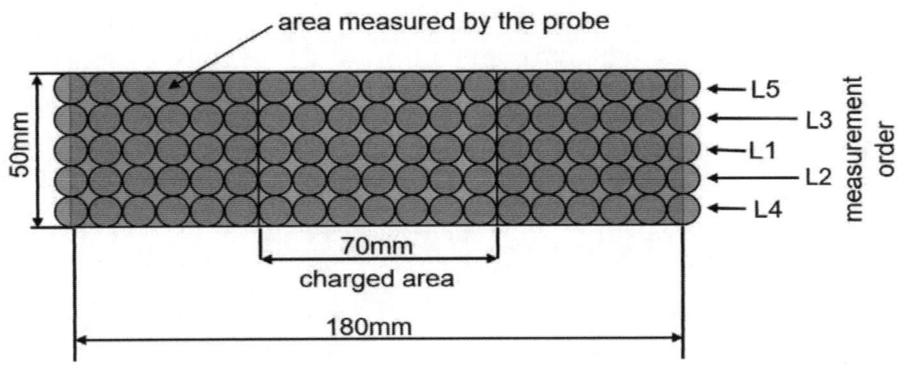


**Figure 2.** Potential measurement mapping.

## 3. Results and discussions

Due to the low number of back and forth cycles (ten, in this experiment), the temperature increase on the sample is limited at a maximum of 1.5 °C. The level of temperature is studied to ensure that a secondary charging effect generate by the heating will not appear. The temperature variation on the surface of the sample is lower than that of air temperature. Therefore, it can be considered that sample heating does not significantly affect the charge.

The initial roughness of the three samples is: $R_{a\,\mathrm{ABS}} = 0.5 \pm 0.1$ μm, $R_{a\,\mathrm{PP}} = 1.2 \pm 0.1$ μm and $R_{a\,\mathrm{PP}} = 0.1 \pm 0.02$ μm. These values are in the same class of roughness (0.1-2 μm). As a consequence, the effect of initial roughness can be considered as non-significant.

The two pairs of samples under study, ABS-PVC and PP-PVC, are expected to have a different tribocharging behavior. The first experiment points out that the average value of the surface potential is the same, at different sliding speeds (figure 3). The average value of the electric potential measured for the pair ABS-PVC is higher than for PP-PVC, in good agreement with ABS and PP in the triboelectric series [25].

In the second experiment, The increase of the normal load between the two samples produce a linear increase of the average surface potential of sample (B), as shown in figure 4. The tendency slope of the electric potential measured for the ABS-PVC pair is steeper than that obtained with the PP-PVC samples. This means that, in contact with the PVC, the ABS can generate a higher charge than the PP, at a given normal load between the bodies in contact.

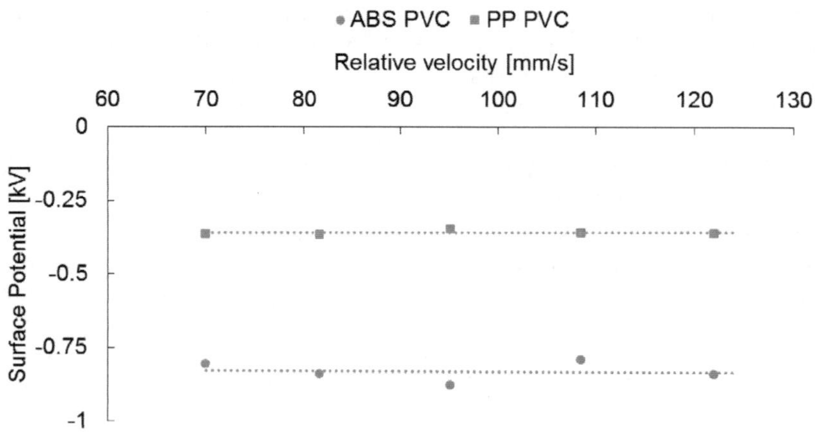


**Figure 3.** The average surface potential of contact area on sample (B) vs. the relative velocity between samples.

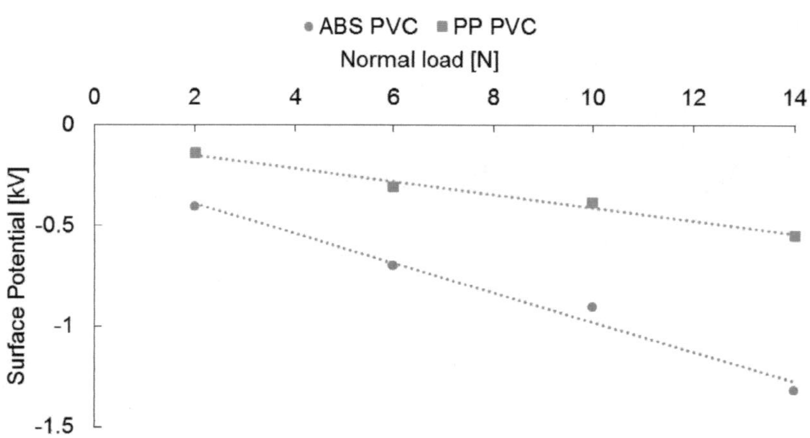


**Figure 4.** The average surface potential of contact area on sample (B) vs. the normal load between samples.

In figures 5 and 6, the repartition of the surface potential on sample (B) is represented for the limit values of the sliding speed, 70 mm/s and 122 mm/s, and of the normal load, 2 N and 14 N, for which the experiments have been conducted. From figure 5 can be seen that with the increase of the sliding speed the charged area is changing the position but the uniformity of potential repartition is similar. Figure 6 indicates that the he footprint of the charge is increasing with the normal load. This phenomenon is partly due to the fact that increasing the load causes also the deformation of the samples. The maximum value of the surface potential is increasing from -1.7 kV for the 2 N load, to almost 4 kV for the 14 N load.

## 4. Conclusions

The two sets of experiments conducted to analyze the generation of triboelectric charge by dry rubbing point out the following conclusions:

(1) The sliding velocity has no influence on the tribocharge generation, either for the rubbing of ABS-PVC or PP-PVC.

(2) A linear variation exists between the load and the average surface potential measured after rubbing of ABS-PVC and PP-PVC pairs. The ratio between the average potential of the two rubbing pairs is kept constant when increasing the normal load between the bodies in contact.

(3) The non-zero surface potential area is increasing with the load due to the increase of contact footprint between the two samples.

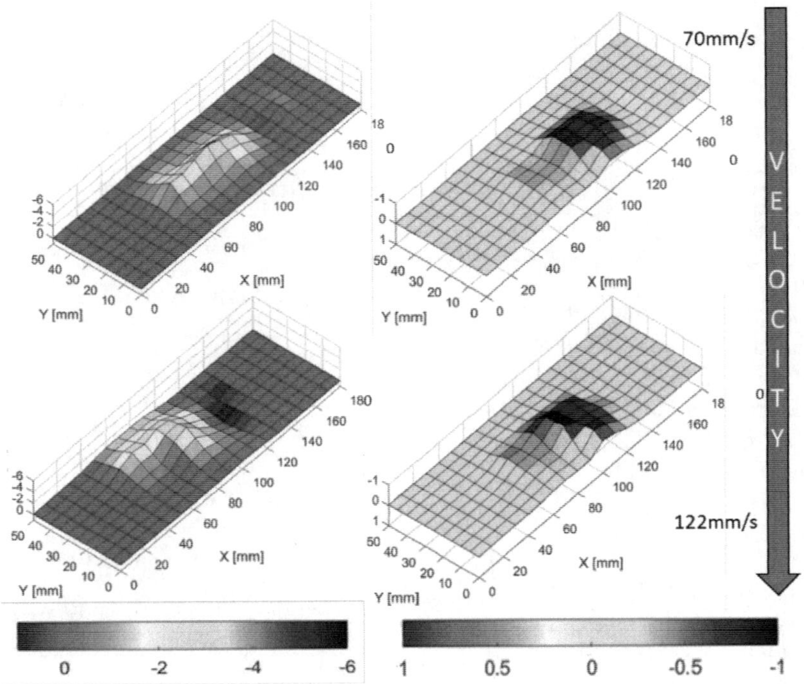


(a) (b)

**Figure 5.** Surface potential mapping for minimum and maximum values of the sliding velocity (70 mm/s and 122 mm/s, respectively); (a) ABS-PVC; (b) PP-PVC.

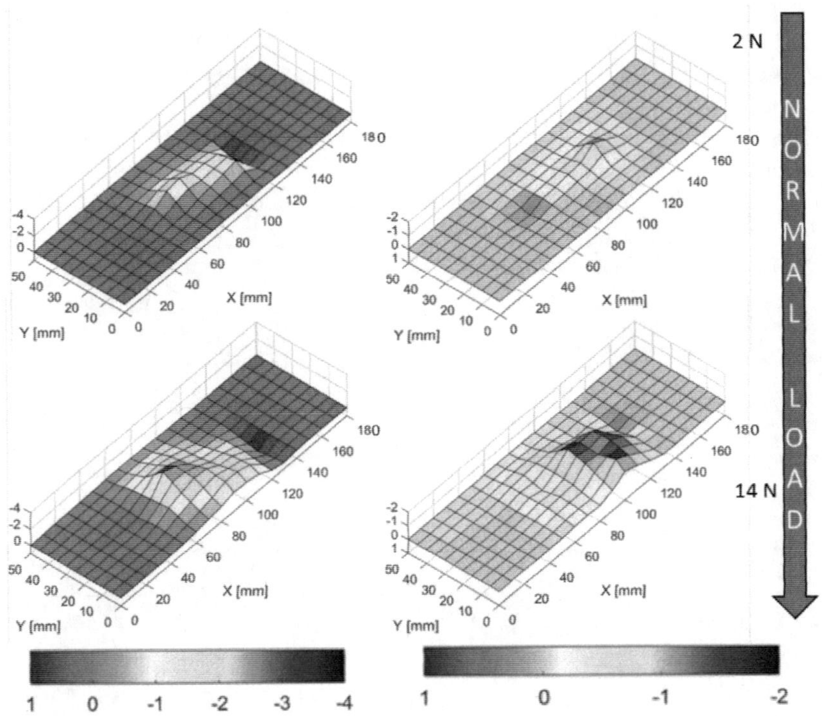


(a) (b)

**Figure 6.** Surface potential mapping for minimum and maximum values of the normal load (2 N and 10 N, respectively); a) ABS-PVC; b) PP-PVC.

**Acknowledgement**
This work was partially funded by the French Government program "Investissements d'Avenir" (LABEX INTERACTIFS, reference ANR-11-LABX-0017-01).

**References**

[1] Zeghloul T, Dascalescu L, Roaugdia K, Fatihou A, Renoux P and Souchet D 2016 *IEEE Trans. Ind. Appl.* **52** 1808-13
[2] Neagoe B, Prawatya Y, Zeghloul T, Souchet D and Dascalescu L 2015 *J. Phys.: Conf. Series* **646** 012058 doi: 10.1088/1742-6596/646/1/012058
[3] Prawatya Y, Neagoe B, Zeghloul T and Dascalescu L 2015 *Conf. Rec. IEEE/IAS Ann. Meet.* doi: 10.1109/IAS.2015.7356756
[4] Unal H and Mimaroglu A 2013 *Surf. Eng.* **29** 455-61
[5] Hosseinia S M and Stolarskia T A 1988 *Surf. Eng.* **4** 322-6
[6] Bailey A G 2001 *J. Electrostat.* **51-52** 82-90
[7] Williams M W 2012 *J. Electrostat.***70** 233-4
[8] Williams M W 2013 *J. Electrostat.***71** 53-4
[9] Loveland R J 1981 *J. Electrostat.***11** 3-11
[10] Mekhalef Benhafssa A, Medles K, Boukhoulda M F, Tilmatine A, Messal S and Dascalescu L 2015 *IEEE Trans. Ind. Appl.* **51** 1-7
[11] Liu L, Oxenham W, Seyam A M and Theyson T 2011 *J Textile Instit.* **102** 1075-85
[12] Liu L, Seyam A M and Oxenham W 2013 *J. Engineered Fibers &Fabrics* **8** 126-36
[13] Komatsu T S , Hashimoto M, Miura T, Arakawa I and Nasuno S 2004 *Appl. Surf. Sci.* **235** 60-4
[14] Ohara K, Nakamura I and Kinoshita M 2001 *J. Electrost.* **51-52** 351-8
[15] Alahmadi A 2014 *Int. J. Sci. Eng. Res.* **5**, 22-9
[16] Guerret-Piecourt C, Bec S and Tréheux D 2001 *C. R. Acad. Sci. Paris* **IV 2** 761–74
[17] Nakayama K 1996 *Wear* **194** 185-9
[18] Harvey T J, Wood R J K, Denuault G and Powrie H E G 2002 *Tribol. Int.* **35** 605–14
[19] Lee D-Yo, Lee J, Hwang J and Choa S-H 2007 *Tribol. Int.* **40** 1253–7
[20] Ning L, Jian L Yang S Wang J Ren J and Wang J 2010 *Tribol. Int.* **43** 568–76
[21] Changa Y-P, Yur J-P, Chou H-M and Chub H-M 2006 *Wear* **260** 1209–16
[22] Kchaoua B, Turki C, Salvia M, Fakhfakh Z and Tréheux D 2008 *Wear* **265** 763–71
[23] Hiratsuka K and Hosotani K 2012 *Tribol. Int.* **55** 87-99
[24] Antoniu A, Dascalescu L, Vacar I V, Plopeanu M C, Tabti B and Teodorescu H N 2011 *IEEE Trans. Ind. Appl.* **47** 1118-25
[25] Diaz A F and Felix-Navarro R M A 2004 *J. Electrostat.* **62** 277-90

# Influence of surface roughness on the tribo-electric process for a sliding contact between polymeric plate materials

**M B Neagoe[1], Y E Prawatya[1, 2], T Zeghloul[1] and L Dascalescu[1]**

[1] Institut P', UPR 3346 CNRS - Université de Poitiers - ENSMA, IUT, 4 avenue de Varsovie, 16021 Angoulême, France
[2] Department of Industrial Engineering, Tanjungpura University, Jl. Ahmad Yani, 78124 Pontianak, Indonesia

E-mail: marian.bogdan.neagoe@univ-poitiers.fr

**Abstract.** Polymers are used in many industrial applications due to their good mechanical and thermal properties. Studies have shown that electrical insulators, such as the polymers, are generating electrostatic charge by frictional contact. This tribo-charging effect influences de sliding conditions, and the level of charge depends of several factors. The aim of this paper is to validate two paradigms: (I) the electrostatic charge increases with contact pressure; (II) the level of charge is influenced by the number of contact points on the surface. The tribocharging experiments were carried out with samples cut from two polymers: Acrylonitrile-Butadiene-Styrene, 5 mm x 15mm x100 mm, roughness: $R_a = 0.4$ µm; and Polyvinyl Chloride, 5 mm x 50 mm x 180 mm, with various roughness values: $R_a$ = 3-12 µm and $R_{Sm}$ = 150-350 µm. The modification of asperity density was accompanied by changes in the electric potential measured at the surface of the PVC samples. Thus, its average value $V$ increased from -0.33 to -0.85 kV, and its maximum value $Vmax$, from -1.56 to -4.83 kV.

## 1. Introduction

Polymers are more and more used as materials for sliding machine components, due to their low cost, ease of manufacturing, as well as good mechanical and thermal properties [1-8]. Their triboelectric charging (i.e., acquiring an electric charge after they come in frictional contact with a different material) is a phenomenon which is not totally understood [8].

Recent studies have shown that the electrical charge generated by such contact between insulating materials has an important effect on the rubbing process, especially in the case of lubricated sliding [9-12]. The charge generation is influenced by many factors: nature and surface state of the materials that are rubbed, ambient conditions, relative velocity of the surfaces in motion, contact pressure, etc [13]. The contact between surfaces is usually made on a number of points related to the surface roughness. When the surfaces slide with respect to each other, a point from one of the surfaces may remain in contact with the other surfaces for a while. After a short displacement the point of contact will give way to a new one. In the course of time multiple points of contact will be involved [13-14]. This consideration has given rise to two paradigms: one (I) according to which the electric charge increases with pressure and the second (II) that says that the electric charge increases with the multiplication of contact points, the Volta-Helmholtz hypothesis [13-14]. A method to change the number of contact points is to modify the roughness of the surfaces involved in the tribocharging process. The aim of this paper is to validate the two paradigms, by investigating the influence of surface roughness on tribocharge generation in the case of polymeric materials.

## 2. Methods and materials

Two types of samples (A): 5 mm x 15 mm x 100 mm and (B): 5 mm x 50 mm x 180 mm are employed for the electric tribo-charging experiments. They are cut from two polymers: Polyvinyl chloride (PVC) and Acrylonitrile Butadiene Styrene (ABS). The samples are chosen in conformity with the tribo series to provide a good triboelectrification [13]. The residual charge at the surface of these samples is neutralized using an ionizing system (electrode ECA 88 BS and high-voltage supply SC 04 B, 5 kV, 7 mA, manufacturer: ELCOWA, Mulhouse, France).

Tribocharging process is carried out using a linear tribometer (figure 1) to generate the electric charge by friction between a pair of samples (A) and (B). The A-type sample is connected to a top holder (2) fixed in a vertical guiding system and using an adjustable arm (3) to vary the normal force. The B-type sample is placed in a holder attached to a back-and-forth sliding motion system (1), actioned electric actuator (4). The tangential and normal forces, as well as the relative displacement of the samples are continuously measured. This triboelectric laboratory bench enables the control of several factors that affect the electrostatic charge generation: relative sliding speed, normal contact force, and number of back-and-forth cycles.

Room temperature and relative humidity are also important factors in electric surface charging experiments, and they were read with a thermo- hygrometer. Their values were respectively set in the intervals 20 – 24°C and 40 to 50%, using a dehumidifier.

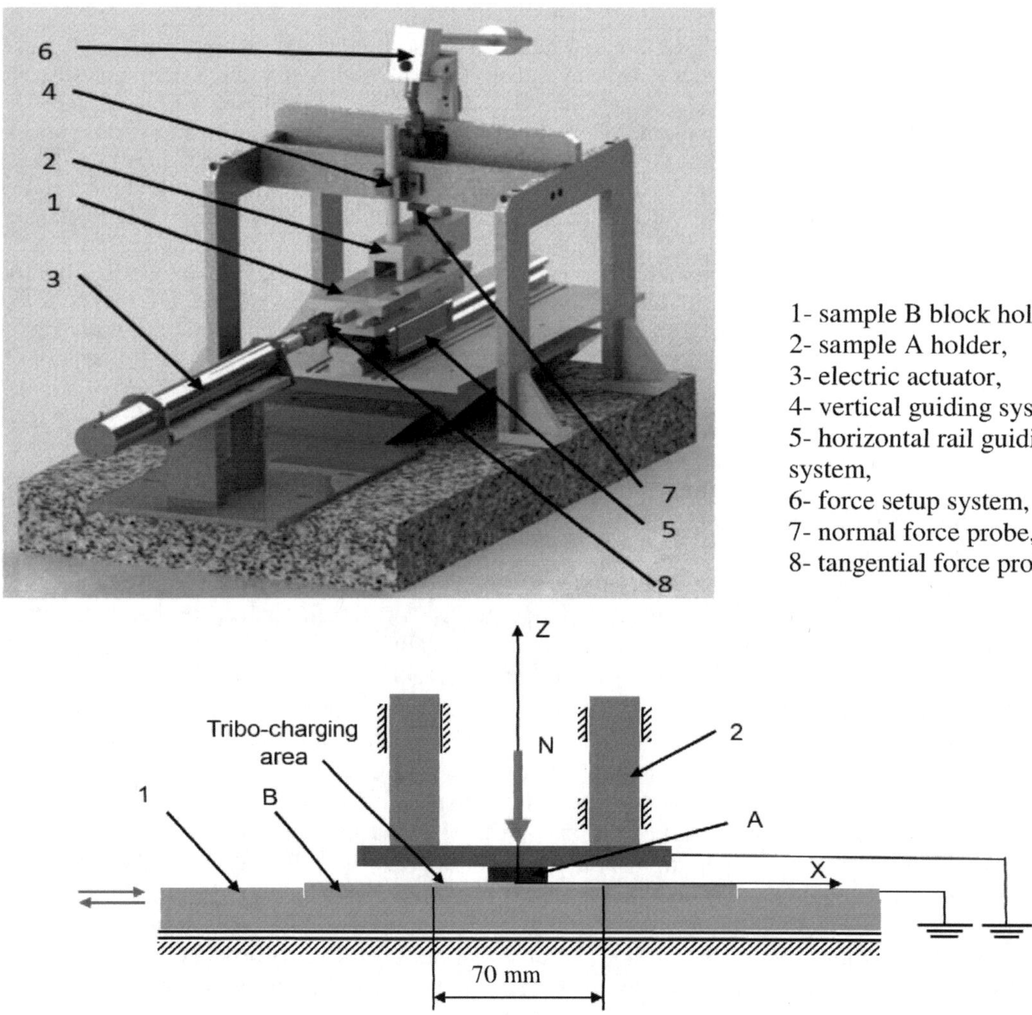


1- sample B block holder,
2- sample A holder,
3- electric actuator,
4- vertical guiding system,
5- horizontal rail guiding system,
6- force setup system,
7- normal force probe,
8- tangential force probe.

**Figure 1.** Schematic representation of the laboratory bench for triboelectric charging

The electric potential mapping techniques can give an image of the tribo-electrically generated charge at the surface of the samples. A proportional relationship exists between the electric potential and the electric charge. Thus, the distribution of electric charge at the surface of the B-type samples was measured with the induction probe of an electrostatic voltmeter.

After charging, the sample is transferred in a 2-axis device that facilitates the cartography of the surface potential. The measurement is performed in 5 x 19 points beginning with line L1 (figure 2). For each experiment, the average and the maximum absolute values of the electric potential were used to evaluate the level electric charge across the rubbing area.

The material pair ABS-PVC was chosen for this experiment. To highlight the correlations of the number of contact points with the level of electric charge generated by friction, the ABS sample A had a smooth initial surface, characterized by a roughness $R_a = 0.4$ µm. The surface state of PVC sample B was modified using the P120, P80, and P40 abrasive discs. The roughness obtained was in the interval $R_a = 3\text{-}12$ µm.

To verify the first paradigm, a set of experiments was made with samples B of similar roughness, prepared with the P40 abrasive disc, at three values of the normal force: $F = 2$ N, 6 N and 10 N, using a sliding speed $v = 75$ mm/s, for a number of sliding cycles $n = 10$. The experiments for verifying the second paradigm were performed with samples B of different roughness, at constant normal force $F =$ 2 N, sliding speed $v = 75$ mm/s, and number of back-and-forth cycles $n = 10$. The variables measured during the process of rubbing were the normal and tangential forces, as well as the relative displacement of the samples in contact.

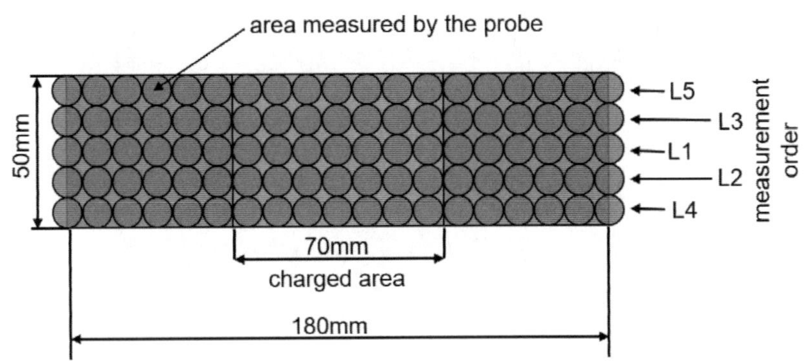


**Figure 2.** Mapping of the electric potential at the surface of the bottom sample.

The roughness measurements were performed using the roughness meter Mitutoyo SJ210 according to the standard ISO-1997. The measurements were repeated three times, for a sampling length of 15 mm. The roughness test was made in the same direction with the sliding. Two roughness coefficients were measured: $R_a$ which is defining the average of the peaks and valleys, and $R_{Sm}$ defining the average distance between two consecutive peaks.

## 3. Results and discussions

In the first set of experiments, carried out with samples of similar roughness $R_a \approx 10$ µm, the surface potential (SP) was found to increase is increasing with the normal load (figure 3). The contact surface between sample A and B being constant, it can be inferred from this that the SP increases with the average pressure applied, which confirms the first paradigm.

Two typical cartographies of the electric potential measured at the surface of B-type PVC samples can be examined in figure 4. The electric potential is less uniform at the surface of the smoother surface. For surfaces characterized by a high roughness, the peaks are higher and less dense. The higher deformation per peak gives a more uniform distributed contact with the smooth surface. For a low $R_a$, the surface undulations generate local high pressures.

In figure 5 is represented the contact between surfaces characterized by different roughness coefficients. The contact point density decreases with the roughness. The global pressure for every sample is constant and is given by the normal force of 10 N divided by the nominal area of contact.

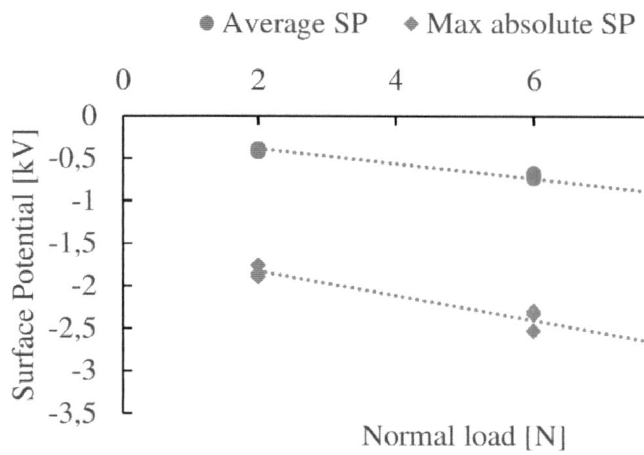


**Figure 3.** Relation between the surface potential and the normal load.

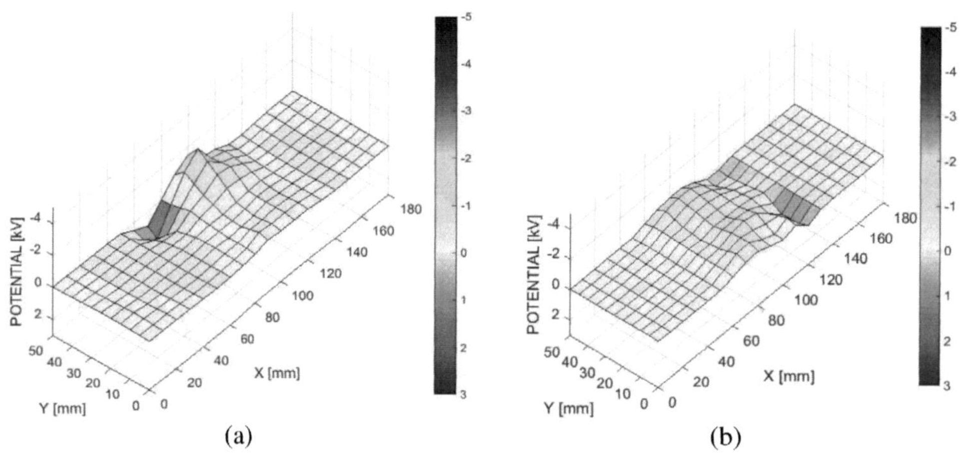


**Figure 4.** Surface potential mapping: (a) $R_a$ =3.7, $R_{Sm}$ = 152.6μm
(b) $R_a$=10.9, $R_{Sm}$= 350.2 μm.

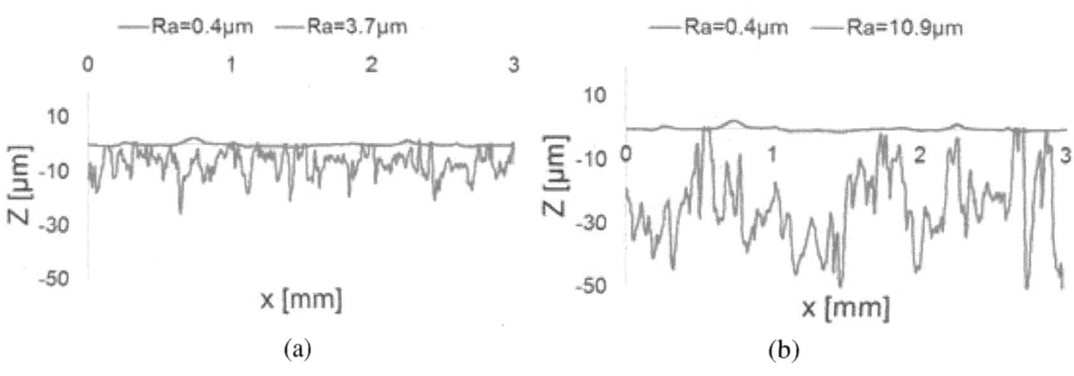


**Figure 5.** Contact due to the roughness asperities: (a) $R_a$=3.7 μm; $R_{Sm}$ = 152.6 μm;
(b) $R_a$ = 10.9 μm; $R_{Sm}$ = 350.2 μm.

The results of roughness measurements point out that $R_{Sm}$ increases with $R_a$ (figure 6), due to the aspect of the asperities of the abrasive disks used for surface preparation. For known $R_{Sm}$ (i.e., the mean distance between two consecutive peaks), the density of peaks can be computed as follows: $\rho = 1/R_{Sm}$ [peaks/mm]. This means that roughness $R_a$ increases when the density of the peaks $\rho$ decreases.

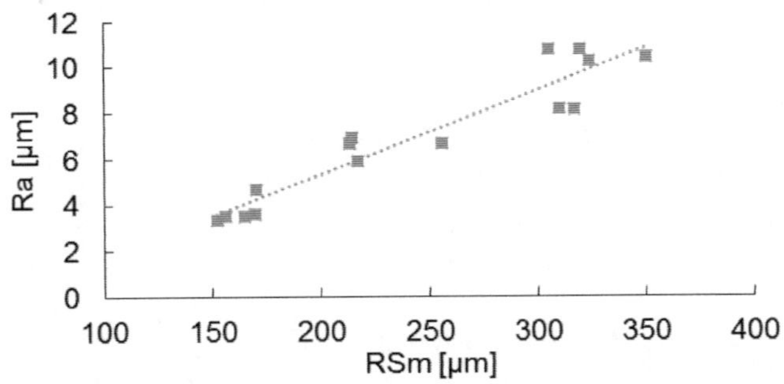


**Figure 6.** $R_{Sm}$ vs $R_a$.

Heating of rubbed surface has been monitored with a thermal camera (figure 7). As the sliding between the two samples was set up to 10 cycles per run, the rubbing process produces only a 1.5°C temperature rise of the sample B, and the secondary charging mechanism generated by heating can be considered not important when compared with the multipoint mechanism.

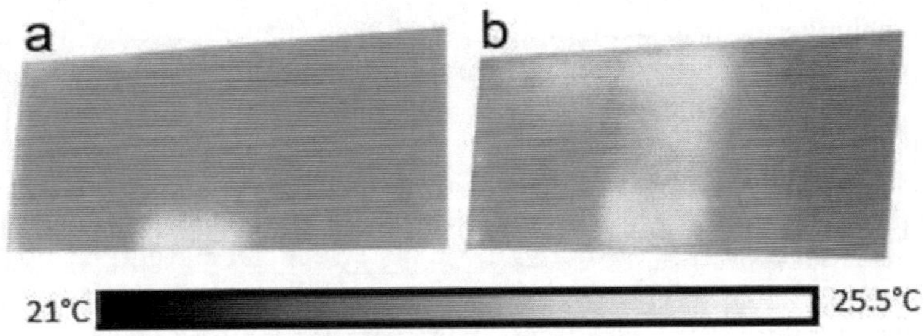


**Figure 7.** Thermal fields: (a) $R_a$ =3.7, $R_{Sm}$ = 152.6 µm;
(b) $R_a$ =10.9, $R_{Sm}$ = 350.2 µm.

The non-uniformity of the thermic field is due to non-uniform pressure contact on the sample. From figures 4 and 7, it can be seen that the surface potential, which is proportional to the charge, follows the temperature pattern. With heating assumed to be co-related to the local pressure exerted between the two bodies in contact, this result validates once more the first paradigm: higher pressure is producing higher electric charge.

The average and maximum absolute values of the surface potential increase with the density of peaks (figure 8). Thus, the charge, which is proportional with the electric potential, increases with the number of contact points between the two bodies. This confirms the second of the two paradigms.

Keeping the same normal load and increasing the number of peaks implies a lower contact pressure per peak, but a constant average pressure. The results plotted in figure 8, obtained at constant average pressure, show that the surface potential is increasing with the density of the peaks. This means that the electric charge generation is more influenced by the number of peaks that enter in contact, than by the local contact pressure at each peak.

## 4. Conclusions

The increasing of the roughness $R_{Sm}$ and of the distance between peaks is accompanied by a more uniform repartition of the normal force across the surfaces in contact, generating a uniform electrostatic charging.

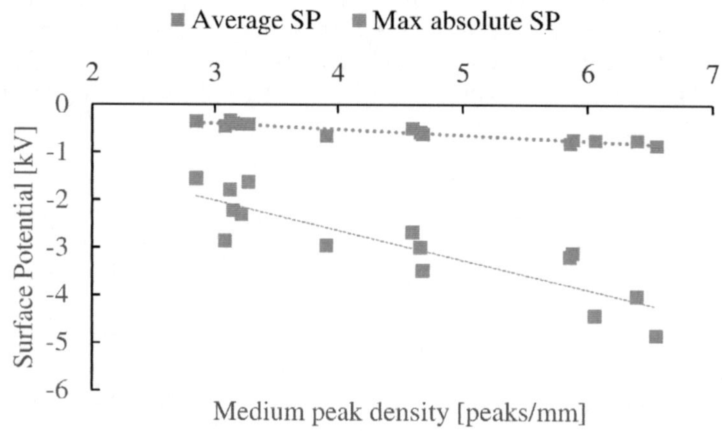


**Figure 8.** Correlation between the surface potential and the medium peak density.

The electrostatic charge generated by triboelectric effect increases with the pressure between the two surfaces that are rubbed together, which confirms the first paradigm.

The electrostatic charge generated increases with the peak density, as predicted by the second paradigm, the Volta-Helmholtz hypothesis.

The effect of the number of peaks is more important than that of pressure, but this phenomenon should be further investigated, to get a better understanding of the physical mechanisms involved.

**Acknowledgement**
This work was partially funded by the French Government program "Investissements d'Avenir" (LABEX INTERACTIFS, reference ANR-11-LABX-0017-01).

**References**

[1] Unal H and Mimaroglu A 2013 *Surf. Eng.* **29** 455-61
[2] Hosseinia S M and Stolarskia T A 1988 *Surf. Eng.* **4** 322-6
[3] Zeghloul T, Dascalescu L, Roaugdia K, Fatihou A, Renoux P and Souchet D 2016 *IEEE Trans. Ind. Appl.* **52** 1808-13
[4] Rymuza Z 2007 *Arch. Civil Mech. Eng.* **7** 177-84
[5] Guerret-Piecourt C, Bec S and Tréheux D 2001 *C. R. Acad. Sci. Paris IV* **2** 761-74
[6] Nakayama K 1996 *Wear* **194** 185-9
[7] Hiratsuka K and Hosotani K 2012 *Tribol. Int.* **55** 87-99
[8] Bailey A G 2001 *J. Electrostat.* **51-52** 82-90
[9] Williams M W 2012 *J. Electrostat.* **70** 233-4
[10] Williams M W 2012 *J. Electrostat.* **71** 53-4
[11] Hogue M D, Buhler C R, Calle C I, Matsuyama T, Luo W and Groop E E 2004 *J. Electrostat.* **61** 259-68
[12] Alahmadi A 2014 *Int. J. Sci. Eng. Res.* **5** 22-9
[13] Chang J S, Kelly A J and Crowley J M 1995 *Handbook of Electrostatic Processes* (New York: Marcel Dekker) p 32
[14] Harper W R 1998 *Contact and Friction Electrification* (Morgan Hill, CA: Laplacian Press) pp 3-4
[15] Diaz A F and Felix-Navarro R M A 2004 *J. Electrostat.* **62** 277-90

# Experimental analysis of axial and radial stress distribution in soft materials used for petrochemical valve stem sealing package

**R G Ripeanu[1], A Ispas[2] and D Ispas[1]**

[1]University Petroleum-Gas of Ploiesti, Mechanical Engineering Department, 39 Bucharest Blvd., 100680 Ploiesti, Romania
[2]S.C. Etanşări GRAFEX S.R.L., 1 Dumbravei Street, 100197 Ploiesti, Romania

E-mail: rrapeanu@upg-ploiesti.ro

**Abstract.** Paper presents experimental results obtained by authors regarding pressure values and distribution, by using a special pressure sensitive paper, type FujiFilm and adequate hardware equipment and a soft program FPD810 Win. Was obtained the proper axial force related for different materials in order to respect the valves sealing demands. Were showed that, pressure maps obtained depends of the construction of sealing package, density of the preformed band rings and corrected sealing replacing on location in case of square braided cord used.

## 1. Introduction

Refineries are industrial sites that manage huge amounts of raw and processed materials and in the same time being intensive consumers of energy and water. That is the reason for refineries to be on the 2$^{nd}$ place [1] (after combustion plants with a power greater than 50 MW) in energy industry domain in Directive 2008/1/EC of 15 January 2008 concerning integrated pollution prevention and control of pollution, IPPC (was included, together with the late six directives, directive 2010/75/I of 24 November 2010) referring industrial emissions.

On the total refineries fugitive emissions (specific in case of sealing's), 75% are generated by valves [1], so diminishing pollution is an important problem that must be solved.

There are many norms [2] (API 589, API 607, API 622ASME B73.1, ASME B73.2, ASTM F 2087-01, ASTM F2168-02, ASTM F 2191-02, ASTM F 1277-02, BS 4371, ISA-SP-93, ANSI/FCI 91-1, MSS SP – 120, MSS SP – 121, TA Luft, VDI 2440, EN ISO 15848-1) which impose very restrictive conditions regarding leakage rate. In Europe are applied especially TA – Luft (Technische Anleitung zur Reinhaltung der Luft, 2002), VDI 2440 (Verein Deuscher Ingenieure) and EN ISO 15848-1, 2. The "Technische Anleitung zur Reinhaltung der Luft" (Technical Guidelines for Air Pollution Control, or TA Luft for short and referred to hereinafter as TA Luft) are like a law in character [3]. Among other things, this norm pursues the objective of reducing emissions with gasket systems which are normally used in the chemicals and petrochemical industries. TA Luft only gives guidelines in respect of compliance with permissible leakage limits and refers to regulations which define basic conditions for inspections [3]. For shutoff and regulating valves, will be applied VDI (Association of German Engineers) Guideline 2440 which may be regarded as state-of-the-art from a legal perspective [3]. Guideline VDI 2440 refers to average gas-like emissions for differing gasket systems but these are not upper limits (see table 1). VDI 2440 refers to the use of high quality metallic bellows with a downstream safety stuffing box or "equivalent gasket systems" as a particularly effective means of reducing emissions. In other words, metallic bellows and so-called "equivalent gasket systems" are established as conforming to TA Luft [3]. This means that only these kinds of gasket systems may be used in Germany on shutoff and regulating valves when a TA Luft medium is subsidized.

**Table 1.** Average gaseous emissions (leakage) for valve gaskets [3].

| Gasket system | Leakage related to the average size of the gasket (mg/sxm) |
|---|---|
| Stuffing box with packing | 1.0 |
| Stuffing box with cup leather, O-ring | 0.1 |
| Stuffing box with packing, stuffing box with cup leather, O-ring (with "TA Air Certificate" according to VDI 2440, Section 3.3.1.3) | 0.01 |
| Metallic bellows, sealed | 0.01 |
| Metallic bellows, sealed (with flat gasket possessing a TA Luft Certificate according to VDI 2440) | 0.001 |
| Stuffing box with packing and sealing medium/suction, metallic bellows, welded on both sides | No emission (technically leak-proof) |

The norm EN ISO 15848-1 has been valid since 2006. These ISO norm describes detailed tests for determining leakage and the creep behavior of valves. The norm was developed under the overall control of the CEN/TC 69 "Industrial Valves" Committee. The norm distinguishes between the following parameters [3]:

- 3 grades of imperviousness for the spindle/shaft;
- a grade of imperviousness for the body gasket;
- 2 testing media (helium and methane);
- 3 grades of firmness;
- 5 grades of temperature;
- 1 distinction between shutoff and regulating valves.

According to EN ISO 15848-1, table 2 shows the grades of imperviousness for die spindle and the shaft.

**Table 2.** Grade of imperviousness/leakage areas [3].

| Grade | Measured leakage rate (mg/sxm) | Remarks |
|---|---|---|
| A (helium only) | $\leq 10^{-6}$ | Typically achieved with bellows gasket or an equivalent spindle/shaft gasket system for swivel valves. |
| B | $\leq 10^{-4}$ | Typically achieved with packing on a PTFE basis or elastomer packing. |
| C | $\leq 10^{-2}$ | Typically achieved with packing on the basis of flexible graphite. |

The majority of the refinery valves, newer or older manufactured have the sealing with atmosphere- to avoid volatile organic compounds emissions, placed on the valve stem of type stuffing box with soft sealing material package. Expanded graphite (flexible) is almost generalized used as soft material [1] [2], [4], in various forms: square braided cord, preformed band rings and combined packages in various densities and construction. Sealing depends on radial pressure developed in package due to axial pressure applied on the gland [2], [5], [6].

The aim of this paper is to present a method of testing new sealing soft materials and to analyze axial and radial stress distribution in these materials used for petrochemical valve stem sealing package, in order to determine the transforming pressure coefficients that are used in dimensioning calculus of sealing package, with respecting fugitive emissions imposed by the norms presented above.

## 2. Stuffing box sealing calculus

Depending of working conditions, stuffing box sealing requires of us to differentiate between static or dynamic use of packings. Very small relative movements of the components are characteristic of seals under static or semi-static load [2]. At valves, but not only, sealing package must have elasticity, small friction coefficient and mainly low leakage. When a packing ring is compressed, it will be subjected to both plastic and elastic deformation depending on the circumstances. In exclusively static sealing applications, such as housing covers, flanges etc., the seal is required to give an elastic reaction in order to adapt itself to the dimensional changes of the seal cavity resulting from temperature and stress variations [2]. Reliable seals should neither extrude at elevated temperatures nor release volatile constituents. Stuffing box sealing calculus respects the equilibrium state of sealing packing within stuffing box as presented in figure 1.

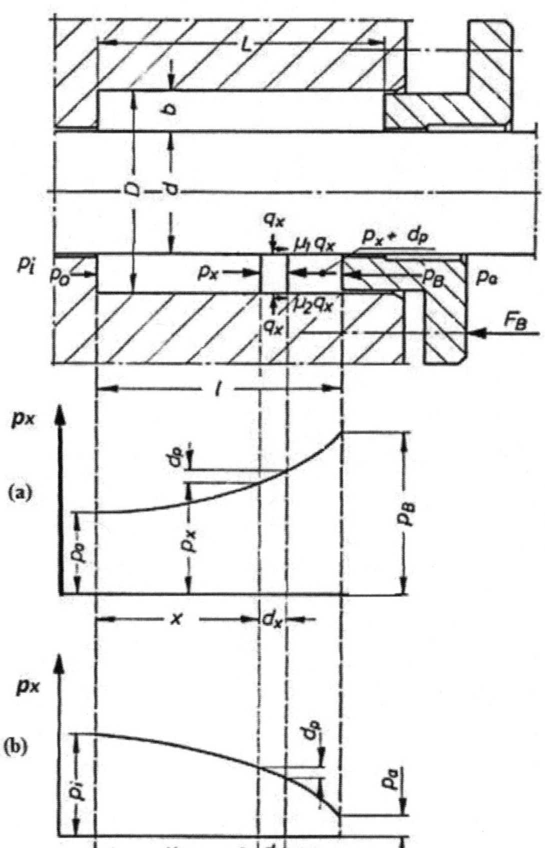


**Figure 1.** Equilibrium state of the internal pressures of sealing packing within the stuffing box.
(a) Tightening pressure distribution;
(b) Sealing pressure distribution.

To obtain the pressure equilibrium at $x$ distance we could obtain the $p_x$ [4], [5], [6]:

$$p_x = p_B \cdot e^{-(\mu_1+\mu_2)\frac{K\cdot(l-x)}{b}} \tag{1}$$

where: $\mu_1$ and $\mu_2$ are friction coefficients between stem and sealing ring respectively between sealing ring and stuffing box.

$$q_x = K \cdot p_x \tag{2}$$

$$F_B = \frac{\pi}{4}(D^2 - d^2) \cdot \frac{p_i - p_a}{K} \tag{3}$$

We could observe that to solve equations (1), (2) and (3) we have to know geometrical parameters, [4, 5, 6]:

$$b = (1.2 \ldots 1.4)\, d^{1/2} \tag{4}$$

$$l = (0.8 \ldots 1.5)\, d \tag{5}$$

and also $K$ is a factor which is specific to the packing, normally ranging from 0.6 to 1.0. $K$ represents transforming of axial pressure which has an exponential distribution on $l$-high of the package, in radial pressure.

$$K = \min(K_1, K_0) \tag{6}$$

where: $K_1$ is the pressure transforming coefficient on stuffing box;
$K_0$ - the pressure transforming coefficient on stem/shaft.

## 3. Experimental tests

### 3.1. Working procedure

In order to experimentally establish $p_x$ -axial pressure distribution of package height and radial pressure on stuffing box and on stem respectively was used the device presented in figure 2 and pressure sensitive paper fixed on stem, on stuffing box, between the rings, on gland and on seat. In figure 3 it is shown the pressure sensitive paper fixed as mentioned above.

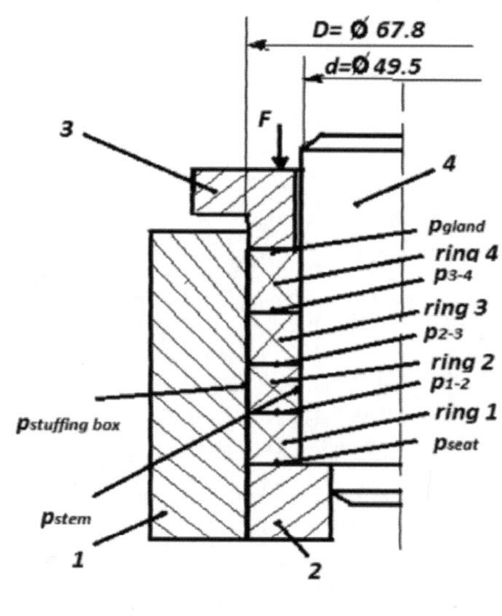


**Figure 2.** Testing device.
1-stuffing box; 2-seat; 3-gland; 4-stem/shaft.

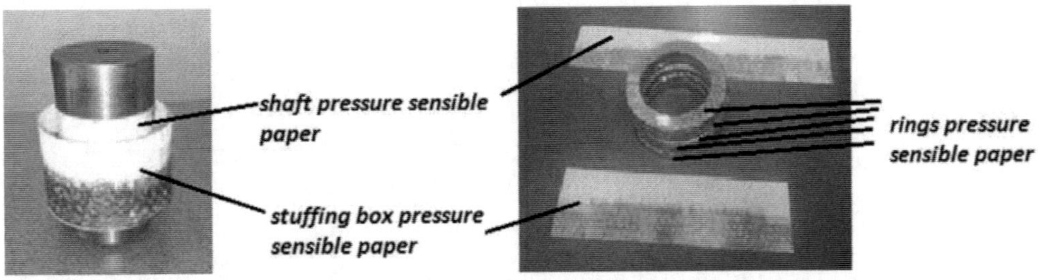


**Figure 3**. Positions of pressure sensitive papers.

Appling load $F$ was made controlled with a dynamometer and loading speed as well as unloading speed was also controlled. The maximum load on gland of 40kN, 50kN and 60kN respectively were maintained 15 seconds and then set was unloaded. FujiFilm pressure sensitive paper registers the pressure depending on temperature and humidity. Tests were performed at $20^0$C and RH=33%, in dry conditions (without lubricant). Also at each test, to avoid possible errors, the calibration was made without stuffing box, pressure sensitive paper being placed between two sealing plates type AF and clearance between stem.

In figure 5 is presented a result obtained with a specific software type FPD 8010 Win.

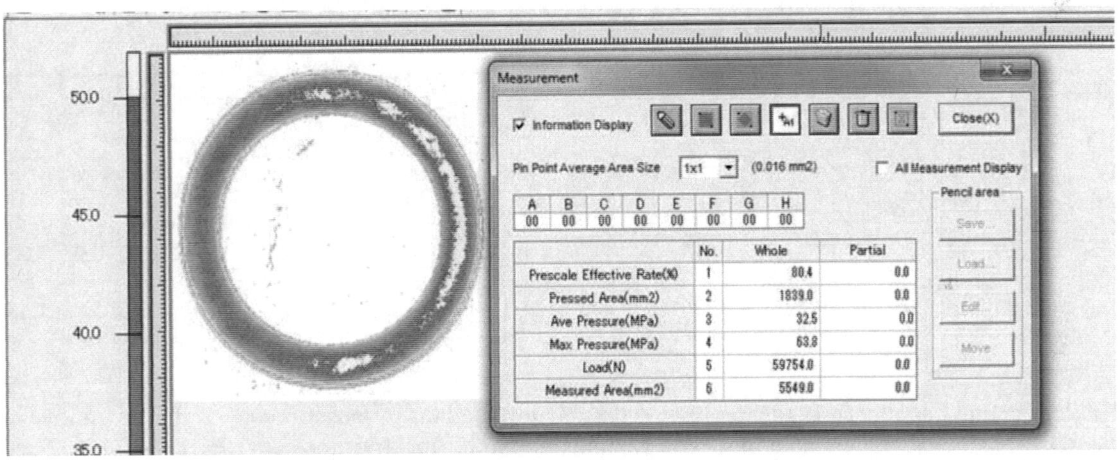


**Figure 4**. FPD 8010 Win pressure sensitive paper results.

From figure 5 we could remark the confidence between measured load of 60000 N and obtained load on surface of pressure sensitive paper of 59754 N , so an error of 0.41%.

### 3.2. *Experimental results*

*3.2.1. Package of expanded graphite rings string type GS.* For experimental tests were made string rings of expanded graphite with dimensions Φ67.8 x Φ49.5 x 10.4 mm, (outer diameter x inner diameter x height) with a mass of 25 g and density of 1.3 g·cm$^{-3}$. The illustrated in figure 2 arrangements was compressed with a dynamometer measured load of 40, 50 and 60 kN. After disassembling each paper pressure sensitive rings were analyzed with soft program FPD 8010 Win. Average axial pressure, $p_x$ on each ring was considered as arithmetic average of two values obtained on each ring face.

Radial pressure, $q_x$, on stuffing box and on stem was established for each ring taking account the height of rings on paper records.

Experimental results obtained for package of expanded graphite rings string are synthetic presented in table 3.

**Table 3.** Results obtained for package of expanded graphite rings string.

| *F* X10³ (N) | $p_B$ (MPa) | Average $p_x$, (MPa) | | | | Stem: Average $q_x$, (MPa) | | | | Stem: $K_0$ | | | | Stuffing box: Average $q_x$, (MPa) | | | | Stuffing box: $K_1$ | | | |
|---|---|---|---|---|---|---|---|---|---|---|---|---|---|---|---|---|---|---|---|---|---|
| | | 1 | 2 | 3 | 4 | 1 | 2 | 3 | 4 | 1 | 2 | 3 | 4 | 1 | 2 | 3 | 4 | 1 | 2 | 3 | 4 |
| 40 | 21.75 | 13 | 14 | 16 | 19 | 10 | 11 | 15 | 16 | 0.77 | 0.79 | 0.94 | 0.84 | 9 | 11 | 14 | 16 | 0.69 | 0.79 | 0.88 | 0.84 |
| 50 | 27.19 | 14 | 15 | 17 | 22 | 13 | 13 | 16 | 20 | 0.93 | 0.87 | 0.94 | 0.91 | 11 | 13 | 15 | 19 | 0.79 | 0.87 | 0.88 | 0.86 |
| 60 | 32.63 | 22 | 23 | 24 | 30 | 19 | 20 | 21 | 29 | 0.86 | 0.87 | 0.88 | 0.97 | 20 | 21 | 22 | 26 | 0.91 | 0.91 | 0.92 | 0.87 |

*3.2.2. Package of preformed rings of expanded graphite tape.* Experimental tests were identical conducted as in case of package of expanded graphite rings string. Were realized preformed rings of expanded graphite tape with Φ67.8 x Φ49.5 x 9.7 mm, (outer diameter x inner diameter x height), with a mass of 22 g and density of 1.1 $g \cdot cm^{-3}$.

Experimental results obtained for package of preformed rings of expanded graphite tape are shown in table 4.

**Table 4.** Results obtained for package of preformed rings of expanded graphite tape.

| *F* X10³ (N) | $p_B$ (MPa) | Average $p_x$, (MPa) | | | | Stem: Average $q_x$, (MPa) | | | | Stem: $K_0$ | | | | Stuffing box: Average $q_x$, (MPa) | | | | Stuffing box: $K_1$ | | | |
|---|---|---|---|---|---|---|---|---|---|---|---|---|---|---|---|---|---|---|---|---|---|
| | | 1 | 2 | 3 | 4 | 1 | 2 | 3 | 4 | 1 | 2 | 3 | 4 | 1 | 2 | 3 | 4 | 1 | 2 | 3 | 4 |
| 40 | 21.75 | 10 | 10 | 12 | 14 | 13 | 13 | 14 | 17 | 1.3 | 1.3 | 1.2 | 1.2 | 5 | 6 | 11 | 16 | 0.5 | 0.6 | 0.92 | 1.14 |
| 50 | 27.19 | 12 | 12 | 15 | 18 | 10 | 12 | 15 | 22 | 0.83 | 1 | 1 | 1.22 | 11 | 11 | 14 | 19 | 0.92 | 0.92 | 0.93 | 1.05 |
| 60 | 32.63 | 19 | 20 | 22 | 29 | 23 | 24 | 26 | 29 | 1.2 | 1.2 | 1.18 | 1 | 19 | 21 | 25 | 33 | 1 | 1.05 | 1.13 | 1.14 |

## 4. Conclusions

Results obtained at experimental tests show that testing method proposed gives us the possibility to characterize the packing materials, by finding pressure transforming coefficients $K_1$, $K_0$ and $K=min(K_1,K_0)$. Pressure transforming coefficients are necessary to solve sealing pressure equations (1), (2) and (3) and to establish the dimensions of sealing package depending of sealed pressure and also to establish the compression load/ torque necessary to obtain sealing in conditions of respecting fugitive emissions limits.

The results obtained for package of expanded graphite rings string and for package of preformed rings of expanded graphite tape presented in table 3 and in table 4 shows that these materials are even better than similar sealing materials recommended by major sealing manufactures (Garlock [7], Klinger et.a.).

It could be observed, also from table 3 and table 4, that axial pressure $p_x$ at contact between gland and ring 4, $p_{gland}$ is smaller than theoretical pressure $p_B$ due to package friction with stem and stuffing box surfaces during compression. From axial and radial pressures records could also be established the average friction coefficients, but at experimental tests friction are between package and pressure sensitive paper, so these friction coefficients are not important.

From table 3 and table 4 could be also observed that with increasing compression load the axial pressure $p_x$ for package of expanded graphite rings string and for package of preformed rings of expanded graphite tape became similar, and this means that could be used a smaller high of package of preformed rings of expanded graphite tape.

**Acknowledgments**
Authors would like to express their gratitude to company SC Etansari Grafex SRL for providing materials and devices used for experimental tests.

**References**

[1] European Sealing Association 2009 *Sealing Technology-BAT Guidance Notes, Guidance Notes to the Best Available Techniques for Sealing Technology Used in Equipment on Industrial Installations Covered by the EU IPPC Directive* ESA Publication No 014/09 rev.1 (Tregarth,UK) pp 24-45

[2] Fluid Sealing Association and European Sealing Association 2006 *Compresion Packing Technical Manual* 3rd ed. FSA and ESA (Wayne, USA) pp 37-83

[3] Riedl A 2007 Emission measurements of industrial valves acording to TA Luft and EN ISO 15848-1 *Valve World* June 51-55

[4] Falticeanu C, Ciortan S, Falticeanu C L, Manea M and Mircea I 2001 *Etansari cu contact mobil. Ghid pentru proiectarea etansarilor cu presgarnitura* (Galati, Romania: Ed. Evrika) pp 63-122

[5] Pengyun S, Kuangmin C and Zongyun D 1997 A theoretical analysis of the lateral pressure coefficients in a soft-packed stuffing-box seal *Tribol. Int.* **30** (10) 759-765

[6] Falticeanu C, Ciortan S and Falticeanu C L 2003 Modern research regarding the soft material of stuffing box packings- tribological behavior *Proc. of 8th Int. Tribology Conf.* ITC'03 (Belgrade, October 8-10) pp 300-304

[7] Garlock Sealing Technologies 2016 *Compression Packing Technical Manual* Garlock (Palmyra NY, USA) pp 1-34

# Influence of tooth profile on the noncircular gear tooth contact

**A Cristescu, L Andrei and B Cristescu**

Department of Mechanical Engineering, "Dunărea de Jos" University of Galaţi, Domneasca 47, Romania

E-mail: landrei@ugal.ro

**Abstract.** With noncircular gears, the continuous modification of the tooth meshing, in terms of variation of the tooth profiles and the line of action position and inclination, makes difficult the implementation of a general standard procedure for the analysis of the noncircular gears tooth contact. In this paper, the authors present a graphical approach that enables the tooth contact static pattern to be produced and evaluated in case of a noncircular gear with complex geometry of the pitch curve. The study is virtually developed, in AutoCAD environment, by animating and investigating the gear solid models in mesh. The tooth static contact analysis enables the path of contact area and distribution to be evaluated in correlation with the following variable initial data: gear pitch curve geometry, tooth profile geometry, as a consequence of different generating procedures, and the gear pressure angle. It was found out that the noncircular gear tooth contact could be improved by choosing different procedures for the tooth flank generation in concave and convex zones and by increasing the gear pressure angle.

## 1. Introduction

The design of the circular gears is controlled through specific standardized calculation. By comparison, the design of the noncircular gears is a complex process based on two main steps: (i) noncircular centrodes modelling and (ii) teeth generation. Considering the design input data, the approaches to noncircular centrodes modeling are mainly based on the transmission ratio variation [1 - 3], the law of motion for the driven element [4] or the driving centrode geometry [5, 6]. The teeth generation is developed from either analytical procedures that consider the particularities of gear line of action [7, 8], the traditional theory of the enveloping surfaces [9, 10], or from the simulation of the gear generation [11, 12]. Basically, the noncircular gears teeth generation methods presented by specialists are based on a constant pressure angle of a the standard value of 20°, but there are certain researches who, in order to improve the meshing conditions, consider different values of the pressure angle: either for all the teeth [11] or from one tooth to another [13, 14].

While the circular gears meshing process is well known and controlled, due to the variable and complex geometry of the noncircular gears, their meshing exhibits specific behavior which is difficult to predict. Unlike the circular gears meshing as approached by standardized procedures and calculations, the teeth contact analysis for the noncircular gears is developed through experimental methods based both on specific elements of the standard gear geometry and on virtually developed models of the noncircular gears [15 - 17].

The authors of this paper have developed a graphical procedure to analyze the static noncircular gear contact by taking into account the characteristics of the noncircular gears meshing process, i.e. the variable position of the instantaneous contact point of the mating centrodes along the gears center line and the variable inclination of the line of action that is related, by a constant pressure angle, to the common tangent of the conjugate noncircular centrodes. All the required rolling parameters are calculated by Hypertext Preprocessor (PHP) codes and imported into the AutoCAD environment. The centrodes geometry is automatically investigated and the teeth placed on convex and concave zones, with minimum and maximum curvature, are picked up. An analytical procedure for the gear teeth profiles generation will provide the best accuracy for all the geometrical and dimensional tooth parameters required by further investigations on gears performance. The virtual gears are built up combining PHP calculus and computer graphics operations specific to the AutoCAD. Finally, the virtual solid gears models are manipulated to simulate the gears meshing and to produce the teeth contact patterns. The distribution and area of the contact pattern are analyzed in correlation with the variable gears design data such as tooth profile geometry provided by the generation process and the gear pressure angle.

## 2. Noncircular gear teeth generation

In order to generate noncircular gear teeth, the specific kinematics of the rolling of the rack-cutter over the noncircular gear blank is considered. To enable generation of noncircular gears with complex geometries, that exhibit also concave zones, without any undercutting, a single tooth rack-cutter of standard geometry is considered. Each flank generating process is locally analyzed in the proximity of the intersection point $E_i$ between the current tooth flank and the gear noncircular pitch curve (figures 1, 2). The analytical procedure of the tooth flank generation process consists in expressing the intersection point between the rack cutter tooth flank and the current line of action during the rolling motion and is presented in detail in previous publications [17, 18]. As the main objective of the paper is to analyze the influence of the noncircular gear tooth geometry on the tooth contact pattern, variations in tooth geometry are achieved by taken into account the following hypotheses: (I) the tooth flank generation is developed by the pure rolling of the rack cutter on both the noncircular pitch curve and on the local osculating circle; (II) the tooth flank generation is developed by the pure rolling of the rack cutter on the noncircular gear pitch curve, considering the pressure angle at different values.

### 2.1. *Using rolling on the noncircular pitch curve*

The first hypothesis of generating the noncircular gear tooth flanks considers the rolling motion of the rack cutter tooth on the correct gear pitch curve. Figure 1 illustrates the geometry and the dimensional parameters taken into account while generating the tooth operating flank addendum, showing:

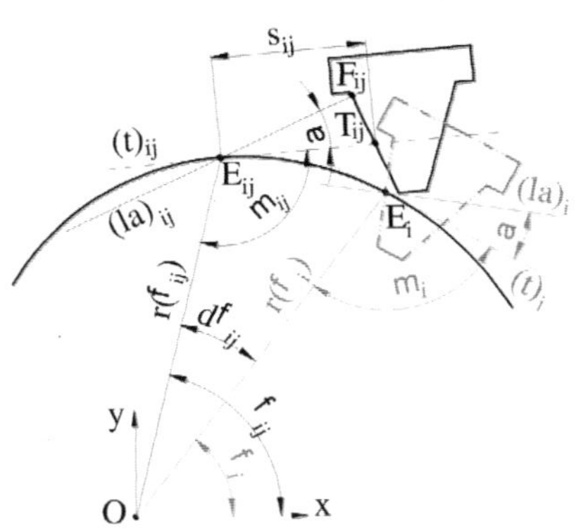


**Figure 1.** Geometric and kinematics of tooth flank generation using rolling on the noncircular pitch curve.

(i) the starting position of the rack cutter tooth and its orientation specific to point $E_i$. The rack cutter pitch line is aligned to the local tangent $(t)_i$ that is inclined by $\mu_i$ angle, relative to the current radius $r(\varphi_i)$. The first point of the tooth flank profile is $F_i \equiv E_i$;

(ii) an arbitrary position of the rack cutter tooth, after the rolling motion defined by the rotational angle $d\varphi_{ij}$ and the displacement $s_{ij}$ equal to the length of the noncircular arc $E_iE_{ij}$; the rack cutter is firstly rotated about $E_{ij}$, so that its pitch line is aligned to the current tangent $(t)_{ij}$ to the noncircular centrode, and is translated by the rolling distance. The generated point $F_{ij}$ of the tooth flank is picked up at the intersection between the rack cutter tooth operating flank and the current line of action $(la)_{ij}$ that is inclined by the constant pressure angle $\alpha$ with respect to $(t)_{ij}$ line. Analyzing the generating process geometry and kinematics from figure 1, the tooth operating flank for the addendum zone is described by the equations:

$$x_{ij} = r(\varphi_{ij}) \cdot \cos\varphi_{ij} - s_{ij} \cdot \cos\alpha \cdot \cos(\mu_{ij} + \alpha + \varphi_{ij}),$$
$$y_{ij} = r(\varphi_{ij}) \cdot \sin\varphi_{ij} - s_{ij} \cdot \cos\alpha \cdot \sin(\mu_{ij} + \alpha + \varphi_{ij}) \tag{1}$$

where $\varphi_{ij}$ and $r(\varphi_{ij})$ are the polar coordinates of the instantaneous center of rotation $E_{ij}$, $s_{ij}$ - the rack displacement during the rolling motion, $\mu_{ij}$ - the inclination angle of the tangent $(t)_{ij}$ with respect to the current radius, $\alpha$ - the constant pressure angle.

Similar equations are written for the tooth dedendum profile of the operating flank and for the non-operating flanks, considering the proper rolling and rack cutter flank [17, 18].

### 2.2. *Using rolling on the local osculating circle*

Analyzing the generating process geometry and kinematics from figure 2, when the rolling motion is performed relative to the local osculating circle, specific to point $E_i$, the tooth operating flank for the addendum zone is described by equations:

$$x_{ij} = r(\varphi_i) \cdot \cos\varphi_i - s'_{ij} \cdot \cos\alpha \cdot \cos(\mu_i + \alpha + \varphi_i),$$
$$y_{ij} = r(\varphi_i) \cdot \sin\varphi_i - s'_{ij} \cdot \cos\alpha \cdot \sin(\mu_i + \alpha + \varphi_i) \tag{2}$$

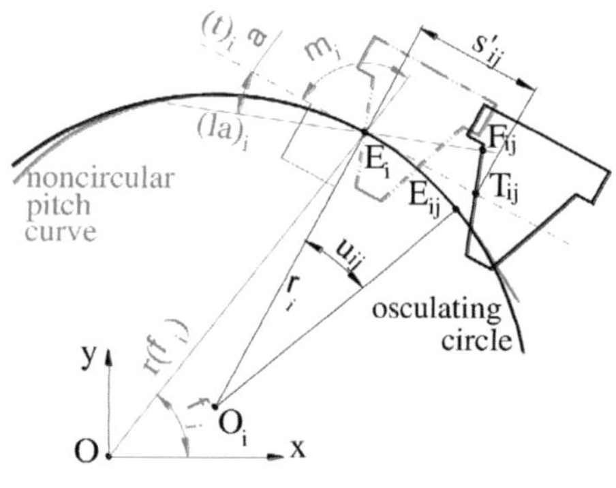


**Figure 2.** Geometric and kinematics of tooth flank generation using rolling on the local osculating circle.

where $\varphi_i$ and $r(\varphi_i)$ are the polar coordinates of the instantaneous center of rotation $E_i$, $s'_{ij}$ - the rack displacement along the tangent $(t)_i$, $\mu_i$ - the inclination angle of the rack cutter pitch line with respect to the radius $r(\varphi_i)$. The rack cutter displacement, $s'_{ij}$, is calculated by:

$$s'_{ij} = \rho_i \cdot u_{ij} \tag{3}$$

where $\rho_i$ is the local osculating circle radius, $u_{ij}$ – the incremental rotational angle of the rolling process relative to the osculating circle.

The tooth dedendum profile of the operating flank and the non-operating flank are described by similar equations [18].

For a pair of noncircular gears defined by the transmission ratio $m_{21}(\varphi)$ and the gear center distance $D$, considering the gears meshing and the coordinate transformation operations, the tooth flanks for the mating noncircular gears are also analitically described [17]. The original PHP codes enable the teeth flanks profiles generation for the noncircular gears pairs. To complete the gear representations, the numerical data is imported to AutoCAD environment and editing operations are applied.

## 3. Tooth contact analysis

The gears static tooth contact is computed based on a simple geometric approach that are using the noncircular gears solid models and the AutoCAD facilities, as presented in other previous publications [19, 20]. The tooth contact pattern is considered as the composed solid that results by intersecting the main and adjacent pairs of meshing teeth whose elastic deformations were not taken into account. The static analysis refers to the gears unload simulating meshing, although an initial extremely reduced torque and rotation, respectively, should be applied to the pinion in order to enable the solids intersection. The paper aim is just to compare the teeth meshing for the above two gear teeth geometries induced by the generation method; a comparison with standard gears meshing is difficult to be made while the tooth flank geometry differs to one tooth to another, there is a different module that defines the tooth geometry along the noncircular gear pitch curves etc., but the authors would consider it for further researches.

Investigations on the noncircular gears static contact is particularly developed for a gears train that is defined by the following design data: gear center distance $D$ = 200 mm, gears number of teeth $z_1 = z_2 = 36$, gear facewidth is $B$ = 20 mm, pressure angle $\alpha \in \{15°, 20°, 25°\}$, tooth position chosen so that a constant angular pitch is met ($\varphi_i = i \cdot 2\pi/z_1$, where $i = 0...(z_1 - 1)$) and the gear transmission ratio is defined by:

$$m_{21}(\varphi) = 1 + \frac{1}{4} \cdot \cos\varphi + \frac{1}{3} \cdot \sin 3\varphi \tag{4}$$

The gear transmission ratio is chosen so as complex pitch curve geometries are achieved, that exhibit convex, concave and almost linear zones. Two pairs of teeth chosen for the analysis (figure 3) are placed in the convex zone, with the highest radius of curvature (tooth $D_{33}$) and in the concave zone, with the smallest radius of curvature (tooth $D_{22}$).

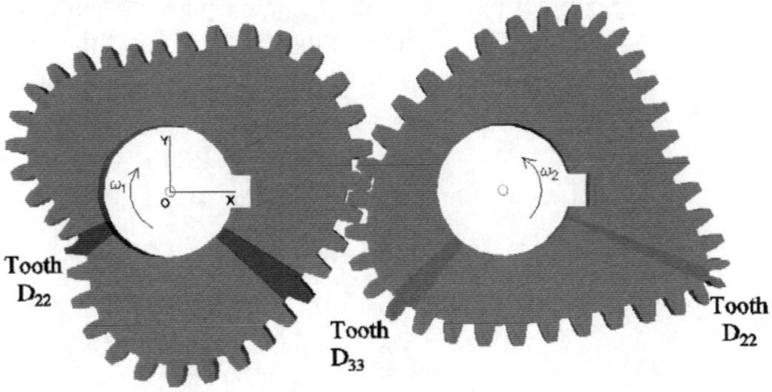


**Figure 3.** Virtual solid models of noncircular gears.

The algorithm proposed for the tooth contact pattern investigation consists of (figure 4):

*Step1:* The gear teeth being analyzed are selected and identified by positioning and geometrical parameters. The main tooth and other four teeth are taken into account (figure 4a);

*Step2:* The tooth being analyzed is placed on the center line by a proper rotational angle of pinion ($\varphi_1$) and driven gear ($\varphi_2$), respectively. To produce the contact pattern, the static meshing is considered and a minimum torque is applied, which is further reflected by a controlled interference induced by an initial pinion rotational angle of 0.005° (figure 4b);

*Step3:* Incremental rolling by $\Delta\varphi_1 = 1^{\circ}$ is applied in both counter clock and clockwise in order to highlight the tooth contact during all the meshing of the conjugated teeth (figure 4c);

*Step4:* The solids are intersected to obtain the static path of contact for the controlled interference. The tooth contact pattern distribution is illustrated on the driving gear teeth (figure 4d) and its area is measured along the main tooth and on the other teeth engaged in the mesh.

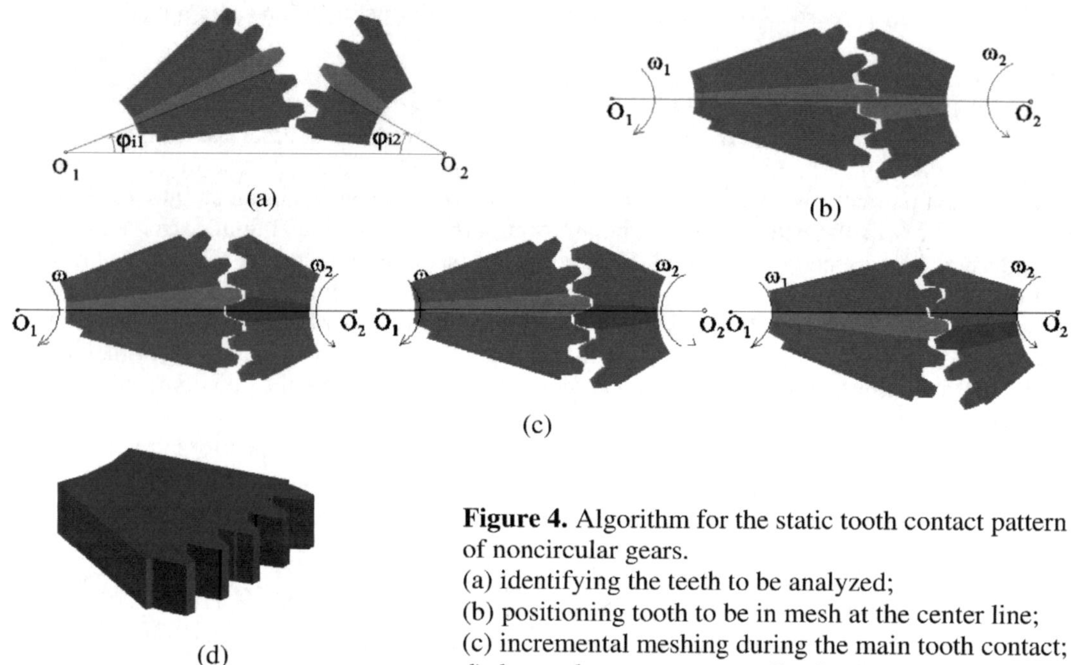


**Figure 4.** Algorithm for the static tooth contact pattern of noncircular gears.
(a) identifying the teeth to be analyzed;
(b) positioning tooth to be in mesh at the center line;
(c) incremental meshing during the main tooth contact;
d) the tooth contact pattern distribution.

*3.1. Influence of tooth generation process on the static tooth contact pattern*
Figure 5 illustrates details of the selected teeth geometry of the noncircular driving gear generated by the rack cutter rolling on the noncircular pitch curve (named as AnCG20) and on the local osculating circle (AnAG20), respectively. The study of the static contact pattern of the selected pairs of teeth, for both noncircular gears trains, provides data regarding the number of teeth in mesh, the area of the path of contact on the main analyzed tooth (APT), the value of the total contact area (APG) and the value of the gear rotational angle while the analyzed tooth is in mesh.

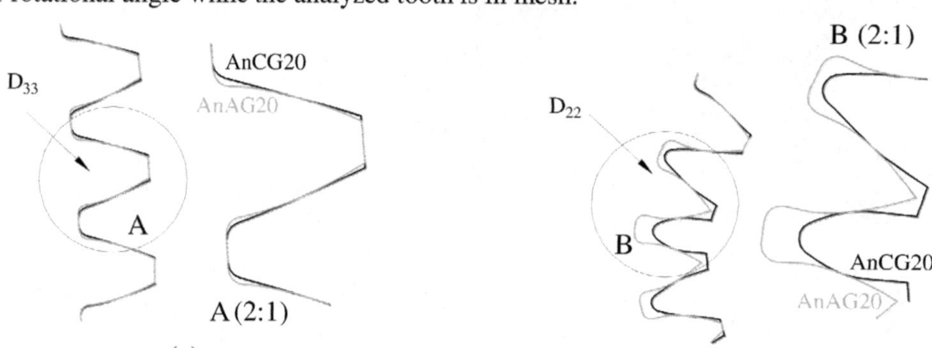


**Figure 5.** Details of noncircular driving gear generated by accurate (AnCG20) and approximate (AnAG20) generation process on convex zone, with the highest curvature radius (a) and on concave zone, with the smallest curvature radius (b).

Table 1 and figure 6 present a selection of data recorded while the specified teeth are in mesh. The global area of contact (APG), presented in table 1, is the sum of the total areas of the solids that result by intersecting the main teeth and the adjacent pairs of meshing teeth; the tooth contact area (APT), shown in figure 6, is the total area of the solid that results by intersecting the main teeth ($D_{33}$, $D_{22}$, respectively). It is noticed that for both gear trains, the teeth $D_{33}$ and $D_{22}$ are in mesh while the driving gear is rotated by 15° and 14°, respectively. Comparing the total area of contact (APG) in the selected convex zone, decreases in the area are recorded in case of ANAG20, by 25% while tooth D 33 gets into mesh, by 59% when contact occurs on the gears center line and by 15% when tooth gets out of mesh. This could be explained by the higher curvature of the correct tooth flank due to the rolling on the noncircular pitch curve arc whose curvature decreases in the vicinity of teeth $D_{33}$, while the approximate tooth flank is an involute of a circular arc with constant high radius, almost a rectilinear segment. The analysis of the tooth contact pattern in the vicinity of tooth $D_{22}$ shows a slightly better behavior for the concave zones of the AnAG20 gear compared to the accurately generated gears, the total contact area being increased by 9% when tooth $D_{22}$ gets into mesh, by 8.2% at center line and by 9.6% when tooth gets out of the mesh. In particular, the comparison of the $D_{22}$ tooth contact pattern area is illustrated in figure 6.

**Table 1.** Contact parameters for noncircular gears generated by accurate and approximate rolling.

| Tooth | | $D_{33}$ | $D_{33}$ | $D_{33}$ | $D_{22}$ | $D_{22}$ | $D_{22}$ |
|---|---|---|---|---|---|---|---|
| Tooth position | | Gets into mesh | On center line | Out of mesh | Gets into mesh | On center line | Out of mesh |
| AnCG20 | No.of teeth | 3 | 3 | 3 | 2 | 3 | 3 |
| | APG [mm$^2$] | 393.494 | 435.432 | 212.065 | 99.644 | 351.419 | 184.228 |
| AnAG20 | No.of teeth | 2 | 3 | 3 | 2 | 3 | 3 |
| | APG [mm$^2$] | 295.183 | 179.988 | 179.079 | 108.815 | 380.354 | 201.912 |

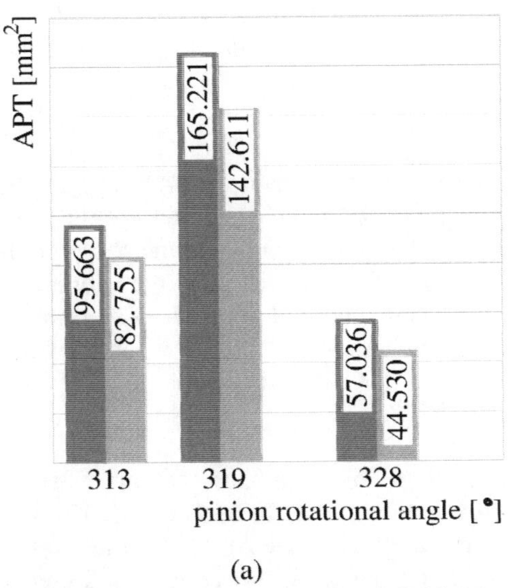


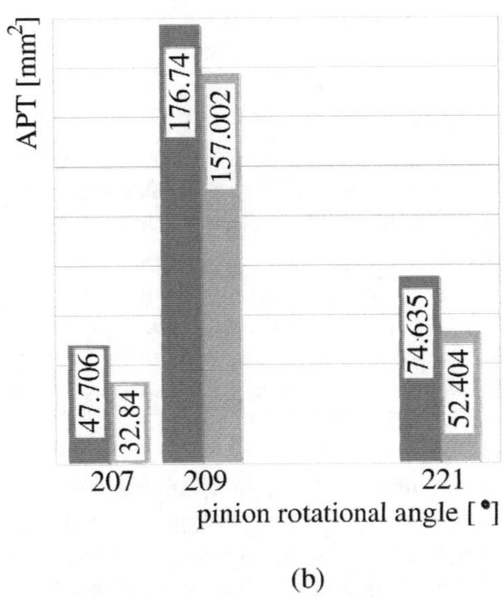


**Figure 6.** Contact area on teeth D33 (a) and $D_{22}$ (b), for the AnCG20 and AnAG20 gear trains.

*3.2. Influence of pressure angle on the static tooth contact pattern*

To analyze the influence of the pressure angle on the tooth contact pattern, three gear trains are compared, generated by using an accurate rolling process, making use of the above mentioned gears design data and three values of the pressure angle, resulting the AnCG15 gear pair for $\alpha$ = 15°, the AnCG20 gear pair for $\alpha$ = 20° and the AnCG25 gear pair for $\alpha$ = 25°, respectively (figure 7).

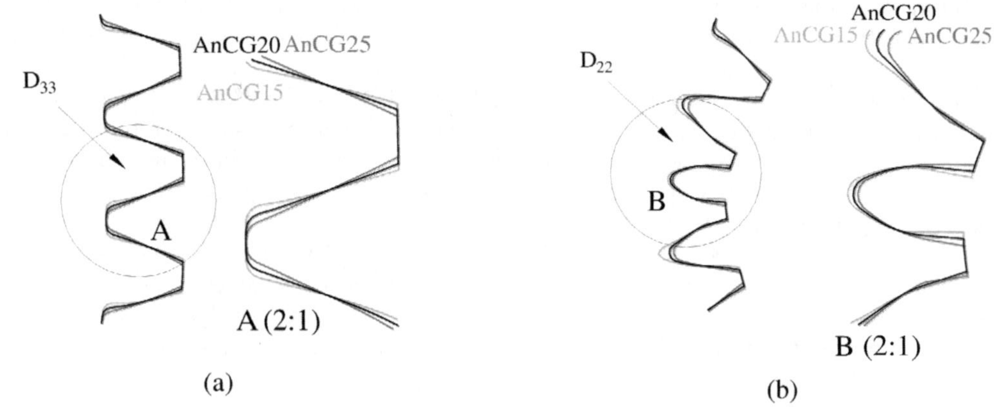


**Figure 7.** Details of noncircular driving gear generated using $\alpha$ = 15° (AnCG15), $\alpha$ = 20° (AnCG20) and $\alpha$ = 25° (AnCG25) for convex zone, with the highest curvature radius (a) and for concave zone, with the smallest curvature radius (b).

**Table 2.** Contact parameters for noncircular gears designed with variable pressure angle.

| Tooth | | $D_{33}$ | $D_{33}$ | $D_{33}$ | $D_{22}$ | $D_{22}$ | $D_{22}$ |
|---|---|---|---|---|---|---|---|
| Tooth position | | Gets into mesh | On center line | Out of mesh | Gets into mesh | On center line | Out of mesh |
| AnCG20 | No.of teeth | 3 | 3 | 3 | 2 | 3 | 3 |
| | APG [mm$^2$] | 393.494 | 435.432 | 212.065 | 99.644 | 351.419 | 184.228 |
| AnCG15 | No.of teeth | 3 | 3 | 2 | 2 | 3 | 3 |
| | APG [mm$^2$] | 312.494 | 402.714 | 198.451 | 99.644 | 351.419 | 184.228 |
| AnCG25 | No.of teeth | 3 | 3 | 2 | 2 | 3 | 3 |
| | APG [mm$^2$] | 398.205 | 448.032 | 230.119 | 118.004 | 398.212 | 231.509 |

The pressure angle influence on the static contact conditions is determined by analyzing the data from table 2 and figure 8. It appears that for both the convex and concave areas, greater pressure angle values determine improved meshing conditions by increasing the contact area for the tooth contact pattern. Thus, for tooth $D_{33}$, the average area is with 9.79% higher for AnCG20 than the average area for gear trainAnCG15 and with 10.26% lower than the values recorded for AnCG25 (table 2). Similarly, for tooth $D_{22}$, the growth in the average area of the contact path is 14.03% when comparing the AnCG20 and AnCG15 gear trains and 12.60% for AnCG25 to AnCG15 (figure 8).

## 4. Conclusions

Conditions of noncircular gear trains meshing are studied in terms of tooth static contact pattern. A virtual graphical method is developed in order to produce the tooth contact and to enable analysis of the influence of the gear generation method and pressure angle, respectively, on the meshing characteristics. The analysis is focused on the contact of the teeth situated in convex and concave zones of the noncircular gears, with extreme values of the gear pitch curve radius of curvature.

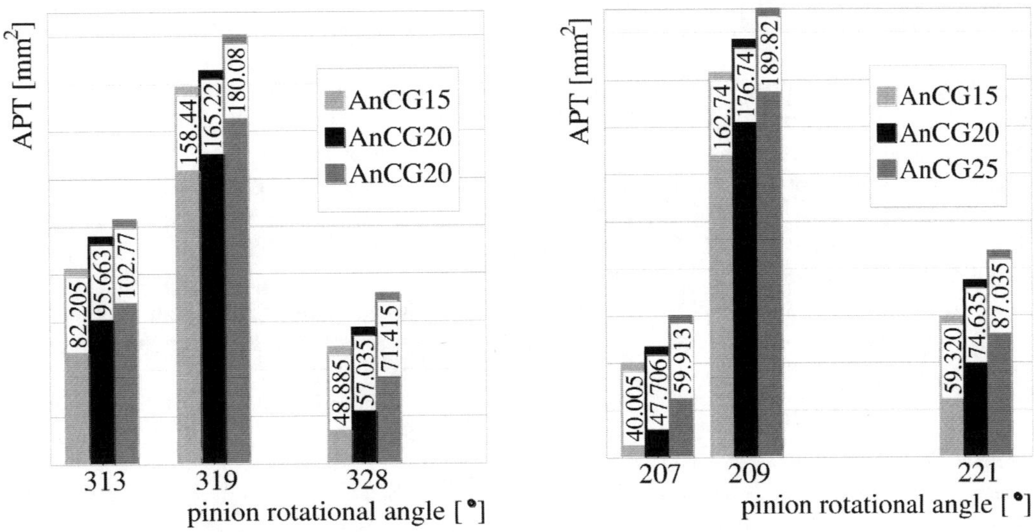


**Figure 8.** Contact area on teetth $D_{33}$ (a) and $D_{22}$ (b) measured on the AnCG15, AnCG20 and AnCG25 gear trains.

Original PHP codes provide neccessary data base to generate the noncircular mating centrodes and the teeth flanks profiles that are further imported in AutoCAD environment where the virtual solid models of the gear trains are produced and incrementally got into the mesh. The teeth contact pattern size and distribution are investigated and compared for noncircular gear trains with variable tooth flank geometries, generated while the rach cutter is rolling on the correct noncircular pitch curve and on the local osculating circle, approximating the tooth profile by an involute. It appeared that gears accurately generated, by performing the rolling process over the noncircular pitch curve, led to better meshing characteristics in case of convex zones, compared to gears generated by rolling over local osculating circle, the contact area of which was improved into concave zones, with high curvature. Modifying the gear tooth geometry by varying the gear pressure angle, it was noticed that high pressure angle provided, for both the convex and concave zones, improved meshing conditions.

**Acknowledgements**
The work has been funded by the Sectorial Operational Programme Human Resources Development 2007-2013 of the Ministry of European Funds through the Financial Agreement POSDRU/159/1.5/S/132397.

**References**

[1] Tsay M F and Fong Z H 2005 Study on the generalized mathematical model of noncircular gears *Math. Comp. Model.* **41** 555–569
[2] Mundo D 2006 Geometric design of a planetary gear train with non-circular gears *Mech. Mach. Theory* **41** 456–472
[3] Cristescu A, Cristescu B and Andrei L 2014 Generalization of Multispeed Gear Pitch Curves Design *Appl. Mech. Mater.* **659** 559–564
[4] Laczik B 2007 Design and Manufacturing of Non-Circular Gears by Given Transfer Function *Proc. ICT* 101–109
[5] Bair B W, Chen C F, Chen S F and Chou C Y 2007 Mathematical Model and Characteristic Analysis of Elliptical Gears Manufactured by Circular-Arc Shaper Cutters *J. Mech. Design* **129** 210-214

[6] Vasie M and Andrei L 2012 Design and generation of noncircular gears with convex-concave pitch curves *The Annals of "Dunarea Jos" University of Galati, Mathematics, Physics, Theoretical Mechanics* 55–60
[7] Danieli A, Mundo D 2005 New developments in variable radius gears using constant pressure angle teeth *Mech.Mach Theory* **40** 203-217
[8] Gao S, Igarashi S, Takayama F, Ohomote R, Tang X, Chen X and Li J 2006 Design and analysis of shapes of elliptic gears *12th Int. Conf. on Geometry and Graphics*
[9] Litvin F L 2009 *Noncircular Gears Design and Generation* (Cambridge University Press)
[10] Figliolini G and Rea P 2014 Effects of the design parameters of involute gears generated by rack-cutters *International Gear Conference 2014: 26th–28th August 2014, Lyon* pp 294-302
[11] Li J, Wu X and Mao S 2007 Numerical computing method of noncircular gear tooth profiles generated by shaper cutters *The International Journal of Advanced Manufacturing Technologies* **33** 1098–1105
[12] Vasie M and Andrei L 2011 Noncircular gear design and generation by rack cutter *The Annals of "Dunarea Jos" University of Galati, Mathematics, Physics, Theoretical Mechanics* **1** 81–86
[13] Kitano A 1977 Elliptic gear wheel *United States Patent 4036073A*
[14] Danieli G A 2007 Toward a greater industrial application of variable radius gearing *12th IFToMM World Congress*
[15] Khan S 2015 Simulation and analysis of transmission error in helical non circular gear model *Int. International Journal of Mechanical Engineering and Technology* **6** 128–136
[16] Barkah D, Shafiq B and Dooner D 2000 3D Mesh generation for static stress determination in spiral noncircular gears used for torque balancing *J. Mech. Design* **124** 313-321
[17] Cristescu B, Cristescu A and Andrei L 2014 Analytical generation of involute flanks of noncircular gear tooth *The Annals of "Dunărea de Jos" University of Galati, Mathematics, Physics, Theoretical Mechanics* **1** 36-43
[18] Cristescu A, Cristescu B and Andrei L 2015 Finite element analysis of multispeed noncircular gears *Appl. Mech. Mater* **808** 244-251
[19] Andrei L, Andrei G, Walton D and Dean K 2008 Analysis of gear quality criteria and performance of curved face width spur gears *The Annals of "Dunărea de Jos" University of Galați*, Fascicle VIII Tribology 30–34
[20] Vasie M and Andrei L 2012 Analysis of noncircular gear meshing *Mechanical Testing and Diagnosis* **4** 70-78

# Dynamic analysis of six-bar mechanical press for deep drawing

**S Mitsi, I Tsiafis and K D Bouzakis**

Laboratory for Machine Tools and Manufacturing Engineering, Mechanical Engineering Department, Aristotle University, University Campus, 54124 Thessaloniki, Greece

E-mail: mitsi@eng.auth.gr

**Abstract.** This paper analyzes the dynamical behavior of a six-bar linkage used in mechanical presses for metal forming such as deep drawing. In the under study mechanism, a four-bar linkage is connected to a slider through an articulated binary link. The motion of the six-bar linkage is studied by kinematic analysis developing an analytical method. Furthermore, using an iterative method and d' Alembert's principle, the joint forces and drive moment are evaluated considering joint frictions. The simulation results obtained with a MATLAB program are validated by comparing the theoretical values of the input moment with the ones obtained from the conservation of energy law.

## 1. Introduction

The mechanical press is widely employed for metal-forming processes. Many different types of linkage drives have been proposed to satisfy the load – stroke characteristics for the specific metal-forming process. The metal forming operations like shearing requires very short stroke of the ram and deep drawing requires a slow and long stroke of the punch. Yan and Chen [1, 2] proposed the novel approach of varying the input speed such that the ram's motion was suitable for both deep-drawing and precision-cutting processes. Hsieh and Tsai [3] developed a novel press system with six links for precision deep drawing so that can generate complicated and flexible output motion. Soong [4] proposed a new design method for single DOF mechanical press with variable speeds and length-adjustable driving links. Yossifon and Shivpuri [5] investigated the design, analysis and construction of a double knuckle press for precision forming driven by a servo motor. The dynamic analysis of the double-toggle press, performed in [6], included the evaluation of bearing forces and the required input torque without considering joint friction. Tso and Liang [7] proposed a nine-bar linkage for mechanical presses in order to obtain slow forming speed and a long period of dwell time in the bottom of the ram stroke. The dynamic analysis and simulation of a hybrid-driven two-degrees-of-freedom seven-bar press using Lagrangian's formulation have been developed by Li and Zhang [8]. The force analysis of a six-linkage used in mechanical presses for metal forming such as coining and blanking, considering joint frictions is presented in [9].

In the present paper the dynamic analysis of a six-bar linkage for mechanical press used for the deep drawing, is investigated. First, the kinematic analysis, dealing with the displacement, velocity and acceleration of the mechanism links, is performed. Then, the force analysis, taking into account the external force and joint frictions, is presented.

## 2. Kinematic analysis of six-bar linkage

The investigated mechanical press contains a six-bar linkage (figure 1). An electrical motor provides the input torque to the crank through a gears transmission. The six-bar linkage is composed of the $A_0ABB_0$ four-bar linkage connected to the slider through the articulated binary link CD. Structural, this mechanism includes the frame 0, the input link 1 and the second-class Assur group RRR (links 2 and 3) connected to the second-class Assur group RRP (links 4 and 5) through the revolute joint C, where R denotes a revolute pair and P a prismatic pair. The joints $A_0$ and $B_0$ are fixed and the input link 1 rotates with a constant known angular velocity $\omega_{10}$. Also the link lengths $l_0=A_0B_0$, $l_1=A_0A$, $l_2=AB$, $l_{AC}=AC$, $l_3=B_0B$, $l_4=CD$ and the distance $a$ are given.

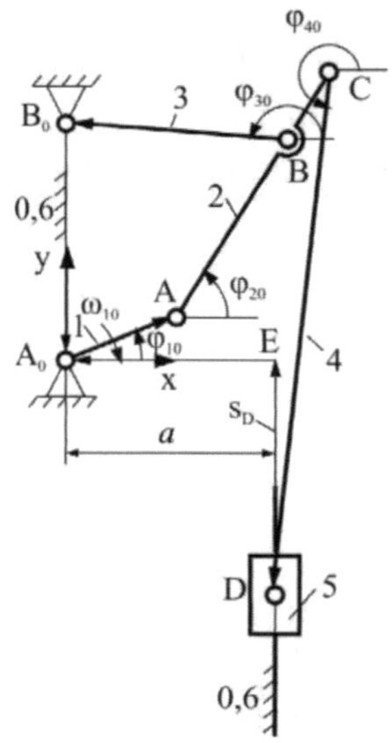


**Figure 1.** Kinematic model of six-bar mechanical press.

The equations for kinematic analysis are written and solved considering a Cartesian coordinate system, with origin the joint $A_0$. The mechanism has two independent closed contours $A_0ABB_0A_0$ and $A_0ACDEA_0$. Every link length and slider displacement (from a convenient reference point E) are represented by suitable planar vectors. The vector equation corresponding to each above independent closed loop is:

For closed contour $A_0ABB_0A_0$:

$$\boldsymbol{l_1} + \boldsymbol{l_2} + \boldsymbol{l_3} + \boldsymbol{l_0} = \boldsymbol{0} \tag{1}$$

and for closed contour $A_0ACDEA_0$:

$$\boldsymbol{l_1} + \boldsymbol{l_2} + \boldsymbol{l_4} + \boldsymbol{s_D} + \boldsymbol{a} = \boldsymbol{0} \tag{2}$$

The vector equations (1) and (2) are projected on the Cartesian reference system. It is obtained:

$$l_1 cos\varphi_{10} + l_2 cos\varphi_{20} + l_3 cos\varphi_{30} = 0 \tag{3}$$

$$l_1 sin\varphi_{10} + l_2 sin\varphi_{20} + l_3 sin\varphi_{30} - l_0 = 0 \tag{4}$$

$$l_1 cos\varphi_{10} + l_{AC} cos\varphi_{20} + l_4 cos\varphi_{40} - a = 0 \tag{5}$$

$$l_1 sin\varphi_{10} + l_{AC} sin\varphi_{20} + l_4 sin\varphi_{40} + s_D = 0 \tag{6}$$

Equations (3) and (4) constitute a set of two nonlinear equations in two unknowns, namely $\varphi_{20}$ and $\varphi_{30}$. To determine the angle $\varphi_{20}$, the equations (3) and (4) are rearranged as:

$$l_3 cos\varphi_{30} = -l_1 cos\varphi_{10} - l_2 cos\varphi_{20} \tag{7}$$

$$l_3 sin\varphi_{30} = -l_1 sin\varphi_{10} - l_2 sin\varphi_{20} + l_0 \tag{8}$$

Squaring the equations (7) and (8) and adding them, the following equation is obtained:

$$A\, cos\varphi_{20} + B\, sin\varphi_{20} = C \tag{9}$$

where:

$$A = -2l_1 l_2 cos\varphi_{10}$$

$$B = 2\left(l_0 - l_1 sin\varphi_{10}\right) l_2$$

$$C = l_2^2 - l_3^2 + l_1^2 cos^2\varphi_{10} + \left(l_0 - l_1 sin\varphi_{10}\right)^2 \tag{10}$$

Equation (9) has the solutions [9]:

$$\varphi_{20} = atan(C/(\pm\left(A^2 + B^2 - C^2\right)^{1/2}) - atan\left(A/B\right) \tag{11}$$

The double sign "±" in equation (11) corresponds to the two different assembly modes of RRR Assur group.

The angle $\varphi_{30}$ is obtained using a similar procedure eliminating the angle $\varphi_{20}$ from equations (3) and (4).

For each value of the angle $\varphi_{20}$, using equation (5), the angle $\varphi_{40}$ is determined [9]:

$$\varphi_{40} = atan(\pm\left(1 - D^2\right)^{1/2} / D) \tag{12}$$

with $D = \left(a - l_1 cos\varphi_{10} - l_{AC} cos\varphi_{20}\right) / l_4$. The double sign "±" in equation (12) corresponds to the two different assembly modes of RRP Assur group.

Finally, once $\varphi_{20}$ and $\varphi_{40}$ are known, from equation (6) the slider displacement $s_D$ is obtained:

$$s_D = -l_1 sin\varphi_{10} - l_{AC} sin\varphi_{20} - l_4 sin\varphi_{40} \tag{13}$$

It has been mentioned that from the four configurations of the six-bar linkage, the one that corresponds to the working specifications of the mechanical press is chosen.

The results of displacement analysis are used for velocity analysis. Differentiating equations (3) and (4) with respect to time and after rearranging terms it is obtained a set of two linear equations in the unknown angular velocities $\omega_{20}$ and $\omega_{30}$. By solving this system the unknown velocities are determined as [9]:

$$\omega_{20} = \left(\omega_{10} l_1 sin\left(\varphi_{10} - \varphi_{30}\right)\right) / \left(l_2 sin\left(\varphi_{30} - \varphi_{20}\right)\right) \tag{14}$$

$$\omega_{30} = \left(\omega_{10} l_1 sin\left(\varphi_{20} - \varphi_{10}\right)\right) / \left(l_3 sin\left(\varphi_{30} - \varphi_{20}\right)\right) \tag{15}$$

where $\omega_{i0} = d\varphi_{i0} / dt \; \left(i = 1,\ 2,\ 3\right)$.

Differentiating equation (5) with respect to time, a linear equation in unknown angular velocity $\omega_{40}$ is obtained and the solution is:

$$\omega_{40} = \left(-l_1 \omega_{10} sin\varphi_{10} - l_{AC} \omega_{20} sin\varphi_{20}\right) / \left(l_4 sin\varphi_{40}\right) \tag{16}$$

The slider velocity $v_D$ is determined by solving the linear equation obtained by differentiating equation (6) with respect to time:

$$v_D = -l_1 \omega_{10} cos\varphi_{10} - l_{AC} \omega_{20} cos\varphi_{20} - l_4 \omega_{40} cos\varphi_{40} \tag{17}$$

Acceleration analysis is performed by differentiating the equations derived in velocity analysis with respect to time. From the obtained equations are determined the angular accelerations $\varepsilon_{20}$, $\varepsilon_{30}$, $\varepsilon_{40}$ and the slider acceleration $a_D$:

$$\varepsilon_{20} = \left(A_1 cos\varphi_{30} + B_1 sin\varphi_{30}\right) / \left(l_2 sin\left(\varphi_{30} - \varphi_{20}\right)\right) \tag{18}$$

$$\varepsilon_{30} = -\left(A_1 cos\varphi_{20} + B_1 sin\varphi_{20}\right) / \left(l_3 sin\left(\varphi_{30} - \varphi_{20}\right)\right) \tag{19}$$

$$\varepsilon_{40} = \left(a_{xC} - l_4 \omega_{40}^2 cos\varphi_{40}\right) / \left(l_4 sin\varphi_{40}\right) \tag{20}$$

$$a_D = -a_{yC} - l_4\varepsilon_{40}cos\varphi_{40} + l_4\omega_{40}^2 sin\varphi_{40} \quad (21)$$

where:

$$A_1 = l_1\omega_{10}^2 cos\varphi_{10} + l_2\omega_{20}^2 cos\varphi_{20} + l_3\omega_{30}^2 cos\varphi_{30}$$
$$B_1 = l_1\omega_{10}^2 sin\varphi_{10} + l_2\omega_{20}^2 sin\varphi_{20} + l_3\omega_{30}^2 sin\varphi_{30}$$
$$a_{xC} = -l_1\omega_{10}^2 cos\varphi_{10} - l_{AC}\varepsilon_{20} sin\varphi_{20} - l_{AC}\omega_{20}^2 cos\varphi_{20}$$
$$a_{yC} = -l_1\omega_{10}^2 sin\varphi_{10} + l_{AC}\varepsilon_{20} cos\varphi_{20} - l_{AC}\omega_{20}^2 sin\varphi_{20}$$

The results of kinematic analysis are furthermore used to calculate the accelerations of the link mass centers needed to perform the kinetostatic analysis of the mechanism.

## 3. Forces analysis

The kinetostatic force analysis takes into account the inertia forces and moments imposed on the links and the joint friction forces considering known the ram force $F_{ext}$. The ram external force is opposed to the motion of the link 5. The friction losses are due to the sliding friction at the contact area between the ram 5 and the frame and the friction moment in revolute joints. In the calculation of friction forces and moments, the joint reaction forces, the characteristic property of the joint materials in contact and the relative velocities are considered.

The determination of the joint reaction forces and the input moment is performed using an iterative procedure [9]. After kinematic analysis, the joint reaction forces and the input moment are calculated without friction. At the solution step m+1 of the iterative procedure, friction forces and moments are assumed to be external loads that have been defined during the determination of the joint reaction forces at the previous iteration step m. The iterative procedure converges when the difference between two successive values of the obtained joint reaction forces is smaller than a prescribed criterion ε that indicates the desired accuracy of the calculation. The solution converges rapidly due to the low values of the friction forces and moments.

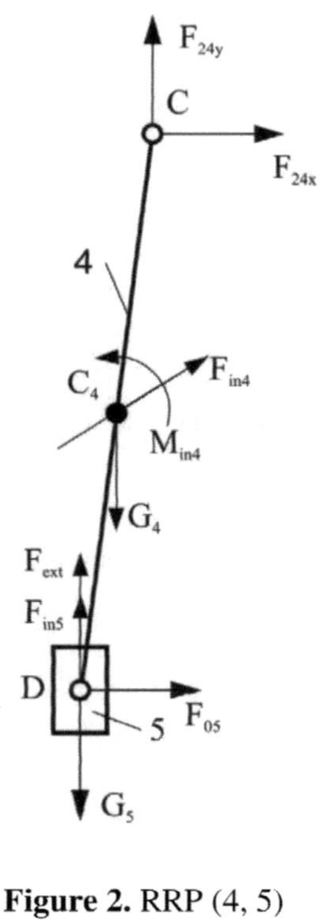


**Figure 2.** RRP (4, 5) dyad forces.

Due to decupled structure of the mechanism, the equations for the force analysis are written and solved separately for each Assur group (dyad) and input link 1 considering D' Alembert principle. In each link k, the inertia force $F_{ink}$ (k=1,n, n=links number) is assumed to act at the mass center $C_k$ and an inertia moment $M_{ink}$ is added.

The dynamic analysis starts with the RRP (4, 5) dyad because the external force $\mathbf{F_{ext}}$ is known. Figure 2 shows the forces and the moments that act on the RRP (4, 5) dyad. The unknown joint reaction forces are $\mathbf{F_{45}} = -\mathbf{F_{54}}$, $\mathbf{F_{05}}$ and $\mathbf{F_{24}}$. The joint reaction $\mathbf{F_{05}}$ is perpendicular to the sliding direction of the ram 5. To calculate the reactions forces, a force equation for links 4 and 5 and a moment equation on link 4 are written.

The vector sum of all the forces that act on the links 4 and 5 is zero:

$$(4,5) - \Sigma \boldsymbol{F} = \boldsymbol{F}_{24} + \boldsymbol{G}_4 + \boldsymbol{G}_5 + \boldsymbol{F}_{in4} + \boldsymbol{F}_{in5} + \boldsymbol{F}_{ext} + \boldsymbol{F}_{05} = \boldsymbol{0} \quad (22)$$

and the vector sum of all the moments on link 4 about D is zero:

$$4 - \Sigma \boldsymbol{M}_D = \boldsymbol{r}_{DC} \times \boldsymbol{F}_{24} + \boldsymbol{r}_{DC4} \times (\boldsymbol{G}_4 + \boldsymbol{F}_{in4}) + \boldsymbol{M}_{in4} = \boldsymbol{0} \quad (23)$$

where $G_4$ and $G_5$ are gravitational forces of links 4 and 5 respectively. The vectorial equations (22) and (23) give three scalar equations on x, y and z, that form a linear system of three equations with three scalar unknowns $F_{24x}$, $F_{24y}$ and $F_{05x}$. For solving this system of equations the MATLAB program is used. The joint reaction force $\mathbf{F_{45}}$ is calculated from the force equation for link 5:

$$5 - \Sigma \boldsymbol{F} = \boldsymbol{F}_{45} + \boldsymbol{F}_{in5} + \boldsymbol{G}_5 + \boldsymbol{F}_{ext} + \boldsymbol{F}_{05} = \boldsymbol{0} \quad (24)$$

The application point of the reaction force $\mathbf{F_{05}}$, calculated from a moment equation about D for link 5, is at D.

Figure 3 shows the forces and the moments that act on the RRR (2, 3) dyad. The unknown joint reaction forces are $\mathbf{F}_{03}$, $\mathbf{F}_{32}$ and $\mathbf{F}_{12}$. The joint force $\mathbf{F_{42}} = -\mathbf{F_{24}}$ was calculated from the previous RRP (4, 5) dyad.

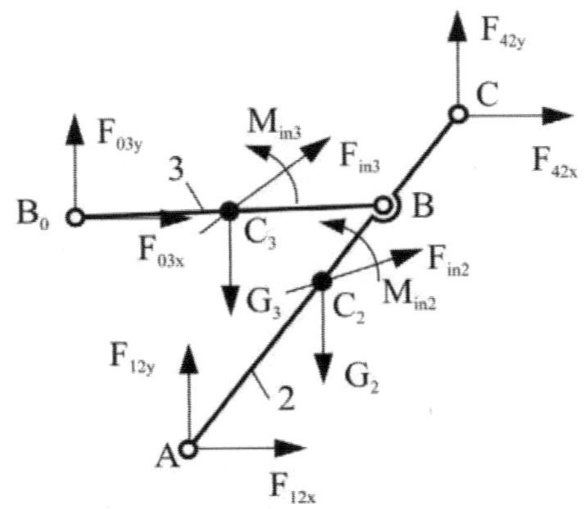


**Figure 3**. D' Alembert principle for RRR (2, 3) dyad.

The vector sum of all the forces that act on links 2 and 3 is zero:

$$(2,\ 3) - \Sigma \boldsymbol{F} = \boldsymbol{F}_{12} + \boldsymbol{F}_{42} + \boldsymbol{G}_2 + \boldsymbol{G}_3 + \boldsymbol{F}_{in2} + \boldsymbol{F}_{in3} + \boldsymbol{F}_{03} = \boldsymbol{0} \tag{25}$$

The vector sum of all the moments on link 2 about B is zero:

$$2 - \Sigma \boldsymbol{M}_B = \boldsymbol{r}_{BA} \times \boldsymbol{F}_{12} + \boldsymbol{r}_{BC2} \times (\boldsymbol{G}_2 + \boldsymbol{F}_{in2}) + \boldsymbol{r}_{BC} \times \boldsymbol{F}_{42} + \boldsymbol{M}_{in2} = \boldsymbol{0} \tag{26}$$

and the vector sum of all the moments on link 3 about B is zero:

$$3 - \Sigma \boldsymbol{M}_B = \boldsymbol{r}_{BB0} \times \boldsymbol{F}_{03} + \boldsymbol{r}_{BC3} \times (\boldsymbol{G}_3 + \boldsymbol{F}_{in3}) + \boldsymbol{M}_{in3} = \boldsymbol{0} \tag{27}$$

The vectorial equation (25) gives two scalar equations on x and y and equations (26) and (27) give one scalar equation on z, that form a linear system of four equations with four scalar unknowns $F_{03x}$, $F_{03y}$, $F_{12x}$ and $F_{12y}$. For solving this system of equations the MATLAB program is used.

The joint reaction force $\mathbf{F_{32}}$ is calculated from the force equation for link 2:

$$2 - \Sigma \boldsymbol{F} = \boldsymbol{F}_{12} + \boldsymbol{F}_{in2} + \boldsymbol{G}_2 + \boldsymbol{F}_{42} + \boldsymbol{F}_{32} = \boldsymbol{0} \tag{28}$$

The force analysis without friction ends with the input link 1. Figure 4 shows the forces and moments that act on the input link. The unknowns are the joint reaction force $\mathbf{F}_{01}$ and the input moment (motor moment) $\mathbf{M_{mot}}$. The joint force $\mathbf{F_{21}} = -\mathbf{F_{12}}$ was calculated from the previous dyad RRR.

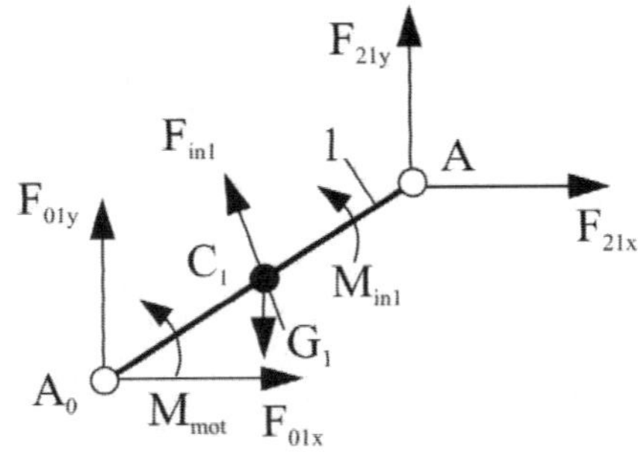


**Figure 4**. D' Alembert principle for input link 1.

The joint reaction force $\mathbf{F_{01}}$ is calculated from the force equation for link 1:

$$1-\Sigma \boldsymbol{F} = \boldsymbol{F}_{01} + \boldsymbol{F}_{in1} + \boldsymbol{G}_1 + \boldsymbol{F}_{21} = \boldsymbol{0} \tag{29}$$

and the input moment $\mathbf{M}_{mot}$ is determined from the equation of moments with respect $A_0$:

$$1-\Sigma \boldsymbol{M}_{A0} = \boldsymbol{M}_{mot} + \boldsymbol{r}_{A0A} \times \boldsymbol{F}_{21} + \boldsymbol{r}_{A0C1} \times (\boldsymbol{G}_1 + \boldsymbol{F}_{in1}) + \boldsymbol{M}_{in1} = \boldsymbol{0} \tag{30}$$

The joint reaction forces without friction are used furthermore in the first step of the iterative procedure to analyze the forces considering the joint frictions. At the solution step m+1 of the iterative procedure, friction forces and moments are calculated with the aid of reaction forces $F_{ij}^{m}$ determined at the step m.

The friction moment $M_{fij}^{m+1}$ at the revolute joint ij, opposed to relative angular velocity $\omega_{ji}$, is:

$$M_{fij}^{m+1} = -sign\,(\omega_{ji})\,\mu\, r_{ij} F_{ij}^{m} \qquad \left(i,\ j = 0, \ldots, 5\right) \tag{31}$$

and the friction force $F_{f05}^{m+1}$ at the prismatic joint D is:

$$F_{f05}^{m+1} = -sign\left(v_{50}\right)\mu\, F_{05}^{m} \tag{32}$$

where $\mu$ is friction coefficient and $r_{ij}$ is joint shaft radius.

These friction moments and friction force, considered to be external loads, are introduced in the corresponding equations of the method above described to calculate the joint reaction forces and input moment at step m+1.

## 4. Numerical application

The above equations have been implemented in a computer program using MATLAB software to generate the simulation presented throughout the paper. The main dimensions of the six-bar linkage (figure 1) used for kinematic analysis are inserted in table 1. These dimensions have been chosen in order to assure an approximately constant and small velocity and, accordingly, a quick return suitable for deep drawing. The driver link 1 rotates with a constant angular velocity of $\omega_{10}$=-4 rad/s. All five moving links are rectangular prism with depths $d_1$=5 mm, $d_2$=$d_3$=8 mm, $d_4$=10 mm, $d_5$=20 mm and heights $h_1$=2 mm, $h_2$=$h_3$=4 mm and $h_4$=5 mm, $h_5$=16 mm. The density of the material is $\rho_{steel}$= 8000 kg/m$^3$. The joint shaft radii are: $r_{A0}$=$r_D$= 20 mm, $r_A$=$r_B$= $r_{B0}$= $r_C$= 30 mm and the friction coefficient is $\mu$=0.1. The center of mass of links 1 and 5 is located in the joint $A_0$ and joint D respectively.

**Table 1**. Main dimensions of six-bar press.

| Dimensions | $l_0$ | $l_1$ | $l_2$ | $l_3$ | $l_{AC}$ | $l_4$ | $l_5$ | $a$ |
|---|---|---|---|---|---|---|---|---|
| (mm) | 128 | 64 | 112 | 112 | 160 | 290 | 80 | 110 |

The ram external force is opposed to the motion of the link 5: $F_{ext}$ = - sign($v_D$)$F_e$, where $F_e$=$10^6$ N throughout the working stroke.

The results for the kinematic simulation are shown in figures 5 and 6. Figure 5 shows the ram displacement (a), velocity (b) and acceleration (c) of the forming mechanical press. It is observed, that in the working zone, the ram velocity is approximately constant and has small values (figure 5b). The links positions for a complete rotation of the driver link 1 (the step of the angle is $10^0$) are given in figure 6.

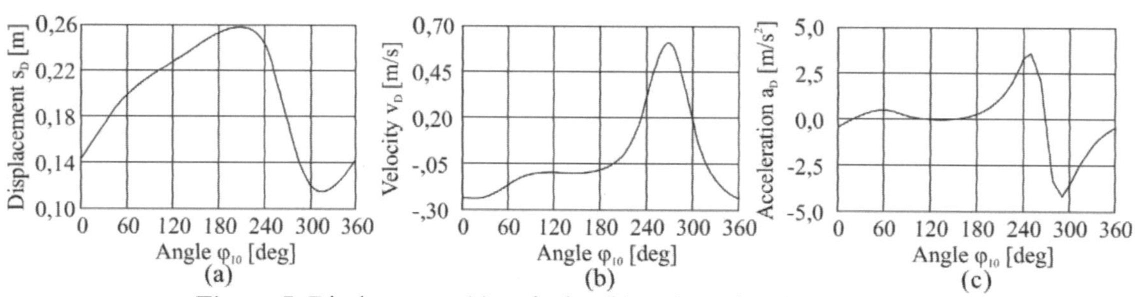


**Figure 5**. Displacement (a), velocity (b) and acceleration (c) of ram.

Some results for the dynamic simulation are shown in figures 7 and 8. Figure 7 illustrates the x, y components of the joint reaction forces $F_{01}$ (a) and $F_{03}$ (b) respectively versus crank angle with and without friction. In figure 8a is given the input torque versus crank angle with and without friction.

In order to validate the simulation results obtained with the MATLAB program, the values of the input torque without friction (method 1 in figure 8b) are compared with the ones obtained from the conservation of energy law (method 2 in figure 8b). It is remarked a very good agreement of the results.

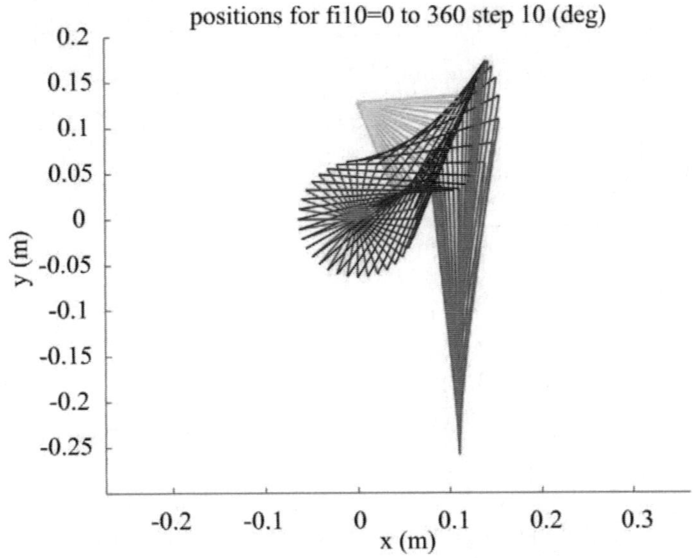


**Figure 6.** The link positions of six-bar press along the entire ram stroke.

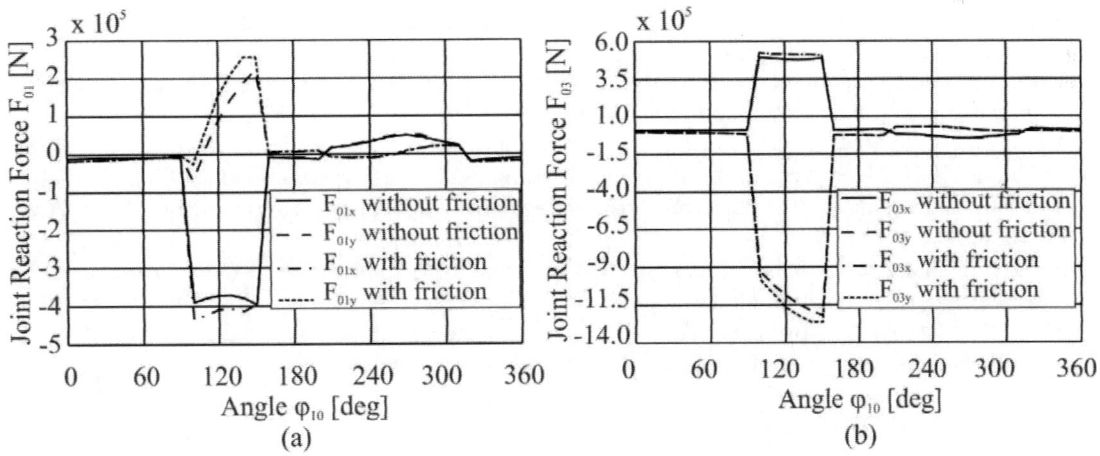


**Figure 7**. Joint reaction forces $F_{01}$ (a) and $F_{03}$ (b).

## 5. Conclusions

In the current study, the kinematic and kinetostatic analysis of a six-bar linkage of a mechanical press for deep drawing has been investigated developing a MATLAB program. An analytical method has been used for the determination of the displacement, velocity and acceleration of the links and the simulation of mechanism motion. The force analysis considering the joint friction is performed with an iterative procedure, applying the D'Alembert principle. It is observed that due to the low values of the friction forces and moments, the solution converges after one iteration step. The developed program can be useful for the optimization of the press design considering different constraints.

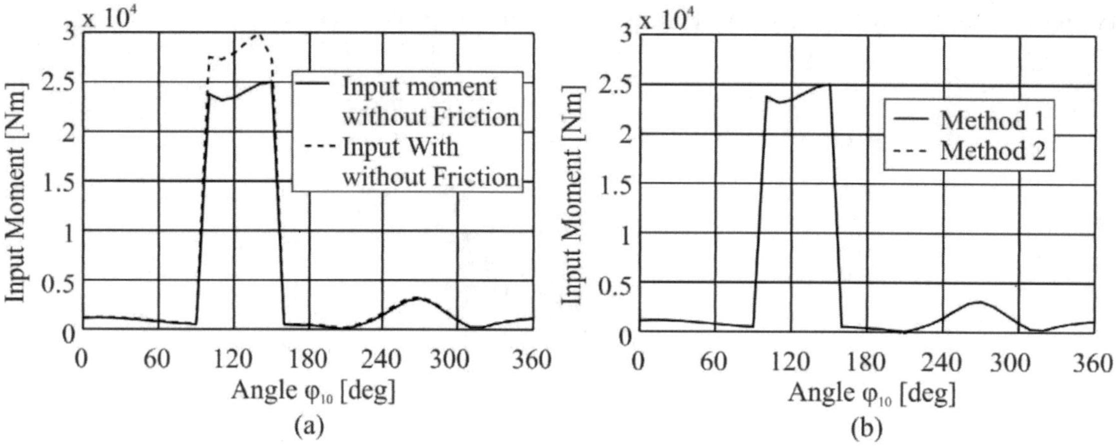


**Figure 8**. Input moment versus crank angle.

**References**

[1] Yan H S and Chen W R 2000 A variable input speed approach for improving the output motion characteristics of Watt-type presses *Int J Mach Tools Manuf* **40** 675

[2] Yan H S and Chen W R 2002 Optimized kinematics properties for Stevenson-type presses with variable input speed approach *Trans ASME J Mech Des* **124** 350

[3] Hsieh W H and Tsai C H 2011 On a novel press system with six links for precision deep drawing *Mech Mach Theory* **46** 239

[4] Soong R Ch 2010 A new design for single DOF mechanical presses with variable speeds and length-adjustable driving links *Mech Mach Theory* **45** 496

[5] Yossifon S and Shivpuri R 1993 Analysis and comparison of selected rotary linkage drives for mechanical presses *Int J Mach Tools Manuf* **33** 175

[6] Yossifon S and Shivpuri R 1993 Design considerations for the electric servo-motor driven 30 ton double knichle press for precision forming *Int J Mach Tools Manuf* **33** 209

[7] Tso P L and Liang K C 2002 A nine-bar linkage for mechanical forming presses *Int J Mach Tools Manuf* **42** 139

[8] Li H and Zhang Y 2010 Seven-bar mechanical press with hybrid-driven mechanism for deep drawing Part 2 Dynamic modelling and simulation *J Mech Sci Technol* **24** 2161

[9] Mitsi S, Tsiafis I and Bouzakis K D 2015 Force analysis of six-bar linkage for mechanical presses considering joint frictions *J Balk. Tribol. Assoc.* **21** 281

IOP Conf. Series: Materials Science and Engineering **174** (2017) 012007 doi:10.1088/1757-899X/174/1/012007

# The effects of mechanical properties and sintering duration on the wear behaviour of silicon nitride

**A Brusilová[1], P Švec[1], Z Gábrišová[1] and M Pokusová[2]**

[1]Institute of Technologies and Materials, Faculty of Mechanical Engineering, Slovak University of Technology in Bratislava, Pionierska 15, 831 02 Bratislava, Slovakia
[2]Institute of manufacturing systems, environmental technology and quality management, Slovak University of Technology in Bratislava, Nám. Slobody 17, 812 31 Bratislava, Slovakia

E-mail: alena.brusilova@stuba.sk

**Abstract.**The effects of hot pressing parameters and mechanical properties on wear of silicon nitride ceramics were analyzed. The weakest microstructure constituent was the brittle crystal boundary phase that broke and chipped off during the wear tests. The volume loss of separate ceramic materials during the wear tests depended mainly in inverse proportion to the hardness of the ceramics. Both microcutting and microcracking mechanisms played a part in the wearing of $Si_3N_4$ ceramics. Based on these experimental results the volume loss and also the wear resistance of the ceramics under study can be described by the model $V \sim H_V^{-1}$.

## 1. Introduction

$Si_3N_4$ based ceramics proves to be one of the advanced materials. An important position of silicon nitride in the industry is given by its properties, especially by its hardness and wear resistance, room and high-temperature strength (up to 1400°C) and by its resistance to corrosion and creep [1]. $Si_3N_4$ has been successfully used as wear resistant material in a wide variety of engineering applications involving contact with metallic surfaces such as drawing dies, roller bearings, cutting tools, and automotive or aerospace engine parts. The friction coefficient of 0.17 in a couple with $Si_3N_4$ and 0.15 with a bearing steel were achieved, using a hydrocarbon binder resulted in 0.13 and/or 0.11. Abrasive wear is the most common mechanism of ceramic material removal. Two basic mechanisms of surface damage are applied during the abrasive wear of ceramics: microcutting and microcracking [2-4].

The mechanisms of microcutting can be described according to a model that comes from Rabinowitz´s conception of abrasive wear mechanism description [5]. In this model the abrasive particle in the cone shape makes grooves on the surface of the solid body. The removed material volume $V$ can be described by equation (1):

$$V = \frac{F.l}{\pi.tg\alpha.H_V} \tag{1}$$

where $F$ is the force necessary to get the abrasive particle into the abraded material, $l$ is the length of a groove on the surface if the cone moves during relative motion in parallel with the worn material surface, $H_V$ is the hardness of ceramics, and $\alpha$ is the angle of the cone which participates in the grooving of the surface. It follows from equation (1) that the removed material volume is dependent only on one material property, hardness $H_V$.

If we consider the mechanism of microcracking, the abrasive particle creates a crack in the plane of the load axis after the overrun of the specific limit value of the load. Cracks spread to the sample surface, where they can develop into a fracture [3, 6]. The volume of wear $V$ can be expressed by means of equation (2).

$$V = \frac{F^{9/8}}{K_{IC}^{1/2} . H_V^{5/8}} \cdot \left( \frac{E}{H_V} \right)^{4/5} . l \tag{2}$$

In this equation (2) $l$ is length of path of abrasive particle, $F$ is load on the abrasive particle, $K_{IC}$ is fracture toughness of the ceramics, $H_V$ is the hardness of the ceramics and $E$ is the modulus of elasticity of the worn ceramic material. According to equation (2) the removed material, volume $V$, is dependent on three material properties: the modulus of elasticity $E$, the fracture toughness $K_{IC}$ and hardness $H_V$.

According to some works [5,7] the ratio of $K_{IC}/H_V$ values appears as the main characteristic determining dominant wear mechanisms of brittle materials during abrasive wear. This parameter defines the dominant wear mechanism at the point of contact. The mechanism of microcutting is dominant at a high value of this rate and simultaneously at a high value of fracture toughness, i.e. wear volume will be depended on hardness ($V \sim 1/H_V$). Brittle fractures dominant at a low value of the rate $K_{IC}/H_V$, i.e. wear will increase with decreasing fracture toughness ($V \sim K_{IC}/H_V$). This influences the growth of wear intensity. According to equations (1) the intensity of microcutting decreases with the hardness of ceramics, and according to equation (2) the intensity of microcracking decreases with fracture toughness of the worn surface. This can lead to a transition from plastic microcutting to brittle microcracking during abrasive wear [2]. Besides the mechanical properties of ceramics, the character of the structure like grain size and $\beta$-$Si_3N_4$ phase ratio plays an important role in determining how the ceramics will react to specific states of stress which arise under specific conditions of wear [3, 8].

## 2. Materials and experiments

The wear resistance test specimens were prepared from powdered silicon nitride, hot-pressed with sintering additives of yttrium oxide and aluminum oxide. The amounts of sintering additives $Y_2O_3$ and $Al_2O_3$ were different in separate specimens, but the proportion of $Y_2O_3$ and $Al_2O_3$ was the same for all the prepared specimens. The $Y_2O_3$ and $Al_2O_3$ powders were added in concentrations that could set creating $Y_3Al_5O_{12}$ phase (YAG - yttrium aluminium garnet) in ceramic materials. This phase contributed to the sintering ability of the ceramics [8]. The designations and compositions of the studied materials are given in table 1.

**Table 1.** Chemical composition of ceramic samples.

| Ceramic sample | Concentration (wt.%) | | |
|---|---|---|---|
| | $Si_3N_4$ | $Y_2O_3$ | $Al_2O_3$ |
| 5 % YAG | bulk | 7.75 | 2.14 |
| 10 % YAG | bulk | 10.34 | 4.30 |

Initial powder compositions were wet mixed in alcohol. After drying and sieving, the powder was compacted in steel dies. The final densification was accomplished using hot pressing techniques in a nitrogen atmosphere with the purity of 99.99 % and an overpressure of 75 kPa. All samples were hot pressed at a temperature of 1680 °C and a pressure of 34 MPa. Three various sintering times 5, 15 and 30 min were applied.

Densities of the hot pressed ceramics were measured by the Archimedes's method. Hardness and fracture toughness were determined by means of Vickers indentation method with a testing load of 98 N. The fracture toughness was also determined by means of Vickers indentation. This method is based on the measurement of crack lengths that rise from the indentation corners. The more brittle the tested material is, the longer are initiated cracks. The wear resistance was evaluated by means of grinding the sample using a pin on disk method. Test samples with diameters of 8.4 mm and a height of 10 mm were placed in contact with corundum grinding paper with a graininess of 120 μm. The grinding trajectory was 125 m and the pressure was 1.5 MPa. Sliding speed max. 0.5 m.s$^{-1}$, radial movement 1.5 mm/ot, dry friction. For experiments was used the equipment for abrasive wear testing (figure 1). The wear resistance was determined based on the volume loss of the samples relative to the grinding trajectory according to equation (3):

$$V_{V/s} = \frac{\Delta m}{\rho . l} \tag{3}$$

where $V_{V/s}$ is volume loss of the samples, $\Delta m$ is weight loss of the samples, $\rho$ is density of the sample and $l$ is grinding path of the sample.

The microstructures of the hot pressed ceramics were observed using a scanning electron microscope. In order to identify the microstructure created, the hot pressed specimens were subjected to XRD analysis.

**Figure 1.** The test equipment.

## 3. Results

The effects of densification, mechanical properties and the microstructure on the wear resistance of the prepared ceramics samples were evaluated.

### *3.1 Densification and microstructure*

Average values from measured densities in particular ceramic samples are included in table 2. The densities of the ceramic samples were influenced by the amount of additives ($Y_2O_3$ and $Al_2O_3$) and by the sintering time. They increased with increasing the concentration of additives and sintering time from 3.21 to 3.27 g.cm$^{-3}$. These values corresponded to the relative density from 97.2 to 98.5 %, which indicate good densification of samples. The smallest density, 3.21 g.cm$^{-3}$, was measured in the 5 % YAG sample with the smallest concentration of the additives, sintered for 5 min. The highest value of 3.27 g.cm$^{-3}$ was achieved in the 10 % YAG sample sintered for 5 and 15 min.

**Table 2.** Density of ceramic samples.

| Ceramic sample | Density (g.cm$^{-3}$) | | |
|---|---|---|---|
| | 5 min | 15 min | 30 min |
| 5 % YAG | 3.21 | 3.24 | 3.25 |
| 10 % YAG | 3.27 | 3.27 | 3.26 |

Each of the samples with different compositions and pressing time were analyzed. The phase composition of samples was identified by the XRD method (Tab. 3). Only two phases were found in all specimens $\alpha$-$Si_3N_4$ and $\beta$-$Si_3N_4$ phases. Phases with a concentration below 5 % couldn't be identified using the XRD method.

**Table 3.** $\beta$-$Si_3N_4$ phase ratio of ceramic samples.

| Ceramic sample | $\beta$-$Si_3N_4$ phase ratio (wt.%) | | |
|---|---|---|---|
| | 5 min | 15 min | 30 min |
| 5 % YAG | 45 | 58 | 75 |
| 10 % YAG | 43 | 62 | 100 |

During hot-pressing, the initial $\alpha$-$Si_3N_4$ powder was transformed to $\beta$-$Si_3N_4$ phase, which can be seen from table 3. The transformation stage increases with both the additive concentrations and the sintering time. The portion of $\beta$-$Si_3N_4$ phase from 43 to 45 % was measured by XRD at a shorter sintering time of 5 min for all prepared compositions. There were small differences between separate compositions, but these were within the measurement error, which was 5 %. The complete 100 % transformation of $\alpha$-$Si_3N_4$ powder into $\beta$-$Si_3N_4$ phase was achieved only in the 10 % YAG samples with the highest concentration of additives and sintered for 30 min.

The study of the microstructure confirmed the effect of the composition on the formed phases. The microstructure consisted of $\alpha$- $Si_3N_4$ and $\beta$-$Si_3N_4$ phase, figure 2 and 3. The amount of overextended $\beta$-$Si_3N_4$ grain increased with both the additive content and at the sintering time and reached the full $\beta$-$Si_3N_4$ microstructure at the highest additive content and longest time, which can be seen in figure 2a) (ceramic sample 10 % YAG sintered for 30 min). This can be explained by the increase in the transformation velocity at higher additive contents and longer times. With the increase of the additive amount, both the grain size and the ratio of a binding phase at the grain boundary increased. This crystal boundary phase is relatively brittle and can by indicated as the weakest component of the microstructure [8, 9]. The grain size also grew with the sintering time.

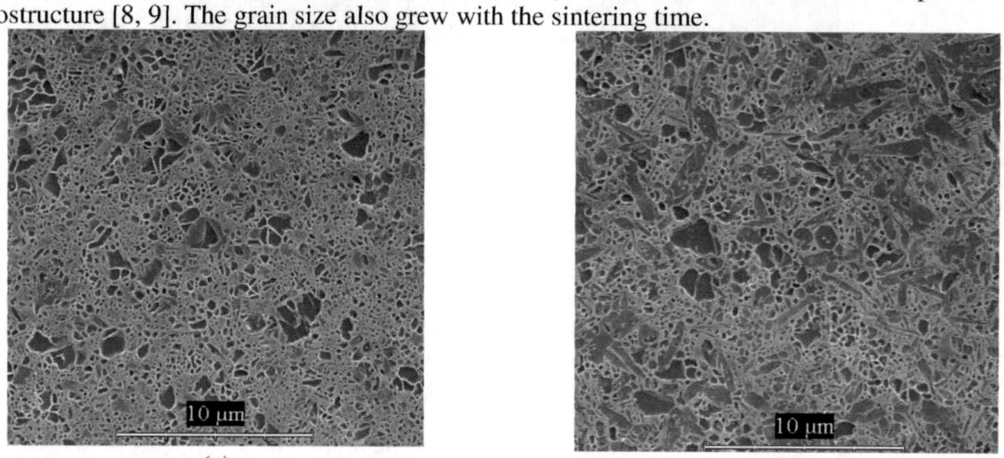


(a) (b)

**Figure 2.** Microstructure of ceramics sample with 10% YAG sintered for a) 5 min b) 30 min.

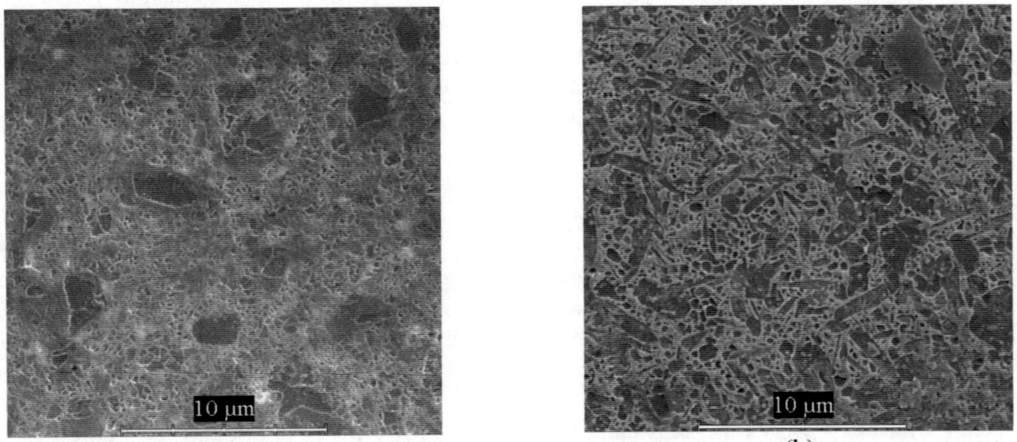


(a) (b)

**Figure 3.** Microstructure of ceramics sample with 5% YAG sintered for a) 5 min b) 30 min.

The effect of the additional concentration and sintering time on the wear of ceramic samples can be seen in figure 4. From this figure it can be seen that the volume change during the wear test decreased with the increasing of the additions and with sintering time. The highest wear resistance was achieved in the 10 % YAG sample, which was pressed for only 5 min, and the least wear resistance was with the 5 % YAG specimen with the smallest additional content, pressed for the longest time of 30 min.

*3.2 Effects of sintering parameters and mechanical properties on wear*

In work [10] was found the highest hardness at each prepared composition was always achieved at shorter sintering times of 5 min. When prolonging the time to 30 mins, the hardness slightly decreased, which can be explained by the increasing of grain size and increasing of β-$Si_3N_4$ phase amount with sintering time that was described above. Thus the hardness decreased with increasing of grain size and also by increasing of the β-$Si_3N_4$ phase amount what could be explained by lower hardness of prismatic β-$Si_3N_4$ crystals in comparison to narrow α-$Si_3N_4$ crystals [11].

The effects of the measured mechanical properties on the wear properties of the ceramic samples were studied in detail and are presented in the figures 5, 6 and 7.

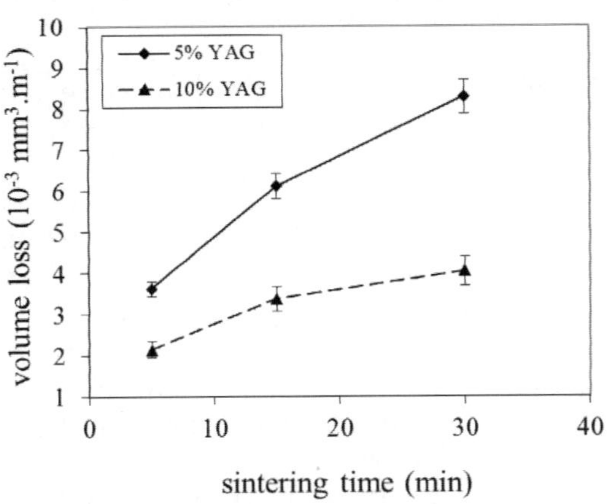


**Figure 4.** Effect of pressing time and additive concentration on wear of ceramics.

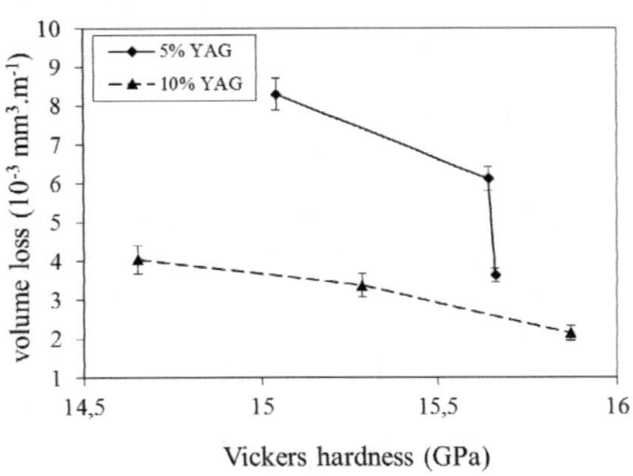


**Figure 5.** Effect of hardness on wear of ceramics.

Hardness had a positive effect on wear resistance (figure 5). Higher hardness resulted in less wear. As the highest hardness was measured in the higher concentration of additives (ceramics 10 % YAG), these samples had the smallest volume losses. The highest volume losses were measured in the 5 % YAG samples with the least additives that was pressed for 5 min. The results in figure 5 correspond well with model ($V \sim H_V^{-1}$), where the volume losses $V$ during the wear tests vary inversely in proportion to the hardness $H_V$ of the ceramics.

A very interesting development was noted when the effect of fracture toughness on wear was measured (figure 6). All compositions showed the same progress. At first wear increased up to the maximum value. After reaching the maximum value wear decreased slightly. This can be explained by the relationship between the $\beta$-$Si_3N_4$ phase and grain size, but the differences between the separate samples within each composition were relatively small. The effect of grain size was dominant. The highest grain size was always in the sample with a middle value of fracture toughness. That means the wear behaviour can't be described only by the effect of fracture toughness on wear.

Wear behaviour may be better described by the model which reflects the effect of ratio fracture toughness / hardness on wear rate. This can be seen in figure 7. These relations fit very well for each separate composition. The higher the value of the calculated ratio of fracture toughness to hardness is, the higher the wear rate is. The highest wear rate was noticed in the 5 % YAG specimen that contained the smallest concentration of additives that was pressed for 30 min. The ratio of fracture toughness to hardness accurately describes the relationship between the decrease of wear resistance in spite of the transformation progress $\alpha$-$Si_3N_4$ phase to the $\beta$-$Si_3N_4$ phase, and the increase in the rate of wear in spite of grain growth.

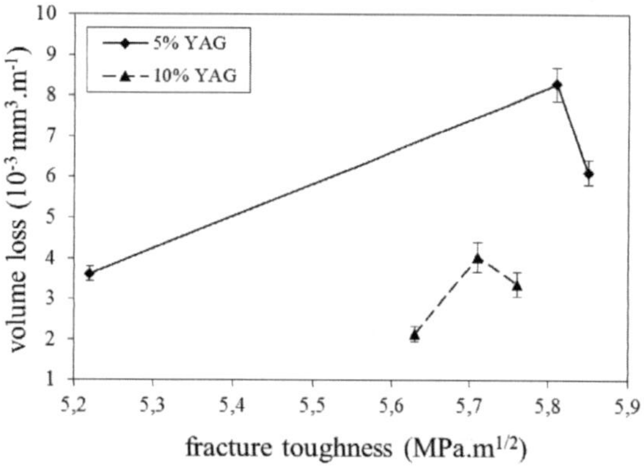


**Figure 6.** Effect of fracture toughness on wear of ceramics.

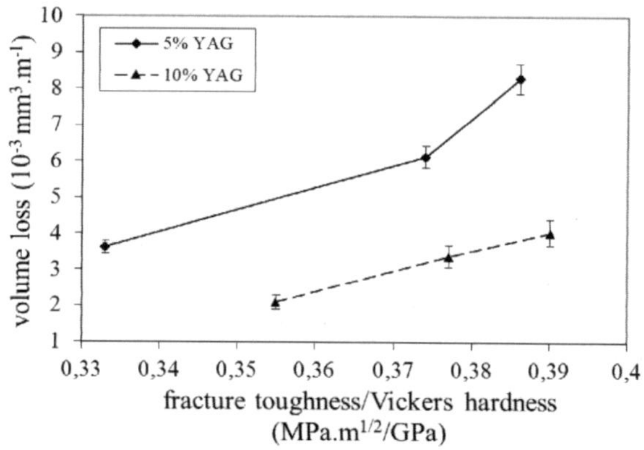


**Figure 7.** Effect of ratio fracture toughness / Vickers hardness on wear of ceramics.

The ceramic surfaces after the wear tests confirmed the effect of additive concentration and sintering time on the wear resistance of specimens that were described above. The specimens which were pressed for a longer time showed greater damage than the specimens pressed for a shorter time. The specimens with higher contents of additives showed less damage than the specimen with smaller additive concentrations. These specimens have higher portions of binding phases, which is present on the grain boundary. The binding phase has a positive effect on the densification, which generally improves the mechanical properties of the ceramic materials. On the other hand, the binding phase is the weakest component of the microstructure. This brittle crystal boundary phase broke and chipped off during the wear tests, which can be seen in figure 8 and 9. Prolongation of pressing time is connected to grain growth, which caused less wear resistance, thus the larger volumes of the surface were extracted.

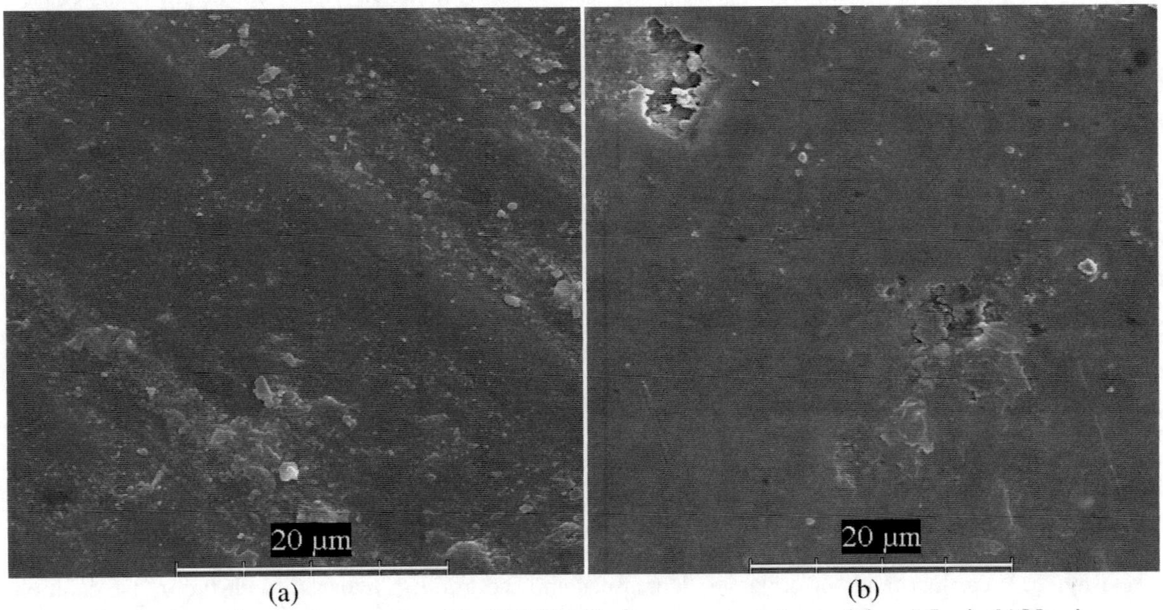


**Figure 8**. Surface of ceramic sample with 10% YAG after wear test sintered for a) 5 min b)30 min.

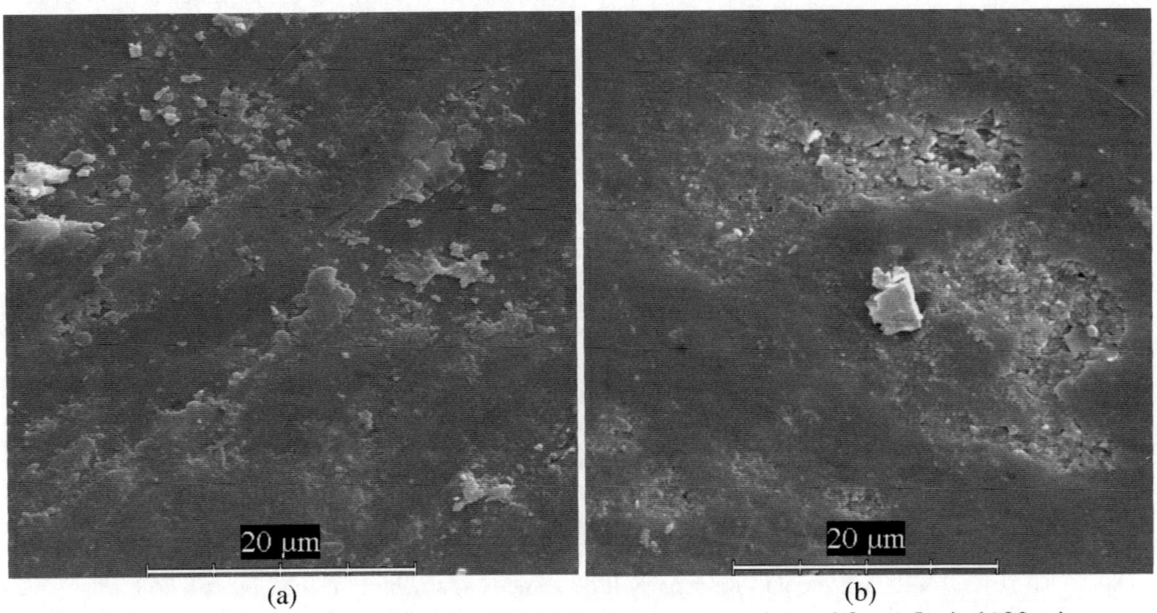


**Figure 9**. Surface of ceramic sample with 5% YAG after wear test sintered for a) 5 min b)30 min.

Consistent with the theoretical background, the volume loss depends on hardness ($V \sim H_V^{-1}$) when the dominant mechanism of wear is microcutting. If the dominate mechanism is microcracking, the volume loss depends on the ratio hardness / fracture toughness ($V \sim K_{IC}/H_V$). All ceramic surfaces were damaged by microcutting and also by microcracking. This can be clearly seen from the observed surface after the wear tests (figures 8, 9). Several scratches and also damaged areas are visible. These areas are characteristic with dropped out material and many microcracks. These characteristic scratches and pits with microcracks were presented on all specimens. Thus both mechanisms caused some wear on all specimens. However the extent of ceramic surface damage increased with pressing time.

Both models could be used to interpret the experimental data during the wearing process of $Si_3N_4$ ceramic with alumina abrasive. In our case, the experimental data corresponded better with the model, which assumes that microcutting as the dominant mechanism. Since there were only very small changes in the fracture toughness of separate specimens, the effect of ceramic hardness can be consider as dominant. Thus the volume loss during wear can be described by the first model ($V \sim H_V^{-1}$).

## 4. Conclusions

This paper concentrated on the analysis of the wear mechanism of $Si_3N_4$ based ceramics fabricated by hot pressing. The effect of preparation parameters chosen, such as chemical composition, sintering conditions in the microstructure, mechanical properties and wear resistance, was evaluated. It was found that wear resistance of hot pressed silicon nitride ceramics depends on the amount of a binder as well as on pressing time. Changes in wear resistance are associated with changes in material structure occuring on compacting the material of concern. The direct proportion between wear resistance and hardness and fracture toughness of ceramics has been demonstrated. Wear was mostly influenced by the hardness of ceramic materials. The specimen with the highest hardness achieved the highest wear resistance. Wear resistance of ceramics decreased with the grain growth and with the transformation progress of narrow $\alpha$-$Si_3N_4$ phase to prismatic $\beta$-$Si_3N_4$ phase.

The weakest microstructure constituent was the brittle crystal boundary phase that broke and chipped off during the wear tests. Two damage mechanisms were taken into consideration on ceramic surfaces during the wear of $Si_3N_4$ ceramic pin samples with alumina grinding disks. Specimen surfaces were damaged by microcutting and also by microcracking mechanisms. The volume loss $V$ of separate ceramic materials during the wear tests depended mainly inversely in proportion to the hardness $H_V$ of the ceramics. Based on these experimental results the volume loss and also the wear resistance of the studied ceramics can be described by model $V \sim H_V^{-1}$.

## Acknowledgements

The authors appreciate the financial support provided by the Slovak Research and Development Agency for the project APVV-0857-12 and APVV-0281-12.

## References

[1] Hvizdoš P, Kašiarová M, Hnatko M and Šajgalík P 2004 Kovove Mater. **42** 51
[2] Medvedovski E 2001 Wear resistant engineering ceramics *Wear* **249** 821
[3] Bocanegra-Bernal M H and Matovic B 2009 Dense and near-net shape fabrication of $Si_3N_4$ ceramics *Mat. Sci.Eng. A – Struct* **500** 130
[4] Dogan C P and Hawk J A 2001 Microstructure and abrasive wear in silicon nitride ceramics *Wear* **250** 256
[5] Rabinowicz E 1995 *Friction and wear of materials* 2th ed (New York :John Wiley and Sons)
[6] Hsu S M and Shen M 2004 Wear of ceramics *Wear* **256** 867
[7] Williams J A 2005 Wear and wear particles – some fundamentals *Tribol. Int.* **38** 863
[8] Lee B T, Han B D and Kim H D 2003 Comparison of fracture characteristic of silicon nitride ceramics with and without second crystalline phase *Mater. Lett.* **58** 74
[9] Emoto H and Mitomo M 1997 Control and characterization of abnormally grown grains in silicon nitride ceramics *J. Eur. Ceram. Soc.* **17** 797
[10] Švec P, Brusilová A and Kozánková J 2009 Effect of microstructure and mechanical properties on wear resistance of silicon nitride ceramics Mater. Eng. **16** 34
[11] Nakamura M and Hirao K 2003 Wear behaviour of a-$Si_3N_4$ ceramics *Wear* **254** 94

# The effect of transport velocity upon spin torque

**S Alaci, F C Ciornei, D Amarandei, L Irimescu, I C Romanu and M A Rotar**

Department of Mechanics and Technologies, "Stefan cel Mare" University of Suceava, Str. Universitatii 13, Suceava, 720229, Romania

E-mail: alaci@fim.usv.ro

**Abstract**. The paper analysis the effect produced by superposition of a rotation motion from contact area on a transport translational motion, having the same value for all contact points. A theoretical model is proposed for the calculus of the total friction torque. The deduced expression shows that the friction torque has a maximum value when the translational velocity is zero and rapidly decreases with increasing the translational velocity. The diminishing is not uniform, but is more evident at the beginning, when the ratio between the transport velocity and angular velocity is comparable to the radius of contact area and, afterwards, the decline of the friction torque is slower. The last part of the paper proposes a solution for an experimental validation of the theoretical results.

## 1. Introduction

The Hertzian point contact assumes that, between the boundary surfaces of two bodies, a theoretical contact point exists. Under normal force loading, in the region of the initial contact point, a contact surface occurs of elliptical shape and having the dimensions as found by Hertz in 1883 [1]. In the same paper, Hertz proves that on the contact area, a contact pressure occurs, having the maximum value in the centre of the contact ellipse and being zero on the boundary of the ellipse. The point contact between two bodies supposes that a class 1 kinematical joint is formed, namely, if one of the bodies is maintained motionless, the other one, from the six elementary motions theoretically possible, has a suppressed motion, the translation along common normal. The other simple motions determine, for an actual case, an extremely complex relative motion on the contact area.

The dynamic study of a body contacting a surface becomes substantially intricate when the hypothesis of smooth bodies is neglected and friction forces are considered between connections. Between the two surfaces, dry or fluid friction can happen. The dry friction case makes the dynamical study of a system complex because the characterisation is made using inequalities and, thus, working with unilateral links [2],[3].. Even under the hypothesis of perfect rigid bodies, the dynamical study with dry friction is difficult. To support this affirmation, the famous problem of the rolling coin can be mentioned [4],[5]. Considering that the relative motion between the two contacting bodies is described by the sliding velocity, positioned in a common tangent plane and by the relative angular velocity. The angular velocity is composed by the spin angular velocity, parallel to the common normal and by the rolling velocity, positioned in the common tangent plane. After setting equal to zero one or the other of the two component of angular velocity, with the hypothesis that the linear velocity is different from zero, there can be examined the rolling motion with sliding and the spinning motion with sliding.

Next, the axi-symmetric contact with spinning and sliding motion is considered, aiming to emphasise the influence of the transport velocity upon the spinning torque.

## 2. Theoretical considerations

As regarding an axi-symmetric contact, the contact area is a circle of radius $r_{max}$, figure 1. A current point is considered on the contact area, as the centre of elementary surface $dA$. The position of a current point from the contact area is set by the vector of position $\boldsymbol{r}$ of point $M$. The position of a current point $M$ can be established in Cartesian coordinates, $M(x,y)$, or in polar coordinates, $M(\rho,\theta)$.

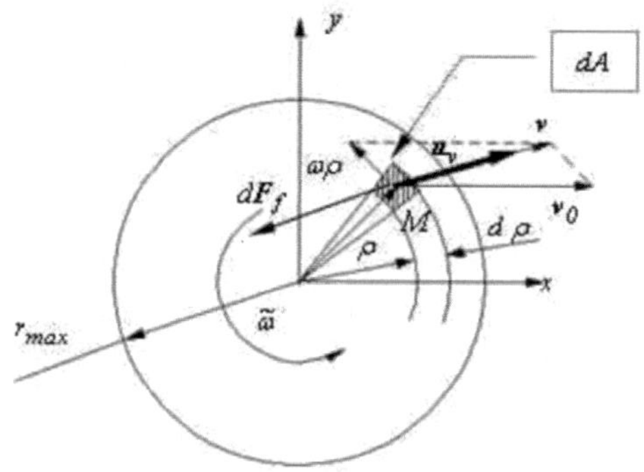


**Figure 1**. Relative motion on contact area.

The absolute velocity of a point from the contact area, $M(x,y)=M(\rho,\theta)$ is, according to [6],[7]:

$$\boldsymbol{v} = \boldsymbol{v}_0 + \boldsymbol{\omega} \times \boldsymbol{r} = \begin{bmatrix} v_0 - \omega y \\ \omega x \\ 0 \end{bmatrix} \tag{1}$$

where $\omega$ is the angular velocity directed, in this case, along the common normal, and $v_0$ is the transport velocity, identical for any contact surface point, with the direction of the horizontal. The unit vector of velocity $\boldsymbol{v}$ is:

$$\boldsymbol{u}_v = \frac{\boldsymbol{u}_v}{|\boldsymbol{u}_v|} \tag{2}$$

By calculating the absolute value of velocity (1) it results that, in any point of contact area, it is different from zero and as a consequence, the elementary friction force from the elementary surface is:

$$d\boldsymbol{F}_f = -\mu\, p(x,y)dA\, \boldsymbol{u}_v \tag{3}$$

where $p(x,y)$ is the pressure acting in point $M(x,y)$ and $\mu$ is the dynamic coefficient of friction. Between all contacting points, except for origin, relative motion exists. The elementary friction torque is:

$$d\boldsymbol{M}_f = \boldsymbol{r} \times d\boldsymbol{F}_f \tag{4}$$

By substituting in relations (1), (2), (3) and (4), after a series of calculus one obtains:

$$d\boldsymbol{M}_f = -\mu\, p(x,y) \frac{\omega\rho^2 - v_0\rho\sin\varphi}{\sqrt{v_0^2 - 2v_0\omega\rho\sin\varphi + \omega^2\rho^2}} dA\boldsymbol{k} \tag{5}$$

where $\boldsymbol{k}$ is the versor normal to the contact surface in the theoretical contact point. To estimate the total torque, the integration of relation (5) on the entire contact area is required. Assuming a Hertzian pressure distribution, the integration can be made straightforwardly if the transition to polar coordinates is done:

$$p(\rho) = p_{max}\sqrt{1-\frac{\rho^2}{r_{max}^2}} \tag{6}$$

where $p_{max}$ and $r_{max}$ are the maximum contact pressure and the maximum contact radius, respectively.

For the spinning torque the final expression is obtained:

$$M_f = -\mu\, p_{max}\int_0^{2\pi}\int_0^{r_{max}}\left(1-\frac{\rho^2}{r_{max}^2}\right)^{1/2}\left(\rho^2-\frac{v_0}{\omega}\rho\sin\varphi\right)\left(\frac{v_0^2}{\omega^2}-2\frac{v_0}{\omega}\rho\sin\varphi+\rho^2\right)^{-1/2}\rho\, d\rho\, d\varphi \tag{7}$$

Relation (7) shows that the friction torque does not depend separately on angular velocity and on transport velocity $v_0$ but depends on the ratio $v_0/\omega$, fraction that has the dimension of a length. For the integral from the right side it cannot be given an analytical expression.

By denoting the characteristic dimension:

$$r_0 = \frac{v_0}{\omega} \tag{8}$$

the relation (7) can be given under the final form:

$$M_f = -\mu\, r_{max}^3 p_{max}\int_0^1 t^2\sqrt{1-t^2}\int_0^{2\pi}\left(t-\frac{r_0}{r_{max}}\sin\varphi\right)\left(t^2-2t\frac{r_0}{r_{max}}\sin\varphi+\frac{r_0^2}{r_{max}^2}\right)^{-1/2} d\varphi\, dt \tag{9}$$

To exemplify the effect of transport velocity upon spin velocity, it was considered the contact between a bearing ball $20mm$ in diameter placed on an horizontal steel plate. One considers the rotational speed of the ball $200rot/min$ to which corresponds an angular spin velocity Although in literature there are models for the description of the many complex phenomenon of dry friction , below it is considered that the entire contact area that is constant . So, $\omega = 20.944 rad/s$ . the friction coefficient is thought to have the constant value $\mu = 0.2$ [8]. For the considered contact case, using the relations from the contact mechanics theory [9] for the normal force $Q = 100N$ it resulted the maximum contact radius $r_{max} = 1.866\cdot 10^{-4} m$ and the maximum contact pressure $p_{max} = 1.371\cdot 10^9 Pa$ .

Applying relation (9), the variation of friction torque with respect to transport velocity was traced as seen in figure 2. The abscissa of the plot is dimensionless by dividing to relative rotational speed of a point from the boundary of contact area, $\omega r_{max}$. It can be noticed that, for the ratio $(v/\omega\cdot r_{max}) < 1.5$ there is a strong dependency. External to this interval, the influence is much more diminished. For very high transport velocities, the total friction torque tends asymptotically to zero.

To obtain an intuitive representation, it was considered the ratio between friction torque and normal load, resulting a term with length dimension, represented in the ordinates of next figures.

In figure 3 there are traced, for several values of transport velocity, the plots of total friction torque versus normal load force. From these graphs it can be remarked that the influence of normal force upon spinning torque is much stronger for reduced values of the transport velocity.

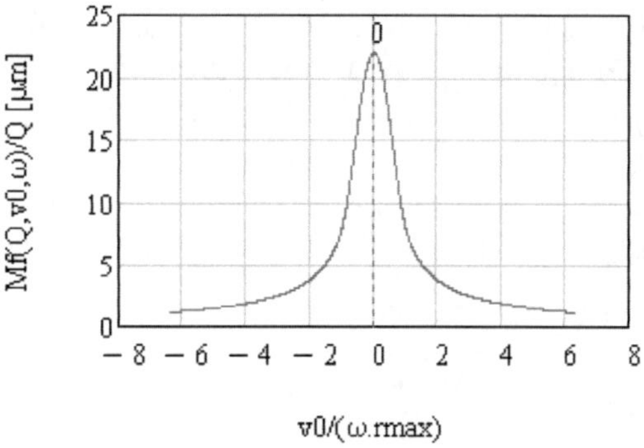


**Figure 2**. Total friction torque variation with transport velocity.

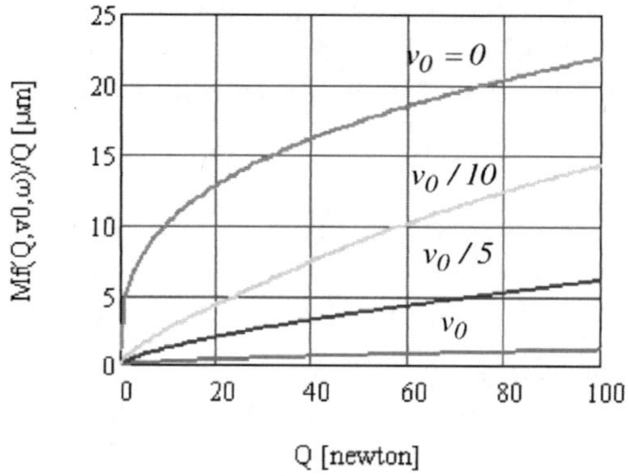


**Figure 3.** Normal force effect upon friction torque for different values of transport velocity.

The expression of the elementary friction torque can be obtained considering the integrant from surface integral (7):

$$M_{elem_f} = -\mu\, p_{max}\left(1-\frac{\rho^2}{r_{max}^2}\right)^{1/2}\left(\rho^2-\frac{v_0}{\omega}\rho\sin\varphi\right)\left(\frac{v_0^2}{\omega^2}-2\frac{v_0}{\omega}\rho\sin\varphi+\rho^2\right)^{-1/2} \tag{10}$$

Another interesting dependency is presented in figure 4. The contact area was divided into concentric circular rings having the same surface area.

$$M_{rad_i} = -\int_{R_{i-1}}^{R_i}\int_0^{2\pi} M_{elem_f}\, dA \tag{11}$$

The friction torque corresponding to each ring was found and it was represented the friction torque as function of external radius of each ring.

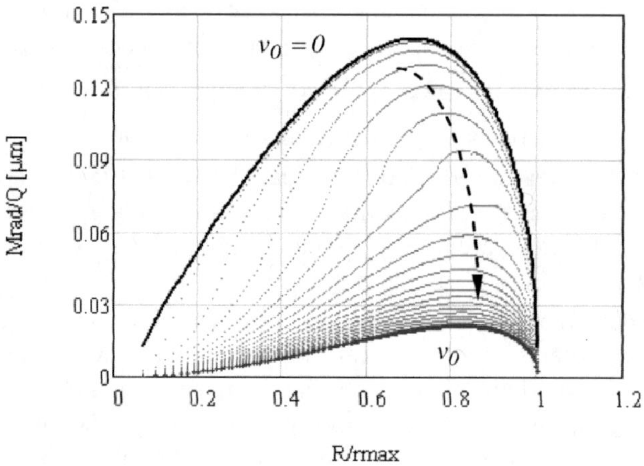


**Figure 4.** Friction torque (divided by normal load) variation with relative radius.

IOP Conf. Series: Materials Science and Engineering **174** (2017) 012008 doi:10.1088/1757-899X/174/1/012008

In table 1 there are presented the spatial plots, level curves and sections on the direction of transport motion and normal to it, for different values of transport velocity. Here, $x$ $y$ are the coordinates of a moving point which is placed inside a rectangle which envelopes the contact area. It can be noticed that, for zero transport velocity, the elementary friction torque is positive on the entire surface and has an axi-symmetric distribution. With increasing transport velocity, regions where the elementary friction torque is negative occur, and thus, the total friction torque decreases. For greater values of transport velocity, the spatial graph of elementary friction torque has the centre of contact area as anti-symmetry point, and therefore the elementary friction torque from a point would be compensated by the one from the symmetrical point with respect to the contact centre and, globally, the friction torque is zero.

**Table 1.** The effect of transport velocity upon elementary friction torque

| $\frac{v_0}{\omega r_{max}} = 0$ | $\frac{v_0}{\omega r_{max}} = 0.643$ | $\frac{v_0}{\omega r_{max}} = 6.43$ |
|---|---|---|
| 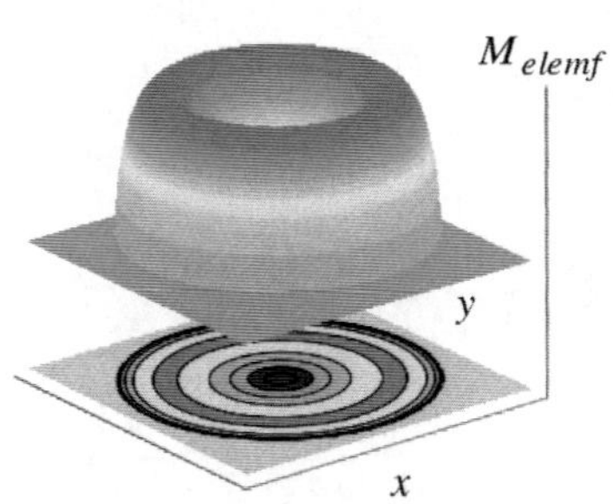 | 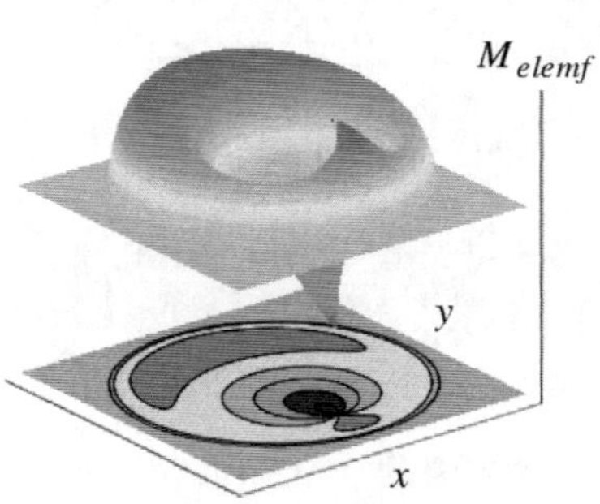 | 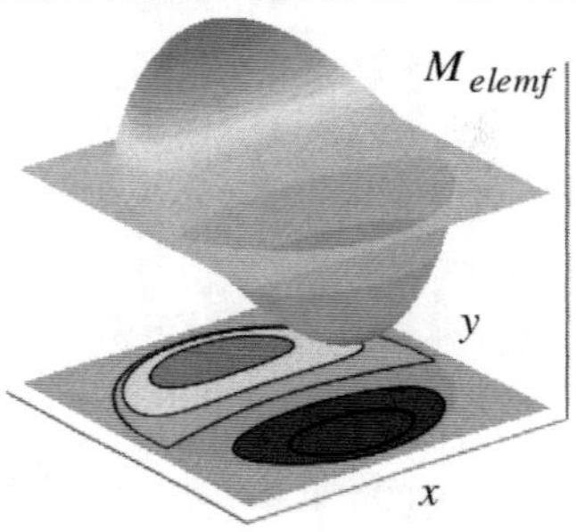 |
| 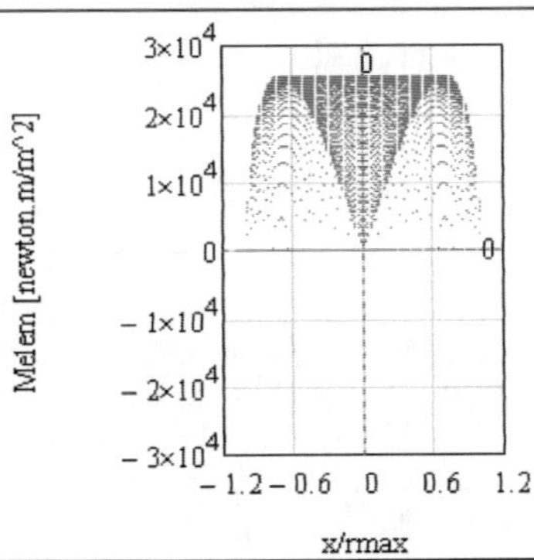 | 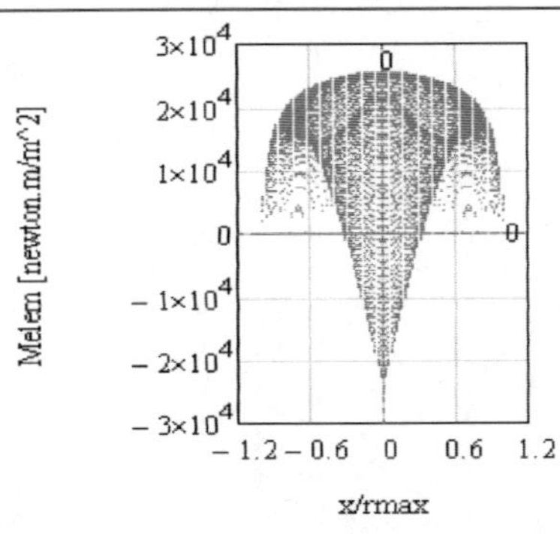 | 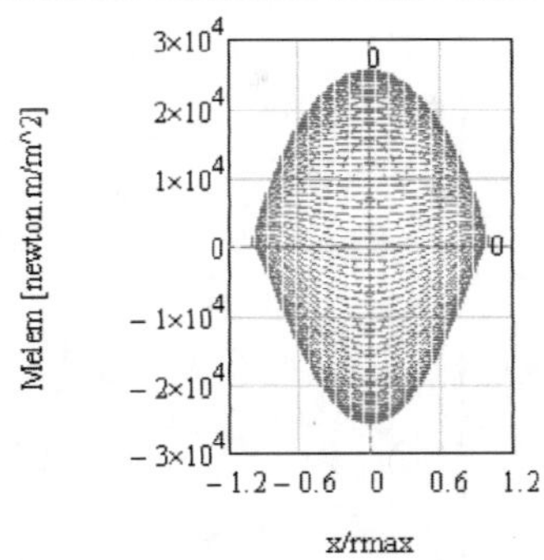 |
| 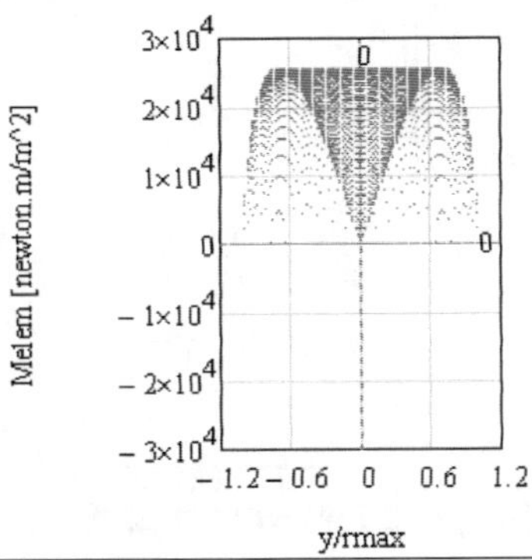 | 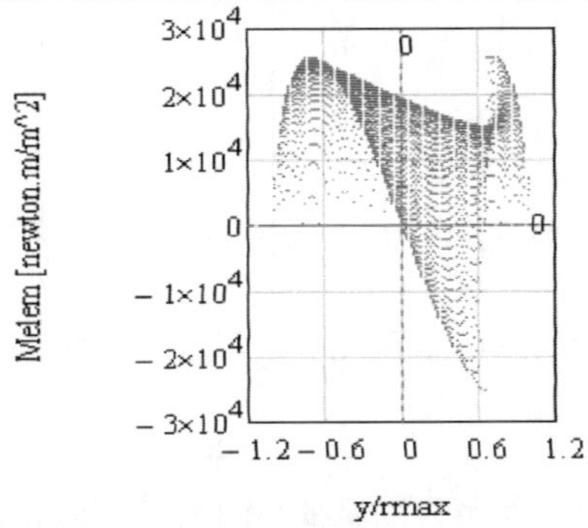 | 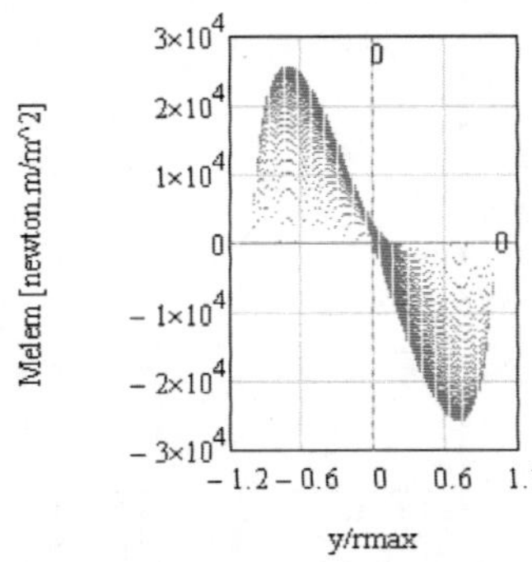 |

IOP Conf. Series: Materials Science and Engineering **174** (2017) 012008 doi:10.1088/1757-899X/174/1/012008

## 3. Experimental validation

Next, an experimental qualitative justification of the influence of transport velocity upon spinning friction torque is presented. It is considered a bearing ball, 60 mm diameter, placed on the surface of a biconcave lens (figure 5). The ball is manually run into rotation motion and it is observed that when the velocity of the ball, after launching, exceeds a certain value, the motion of the ball is spatial and the contact point moves on the surface of the lens. The angular velocity of the ball decreases due to friction and finally, the contact point becomes immobile and the ball has a motion around the vertical axis passing through the contact point. The rotation motion carries on till resting is settled.

When following launching, the angular velocity of the ball was not high enough, the motion of the ball was from the beginning a rotation around a still axis. Applying a reflective stamp on the ball surface in the equator region, it can be found the rotation velocity of the ball using a non–contact tachometer. In figure 6 there are presented the plots of ball rotational speed for a series of successive launchings. For all launches one can notice the existence of two regions: a zone where a noise exists, corresponding to spatial motion and a region much smoother, corresponding to the motion around an immobile axis.

**Figure 5**. Experimental set-up.

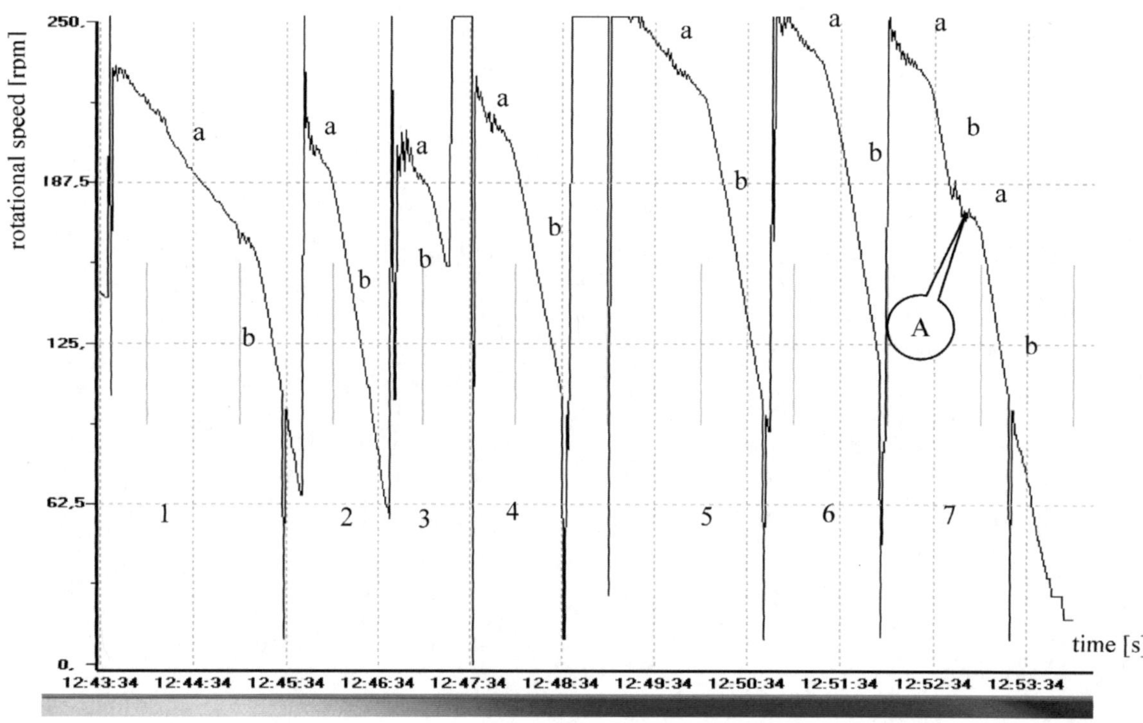


**Figure 6.** Experimental plots of rotational speed versus time.

The slope of the region with noise is smaller than the slope for the smooth region, fact that confirms that when the contact point is mobile the friction torque is smaller than for the case when it is immobile. Additionally, the two regions are connected, fact that confirms that a continuous variation with velocity of the friction torque. On the smooth regions the plots are straight lines and parallel to each others and, thus, they prove the presence of a constant torque, having the same value for all launches. In the A point from the curve corresponding to the last launching it can be noticed a region with noise intercalated within a smooth region. This region was produced by applying a small shock upon the lens after the rotation about the immobile axis was installed. After the shock was applied, the motion of the ball becomes spatial but quickly diminishes and the rotation about the axis restores.

## 4. Conclusions

The paper considers the Hertzian contact of two axi-symmetric bodies. In a first phase, one considers that the relative motion between the two bodies is rotation about common normal. The distribution of elementary friction forces on contact area generates a spinning friction torque. The work studies the effect upon relative motion from contact area of superposition of a relative translation having the same values in all points of contact area.

A theoretical model for the estimation of the total friction torque is proposed. The expression obtained shows that the friction torque reaches maximum for zero translation velocity and decreases rapidly with increasing the translation velocity. However the decrease is not uniform, being more evident at the beginning when the ratio between the transport velocity and angular velocity is comparable to the radius of contact area and afterwards, the decrease of friction torque is much slower.

The last part of the paper proposes a solution for the experimental corroboration of theoretical results. The contact between a ball and a spherical cavity is considered and after launching the ball into rotation motion, two stages can be identified: in a first phase, the ball has a spatial motion and the contact point is mobile while during the second phase the motion of the ball stabilizes and the direction of angular velocity remains still.

**Acknowledgement**

The infrastructure used for this work was partially supported by the project "Integrated Center for research, development and innovation in Advanced Materials, Nanotechnologies, and Distributed Systems for fabrication and control", Contract No. 671/09.04.2015, Sectorial Operational Program for Increase of the Economic Competitiveness co-funded from the European Regional Development Fund.

**References**

[1] Hertz H R 1881 *Ueber die Beruehrung elastischer Koerper (On Contact Between Elastic Bodies)* Journal für die reine und angewandte Mathematik 92 pp 156-171

[2] Glocker C 2001 Set-Valued Force Laws, Dynamics of Non-Smooth Systems. Vol. 1 *Lecture Notes in Applied Mechanics* (Springer-Verlag Berlin) chapter 10 pp 131-154

[3] Leine R I, Glocker C 2003 A set-valued force law for spatial Coulomb–Contensou friction. *Eur. J. Mech. A/Solids* **22** 193–216

[4] Pennestrì E, Rossi V, Salvini P, Valentini P P 2016 Review and comparison of dry friction force models *Nonlinear Dyn.* **83** (4) 1785–1801

[5] Marques F, Flores P, Lankarani H M 2016 On the frictional contacts in multibody system dynamics, *Multibody Dyn. Comput. Methods Appl. Sci.* **42** 67–91

[6] O'Reilly O M 1996 The dynamics of rolling disks and sliding disks. *Nonlinear Dynamics* **10 (3)** 287– 305

[7] Kane T R and Levinson D A 1985 *Dynamics. Theory and Applications* (McGraw-Hill Book Company) pp 15-24

[8] Friction and Friction Coefficients available at http://www.school-for- champions.com/science/ friction_coefficient_greased.htm#.V625nNR94_4

[9] Johnson K L 1985 *Contact Mechanics* (Cambridge University Press) chapter 4 pp 84-106

IOP Conf. Series: Materials Science and Engineering **174** (2017) 012009 doi:10.1088/1757-899X/174/1/012009

# Tribological behaviour of unidirectional carbon fibre-reinforced epoxy composites

**Y Şahin[1] and P De Baets[2]**

[1] Department of Manufacturing Engineering, Gazi University, Besevler-06500, Ankara, Turkey
[2] Soete Laboratory, Ghent University, Technologiepark 903, Zwijnaarde 903, B-9052 Gent, Belgium

E-mail: ysahin@gazi.edu.tr

**Abstract.** Tribological behaviour of unidirectional carbon fibre-reinforced epoxy composites containing 42wt.% (CU42) and 52wt.% (CU52) carbon fibres fabricated by moulding technique was investigated on a pin-on-flat plate configuration. It is the first time to measure static and dynamic coefficient of frictions and wear rates of epoxy composites under heavy loading conditions. Microstructures of composites were examined by scanning electron microscopy (SEM). The experimental results indicated the carbon fiber improved the tribological properties of thermoset epoxy by reducing wear rate, but increased the coefficient of friction. At higher load, average wear rates were about $10.8 \times 10^{-5}$ $mm^3/N.m$ for composites while it was about $38.20 \times 10^{-5}$ $mm^3/N.m$ for epoxy resin. The wear rate decreased with decreasing load while friction coefficient increased with decreasing load. Moreover, friction coefficient of composites of CU42 tested at 90 N load was measured to be in the range 0.35 and 0.13 for static and dynamic component, respectively.

## 1. Introduction

Fibre-reinforced composites have been used successfully for many decades as engineering applications such as aerospace, automobile, chemical, construction and ship/marine industries and sport equipment due to high specific strength and high modulus, wide varieties of availability and design flexibility as compared to metal based counterparts [1, 2]. The application of carbon fibers as reinforcement in polymer composites has been continuously growing during the last few years. The main advantages of such fibers are their specific strength and specific modulus, excellent thermal stability, simplicity of production, low specific gravity, low coefficient of friction and wear, weight savings and performance [3].

The mechanical properties of polymer based composites are greatly influenced by many factors such as the effect of fiber loading, fiber length, fiber, matrix, fibre diameter, processing, interface between the fibre and matrix and orientation having a significant influence on mechanical behavior of polymer composites [4, 5]. The mechanical properties such as hardness, tensile strength, tensile modulus, ductility are improved as the fibers reinforcement content increase in the polymeric matrix materials [6-11]. Over the last two decades, the investigation on wear and friction characteristics of many types of polymer composites were reported, such as carbon fibre reinforced polyamide polymer composite [12,13,14], carbon fibre epoxy composite [15,16,17], carbon fibre/glass epoxy composite [18-23] and fibre length and particle sizes on epoxy composite [24,25] and nano-size effect [26,27].

Also, influence of parameters such as fibre content, fibre angle, type of resin/fibre, process parametres and, especially fibre orientation on the wear rate is very important [5,28-30]. Thus, the aim of the present work is to study the friction and dry sliding wear behaviour under two fiber contents and different loads at a fixed sliding speed for carbon unidirectional-fibre-reinforced epoxy composites using a pin-on-flat wear tester.

## 2. Experimental

### *2.1. Materials*

Both unfilled and carbon-reinforced epoxy composites were manufactured by hand lay-up technique. The carbon fibers were used as a reinforcement and the resin used in this work is commercial SR 8500 epoxy resin and hardener is SD 860x supplied by MCtechnic Ltd., Neitherland. The filler materials used in this study is carbon fabrics supplied by MCtechnic Ltd. The carbon plain weave fabrics is 200 $g/m^2$ and its diameter is about 7-10 μm. The manufacturing process involves the mixing of the epoxy resin with a hardener at a ratio of 100:28. Then the catalyzed resin mixtures are spatulated on composite panel of 150x150 $mm^2$ in size with a thickness of 4.0 mm mold plate. The reinforced of 7 and 10 layers are soaked by the resins. A metal roller is used to compact the laminate so that a uniform thickness could be obtained. A bottom layer and top layer of thee laminate was covered by the mold sheets, which were coated with a release agent for easy separation after curing.

The specimens required for adhesive wear studies (5x5x4 $mm^3$) were cut from the laminated composites by using a water jet machining. Three different types of specimens were prepared for this study. The unidirectional carbon fibres used as filler in range of 0 wt.%, 42 wt.% to 52 wt.%, respectively. Hereafter these composites and epoxy were referred as a CU42, CU52 coded number, and EP coded number, respectively. Microstructure of composites was investigated by SEM of their cross-sections mounted in a transparent resin and polished up to 6 μm by diamond paste.

### *2.2. Friction and wear tests*

The experimental apparatus is a pin-on-flat wear-testing machine with a reciprocating motion. A pin specimen is fixed to a reciprocating stage or to a pin specimen holder by setting screws. The pin is then mounted in a steel holder in the wear machine so that it is held firmly perpendicular to that of the flat surface. The normal load is applied through a spring and lever system that pulls down a transverse bridge structure over composite specimen. Friction force is measured with a strain-gauge detector installed on the wear testing machine. The coefficient of static friction is the friction force that exists between two objects when neither of the objects is moving [5]. The coefficient of kinetic friction is the force between two objects when one object is moving, or if two objects are moving against each other. The material loss of testing samples is calculated from weight loss of composites. Carbon fibre reinforced composite (CFRC) is slid in a reciprocating motion against cold rolled steel AISI 42CrMo4 grinded to perpendicular to the sliding direction. The steel counter face is fixed to a base plate sizes of 58 x 38 x 4 mm with an average surface roughness of 0.42 μm perpendicular to the sliding direction. The sliding stroke is 15 mm. Both composite specimens (CU42, CU52) are tested fibres parallel to sliding direction (p-orientation). The total sliding distance is about 2016 m/3024 m. The wear pin is cleaned in acetone prior to and after the wear tests, and then weighed on a microbalance with 0.1 mg sensitiveness. The wear rate is calculated by measuring the mass loss, density and known sliding distance and load. The specific wear rate (Ks) is then expressed on volume loss basis.

During sliding, the mechanical energy is mainly dissipated as heat. This may lead to deterioration of mechanical properties, which turn often decrease of friction and wear performance [31]. This limits the life time of components made from polymer composites. This is especially a problem in applications involving high contact pressure (p) and sliding speed (v) since this results in a high specific wear rate of energy disipation ($Q_d$), i.e. the rate of dissipated energy per unit apparent contact area:

$$Q_d = \mu * p * v \qquad (J/m^2\ s) \qquad (1)$$

It is seen that $Q_d$ is directly proportonal to $\mu$ and pv condition. Therefore, for polymer composites tested under heavy conditions, the thermal properties must be improved or frictional heating must be reduced by e.g. applying lubricating resins or fillers, which is the case of this study.

## 3. Results and discussion

### *3.1. Microstructure*

Figure 1 shows a typical microstructure of unidirectional fibre reinforced epoxy composites, which are included 42 wt.% and 52wt. % carbon fibres under SEM, respectively. This SEM image at lower magnification indicates a cross-sectional of longitudinal fibres and epoxy matrix, but it does not contain any porosity in the structure. Figure 1 (a) shows the SEM of the microstructure for a CU42 composite. The distributions of carbon fibres in the matrix seemed not to be very uniform. Its dimensions are about 7-10 µm. SEM image in Figure 1 (b) indicates that the distributions of fibres in the matrix seemed to be more uniform than that of CU52 fibre-reinforced composites. Its dimensions are the same with the previous one.

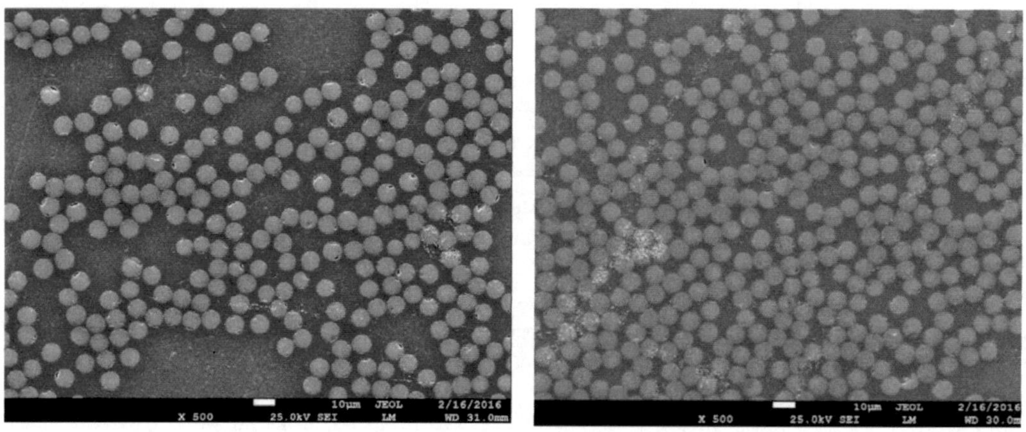

(a) (b)

**Figure 1**. Cross-sectional view of unidirectional carbon fibre-reinforced epoxy composites taken by SEM. (a) Micrograph of CU42 composite, showing fibre distributions in the microstructure, (b) Micrograph of CU52 composite, indicating a more uniform fibre distributions.

### *3.2. Coefficient of friction*

The normal force and the tangential friction force of pin-on-flat for composites and its polymers against the smooth steel were measured in real time using a load cell. Typical plots of variations in the static and dynamic coefficient of friction as a function of the applied load for CU42 fibre-reinforced composites at a fixed speed are shown in Figure 2. In this figure, the symbols used as CU42-us and CU42-udy indicates the static COF and dynamic COF, respectively. The friction coefficient of the studied epoxy composite/steel tribo-pairs at lower load was measured to be in the range 0.45 and 0.17 for static and dynamic component, respectively. Also, the static and dynamic friction coefficients appeared to vary similarly as a function load, but the dynamic friction coefficient exhibited lower values than the static component.

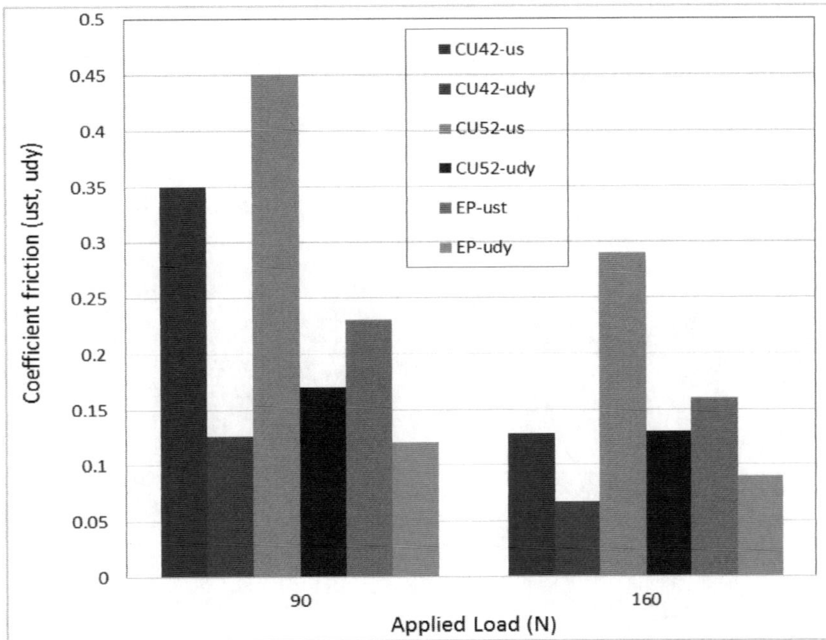


**Figure 2.** Variations in static and dynamic coefficient of friction as a function of applied load for CU42 fibre-reinforced composites, sliding at 0.42 m/s against the smooth steel under two different loads.

Figure 3 shows the variations in static and dynamic coefficient of friction as a function of number of cycles for CU42 fibre-reinforced composites, sliding at 0.42 m/s against the smooth steel under 90 N load and 160 N load. The dynamic friction coefficient exhibited lower values than the static component. However, the static and dynamic friction coefficients appeared to vary similarly as a function sliding distance. The steady state friction coefficient was obtained at $2.5x10^4$ when tested at 90 N load. The static and dynamic friction coefficient obtained was about 0.40, 0.17, respectively. It decreased considerably with increasing the load (Fig. 3b). This might be due to decreasing the ploughing component and thermal softening the surface. In addition, variations in static and dynamic friction coefficient for CU52 fibre-reinforced composites is plotted as a function of time in Fig. 4 under different loads. The friction coefficient increased rapidly throughout the first meters of sliding and subsequently decreased because the running-in period decreased considerably. It was observed for the carbon and glass fibre reinforced of PEEK composite against steel at long sliding [22]. After the initial running in period the friction coefficient stabilized as a friction of sliding distance, which is the case for running time of present results. PEEK CF30 presented the less COF (0.18), followed by PEEK (0.21). However, when the conditions changed the COF increased. COF and wear rate of carbon fibre epoxy matrix composites measured [32]. Initial COF recorded was 0.65 that for the peak after wear in was 0.74. Thus, the COF for composite/composite contact should be considered as being high. Friedrich et al.[26] studied COF for two PA66 composites with and without nano-$TiO_2$. At the beginning of the running in stage, the sliding performance was similar to both. However, the COF and contact temperature for nanocomposites were abruptly reduced after an initial stage of the contact period, lasting about an hour.

The static friction decreased at higher normal load, which was the same trend for the dynamic friction because the time to establish a steady-state friction shortened. This might be due to a reduced contribution of ploughing component and frictional heat for the polymer, which increased the surface temperature, inducing the adhesive, abrasive wear, plastic deformation and fibre breakage.

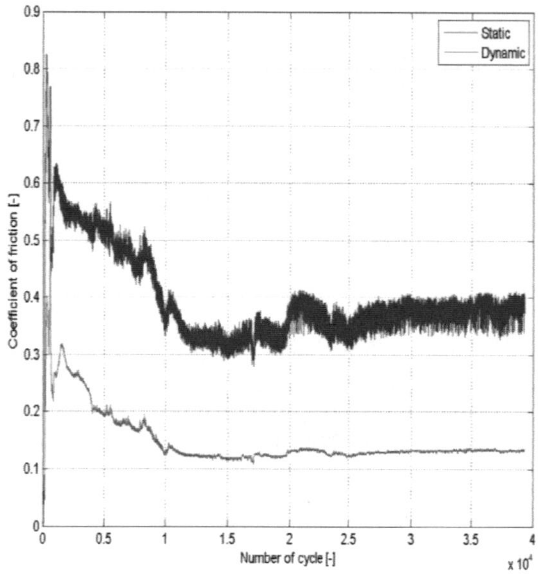


(a) 90 N

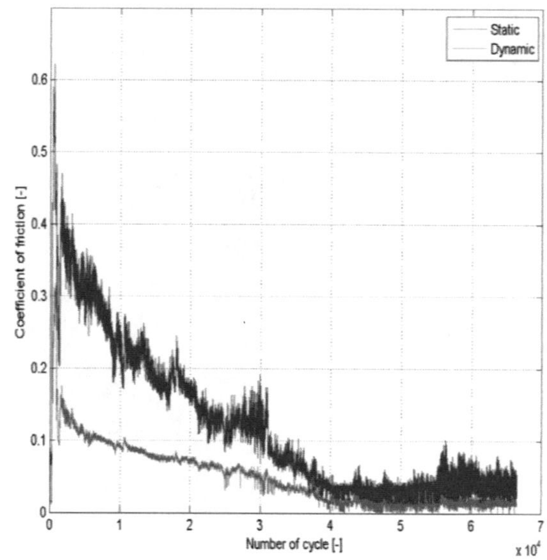


(b) 160 N

**Figure 3.** Variations in static and dynamic coefficient of friction as a function of number of cycles for CU42 fibre-reinforced composites, sliding at 0.42 m/s against the smooth steel under (a) 90 N load, (b) 160 N load.

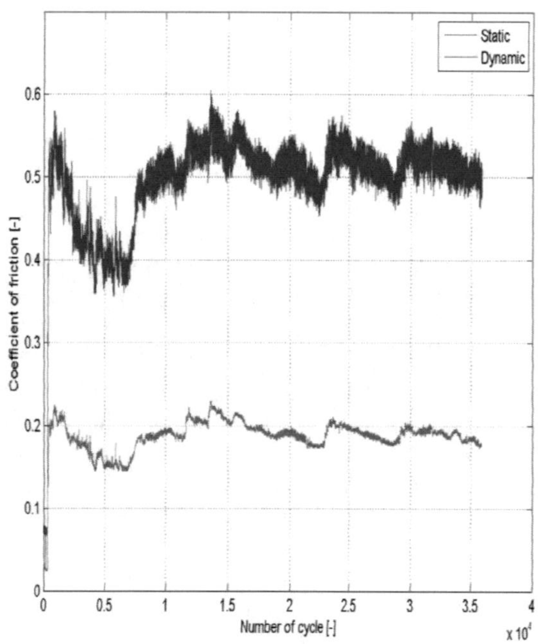


(a) 90 N

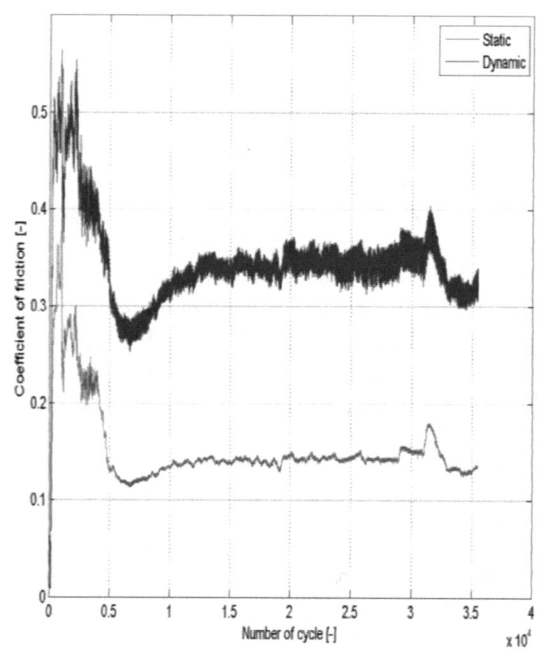


(b) 160 N

**Figure 4**. Variations in static and dynamic coefficient of friction as a function of number of cycles for CU52 fibre-reinforced composites, sliding at 0.42 m/s against the smooth steel under (a) 90 N load, (b) 160 N load.

In addition, large fluctuations occurred for lower load and the initial stage of friction. Wang et al.[17] found the $MoS_2$ filler was effective in reducing the friction coefficient of nylon at 100 N load because transfer film formed by nylon and $10MoS_2$ plus naylon composite. The COF was reduced by 35% by substituting the glass fibre weave with the carbon/aramid weave, epoxy resin (EP), EP/Glass, Carbon+Aramid (C/A) [31]. The best tribological properties were obtained when the fibres were parallel and anti-parallel to the sliding direction due to lubricating effect of carbon fibre. Average COF for G/EP and CA/EP was about 0.63 and 0.41, but EP was about 0.69. Matsunage et al.[33] measured the COF for carbon fibre/epoxy matrix composites in contact with stainless steel under reciprocating sliding. The COF was varied 0.10 – 0.25 with a maximum for transverse fibre orientation. However, the COF varied between 0.45 and 0.65 for graphite/epoxy composite and carbon/PEEK composites, respectively [34].

### *3.3. Wear rate*

The experimental results of the dry wear of epoxy composites at different conditions are shown in Table 2. The tests relevant to this table were carried out at 0.42 m/s speed, but indicated loads. The specific wear rate of composites and its matrix experiments are plotted as a function of the load in Fig.5. There were no significant difference in wear rate between two volume fractions of composites because of the similar alignment of fibres in the matrix. It is evident from the figure that the specific wear increased considerably with increasing the load for the epoxy matrix, but it decreased slightly with increasing the load for CU42 composite. The average wear rate of tested composites at 90 N load was about $12.36x10^{-5}$ $mm^3/N.m$, which was decreased with increasing the fibre content. Typical wear data, obtained throughout reciprocative sliding experiments for composite/steel combinations, are plotted as a function of the applied load in this figure for composites.

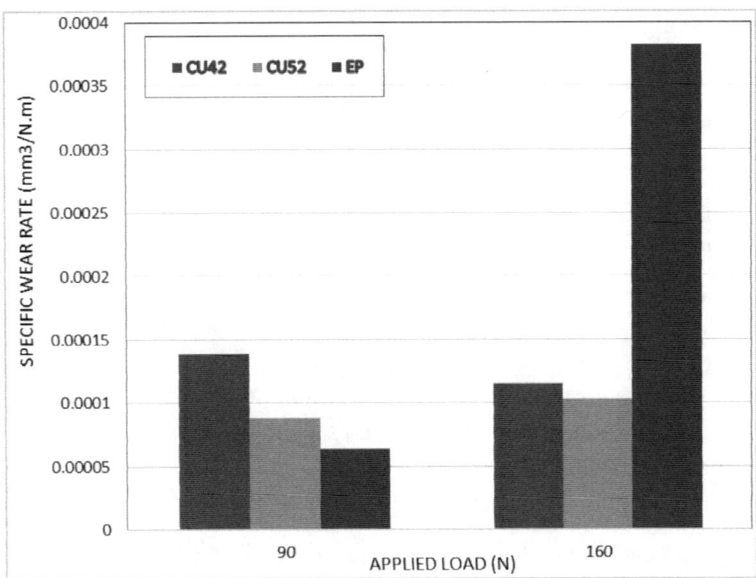


**Figure 5.** Specific wear rate as a function of load for unidirectional carbon fibre-reinforced composites, sliding at 0.42 m/s tested against the smooth steel.

Other reason was that adhesion component became a more effective due to thermal effects. The temperature at the frictional surfaces increased with increasing the load and the frictional heat on composites could not be distributed in time due to the poor ability of heat transfer. It is found that carbon fibre decreased the wear rate for nylon while wear rate increased and transfer film was formed by nylon and $10MoS_2$ plus nylon composite [17]. The aramid fibres did not show exposed fibre ends as opposed to the carbon and glass fibres. It seemed to prevent the micro-cracking of the resin due to their toughness.The carbon fibres seemed to be worn thinner and then ultimately fracture into smaller pieces [31]. The main wear mechanisms were slow mean of the matrix on the original surface and cracking of the fibre/matrix interface causing matrix and fibres to be broken off [32].

## 4. Conclusions

The following conclusions related to tribology of experimental results on carbon fibre-reinforced epoxy composites were drawn:

The microstructure is examined by SEM to see the distributions of fibres in the epoxy matrix. The experimental results indicated the unidirectional carbon fibers improved the tribological properties of the epoxy resin by reducing the wear rate, but increased the coefficient of friction. The wear rates of composites and epoxy under the loads applied varied from $10.8x10^{-5}$ $mm^3/(N.m)$ and $38.2x10^{-5}$ $mm^3/(N.m)$, respectively. The friction coefficient of CU42 and CU52 composites/steel tribo-pairs when tested at 90 N load was measured to be in the range 0.35 and 0.13, 0.45 and 0.17 for static and dynamic component, respectively. Also, the coefficient of friction decreased with increasing the load.

## Acknowledgements

This research is financially supported by the Turkish Scientific and Technical Council Foundation of Turkey (Grant No. 1059B191401938).

## References

[1] Paiva, J M F, Santos A D N and Rezende M C 2009 *Mat. Res.* **12 (3)** 367-374
[2] Paiva J M F , Mayer S and Rezende M C 2005 *Mat. Res.* **8** 91-97
[3] Su F H, Zhang Z Z, Wang K and Jiang W 2006 *Compos. Part A-Appl S* **37** 1351-57
[4] Wonderly C, Grenestedt, Fernlund G and Cepus E *2005 Compos. Part B-Eng* **36** 417-426
[5] Şahin Y 2015 *Kompozit Malzemelere Giriş (Introduction to Composite Materials)*, 3rd Ed. (Ankara, Turkey: Seçkin Yayıncılık)
[6] Rahmani H, Najafi S H M, Matin S S and Ashori A 2014 *Polym. Eng. Sci.* **54** 2676–82
[7] Ary Subagia I D G, Kim Y, Tijing L D, Kim C S and Shon H K 2014 *Compos. Part B-Eng* **58** 251–258
[8] Zhou Y, Pervin F, Jeelani S and Mallick P K 2008 *J. Mater. Proces. Technol.* **198** 445–453
[9] Han S H, Oh H J, Lee H C and Su S S 2012 *Compos. Part B-Eng.* **45** 172-177
[10] Tarpani J R, Milan M T, Spinelli D and Bose W W 2006 *Mat. Res.* **9** 115-120
[11] Jagannatha T D and Harish G 2015 *Int. J. Mech. Eng. & Rob. Res.* **4** 131-136
[12] Kurokawa M, Uchiyama Y, Iwai T and Nagai S 2003 *Wear* **254** 468–473
[13] Kukureka S N, Hooke C J, Rao M, Liao P and Chen Y K 1999 *Tribol. Inter.* **32** 107–116
[14] Wang Q H, Zhang X R and Pei X Q 2010 *Mater. Design* **31** (8) 3761–3768
[15] Dhieb H, Buijnster J G and Celis J P 2011 *Sustainable Construction and Design* **2 (1)** 53-58
[16] Zhou X H, Sun Y S and Wang W S 2009 *J. Mater. Proces. Technol.* **209** 4553-57
[17] Wang J, Gu M and Songhao S G 2003 *Wear* **255** 774-779.
[18] Suresha B, Chandramohan G, Samapthkumaran P, Seetharamu S, Vynatheya S 2006 *J. Reinf. Plast. Comp.* **25** (7) 771-782
[19] Suresha B, Sidaramainh K, Seetharamu S and Kumaran P S 2009 *Wear* **267** 1405-14
[20] Jia J, Chen J, Zhou H, Hu L and Chen L 2005 *Compos. Sci. Technol.* **65** 1139-47
[21] Larsen T andersen T L, Thorning B, Horsewell A and Vigild M E 2007 *Wear* **262** 1013-20
[22] Davim J P and Cardoso R 2009 *Wear* **266** 795-799
[23] Furey B S and Michael J 1993 *Wear* **162** 385-396
[24] Zhang H, Zhang Z and Friedrich K 2007 *Compos. Sci.Technol.* **67** 222-230
[25] Akshatha B S and Gowda DrT 2015 *Int. J. of Eng. Res. & Technol. (IJERT)* **4** 282-286
[26] Friedrich K, Zhang H, Lim L H and Yang J 2005 *Compos. Sci.Technol.* **65** 2329–43
[27] Zhang G, Chang L and Schlarb A K 2009 *Compos. Sci. Technol.* **69** 1029–35
[28] Dwiedi U and Chand N 2009 *Appl. Compos. Mater.***16** 93-100

[29] Prasad M P, Vinod B D and Suder R L J 2014 *Intern. Journal of Engineering Research and General Science* **2** 809-814

[30] Zhang G, Rashea Z and Schlarb A K 2010 *Wear* **268** 893-899

[31] Larsen Q T andersen T L, Thorning B and Vigild M E 2008 *Wear* ***264*** 857-868

[32] Schön J 2004 *Wear* ***257*** 395-407

[33] Matsunaga S, Matsubara T, Wang W X and Takao Y 2001 Effects of reciprocation number on the friction behaviors of carbon/epoxy for various fiber orientations and high contact pressures, *Proc. of the ICCM-13* Beijing China pp ID. 1446

[34] Tripathy B S and Furey M J 1993 *Wear* ***162-164*** 385-396

# Abrasive wear of high velocity oxygen fuel (HVOF) superalloy coatings under vibration load

**M Kandeva[1], B Ivanova[1], D Karastoyanov[2], T Grozdanova[3] and E Assenova[4]**

[1]Faculty of Industrial Technology, Technical University of Sofia, Bulgaria
[2]Institute for Information and Communication Technologies, Bulgarian Academy of Sciences, Bulgaria
[3]Space Research and Technologies Institute, Bulgarian Academy of Sciences, Bulgaria
[4]Society of Bulgarian Tribologists, Sofia, Bulgaria

E-mail: kandevam@gmail.com

**Abstract**. The present paper considers wear of coatings deposited by HVOF (High velocity oxy-fuel) technology, under conditions of dry friction against abrasive surface accompanied with the action of vibrations perpendicular to the sliding axis. Results are obtained with four type coatings: two types with Ni matrix of composition 602P – without preliminary heating of the basic surface (the substrate) and after substrate heating up to 650°C in a chamber; coating WC-12Co with tungsten matrix and coating obtained by 1:1 proportion powder mixture of both compositions 602P and WC-12Co. Results about the thickness, hardness and coating' morphology are presented, as well as dependences of the wear and the relative wear resistance on vibration speeds in the interval 3.03 to 21.08 mm/s. New results are obtained about the nonlinear relationship between abrasive wear and vibration speed showing minimal wear for all specimens by 6.04 mm/s. It is found that lowest wear shows WC-12Co coating in the entire interval of vibration speed variation: 3.03 to 21.08 mm/s.
The obtained results are new in the literature; they are not presented and published by the authors.

## 1. Introduction

Modern equipment operates under impact of dynamical load (collision and vibrations) of various intensities.

Mechanical impact and the resulted vibrations involve machine functional characteristics distortion and failures. Variation in vibration parameters influence friction and wear intensity. The strength and wear resistance of materials under vibration conditions are of crucial importance for their functionality and operational resource. Vibration characteristics vary in large interval in the non-stationary operating regime of tractor engines and gas turbines, the variations depending on the stage of wear. Vibration frequency varies from 180 to 6500 Hz, and the amplitude from 0.5 to 10 μm in faulty engines. It is possible that resonance effects emerge in the case of quick change of operation regime, which results in acceleration of fatigue and other wear mechanism [1÷8].

Abrasive wear mostly accompanied by vibration impacts is the basic reason for putting out of normal service of building, agriculture and mining machinery and equipment [5÷7].

One of the ways for improvement of the exploitation resource of technical systems working under conditions of abrasive-vibrational impact is the improvement of surface wear resistance by coating deposition through the HVOF technology.

HVOF (High velocity oxy-fuel) technology is a modern tribotechnology for high wear resistant coatings applied in components, contact joints and surfaces operating under heavy-loaded conditions of abrasion, corrosion, erosion, cavitations, high dynamic loads, vibrations, etc. in transport, mining and power industry, oil and gas industry, road machinery, etc. [7÷25]

HVOF technology for coating deposition consists in impact interaction between the supersonic speed powder particles in the flame stream and the material surface (the substrate) to be coated [10÷17].

The fuel gas (kerosene, acetylene, propylene and hydrogen) and the oxygen are supplied to the combustion chamber, and the flame of temperature higher than 3000°C flows out with high velocity under high pressure though the nozzle. The powder composition is supplied axially under high pressure to the combustion chamber or to the nozzle from the side of the lower pressure (figure1).

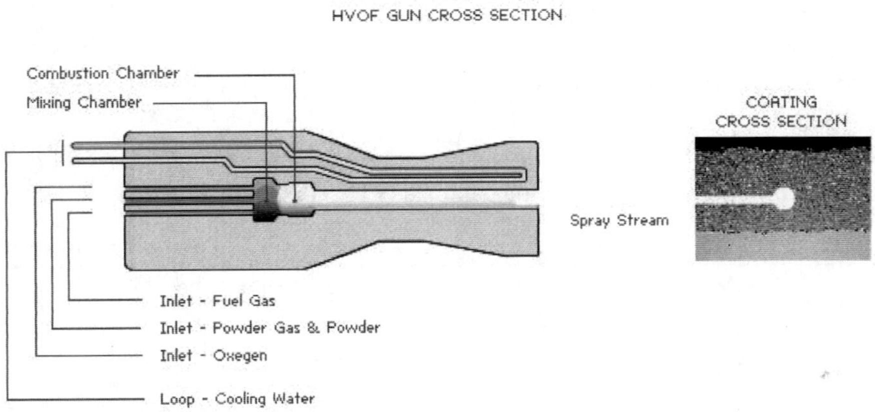


**Figure 1.** Scheme of the system for coating by HVOF (High velocity oxy-fuel) process.

The powder compositions called also powder super alloys, are mixtures of particles of micrometer size composed of different metals, alloys, ceramics, and polymers. The particles brought by the flame steam change into plastic state in the form of a drop, and after the contact with the substrate deform into thin lamellas. The particles-drops cool at the collision with the roughness forming adhesion and cohesion contact bonds with the substrate surface and between each-other structuring the laminar configuration of the composite coating [13÷22].

The present paper sets out the study on the influence of load vibration speed value upon abrasive wear parameters of HVOF-coatings.

## 2. Materials and technology

Four kind coatings are studied obtained by the HVOF process technology, which are deposited on a substrate of equal material, namely steel of composition given in table 1. The substrate hardness is between 193.6 and 219.5 HV.

**Table 1.** Chemical composition (wt. %) of the coated material (substrate).

| Element | C | S | Mn | P | Si | Cr | Ni | Fe |
|---|---|---|---|---|---|---|---|---|
| Percentage | 0.15 | 0.025 | 0.8 | 0.011 | 0.21 | 0.30 | 0.30 | Balance |

Two type powder mixtures of trade names 602P and WC-12Co are used for the coating deposition; their chemical composition, physical and mechanical properties are given in tables 2 and 3.

The particles are of average size 45 ± 2.5 μm; they are obtained by agglomeration process in clouding sintering.

The coatings are deposited by the system of MICROJET+Hybrid model with technological parameters shown in table 4.

**Table 2.** Coating 602P.

| Chemical composition,wt.% | Melting point, °C | Mohs' hardness |
|---|---|---|
| Cr: 13.2 | 1907 | 8.5 |
| Si: 3.98 | 1414 | 7 |
| B: 2.79 | 2076 | 9.5 |
| Fe: 4.6 | 1538 | 4 |
| Co: 0.03 | 1495 | 5 |
| C: 0.63 | 3550 | - |
| Ni: balance | 1455 | 4 |

Surface temperature during the deposition at 120 mm distance: 263°C
Coating adhesion: 42-43 MPa

**Table 3.** Coating WC-12Co.

| Chemical composition,wt.% | Melting point, °C | Mohs' hardness |
|---|---|---|
| Co: 12 | 1495 | 5 |
| C: 2.4 | 3550 | - |
| Fe < 0.1 | 1538 | 4 |
| Ni < 0.1 | 1455 | 4 |
| W:balance | 3422 | 7.5 |

Surface temperature during the deposition at 120 mm distance: 95°C
Coating adhesion: 63 – 69 MPa

**Table 4.** Technological regime parameters for HVOF coating deposition.

| № | Parameter | Technological regime |
|---|---|---|
| 1. | Propylene/oxygen ratio, % | 55/100 |
| 2. | Jet velocity, m/s | 1000 |
| 3. | Distance „nozzle-coating" L, mm | 120 |
| 4. | Angle between orifice axis and surface, $\alpha$ ° | 90 |
| 5. | Air pressure from compressor, bar | 5 |
| 6. | $N_2$ pressure in the proportioning device, bar | 4 |
| 7. | Velocity of powder material feeding, tr/min | 1.5 |
| 8. | Mass flow rate of the powder material, g/min | 22 |

The following coatings are produced: № 1 – with powder compsition 602P on a substrate without preliminary heating; № 2 – with powder compsition 602P on a substrate with 5 min preliminary heating up to 650°C in a chamber; № 3 – with powder compsition WC-12Co without preliminary heating of the substrate; № 4 – with powder compsition of 1:1 mixture of both compositions 602P and WC-12Co, without preliminary heating of the substrate. After the mechanical treatment all coatings obtain equal thickness of 680 μm.

## 3. Experimental procedures

Vibro-abrasive wear tests were carried out on the modified pin-on-drum tribometer which is used for abrasive wear tests without vibrations. The tests were performed in the ambient air at room temperature. A schematic diagram of modified pin-on-drum tribometer is presented in figure 2 [4, 6, 7].

Vertical cylindrical specimen (3) with the deposited coating is positioned perpendicular to the impregnated corundum abrasive paper (5) with grain size of (P120 grit). The abrasive paper is fastened to, and supported by a horizontal cylinder (2). The cylinder driven by AC motor (1) rotates with given constant rotational speed about its horizontal axis. Specimen (3) is fixed in the loading

IOP Conf. Series: Materials Science and Engineering **174** (2017) 012010 doi:10.1088/1757-899X/174/1/012010

head (4) through elastic connection, which allows self-adjustment of the specimen to the abrasive paper (5) and provides possibility for rotation of the specimen around its vertical axis. The loading head (4) engages the static rack bar (7) through the horizontal gear ring (6), providing worm drive and rotation of the specimen around its vertical axis. Static rack bar also provides horizontal movement of the loading head and test specimen (3). In this way a helical wear track is formed. Rotation of the specimen around its vertical axis provides the homogeneous wear in all points of the contact area. Cleaning of the abrasive paper (5) from wear debris is done by an appropriate brush fixed to the loading head (4), together with a vacuum pump.

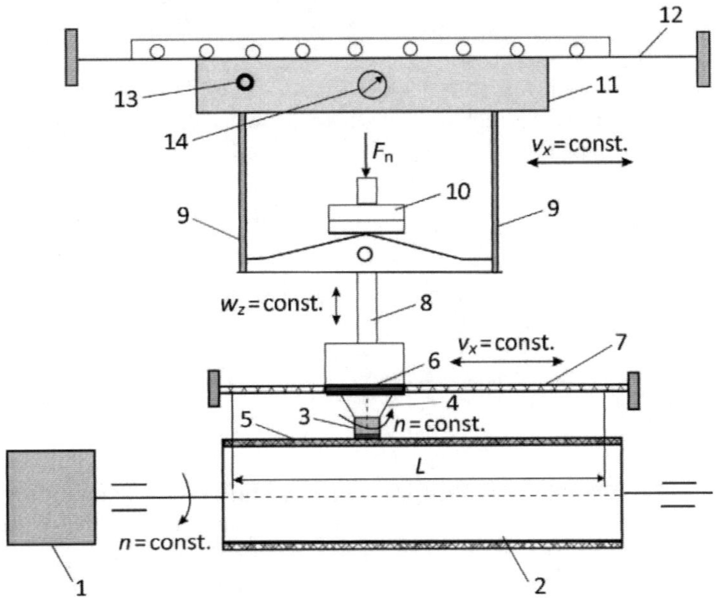


**Figure 2.** Schematic diagram of vibro-abrasive wear testing on modified pin-on-drum tribometer: 1 – AC motor; 2 – drum (horizontal cylinder); 3 – pin (specimen) with coating; 4 – loading head; 5 – abrasive surface (abrasive paper); 6 – gear ring; 7 – static rack bar; 8 – specimen fastening mechanism; 9 – vibrating frame; 10 – loading weights; 11 – vibrator; 12 – vibrator supporting structure with driving mechanism; 13 – on/off button of the vibrator; 14 – regulator of vibrations parameters.

The vibration parameters: vibration displacement [mm], vibration speed [mm/s] and vibration acceleration [mm/s$^2$] are measured by the „PCR-VT 204" vibrometer (figure 3 a,b).

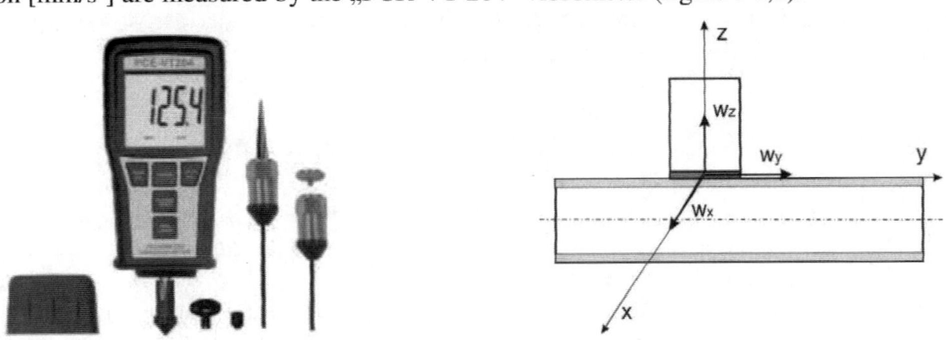


(a) (b)

**Figure 3.** (a) „PCR-VT 204" vibrometer; (b) directions of measured velocity components.

Table 5 shows the vibration components values in three directions: vertical vibration - $w_z$; vibration in sliding direction - $w_y$ and transversal vibration - $w_x$ in x direction (figure 3b) for each indication of the regulator of vibrations parameters 14 (figure 2).

**Table 5.** Values of the vibration velocity components and total vibration velocity.

| **Regulator calibration** | **1** | **5** | **7** | **9** | **10** |
|---|---|---|---|---|---|
| $w_z$, mm/s | 3 | 6 | 9 | 16 | 20 |
| $w_x$, mm/s | 0.2 | 0.35 | 0.8 | 3.8 | 5 |
| $w_y$, mm/s | 0.4 | 0.6 | 0.9 | 3.8 | 4.4 |
| w, mm/s | 3.03 | 6.04 | 9.08 | 16.88 | 21.08 |

Different values of the vibration speed in the interval 3.03 to 21.08 mm/s are given by means of the vibration regulator. The interval is chosen in accordance to the requirements for assessment and classification of the four groups (K, M, Gand T) machines and equipment in ISO 2372 and VDI 2056.

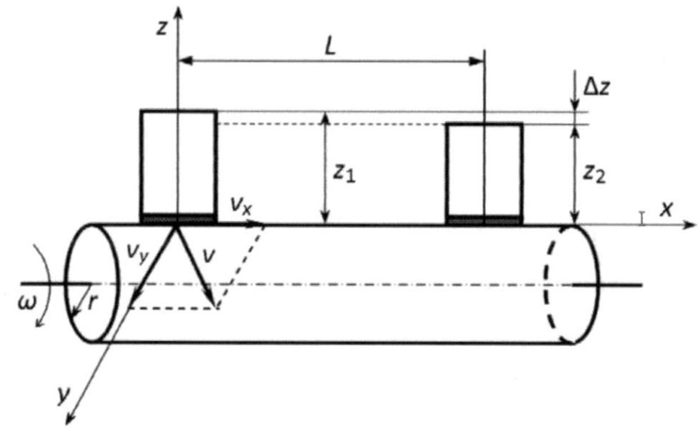


**Figure 4.** Scheme of the sliding speed of the contact area between the specimen and the rotating cylinder.

The two components $V_y$ and $V_x$ of specimen's sliding speed $V$ are determined according to figure 4 by the equations:

$$V_x = \frac{L}{t};\; V_y = r.\omega = r\frac{\pi.n}{30};\; V = \sqrt{V_y^2 + V_x^2} \tag{1}$$

where: L is the friction way passed by the specimen along the generating line of the cylinder (figure 4). In the conducted experiment experiment L = 0,6 m, passed by the specimen in t = 2,28 min. The cylinder radius is r = 0,075 m, and the revolutions of the cylinder are $n = 40$ rpm. The total friction way on the spiral is calculated by the formula:

$$S = v.t.N \tag{2}$$

where $N$ is the number of sliding cycles. The sliding cycles reveal how many times is passed the friction way L.

Each coated specimen is tested at different vibration speeds in the interval 3.03 to 21.08 mm/s for equal friction conditions. The specimens are of cylindrical form with diameter d = 9 mm, height – 30 mm and of equal contact surface roughness Ra = 0.425 μm.

The study was carried out at constant friction parameters: load 3.92 N (0.4kg), nominal contact area $A_a = 0.64.10^{-4}$ m$^2$, nominal contact pressure $p_a = 6.12$ N/cm$^2$, sliding speed v = 0.31 m/s and friction way S = 49.01 m.

The procedure of wear characteristics study consists in measurement of mass wear for each specimen at given constant values of the parameters: load, friction way, sliding speed, abrasive surface type and a given vibration speed. Then a different value of the vibration speed is given under the same sliding conditions, and the mass wear is measured again. The tests are repeated with each specimen for several vibration speed values.

The comparison of the coatings is made based on the parameters: mass wear *m*, relative wear resistance $q_m$ and vibration factor of the abrasive wear $\lambda_w$.

The relative wear resistance is the ratio between friction way *S* and mass wear *m*, i.e.

$$q_m = \frac{S}{m}, \text{ [m/mg]} \quad (3)$$

The vibration factor of the abrasion $\lambda_w$ is dimensionless and is given by the ratio between the specimen's mass wear $m_w$ at the given vibration speed and the mass wear of the same specimen without vibrations - $m_o$:

$$\lambda_w = \frac{m_w}{m_o} \quad (4)$$

## 4. Experimental results

Results about the mass wear, the relative wear resistance and the vibration factor of wear are obtained for the studied coatings at different vibration speed values.

Table 6 shows the results for the mass wear of the four type coatings and the base surface at abrasive friction without vibration and five values of the vibration speed in the interval 3.03 to 21.08 mm/s.

Table 7 gives the interval of variation and the average value of coatings hardness before the wear, after the abrasive friction without vibration and after the friction with vibration at vibration speed 21.08 mm/s.

**Table 6.** Abrasive wear of tested coatings.

| № | **Coatings** | **Vibration velocity, mm/s** | 0 | 3.03 | 6.04 | 9.08 | 16.88 | 21.08 |
|---|---|---|---|---|---|---|---|---|
| 1 | 602 P | Mass loss, mg | 7.2 | 5.8 | 4.9 | 7.2 | 9.6 | 11.5 |
| 2 | 602P+HT 650°C | Mass loss, mg | 4.6 | 3.5 | 2.5 | 4.8 | 6.2 | 7.4 |
| 3 | WC-12Co | Mass loss, mg | 0.5 | 0.4 | 0.4 | 0.5 | 0.6 | 0.6 |
| 4 | 602P+WC-12Co1:1 | Mass loss, mg | 5.5 | 4.1 | 3.2 | 5.6 | 7.3 | 8.2 |
| 5 | coated material | Mass loss, mg | 20.6 | 16.5 | 12.4 | 18.5 | 20.6 | 25.4 |

**Table 7.** Coatings hardness before and after abrasive friction without and with vibration.

| № | **Coating** | Hardness before friction, HV | Hardness, HV w = 0 mm/s | Hardness, HV w = 21.08 mm/s |
|---|---|---|---|---|
| 1 | 602P | 168÷196<br>182 | 165÷193<br>179 | 141÷163<br>152 |
| 2 | 602P+HT650°C | 310÷328<br>319 | 301÷325<br>313 | 290÷309<br>300 |
| 3 | WC-12Co | 405÷441<br>423 | 398÷432<br>415 | 381÷409<br>395 |
| 4 | 602P+WC-12Co 1:1 | 480÷508<br>494 | 460÷498<br>479 | 449÷490<br>470 |

Figures 5, 6, 7 and 8 graphically show the mass wear variation with different vibration speeds for all tested coatings and the substrate.

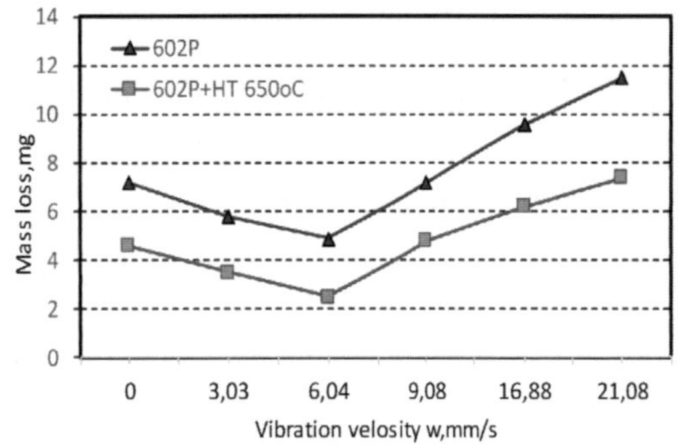


**Figure 5.** Mass loss variation with the vibration speed for coating 602P without and with preliminary heating of the substrate.

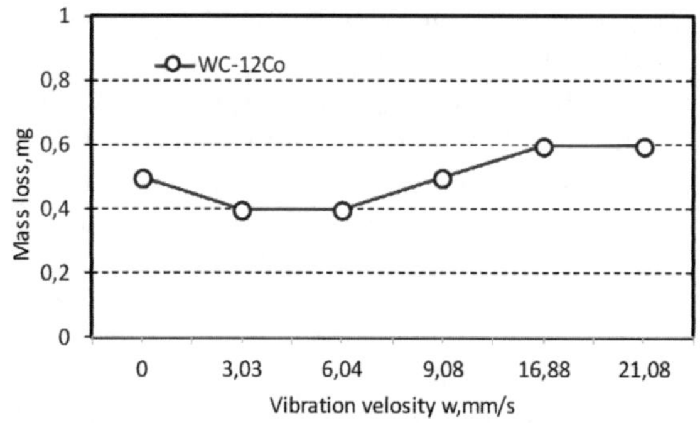


**Figure 6.** Mass loss variation with the vibration speed for coating WC-12Co without preliminary heating of the substrate.

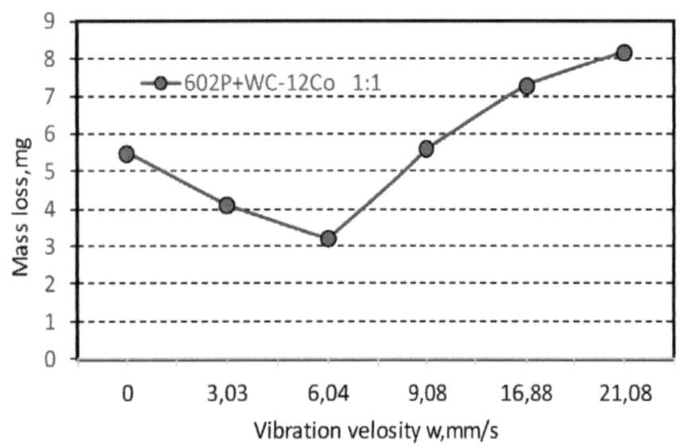


**Figure 7.** Mass loss variation with the vibration speed for coating 602P+WC-12Co 1:1 without preliminary heating of the substrate.

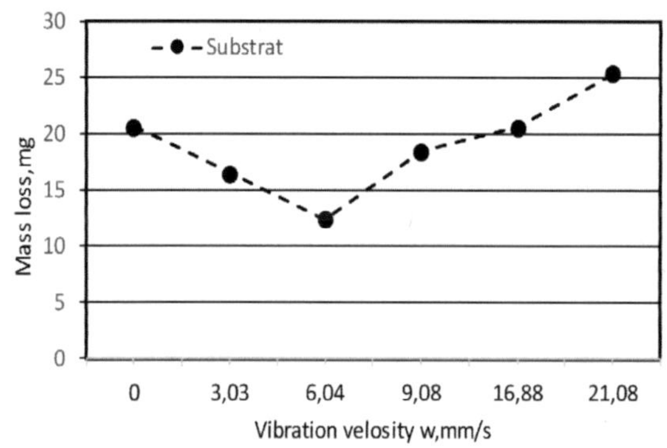


**Figure 8.** Mass loss variation with the vibration speed for the substrate.

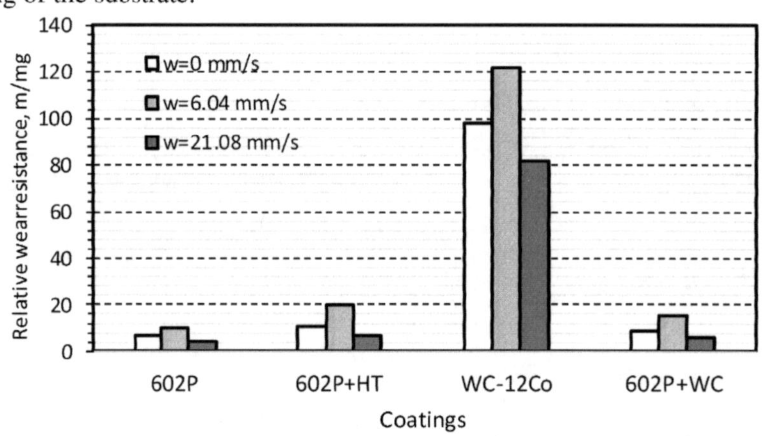


**Figure 9.** Diagram of the relative wear resistance of coatings with and without vibration.

Figure 9 shows the diagram of the relative wear resistance of the coatings for the cases without vibration and for two values of the vibration speed.

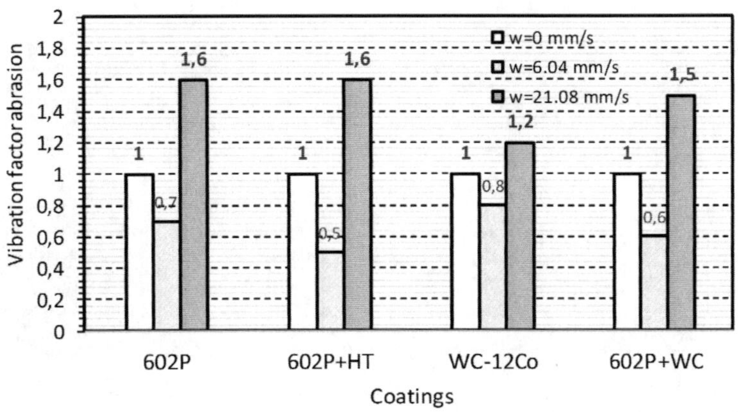


**Figure 10.** Diagram of the vibration factor at abrasion for all coatings.

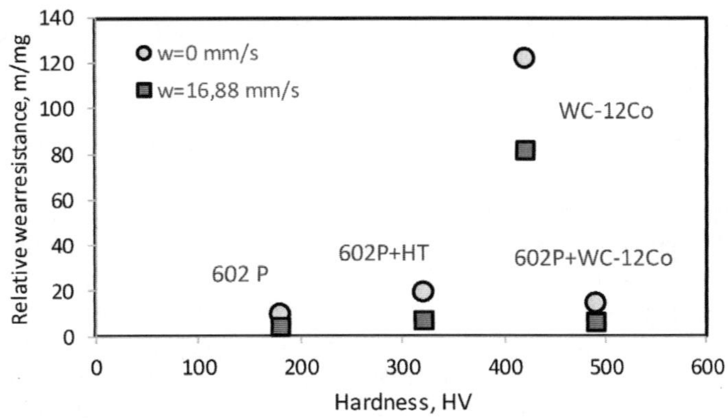


**Figure 11.** Variation of the relative wear resistance with coatings hardness before friction and after friction without and with vibration.

Figure 10 shows the diagram of the vibration factor during abrasion for all coatings, and figure 11 shows the relationship between the relative wear resistance and the hardness of the coatings before friction, after friction without and with vibration and the hardness of the coatings measured before friction.

**Figure 12.** Morphology of 602P coating on steel.

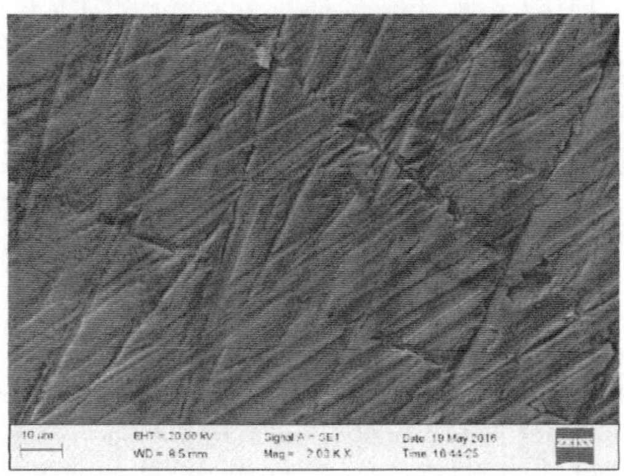

**Figure 13.** Morphology of 602P+HT 650°C coating on steel

**Figure 14.** Morphology of WC-12Co coating on steel.

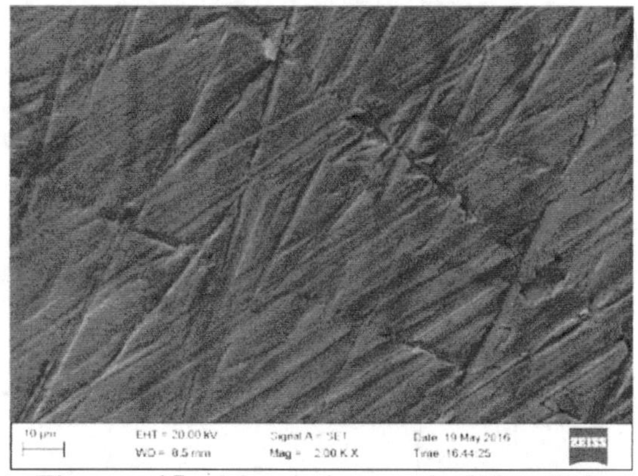

**Figure 15.** Morphology of 602P+WC-12Co coating on steel.

Figures 12 - 15 give the morphology of the coatings after abrasive friction at vibration speed of 21.08 mm/s.

## 4. Discussion and conclusions

The basic interpretations and conclusions of the present study are as follows:

- The value of the vibration speed influences the value and the character of variation of coatings mass wear. The wear decreases up to a given vibration speed value, then increases and for w > 16,88 mm/s the amount of wear attains higher values than those without vibration (figures 5,6,7);
- The relationship between wear and vibration speed shows nonlinear character with a minimum value, which appears for all specimens at vibration speed 6.04 mm/s (figures 5,6,7,8);
- Lowest wear for all specimens in the whole interval of vibration speed shows the coating WC-12Co, the wear of which is a whole order of magnitude lower than those of the other coatings. The relation between wear and vibration speed also shows a minimum at w = 6.04 mm/s, however at higher speeds w > 16.88 mm/s the wear exhibits a constant value close to the value under conditions without vibration (figure 6);
- The preliminary heating of the substrate to 650°C during 5 min at the deposition of 602P coating results in decrease of wear with 55% for the entire interval of vibration speed variation (figure 5). With the change of the vibration speed, the relative wear resistance varies less compared with the other coatings, its value being close to that for the case without vibration (figures 9,10);
- The wear and the relative wear resistance of the combined coating 602P+WC-12Co(1:1) are not much different from those of coatings 602P and 602P+HT650°C (figures 9,10);
- Highest hardness shows the coating 602P+WC-12Co (1:1) before friction, and after friction without and with vibration speed 21.08 mm/s (table 7). This coating however does not exhibit the highest relative wear resistance (figure11). This fact confirms the idea that for composite coatings and materials there is not always a correlation between the hardness and the wear resistance;
- The analysis of the morphology of the coatings after abrasive wear at vibration speed 21.08 mm/s shows that in the case of the coating WC-12Co the basic wear mechanism is mechanical scratch by the abrasive particles during the relative motion. Figure 14 shows that the worn coating has a homogeneous texture in the shape of plane network of scratches obtained by the plane motion of the specimen. The high wear resistance of the coating is related to the specific individual functions of the components in its structure: the WC particles assure high resistance of the coating against the cutting tangential action of the abrasive particles and the fatigue strength to vibrations of the coating; the cobalt (Co) plays the contact role in coating structure building internal boundary network between the WC particles and guaranties the wholeness of the coating. The wear resistance of composite coatings is mostly determined by the thickness and the physical and mechanicals properties of this network. The internal contact network defines in many cases the damping properties of the coating under conditions of pulse and periodical perturbations;

- The basic mechanism of wear for coatings 602P and 602P+WC-12Co (1:1) is fatigue destruction of the surface due to the periodically varying load with high vibration speed (figures 12 and 15).

The overall conclusion is that the operation of machinery and equipment under extreme conditions of vibration load and abrasive friction is optimal in the case of WC-12Co coatings deposited through HVOF technology. The preliminary heating of the substrate during 602P coating deposition leads to 50% wear resistance improvement.

**Acknowledgments**
This study is connected with of the tasks under the Project **DUNK-01/3** "University R&D Complex for innovation and transfer of knowledge in micro/nano-technologies and materials, energy efficiency and virtual engineering" funded by the Bulgarian Ministry of Education and Science, and under CEEPUS III Network: CIII BG 0703-04-1516 "Modern Trends in Education and Research on Mechanical Systems – Bridging Reliability, Quality and Tribology*"*.

**References**
[1] Chen G 2014 *Handbook of Friction-Vibration Interactions (1st ed.)* (Woodhead Publishing: Elsevier)
[2] Stachowiak G and Batchelor A W 2013 *Engineering Tribology (4th ed.)* (Butterworth-Heinemann: Elsevier)
[3] Oskolkov V 2000 *Mechanical contact in conditions of forced oscillations in the range of static friction* Doctoral Thesis (Russia)
[4] Kandeva M, Karastoyanov D and Vencl A 2016 *Advanced Tribological Coatings for Heavy-Duty Applications: Case Studies* (Sofia: Marin Drinov Academic Publishing House)
[5] Kandeva M, Karastoyanov D, Ivanova B and Assenova E 2016 *Tribological Interactions of Spheroidal Graphite Cast Iron Microalloyed by Tin* (Sofia: Marin Drinov Academic Publishing House)
[6] Kandeva M, Grozdanova T, Karastoyanov D, Ivanova B, Jakimovska K and Vencl A 2016 Abrasive wear under vibrations of the spheroidal graphite cast iron microalloyed by tin *J Balk Tribol Assoc* **22** (2A) 1769- 80
[7] Kandeva M and Grozdanova T 2015 *Proc.14th Int. Conf. on Tribology SERBIATRIB'15* Belgrade - Serbia pp 272-278
[8] Pawlowski L 2008 *The Science and Engineering of Thermal Spray Coatings* (2nd ed.) (Chichester: John Wiley&Sons)
[9] Vencl A, Manić N, Popovic V and Mrdak M 2010 Possibility of the abrasive wearresistance determination with scratchtester *Tribol. Lett.* **37** (3) 591-604
[10] Oksa M, Turunen E, Suhonen T, Varis T and Hannula S 2011 Optimization and characterization of high velocity oxy-fuel sprayed coatings: Techniques, materials and applications *Coatings* **1** (1) 17-52
[11] Vencl A 2012 Optimization of the deposition parameters of thick atmospheric plasma spray coatings *J Balk Tribol Assoc* **18** (3) 405-414
[12] Mrdak M R, Vencl A, Nedeljković B D and Stanković M 2013 Influence of plasma spraying parameters on properties of the thermal barrier coatings *Mater Sci Tech* **29** (200) 559-567
[13] Cabral-Miramontes J A, Gaona-Tiburcio C, Almeraya-Calderón F, Estupiñan-Lopez F H, Pedraza-Basulto G K and Poblano-Salas C A 2014 Parameter studies on high-velocity oxy-fuel spraying of CoNiCrAlY coatings used in theater on autical industry *International Journal of Corrosion* **2014** Article ID 703806
[14] Wood R J K and Roy M 2013 Tribology of thermal-sprayed coatings in *Surface Engineering for Enhanced Performance against Wear* ed M Roy (Wien: Springer) pp. 1-43
[15] Chivavibul P, Watanabe M, Kuroda S, Kawakita J, Komatsu M, Sato K and Kitamura J 2010 Effect of powder characteristics on properties of warm-sprayed WC-Co coatings *J Therm Spray Techn* **19** (1) 81-88

[16] Liu Y, Fischer T E and Dent A 2003 Comparison of HVOF and plasma-sprayed alumina/titania coatings – Microstructure, mechanical properties and abrasion behaviour *Surf Coat Tech* **167** (1) 68-76

[17] Peichev I 2010 Application of high-speed technological processes /HVOF and plasma/ for the laying of wear-resistant coatings. Protection of corrosion and erosion in electric power plants *Tribological Journal BULTRIB* **1** (01) ed M. Kandeva (Sofia: TEMTO) 101-107

[18] Peichev I 2010 in *Proc. of the 24th Int. Conf. 65 Years Faculty of Machine Technology* (Sozopol-Bulgaria) pp 561-564

[19] Peichev I 2010 in *Proc. 24th Int. Conf. 65 Years Faculty of Machine Technology* (Sozopol-Bulgaria) pp 565-571

[20] Peichev I, Kandeva M, Assenova E and Pojidaeva V *2011* About the deposition of super alloys by means of supersonic HVOF in *Proc. 12th Int. Conf. on Tribology SERBIATRIB'11* (Kragujevac, Serbia) pp 91-97

[21] Peichev I and Kandeva M 2012 Characterizing of composite coatings deposited through gas-thermal supersound stream (HVOF-process) at different technological regimes *Tribological Journal BULTRIB* **2** ed M. Kandeva (Sofia: TEMTO) 187-197

[22] Peichev I, Kandeva M and Dimitrova A 2012 Wear of composite coatings deposited through gas-thermal supersound stream (HVOF-process) *Tribological Journal BULTRIB* **2** ed M. Kandeva (Sofia: TEMTO) 198-212

[23] Kandeva M, Ivanova B, Pozhidaeva V, Karastoyanov D and Javorova J 2013 in *Proc. of the 5th World Tribology Congress* (Turin, Italy) Paper ID 1251

[24] Kandeva M, Karastoyanov D, Vencl A, Peichev I, Ivanova B and Assenova E 2014 in *Proc. of the Int. Conf. on Tribology and Reliability* (Saint Petersburg, Russia) pp 307-321

[25] Kandeva M and Peichev I 2012 in *Proc. of the 9th Int. Congress on Machines, Technologies, Materials* **2** (Varna, Bulgaria) pp 131-135

# Experimental research on energy losses through friction in order to increase lifetime of hydraulic cylinders

**C Cristescu, R Radoi, C Dumitrescu and L Dumitrescu**

Hydraulics and Pneumatics Research Institute, 14 Cutitul de Argint Street, 040558, Bucharest, Romania

E-mail: cristescu.ihp@fluidas.ro

**Abstract**. Starting from some theoretical considerations, the article presents some experimental results on the assessment of dry and viscous friction forces that occur in the operation of hydraulic cylinders, the basic elements from the structure of hydraulic drive systems. It also presents some specific aspects regarding the durability of the seals of hydraulic cylinders, in order to increase lifetime of hydraulic cylinders, respectively lifetime of hydraulic drive systems.

## 1. Introduction

As it is known, improving energy efficiency of industrial machinery and equipment is a constant direction of activity for researchers and manufacturers of such products. On this line, a couple of theoretical and experimental research activities have been developed in order to find new technical solutions for driving, and also to assess/quantify their effects in terms of energy, with a goal to improve efficiency of various drive systems. Such research activities have also been conducted for hydraulic and pneumatic drive systems (Fluid Power systems), and the results obtained were very good, especially in cases when electronic control and adjustment systems have been comprised in the hydraulic actuators, this meaning also applying information technology in drive processes and developing intelligent hydraulic drive systems. To improve energy efficiency of hydraulic drive systems, an important direction of action was to increase energy efficiency of each individual component, targeting a decrease in fluid flow and pressure losses, in energy losses, generally speaking, and correspondingly targeting increased efficiency rates and prolonged life span of components.

One of the most important components of hydraulic drives is represented by hydraulic drive motors, which convert hydrostatic energy into mechanical energy. There are three types of hydraulic motors, namely: linear, rotary and reciprocating hydraulic motors.

This paper discusses issues concerning the **linear hydraulic motors**, also generically known as "***hydraulic cylinders***" and presents some experimental research activities conducted in the institute INOE 2000-IHP in Bucharest, Romania, aiming to assess the total energy losses in the operation of a hydraulic cylinder and increase the life span of these components.

These total energy losses include viscous friction losses occurring on fluid flow [1], friction losses that occur in the hydraulic cylinder rod seal [2], friction losses that occur in the hydraulic cylinder piston seals [3] and [4], and also dry (or Coulomb) friction energy losses in rod and piston guides.

An important part in increasing the life span of the hydraulic cylinders is played by durability of sealing systems, which are the first causing decommissioning of hydraulic cylinders.

On this line, the famous manufacturers of hydraulic cylinders, figure 1, have designed, tested and adopted new materials and novel constructive and sealing solutions, figure 2, high efficiency [5].

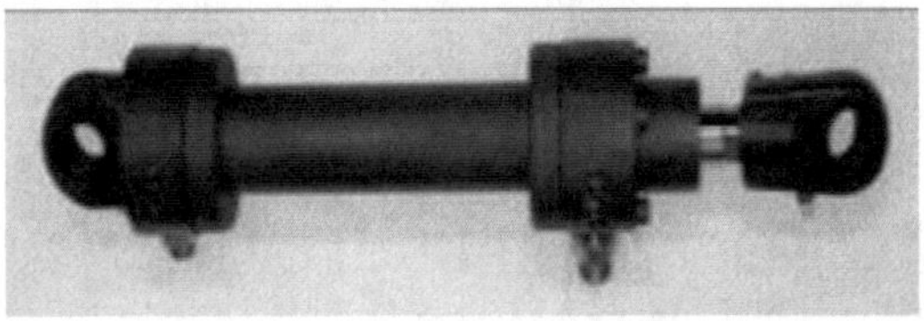

**Figure 1.** Hydraulic cylinder developed by REXROTH.

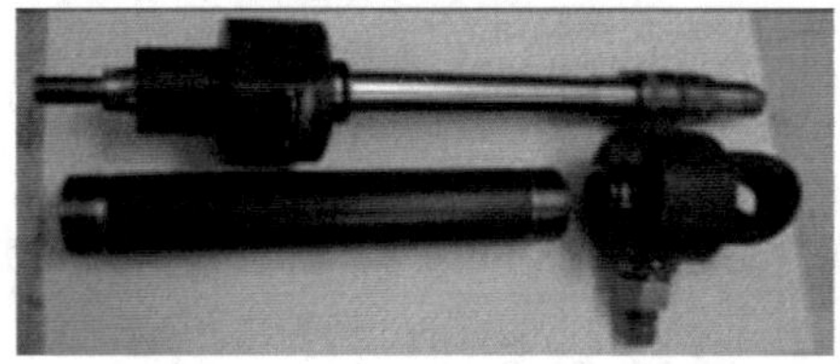

**Figure 2.** Novel constructive solutions.

By using new materials, which have low friction coefficients, companies have abandoned the old sealing systems, based on **U** or **V**-shaped gaskets / sleeves, figure 3, and they have developed novel innovative sealing solutions, which on the one hand decrease friction losses, and on the other hand lead to prolonged life span of seals, and implicitly prolonged life span of hydraulic cylinders, figure 4.

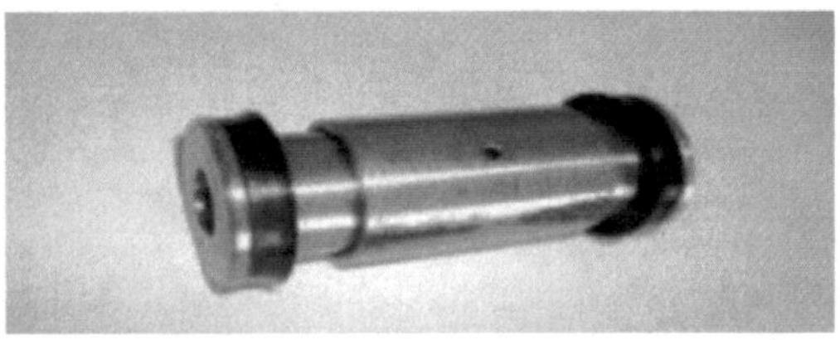

**Figure 3**. Piston seal with U-shaped gaskets.

**Figure 4.** Novel sealing and guidance solutions.

This was the reason why Hydraulics and Pneumatics Research Institute in Bucharest has developed a serious activity of theoretical and experimental research, in order to know the quantitative and qualitative evolution of friction forces during the working cycles, and thus to know all the resistance forces from inside of hydraulic cylinders. For experimental determination of all resistance forces occurring inside of hydraulic cylinders, during the working cycles, there has been designed and developed an experimental device and a test bench, equipped with modern "on-line" system for measuring the evolution of the parameters of interest, which will be presented in the next chapter.

## 2. Presentation of experimental research infrastructure

In order to determine the resistance forces occurring during operation of hydraulic cylinders, there has been designed and developed an experimental device able to simulate in the laboratory the real operating conditions of a hydrostatic drive system.

This device has been installed on a test bench already existing in the laboratory, capable of generating the alternative forces for driving the hydraulic cylinder under tests.

### *2.1. Principle of the experimental device*

The principle at the basis of developing the experimental device has consisted in using a hydraulic cylinder in the current manufacturing, vertically mounted by means of spherical joints, in order to remove side forces of gravity and bending forces. Actuation of the cylinder rod, during the upward and downward stroke, is done by the existing test bench the cylinder is mounted on. Creating the working pressures inside the cylinder, alternatingly on the two piston sides, for the two directions of work, has been performed by means of hydraulic circuits controlled by pressure valves, which allow generation of pressure stages.

### *2.2. Determination of resistance forces*

Determination of total resistance forces is done separately for each of the two strokes, the upward stroke, figure 5, and the downward stroke, figure 6, of the hydraulic cylinder under tests. For each stroke, one shall write the force balance equation, both on the piston and the cylinder liner.

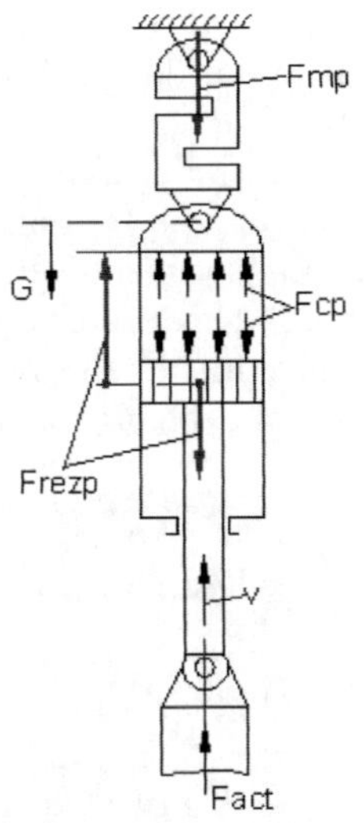


**Figure 5.** Upward stroke.

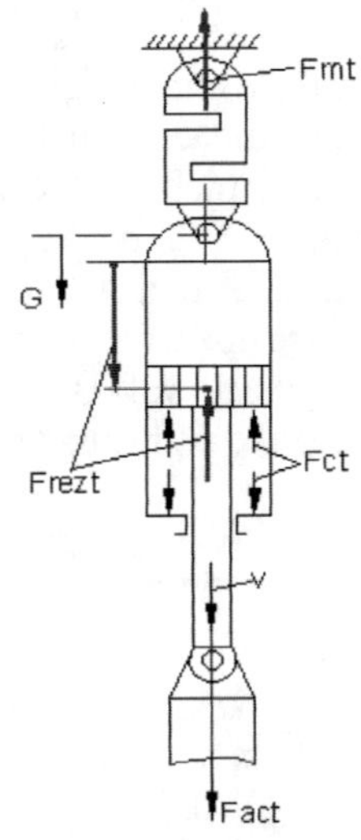


**Figure 6.** Downward stroke.

In the table 1, are presented the forces balance equations on the upward and downward strokes, both for piston and liner of the cylinder on which the transducer measuring the forces $F_{mp}$, and respectively $F_{mt}$, is installed.

**Table 1.** The forces balance equations on the upward and downward strokes.

| Upward stroke, figure 5 | Downward stroke, figure 6 |
|---|---|
| - for piston: | - for piston: |
| $F_{act} = F_{cp} + F_{rezp}$ (1) | $F_{act} = F_{cp} + F_{rezt}$ (3) |
| - for cylinder liner: | - for cylinder liner: |
| $F_{mp} + G = F_{cp} + F_{rezp}$ (2) | $F_{mt} + G = F_{ct} + F_{rezt}$ (4) |

As one can see in figures 5 and 6, in the above equations, forces $F_{cp}$ and $F_{ct}$ are the hydraulic forces generated by the pressure step (p) at which testing is conducted, applied on the large side of the piston, respectively on its small side. They are calculated automatically by the computing system of the data acquisition and processing system. The forces $F_{mp}$ and $F_{mt}$ are experimentally measured by the force transducer that cylinder liner is suspended to and they are sent to the computing system. The force of gravity G is the total weight of hydraulic parts (cylinder liner, fittings, pressure transducers, gauges and hydraulic hoses) mounted on the cylinder liner. The unknown forces are the resistance forces $F_{rezp}$ and $F_{rezt}$, which decrease performance of hydraulic cylinders and represent losses in the system.

Experimental research aim is exactly to determine these resistance forces ($F_{rezp}$ and $F_{rezt}$) and relate them to the calculated nominal forces, produced by the working pressures.

From the above equations (2) and (4), by removing indices p and t, we obtain the calculation relation for resistance forces, in the form:

$$F_{rez} = F_m - F_{cp} \pm G \quad (5)$$

In equation (5), for the upward stroke the force of gravity G enters with minus (-) sign, while for the downward stroke, it enters with plus (+) sign. The computerized system acquires all data, automatically calculates the resistance forces and plots the variation charts for all parameters.

### *2.3. Presentation of the experimental device*

Based on the principle of measuring the forces, shown in figures 5 and 6 and quantified by equation (5), there has been designed the experimental device shown in figure 7, which has been physically developed and installed on the test bench, as one can see in figure 8.

Structure of the experimental device is as follows: hydraulic cylinder under tests HC, spherical joint AS, force transducer FT, bench actuation rod AR, ensuring performing of work strokes, manometer M1 and pressure transducer TP1, on the large piston side, manometer M2 and pressure transducer TP2, on the small piston side, and also hydraulic hoses needed for connecting to the oil tank, by means of two check valves and respectively for connecting to the proportional pressure valves. The pressure steps are controlled directly from the computer keyboard and are kept constant.

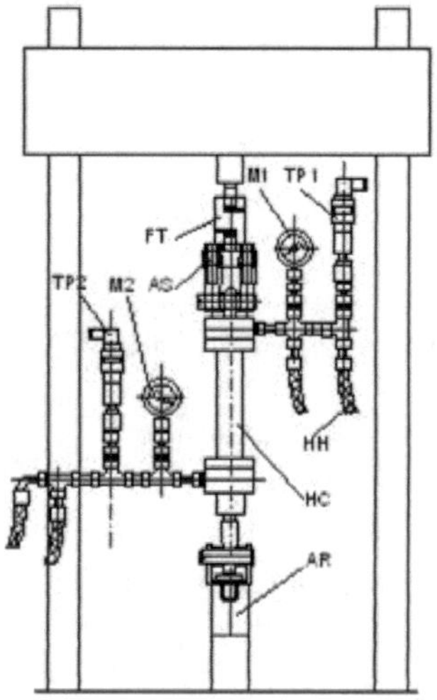


**Figure 7.** Layout of the experimental device.

**Figure 8.** The experimental device installed on the test bench.

When performing the both working strokes, the space generated by piston is filled with oil sucked from the tank by means of check valves, the oil being kept at the set pressure, controlled by proportional pressure valves, so are maintained constant the working forces.

## 3. Presentation of the experimental stand

To conduct experimental determinations of resistance forces occurring in operation of hydraulic cylinders, there has been necessary to develop a complex experimental stand, comprising both the experimental device and the test bench, and also IT system. Figure 9 presents an overview of the experimental stand, and in figure 10 one can see in detail the data acquisition and IT system.

The existing testing bench, presented in figure 11, by its own hydraulic unit, provides vertical movement in either direction; the movable rod (AR) of the hydraulic cylinder of the existing stand ensures the reciprocating movement to the rod and piston of the experimental device.

In figure 12 one can see the pressure transducer and manometer for the pressure above the piston, and figure 13 shows the pressure transducer and manometer for the pressure below the piston.

**Figure 9.** Overview of the experimental stand.

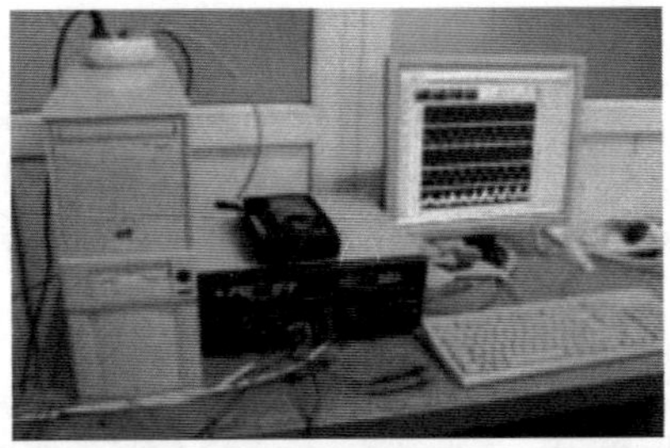

**Figure 10.** View of the IT system.

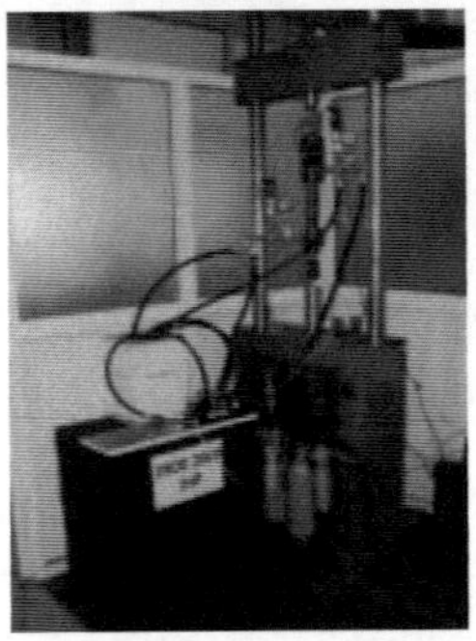

**Figure 11.** The experimental device on the test bench.

**Figure 12.** The pressure transducer and manometer above piston.

**Figure 13.** The pressure transducer and manometer below piston.

Measuring of the piston working stroke inside the cylinder body is done by the stroke transducer presented in figure 14. Figure 15 presents the force transducer used to measure the resistance forces in the piston seals. The oil temperature can be seen on the temperature transducer shown in figure 16.

**Figure 14.** The special stroke transducer of the existing bench.

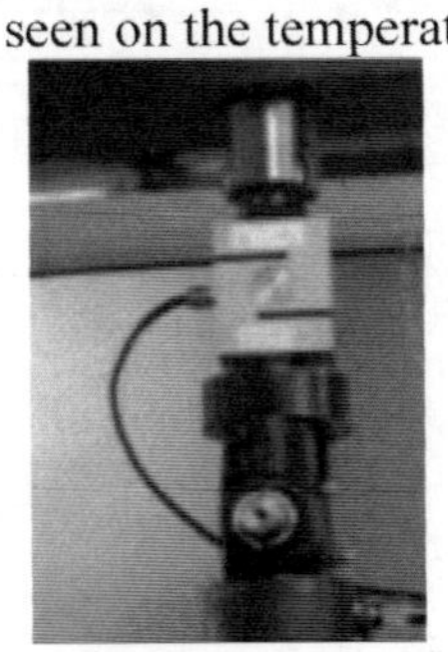

**Figure 15.** The force transducer.

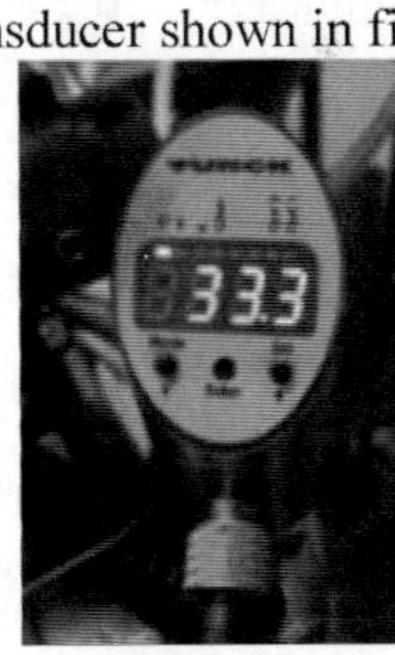


**Figure 16.** The temperature indicator.

The experimental stand has other transducers as well, such as a digital thermometer for the ambient temperature and also a flow transducer which can measure the oil flow in the main pump of the stand.

By means of special electric cables all signals provided by transducers reach the acquisition board installed on the computer, and this one, based on specialized software, allows capturing, storage and processing of the measured data.

## 4. Some experimental results

In order to measure and record variation of resistance forces which occur within hydraulic cylinders there has been conducted experimental research which led to obtaining a lot of graphical experimental results; some of them are presented bellow.To this end, there has been necessary to set the parameters of interest and define a testing methodology. The object subjected to an experimental research was a hydraulic cylinder manufactured by Rexroth, having the piston diameter of 40 mm, rod diameter of 25 mm and stroke of 160 mm.

The dimensions of the tested hydraulic cylinder are identical with the dimensions of the main elements of the research devices used to determine the friction forces in rod seal [2] and piston seal, [3] and [4], precisely in order to create the possibility for a comparison of the obtained results.

The methodology for testing the hydraulic cylinder in order to determine total resistance forces occurring during its operation consists in making a series of strokes (1-3 strokes), with certain pressure on the large piston side, during the upward stroke, respectively on the small piston side (on the rod), during the downward stroke. Pressure steps have been created by means of proportional pressure valves (SPP1 and SPP2), with values: 25 bar, 50 bar, 100 bar, 150 bar and 200 bar - the maximum working range of valves. For each pressure step there have been carried out, as a rule, three determinations with three consecutive cycles.

Following the experimental research conducted, there has been obtained a complete set of experimental results, for 5 steps of pressure, with 3 determinations for each step, each measurement having 3 consecutive working cycles. There have been carried out 45 experimental determinations.

For each measurement, there were obtained the complex variation diagrams for the following parameters: the measured forces ($F_{mp}$ and $F_{mt}$); the working pressure on the large piston side ($p_1$) and also on the small piston side ($p_2$); the working stroke (s) and working speed (v). Automatically, by means of the computerized system, there are obtained the values for pressure forces calculated, $F_{cp}$ and $F_{ct}$, resistance forces $F_{rezp}$ and $F_{rezt}$, and reports of resistance forces and calculated forces.

The obtained complex graphical variations create a complete picture over the variation of resistance forces during pressure increase.

An example of numerical values of the parameters of interest, obtained at a set of experimental determinations, can be seen for the upward stroke, with pressure on the large piston side, in table 2, and for the downward stroke, with pressure on the small piston side, in table 3.

**Table 2.** Parameters on the upward stroke.

| $P_1$ [bar] | $F_{mp}$ [daN] | $F_{cp}$ [daN] | $F_{rezp}$ [daN] | $F_{rezp}$ / $F_{cp}$ [-] | $F_{rezp}$ / $F_{cp}$ [%] |
|---|---|---|---|---|---|
| 25 | 507.12 | 491.72 | 27.40 | 0.056 | 5.6 |
| 50 | 819.44 | 801.93 | 29.51 | 0.037 | 3.7 |
| 100 | 1407.53 | 1390.00 | 33.36 | 0.021 | 2.1 |
| 150 | 2012.60 | 1996.20 | 43.98 | 0.017 | 1.7 |
| 200 | 2654.60 | 2599.72 | 66.88 | 0.026 | 2.6 |

**Table 3.** Parameters on the downward stroke.

| $P_2$ [bar] | $F_{mt}$ [daN] | $F_{ct}$ [daN] | $F_{rezt}$ [daN] | $F_{rezt}$ / $F_{ct}$ [-] | $F_{rezt}$ / $F_{ct}$ [%] |
|---|---|---|---|---|---|
| 25 | 233.20 | 238.10 | 4.90 | 0.020 | 2.0 |
| 50 | 411.92 | 423.52 | 21.00 | 0.027 | 2.7 |
| 100 | 730.06 | 784.31 | 54.25 | 0.069 | 5.3 |
| 150 | 1062.16 | 1151.25 | 89.08 | 0.077 | 7.7 |
| 200 | 1386.81 | 1537.41 | 150.62 | 0.098 | 9.8 |

There are obtained a lot of complex graphs resulted for each pressure steps and a certain speed step. One experimental result is given in figure 17, where there are plotted the complex characteristic graphs for pressure step values of 200 bar and high speed (theoretically about 100 mm/s). The graphs shown in figure 17a are taken directly from the PC screen, while the graphs shown in figure 17b are the same, but delivered by the computer. From the above graphs one can notice **the repeatability in value** for the 3 consecutive cycles, which substantiates the accuracy of experimental determinations for all parameters of interest (stroke, speed, pressures, and measured forces).

From tables 2 and 3 presented above one can notice that **the total resistance forces** occurring in the operation are proportional to the working pressure of the hydraulic cylinders.

Based on the experimental data listed in tables 2 and 3, there have been plotted the variation graphs for measured forces $F_{mp}$ and $F_{mt}$, shown in figure 18, for calculated forces $F_{cp}$ and $F_{ct}$, in figure 19, and also for resistance forces, as quantities in figure 20 and as a percentage in figure 21.

These graphs obtained have been analyzed in detail and there have been drawn some particularly interesting conclusions.

From the graphs in figures 18 and 19 one can see that the measured forces ($F_{mp}$) and the calculated forces ($F_{cp}$) are higher on the large piston side, compared to those on the small piston side ($F_{mt}$ and $F_{ct}$), which is normal, given the different piston surfaces.

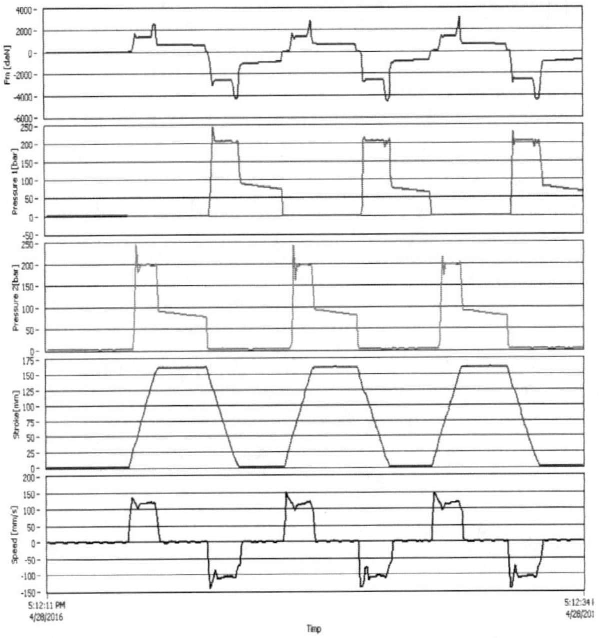


**Figure 17.** The complex graphs for pressure step of 200 bar.

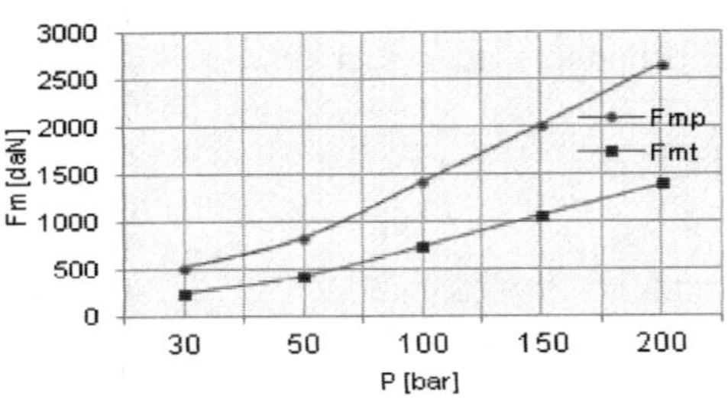


**Figure 18.** Variation of measured forces.

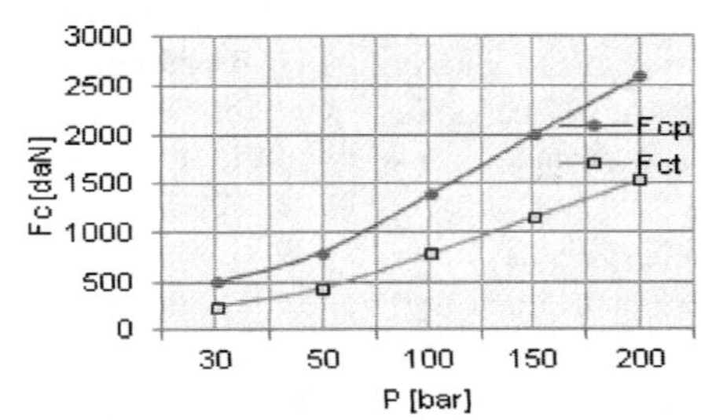


**Figure 19.** Variation of calculated forces.

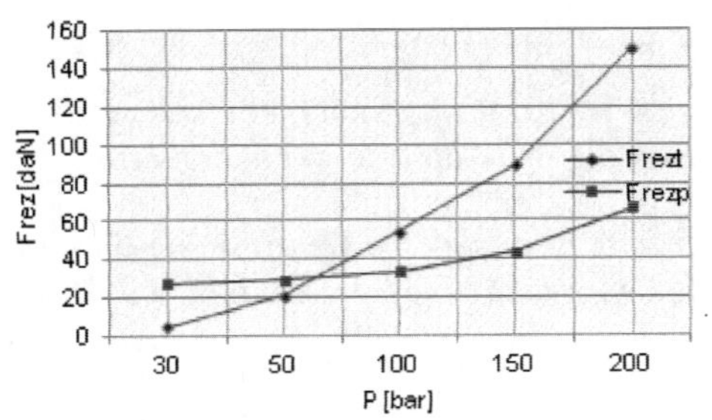


**Figure 20.** Variation of resistance forces.

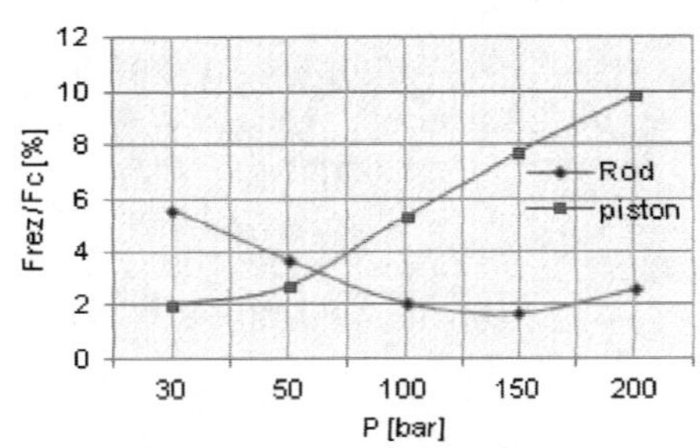


**Figure 21.** Percent variation of the $F_{rez}/F_c$ ratio.

In figure 20 one can notice that, for the pressures greater than 50-60 bar, the resistance forces measured on the small piston side ($F_{rezt}$) are higher than the resistance forces on the large piston side ($F_{rezp}$). This can be explained by the fact that, when pressing ia made on the piston, the resistant forces are generated by friction of the seal and the guidance of the piston, and when the pressing is on the rod, the resistance forces are generated by the friction of the seal and guide on rod and, also, by the friction on the seal and guide of the piston. At pressure under 50-60 bar, the friction forces on seals are most important, compared to the guides forces and, therefore, the forces on the piston are largest.

As a matter of fact, the graphs in figure 21 show that, as a percentage, the ratio of resistance forces ($F_{rez}$) and theoretically calculated forces ($F_c$), which is **the expression of energy losses** over the two drive strokes of hydraulic cylinders, is higher on the retraction stroke of the rod than on the extension stroke, and active useful forces are higher on the extension stroke.

In these conditions, energy efficiency is much lower in the retraction stroke of the rod, where losses are (3 – 10) %, and higher in the extension stroke, where losses represent only (3 - 5) %. These values are for innovative materials and novel sealing and guidance solutions, which are much lower than those indicated in the literature, based on traditional solutions, generically estimated at about (10 – 15) %, but no difference is made in terms of losses over each of the two strokes [6].

## 5. Conclusions

This paper presented the infrastructure and the results obtained following the development of an experimental research, on assessment of total resistance forces occurring in operation of hydraulic cylinders, in order to determine energy losses and efficiency for the two working strokes of cylinders.

Following the design and development of an experimental device and test bench, there have been obtained several complex numerical data and variation graphs for the parameters of interest, which enabled assessment of resistance forces, and respectively assessment of energy losses.

By using a high performance hydraulic cylinder, it has been demonstrated not only that resistance forces are **different as percentage for the two strokes** of the hydraulic cylinder, but also that they are higher on the small side of the piston, where active force developed by the hydraulic cylinder is lower and, therefore, energy efficiency, when performing the retraction stroke, is much lower than the efficiency when performing the extension stroke. On the extension stroke, the losses are much lower, and on the retraction stroke, the losses are higher, but not as high as reported in the literature.

These results and conclusions represent an original contribution of the paper authors.

Decrease of energy losses by adopting novel materials and innovative guiding and sealing solutions results in **increased life span** of the hydraulic cylinders used in hydrostatic drive systems.

**Acknowledgment**

Paper has been made with financial support of ANCSI, under the research NUCLEU Program -2016, project *Physics of processes for reducing energy losses and developing renewable energy resources by use of high-performance equipment*, project no. PN 16.40.03.01, financial agreement no. 5N/2016.

## References

[1] Cristescu C, Dumitrescu C and Vrânceanu G 2016 Consideration on energy loss in hydraulic drive systems *Hidraulica* **1 (2016)** 36 – 46

[2] Cristescu C, Drumea P, Guta D I, Anghel S and Crudu M 2011 Experimental measurements for determination of frictional forces within the rod seals of hydraulic cylinders *Proc. 7th Int. Conf. on Tribology* (Thessaloniki) pp 163-170

[3] Cristescu C, Drumea P and Heilp O 2012 Experimental Researches for Determining the Friction Forces in the Piston Seals of the Hydraulic Cylinders *Proc. 17th Int. Sealing Conf. ISC-2012* (Stuttgart) pp 473-482

[4] Cristescu C, Dumitrescu C, Radoi R and Dumitrescu L 2014 Experimental Research for Measuring Friction Forces from Piston Sealing at the Hydraulic Cylinders *Proc. 8th Int. Conf. on Tribology-BALKANTRIB'14* (Sinaia) pp 341-349

[5] Rexroth Bosch Group, Industrial Hydraulic Cylinder available at https://www.boschrexroth.com/en/xc/products/product-groups/industrial-hydraulics/cylinders/index (accessed on 12.05.2016)

[6] Marin V, Moscovici R and Teneslav D 1981 *Sisteme hidraulice de acţionare şi reglare automată* (Bucureşti: Editura Tehnică) pp 46-61

# Structural-energetic regularities of tribocontact evolution*

**S Fedorov**
Kalilingrad State Technical University, Kaliningrad, Russia

E-mail: fedorov@klgtu.ru

**Abstract.** The contact of friction lives with its own very interesting life. It gets evolved. This evolution of contact changes the states and properties of it. How can we describe the tribocontact evolution? Any nature systems are very well described by thermodynamic method. The contact of friction has the system properties and therefore one may use thermodynamic approach. From thermodynamic point of view friction is a competition of two simultaneous, interconnected and opposite tendencies of accumulating latent (potential) energy of various kinds of defects and damages of contact volumes structures and releasing (dissipation) energy due to various relaxation processes. This friction adaptive-dissipative model for analysis of evolution of tribocontact as elementary tribosystem is examined.

## 1. Introduction

This paper in its basis is a logical completion of axiomatic analysis of sliding friction (rolling) within the framework of triboergodynamics, a scientific trend suggested by the author [1].

The general evolution regularities of states and properties of tribosystem in the frame of triboergodynamics are analysed. Triboergodynamics is based on our modern knowledge of friction: 1. friction is a phenomenon of resistance to relative motion between two bodies, originating at their surfaces of contact area; 2. friction is the process of transformation and dissipation of energy of external movement into other kinds of energy; 3. friction is the process of elasto-plastic deformation localized in thin surface layers of rubbing materials.

Methodology of triboergodynamics [1] is based on the analysis method to plastic deformation of ergodynamics of deformed solids [2-4]. Ergodynamics is a synthesis to the problem of deformation most general laws of thermodynamics for non-reversable processes, molecular kinetics and dislocation theory in their mutual, dialectical tie on the basis of a most general law of nature – the law of energy conservation at its transformations.

Within the framework of triboergodynamics the model of elastic-plastic deformation of contact volumes is examined as a generalized mechanism of transformation and dissipation energy and determines essence of resistance to surfaces displacement.

Friction is regarded as a global (energetical) phenomenon of relative movement transformation. It strongly obeys equation of energy balance and from thermodynamic point of view it is a competition of two simultaneous, interconnected and opposite tendencies of accumulating latent (potential) energy $\Delta U_e$ of various kinds of defects and damages of contact volumes structures and releasing (dissipation) energy $Q$ due to various relaxation.

The major distinction of triboergodynamics from general ergodynamics of deformed solids is «scale factor» which exhibits itself in existence of critical friction volume. This volume determines the

* This paper was accepted for publication in Proceedings after double peer reviewing process but was not presented at the Conference ROTRIB'16

limit friction parameters and separate, in essence, the surface deformation from the traditional volume deformation.

## 2. Structural-energetic interpretation of friction process

As it is known friction is characterized as a product of frictional forces $F$ by friction distance $\ell$, that is, the work $\omega_f$, expended on overcoming frictional forces

$$\omega_f = F\ell, \tag{1}$$

According to the energy balance scheme (figure 1) for plastic deformation and fracture [4] presented below, equations for elementary friction work $\omega_f$ (without lubrication) has the view:

$$\omega_f = \Delta u_e + q, \tag{2}$$

$$\dot{\omega}_f = \dot{u}_e + \dot{q}. \tag{3}$$

Here $\omega_f = d\omega_f / dt$ is a power of friction dissipation of energy; $\dot{u}_e = du_e / dt$ is the rate of storing latent energy in deformed (contact) volumes; $\dot{q} = dq/dt$ the power of thermal effect of plastic deformation (friction).

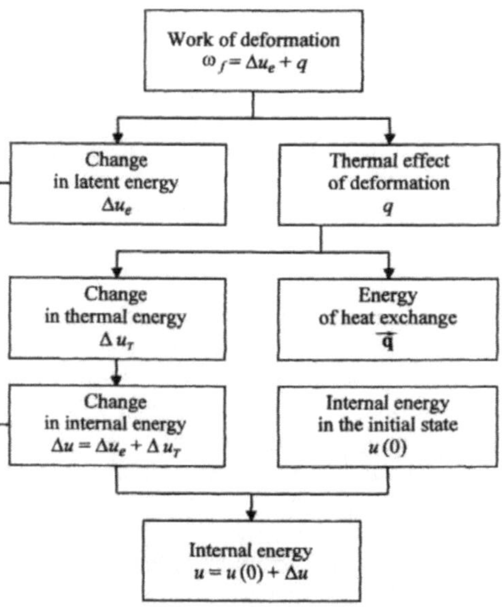


**Figure 1.** Scheme of the energy balance for the plastic deformation of a solid body [1-4].

Since the contact volumes of both materials of the friction pair (figure 2) are deformed, Equations (2) and (3) should be rewritten as:

$$\omega_f = \Delta u_{e1} + \Delta u_{e2} + q_1 + q_2, \tag{4}$$

$$\dot{\omega}_f = \dot{u}_{e1} + \dot{u}_{e2} + \dot{q}_1 + \dot{q}_2. \tag{5}$$

These equations show, that from thermodynamic point of view, the work $\omega_f$ of friction forces, (friction power $\dot{\omega}_f$) may be divided conventionally into two specific parts.

The first part is related to variation of the latent (potential) energy ($\Delta u_{e1}$ and $\Delta u_{e2}$) in deformed (contact) volumes. This is the energy of various simple defects and damages which are generated and accumulated in the bulk. This energy is unique and the total characteristic of submicro-and

IOP Conf. Series: Materials Science and Engineering **174** (2017) 012012 doi:10.1088/1757-899X/174/1/012012

microstructural variations occurring in plastically deformed volumes [2-4].This is a measure of strain hardening and damage of material.

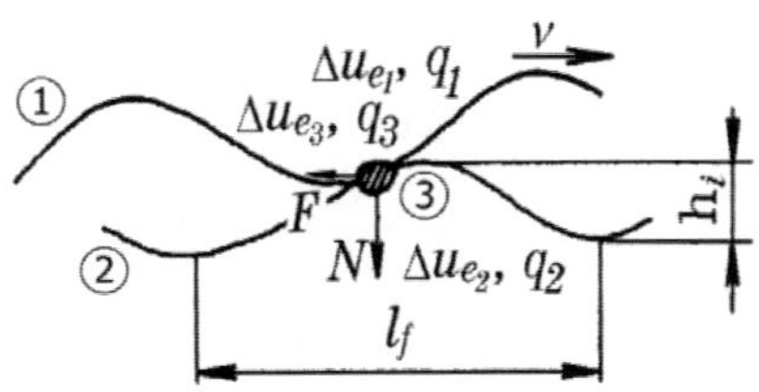


**Figure 2.** Schematic view of friction's contact [1].

The second part of the friction work $\omega_f$ is related to dynamic recovery which is accompanied by releasing latent energy and thermal effect of friction ( $q_1, q_2$ ).This energy involves displacement and annihilation of various simple defects of opposite sign terminating at the free surface, healing reversible submicroscopic discontinuities, etc.

The relations between $\Delta u_{e1}$ and $\Delta u_{e2}$, as well as $q_1$ and $q_2$ are defined by physico-chemical properties of the materials of the friction pair, their structure and friction conditions.

Since the contact volumes (not unit sizes) of the materials forming a friction couple become strained by friction (figure 2), equations (4) and (5) can be rewritten in the form:

$$W_f = \Delta U_e + Q = \Delta U_{e_1} + \Delta U_{e_2} + Q_1 + Q_2, \tag{6}$$

$$\dot{W}_f = \dot{U}_e + \dot{Q} = \dot{U}_{e_1} + \dot{U}_{e_2} + \dot{Q}_1 + \dot{Q}_2 \ . \tag{7}$$

Consequently, the equations for friction work $W_f$, frictional force $F$ and friction coefficient $\mu$ (without lubrication) has the view:

$$W_f = \Delta U_e + Q = \Delta U_{e_1} + \Delta U_{e_2} + \Delta U_{T_1} + \Delta U_{T_2} + \vec{Q}_1 + \vec{Q}_2, \tag{8}$$

$$\dot{W}_f = \dot{U}_e + \dot{Q} = \dot{U}_{e_1} + \dot{U}_{e_2} + \dot{U}_{T_1} + \dot{U}_{T_2} + \dot{\vec{Q}}_1 + \dot{\vec{Q}}_2 \ , \tag{9}$$

$$F_l = \frac{\Delta U_e}{l} + \frac{Q}{l} = \frac{\Delta U_{e_1} + \Delta U_{e_2}}{l} + \frac{Q_1 + Q_2}{l} \ , \tag{10}$$

$$F_v = \frac{\dot{U}_{e_1} + \dot{U}_{e_2}}{v} + \frac{\dot{Q}_1 + \dot{Q}_2}{v} = F_{mechanical} + F_{molecular} , \tag{11}$$

$$\mu_l = \frac{\Delta U_{e_1} + \Delta U_{e_2}}{Nl} + \frac{Q_1 + Q_2}{Nl} = \mu_{adapt} + \mu_{dis} = \mu_{adapt} + \mu_{T(dis)} + \mu_{\vec{Q}(dis)} , \tag{12}$$

$$\mu_v = \frac{\dot{U}_{e_1} + \dot{U}_{e_2}}{Nv} + \frac{\dot{Q}_1 + \dot{Q}_2}{Nv} = \mu_{deformation} + \mu_{adhesion} , \tag{13}$$

where $\Delta U_e = V_f \Delta u_e$; $Q = V_f q$; $\vec{Q} = V_f \vec{q}$; $\dot{U}_e = V_f \dot{u}_e$; $\dot{u}_e = \mathrm{d}u_e / \mathrm{d}t$ - is the rate of latent energy density change in the contact volumes; $V_f$ - is the deformable (friction) volume; $\mu$ - friction coefficient; $\mu_{adapt}$ - adaptive friction coefficient; $\mu_{T(dis)}$ and $\mu_{\vec{Q}(dis)}$ - statical and dynamical components of dissipative friction coefficient; $\Delta U_T$ - thermal component of internal energy; $N$ - normal load; $l$ - distance of friction; $v$ -sliding velosity. The latent energy density $\Delta u_e$ is an integral parameter of tribostate and damageability (failure ( $\Delta u_e^*$ )).

Thus, viewed thermodynamically, the work done by friction forces $W_f$ (the friction power $\dot{W}_f$), the friction force $F$ and the friction coefficient $\mu$ may be classified conventionally into two specific components with different kinetic behavior [3]. The first component is associated with microscopic mechanisms of adaptive type and relates to the change of latent (potential) energy ($\Delta u_{e_1}, \Delta u_{e_2}$) of various elementary defects and damages that are generated and accumulate in the deformable volumes of materials friction pair. This energy is a unique and integral characteristic of the submicro- and microstructural transformations that occur in plastically strained materials [2]. This energy is a measure of strain hardening and damageability of materials. The second component is associated with microscopic mechanisms of dissipative type and relates to dynamic recovery processes in which latent energy is released and heat effect of friction ($q_1, q_2$) take place. This energy originates in the motion and destruction of various elementary defects of opposite signs, the egress of these defects to the surface, the healing of reversible submicroscopic discontinuities, etc. The ratios of the components $\Delta u_{e_1}$ and $\Delta u_{e_2}$ as well as $q_1, q_2$ of the balance vary over a wide range, depending on the physical, chemical, and structural properties of the materials that comprise the friction couple and the friction process conditions.

Thus, the thermodynamic analysis of friction (plastic deformation and fracture) has led to generalized (two-term) relations for the force $F$ and coefficient of friction $\mu$, which agrees with current concepts [5,6] of the nature of friction – molecular-mechanical theory (11) and deformable-adhesion theory (13). But it is more correct to speak about the adaptive-dissipative nature (model) of friction (12).

Relationships (6)-(13) which generalize the mechanism of energy dissipation at friction allow to classify the tribosystem states. According to ergodynamics of deformed solids (relationships $\Delta u = \Delta u_e + \Delta u_T$ and $q = \Delta u_T + \vec{q}$) and equations (6)-(7) may be transformed to:

$$W_f = \Delta U_{e1} + \Delta U_{e2} + \Delta U_{T1} + \Delta U_{T2} + \vec{Q}_1 + \vec{Q}_2, \quad (14)$$

$$\dot{W}_f = \dot{U}_{e1} + \dot{U}_{e2} + \dot{U}_{T1} + \dot{U}_{T2} + \dot{\vec{Q}}_1 + \dot{\vec{Q}}_2. \quad (15)$$

As follows from equations of energy balance (14), (15), all exhibitions of friction and wear may be reduced conventionally at least to two basically different states: the first state defines all types of damage and wear, the second - the so-called "wearless" condition.

The state of damage and wear is characterized by the components of energy balance (14), (15), which are responsible for accumulation of internal energy in deformed volumes $\Delta u = \Delta u_{e1} + \Delta u_{e2} + \Delta u_{T1} + \Delta u_{T2}$, i.e. the process is irreversible. The "wearless" state is characterized by the components responsible for dynamic dissipation (reversibility) of strain energy into elastic and structural dissipated energy $\vec{q} = \vec{q}_1 + \vec{q}_2$ of friction contact.

In its turn, the first state may be classified depending on the relation between potential $\Delta u_e$ and kinetic $\Delta u_T$ components of internal energy. It is subdivided conventionally into mechanical damage and wear (due to so-called structure activation) and thermal damage and wear (due to thermal activation). For instance, let the thermal component of internal energy $\Delta u_T$ be equal to zero ($\Delta u_T = 0$) and the internal energy variation at damage and wear be defined only by variation of the potential component $\Delta u_e (\Delta u = \Delta u_e)$. Then, the mechanical damage and wear with brittle fracture of surfaces take place. On the contrary, if we have $\Delta u_e = 0$ ($\Delta u = \Delta u_T$), then the thermal damage and wear with ductile fracture of surfaces take place. All the intermediate values of the components are associated with quasi-brittle or quasi-ductile fracture of solids.

In the most general case, the energy balance at dry friction (6) should be written as:

$$W_f = \Delta U_{e1} + \Delta U_{e2} + \Delta U_{e3} + Q_1 + Q_2 + Q_3. \quad (16)$$

In the special case, where the friction is localized into volume of the "third body" (figure 2) equation (16) develops into

$$W_f = \Delta U_{e3} + \vec{Q}_3 . \tag{17}$$

Here $\Delta U_{e3} = V_3 \Delta u_{e3}$.

## 3. Energy interpretation of Leonardo da Vinci (Amonton's) friction coefficient

According to thermodynamic theory of strength [2], the structure parameter should be related to the portion of the accumulated plastic deformation that is responsible for strain hardening. This portion is uniquely and integrally defined by the density of the potential component of internal energy (that is, the latent energy density $\Delta u_e$) of various defects and damages that accumulate in a plastically strained material. With this in mind, if we neglect the heat effect $Q$ of friction, one will infer from the thermodynamic analysis of friction of equations (12)-(13) that the Amonton (Leonardo da Vinci) friction coefficient is:

$$\mu = \frac{\Delta U_e}{\mu^* N l} = \frac{F}{N}; \qquad F = \frac{\Delta U_e}{l}; \qquad Q \cong 0, \qquad \mu^* = 1 . \tag{18}$$

Consequently, the coefficient of friction has a very deep physical sense. On the one hand, it is the parameter which generally characterizes the resistance of relative displacement (movement) of surfaces, for it reflects the portion of energy, which «is done by friction away» as accumulated latent energy $\Delta U_e$, by relation to parameter of external forces work $\mu^* N l$ (energy of external relative movement). On the other hand, it is the generalized characteristic of damage, for it is defined of the latent energy density $\Delta u_e$ as integral characteristic of the structure defectiveness measure, because this energy is the generalized parameter of damage. Here too, coefficient of friction generally reflects the structural order (disorder) of deforming contact volume, since the parameter $\Delta U_e = \Delta u_e V_f$ is defined by the energy of defects and damages of different types, that are accumulated into contact volumes $V_f$ solids.

Therefore, coefficient of friction is a true and generalized parameter of tribosystem state. From this conclusion we can say that the analysis of the evolution of the states of a tribosystem is primarily an analysis of the latent deformation energy accumulated within the contact friction volumes.

## 4. Structural-energy regularities of rubbing surfaces evolution

We propose an energetic interpretation of the experimental friction curves $\mu = \mu(N, v)$ [5,7] in view of structural-energy diagram (figure 3) for evolution of rubbing surfaces. According to our concept [1], the ascending portion of the friction coefficient curve $\mu$ is mainly controlled by processes associated with the accumulation of latent energy $\Delta U_e$ in various structural defects and damages. Here the increase in $\mu$ is due to the increasing density of latent (potential) energy $\Delta u_e$ and the increasing adaptive friction volume $n_*$. The descending portion of the friction curve is mainly controlled by processes associated with the release and dissipation of energy $Q = \Delta U_T + \vec{Q}$. Here the decrease in $\mu$ is due to the decrease in latent energy density within the friction volume $V_f$ or (which is virtually the same) to the decrease of the adaptive friction volume $V_{adapt}$ ($u_e = u_e^*$) and to the increase of the dissipative volume $V_{dis}$ ($\vec{q}^* = u_e^*$).

Evolution of tribosystem, presented as a diagram view (figure 3), has an adaptive-dissipative character (6)-(13) and reflects the competitive (dialectical) nature of friction. Evolution curve has the row of principal points (1-5) of transitional tribosystem states, which strictly obeys the balance principle of friction; there are more characteristic areas of tribosystem behavior between these points. These areas reflect the common properties of nonlinear dynamic of evolution.

So, in figure 3 it is possible to see the following conventionally designated points and stages: 0-1 – a stage of static friction and deformational strengthening; 1 – a point of limit for deformational

IOP Conf. Series: Materials Science and Engineering **174** (2017) 012012 doi:10.1088/1757-899X/174/1/012012

strengthening; 1-2 – a stage of pumping of excess energy; 2 – a point of gripping (adhesion) and transition of outer friction into internal (critical non-stability): 2-3 – a stage of forming dissipation structures (formation of heat fluctuation in friction volume); 3 – a point of minimum compatibility (maximum frictionness); 1-2-3 - a stage of selforganization; 3-4 – a stage of compatibility; 4 – a point of wearlessness (anormal-low friction); 5 – a point of thermal adhesion.

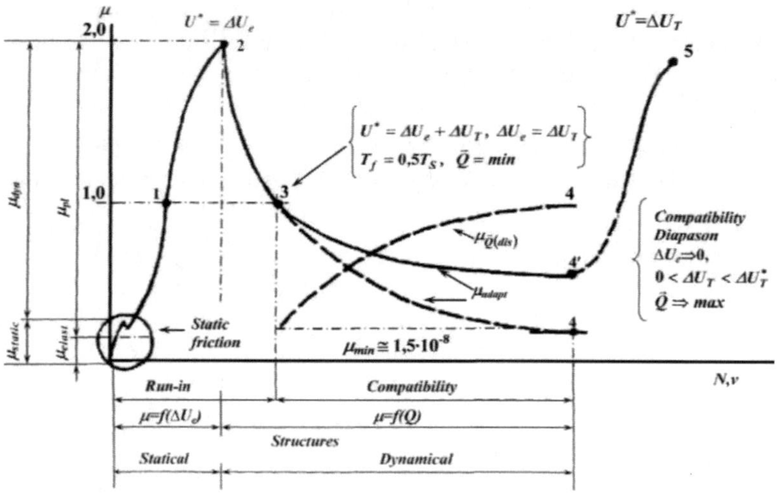


**Figure 3.** Structural-energy diagram for evolution of rubbing surfaces [1].

An ideal evolution of tribosystem is symmetrical. The process starts and finishes within areas of elastic behavior. A plastic maximum (a superactivated condition) exists between them as a condition of selforganisation and adaptation.

In the most general case evolution (adaptation) regularities of tribosystems may be presented as a 2-stage (figure 3). At the first stage (0-2) of adaptation the evolution of friction contact rushes to form some critical volume of friction $V_f^*$ (point 2). It is elementary tribosystem that is the elementary and self-sufficient energy transformer. The first stage - latent energy density growth $\Delta u_e$ to a limited magnitude $\Delta u_e^*$ within critical friction volume $V_f^*$.

This friction volume $V_f^*$ is constant at the second stage of evolution, but here it is evolutionary developed owing to structural transformation; by this one may realize wide spectrum of compatibility friction structures (figure 3). The second stage (2-4) – structural transformation of critical friction volume (elementary tribosystem) $V_f^*$ into adaptive $V_{adapt}$ and dissipative $V_{dis}$ volumes (figure 4). The limit (point 4) of this stage is characterized by a full transformation of adaptive critical volume $V_{adapt}^*$ into $V_{dis}^*$ dissipative.

The volumes mentioned above characterize different regularities of transforming energy of outer mechanical movement at friction. Adaptive volume $V_{adapt}$ is connected with non-reversible absorption of deformation energy. It is in this volume where latent deformation energy $\Delta u_e$ accumulates and where the centres of destruction initially emerge (birth). Dissipative volume $V_{dis}$ is capable of reversible transformation (dissipate) of outer movement energy. It doesn't accumulate latent deformation energy owing to reversible elastic-viscous-plastic deformation.

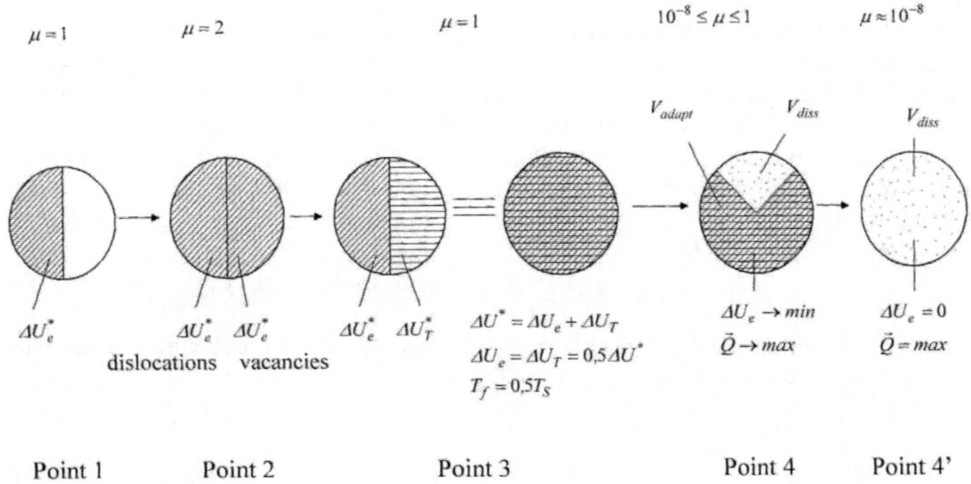


**Figure 4.** A schematic evolution of the contact volume of friction in diagram's points 1-5 [1].

Suggested theoretical and calculation estimation [1] showed that dissipative friction volume performs reversible elastic energy transformation of outer mechanical movement with density $\vec{q}^*$ equal to critical density of latent energy $u_e^*$.

Culmination of tribosystem evolution is its final and limited condition of point 4 – a state of anomalously low friction and wearlessness (maximum efficient).

A schematic evolution of the contact volume of friction in diagram's points 1-5 is presented in figure 4.

The calculation [1] shows that at an ideal tribosystem evolution an adaptive (Amontons) friction coefficient $\mu_{adapt}$ in a point 2 of a diagram falls abruptly down, reaching in a point 4 the value of elastic friction coefficient $\mu_{elast}$. For point 4 of compatibility area 3-4 an equation of energy balance (12) should be put in the following way:

$$\mu_{adapt} = \mu^* - \mu_{dis} = 1 - \mu_{dis} = \mu_{plast} = 0 = \mu_{elast}; \qquad \mu^* = 1{,}0. \tag{19}$$

Thus, point 4 stands for an ideal evolution of contact friction volume a condition of ideal elastic-viscous-plastic deformation. Equation (19) shows as a matter of fact exactly it, i.e. Amontons friction coefficient $\mu_{adapt}$ being in its essence plastic friction coefficient $\mu_{plast}$ has a minimum value equal to zero. It follows then, that plastic friction became elastic with friction coefficient $\mu_{elast}$. It means that plastic deformation of contact volume friction is implemented with the maximum dynamic dissipation ($\vec{Q} = \max$) of accumulated latent energy. That is why the value of accumulated energy in point 4 is equal to zero ($\Delta U_e = 0$). This fact proves an ideal condition at full evolution of contact volume. From the physics point of view this condition may be explained by the full dissipation of accumulated energy $\Delta U_e^*$ in point 2 and by newly emerged structures of point 4 in the form of elastic energy of interaction between them (dynamic dissipation energy $\vec{Q}^*$). Here $\mu_{dis} = 1{,}0$. The structural elements themselves are defectless - $\mu_{adapt} = 0$, and friction is elastic - $\mu = \mu_{elast}$.

It has been demonstrated [1] that value of minimum adaptive friction volume $V_{adapt}^{\min}$ corresponding to the zero meaning of plastic friction component $\mu_{adapt}$ is not equal to zero, but is equal to some minimum structural element of deformed solid body.

## 5. Mechanical or nano quantum idea

As the result of more complete evolution of elementary tribosystem (figure 3, point 4) the unique nanostructure is formed and the basis of which is one mechanical (nano) quantum [8]. This mechanical quantum represents the minimum number of atoms that can be arranged in a structure capable of reversibly absorbing and dissipating (recovering) the energy of external mechanical motion. Mechanical quantum represents the least structural form of solid material body in conditions of plastic deformation too and under transition tribosystem across the limit activated state by development of selforganazing processes of tribosystem adaptation are formed. The universal size (volume) of mechanical quantum (2) is equal to:

$$V_Q = (e^3)^3 = (20{,}08553695)^3 = 8103{,}083969 \text{ atom's oscillators.} \quad (20)$$

Mechanical quantum is dynamic oscillator of dissipative friction structure. Linear size of it quantum (3) is equal to diameter of spherical ideal crystal:

$$D_Q = 2 \cdot W \cdot d_a \cdot (3/4 \cdot \pi)^{1/3} = 7{,}177 nm. \quad (21)$$

Here $d_a$ - mean atomic diameter for metals; $W = e^3$ - parameter of state for mechanical quantum.

Mechanical quantum (figure 5) can be examined as the elementary nanostructure of metal's solid body. Calculations have shown the number $N_Q$ of such mechanical «quanta» (subtribosystems) within the elementary tribosystem's volume $V_f^* = V_{dis}^*$ to be $0{,}63 \cdot 10^8$, which is close to the safe number of fatigue cycles.

**Figure 5.** Model of elementary nanostructure of friction (8103 atomic cubical cells) [1,8].

In these terms (point 4) only one mechanical quantum is the lost – standard of wear. The tribosystem (friction contact) has the ideal damping properties – «wearless».

The principle of mechanical quantum determines nanoquantum levels of all friction parameters of compatible (optimal) tribosystems and other.

## 6. Conclusions

1. From the energy balance equations of friction follows that the evolution of tribosystem has an adaptive-dissipative character.
2. Friction is a competition of two simultaneous, interconnected and opposite tendencies of accumulating latent (potential) energy of various kinds of defects and damages of contact volumes structures and releasing (dissipation) energy due to various relaxation.

IOP Conf. Series: Materials Science and Engineering **174** (2017) 012012 doi:10.1088/1757-899X/174/1/012012

3. The most complete evolution of tribosystem has symmetrical view - the friction process is started and finished within elastic area.
4. The unique nanostructure at the basis of which is one mechanical (nano) quantum is formed by elastic-ductile-plastic and dynamic dissipative state of ideal tribocontact.
5. We can suggest the mechanical quantum as the least structural form of solid material body (subtribosystem) and the standard of wear.
6. All parameters of compatibility (optimal) friction have to be in quanta levels - commensurable with the parameters of the one mechanical quantum.

**References**

[1] Fedorov S V 2003 *The Foundations of Triboergodynamics and Physico-Chemical Prerequisits of Compatibility Theory* (Kaliningrad: Kaliningrad State Technical University Press) p 416
[2] Fedorov V V 1979 *Thermodynamic Aspects of Strength and Fracture of Solids* (Tashkent, Science) p 168
[3] Fedorov V V 1985 *Kinetics of Damage and Fracture of Solids* (Tashkent, Science) p 167
[4] Fedorov V V 1991 Ergodynamic concept of failure *J. Streng. of Mat.* (Translated from Russian), **23** 8 883-889
[5] Kragelskii I, Dobychin M and Kombalov V 1977 *Friction and Wear Calculation Methods* (Moscow, Mechanical Engineering) 526
[6] Bowden F P, Tabor D 1973 *Friction. An Introduction to Tribology* (New-York: Anchor Books Edition) 178
[7] Lancaster I 1973 *J. Plastics and Polymers* **41** 296-306
[8] Fedorov S 2009 The Mechanical Quantum of Dissipative Friction Structures is the Elementary Tribonanostructure *Proc. IV WTC2009* Kyoto p 926

IOP Conf. Series: Materials Science and Engineering **174** (2017) 012013 doi:10.1088/1757-899X/174/1/012013

# The dynamic behavior of sliding tribosystems in unstable operating conditions

**A Postaru[1], Ghe Postaru[1], V Ceban[1], C Spânu[2] and P Stoicev[1]**

[1] Technical University of Moldova, Block 1, Bulevardul Ştefan cel Mare şi Sfînt 168, Chişinău 2004, Republic of Moldova
[2] "Dunărea de Jos" University of Galati, Faculty of Engineering, Domnească 111, 800201, Romania

E – mail: constantin.spanu@ugal.ro

**Abstract.** The paper addresses the issue of the influence that the law of friction has on the dynamic behavior of the mechanical system that interacts with a tribosystem. The emergence of certain nonlinearities of higher order into the law of friction leads to an intensification of the dissipative process and to tribosystem destabilization. Consequently, friction excited self-oscillations are generated into the elements of the mechanical system with a wide spectrum of frequencies, sustained from the external source of energy. The theoretical and experimental modelling of the dissipative process and of the generation of is based on the frictional harmonic oscillator that interacts with the tribosystem. The oscillator is used as a sensitive element to the fluctuations of the frictional force and as a measure of the dissipated energy. Starting from the model, the elaboration of a method and of devices for experimental research provided the opportunity to study the behavior of the tribosystem in unstable operating

## 1. Introduction

As functional components of mechanical systems, tribosystems collaterally affect the former's dynamic behavior and have a predominant role in energy dissipation. The evolution character of the dissipative process is influenced and correlatively connected with the friction characteristic (law) occuring at the relative motion of contact surfaces. At present, a series of different laws have been formulated only for dry friction: simple-static; complicated-dynamic. Considering also lubrication (with Stribeck effect), the series of the friction laws diversify [2,5,8]. In fact, the laws of friction are complex and include the influence of a series of factors of a different nature related to working, geometric and micro geometric, to tribosystem structure, to the source, properties, and characteristics of the materials for triboelements, and to the working environment.

When nonlinearities of different orders occur into the friction characteristic (with fluctuations in the frictional force), in the mechanical system elements are generated noises under the form of self-oscillations with a wide spectrum of frequencies [2, 5, 8]. The structure of the spectrum, the amplitude and shape of the oscillations are influenced by charging parameters, the friction regime and energy dissipation factors, the properties and state of the materials for triboelements and the lubricant, the origin and intensity of processes arising in the contact area. For problems of such complexity a reliable research method remains the experimental one. However, experimental modelling should be formalized and executed within the framework of the fundamental equations of nonlinear dynamics.

## 2. Dynamic modelling of mechanical system-tribosystem interaction

A model commonly used to describe and study the oscillatory processes in different systems is the harmonic oscillator. The oscillator is the basis of both mathematical models and necessary technical devices for tests and the experimental research of the studied systems.

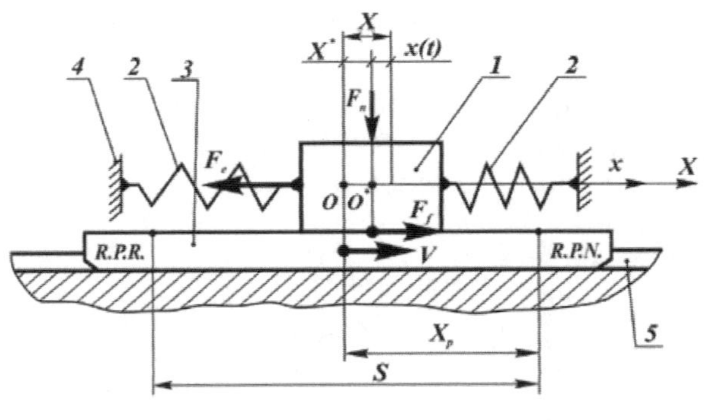


**Figure1.** Mechanical oscillator scheme.

The oscillator has been accepted as a model for studying the interaction between the tribosystem characteristics and the mechanical system (figure 1). It consists of block 1 with mass $\boldsymbol{m}$ linkconnected to housing 4, fixed on both sides by means of two similar elastic elements 2, of low rigidity $\boldsymbol{c}$. The angular frequency of the oscillator is $\omega = \sqrt{c/m}$). The tribological connection between the oscillator and the triboelements is realized through the contact between block 1 and platform 3. Platform 3, driven by a crank-type mechanism, performs a translational reciprocating movement on guide 5 within distances $\boldsymbol{S}$, with speed:

$$V = r\Omega\left(sin\varphi_m + \frac{\lambda}{2}(1 - sin2\varphi_m)\right) \tag{1}$$

where: $\Omega$ - the angular speed of the crank, $r$ - the radius of the crank; $l$ - the length of the rod; $\lambda = \frac{r}{l}$; $\varphi_m$ - the rotation angle of the crank.

Initially, the $X$ coordinate's origin of the gravity center of block 1 is in the stable equilibrium point $O$. When the platform begins to move with speed $V$ on distance $S$ distance, block 1, influenced by the friction force $F_f$, will move in direction $X$ with speed $\dot{X}$. The relative speed between the contact surfaces of the friction bond becomes $v_r = \dot{X} - V$.

Connecting the oscillator to the tribosystem results in a system composed of two subsystems of different nature (mechanical and dissipative) with own dynamic behavior, influencing each other during working. The evolution of the dissipative process (of energetical essence) can be studied only from the perspective of Lagrangean formalism, according to which the generalized dissipative force $Q_d$ derives from a force function called Rayleigh dissipative function [6, 7], defined by the relationship:

$$\Phi_d = \sum_{j=1}^{N} k_j \int_0^{v_j} f_j(u)du \tag{2}$$

where: $k_j$ and $f_j(u)$ – the positive functions defined on spaces $j$ of the contact real elementary areas that are dependent on the $q = X$ coordinate and on the generalized speed $\dot{q} = \dot{X}$ of the oscillator, on speed $V$ of the platform, and on the internal and external parameters of the tribosystem; $v_j$ – the relative local speed of the surfaces on the contact real elementary areas of the spaces; $N$ – the number of real elementary areas within the boundaries of the contact nominal area.

In the sliding tribosystem the role of dissipative generalized force $Q_D$ is played by the total force of friction $F_f$, defined by the gradient of the dissipative force in the direction of the relative motion of the contact surfaces [2, 8 ]

$$Q_d = F_f = -\frac{\partial}{\partial(\dot{X}-V)}\sum_{j=1}^{N} k_j \int_0^{v_j} f_j(u)du = -\frac{\partial \Phi_d}{\partial v_r} \tag{3}$$

where: $\dot{X} - V = v_r$ – the generalized speed is represented by the relative speed of the surfaces when the platform and the block of the oscillator move.

The movement of the oscillator under the action of the dissipative forces is described by the Lagrange equation

$$\frac{d}{dt}\left(\frac{\partial L}{\partial \dot{X}}\right) - \frac{\partial L}{\partial X} = -\frac{\partial \Phi_d}{\partial v_r} \tag{4}$$

$= (T - \Pi)$ – the Lagrange function (kinetic potential); $\left(T = m\frac{\dot{X}^2}{2}\right)$ - the kinetic energy of the oscillator; $\left(\Pi = c\frac{X^2}{2}\right)$ – the potential energy accumulated into the elastic elements of the oscillator.

If the working parameters are maintained at constant level, the oscillator asymptotically stabilizes his position temporarily "*freezing*" in the vicinity of the unstable equilibrium point $O^*$ (figure 1) with coordinate $X^*$. When the position of the block is stabilized, speed $\dot{X} \to 0$, and the contact relative speed becomes $v_r = -V$. In this state the oscillator passes in a steady and stable working mode relative to the point $X^*$, where block 1 will be in the balance of forces in the movement direction of the platform $(F_f + F_e = 0)$, where: $F_f = F_c$ – the constant component of the Coulomb type friction force; $F_e = -cX^*$ - the force of elasticity. On the oscillator's passing in steady state, the dissipative friction force between the block and the platform remains linearly dependent on the generalized coordinate $X^*$and the platform speed $V$.

Lagrange equation for the steady state conditions of the oscillator takes the form.

$$-\frac{\partial L}{\partial X^*} = \frac{1}{2}c\frac{\partial}{\partial X^*}(X^*)^2 = cX^* = -\frac{\partial \Phi_d}{\partial(-V)} = F_c = const \tag{5}$$

According to (5), the friction force $F_c$ and the energy dissipation power $P_d(V)$ for motion with relative speed $(-V)$.

$$F_c = cX^* \qquad P_d(V) = -cX^*V \tag{6}$$

The loss of stability violates the balance of forces, the movement of the oscillator being determined by the variation of the dissipative forces. In the event of some instability in the operation of the tribosystem, with disruptive fluctuations of the friction force, the oscillator passes into a self-oscillation regime maintained from the external source of energy. The nature and evolution of the dissipative process can be efficiently set in the analysis result of the oscillator's motion in the phase space (figure 2) built in the phase coordinates $Y = X$ and $Z = \frac{\dot{X}}{\omega}$. To this aim, the ratio between the motions of the representative point $M$ on the phase trajectories for each two cycles in a row $(i)$ and $(i+1)$ are examined step by step.

In examining the movement on the phase trajectory of cycle $i$ in self-oscillation state (figures.1, 2) the coordinate and speed of the oscillator are $X_i = X_i^* + x_i$, $\dot{X}_i = \dot{X}_i^* + \dot{x}_i$,
where: $X_i^* = (X_i^{max} + X_i^{min})/2$ - average component; $x_i$ - variable component; $\dot{X}_i^* = \Delta X_i^* \omega$ - speed of passage from the previous cycle $(i-1)$ to cycle $i$. On the $i$ cycle path, two types of movement can be identified: 1 - with low speed $\dot{X}_i^*$ , determined by the variation of the mean component coordinate $X_i^*$ of the cycle; 2 -with high speed $\dot{x}_i$, determined by the variation of the $x_i$ coordinate and angular own frequency $\omega$ of the oscillator.

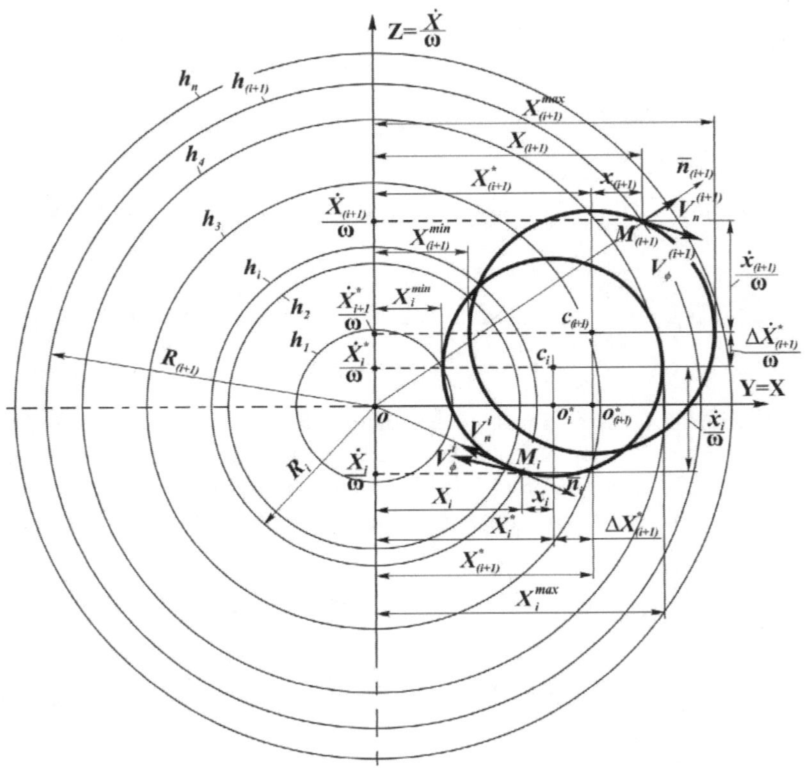


**Figure 2.** Representation of self-oscillatory process in phase space.

Admitting the prime integral $\left(\dot{X}_i \frac{\partial L_i}{\partial \dot{X}_i} - L_i = E_i\right)$ of equation (4) as total energy of the oscillator, represented in the phasic space (figure 2) through the orbit of level $h_i$ drawn with the representative radius $R_i$, is the following is obtained:

$$E_i = \frac{1}{2}c\left[X_i^2 + \left(\frac{\dot{X}_i}{\omega}\right)^2\right] = \frac{1}{2}c(X_i^*)^2 + cX_i^* x_i + \frac{1}{2}c\left[x_i{}^2 + \left(\frac{\dot{x}_i}{\omega}\right)^2\right] = \frac{1}{2}cR_i^2 = h_i \qquad (7)$$

where: $R_i = \sqrt{Y^2 + Z^2} = \sqrt{X_i^2 + \left(\frac{\dot{x}_i}{\omega}\right)^2}$ - the representative radius of the $h_i$ orbit at intersection with the trajectory of the representative point $M_i$trajectory at movement on cycle in the phasic space.

Differentiating equation (7) by time, the energy variation of the oscillator and the evolution of the dissipative process under the influence of the friction force during the $i$ cycle is determined:

$$\frac{dE_i}{dt} = c\left(X_i\dot{X}_i + \frac{\dot{X}_i\ddot{X}_i}{\omega^2}\right) = c\left[X_i^*\dot{X}_i^* + \left(X_i^*\dot{x}_i + \frac{\dot{X}_i^*\ddot{x}_i}{\omega^2}\right) + \left(x_i\dot{x}_i + \frac{\dot{x}_i\ddot{x}_i}{\omega^2}\right)\right] = cR_iV_n^i \qquad (8)$$

where: $V_n^i$ – the projection of the phasic speed $V_\Phi^i$ on the direction of the normal $\bar{n}_i$, drawn through the $M_i$ representative point moving on the phasic trajectory of the cycle when intersecting the representative orbit $h_i$. From a dynamic point of view, $V_n^i$ is the rate of change (dissipation) of the oscillator's energy at movement of the $M_i$ point on the phasic path.

According to relation (8), the variable component of the oscillator's energy includes three groups of factors:

1. $X_i^* \dot{X}_i^*$– for variation of the low speed and low frequency component;
2. $x_i \dot{x}_i + \frac{\dot{x}_i \ddot{x}_i}{\omega^2}$ - for variation of the high speed and high frequency component;
3. $X_i^* \dot{x}_i + \frac{\dot{X}_i^* \ddot{x}_i}{\omega^2}$ – for mutual influence of the low and high speed factors.

Setting the mechanical status of the oscillator by experimental methods, based on expressions (7) and (8), the normal component $V_n^i$ of the phasic speed is determined, which comprises the three groups of influence factors.

$$V_n^i = \frac{X_i \dot{X}_i + \frac{\dot{X}_i \ddot{X}_i}{\omega^2}}{R_i} = \frac{X_i^* \dot{X}_i^* + \left(X_i^* \dot{x}_i + \frac{\dot{X}_i^* \ddot{x}_i}{\omega^2}\right) + \left(x_i \dot{x}_i + \frac{\dot{x}_i \ddot{x}_i}{\omega^2}\right)}{\sqrt{X_i^2 + \left(\frac{\dot{X}_i}{\omega}\right)^2}} = \left(V_n^i\right)^{**} + \left(V_n^i\right)^{*} + \left(V_n^i\right)^{v} \tag{9}$$

At temporary "*freeze*" of the oscillator's mechanical status in the point of coordinate $X_i^*$, the instant steady state functioning conditions are obtained fixed on the average level of the oscillation cycle generated by the stationary friction force $F_c^i$ at the platform movement with speed $V^i$. Based on relations (5) and (6), the solution for friction force and energy dissipation power on the average cycle component is obtained.

$$F_c^i = cX_i^* \tag{10}$$

$$P_d^i\left(V^i\right) = -cX_i^* V^i \tag{11}$$

The total energy dissipation power in the contact area for each movement cycle consists of two basic components: $P_d^i\left(V^i\right)$ – instantly constant, defined by speed $V^i$within cycle limits; variable $P_d^i\left(V_n^i\right)$, defined by speed $V_n^i$.

$$P_d^i = P_d^i\left(V^i\right) + P_d^i\left(V_n^i\right) = -c\left(X_i^* V^i + R_i V_n^i\right) \tag{12}$$

Taking into account relation (9) the following is obtained:

$$P_d^i = F_f^i v_r^i = -\frac{\partial \Phi_d^i}{\partial v_r^i} v_r^i = -\left[cX_i^* V^i + cR_i\left(V_n^i\right)^{**} + cR_i\left(V_n^i\right)^{*} + cR_i\left(V_n^i\right)^{v}\right] \tag{13}$$

The total instantaneous value of the friction force during cycle $(i)$ conditioned by the achievement of the working unstable self-oscillation state:

$$F_f^i = \frac{P_d^i}{v_r^i} = -\left[cX_i^* \frac{V^i}{v_r^i} + cR_i \frac{\left(V_n^i\right)^{**}}{v_r^i} + cR_i \frac{\left(V_n^i\right)^{*}}{v_r^i} + cR_i \frac{\left(V_n^i\right)^{v}}{v_r^i}\right] \tag{14}$$

or

$$F_f^i = \left[F_c^i + \left(F_v^i\right)^{**} + \left(F_v^i\right)^{*} + \left(F_v^i\right)^{v}\right] \tag{15}$$

where: $F_c^i = -cX_i^* \frac{V^i}{v_r^i}$; $\left(F_v^i\right)^{**} = -\,cR_i \frac{\left(V_n^i\right)^{**}}{v_r^i}$; $\left(F_v^i\right)^{*} = -R_i \frac{\left(V_n^i\right)^{*}}{v_r^i}$; $\left(F_v^i\right)^{v} = -cR_i \frac{\left(V_n^i\right)^{v}}{v_r^i}$.

The component $F_c^i$ (of Coulomb type) of the friction force is defined by the linear factors of the dissipative function $\Phi_d$ in the vicinity of the $X_i^*$ coordinated point. The variable (fluctuating) components of the friction force $\left(F_v^i\right)^{**}$, $\left(F_v^i\right)^{*}$, $\left(F_v^i\right)^{v}$occur as a result of various dynamic effects from the contact zone with higher order nonlinearities and may vary in a wide range of frequencies and amplitudes.

When condition $v_r = \dot{x} - V < 0$ is accomplished, the $F_c$ component of the friction force has always the same sense as platform speed vector $\boldsymbol{V}$ and determines the energy dissipation level in the contact zone at relative motion of the surfaces. The variable components (depending on the speed signs $(V_n^i)^{**}$, $(V_n^i)^*$, $(V_n^i)^v$sign) can change their sign according to direction within the limits of the same oscillation cycle. On the negative direction, the energy previously accumulated in the oscillator's elements dissipates in the contact area, and, on the positive direction, through the tribosystem, a new portion of energy from the external source is introduced into the oscillator. This behavior relates to the achievement of the dynamic effect of variable dissipation on direction, also called "negative friction" effect, which is the main cause of engendering friction excited self-oscillations in the mechanical system [1, 2, 5, 7].

Accepting cycle $(i)$ as a benchmark, at crossing to the next cycle $(i+1)$, the $X_i^*$ (coordinated point (figure 2) acquires an additional movement $\Delta X_{(i+1)}^*$ with the speed $X_{(i+1)}^*$, where: $\Delta X_{(i+1)}^* = X_{(i+1)}^* - X_i^*$; $\dot{X}_{(i+1)}^* = \Delta X_{(i+1)}^* \omega$; $X_{(i+1)}^* = (X_{(i+1)}^{max} + X_{(i+1)}^{min})/2$. The coordinate, the absolute speed and the relative speed within the cycle limits $(i+1)$ with respect to cycle $(i)$ will be: $X_{(i+1)} = (X_{(i+1)}^* + x_{(i+1)}) = (X_i^* + \Delta X_{(i+1)}^* + x_{(i+1)})$; $\dot{X}_{(i+1)} = (\dot{X}_{(i+1)}^* + \dot{x}_{(i+1)})$; $v_r^{(i+1)} = (\dot{X}_{(i+1)} - V^{(i+1)})$.

The prime integral of equation (4) for the cycle $(i+1)$

$$E_{(i+1)} = \frac{1}{2}c\left[\left(X_{(i+1)}^*\right)^2 + \left(\frac{\dot{X}_{(i+1)}^*}{\omega}\right)^2\right] + c\left[X_{(i+1)}^* x_{(i+1)} + \frac{\dot{X}_{(i+1)}^* \dot{x}_{(i+1)}}{\omega^2}\right] + \frac{1}{2}c\left[x_{(i+1)}^2 + \left(\frac{\dot{x}_{(i+1)}}{\omega}\right)^2\right] = h_{(i+1)} \tag{16}$$

Similarly, the instantaneous frictional force is determined at the motion of the oscillator during the $(i+1)$ cycle

$$F_f^{(i+1)} = -\frac{P_d^{(i+1)}}{v_r^{(i+1)}} = -\left[cX_{(i+1)}^* \frac{V^{(i+1)}}{v_r^{(i+1)}} + cR_{(i+1)}\frac{\left(V_n^{(i+1)}\right)^{**}}{v_r^{(i+1)}} + cR_{(i+1)}\frac{\left(V_n^{(i+1)}\right)^{*}}{v_r^{(i+1)}} + cR_{(i+1)}\frac{V_n^{(i+1)}}{v_r^{(i+1)}}\right] \tag{17}$$

$$F_f^{(i+1)} = \left[F_c^{(i+1)} + \left(F_v^{(i+1)}\right)^{**} + \left(F_v^{(i+1)}\right)^{*} + \left(F_v^{(i+1)}\right)^{v}\right] \tag{18}$$

The loss of system stability can occur for both types of movements: of low frequency and low speed, and with high frequency and high speed. As an experimental criterion for assessing the movement regime, the deviations of the displacements $\Delta X_{(i+1)}^*$ and $\Delta x_{(i+1)}$between each pair of consecutive cycles are used:

under steady state oscillatory motion

$$\begin{cases} \Delta X_{(i+1)}^* = X_{(i+1)}^* - X_i^* = 0 \\ \Delta x_{(i+1)} = \left(x_{(i+1)}^{max} - x_{(i+1)}^{min}\right) - \left(x_i^{max} - x_i^{min}\right) = 0 \end{cases} \tag{19}$$

under unstable state oscillatory motion

$$\begin{cases} \Delta X_{(i+1)}^* = X_{(i+1)}^* - X_i^* \neq 0 \\ \Delta x_{(i+1)} = \left(x_{(i+1)}^{max} - x_{(i+1)}^{min}\right) - \left(x_i^{max} - x_i^{min}\right) \neq 0 \end{cases} \tag{20}$$

## 3. Experimental results

Based on the above theoretical model, an original model of tribometer was made, with a cyclic translational motion and equipped with proper systems for measuring the dynamic characteristics (variables) of the system. As dynamic variables the following are used:

1. displacement $X_i$ $(mm)$i in the movement direction of the platform of the mass center of the oscillating part against point "$O$" of the oscillator's stable equilibrium;
2. linear speed $v_i = \dot{X}_i(^{mm}/_{s})$ of the mass center of the oscillating part;
3. linear acceleration $a_i = \ddot{X}_i\left(^{mm}/_{s^2}\right)$ of the mass center of the oscillating part;
4. linear speed $V^i(^{mm}/_{s})$ of the platform in the movement direction within the limits of cycle $i$ of the oscillator's oscillation;
5. cyclic frequency of the platform $n_c$ $(min^{-1})$;
6. experimental average temperature $\theta_k$ $(^oC)$ in the contact area during cycle $k$ of the platform movement.

The dynamic variables are recorded, step by step, in the form of time series for each cycle$(i)$ of the oscillator's movement and over each cycle $(k)$ of the platform movement.

Computerized technologies for recording the dynamic variables and for processing of the time series were used to study the behavior of the tribosystem in unstable operating conditions. The evolution of the friction process was estimated through the variation characteristic (law) of the friction force $F_f^k = F_f^k(v_r)$ for relative speed variation; the evolution of the dissipative process (energy dissipation) was estimated through the work of the friction force $W_f^k$ cumulated for each cycle $(k)$ of the platform movement.

In the case of the translational cyclical movement, the accuracy of local friction force determination depends on the resolution $s = \frac{T_k}{T}$s of the system, established within the limits of the platform movement cycle, where $T = \frac{2\pi}{\omega}$ - the period of the oscillator's self-oscillation, $T_k = \frac{2\pi}{\Omega}$ - the period of cycle $(k)$ of the platform movement. Resolution $s$ represents the oscillator's number of cycles included in a platform movement cycle.

As a benchmark characteristic for establishing the experimental friction law, the integrated mean within the limits of cycle $(i)$ period of the local friction force is used

$$F_f = \frac{\omega}{2\pi}\int_{-\frac{\pi}{\omega}}^{+\frac{\pi}{\omega}} F_f^i\,dt \tag{21}$$

where: $F_f^i$ - the instantaneous value of the friction force within the cycle $(i)$ limits.

The level of energy dissipation (represented by the work of friction forces integrated on period $T$ of cycle $(i)$

$$W_f^i = \int_{-\frac{\pi}{\omega}}^{+\frac{\pi}{\omega}} F_f^i\, v_r^i dt \tag{22}$$

The work of friction forces (dissipated energy) $W_f^k$ during the cycle $(k)$ period of the platform movement totalled

$$W_f^k = \sum_{i=1}^{s} W_f^i \tag{23}$$

The results of the experimental data analysis for a number of couples of materials and lubricants revealed a different and complex behavior of the frictional force (figure 3) on different portions and areas of the characteristics points, at relative movement of the contact on the platform cycle strokes $(S)$. Within a cycle the following are identified: the DSM stroke of the platform motion on the direct sense of the movement and the OSM stroke for opposite movements; portions with acceleration movement (AM) per stroke and deceleration movement zones (DM); (ZRP) zones of return points of the contact per stroke and (PMS) zones of points of maximal speed of the cycle. If the platform is actioned by the crank mechanism, return points with different kinematic characteristics are obtained at the end of the strokes: (RPN) - return point near; (RPR) - return point removed. A pronounced dynamic behavior of the friction force occurs when changing speed direction at entry into and exit from the areas of return points.

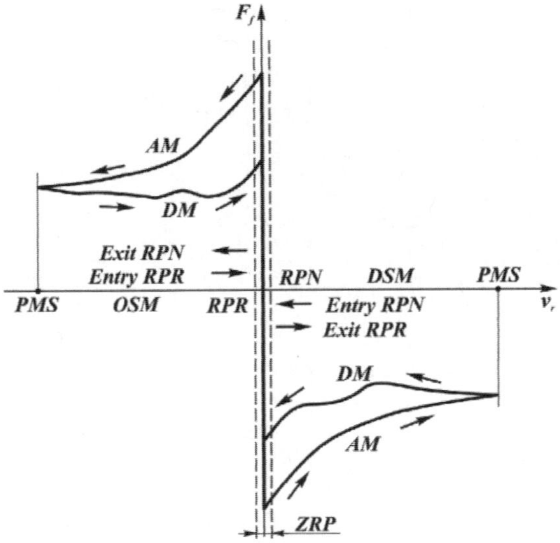


**Figure 3.** Law of friction determined experimentally for cyclical translatory movement.

During the experimental research, the following conditions were set: oscillator mass $m = 0{,}2kg$; rigidity of the elastic element $\boldsymbol{c} = 100N/mm$; angular frequency of the oscillator $\omega = 6280s^{-1}$; angular frequency of the crank $\Omega = 31{,}141s^{-1}$; cyclic frequency of the platform $n_c = 300min^{-1}$; system resolution $s = 200$; contact load in two ways (1 – with constant normal loading during the testing period, 2 - loading in consecutive steps); drop lubrication. The experimental temperature $\boldsymbol{\theta}$ in the contact area was measured with a K type mini K thermocouple. Temperature evolution in the contact area occurs due to heat produced at the exhaust of mechanical energy. Contact form - flat with dimensions: width $b = 2mm$; length $l = 40mm$. Stroke length $S = 100mm$.

Under study was the tribological behavior of the materials of the pair: block triboelement - electrolytic chromium on a steel surface; platform triboelement - 38KH2MYUA steel (similar to 41CrAlMo7). The contact was lubricated with MT- 16P GOST 6360-83 oil (similar to SAE 40 API CB).

Experimental results are shown in figures 4 and 5. For temperature increase in the contact area (figure 4) for a load $F_n = 2{,}0kN$, the energy dissipation level per cycle $(k)$ of platform movement will vary non-linearly with a trend of asymptotic stabilization, between the work values of the friction forces of $W_f^k = (40 \dots 50)\boldsymbol{J}$. If the operating parameters are maintained constant ($F_n$ load and cyclic speed of the platform $n_c$), process stabilization is complete at temperature values $\theta = (240 \dots 250)^{\circ}C$.

Of special importance for determining the dynamic behavior of mechanical systems is the evolution of the law of friction in concrete working conditions. Figure 4 shows a significant influence of temperature on friction force for relative speed $v_r$ variation per cycle strokes of platform movement. When temperature increases, the friction force changes both values and the variation manner per stroke.

Different behavior is identified if the variation of the contact load force $F_n$ takes place at constant temperatures $\boldsymbol{\theta}$. To achieve a boundary lubrication regime, during the experiment the temperature was maintained at high-level in the contact zone, with a permissible variation within $\theta = (200 \dots 210)^{\circ}C$. When the contact was loaded (figure 5), a practically linear dependence of the level of energy dissipation $W_f^k$ on force $F_n$. was obtained. In this case. the evolution manner of the force in the experimental friction law changes.

IOP Conf. Series: Materials Science and Engineering **174** (2017) 012013 doi:10.1088/1757-899X/174/1/012013

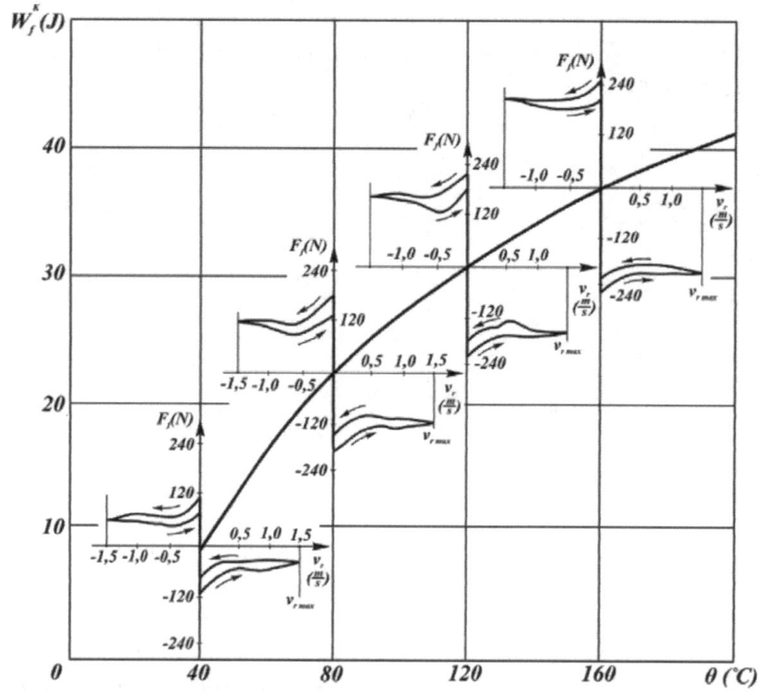


**Figure 4.** Character of the dissipative process and evolution of the friction law with experimental temperature variation θ in the contact area for normal force loading Fn=2,0kN.

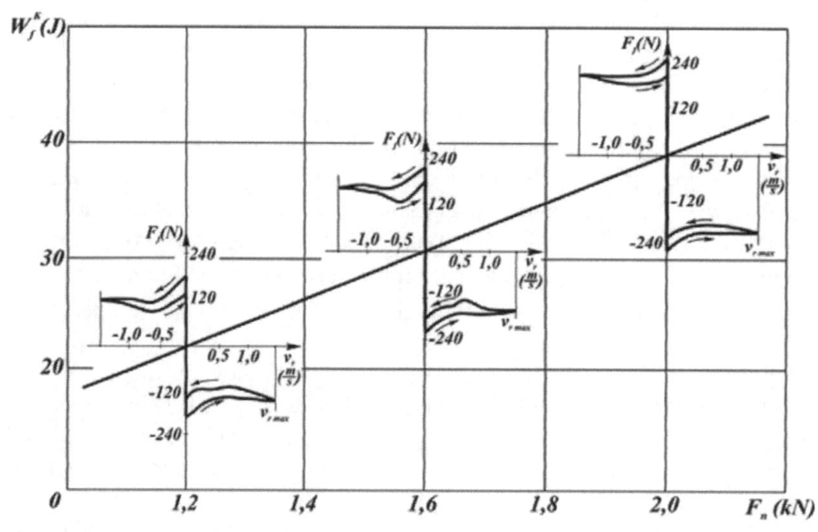


**Figure 5.** Nature of the dissipative process and evolution of friction law for variation of the load force F_nat constant temperature θ into contact area.

IOP Conf. Series: Materials Science and Engineering **174** (2017) 012013 doi:10.1088/1757-899X/174/1/012013

## 4. Conclusion

In cases when the friction law (for the couple of materials used in the construction of the triboelements) higher order nonlinearities occur, the friction forces generate in the mechanical system elements noises as self-oscillations in a wide range of frequencies.

Based on Lagrange equation, the dynamic model of the interaction between the mechanical system and the tribosystem was developed. The harmonic oscillator with elastic elements was accepted as mechanical system for modeling.

The examination of the dynamic model identified the structure of the friction force in unstable operating conditions of the mechanical system. In the structure of the total friction force $F_f$ four possible components appear: a component $F_c^i$ (of Coulomb type), defined by the linear factors of the dissipative function $\Phi_d$; three variable (fluctuating) components within the limits of each oscillation cycle $(F_v^i)^{**}$, $(F_v^i)^{*}$, $(F_v^i)^{v}$ (occur as a result of various dynamic effects in the contact zone with higher order nonlinearities and can vary over a wide range of frequencies and amplitudes.

Based on the model with harmonic oscillator, an original model of tribometer has been made, with cyclical translational movement, equipped with proper measuring systems for the dynamic characteristics of the oscillator, and the method for experimental determination [3] of the dynamic characteristics of the sliding tribosystem in unstable operating regime was developed.

## References

[1] Andronov A A, Vitt A A and Khaikin S E 1981 *Teoriya kolebany* (Moskva: Nauka)
[2] Armstrong-Helouvry B, Dupont P and Canudas De Wit C 1994 A Survey of Models, Analysis Tools and Compensation Methods for the Control of Machines with Friction *Automatica*, **30**(7) 1083-138
[3] Crudu I 2011 *Tribomodelarea* (Bucureşti: AGIR)
[4] Kragelskii I V 1968 *Treniye i iznos* (Moskva: Izdatelstvo Mashinostroyeniye)
[5] Kragelsky I V and Gitis N V 1987 *Friktsionnye avtokolebaniya* (Moskva: Nauka)
[6] Landau L D and Lifshits Ye M 1988 *Teoreticheskaya fizika: Mekhanika Tom1* (Moskva: Nauka)
[7] Lurye A I 1961 *Analiticheskaya mekhanika* (Moskva: Izdatelstvo fiziko-matematicheskoy literatury)
[8] Wojewoda J, Stefanski A, Wiercigroch M and Kapitaniak T 2008 Hysteretic effects of dry friction: modelling and experimental studies *Phil. Trans. R. Soc. A* **366** 747-65

IOP Conf. Series: Materials Science and Engineering **174** (2017) 012014 doi:10.1088/1757-899X/174/1/012014

# Experimental studies regarding wear processes through dry friction of the superficial layer for an unconventional treated steel

**C P Papadatu**

"Dunarea de Jos" University of Galati, Faculty of Engineering, Str. Domneasca 47, Galati, Romania

E-mail: papadatu.carmen@yahoo.com

**Abstract.** The aim of the studies on tribomodels regarding the dry friction wear and the processes of the wear in this case is to determine the wear intensity for a certain type of tribomodel and for a certain type of material. An Amsler stand was used. The experimental study uses two distinct values of task of loading (Q). The rolls with different diameters were used in order to obtain different sliding degrees ($\xi$). Studies have been made for the improvement of the mechanical properties of a Chromium – Molybdenum alloyed steel, unconventional treated in a magnetic field. The Thermo-magnetic treatment was applied before a thermo-chemical treatment. The mechanical properties of the material have been improved, particularly in the case of a great content of aluminum and chromium. The hardness values of the superficial layers which have been obtained after a complex thermo-magnetic and thermo - chemical treatments, the superficial layers content and the behavior of the steels at the wear tests were used as criterion. Diffractometric analysis and a statistical modeling completed this study.

## 1. Introduction

An important quantity of the aluminum in the structure of the steel increases the thermo-magnetic treatment power and the results are the best. At the same time, the existence of aluminum content in the structure of the steel causes some hardening problems which are countered by the Chromium existence.

There were considered different thermo-magnetic treatments as improvement treatments with cooling in water in magnetic field applied before ionic nitriding treatment (plasma nitriding) at 530 ° C. This temperature of the thermo-chemical treatment was considered at 530° C, being specific for this kind of the improvement steel. The influence of such parameters affecting the nitriding layers' thickness, hardness, composition and residual stress was evaluated [1]. In the literature, taking into account the nitriding process, at all temperature below 510-520 ° C substantial quantity of the "S-phase" was found to be present, especially in the case of the austenitic stainless steel [2]. In case of low temperature (100° C – 510° C), plasma nitriding produces the expanded austenite (the S-phase) with good behavior at friction. Phase $\gamma$ ($Fe_4N$) appear at higher temperatures than 500° C and reduces the thickness of the S-phase. In this paper it was considered an improvement steel grade alloyed with Aluminum and Chromium and the treatments temperatures have been chosen for this case. The increasing of the depth of the superficial layers in the case of unconventional treatment applied has been reported in accordance with the depth of the superficial layer for the same steel which suffered a classic improvement treatment before a thermo-chemical treatment. The magnetic field modifies the residual stresses which were obtained by treatment of hardening/tempering. This process depends by the content of the carbon from the structure of the steel. In all the cases, the cooling in magnetic field has been made during the improvement treatment of these steels, the residual stresses by hardening decreases, the residual austenite quantity decreases too and – as a result - the magnetic field has positive effect on the mechanical properties because the hardness of the steels and the wear resistance increase.

If Aluminum and Chromium contents increase in the structure of the steel the residual austenite quantity decreases more rapidly. The martensitic quantity and the hardness of the steel increases significantly, more than in the case of the steels with approx. the same content of Carbon but with lower quantity of Aluminum. As a consequence, the magnetic field intensity, the content of the Carbon and the content of the Aluminum from the steels have an important influence. Because of these aspects, the tendency of breaking decreases and the probability of the fragile breakage no longer exists. Magnetostriction determines local oscillations resulting local plastic deformations [3, 5, 8, 14].

Magnetostriction determines a reduction of the quantity of the residual austenite. Furthermore, this situation implies a higher hardness of the material and for many applications - good endurance characteristics.

The control of the ion (plasma) nitrided layer using micro-hardness curves involves the determining of the micro-hardness from the superficial layer on the depth. Following this influence of the magnetic field can be seen the effects of the aluminum (1.18%) and chromium (1.38%) – as alloying elements of the steel - and the influence of the thermo-magnetic treatment on the micro-hardness of the superficial layers obtained through thermo-chemical treatment applied after thermo-magnetic treatments. This influence is determined by the increasing of the thickness of the superficial layer. This superficial layer thermo-magnetic and thermo-chemical treated has higher micro-hardness values than the superficial layer classical treated.

To complete this study, it was considered a statistical model. It is necessary to prove the causation between a normal load (Q) and the worn-out depth of the layer (Uh). I resort to statistical methods for checking the next assumption: I considered that there is a causal relationship between the independent variable Q and dependent variable Uh. Between these variables there is a direct connection. Application of analysis models of regression and correlation implies some steps [10].

## 2. Experimental Procedure

For the experimental program, were considered the samples (rollers) from the material which is a steel grade of improvement for a machine part construction. This material has the following principal components: 0.38 % C, 1.18 % Al, 1.38 % Cr, 0.17 % Mo, 0.5 % Mn, 0.058 % Cu, 0.25 % Si, 0.26 % Ni, 0.026 % P, 0.026 % S. The existence of the Molybdenum content in the composition of the steel decreases the stiffening phenomenon. The outer diameter of the rollers has 40 mm and the inner diameter of the rollers has 16 mm. [2, 3, 4, 6].

The first stage from the complex program of treatments consists in thermo-magnetic treatments.

The treatment t1 represents a Martensitic hardening process (at 920 °C) and high tempering (at 620°C), a classic treatment of improvement (Magnetic field intensity is H =0).The other treatment, t3, represents a hardening process (just cooling in water in strong alternative current (A.C.) of magnetic field) and high tempering process (just the cooling in water in strong A.C. magnetic field). The treatment t4 represents a hardening process (just cooling in water, in direct current of magnetic field) and high tempering process (just cooling in water, in D.C. magnetic field).

The second stage from the complex program of the treatments consists in applying the thermo-chemical treatment: a plasma (ion) nitriding at 530 °C, after thermo-magnetic treatment, applied at the different samples from the same steel grade considered. The treatments were considered such as: Tca = T3'= t3 + plasma nitriding; Tcc = T4'= t4 + plasma nitriding, T1'= Tclassic.

Micro-hardness (Vickers) was measured on the treated surface layer obtained by thermo-chemical treatment regimes shown above. Were performed a minimum eight determinations for each case.

The wear tests have been made using an Amsler machine, roller on roller, taking two sliding degrees ($\xi$ = 10% or, 20%), testing in time (3 hours). After each hour of wear tests the external diameter was measured. It were determined the wear resistance of the rollers through dry friction and the surface structure evolution for different parameters of testing regimes. Some factors which influence this kind of wear process are: the contact geometry of the friction couple (roller on roller), the technological parameters (surface quality, thermo-chemical treatment) and the exploitation conditions (for example: the thermal solicitation). Wear tests were carried out on an Amsler machine, using several couples of rollers, each couple corresponding to different sliding degrees $\xi$, defined as [3, 4, 7, 14]:

$$\xi = (1 - v_2/v_1) \cdot 100, [\%] \quad (1)$$

where v1 and v2 are the peripheral velocities of the rollers in contact, each one having their specific peripheral velocity. The rollers are in a direct contact (a linear contact) and the friction process created through a particular combination of angular speeds (n1, n2) and diameter sizes (d1, d2) exists. Index 1 or 2 are added for the roller 1 or 2, respectively, both of the same tested friction couple.

For instance, $\xi$=*10 %* is obtained for a pair of tested rollers having d1= 40 mm, n1=180 rpm and d2= 40 mm, n2=162 rpm; $\xi$=*20 %* is obtained for a pair of tested rollers having d1= 44 mm, n2=180 rpm and d2=40 mm, n2=62 rpm; the level of the stress is corresponding to a normal load (Q) of 150 daN and the linear contact between rollers is *b*=10 mm ( represents the width of the rolls) [5,8].

## 3. Results and discussion

After plasma (ion) nitriding, the white superficial layers had a higher depth in the case of applying the magnetic field. In figure 1 was presented the evolution of the micro-hardness values (Vickers, $HV_{0.1}$) in the plasma nitrided layer, depending on the treatments**,** measured in depth of the samples (of the leading rolls of wear)**.** It has been noted that: HVa – microhardness of the samples treated in magnetic field (alternative current); HVc - microhardness of the samples treated in magnetic field (direct current); DGR – The thikness of the white layer (the superficial layer).

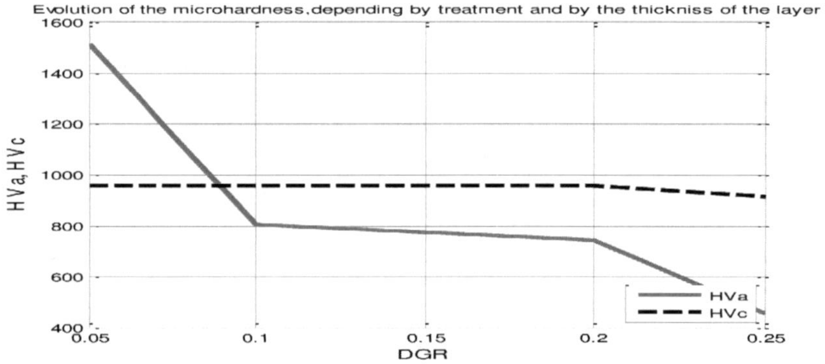


**Figure 1**.The evolution of the microhardness (HV0.1) depending by the treatment and the thickness of the layer.

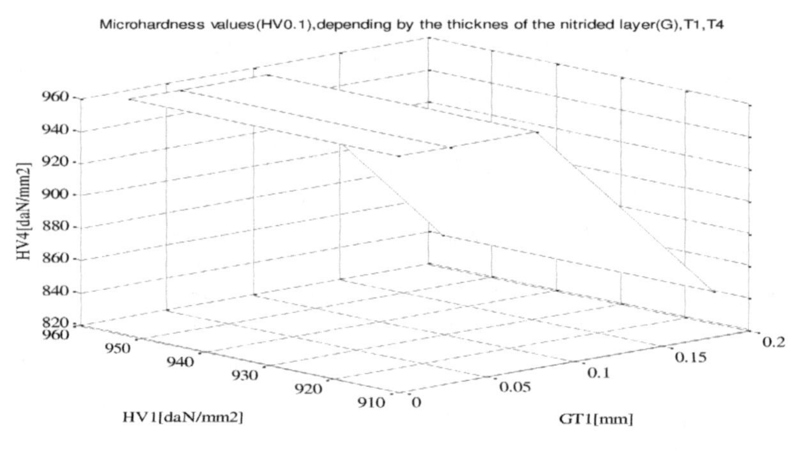


**Figure2.** A comparison between the microhardness of the layer (HV1) obtained by classic treatment and the microhardness of the layer (HV4) obtained through the unconventional treatment T4

In the figures 2 and 3 were presented a comparison between the microhardness of the layer obtained through the classic treatment and the microhardness of the layer obtained by unconventional treatments (T3 and T4). The study was continued with diffractometric analysis.

In the figures: 4, 5,...,until 11, were presented the variation of the diffractometric characteristics, the distribution of the phases from the superficial layers and the internal tensions evolution depending by duration of the wear process through the dry friction.

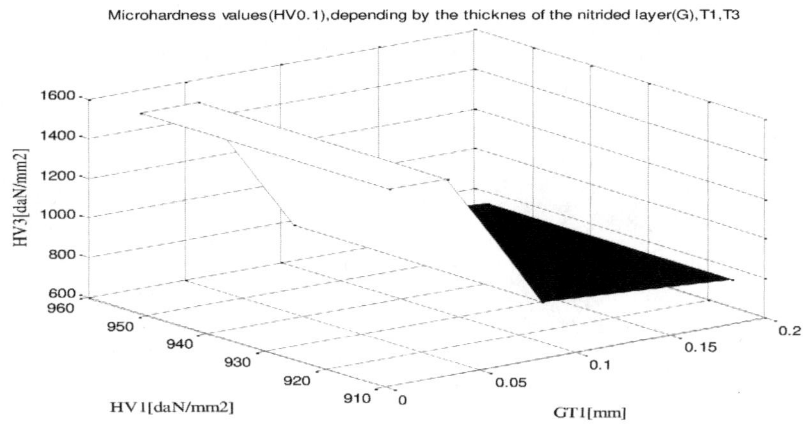


**Figure 3**. A comparison between the microhardness of the layer obtained by classic treatment (HV1) and the microhardness of the layer obtained by unconventional treatment T3 (HV3), taking into account the thikness of the superficial layer (GT1).

It was observed an improvement of the mechanical properties (hardness) in the superficial layer because of the distribution of the γ'-$Fe_4N$ phase, especially in the case of the incresing of the normal load (Q) and for the increasing of the sliding degrees (ξ). The structural and magnetic properties of epitaxial γ'-$Fe_4N$ iron nitrides films have been investigated by Costa – Krämer J L, Borsa D M, and others, in [8], and it was explained why this phase is so important. According with [7], the magnetization properties described are so far are consistent with a single phase of epitaxial film, having a cubic structure and a positive anisotropy constant, i.e., the [100] directions are easy magnetization axes. All the magnetic characteristics presented so far are then dictated by the value of the anisotropy constant. This value can be estimated from the hysteresis loops obtained applying the field along a hard magnetization axis, if the value of the saturation magnetization is known [8].

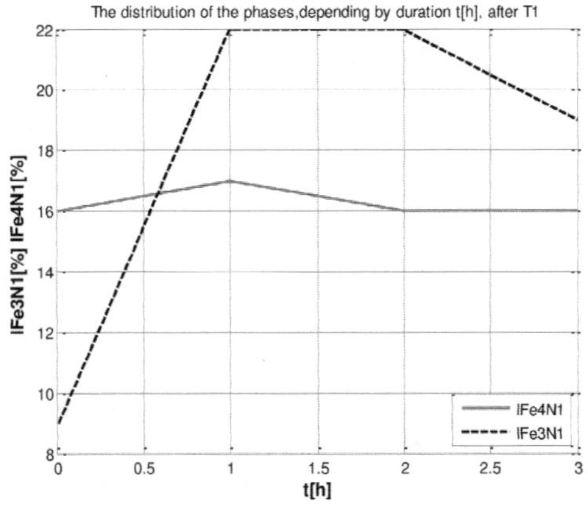


**Figure4.** Distribution of $I_{Fe4N}$ and $I_{Fe3N}$ in the superficial nitrided layer depending by the duration of the wear process, for Q = 75 daN and ξ =10 % (T1).

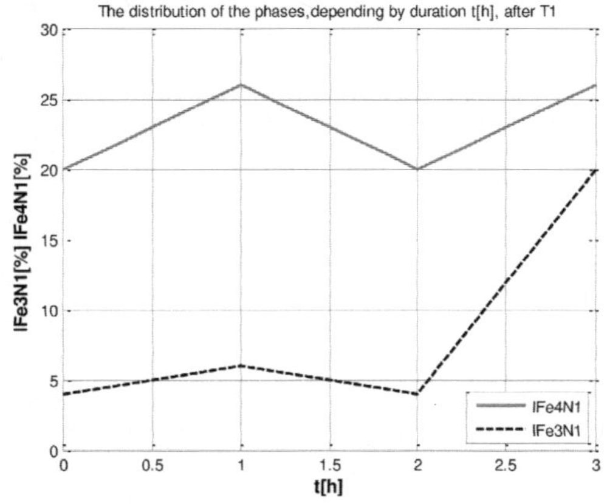


**Figure 5.** Distribution of $I_{Fe4N}$ and $I_{Fe3N}$ in the superficial nitrided layer depending by the duration of the wear process, for Q = 150 daN, ξ = 20 % (T1).

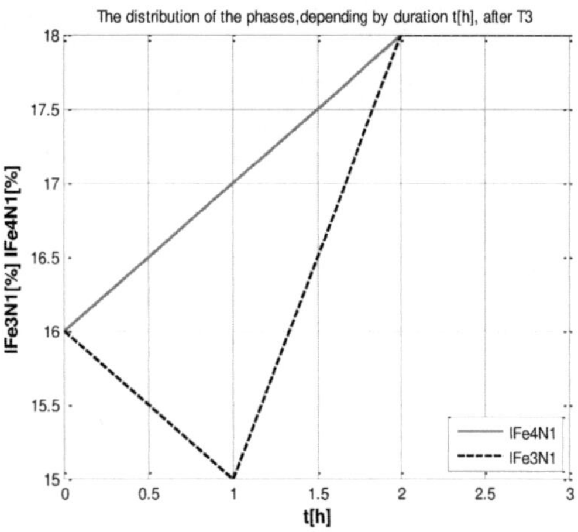


**Figure 6.** Distribution of $I_{Fe4N}$ and $I_{Fe3N}$ in the superficial nitrided layer depending by the duration of the wear process, for Q =150 daN, ξ = 20 % (T3).

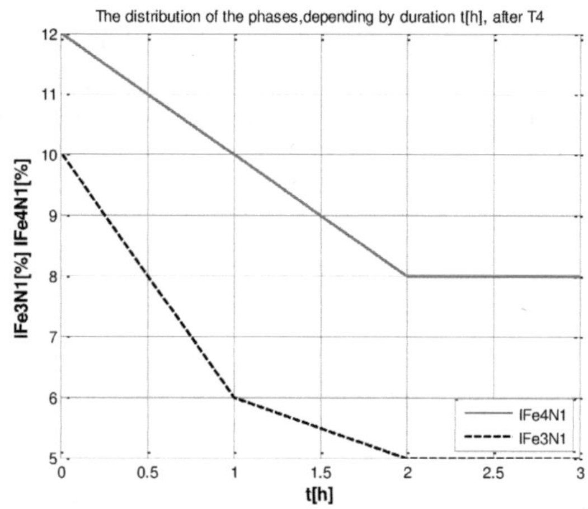


**Figure 7**. Distribution of $I_{Fe4N}$ and $I_{Fe3N}$ in the superficial nitrided layer depending by the duration of the wear process, for Q =150 daN, ξ = 20 % (T4).

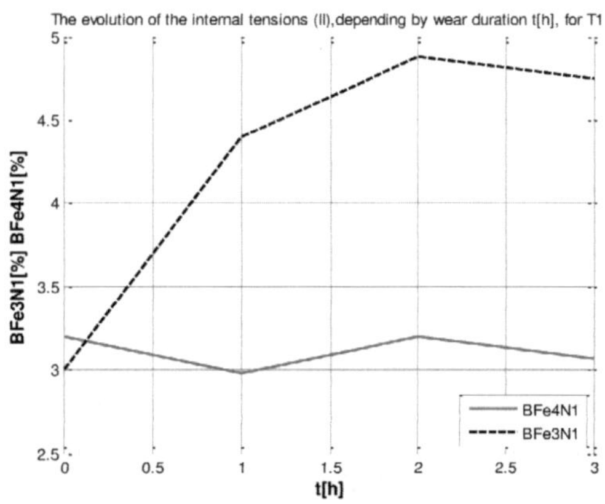


**Figure 8**. Distribution of $B_{Fe4N}$ and $B_{Fe3N}$ in the superficial nitrided layer depending by the duration of the wear process, for Q =75 daN, ξ =10 % (T1).

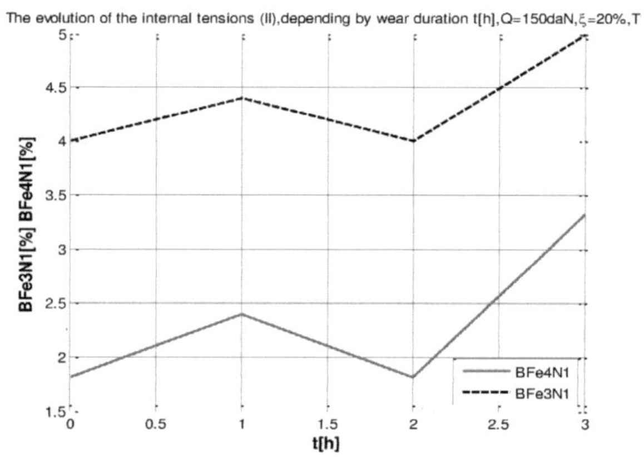


**Figure 9**. Distribution of $B_{Fe4N}$ and $B_{Fe3N}$ in the superficial nitrided layer depending by the duration of the wear process, for Q =150 daN, ξ = 20 % (T1).

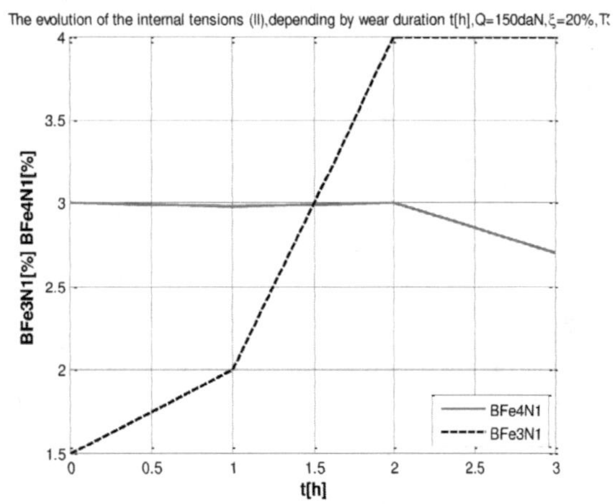


**Figure 10**. Distribution of $B_{Fe4N}$ and $B_{Fe3N}$ in the superficial nitrided layer depending by the duration of the wear process, for Q =150 daN, ξ = 20 % (T3).

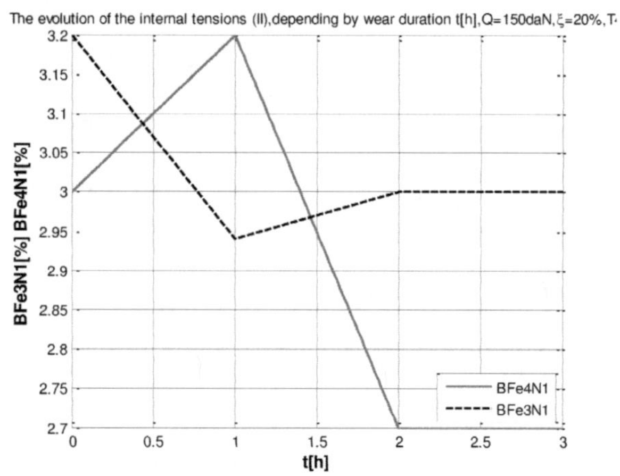


**Figure 11**. Distribution of $B_{Fe4N}$ and $B_{Fe3N}$ in the superficial nitrided layer depending by the duration of the wear process, for Q =150 daN, ξ = 20 % (T4).

In figures 8 - 11, one could observe that the internal tensions of the phases determine a lower resistance of the $Fe_3N$ phase during the cyclical fatigue.This fact was determined by machanical oscilations created by the magnetic field (A.C. current) through the permanently changes of the field lines directions, because of the magnetostriction existence. This magnetostriction from thermo-magnetic treatment changes the re-crystallization conditions, the speed of germination. That's why the hardness of the superficial nitrided layers increases. The degree of the Martensitic Tetragonalitate (c/a) is higher in the case of $\gamma$'-$Fe_4N$ (see figure 12).

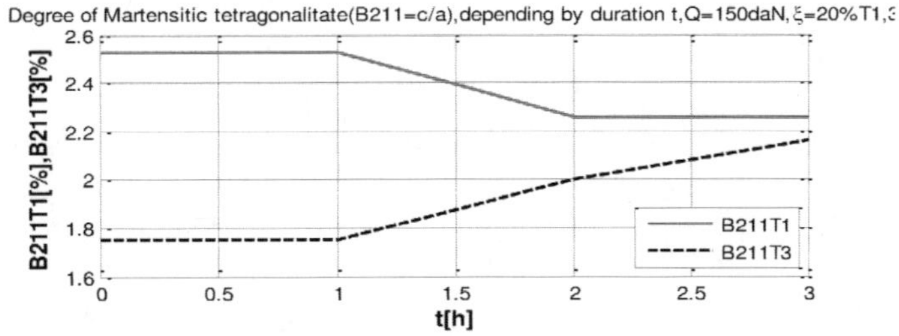


**Figure12**. The evolution of the c/a characteristic, depending by duration, in case of the treatments: T1 (classic) and T3 (unconventional treatment)

In figures 13 - 15, were presented some micro-structural aspects of the superficial nitrided layer, in case of T1 (classic treatment) and in cases of T3 and T4 (unconventional treatment regimes), considering a Nital attack 2%. From the structural point of view, applying a magnetic field (A.C. or, D.C.), there is a noticeable finishing grain size, from 9 value until 7-8 value, This situation will enable a better diffusion all along the grain boundary, when applying a thermo-chemical diffusion treatment such as ion nitriding.

**Figure 13.** Superficial layer thickness, in the case of T1 treatment (x100).

**Figure 14.** Superficial layer thickness, in the case of T3 treatment (cooling in A.C. Current –magnetic field), (x100).

**Figure 15.** Superficial layer thickness, in the case of T4 treatment (cooling in D.C. Current –magnetic field), (x200).

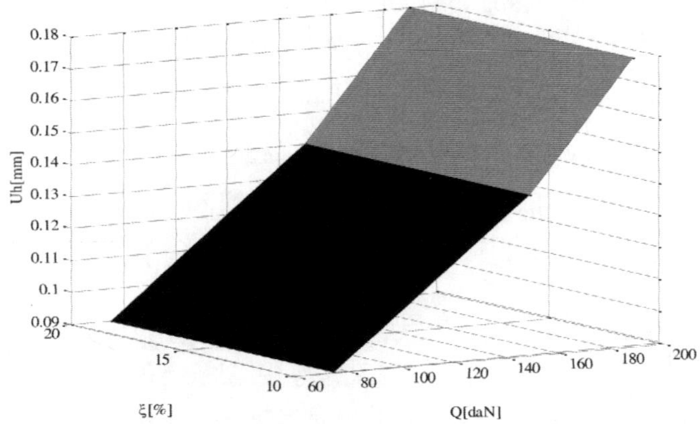


**Figure16**. Worn-out depth of the superficial layer (Uh) evolution, depending by sliding degrees ($\xi$ =10% or 20%) and normal force (Q), for the treatment T1 case.

It was considered the evolution of the worn-out layer depth (Uh) during three hours of the wear process through dry friction process and the values are presented in Table 1.

**Table 1.** The worn-out layer depth evolution

| Q [daN] | 75 | 150 | 190 |
|---|---|---|---|
| Uh [mm] | 0.09 | 0.14 | 0.18 |

In figure16, the worn-out depth (Uh) of the superficial layer increases simultaneously with the increase of the normal load (the down force, Q).

## 4. Statistical model

It can be said that the linear model of uni-factorial regression can describe with success the relationship between the two indicators analyzed.

The main problem of any regression model is the determination of the model parameters, operation which can be performed using the method of least squares. It starts from the regression equation of a simple linear model:

$$y = a + bx + u\ t = 1 \tag{3}$$

where: t, u are theoretical values of variable y obtained only function by the values of essential factor x and by the estimators values of the parameters "a" and "b", respectively, "â" and "$\hat{b}$".

To prove the causation between a normal load (Q) and the worn-out depth of the superficial layer (Uh), resort to statistical methods for checking the next assumption: it is assumed that there is a causal relationship between the independent variable (Q) and a dependent variable (Uh). Between these variables there is a direct connection. Application of analysis models of regression and correlation implies the following steps.

First step is represented by a correct identification of the variables X and Y. In this case, Q will be considered the factorial variable X (the cause) and Uh will be the dependent variable y (the efficient variable).

The second step is represented by a verification of the existence of the connection between X and Y. For this method, the graph of the correlation called *the correlogram* will be used. Based on the graph of correlation we can say that between the two variables, there is a direct connection. Because of the tendency towards the concentration is around a straight line, the graph suggests a linear connection.

The third step is the establishment of the mathematical form of the connection: the development of the regression equation corresponding to this regression model. In the case of regression unifactorial, it is considered that on a resulting characteristic "***y***" acts the only variable factor "***x***" and the others can have a constant and negligible action: y = f(x). Considering that the graphic of the function f (x) is a straight line, results a linear model of the simple regression [10]. The function of estimation is given by the expression:

$$\hat{y} = a + bx \tag{4}$$

Modeling function is given by the following expression:

$$y_{xi} = a + bx_i, \tag{5}$$

for the linearity of connection and it means an additive interaction scheme of variables, in conditions by uniform changing with constant amounts of characteristic "x" [11].

The parameter "a" is the value of the regression function at the point x = 0. This parameter "a" is the ordinate by origin, meaning the point where the regression right intersects the 0y axis.

The parameter "b" called the regression coefficient, shows the amount with which you can modify the variable "y", to change the variable with one unit.

The Method of least squares assumes the minimization of the following function:

$$F(\hat{a},\hat{b}) = \min\sum_{t=1}^{n}(y_t - \hat{y}_t)^2 = \min\sum_{t=1}^{n}(y_t - \hat{a} - \hat{b}x_t)^2 \tag{6}$$

The minimum condition of this function implies the formation of the system of two equations with two unknowns:

$$n\hat{a} + \hat{b}\sum_{i=1}^{n} x_i = \sum_{i=1}^{n} y_i\ , \tag{7}$$

$$\hat{a}\sum_{i=1}^{n} x_i + \hat{b}\sum_{i=1}^{n} x_i^2 = \sum_{i=1}^{n} x_i y_i, \tag{8}$$

For this case, n =3.

One can write the equations from the following system of equations:

$$3a + b\,(75 + 150 + 190) = 0.09 + 0.14 + 0.18$$
$$a\,(75 + 150 + 190) + b\,(75^2 + 150^2 + 190^2) = 75 \cdot 0.09 + 150 \cdot 0.14 + 190 \cdot 0.18$$

Final results are:

b = 0.00076 > 0; a = 0.0315

Because the coefficient "b", also called *regression coefficient,* has a positive value, it shows a direct relation between the two variables. In the graph of correlation the regression coefficient (b) represents the straight line slope. Replacing the values of "a" and "b" in the regression equation, it was obtained the new regression equation:

$$\hat{y} = 0.00076\,x + 0.0315 \quad (9)$$

$$\text{or,}\ \ y_{xi} = 0.00076\,x_i + 0.0315 \quad (10)$$

It is estimated that the connection between the two variables (Q and Uh) is the straight line equation. Using the coefficients "a" and "b" it is calculated the regression equation value of each 'x" characteristic size. These values of the regression equation are also called theoretical values of y according to x. The operation of replacing of the real terms y with the values of regression equations (theoretical values) is called adjustment.

To check the accuracy of parameters calculation of a regression function it is used the following relation:

$$\sum \hat{y} = \sum y \quad (11)$$

This relation is based on the fact that by calculating the regression equation it is obtained a redistribution of the degree of factors influence. To emphasize the logical connection between phenomena is necessary to investigate a large number of individual cases in which deviations (in either direction) is offset each other [13].

Comparing the real values of **y** with the adjusted values for resulting feature **ŷ**, there are small differences, presented in table 2. It is estimated that the bond between these two variables (Q and Uh) is represented by the equation of straight line (Figure 17). Using Matlab software, it was obtained the correlation graphic corresponding to the equation (9), for the values of correlation from table 2 (figure 17).

**Table 2.** Adjusted values

| x | ŷ |
|---|---|
| 75 | 0.0885 |
| 150 | 0.1455 |
| 190 | 0.1759 |

**Table 3.** Values compared

| ŷ | y- ŷ |
|---|---|
| 0.0824 | 0.09-0.0885=0.0015 |
| 0.170 | -0.0055 |
| 0.216 | 0.0041 |

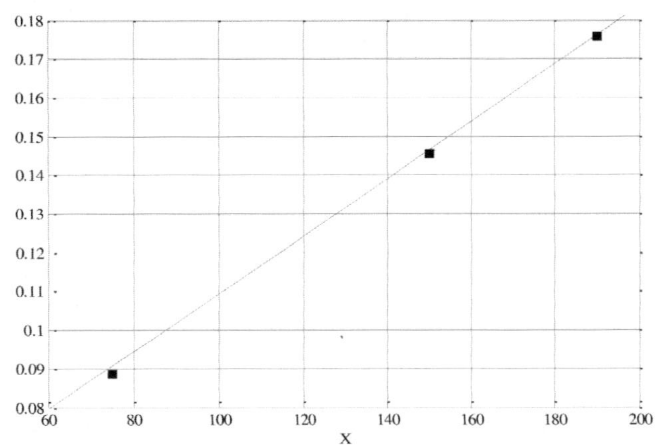


**Figure 17.** Representation of the equation: ŷ = 0.00076 x + 0.0315

The fourth Step represents the characterization of the connection intensity between these two variables using the correlation coefficient "$r_{x,y}$" and the correlation report [10, 11, 13].

$$\bar{y} = \frac{1}{n} * \sum_{i=1}^{n} x_i\, y_i \quad (12)$$

The above relation, for *covariance coefficient*, can be written [10, 11, 13]:

$$M(x \cdot y) = n^{-1} \sum_{i=1}^{n} x_i y_i, \tag{13}$$

For n =3, $\bar{y}$ = M(x y) = 61.95/3 = 20.65
The simple correlation coefficient "r" can be written [13] :

$$r = \frac{n\sum_n x_i y_i - \left(\sum_n x_i\right)\left(\sum_n y_i\right)}{\sqrt{n\sum_n x_i^2 - \left(\sum_n x_i\right)^2} * \sqrt{n\sum_n y_i^2 - \left(\sum_n y_i\right)^2}} \tag{14}$$

$$r = (3 \cdot 61.95 - 415 \cdot 0.41) / 20450^{1/2} \cdot 0.0123^{1/2} = 15.70/15.85 = \mathbf{0.99} > 0$$

In this case, $r_{x,y} = 0.99 > 0$

This result indicates the correlation, a direct connection between these two variables (Q and Uh).Theoretically, if the value is closed by 1, we have a very strong connection between the variables. In this case, it can be talk about a relative deterministic connection because $r_{x,y} = 0.99$ and the following relation: $0.95 \leq r \leq 1$ was demonstrated. For the interpretation of nonzero values for coefficients of correlation, an explanation graphics is much more suggestive in mathematical statistics. The value of the correlation coefficient is in the dependency pairs ($x_i$, $y_i$) with the distribution of values in a rectangular XOY references.

Regarding the geometric configuration of the corresponding distribution points, the distinction is made between the following cases:
a). The points are alignment along a line: it can be a right ascending line ($r_{xy}$=1), or it can be a right line downward ($r_{xy}$= -1). These situations indicate a dependence relation between the two variables.
b) The points are dispersed random, the cloud of the points hasn't orientation. The two variables are independent or uncorrelated ($r_{xy} = 0$).

It is calculated the simple ratio of correlation (Rx,y) taking into account the multiple ratio of correlation according to [10,11,12,13].

It is find that Rx,y is approximately equal with the simple coefficient of the correlation "r". To confirm the linearity of the connection, the following relation must be met:

$$| r_{x,y} | = R_{x,y} \tag{15}$$

In this case,

$$| r_{x,y} | = R_{x,y} = 0.99$$

It was confirmed statistically the direct linear connection between these two variables: Q and Uh.

The correlation is positive because the graph of the correlation is linear ascending and is strong enough because the points are close. It is a linear correlation. Testing suitability of the model is the formula for the correlation coefficient obtained ($0.95 \leq r \leq 1$).

## 5. Conclusions

The first research direction was to increase the mechanical characteristics (hardness) of the steel through thermo-magnetic treatment in order to applying the thermo-chemical diffusion treatment under thermo-magnetic treatment temperature. Another line of the research was to study the influence of the thermo-magnetic treatment on the superficial layer thermo-chemically treated.

The originality consists of applying the thermochemical diffusion treatment after thermo-magnetic basic treatment, with the mention that the thermo-chemical treatment temperature is lower than the temperature of thermo-magnetic treatment. This condition has been mentioned in order to not modify the properties of the material after the thermo-magnetic treatment during the thermo-chemical one.

A mathematical model has been made to confirm the connection between some characteristics of the material according with the parameters of treatments.

A statistical model from Economics which demonstrated the causal relation between the normal loading (Q) and the depth of the worn-out nitrided layer (Uh) after three hours of wearing process, has been proposed in this study.

During this work, a correlation, a regression line through the linear model of uni-factorial regression using the least squares *approximation* method has been proposed. It was demonstrated that a direct connection exists between these two variable.

In the cases presented up, it can be talk about a relative deterministic connection because $r_{x,y} = 0.99$ and the following relation: $0.95 \leq r \leq 1$ was demonstrated.

## References

[1] Cavaliere P, Perrone A and Silvello A 2016 Multi-objective optimization of steel nitriding *Engineering Science and Technology, an International Journal* **19** (1) 292-231

[2] .Lubrecht A A and Dalmaz G (Eds) 2004 *Transient Processes in Tribology: Proceedings of the 30th Leeds-Lyon Symposium on Tribology* (Amsterdam: Elsevier)

[3] Papadatu C P 2005 *Researches on improving the properties and reliability of some steel grades used for manufacturing metallurgical equipment*, PhD.Thesis, Dunarea de Jos University of Galati, Romania.

[4] Papadatu C P 2013 Study regarding the influence of tribological factors on the superficial layers of steels treated with plasma nitriding *Analele Universitatii Maritime Constanta (Annals of Constanta Maritime University)* **19** 151-156

[5] Papadatu C P 2012 Monitoring of the sustainability for steels subjected to unconventional treatments *The Annals of "Dunarea de Jos" University of Galati*, Fascicle **VIII (2012)** 98-102

[6] Rus V, Vermeşan G,Vermeşan H and Rus S, 2003 Ion nitriding: present and future *Intern. Conf. on Materials Science and Engineering BRAMAT 2003, Braşov*

[7] Gheorghies C, Papadatu C P , Stefanescu I 2004 Study on the model of behaviour of some non-conventional treated steels, during friction process, *Proc. of the Intern. Metallurgical and Materials Conf. CENTIMFE, Marinha Grande, Portugal*

[8] Papadatu C P 2007 *Posibilitati de imbunatatire a calitatii unor oteluri utilizate in industria metalurgica (Posibilities to improve the quality of some steel grades from metallurgical industry) (*Galati: Ed. Fundatiei Universitare Dunarea de Jos)

[9] Kosta-Krämer, Borsa D M at al. 2004 Structure and magnetism of single-phase epitaxial γ'-Fe4N, *Physical Review* B **69 (14)** 144402-1 - 144402-8

[10] Oroian M, Radnoti-Szakacs A, Simofi A 2013 Studiu empiric asupra relatiilor de cauzalitate dintre rata inflatiei si unele variabile macroeconomice, *Academica Science Journal*. Studia Series **(2) 1** 21-28.

[11] Scradeanu D 2015 *Modele cantitative statistice* available at http://www.unibuc.ro/prof/scradeanu_d/docs/2013/noi/08_21_27_123_Modele_CANTITATIVE_STATISTICE_DSCRD.pdf

[12] Breaz N, Craciun M, Gaspar P, Miroiu M, Paraschiv-Munteanu I, 2011, *Modelarea matematica prin Matlab,* available at http://www.edumanager.ro/community/documente/modelare_matematica_prin_matlab.pdf

[13] Baron T, Korka M, Pecican E, Stanescu M 1981 *Statistica* (Bucuresti: Ed. Didactica si Pedagogica)

[14] Papadatu C P, Coteţ A, Vasilescu E, Andrei G 2016 Experimental researches regarding the possibility to increase the durability of a CrMo alloyed steel subjected to some non-conventional treatments, applied in magnetic field *Proc. of the conference METAL 2016* pp 1133-38

# A comparative study on tribological behavior between metal and polymeric composites used to repair bronze made parts in dry reciprocating sliding tests

**V Iliuţă[1], M Rîpă[1], J Javorova[2] and G Andrei[1]**

[1] "Dunărea de Jos" University of Galaţi, Galaţi, Romania
[2] University of Chemical Technology and Metallurgy, Sofia, Bulgaria

E-mail: virgil.iliuta@ugal.ro

**Abstract.** The paper presents the researches on tribological behaviour of two materials used for parts subjected to friction, under dry sliding conditions. The composite material is a product of the company Diamant Metallplastic GmbH, Germany and the manufacturer recommend it for repairing or reconditioning worn or damaged bronze made parts. The material belongs to Multimetall category, which is a highly resistant 2-component repair system. This polymer composite material was tribologically tested in dry friction reciprocating conditions, in ball-on-flat configuration, using the tribometer UMT-2 (Bruker, former CETR). The counterpiece was a steel ball. Typical test conditions were as follows: normal loads of 20, 30, 40 and 50 N, sliding distance of 100 m, stroke length of 5mm, average sliding speed of 3,5 mm/s, room temperature and relative humidity of 50-60%. After testing, wear tracks were examined by electronic microscopy (SEM), laser profilometry and the profilometric module of the tribometer CETR-UMT-2. The results of the composite material analysis were compared with the similar ones obtained under similar test conditions for the bronze (metal material).

## 1. Introduction

Researches carried out on polimeric composite materials have shown the advantages of using them in several different areas: automotive industry, manufacturing, shipbuilding, oil and gas industry, foundries, chemical industry, pump construction, housing construction etc [1, 2, 3, 4, 5, 6]. By reinforcing plastic materials with fibre or metalic particles, polimeric composites with remarkable properities were obtained, properities such as: low density, good stiffness, higher strength/ density ratio, low friction coefficient, good surface quality, etc. [7, 8, 9, 10, 11, 12, 13, 14, 15, 16, 17].

Due to some advantages such as easy molding processing, good adhesion to metal surfaces and development of functional surfaces without machining, polymer composites reinforced with fibers or particles are used in repairing metal parts [15, 16, 17].

Composite materials are frequently used for replacing expensive and energy consuming metallic materials. Lately, the fields of application for these materials have expanded in areas such as: automotive maintenance and repair, machine manufacturing, steel industry, etc. [1, 2, 3, 4].

Tribological tests of composite materials are carried out using different methods in various test configurations, such as *ball-on-flat test configuration in reciprocating dry sliding* [7, 8, 9, 10, 11]. In the mentioned papers, the variation of friction coefficient and surface morphology are presented, depending on parameters such as: loading force, sliding distance, test duration, movement frequency.

In the papers [12, 13] the results for tribological tests are presented. The tests were carried out on a few polymeric composite materials. In many papers, the topography of wear traces is used as an indicator of wear behavior for the tested material [14]. Surface topography is studied using digital profilometers and microscopic analysis.

## 2. Experimental procedure

### *2.1. Materials' properties*

Tribological research presented in this paper were performed on samples made from two materials: 1. composite material (code SBC) used for the correction and repairing of defects of bronze work pieces and 2. bronze – metallic material (code SB).

The composite material is produced by the German company "Diamond Metallplastic GmbH" [5]. In the epoxy matrix, particles of Cu, Zn, Sn and various allotropic forms of $SiO_2$ are introduced. Composite consists of two components: the epoxy resin and the hardener supplied in the correct quantities recommended by the producer. The components are mixed and then applied with spatula on the specially manufactured metallic surfaces of samples.

Before the testing, the active surfaces were cleaned, washed and dried (approx. 18 h) [5]. After drying, surface of samples was manufactured by machining.

The main characteristics of the composite material are shown in Table 1.

**Table 1** Main mechanical characteristics of the composite material [5, 15].

| Material code | Tensile Strength [MPa] | Compressive Strength [MPa] | Bending Strength [MPa] | Shear Strength [MPa] | E-Modulus DIN 53457 [MPa] | Hardness [Shore D] | Specific Weight [g/cm$^3$] | Temperature Resistance permanent [°C] | Temperature Resistance temporary [°C] |
|---|---|---|---|---|---|---|---|---|---|
| SBC | 62 | 155 | 79.5 | 16.5 | 5800 | 86 | 2.2 | -32 to +160 | +350 |

The metallic material (bronze) has the mechanical characteristics shown in Table 2 and the roughness parameters presented in Table 3.

**Table 2** The mechanical properties of metallic material - Bronze [15].

| Material code | Tensile Strength $R_m$ [MPa] | Yield Strength $Rp_{0.2}$ [MPa] | Elongation at Break $A_5$ [%] | Brinell Hardness HB | Specific weight [g/cm$^3$] |
|---|---|---|---|---|---|
| SB | 506 | 405 | 12 | 134 | 7.6 |

**Table 3** The roughness of samples.

| Material code | $R_a$ [μm] | $R_q$ [μm] | $R_t$ [μm] | $R_{sk}$ [ - ] | $R_{ku}$ [ - ] |
|---|---|---|---|---|---|
| SBC | 1.81 | 6.67 | 32.68 | -3.72 | 25.33 |
| SB | 0.32 | 0.42 | 3.66 | -0.18 | 4.90 |

The final dimensions of the samples were Φ76x7.

The counterpiece was a steel ball (AISI 52100 GCr15, grade SKF like) with a diameter of 6 mm and hardness of 80HRC. The surface roughness was about $R_a$=0.06 μm.

### 2.2. *Experimental setup and procedures*

For carrying out the tests, the reciprocating module of UMT-2 tribometer was used. The reciprocating tests were conducted in accordance with ASTM G133-05 standard [18], at temperatures between 20 and 26 °C and a relative humidity of 50-60%. For the loading force four values were used: 20, 30, 40 and 50 N. The other test parameters were: the stroke length - 5mm, the sliding distance - 100 m and the test duration - 475 min. Before testing, the samples were cleaned with an organic solvent and then dried in hot air (50 °C).

Tribometer software provided the following parameters: coefficient of friction (μ) and the depth of linear wear ($U_h$), measured for the two materials for all load forces.

After testing, the wear traces were studied using the μSCAN profilometer (®NanoFocus)) and analyzed with SPIP software (™ Image Metrology, Hørsholm) and with the UMT-2 tribometer profilometric module.

Data acquisition of the digital profiles for 2D profilometric analysis was made in three cross-sectional profiles at the distance of 2 mm from each other: i – inside, e – outside of the disc and m – middle of the disc (figure 1). Also, the 3D profile for the 40 N and 50 N load force was obtained, in the case of the two materials (figure 9 and 10).

The profilometric analysis was carried out by using following parameters: (a) average cross sectional area and (b) average width of the profile of wear tracks. The measured area is up limited by the highest peaks of adjacent points of the profile and below by the lowest point of the profile.

Investigations on images were carried out with Quanta 200 SEM (®FEI Co., Czech Republic) and the optical microscope Neophot-2 (®Carl Zeiss, Germany).

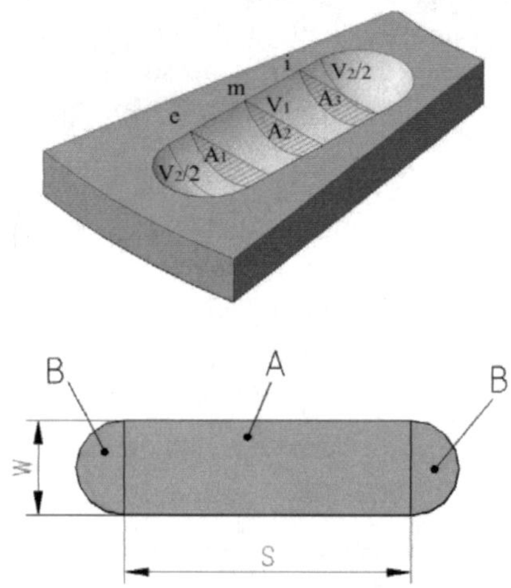


**Figure 1.** Graphical modeling of wear tracks [15].

## 3. Results and discussions

### 3.1. *Variation of the coefficient of friction*

Figure 2 presents the variation of the coefficient of friction (COF) during the tests of the composite material (code SBC) and bronze (code SB). The graphs have been constructed by plotting the acquired data for the sampling interval of 1/10s. In the case of the composite material (code SBC), the variation of the coefficient of friction shows three stages for all four loading forces (figure 2a). In all three stages, the friction force increases with the increasing of the loading force:

- COF increases linearly in the first stage, the rate having higher values at higher load forces; in the second stage there is a rapid increase of COF about 3-4 times, and in the third stage of COF variation one could observe the lowest rate. In this stage the instability of the process is higher comparing with the other stages;

- in contrast to the first and second stages, in the third stage the rates of variation of the coefficient of friction are comparable.

Studying the variation of COF for bronze (code SB) (figure 2b) significant differences comparing to the composite material are observed:

- a sharp increase in first stage of the variation of COF (on a length of approx. 1-2 m) was observed; then COF increases at a rate slightly bigger than for the composite material, regardless of loading force;

- bronze friction behavior is similar for all the range of loading forces.

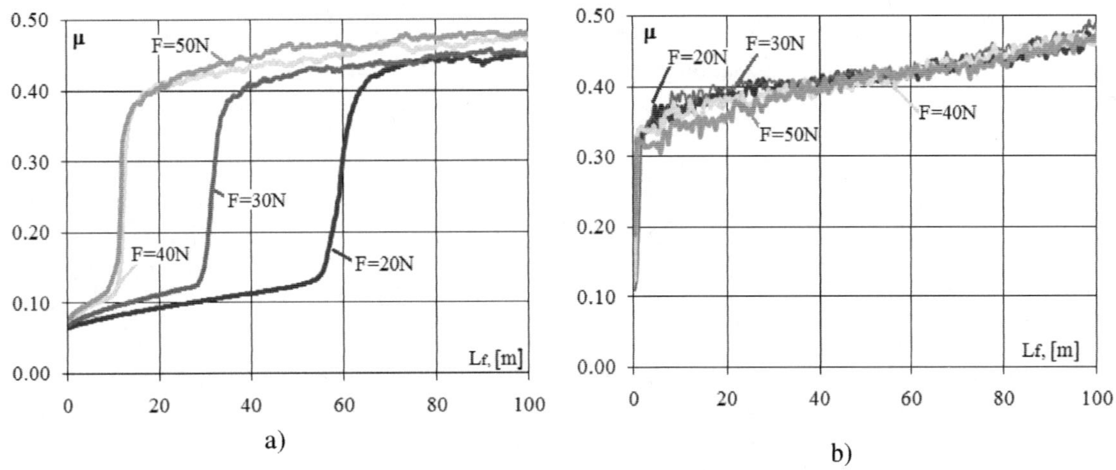


**Figure 2.** Variation of the coefficient of friction (F=20, 30,40 and 50N, $L_f$=100m): a) composite material for bronze (code SBC); b) bronze (code SB).

Figure 3 presents the variation of the average COF (calculated as the average of the values acquired over the last 25 m of the length of friction). One could observe that the two materials (composite material (code SAB) and bronze (code SB)) have the average coefficient of friction about the same. This observation could recommend the using of composite material in reconditioning bronze metal parts.

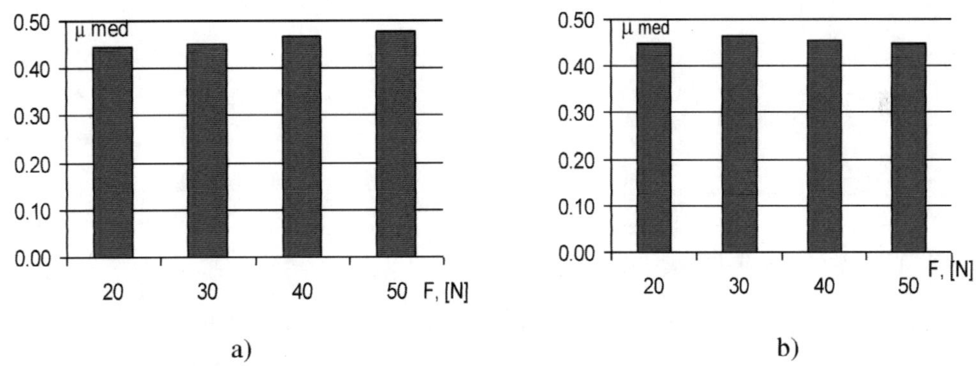


**Figure 3.** Variation of the average coefficient of friction (F=20, 30,40 and 50N, $L_f$=100m): a) composite material for bronze (code SBC); b) bronze (code SB).

*3.2. Variation of the linear wear*

Figure 4 shows the variation of the linear wear (values directly obtained from the testing equipment). For the composite material (figure 4a), two stages of the evolution of linear wear could be observed. In a first stage, linear wear rate has a very small increasing, bigger for the increasing load forces. Also, the length of this first stage decreases with the increasing of the load force. An explanation could be the occurring of more contact points between the reinforcement particles of the composite and the spherical surface of the counterpart.

In the second stage the linear wear has a rapidly increasing, at rates that increase with increasing of the force. It follows a slowly decreasing of the rate of linear wear increasing, due to the flattening of the peaks and the polishing metal particles embedded in the composite material structure.

For the metallic material (code SB) linear wear increases, with relatively constant rates (figure 4b).

In this case the higher rate is observed at the end of the range of applied forces. The values of linear wear for the studied friction length (100 m) are higher in the case of the composite material (code SBC) comparing to the metal (bronze, code SB).

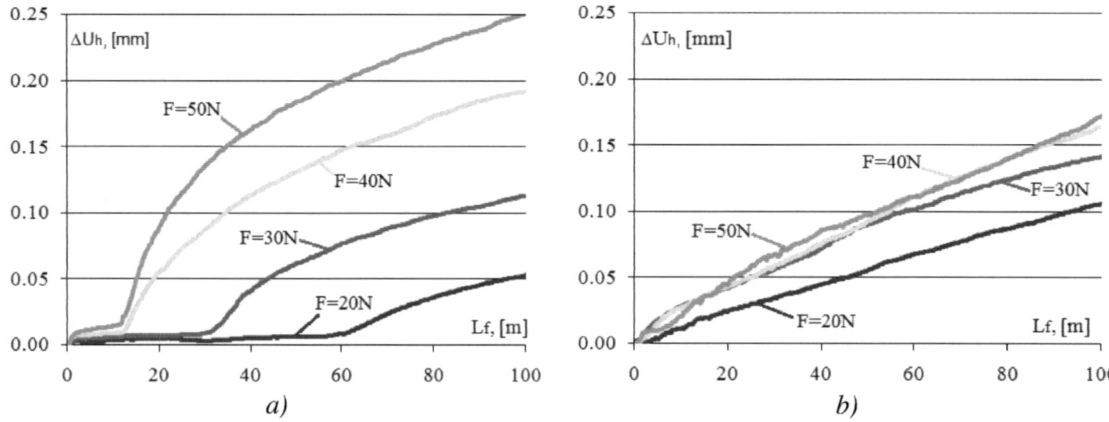


**Figure 4.** Variation of linear wear (F=20, 30,40 and 50N, $L_f$=100m): a) composite material for bronze (code SBC); b) bronze (code SB).

*3.3. Variation of volumetric wear*

Some parameters related to the wear volume were calculated, taking into account the material loss volume, V1 (figure 1) on the cylindrical part of wear track ([18]):

- wear volume (V) [18]:

$$V = A \cdot s \tag{1}$$

- wear intensity ($I_v$) [6,19]:

$$I_v = \frac{V}{L_f} \tag{2}$$

- specific wear rate coefficient (k) [6]:

$$k = \frac{V}{F \cdot L_f} \tag{3}$$

where:

$A_w$- average cross-sectional area of wear track;

$L_f$ - stroke length, mm;

F - normal loading force, N.

The average cross-sectional area of the wear track (A) has been calculated as the average of the three areas of sections $A_1$, $A_2$ si $A_3$. Areas $A_i$ were determined based on measurements made with the laser profilometer:

$$A = \sum_{i=1}^{3} A_i \tag{4}$$

The figure 5 shows that for load forces up to about 43N, composite material has lower values of volume wear than bronze. For repairing bronze parts, composite material can be recommended, but only for forces smaller than 43N.

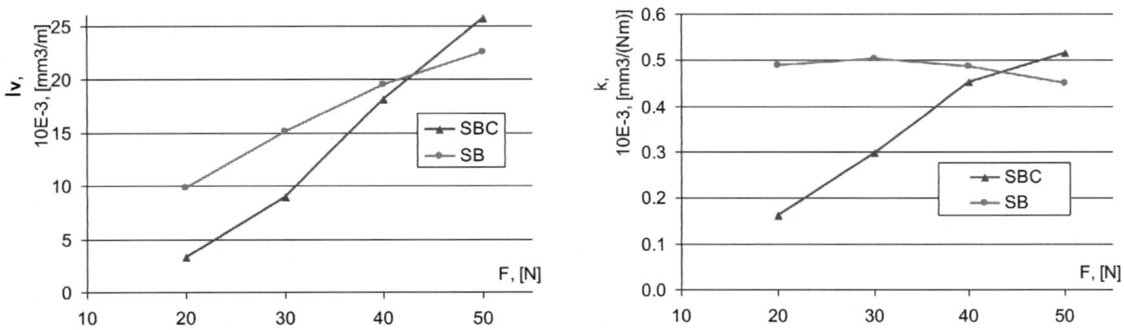


a) b)

**Figure 5.** Variation of wear volume parameters (F=20, 30,40 and 50N, $L_f$=100m):
a) the intensity of wear; b) the specific wear rate.

### *3.4. Profilometric studies*

The results of the processing of 2D digital profiles of wear tracks for the two materials are presented in figure 6.

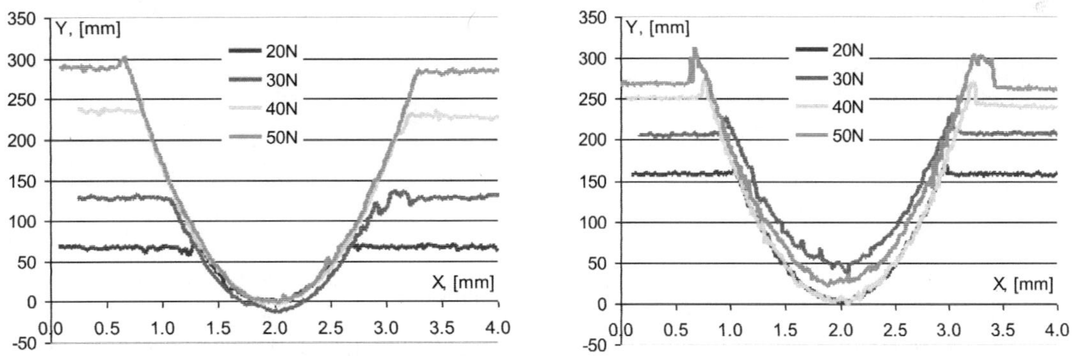


a) b)

**Figure 6.** Profiles (2D) for the loading forces 20÷40N and $L_f$=100m:
a) composite (code SBC); b) brass (code SB).

It can be observed that the composite material behaves better than bronze at lower loading forces (20N and 30N). The width and the depth of wear tracks are larger in the case of bronze. For high load forces (40N and 50N), 2D-profiles wear tracks of the two materials are comparable, previously mentioned parameters having almost similar values. At higher loading force 50N, the both parameters of cross-sectional area are higher in the case of the composite material.

Also, analyzing the values of cross-sectional area and of the width of wear tracks, one could observe that both parameters have increased considerably in the case of the composite material comparing to bronze (figure 7 and figure 8).

Figures 9 and 10 present the 3D profiles of the wear tracks for both materials, for the load forces of 40N and 50N.

Figures 11 and 12 present the 3D plots of the wear track for the composite material and for bronze, respectively, for a scanned area of 0.5x0.5 mm.

In figure 11, for the composite material, it can be observed small and smooth hallows, which can become bigger and bigger, once the load force increases. In the case of bronze, scratches and furrows can be observed (figure 12), once the load force increases. Also, it can be observed that for higher loading forces, particles of material are ripped apart. It can be concluded that the dominant wear mechanism in the case of composite material is the adhesive wear, and for bronze the dominant wear mechanism is the abrasive wear.

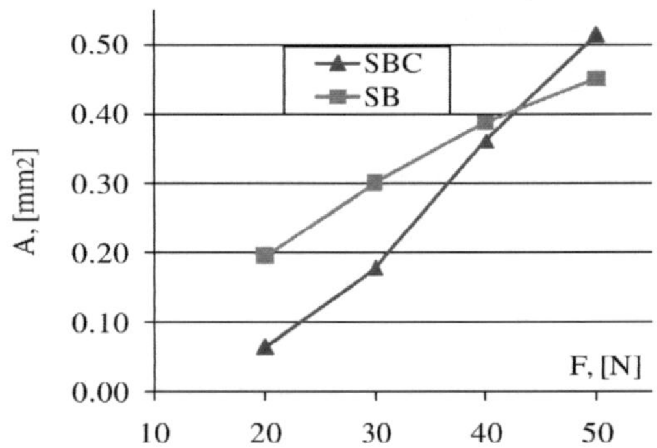


**Figure 7.** Variation of the cross sectional area of the track vs. loading force.

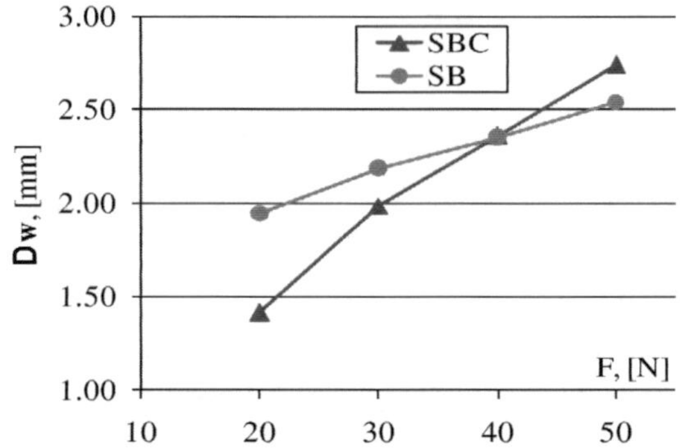


**Figure 8.** Variation of the width of the track vs. loading force.

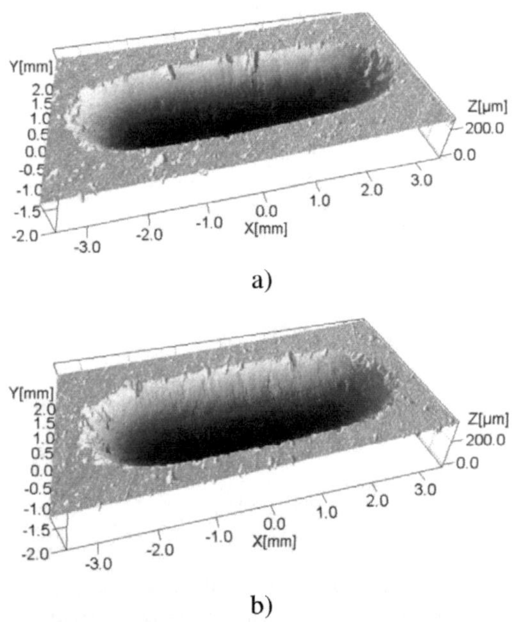


**Figure 9.** Wear track (3D plots) for the composite material, for two values of the loading force: a) F=40N; b) F=50N.

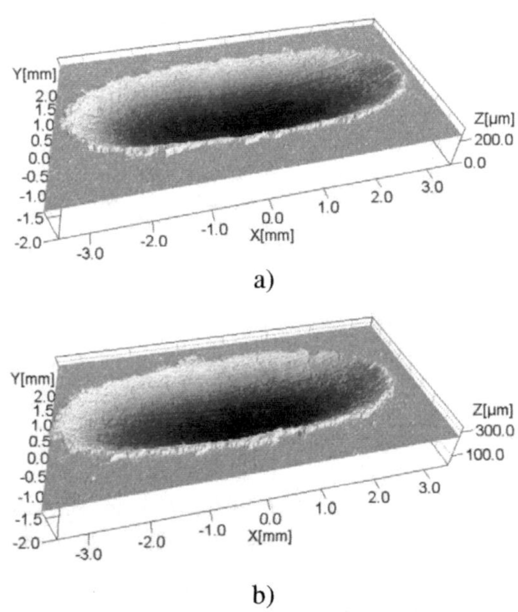


**Figure 10.** Wear track (3D plots) for the bronze, for two values of the loading force: a) F=40N; b) F=50N.

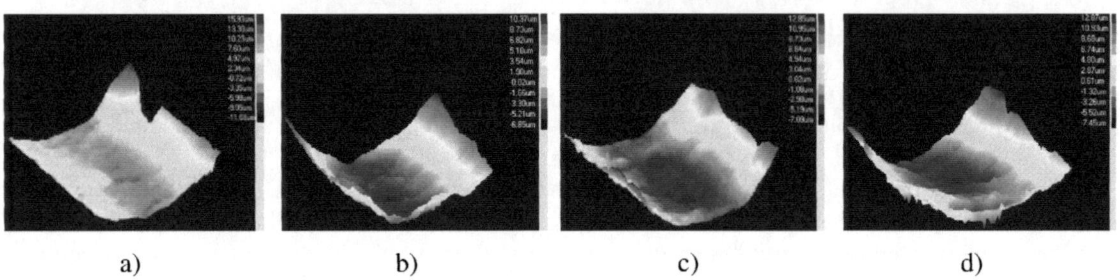


**Figure 11.** Details of the wear track for the composite material (code SBC) for four values of the load force: a) F=20N; b) F=30N; c) F=40N; d) F=50N.

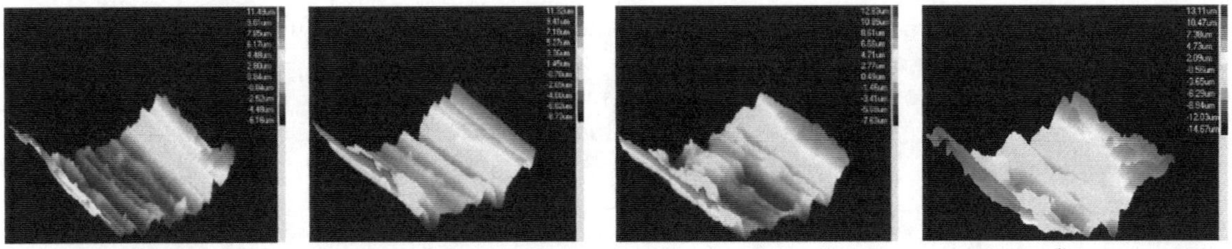

a) b) c) d)

**Figure 12.** Details of the wear track for the bronze (code SB) for four values of the load force: a) F=20N; b) F=30N; c) F=40N; d) F=50N.

To highlight the wear phenomena, the wear traces were studied with the Quanta 200 SEM and the optical microscope Neophot-2. In figure 13 images of wear tracks for composite material (code SBC) and bronze (code SB), obtained with the optical microscope Neophot-2 are presented.

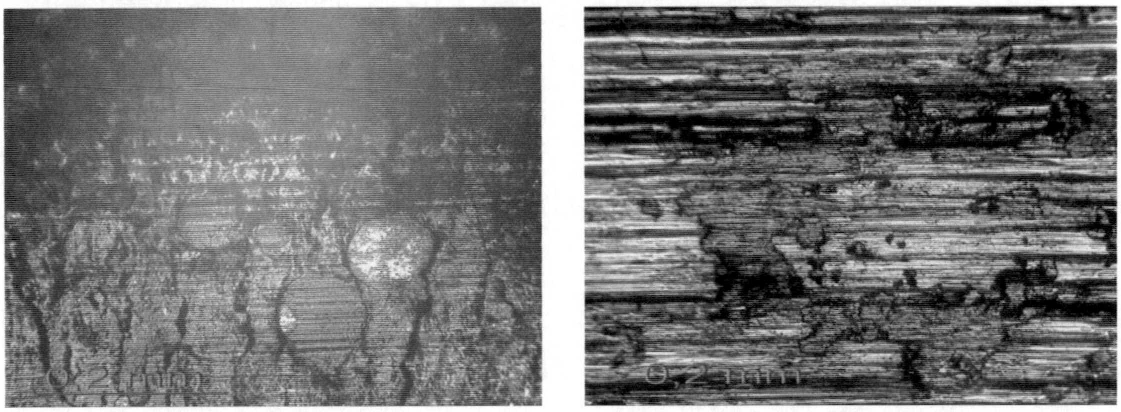

a) b)

**Figure 13.** Microscopic image magnitude x12,5 (F=30 N, $L_f$=100m) a) composite material for bronze (code SBC); b) bronze (code SB).

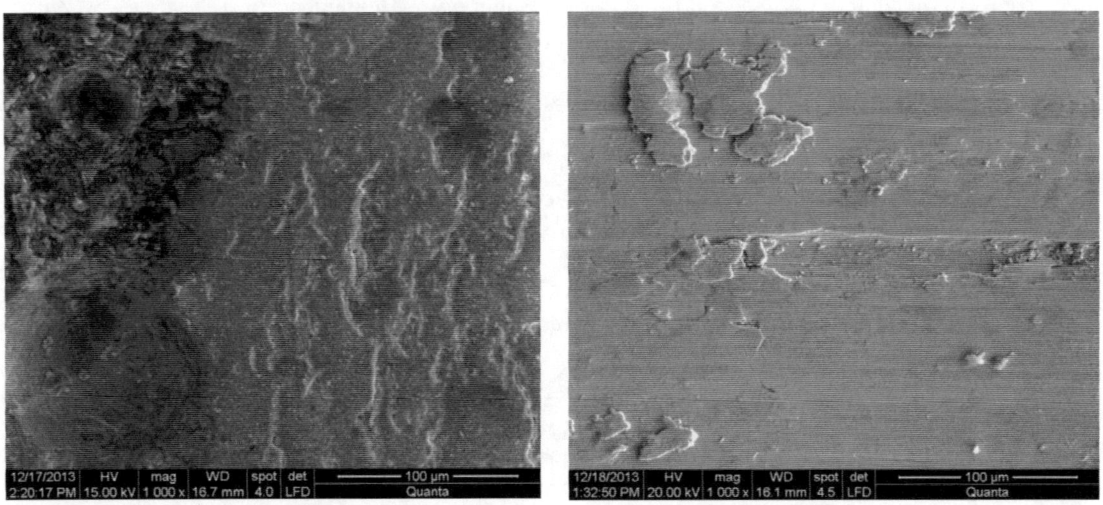


a) b)

**Figure 14.** SEM images obtained with Quanta 200 SEM, magnitude 1000, F=**30** N, $L_f$=100m: a) composite material (code SBC); b) bronze (code SB).

Images emphasize phenomena of particles' conglomeration for the composite material, metal particles and traces of their dislocation (figure 14a).

In the case of bronze, one could observe the effects of the dominant wear mechanism, abrasive wear (figure 13b) but also small area of bonding material, as effects of delamination wear on the metal surface (figure 14b).

## 4. Conclusions

In the present paper, the authors meant to develop a comparison between a metallic particles reinforced polymer composite and a metallic material, both frequently used in technical applications.

The aim of this work was a comparative study of the wear and the topography of wear tracks of two materials, composite material and bronze.

The results of testing both materials, in dry friction reciprocating ball-on-flat conditions, show:

- composite material has tribological characteristics similar to those of bronze and can be used for reconditioning of bronze;
- adhesive wear is the dominant wear mechanism for the composite material, covering also the abrasion wear;
- for bronze, the dominant wear mechanism is abrasive wear, in the presence of some effects of adhesive wear;
- for loading forces up to 40N, the polymer composite material behaves better than bronze, presenting lower values for depth, width and area of wear tracks;
- for loading force 40N, both materials have a similar wear behavior;
- for the highest loading force of 50N, bronze has a better wear behavior than the polymer composite material.

## References

[1] Brydson J A 1999 *Plastics Materials* Seventh Ed. (Oxford: Butterworth-Heinemann)

[2] Crawford R J 1998 *Plastics engineering* Third Ed. (Oxford: Butterworth-Heinemann)

[3] Friedrich K 1993 *Advances in composites tribology* Composite Materials Series 8 (Amsterdam: Elsevier)

[4] Ashby M F and Jones D R H 1998 *Engineering materials 2: an introduction to microstructures, processing and design* Second Ed. (Oxford: Butterworth-Heinemann)

[5] *** Cold welding: top quality connection without thermal stress available at http://diamant-polymer.de/en/products/ultametal/ (14.03.2016)

[6] Stachowiak G W 2005 *Wear – Materials, mechanisms and practice* (Hoboken: John Wiley & Sons)

[7] Li J and Xia Y C 2009 The reinforcement effect of carbon fiber on the friction and wear properties of carbon fiber reinforced PA6 composites *Fiber. Polym.* **10** 519-525

[8] Li J and Cheng X H 2007 Effect of rare earth solution on mechanical and tribological properties of carbon fiber reinforced thermoplastic polyimide composite *Tribol. Lett* **25** 207-214

[9] Li J and Li X Z 2010 Evaluation of the tribological properties of carbon fiber reinforced poly (vinylidene fluoride) composites *J. Mater. Eng. Perform.* **19** 1025-1030

[10] Li J and Sheng X H 2010 The effect of PA6 content on the mechanical and tribological properties of PA6 reinforced PTFE composites *J. Mater. Eng. Perform.* **19** 342-346

[11] Nie W Z, Li J. and Li X Z 2010 The addition of carbon fiber on the tribological properties of poly(vinylidene fluoride) composites *Fiber. Polym.* **11** 559-564

[12] Olea-Mejia O, Brostow W and Buchman E 2010 Wear resistance and wear mechanisms in polymer + metal composites *J. of Nanosci. Nanotechno.* **10** 8524-30

[13] Myshkin N K, Petrokovets M I and Kovalev A V 2005 Tribology of polymers: adhesion, friction, wear, and mass-transfer *Tribol. Int.* **38** 910–921

[14] Lee J H, Xu G H and Liang H 2001 Experimental and numerical analysis of friction and wear behavior of polycarbonate *Wear* **251** 1541–56
[15] Iliuţă V, Rîpă M, Preda A and Andrei G 2005 Friction and wear behavior of moglice polymer composite through dry sliding ball-on-flat reciprocating test *Applied Mechanics and Materials* **808** 137-142
[16] Iliuţă V, Preda A, Andrei G and Bîrsan I 2014 Wear assessment of polymer composite filled with metal particles through ball-on-flat reciprocating test *Mater. Plast.* **51** (4) 359-362
[17] Iliuţă V, Rîpă M, Andrei G, Preda A, Suciu C and Javorova J 2014 profilometric evaluation of the worn surfaces under dry reciprocating wear conditions of a composite material to repair brass made parts *Applied Mechanics and Materials* **657** 437-441
[18] *** ASTM G 133–05 *Standard test method for linearly reciprocating ball-on-flat sliding wear*
[19] Rîpă M and Deleanu L 2008 *Deteriorări în tribosisteme* (Galaţi: Ed. Zigotto)

# Study on friction behaviour of brake shoe materials for mining hoist

**M Ungureanu N S Ungureanu and I Crăciun**

Technical University of Cluj Napoca, North University Center of Baia Mare, Faculty of Engineering, Department of Engineering and Management of Technology, Victor Babes St., 62A, Baia Mare, RO-430083, Romania

E-mail: m_ungureanu@cunbm.utcluj.ro

**Abstract**. The friction coefficient in the brake linkages has an important influence on the braking efficiency and safety of machines. The paper presents a method for the study of the friction coefficient of the friction couple brake shoe-drum for mining hoist. In this context, it is interesting to define the friction coefficient, not just according to the materials in contact, but according to the entire ensemble of tribological factors of the friction couple.

## 1. Introduction

Friction can be defined as a complex process of molecular, mechanic and energetic nature. Dry friction is characterized by direct immediate contact of the surfaces in relative motion. No lubricant film is interfering between the brake-shoe friction areas, except the films absorbed from the ambient gaseous environment. The frictional properties of the brake shoe have very important influence on the braking reliability and safety of mine hoists. [1]

Most researches [1][2][3][4][5] show that the friction depends on speed, pressure, temperature, air humidity, real contact surface, chemical composition of materials etc. In these conditions, the friction coefficient, presented in many tables, has just informative values without presentation of experimental conditions of the measuring process.

In this context it is interesting to define the friction coefficient, not just according to the materials in contact, but also according to the entire ensemble of tribological factors of friction couple brake shoe-drum for mining hoist.

## 2. The analysis of tribological factors of friction couple for mining hoist

By corroboration of the data established by Czihoş [6] and Crudu [7] with the aspects specific to the braking system in case of mining plants, the structuring of the parameters of the shoe – drum represented in figure 1 has been carried out.

The friction materials are equipped with different types of braking system, functioning in different mediums, therefore demanding separate analysis for each particular case, for the tribological factors of friction.

The braking systems are equipped with different friction materials and they function in different environments, therefor they demand separate analysis of the tribological factors for each case.

The brake drum of the mining hoist is made of steel or cast iron. As many theoretical and experimental researches have proved, the most recommended material for the brake drum and the brake disks is cast iron (it has better thermo-physical properties than steel and a better stability of the friction coefficient). However, most of the brake drum are made of steel [8][9]. The materials from which the brake drum of the mining hoist are made must have the following characteristics: good thermal conductivity; high friction coefficient between it and the friction lining; resistance to wear.

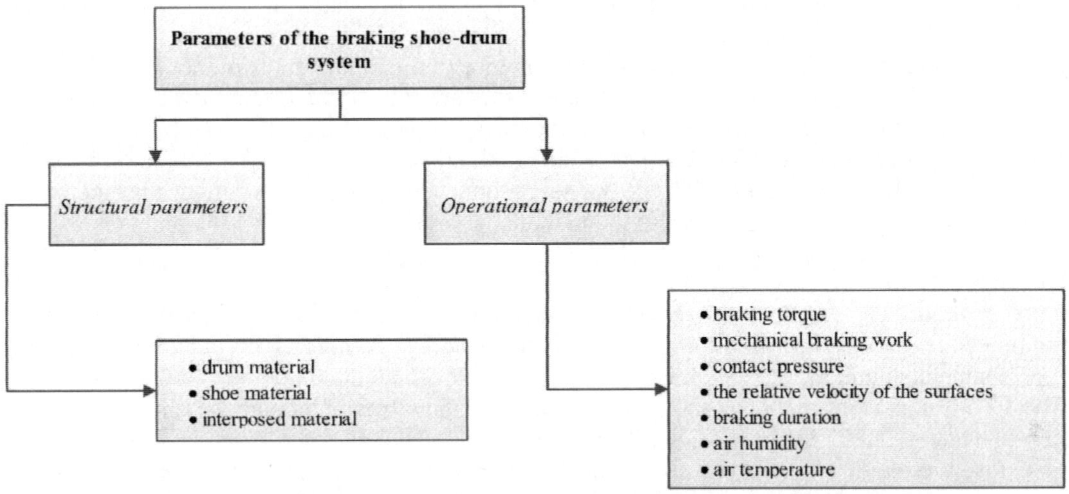


**Figure 1.** Tribological parameters on which the friction coefficient depends.

Brake shoes used for mining hoist are made of composite materials.
The introduction of different fillings in the composition of friction materials influences the working conditions and the friction characteristics in different ways [1].

The filling used in the past was usually asbestos, due to some specific properties: mechanical strength, high temperature resistance, low heat conductivity and good resistance to wear. At present, it is necessary to replace it with other types of materials for environmental reasons. Therefore, the following are used as fillers: glass fibers, mineral fibers, Kevlar, ceramic materials, etc. [10] [11].
Inserting metal dust creates favorable conditions for removing the heat from the friction surfaces and increases the stability of the friction coefficient.

The main characteristics that are imposed on the friction linings are [4][6][10]: high and as stable as possible friction coefficient between them and the brake drum, resistance to wear, low friability, good processability anti-inflammability.

The normal force between brake shoe and drum is variable during braking. Regarding the variation of the force N during braking N (t), researches in the field have established that N varies according to a linear law as follows: [12]:

$$N = c \cdot N_{\max} \cdot t_f , \tag{1}$$

where:

- c – variation (non-uniformity) coefficient of the pressure force of the shoes on drums,
- $N_{max}$ – maximum value of the normal force,
- $t_f$ – braking duration.

It follows that the braking pressure is variable.
**The angular speed** of the drum is also variable during braking.
The variation function is [6]:

$$\omega = \omega_0 \left( 1 - \frac{t}{t_f} \right) \tag{2}$$

In which $t_f$ is the braking duration, $\omega_0$ is the angular speed of the rim at the beginning of braking speed. In conclusion, the sliding velocity is also variable.

**Temperature of the braking surface** influences the coefficient of friction.

During the braking process, the friction work is converted into heat at the level of friction surfaces, thus endangering the quality of braking [1].

Experimental research has shown that the most intensely heated are the brakes of the mining hoists with cages, which descends the weights and having synchronous drive without dynamic braking [8]. The thermal condition of the brakes depends on the physical properties of the materials of the friction torque (shoe - drum) and on the construction of the brake [9].

In the case of shoe brakes about 95% of the heat produced on the friction surfaces is taken over by the drum and only 5% is taken over by the shoes [12].

## 3. Experimentation

The objective of the experimental research is to establish the friction coefficient of the friction material between brake shoe and drum when it is subjected to compressive loads, high temperature and environmental humidity.

The FF-30 non-asbestos brake shoe was selected as the experimental brake shoe material. The drum was made of steel S235JR SR EN 10025-2, 2004.

FF-30 is an ecologic non-asbestos friction material, with a good braking/engagement/clutch release efficiency, good wear characteristics and good mechanical resistance, during the working temperature range. It contains mineral fibers, pulverous inorganic materials, phenol-formaldehyde resin and powder synthetic rubber, complying with the requirements of the legislation in force on environmental and work safety.

The friction experimental stand, figure 3, was created with the target to simulate the operating conditions of drum brakes of mining hoist.

The block diagram for the friction experimental stand is presented in figure 2:

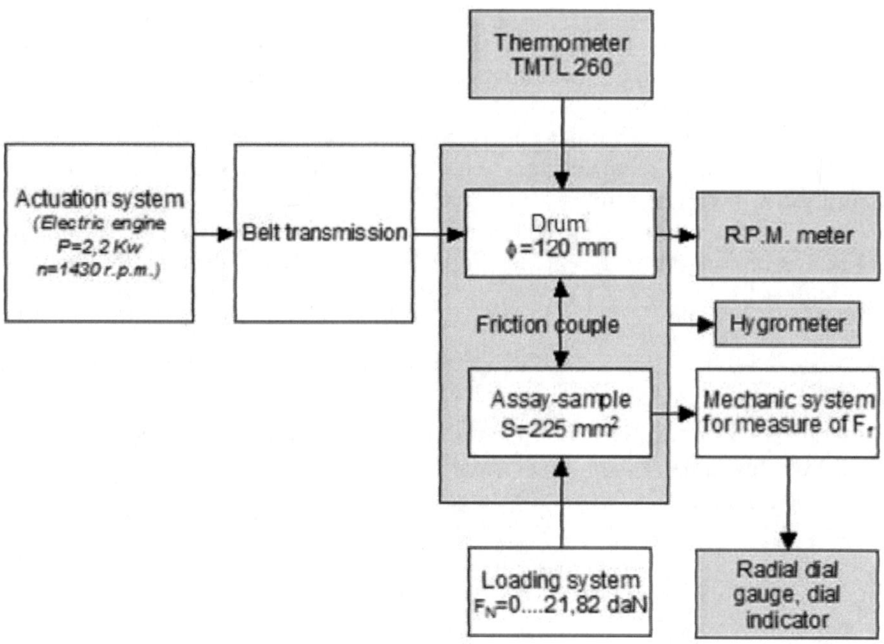


**Figure 2.** Block diagram of the experimental stand.

The drum is 120 mm in diameter and it is in contact with the assay sample (2) made from friction material type F - 30 having a 225 mm$^2$ contact surface. The assay sample is fixed in a vertical device (3). On the vertical device is applied the set weight which is used to adjust the load gradually (Fn).
The device is a charging equipment with standard weights. The movement of the drum is provided by an electric engine P=2.2 kW, n=1430 r.p.m. The motion is transmitted from the engine to the drum via belt transmission.

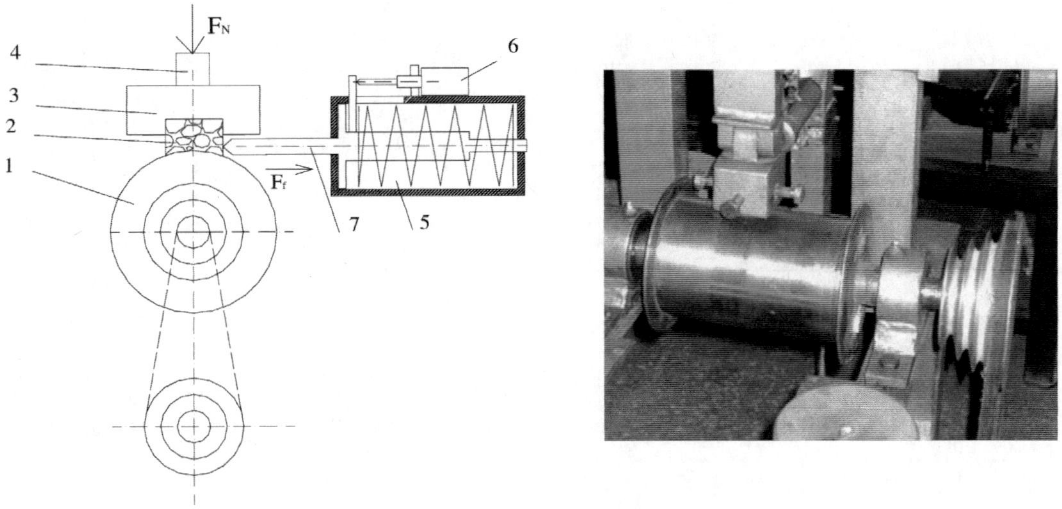


**Figure 3.** Experimental stand.

The belt transmission has four positions, corresponding to four values of speed of the drum. The friction forces between the assay sample and drum move the piece (7) and determine the compression of the spring (5). The compression is measured with the dial indicator (6). Knowing the characteristic of the compression spring depending on the shift, read with the dial indicator, the tangential force is deduced.
A water sprayer with water collected from mine was used for the simulation of humidity of technical environment. The humidity was measured with a hygrometer type testo 615.
Another condition was the heating of the drum surface at the nominal temperature of function (inspected with thermometer TMTL 260).The loads applied to the assay were of 225 N, 112.5 N, 75 N respectively the braking pressure values 1MPa, 0.5MPa, 0.33MPa. The tests were carried out for four peripheral speeds of the drum and for humidity of 60% and 90% (table 1). The time for each experiment was 2 seconds. Also each experiment was done 3 times.

**Table 1** Testing values of braking parameters

| Pressure [MPa] | Temperature [° C] | Speed [m/s] | Humidity [%] |
|---|---|---|---|
| 1 | 30.000 | 5.500 | 60.000 |
| 0.5 | | | 90.000 |
| 0.33 | | 8.000 | 60.000 |
| | | | 90.000 |

| | | | |
|---|---|---|---|
| | 30.000 | 10.500 | 60.000 |
| | | | 90.000 |
| | | 12.000 | 60.000 |
| | | | 90.000 |
| | 90.000 | 5.500 | 60.000 |
| | | | 90.000 |
| | | 8.000 | 60.000 |
| | | | 90.000 |
| 1<br>0.5<br>0.33 | | 10.500 | 60.000 |
| | | | 90.000 |
| | | 12.000 | 60.000 |
| | | | 90.000 |
| | 150.000 | 5.500 | 60.000 |
| | | | 90.000 |
| | | 8.000 | 60.000 |
| | | | 90.000 |
| | | 10.500 | 60.000 |
| | | | 90.000 |
| | | 12.000 | 60.000 |
| | | | 90.000 |

In these conditions, the determination of the friction coefficient was realized. Experiments were performed by a sequence presented in table 1..

After carrying out the experiments we get the variation charts presented in figures 4-9.

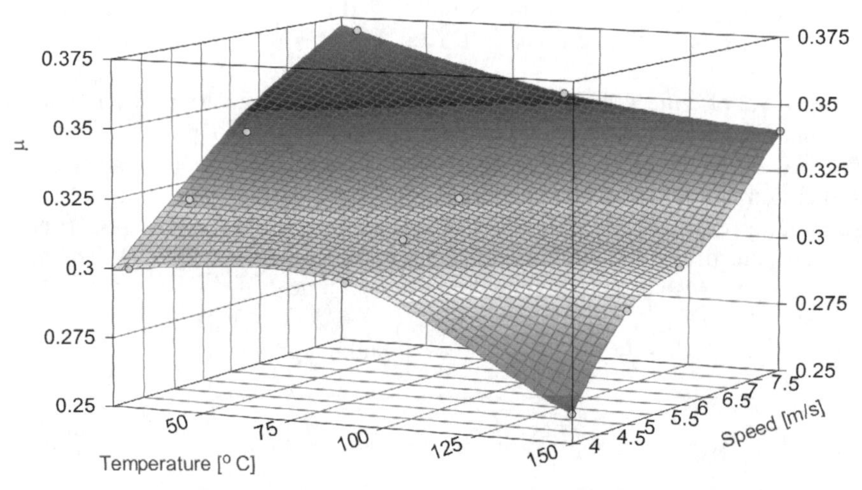


**Figure 4.** Friction coefficient for 1 MPa and 60% humidity

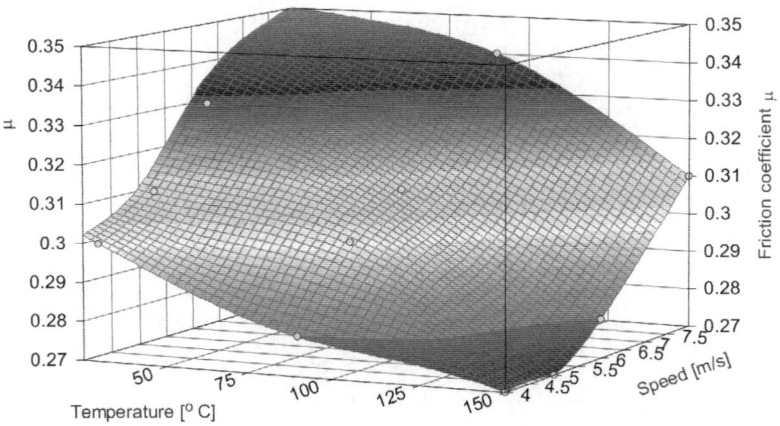


**Figure 5.** Friction coefficient for 1 MPa and 90% humidity

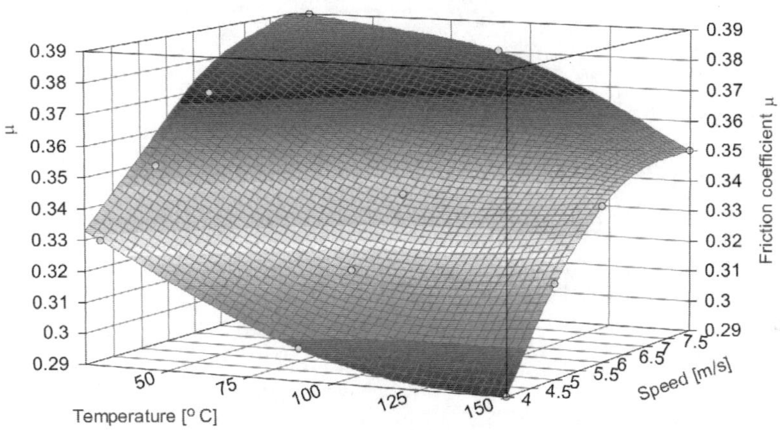


**Figure 6.** Friction coefficient for 0.5 MPa and 60% humidity

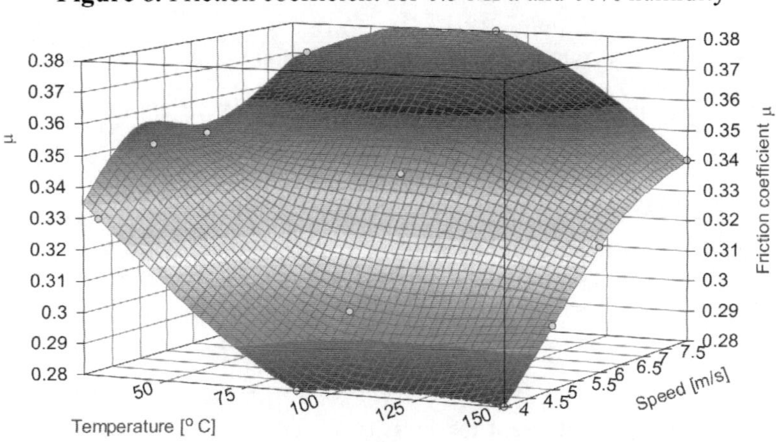


**Figure 7.** Friction coefficient for 0.5 MPa and 90% humidity

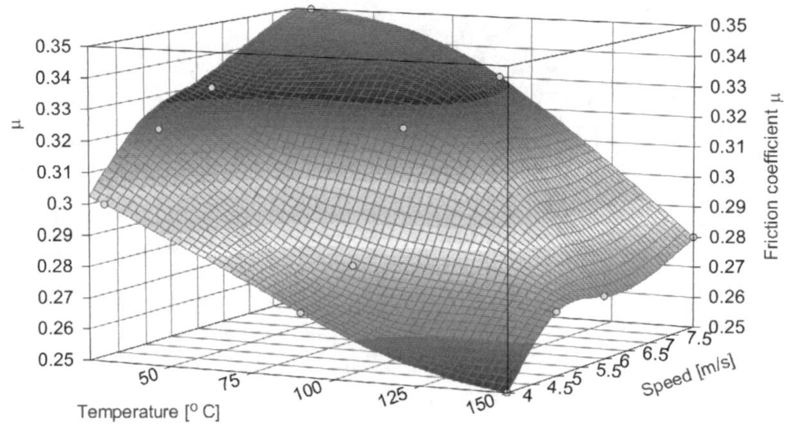


**Figure 8.** Friction coefficient for 0.33 MPa and 60% humidity

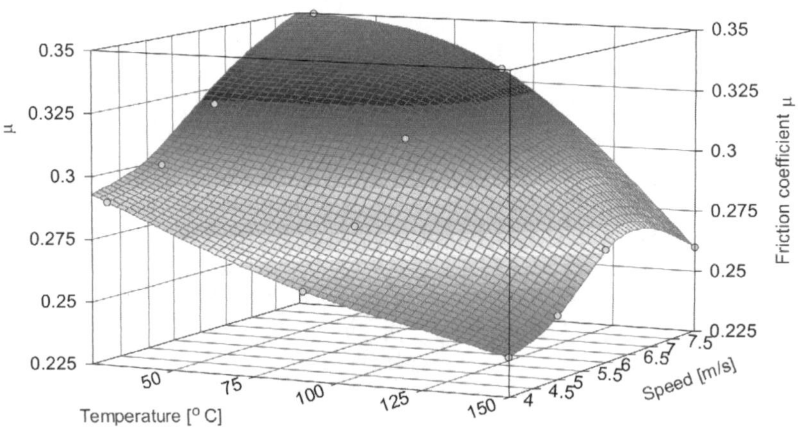


**Figure 9.** Friction coefficient for 0.33 MPa and 90% humidity

In order to assess the friction coupling, the following coefficients have been introduced [12]:

- stability coefficient for the friction coefficient:

$$\alpha_{CT} = \frac{f_{cp}}{f_{\max}} \tag{3}$$

In which $f_{cp}$ is the average value of the friction coefficient during braking, $f_{\max}$ is the maximum value of the friction coefficient in the braking process.

- coefficient of braking efficiency:

$$\beta_{e\varphi} = \frac{\alpha_{CT}}{t_f^2} \tag{4}$$

In which $\alpha_{CT}$ is the stability coefficient for the friction coefficient and $t_f$ is the braking duration.

Using diagrams (3) and (4), we can determine the stability of the friction coefficient and the efficiency of the brake considering the ensemble of tribological factors (different values of speed, temperature and humidity), for different mining hoists.

## 4. Conclusions

This paper presents the parameters that influence the friction coefficient for the shoe brakes of the mining hoist. Then the influence of four parameters (braking pressure, sliding velocity, surface temperature and humidity) on the friction coefficient has been studied using an experimental stand.

Analyzing the resulted charts, the following conclusions occurred:

- The friction coefficient increases with the increase of relative speed;
- The friction coefficient decreases in correlation with the increase of temperature;
- The friction coefficient decreases in correlation with the increase of the contact pressure;
- The friction coefficient decreases in correlation with the increase of humidity.

The presented method allows a better assessment of the friction coefficient in tribological conditions of operation. Also for concrete operation conditions one can determine the stability coefficient of the friction coefficient based on the experimental data.

## References

[1] Bao J, Zhu Z, Tong M and Yin Y 2011 Intelligent Predictions on Frictional Properties of Non-Asbestos Brake Shoe for Mine Hoister Based on ANN Model *The 2nd Intern. Conf. on Inteligent Control and Information Processing* pp 708-712

[2] Chen J D and Chern L J H 1996 Effect of humidity on the tribological behaviour of carbon-carbon composites *Wear* **193** 38-47

[3] Bashir M, Saleem S S and Bashir O 2015 Friction and wear behavior of disc brake pad material using banana peel powder, *IJRET: International Journal of Research in Engineering and Technology* **4** (2) 650-659

[4] Mutlu I, Sugözü I and Keskin A 2015 The effects of porosity in friction performance of brake pad using waste tire dust *Polímeros* **25** (5) 440-446

[5] Abed A A, Saad N A, Matloub F K 2012 Evaluation of resins and fibers content for friction materials industry *The Iraqi Journal For Mech. and Material Eng.* **12** (3) 533-544

[6] Czichos H, Habig K-H 1992 *Tribologie-Handbuch Reibung und Verschleiß* Vieweg-Verlag, Wiesbaden

[7] Crudu I, 1985, On the concept of Tribo-system and Tribo-modeling criterion,. *P. 4th . European Tribology Congress,* EUROTRIB 85 Lyon

[8] Magyari A, Achim M 1998 *Culegere de caracteristici tehnice pentru alegerea şi proiectarea instalaţiilor de extracţie miniere* Litografia I.M.P.

[9] Banciu M, Tudoreanu N 1993 *Normativ de calcule tehnice privind uzura şi oboseala elementelor de rezistenţă ale mecanismelor frânelor de manevră şi de siguranţă ale maşinilor de extracţie* (Bucureşti: IPROMIN)

[10] Bahadur S, Tabor D 1985 Role of fillers in the friction and wear behavior of high density polyethylene, in: *L.-H. Lee Ed. , Polymer Ž . Wear and Its Control*, ACS Symp. Ser. 287, pp 253–268

[11] Qian Y, Zhang W, Ge M, Wei X 2013 Frictional response of a novel C/C–ZrB2–ZrC–SiC composite under simulated braking *Journal of Advanced Ceramics* **2** (2) 157-161

[12] Belobrov V I 1981 *Dinamika, nagrev i iznos tormozov şahtnâh podbemnâh maşin* (Kiev: Naukova Dumka)

IOP Conf. Series: Materials Science and Engineering **174** (2017) 012017 doi:10.1088/1757-899X/174/1/012017

# Micro-scratching tests of a rolled aluminium alloy AA2024-T351 thick plate using a diamond micro-blade

**E Pirva[1], A Tudor[1], A Gavrus[2], G Chisiu[1], N Stoica[1] and A Predescu[1]**

[1]Machine Elements and Tribology Department, Faculty of Mechanical Engineering and Mechatronics, University POLITEHNICA of Bucharest, 313 Independentei Street, Bucharest, Romania

[2]Mechanical and Control System Engineering Departament, Mechanical and Civil Engineering Laboratory LGCGM-EA3913, INSA Rennes, France

E-mail: elisabeta.pirva@yahoo.com

**Abstract.** The present research work is focused on investigating the apparent coefficient of abrasive friction of a rolled thick plate of an AA2024-T351 aluminium alloy, using micro-scratch tests. For this study, specific materials specimens and a particular UMT Micro-Scratch Equipment were used. The test involved the generation of a scratch process at a local scale using a diamond stylus (micro-blade defined by a radius of 0.8 μm) moving along a specified path under a constant normal force (10 N) and with a constant speed (0.2 mm/s). For the characterization of the surface quality, two orthogonal directions were considered: the longitudinal one, along the rolling direction, and the corresponding transversal one. Given the fractal nature of the surface, an investigation was done in order to assess its influence on the coefficient of abrasive friction. The fractal dimension $D_f$, one of the most important parameters in a fractal surface analysis, was used to determine this influence in the global friction and abrasion phenomena. The abrasion factor was calculated using the Zum Gahr method for the data obtained with a specialized Mitutoyo SJ-301 surface tester. Measurements were made at the beginning, middle and at the end of the scratch channel. The obtained value for the abrasion factor was slightly less than zero. Other influences of anisotropic material features on global abrasion effects were also analyzed via comparisons of the coefficients of abrasive friction for both static and kinematic conditions.

## 1. Introduction

The scratch test is an established procedure commonly used as a technique for studying mechanical properties of materials near their surface. This procedure is of great interest to both academic and industrial communities [1, 2]. The micro-scratching test is one of the most widely used methods to obtain an effective and fast assessment of the critical loads related to adhesion properties of coating. There are two types of scratch tests that can be done: the scratch hardness test (scratch with constant normal load on a specimen and on a reference material using a stylus) and the scratch adhesion test where a progressive or a constant load is applied [3].

The scratch hardness and the materials surface deformation mechanisms depend mainly on the material's rheological behavior, on the indenter geometry and on the friction at the interfaces [2].

## 2. Experimental setup

### *2.1 Material properties*
For the micro-scratch tests a rolled aluminium alloy AA2024-T351 was used, which is one of the most popular high-strength aluminium alloys, having a good behavior at high temperatures and a low corrosion resistance. The composition of the 2000 aluminium alloy series has as a distinctive feature a copper content of approximately 4% and almost unnoticeable small impurities of iron. The mechanical proprieties are shown in table 1, from the technical literature and from the previous work concerning anisotropic rheology of AA2024 rolled thick sheets, conducted at National Institute of Applied Sciences of Rennes (INSA Rennes) [4].

**Table 1.** General mechanical proprieties of rolled AA2024-T351 aluminium alloy.

| Propriety | AA2024-T351 |
|---|---|
| Strength | > 420 N/mm$^2$, |
| Yield strength | 0.2%>260N/mm2 |
| HD harshness | 120 Kgf/mm$^2$ |
| Young modulus | 71000-74000 MPa |
| Poisson ratio | 0.33 |

### *2.2 Micro-scratch apparatus*
For this study it was used an UMT Micro-Scratch Equipment (figures 1, 2) from the Machine Elements and Tribology Department (OMTR) – University Politehnica of Bucharest (UPB). The equipment can provide rotational, translational or reciprocating motions with speeds starting from 0.1μm/s up to 10m/s. A constant or a progressive load between 0.05 N and 1000 N can be applied on the sample by the carriage. This normal force $F_z$ is controlled by a close-loop feed-back procedure. The tangential force ($F_x$), normal load ($F_n$), penetration depth and the coefficient of friction, or of abrasive friction, can be measured and recorded at a total sampling rate of 20 kHz [5].

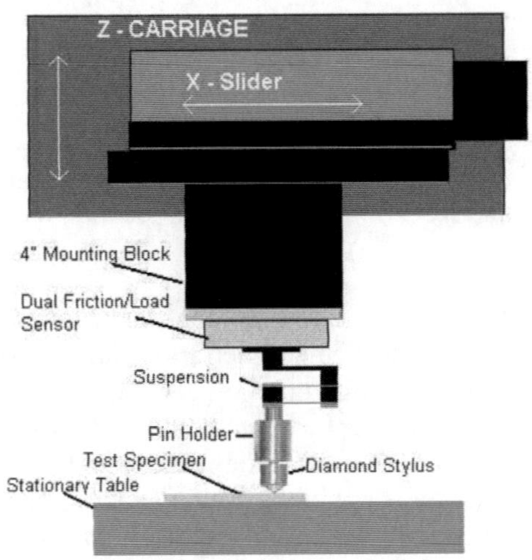


**Figure 1.** UMT Micro-Scratch configuration.

**Figure 2.** UMT Micro-Scratch.

The micro-scratch process is performed by moving a diamond stylus along a specified trajectory, maintaining a constant normal force and a uniform speed (figure 3). The thick plate specimen of 10 mm thickness has been prepared and machining by Mechanical Common Center (CCM) of INSA Rennes. The experimental tests were made at OMTR – UPB using the operating parameters shown in table 2.

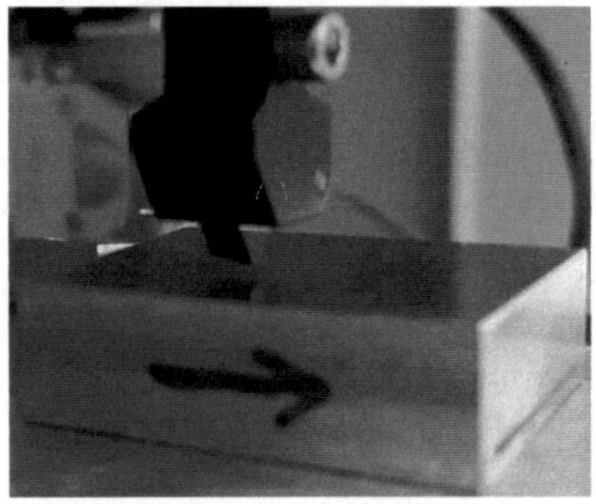

**Figure 3.** Micro-blade (with a radius of 0.8 μm) and the AA2024-T351 specimen used for the scratch tests.

**Table 2.** Micro-scratch parameters.

| | |
|---|---|
| Load type | constant |
| Load value | 10 N |
| Scratching speed | 0.2 mm/s |
| Scratch Length | 5 mm |
| Indenter | micro-blade, radius= 0.8 μm |

In order to analyze the quality of the material surfaces there were made scratch tests in two perpendicular directions. Five scratch tests along the rolling longitudinal direction – LD and other five scratch tests in the transversal direction – TD (figure 4).

**Figure 4.** Schematic illustration of longitudinal and transversal orientation of scratch.

## 3. Experimental results

In figure 5 the variations in time of the tangential forces and of the apparent coefficient of abrasive friction are presented. In figure 6 it is plotted the time variation of the penetration depth. In these figures the results for both scratch directions (LD – rolling longitudinal direction and TD – transversal direction) are shown as a statistic mean of the five measurements done for each case.

Abrasive friction is considered to be one of the main causes of mechanical energy loss. It is known that two forms of friction exist, corresponding to static and kinetic conditions. Static friction represents the frictional force acting between two surfaces which are attempting to move relative to each other, but sliding has not occurred yet, whereas the kinetic friction is the frictional force acting between the two surfaces which are in motion relative to each other.

Starting from the above described scratching tests, an apparent coefficient of abrasive friction can be defined as the ratio between the measured tangential force and the normal one:

$$\mu_s = F_s / F_n \qquad \mu_k = F_k / F_n \tag{1}$$

where: $F_s$, $F_k$- static and kinetic abrasive tangential force; $F_n$- normal force.

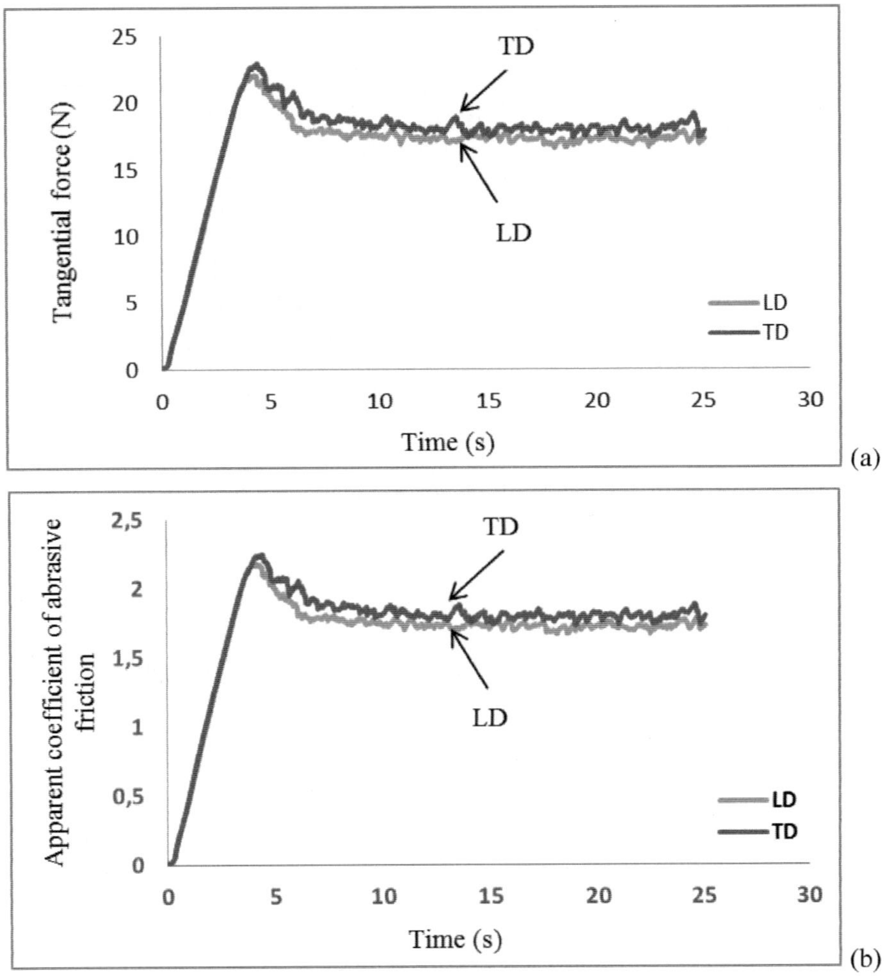


**Figure 5.** (a) The time variation of the tangential force,
(b) The time variation of the apparent coefficient of abrasive friction.

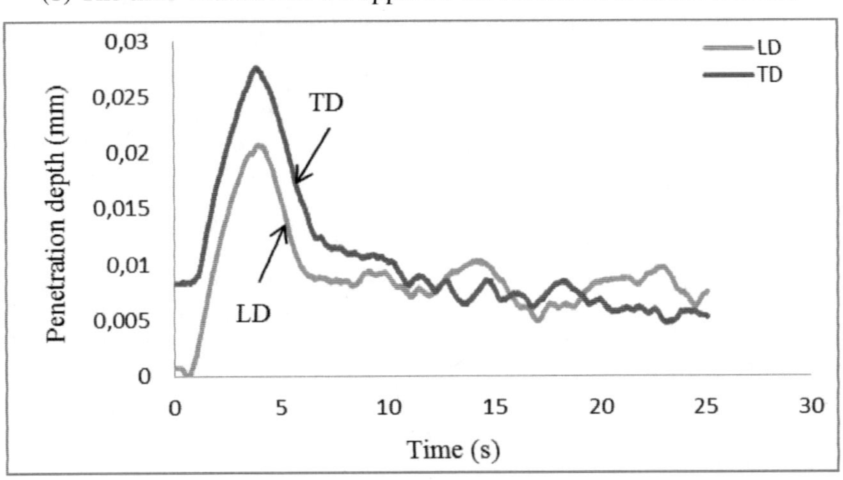


**Figure 6.** The time variation of the penetration depth.

From figure 5 it can be observed an apparent coefficient of static-abrasive friction in the interval 0-4 s and a kinetic one after 4s. Considering that the surfaces have a fractal behavior up to 1.1 μm, it had to be determined if there is any possible influence of this characteristic upon the coefficient of abrasive friction. The fractal dimension $D_f$ was used to describe the fractal quality of the surface [6]. The values of this parameter were obtained for each scratch direction in [6]: along the longitudinal direction $D_f = 1.216$ and along the transversal direction $D_f = 1.113$.

According to the results synthesized in figure 7, it can be noticed, that there is little influence on the coefficients of abrasive friction, both in the static and in the kinetic case, by the fractal surface behavior. It is observed that the loading direction (longitudinal direction or transversal direction) has a small influence of 3% on the apparent coefficient of abrasive friction.

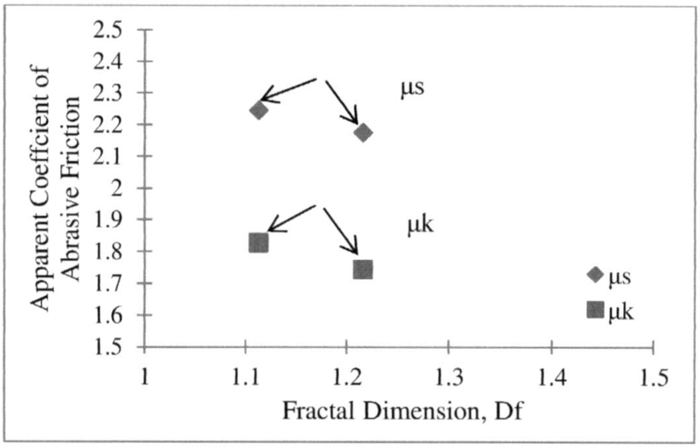


**Figure 7.** The variation of global static and kinetic apparent coefficients of abrasive friction with the fractal dimension $D_f$.

These results must be interpreted taken into account that the determined coefficient of abrasive friction is only an apparent one and not on a Coulomb friction coefficient. The latter can be estimated only if the measured tangential force is close to a real sliding friction force, where its component generating the plastic deformation during the scratching is reduced to a minimum. Generally this condition can be obtained if the applied contact pressure is smaller than one third of the Vickers hardness of material. Further tests along the diagonal direction DD must be added in the future in order to conclude on the influence of the anisotropy and of the fractal dimension.

## 4. The abrasion factor

The abrasion factor is defined as the ratio between the material volume that was moved in the wear process and the wear volume. For ductile materials, the abrasion factor can be estimated using the following formula [7]:

$$f_{ab} = 1 - (A_1 + A_2) / A_v \quad (2)$$

where $A_1$ and $A_2$ are the structured surfaces areas situated above the initial axis of the undistorted surface and $Av$ is the area of the deformed surface structure under the initial axis [8].

The principle of Zum Gahr's formalism of a cross-sectional cut it is shown in the following figure [9].

To compute the abrasion factor, the surface profile of each scratch channel was measured with a specialized Mitutoyo SJ-301 surface tester (figure 9). The measurements were made at the beginning, middle and at the end of the scratch channel.

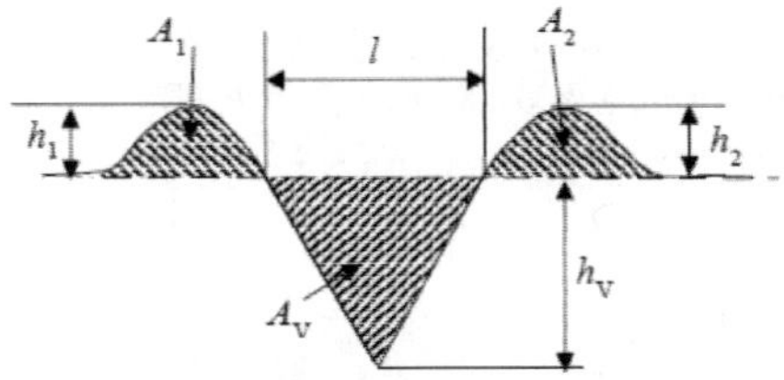


**Figure 8.** The principle of Zum Gahr of a cross-sectional cut.

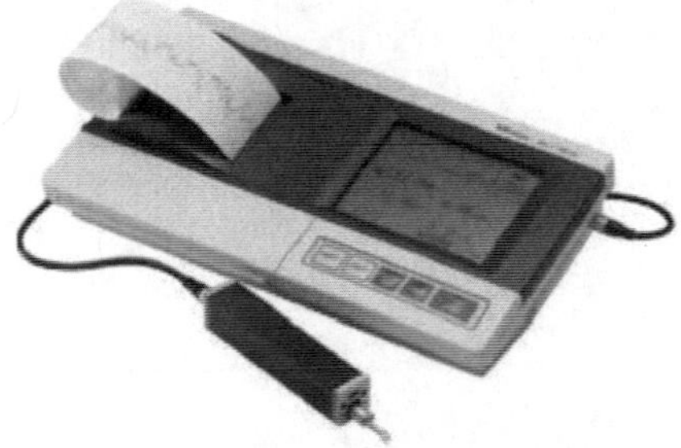

**Figure 9.** Mitutoyo SJ-301 surface tester [8].

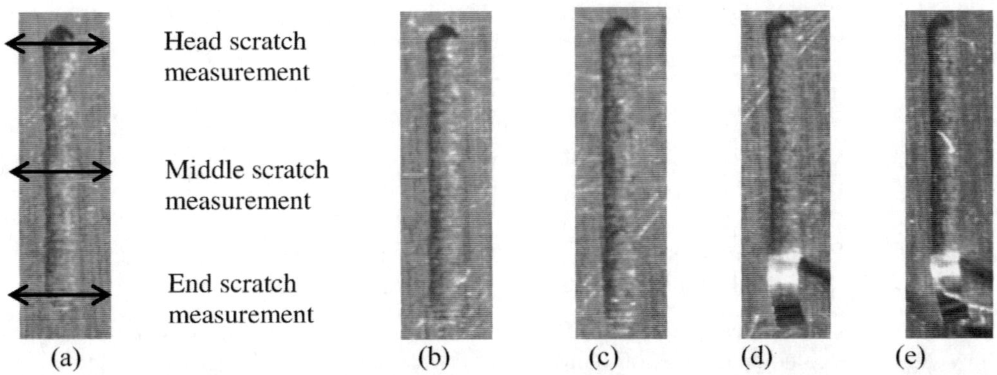


**Figure 10.** Microscopic images of the five micro-scratch tests for LD (rolling longitudinal direction).

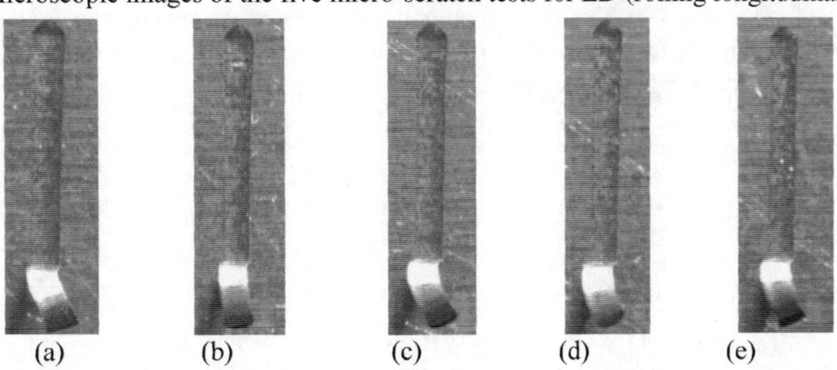


**Figure 11.** Microscopic images of the five micro-scratch tests for TD (transversal direction).

Microscopic images of the scratch channels for all five micro-scratch tests (for longitudinal direction and transversal direction) are presented in figure 10 and figure 11. The profilograms for the first scratch test (longitudinal direction) are shown in figure 12.

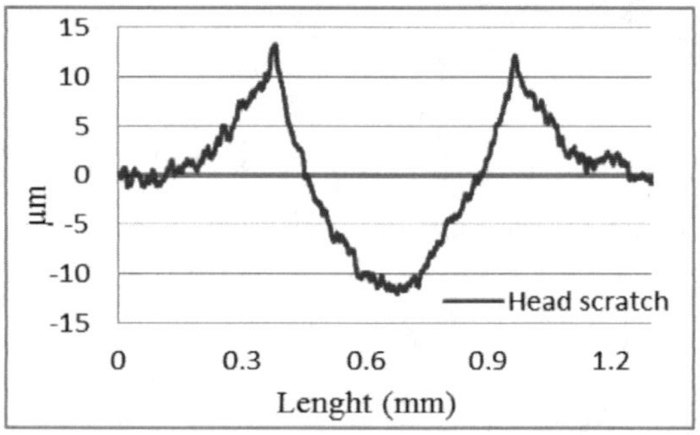


(a)

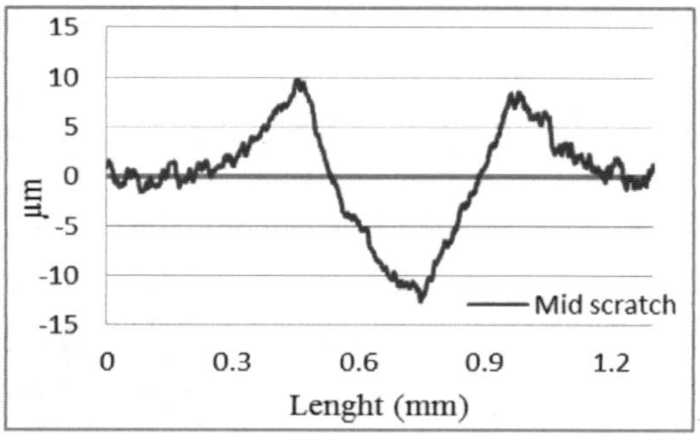


(b)

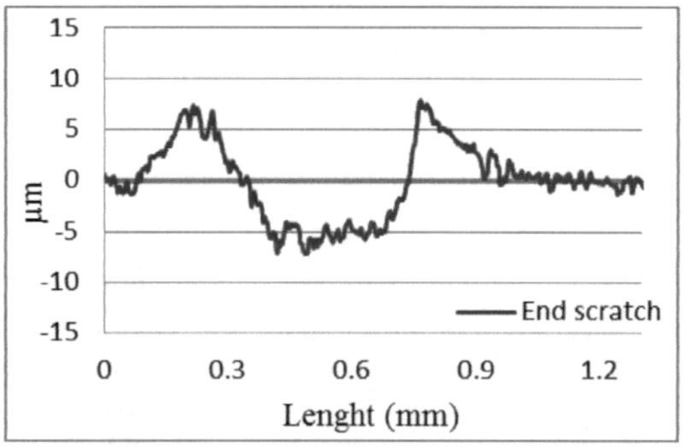


(c)

**Figure 12.** Profilograms for the first scratch test (longitudinal direction) at the beginning, middle and the end of the scratch channel.

The abrasion factor was calculated at the beginning, middle and the end of the scratch, using the principle of Zum Gahr. The calculated abrasion factors, for all tests, and their average are presented in table 3 and table 4.

The obtained values show a relatively important influence of the material anisotropy, especially regarding the ratio factor that has a variation from 4% to 30%.

**Table 3.** Abrasion factors for rolling longitudinal direction (LD).

| Scratch test | Head scratch | Middle scratch | End scratch | Ratio |
|---|---|---|---|---|
| 1 | -0.41892 | -0.11949 | -0.11898 | -0.219 |
| 2 | -0.22165 | -0.42426 | -0.15980 | -0.269 |
| 3 | -0.32195 | -0.60129 | -0.19630 | -0.242 |
| 4 | -0.36294 | -0.41496 | -0.31339 | -0.364 |
| 5 | -0.41745 | -0.38424 | -0.25190 | -0.351 |

**Table 4.** Abrasion factors for transversal direction (TD).

| Scratch test | Head scratch | Middle scratch | End scratch | Ratio |
|---|---|---|---|---|
| 1 | -0.34102 | -0.12499 | -0.16630 | -0.211 |
| 2 | -0.21360 | -0.43523 | -0.35275 | -0.334 |
| 3 | -0.32153 | -0.44334 | -0.30773 | -0.358 |
| 4 | -0.29634 | -0.37073 | -0.40975 | -0.359 |
| 5 | -0.47566 | -0.48230 | -0.36008 | -0.439 |

## 5. Conclusions

The aim of this paper was to determine whether there is a possible influence of the fractal nature of the surface on the apparent coefficient of abrasive friction. The samples used were plates of 10mm thickness, made of rolled aluminium alloy – AA2024-T351. The scratch tests were performed along two directions: a longitudinal one, corresponding to the rolling direction of the sample, and a transversal one. The impact of the anisotropic character of the material, due to the manufacturing process, was thus taken into account along with the desired parameters.

The micro-scratch tests were realized by moving a diamond stylus (micro-blade defined by a radius of 0.8 μm) along a specified trajectory (5mm length), maintaining a constant normal force (10N) and a uniform speed (0.2 mm/s). It can be observed an apparent coefficient of static-abrasive friction in the interval 0-4 s and a kinetic one after 4s.

Starting from a previous work, it was considered that the surfaces have a fractal behavior up to 1.1 μm and it was determined if there is any possible influence of this characteristic upon the coefficient of abrasive friction. It can be noticed that there is small influence of 3% on the coefficients of abrasive friction, both in the static and in the kinetic case, by the fractal surface behavior.

The results must be interpreted taken into account that the determined coefficient of abrasive friction is only an apparent one and not on a Coulomb friction coefficient. In a future work, to estimate the variation of Coulomb friction, smallest normal forces that 10 N will be considered.

Further tests along the diagonal direction DD must be added in the future in order to conclude on the influence of the anisotropy and of the fractal dimension.

The abrasion factor was calculated using the principle of Zum Gahr method. Measurements were made at the beginning, middle and at the end of the scratch channel. The obtained value for the abrasion factor was slightly less than zero, which confirms the good ductility of the aluminium alloy AA2024-T351 even for a micro-plastic process.

## References

[1] Wredenberg F and Larsson P L 2009 Delamination of thin coatings at scratching: experiments and numeric *J Mech Mater Struct* **4** (6) 1041-62

[2] Felder E and Bucaille J-L 2006 Mechanical analysis of the scratching of metals and polymers with conical indenters at moderate and large strains *Tribol Int* **39** 70-87

[3] Kuiry S 2012 Advanced Scratch Testing for Evaluation of Coatings (webinar presentation) available at https://www.bruker.com/fileadmin/user_upload/8-PDF-Docs/SurfaceAnalysis/TMT/Webinars/Advanced-Scratch-Testing-for-Evaluation-of-Coatings-Slides.pdf (2016)

[4] Pirva E 2014 *Etudes mécaniques et multi-échelle du comportement isotrope et anisotrope 3D en grandes déformations plastiques. Application à la modélisation numérique de l'emboutissage profonde des tôles des alliages d'aluminium* Master Dissertation INSA Rennes France

[5] *** UMT Multi-Specimen Test System Hardware Installation & Applications Manual available at http://erc.ncat.edu/Facilities/Manuals/CETR_UMT2_Manual.pdf (Jan. 2016)

[6] Pirva E, Tudor A and Gavrus A 2016 Fractal analysis of surface micro-topography for a rolled anisotropic thick sheet of aluminium alloy AA2024-T351 *IOP Conference Series: Materials Science and Engineering* **147** (1) p. 012017

[7] Tudor A and Vlase M 2010 *Uzarea Materialelor* (Bucuresti: Editura BREN)

[8] Petrescu A M, Paduraru G I, Tudor A and Stoica N A 2015 Mechanical testing of electronic printed circuit boards and soldering alloys *SERBIATRIB '15, the 14th Intern. Conf. on Tribology* pp. 155-158

[9] Ghisiu G and Tudor A 2014 Characteristics of UHMW polyethylene by scratching method *J Balk Tribol Assoc* **20** (1) 138–150

# The adhesion and hysteresis effect in friction skin with artificial materials

**K A Subhi**[1,2], **A Tudor**[1], **E K Hussein**[2] **and H S Wahad**[1]

[1] University Politehnica of Bucharest, Spl. Independentei 313, 060042, Bucharest, Romania
[2] AL-FURAT AL-AWSAT Technical University of Iraq

E-mail: qusay_abunoor@yahoo.com

**Abstract.** Human skin is a soft biomaterial with a complex anatomical structure and it has a complex material behavior during the mechanical contact with objects and surfaces. The friction adhesion component is defined by means of the theories of Johnson-Kendall-Roberts (JKR), Derjaguin-Muller-Toporov (DMT) and Maugis – Dugdale (MD). We shall consider the human skin entering into contact with a rigid surface. The deformation (hysteresis) component of the skin friction is evaluated with Voigt rheological model for the spherical contact, with the original model, developed in MATHCAD software. The adhesive component of the skin friction is greater than the hysteresis component for all friction parameters (load, velocity, the strength of interface between skin and the artificial material).

## 1. Introduction

Tribology science and technology mating surfaces being in relative opposite motion are not limited for improving engineering problems in mechanical engineering applications. Nowadays, the important part of the actual solutions is related to the human skin tribological aspects. The human skin in vivo is commonly one of the famous interacting surfaces which are found in relative motion in many of the daily activities of man. So many questions are typically related to optimizing the friction problems of the skin–product interactions, rather than to optimizing the use [1].

A well-known method for examining and analyzing the tribological performance is based on a general treatment, which is called the systems approach [2].This means that a tribological contact condition is separated from the application studied, by using an envelope system as proposed (contact area specifications). The contact condition separated by this envelope is considered as a system, that is, a set of factors interconnected by structure and function. The human skin is a biomaterial soft with a complex anatomical structure and it has a complex material behavior while coming into mechanical contact with the object and with the end prosthesis. Amonton's empirical rules of friction are not suitable for skin and the theoretical concept of elastomer friction can be used. Thus, the friction of elastomers involves two terms (non-interacting), adhesion and deformation [3].The report of these two components for the skin is the function to state of lubrication (dry friction, limit or mixed lubrication).

The adhesion component is defined by the surface energy of skin and three adhesion theories are used. The deformation component of human skin is determined by a new model, which uses the hysteresis and plowing phenomena.

The aim of this paper is to define quantitatively these two components of friction, for the elastic and viscoelastic properties of skin, with the new model for the hysteresis component of friction between the human skin (Voight material) and the rigid sphere.

## 2. Adhesive contact mechanics theories for skin

The theories explaining the adhesion and separation of two spherical elastic contact surfaces under the action of purely normal forces are investigated [4]. The adhesions between the rigid spheres where the surface forces are controlled by the Lennard–Jones potential equations, which give the total force–separation relations, were analyzed by Bradley in 1932 [5]. The adhesion theories taking elastic deformations into account were presented by the scientists Johnson, Kendall & Roberts (JKR) in 1971 Derjaguin, Muller &Toporov (DMT) in 1975 [5] and Maugis-Dugdale (MD) in 1992 [6]. When two elastic spheres come into contact with each other or with another flat surface, the resulting contact area can be estimated by using single-asperity contact mechanics models, such as those detailed by Hertz (H), Johnson, Kendall, and Roberts (JKR) [7], Derjaguin, Muller and Toporov (DMT) [8] or Maugis and Dugdale (MD).

We shall consider the contact between one rigid sphere (radius $R_r$) and equivalent elastic plane (elasticity $E_r$=(3/4)$K_e$): $K_e = \frac{3}{4}\left(\frac{1-\nu_1^2}{E_1} + \frac{1-\nu_2^2}{E_2}\right)$ where $\nu_1$ and $\nu_2$ are the Poison coefficients, $E_1$ and $E_2$ are elasticity modulus for contact materials.

The contact parameters are: the surface energy of sphere with elastic plane (γ), the equilibrium spatting in the Lennard- Jones potential ($z_o$) and the maximum tension (strength) of Lenard-Jones potential ($\sigma_o$). The Derjaguin (β), Tabor (Z) and Maugis (λ) dimensionless parameters can be used for describing the characteristics of the adhesion contact of skin with the rigid sphere:

$$\beta = \frac{64}{3\pi}\left(\frac{R_r\gamma^2}{\pi K_e^{\ 2} z_o^{\ 3}}\right)^{1/3};\ Z = \left(\frac{16R_r\gamma^2}{9K_e^{\ 2} z_{.o}^{\ 3}}\right)^{1/3};\ \lambda = 2\sigma_o\left(\frac{R_r}{\pi\gamma K_e^{\ 2}}\right)^{1/3} \quad (1)$$

The relations between these parameters are $\lambda = 0.4Z$ *and* $\lambda = 1.157\beta$.

In order to define the adhesion effect about the friction coefficient, it is accepted the dimensionless parameters of Johnson and Maugis, for normal load ($F_{na}$), contact radius ($a_a$) and penetration (Δ) [5, 6]:

$$a_a = a\cdot\left(\frac{K_e}{\pi\cdot\gamma\cdot R_r^{\ 2}}\right)^{\frac{1}{3}} \qquad F_{na} = \frac{F_n}{\pi\cdot\gamma\cdot R_r} \qquad \Delta = \delta\cdot\left(\frac{K_e^{\ 2}}{\pi^2\cdot\gamma^2\cdot R_r}\right)^{\frac{1}{3}} \quad (2), (3), (4)$$

Thus, the Hertz ($a_{aH}$), JKR ($a_{aJC}$), DMT ($a_{aDC}$) and MD ($a_{aMC}$) radius contact for Hertz, JKR, DMT and MD models are

$$a_{aH} = F_{na}^{\ 1/3};$$

$$a_{aJC} = 6^{1/3}\left(\frac{1+\sqrt{1+\frac{2F_{na}}{3}}}{2}\right)^{2/3} \quad a_{aDC} = 2^{1/3}\left(1+\frac{F_{na}}{2}\right)^{1/3} \quad a_{aMC} = \left(\frac{\alpha+\sqrt{1-\frac{F_{na}}{F_{nc}}}}{1+\alpha}\right)^{2/3} a_{oc} \quad (5)$$

To solve contact parameters for Maugis-Dugdale model (MD), it is adopted the numerical solution of Carpick [9]. Thus,

$$\alpha = \frac{1-\exp\left(\dfrac{-\lambda}{0.924}\right)}{1.02}; \; F_{nc} = \frac{-7}{4}+\frac{1}{4}\frac{4.04\lambda^{1.4}-1}{4.04\lambda^{1.4}+1}; \; a_{oc} = 1.54+0.279\frac{2.28\lambda^{1.3}-1}{2.28\lambda^{1.3}+1} \qquad (6)$$

Figure 1 shows the contact radius in Hertz (H), JKR, DMT and MD models, respectively, determined by the Mathcad software. It is observed the effect of the adhesion parameter (λ) is related to the contact radius and the penetration in MD model. When the adhesion parameter of skin is small ($\lambda < 0.1$), the contact radius can be defined by the DMT theory and for greater values ($\lambda > 3$), the radius contact can be determined by JKR theory.

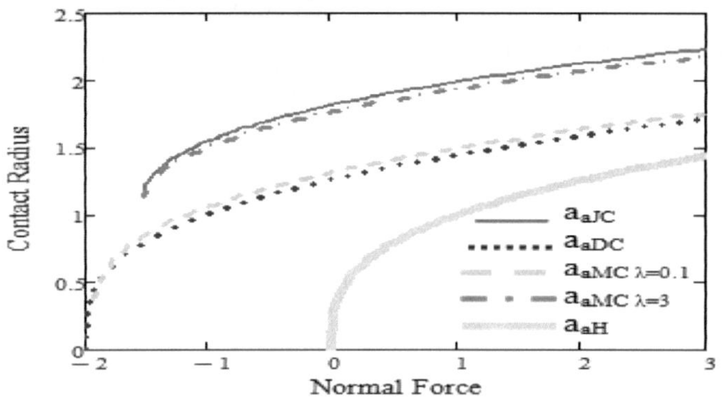


**Figure 1.** Contact radius vs normal load in H, JKR, DMT and MD (λ = 0.1 and λ = 3) models.

The adhesion parameter of skin varies between large limits (λ= 0.01….6), as a function of strength of Lenard – Jones potential, surface energy, equivalent elasticity and equivalent radius.

We accept the constant shearing strength ($\tau_o$) in contact, as a function to friction. In accordance to this hypothesis, the conventional Coulomb friction coefficient can be evaluated directly:

$$\mu_H = \frac{\pi^{2/3} a_{aH}{}^2 \tau_a}{F_{na}}; \; \mu_J = \frac{\pi^{2/3} a_{aJC}{}^2 \tau_a}{F_{na}+1.5}; \; \mu_D = \frac{\pi^{2/3} a_{aDC}{}^2 \tau_a}{F_{na}+2}; \; \mu_M = \frac{\pi^{2/3} a_{aMC}{}^2 \tau_a}{F_{na}-F_{nc}} \qquad (7)$$

where $\tau_a = \tau_o \left(\dfrac{R_r}{\gamma K_e{}^2}\right)^{1/3}$ is the friction shear strength of interfacial surfaces.

Figure 2 shows the conventional friction adhesion coefficient for H, JKR, DMT and MD models as a function to normal load.

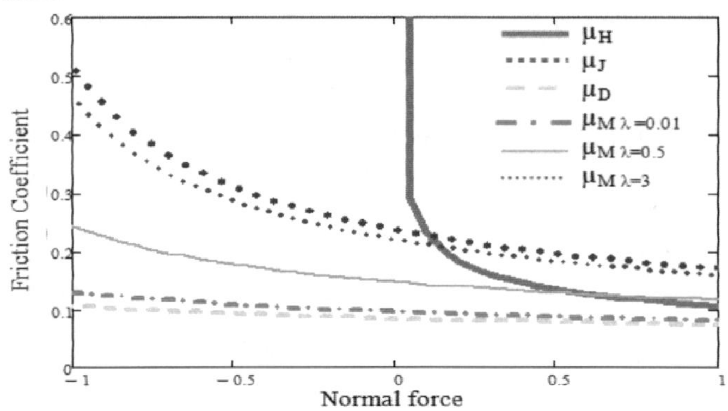


**Figure 2.** Friction coefficient vs normal load in H, JKR, DMT and MD (λ = 0.01, λ = 0.5 and λ = 3).

It is observed that adhesion friction component decreases with normal load for all adhesion models.

## 3. Hysteresis friction component

The human skin can be considered as an elastomeric material. It is applayd the Voigt rheologic model of behavior material properties (elasticity modulus $E_1$ and $E_2$ and viscosity η) [10]:

$$E_p = \frac{E_1 E_2^2 + \omega^2 (E_1 + E_2)\eta^2}{E_2^2 + \omega^2 \eta^2}; \; \eta_p = \frac{\eta E_2^2}{E_2^2 + \omega^2 \eta^2}, \tag{8}$$

where $\omega$ is angular frequency in dynamic model [10].

The asymmetry of pressure distribution of rigid sphere with equivalent Voigt viscoelastic material of plane is characterized by L, $x_o$ and $y_o$ (figure 3) [10].

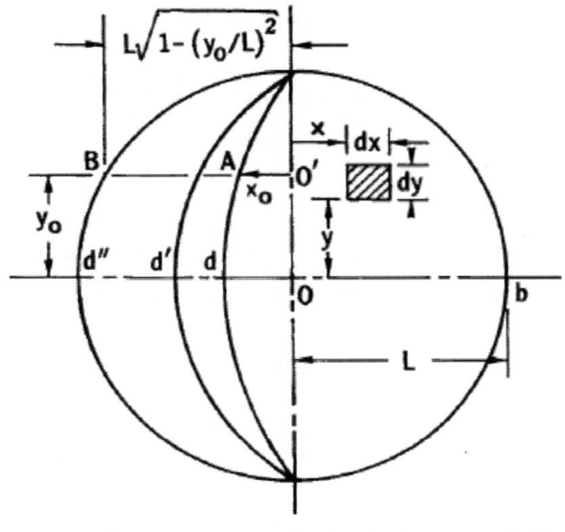


**Figure 3.** Asymmetrical spherical contact [10].

The mechanical equilibrium and boundary condition are

$$F_n = \int_{x_o}^{L} \int_{y_o}^{L} p(x,y)dxdy \text{ and } p(x_o, y_o) = 0 . \tag{9}$$

In the rolling sphere in x direction appears the moment ($M_f$) of all pressure force about the centre O of the sphere. Thus, the hysteresis force in rolling direction is

$$F_{hyst} = \frac{M_f}{R_r} = \frac{\int_{x_o}^{L} \int_{y_o}^{L} xp(x,y)dxdy}{R_r} . \tag{10}$$

The hysteresis friction component ($\mu_h$) can be obtained by the integral (10) ($F_{hyst}$) and the normal load ($F_n$) (9):

$$\mu_h = \frac{F_{hyst}}{F_n} = \frac{F_{hysta}}{F_{na}} . \tag{11}$$

It is used the following dimensionless parameters:

$$F_{na} = \frac{2F_n L}{E_p R_r^3}; \; F_{hysta} = \frac{2F_{hyst} L}{E_p R_r^3}; \; x_1 = \frac{x_o}{L}; \; y_1 = \frac{y_o}{L}; \tag{12}$$

where $L_a = L / R_r$, $\gamma_V = \frac{\beta_V}{L_a}$ *with* $\beta_V = \frac{\eta_p}{E_p}$,

and is obtained the normal load and hysteresis force:

$$F_{na} = L_a^3 (1-x_1)(1-y_1)\left[1+\frac{\beta_V}{L_a}(1+x_1)-\frac{1}{3}(1+x_1+x_1^2)+\frac{1}{4}L_a(1+x_1+x_1^2+x_1^3)\right]. \quad (13)$$

$$F_{hysta} = L_a^5 (1-x_1)(1-y_1)\begin{cases}(1+x_1)\left[\frac{1}{2}-\frac{1}{6}(1+y_1+y_1^2)+\frac{L_a}{8}(1+y_1+y_1^2+y_1^3)\right]+ \\ \frac{2\beta_V}{L_a}(1+x_1+x_1^3)+\frac{\beta_V L_a}{5}(1+x_1+x_1^2+x_1^3+x_1^4+x_1^5)\end{cases}. \quad (14)$$

From the boundary condition, $x_1 = \left(1+\gamma_V^2 - y_1^2\right)^{1/2} - \gamma_V$.

The medium hysteresis friction coefficient from circular contact will be obtained by

$$\mu_{hm} = \frac{1}{L_a}\int_0^{L_a} \mu_h dy_1 . \quad (15)$$

Thus, the figures 4 and 5 show the hysteresis friction components as a function to sliding velocity (V) and the normal load of the rigid spheres with same radius ($R_r$) in contact with the viscoelastic skin for three viscoelastic properties of skin, as a function to age. The rheological properties of human skin are considered and adapted by Zahouani [11]. In the other sections of the paper, it is necessary to extend the experimental results about the rheology properties of the human skin, in some parts of the human body. These results will be used to define the friction between the skin and artificial materials, used in endoprothesys.

The normal load is appreciated by means of the dimensionless Stribeck pressure $p_{Sa} = \frac{F_n}{\pi R_r^2 E_p}$.

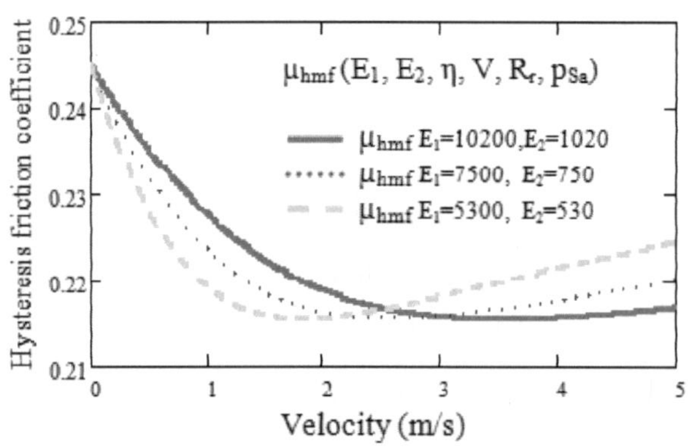


**Figure 4.** Hysteresis friction coefficient vs velocity.

The rolling contact implies a micro slip rather than a gross sliding within a mating surface. During the rolling process, only part of the work expanded elastically compresses, the material is recovered and the remaining one will appear as hysteresis friction and it ultimately become the same as it was when it was generated.

In the theoretical model of the hysteresis component of the skin friction with a rigid sphere we shall use some values of elasticity and viscosity in the Voight rheological behavior of skin. In this model, it is possible to analyze the effect of the velocity and the normal load of the hysteresis friction component for some rheological properties of skin (figures 4 and 5). It is noticed that the deformation

IOP Conf. Series: Materials Science and Engineering **174** (2017) 012018 doi:10.1088/1757-899X/174/1/012018

(hysteresis) friction component is at its minimum with respect to velocity values and increases non-linearly with the normal load, evaluated by Stribeck contact pressure (figure 5).

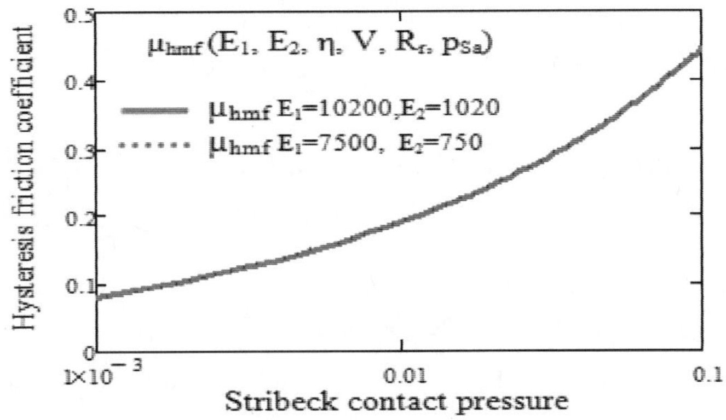


**Figure 5.** Hysteresis friction coefficient vs Stribeck contact pressure.

## 4. Conclusions

Human skin can be treated as viscoelastic material due to its specific mechanical properties. The adhesion with the other contacted surface is also exhibited and it can be complied with by using the intermediate model Maugis-Dugdale, model which represents the approximate solution between the two theories JKR and DMR models. The friction conditions of the human skin can be analyzed by all adhesion theories.

The adhesive component of the coefficient of friction decreases proportionally with the normal force ($F_{na}$) the force increases and so does the hysteresis component non-linearly with normal load.

. The new model of hysteresis friction coefficient of viscoelastic skin with rigid sphere can explain the effect of the velocity in terms of the friction of the skin coming in contact with some artificial materials.

## References


[1] Van Der Heide E, Zeng X and Masen M A 2013 Review Article Skin tribology *Friction* **1 (2)** 130- 142

[2] Czichos H 1978 A Systems Approach to the Science and Technology of Friction Wear and Lubrication W*ear* **54**

[3] Zhou Z R, Jin Z M 2015 Biotribology: Recent progresses and future perspectives *Biosurface and Biotribology* **1** 3-24

[4] Adams, G G and Nosonovsky M 2000 Contact modeling *Trib. Int.l* **33** 431-442

[5] Johnson K L 1997 Adhesion and friction between a smooth elastic spherical asperity and a plane surface *Proc. R. Soc. Lond.* A **453** 163-179

[6] Maugis D 1992 Adhesion of Spheres - the JKR-DMT Transition Using a Dugdale Model *J. Colloid Interface Sci.* **150** 243-269

[7] Johnson K L, Kendall K and Roberts A D 1971 Surface Energy and Contact of Elastic Solids. *Math. Phys. Sci.* **324** 301-313

[8] Derjaguin B V, Muller V M and Toporov Y P 1975 Effect of Contact Deformations on Adhesion of Particles *J. colloid interface sci.* **53** 314-326

[9] Carpick R W, Ogletree D F, and Miquel Salmeron 1999 A General Equation for Fitting Contact Area and Friction vs Load Measurements *J. Colloid Interface Sci.* **211** 395-400

[10] Moore D F 1972 The Friction and Lubrication of Elastomers. *Pergamon Press Oxford*

[11] Zahouani H, Boyer G, Pailler-Mattei C, Ben Tkaya M and Vargiolu R 2011 Effect of human ageing on skin rheology and tribology *Wear* **271** 2364-69

# Tribological properties of the disc brake friction couple materials in the range of small and very small speeds

**N A Stoica[1], A M Petrescu, A Tudor and A Predescu**

Department of Machine Elements and Tribology, University Politehnica of Bucharest, 313 Splaiul Independentei, Bucharest, Romania

E-mail: nicolae.stoica@upb.ro

**Abstract**. The tribological properties of the friction couple materials have a major influence on the brake system operation and its failure. One of the main phenomena associated as a symptom of failure in the brake system are the noises and vibrations produced during braking. The stick-slip phenomenon is attributed as the cause of these noises and vibrations. The stick-slip phenomenon usually appears at low and very low sliding speeds and is described as intermittences in the friction process caused by the differences between the values of the kinetic and the static friction coefficients. The present paper addresses an investigation about the influence of the static and kinetic friction on the occurrence of above mentioned noises and vibrations in the disc brake system. For this, extensive experimental work was performed on a laboratory tribometer in the form of pin-on-disc tests, where the pin was manufactured out of an automotive brake pad and the disc was manufactured out of an automotive grey cast iron brake disc. The results highlight the effects of the sliding speed and contact pressure on the friction coefficient and its influence on the brake noises and vibrations caused by the stick-slip phenomenon.

## 1. Introduction

The tribological properties of the friction couple materials have a major influence on the brake system operation and its failure. One of the main phenomena associated as a symptom of failure in the brake system are the noises and vibrations produced during braking. The noises and vibration produced during braking have led the owners of vehicles to present complaints against the automobile manufacturer, which lead to a large amount of money being spent on the warranty and repair costs [1-4]. A wide array of brake noise and vibration phenomena are described by an even wider array of terminology. Squeal, groan, chatter, judder, moan, hum, and squeak are just a few of the names found in the literature [2]. Researchers attribute the stick-slip phenomenon as one of the causes of these noises and vibrations [1-7].

Typically, for the stick-slip phenomenon to occur, the static friction coefficient ($\mu_s$) between the two contact surfaces of the friction materials must be greater than the kinetic friction coefficient ($\mu_k$) [7, 8]. The stick-slip phenomenon usually appears at low and very low sliding speeds and is described as intermittences in the friction process caused by the differences between the values of the kinetic and the static friction coefficients. It appears in friction couples with dry or limited friction regime, when the sliding speed is in the range of 0.01 – 3 mm/s or when the angular speed is somewhere in the 1 –

25 rad/s [9, 10]. If these speeds have higher values, then the movement takes the form of self-induced vibrations sustained by the friction force itself [11].

The present paper addresses an investigation about the influence of the static and kinetic friction on the occurrence of above mentioned noises and vibrations in the disc brake system. For this, extensive experimental work was performed on a UMT laboratory tribometer in the form of pin-on-disc tests, where the pin was manufactured out of an automotive brake pad and the disc was manufactured out of an automotive grey cast iron brake disc.

## 2. Testing procedure

The experimental research presented in this paper was conducted on a Bruker (former CETR) UMT-2 Tribometer, which can be used effectively for the tribological testing of ferrous and non-ferrous metals, plastics, ceramics, paper, composites, thin and thick coatings, as well as of solid lubricants, lubricating fluids, oils and greases.

For the current tests conducted on the UMT a pin-on-disc system setup was used. This test method involves a pin shaped upper specimen that slides against a rotating disk as a lower specimen under a prescribed set of conditions. The UMT allows for monitoring during the test the actual dynamic normal load, friction force and friction coefficient, depth of wear and contact acoustic emission. In figure 1 we have an overview of the UMT tester and the two specimens used for the experiments. The upper specimen is a pin manufactured out of a semi-metallic automotive brake pad and the lower specimen is a disc manufactured out of an automotive grey cast iron. The pin has a diameter of 12 mm and the disc has a 55 mm diameter. Thus the contact between the pin and the disc is a flat circular surfaces with an area of 113.1 $mm^2$.

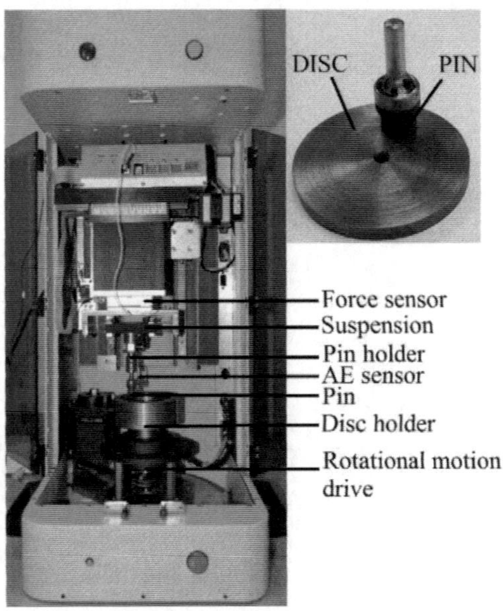


**Figure 1.** Overview of the UMT tester and the two specimens.

The UMT tester was equipped with a model DFH-20 two dimensional force sensor used to measure the sliding and breakaway friction force between the upper and lower test specimens as well as measuring and controlling the loading force. The model DFH-20 dual friction/load sensor has a range from 2 to 200 N and a resolution of 10 mN. In order to maintain a constant load it is sometimes necessary to have a suspension between the force sensor assembly and the upper specimen holder. The suspensions equipped on the UMT tester is a spring device used to compensate for variations in the distance between the force sensors and the surface of the lower specimen when the lower specimen is in motion. The upper specimen holder connects the pin to the suspension, while the lower specimen holder connects the disc to the model S25LE lower rotational motion drive used for the tests. The contact acoustic emission (AE) sensor is mounted on the pin and gives a voltage output.

The tests were performed at a normal load of 50 N, 75 N and 100 N and at three different speeds: 20 rpm, 40 rpm and 60 rpm. The values of the coefficient of friction (COF) and acoustic emission (AE) were monitored during the tests. The testing time for all the nine test was 300 s. Before these tests, a series of running in tests at 100 N and 60 rpm with a total time of one hour were undertaken to ensure that the real contact area is as close as possible to the nominal contact are so that the measured parameters do not vary from one test to another done at the same load and speed.

## 3. Results and discussions

The results were analyzed graphically and they highlight in comparison the evolution of the friction coefficient and acoustic emission during the tests.

From the diameter of the pin we can determine the nominal contact surface between the pin and the disc, which is 113.1 $mm^2$ and with the normal load we can determine the contact pressure between the pin and the disc. So, for 50 N the contact pressure is 0.44 MPa, for 75 N we have a contact pressure of 0.66 MPa and for 100 N the contact pressure is 0.88 MPa. Also, corresponding to the 20 rpm, 40 rpm and 60 rpm rotational speeds and the distance between the center of the pin and that of the disc (21.5 mm) the sliding speed is 0.045 m/s, 0.09 m/s and 0.13 m/s.

All three test conducted at 50 N present almost no acoustic emission as shown in figure 2.

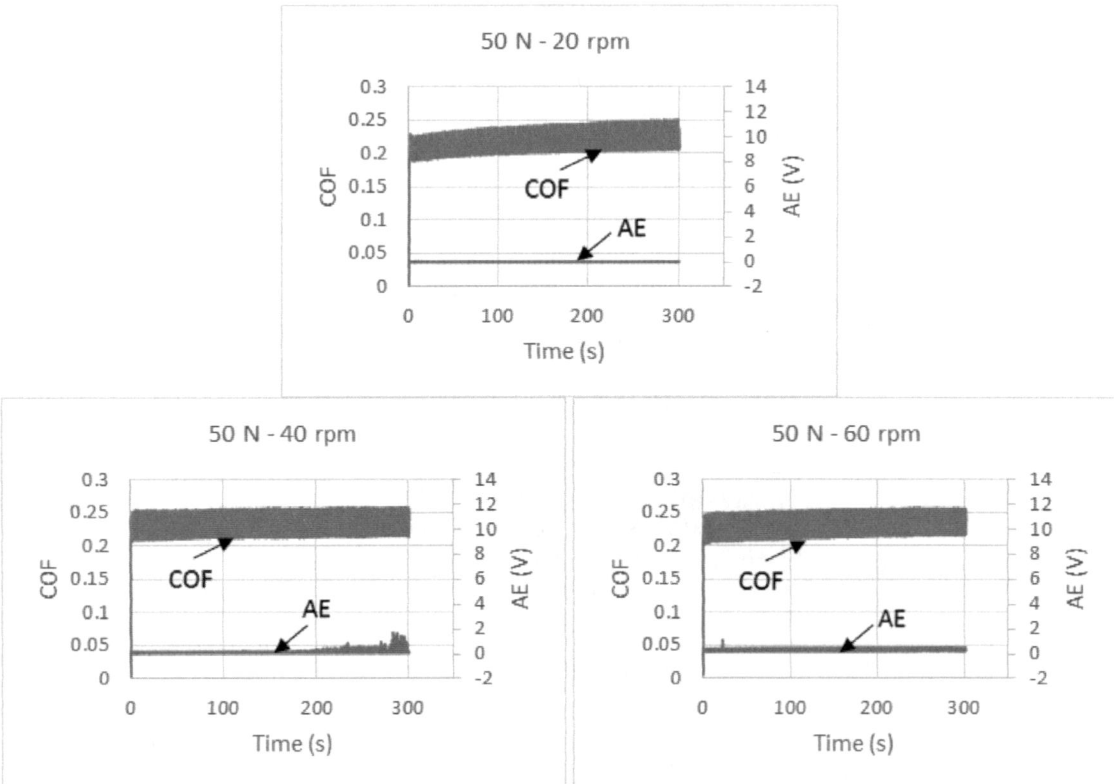


**Figure 2.** The evolution of the friction coefficient and acoustic emission for the three 50 N tests conducted at 20, 40 and 60 rpm.

As seen from in the graphs, only the test conducted at 50 N and 40 rpm the acoustic emission has a slight increase, but that is in the end of the 300 s duration of the test and no noise was audible as in the other two cases of the 50 N tests. The cyclical variation of the friction coefficient observed in the

IOP Conf. Series: Materials Science and Engineering **174** (2017) 012019 doi:10.1088/1757-899X/174/1/012019

graphs is given by the loading force. This is explained by the fact that the misalignment of the surface of the disc determines a cyclical variation of the loading force and the tester is not able to keep the load constant throughout a rotation of the disc. This phenomenon is present in all the pin-on-disc tests.

Figure 3 presents the evolution of the friction coefficient and acoustic emission for the three 75 N tests conducted at 20, 40 and 60 rpm. In this case, when the load is increased to 75 N we can observe an increase of the acoustic emission. Comparing the three graphs we can see that by increasing the speed the acoustic emission will have a higher amplitude.

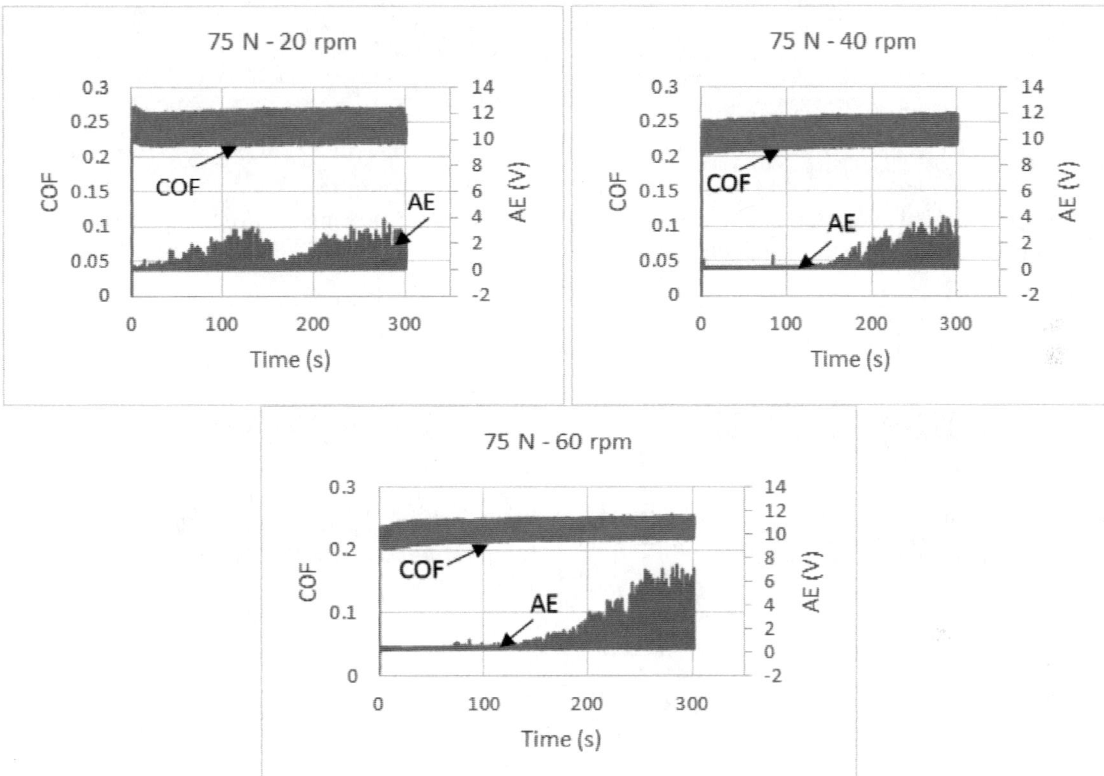


**Figure 3.** The evolution of the friction coefficient and acoustic emission for the three 75 N tests conducted at 20, 40 and 60 rpm.

During the three 75 N tests a short duration and low intensity noise coming from the two samples in contact can be heard. This noise could be heard in the last part of the tests and was cyclical, repeating every second, with just a few intermittences. In the last part of the 75 N and 60 rpm the noise grew louder with time and was louder than in the other two 75 N tests.

For a better understanding of what is happening, we have chosen a 10 seconds period of the experiment to highlight the phenomenon. This period can be observed in figure 4, in which we can see the evolution of the friction coefficient and acoustic emission between second 200 and second 210 of the 300 seconds test conducted at 75 N and 60 rpm. Here, the disc makes a complete rotation every second and the friction coefficient has the almost the same evolution from a cycle to another, but compared with the tests where we don't have acoustic emission there is a sudden drop and rise in the values of the friction coefficient that repeats every cycle with different amplitudes. This can be considered a stick-slip period.

The last tests were performed at a normal load of 100 N. Figure 5 presents the evolution of the friction coefficient and acoustic emission for the three tests conducted at a normal force of 100 N and 20, 40 and 60 rpm speed.

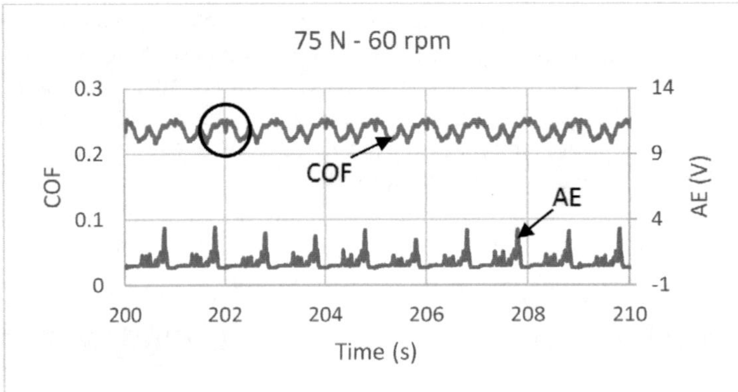


**Figure 4.** The evolution of the friction coefficient and acoustic emission for a 10 seconds period for the 75 N test conducted at 60 rpm.

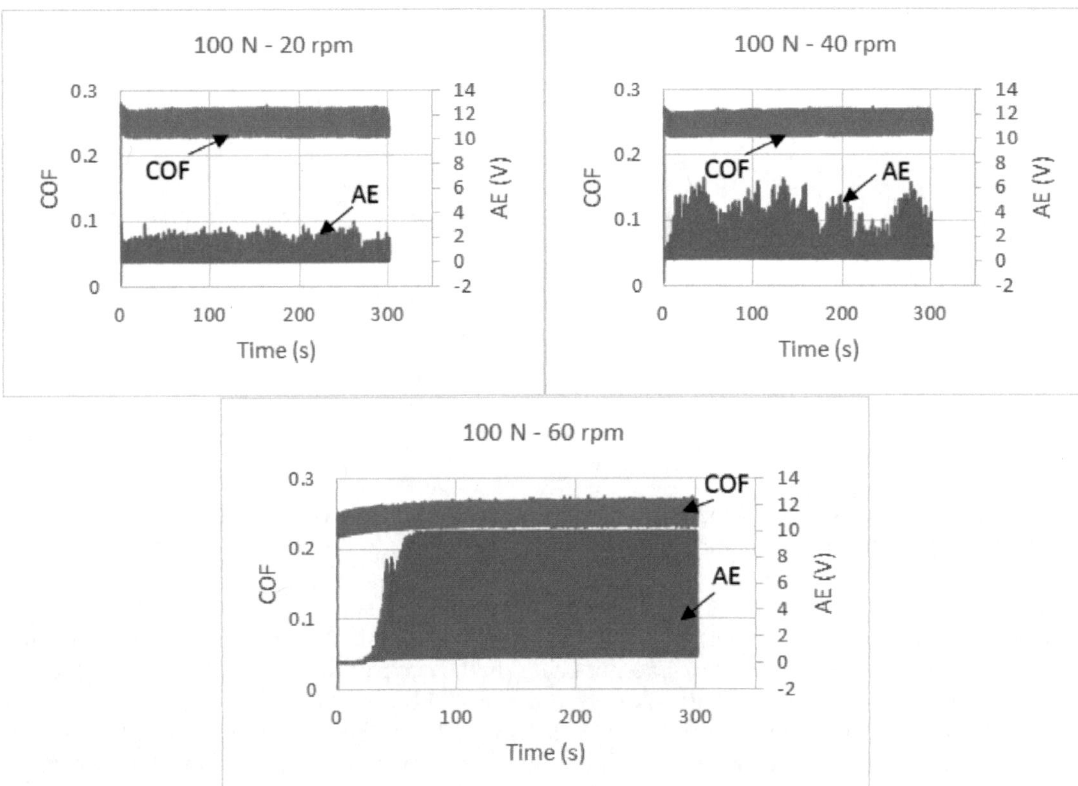


**Figure 5.** The evolution of the friction coefficient and acoustic emission for the three 100 N tests conducted at 20, 40 and 60 rpm.

From the graphs in figure 5 we can see that for all the three tests conducted for the 100 N normal force we have acoustic emission which increases with the rotational speed of the disc. The noise present in the 75 N tests was also present in the 100 N test, but much louder, and it grew in intensity as the speed increased. For the 100 N tests the amplitude of the stick-slip phenomenon described above increases with the increase of the relative speed between the pin and the disc. This can be observed in

IOP Conf. Series: Materials Science and Engineering **174** (2017) 012019 doi:10.1088/1757-899X/174/1/012019

figure 6 where there is a sudden drop and rise in the values of the friction coefficient that repeats every cycle.

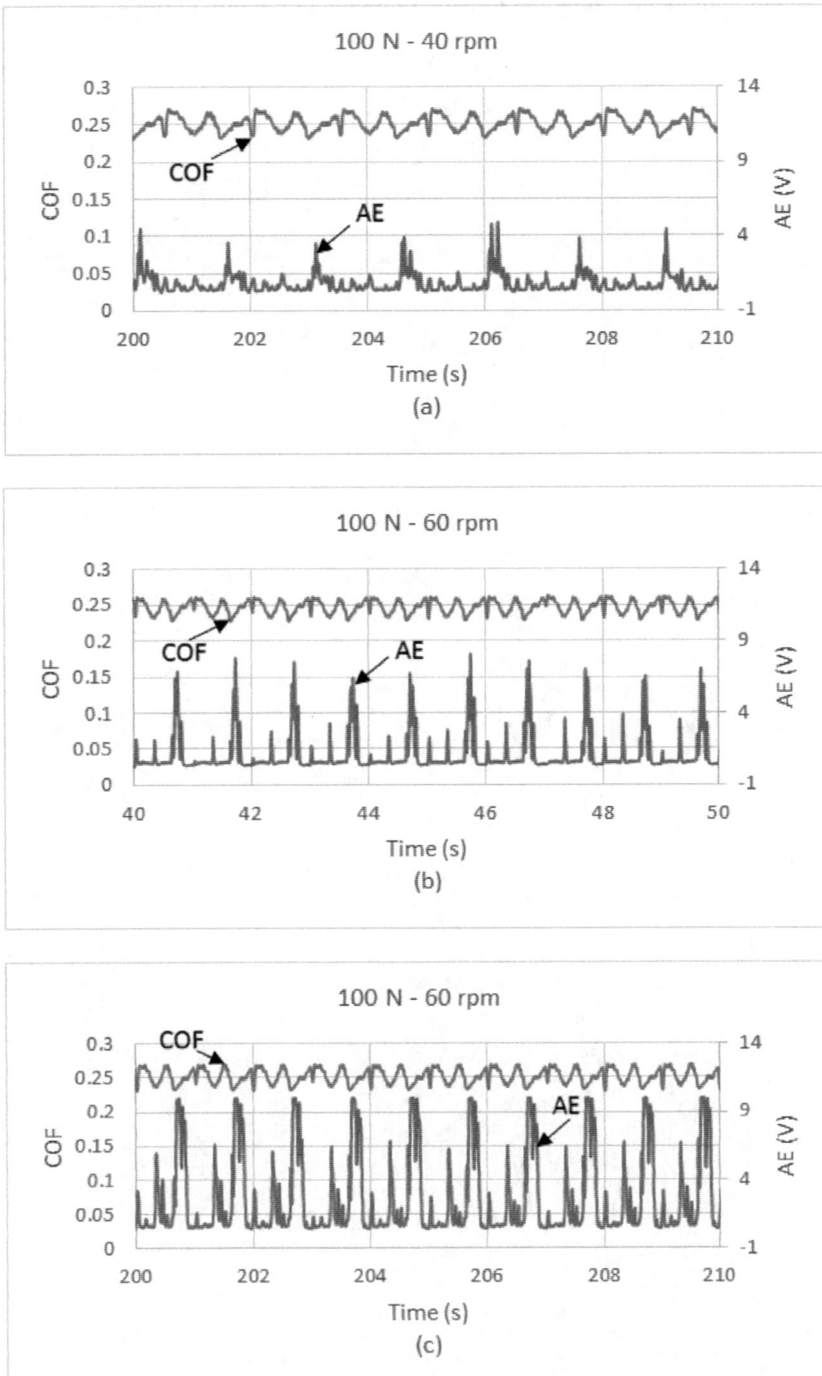


**Figure 6.** The evolution of the friction coefficient and acoustic emission for a 10 seconds period for the 100 N test conducted at: a) 40 rpm; b) 60 rpm (beginning of the test); c) 60 rpm (end of the test).

To better observe the correlation between the stick-slip phenomenon and the acoustic emission a two seconds period was chosen. This period can be seen in figure 7 for the 100 N and 40 rpm test (figure 7 a), the beginning of the 100 N and 60 rpm test (figure 7 b) and the end of the 100 N and 60 rpm test (figure 7 c).

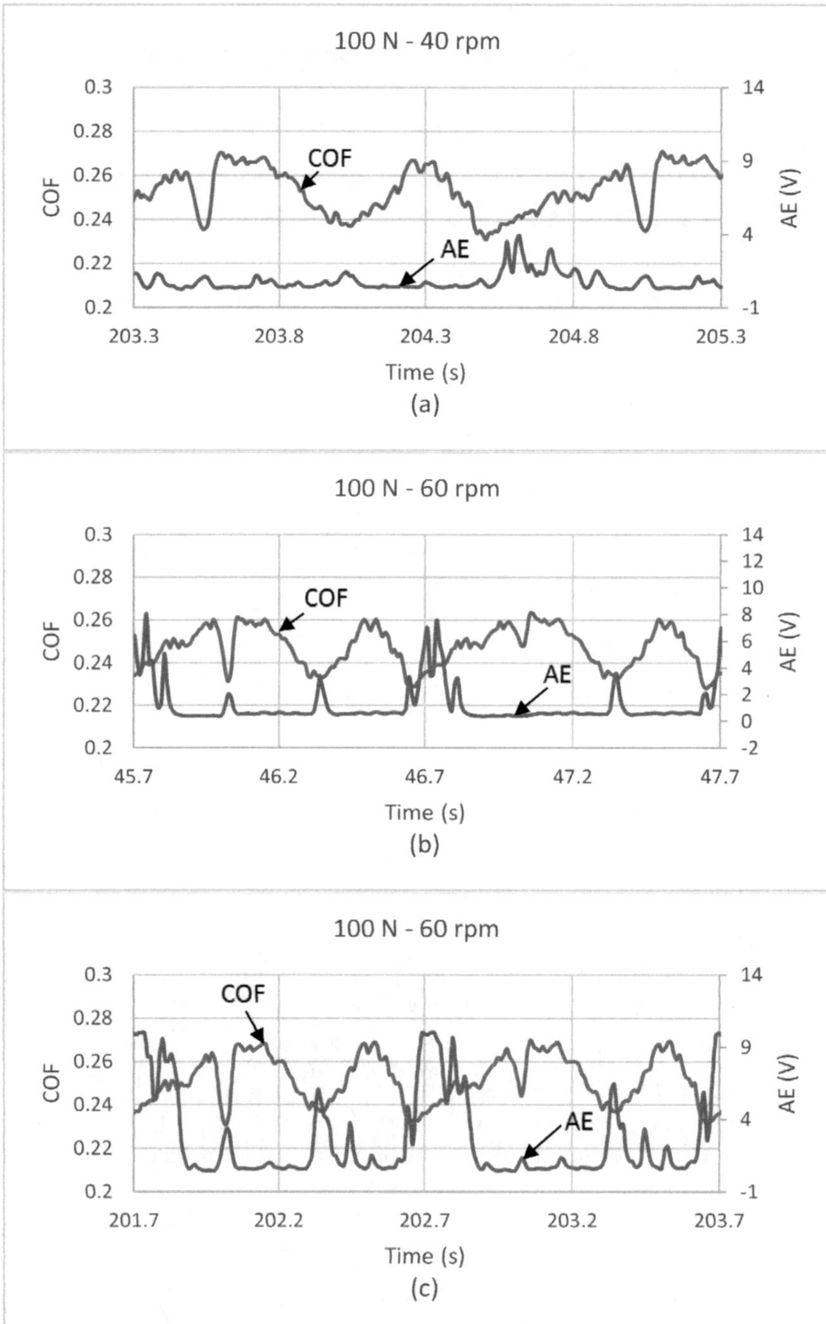


**Figure 7.** The evolution of the friction coefficient and acoustic emission for a 2 seconds period for the 100 N test conducted at: a) 40 rpm; b) 60 rpm (beginning of the test); c) 60 rpm (end of the test).

IOP Conf. Series: Materials Science and Engineering **174** (2017) 012019 doi:10.1088/1757-899X/174/1/012019

The results presented in figure 6 and figure 7 highlight the correlation between the acoustic emission and the sudden drop and rise of the values of the friction coefficient. Thus, we can see that amplitude of the friction coefficient influences the amplitude of the acoustic emission. Regarding the noise produced during the 100 N tests, it was more intense than the 50 N and 75 N tests and it grew in intensity as the rotational speed of the disc sample was increased.

The average value of the acoustic emission was calculated for each test in order to see how the contact pressure and sliding speed influences the level of the acoustic emission. These values have been inserted in the two graphs in figure 8.

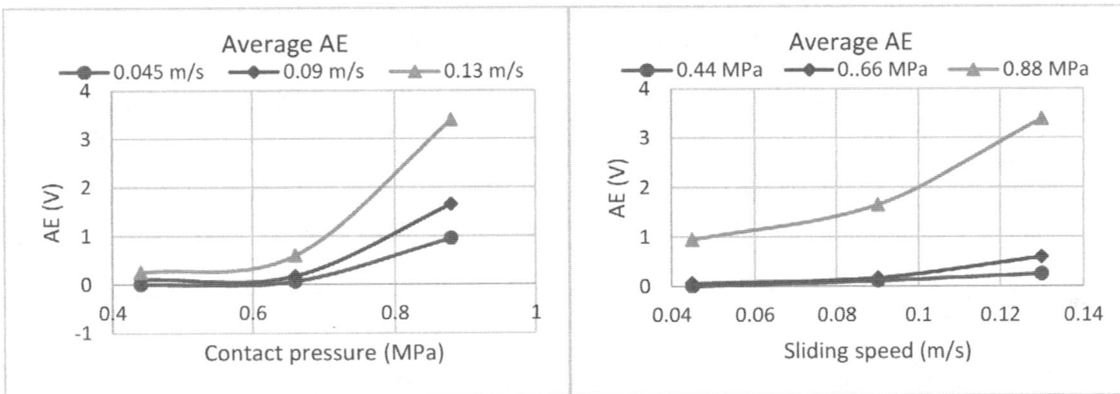


**Figure 8.** The evolution of the average AE versus contact pressure and sliding speed

As clearly seen form the two graphs from figure 8, by increasing either the contact pressure or the sliding speed the level of the acoustic emission increases significantly. For the current tests, when the sliding speed was tripled the average level of the acoustic emission increased from 0 V to 0.25 V for the 50 N test, from 0.06 V to 0.59 V for the 75 N test and from 0.95 V to 3.38 V for the 100 N test. Also, by doubling the contact pressure between the pin and the disc the average level of the acoustic emission increases from 0 V to 0.95 V for the 20 rpm test, from 0.11 V to 1.65 V for the 40 rpm test and from 0.25 V to 3.38 V for the 60 rpm test.

Friction generates heat and a significant temperature rise at the contact between the two sliding surfaces can influence the tribological properties of the brake friction couple materials. Consequently the maximum temperature rise has been estimated for the following conditions:

- maximum sliding speed: 0.13 m/s;
- maximum contact pressure: 0.88 MPa;
- average friction coefficient: 0.25;
- total sliding time: 300 s;
- disc mass: 0.167 g;
- specific heat for grey cast iron: 560 J/(kg·grd).

It is considered that 80% of the heat generated by friction is dissipated through conduction in the grey cast iron disc. In these conditions, the estimated temperature rise at the local contact is 6.8°C. This temperature rise does not significantly alter the mechanical properties or the tribological properties of the materials.

## 4. Conclusions

The results highlight the effects of the sliding speed and contact pressure on the friction coefficient and its influence on the brake noises and vibrations caused by the stick-slip phenomenon.

The results of the pin-on disc tests indicate that in the case of the 50 N (0.44 MPa) tests, where the friction coefficient has no major variations caused other than the misalignment of the surface of the disc, there is no acoustic emission and audible noise.

On the other hand, in the cases of the 75 N (0.66 MPa) and 100 N (0.88 MPa) tests, where the stick-slip phenomenon can be observed a notable amount acoustic emission is also present, accompanied by an audible noise.

The results reveal that by increasing the contact pressure and the sliding speed between the two samples the amplitude of the stick-slip increases, which can be correlated with the increase of the acoustic emission level and the intensity of the audible noise.

**References**

[1] Yoon S W, Shin M W, Lee W G and Jang H 2012 Effect of surface contact conditions on the stick–slip behavior of brake friction material *Wear* **294-295** 305-12

[2] Kinkaid N M, O'Reilly O M and Papadopoulos P 2003 Automotive discbrake squeal *J. Sound. Vib.* **267** 105-66

[3] Kruse S, Stingl B, Hieke J, Papangelo A, Tiedemann M, Hoffmann N and Ciavarella M 2014 The influence of loading conditions on the static coefficient of friction – a study on brake creep groan *Topics in Modal Analysis I* **7** 149-60

[4] Neis P D, Ferreira N F, Matos L T, Masotti D and Poletto J C 2015 Definition of a suitable parameter for characterizing creep-groan propensity in brake pads *SAE Tech. Pap.* 2015-36-0004

[5] Akay A, Giannini O, Massi F and Sestieri A 2009 Disc brake squeal characterization through simplified test rigs *Mech. Syst. Signal Pr.* **23** 2590-2607

[6] Polt G, Sebestyén T, A Pauschitz and Franek F 2005 Investigations of stick-slip effects in wet friction drive elements *Mater. Sci. Forum* **473-474** 249-54

[7] Stoica N A and Tudor A 2015 Some aspects concerning the behaviour of friction materials at low and very low sliding speeds *Tribology in Industry* **37** 374-79

[8] Gao C, Kuhlmann-Wilsdorf D and Makel D D 1994 The dynamic analysis of stick-slip motion *Wear* **173** 1-12

[9] Pavelescu D, Mușat M and A. Tudor 1977 *Tribologie (Tribology)* (Bucharest: Ed. Did. Ped.)

[10] Antoniu S S and Cameron A 1976 The friction-speed relation for stick data *Wear* **36** 235-54

[11] Tudor A and Bo L C 1981 Aspecte teoretice privind influenta vitezei de antrenare si a amortizarii asupra mișcării sacadate *Studii și Cercetări de Mecanică Aplicată (Applied Mechanics Research and Studies)* **40** 741-51

# Analysis of progressive cavity pumps specific wear processes using Finnie models

**M Popescu[1], A I Popovici[2], F N Petrescu[2] and N N Antonescu[1]**

[1]Petroleum-Gas University of Ploiesti, Blvd. Bucureşti no. 39, code 100680, Ploiesti, Romania
[2]Department of Mechanical Technology, Technical University of Civil Engineering Bucharest, 59 Plevnei Way, code 010223, Bucharest, Romania

E-mail: arys_jo@yahoo.com

**Abstract**. Progressive cavitation pumps are designed to work in aggressive environments thus their wear is inevitable. The specific tribological coupling of these pumps is composed of a helical rotor with a single outward helix and a stator with a double inside helix. These elements are in relative motion and direct contact to each other and also in direct contact with the pumped oil. Therefore the main forms of wear of rotor-stator coupling elements are the abrasive wear and the erosion wear. In this paper is presented the analysis of erosion with Finnie models. The results are useful for estimation of progressive cavity pumps specific abrasive-erosive wear.

## 1. Introduction

The specific tribological coupling of progressive cavitation pump is composed of a helical rotor with a single outward helix and a stator with a double inside helix.

The rotor-stator coupling wear resistance and working life are ensured by high-performance solutions concerning the design and manufacture of progressive cavity pumps.

The pumps are designed to work in aggressive environments so their wear is inevitable.

Specific tribological processes of progressive cavity pumps can be studied considering the followings:

- rotor and stator materials;
- the characteristics of pumped fluid;
- contamination of oil pumped under pressure.

The rotor is made of steel (alloy steel or stainless steel with high hardness) and the stator is made of an elastomer with well-defined properties.

The specific couplings components are in relative motion and direct contact with each other and in direct contact with the pumped oil. The wear of the specific tribological coupling of the progressive cavitation pumps (rotor-stator) must be approach in this context.

The manner in which is acchived the sealing between the helix of the rotor and the stator in order to ensure the high presure of the fluid in the cavities formed, thus avoiding the reverse flow of discharge chamber to the aspiration chamber is of great tribological interest.

The fluids circulated by progressive cavitation pumps are characterized by the followings:

- viscosity and specific weight differentiated in a range of relatively high values;
- the content of solid particles (in some cases can reach up to 60% sand);
- water content which can reach up to 90%;
- the temperature of the pumped liquid that can be relatively high: 60 ÷ 120 ° C for fluids without sand and 40 ÷ 90 ° C for fluids containing sand;
- the existence of corrosive agents in the pumped oil ($CO_2$, $H_2S$ or $O_2$).

Abrasive wear and erosion wear are the main forms of wear of the coupling components (rotor and stator). Essentially, abrasion and erosion are caused by the impact of the rigid particle on the active surfaces of the pump. The impact is mainly differentiated by the nature of the material subjected to stress: elastic impact and plastic impact.

According to Johnson [1] the elastic impact occurs at a speed nearly 0.1 m/s, and the full plastic penetration occurs at a speed of 5 m/s in progressive cavitation pumps case. For plastic materials the plastic deformation of the target surface occurs at a very low speed (for hard steel with $\sigma_c$ = 1000 N/mm$^2$, the impact speed is 0.14 m/s [2]).

Pumped oil contains abrasive solid particles in suspention. The particles entrained by the rotor spiral are pressed on the surface of the stator which is made of an elastomer (synthetic rubber) that deforms and fixes the particles during the contact. In this way, it is achived an efficient sealing of pumping chambers and thus the effects of abrasion are reduced. After the removal of the rotor from the stator the particle of sand that was interposed falls under its own weight into the pumping chamber (cavity). In this case the particles circulated by the fluid with certain speeds are directed to the surface of the rotor and the stator and become active in terms of abrasion and erosion.

The arguments previous presented underline the requirement of studying wear processes specific to progressive cavity pumps.

## 2. Analytical models of abrasive erosion

Theoretical and experimental research of erosive wear processes have been and are currently a constant concern of many experts in the field of tribology [3-8]. The research results of erosion were materialized by the development of mathematical models specific to these tribological process. In his study, Meng [9] presented and analyzed 28 such models. Essentially the mathematical models of erosion suggest calculation equations for the intensity of erosive wear.

Development of abrasive erosion models is obviously marked by technical level of theoretical and experimental investigation resources which are currently in an accelerated dynamic of improvement.

It is an unanimous assessment of experts in the field that 1990 marks an important milestone in the research development history of erosive wear [10-11].

The pioneering stage in development of erosion mathematical models is marked by classic models foundation, characterized as the "spine" of the research of erosive processes.

Thus, the classical models of erosion, developed before 1990 are those proposed by Finnie in 1960 [12] and in 1972 [13], Bitter in 1963 [14], Hutchings (in 1974, 1979 and 1981) [15], Johnson and Cook (in 1983 and 1985) [16], and Sundararajan and Shewmon in 1983 [17]. These models have been validated by research performed by specialists in the fiels and undeniably reflects the true image of the essence of erosion mechanism.

Using the numerical calculation methods for describing the mechanism of erosion it opens a new stage of research in this field which is located between 1990-1995 and continued up to present date [10]. Using finite element method bi- and tridimenional models of erosion caused by abrasive particles specified forms (spherical in most cases) and different sizes which interacts with the surfaces of ductile and brittle materials were developed. Thus, based on the finite element method, Wang and Yang [10] model describe the effect of impact angles, impact speed and penetration depth of the particle on the intensity of erosive wear process.

## 3. Results of the erosion wear analysis using Finnie models. Discutions

Finnie models are applied for two erosion processes: the one in which the proportion of abrasive particles with microcutting effect ($p_a$) is 10% and the second one in which the percentage is 50%.

The proposed equations for the first Finnie model [12] (developed in 1960) are as following:

$$E_F = \frac{p_a \cdot \rho_m \cdot v^2 \cdot (1-c_r^2)}{0.9272 \cdot H_s \cdot \psi \cdot K_F}\left(\sin 2\alpha - \frac{6}{K_F}\sin^2\alpha\right), \text{ for } tg\alpha \le \frac{K_F}{6} \quad (1)$$

$$E_F = \frac{p_a \cdot \rho_m \cdot v^2 \cdot (1-c_r^2)}{0.9272 \cdot H_s \cdot \psi \cdot K_F}\left(\frac{K_F}{6} cos^2\alpha\right), \text{ for } tg\alpha \ge \frac{K_F}{6} \quad (2)$$

where: $E_F$ – dimensionless wear rate; $p_a$ – the percentage of abrasive particles with microcutting effects; $\rho_m$ – target material density; $v$ – particle impact speed (m·s$^{-1}$); $\alpha$ – the angle of incidence; $c_r$ – restitution coefficient; $H_s$ – static hardness; $\Psi$ – the ratio of the contact length ($L$) and cutting depth ($\delta$) of impact area; $K_F$ – the ratio of the horizontal component ($F_0$) and the vertical component ($F_v$) of the characteristic impact force.

The equations are inferred by considering the following assumptions:

- abrasive particles are sharp;
- cutting and recoil effects of the particles are significant;
- the material is plastically deformed by the impact of the abrasive particle.

The calculation formulas modified by Finnie (in 1972) [13] have the following structure:

$$E_F = \left(\frac{p_a}{2K}\right)\frac{\rho_m \cdot v^2 \cdot (1-c_r^2)}{0.9272 \cdot H_s}\left(\sin 2\alpha - \frac{8}{K_F}\sin^2\alpha\right), \text{ for } tg\alpha \le \frac{K_F}{6} \quad (3)$$

$$E_F = \left(\frac{p_a}{16}\right)\frac{\rho_m \cdot v^2 \cdot (1-c_r^2)}{0.9272 \cdot H_s}\left(cos^2\alpha\right), \text{ for } tg\alpha \ge \frac{K_F}{6} \quad (4)$$

The involved parameters in the calculation formulas, for both Finnie models, are listed in table 1, together with their associated values used in the analysis performed.

**Table 1.** Finnie models parameters (for plastic deformation).

| parameter | calculation value |
|---|---|
| $p_a$ – the percentage of abrasive particles with microcutting effects for a concentration of 3,3x10³ mg/l sand | • 50% - for the first model: $p_a$=1.65 kg/m$^3$; • 10% - for the second model: $p_a$= 0.33 kg/m$^3$; |
| $\rho_m$ – density of the target area [kg/m$^3$] | 1.19 |
| $\alpha$ – the angle of incidence [°] | 30 ° |
| $v$ – particle impact speed [m/s] | 4.71 |
| $H_s$ – surface hardness [N/m$^2$] | stator I: 80; stator II: 75; stator III: 70 |
| $c_r$ – restitution coefficient | 0.5 |
| $\rho_a$ – density of the particle material [kg/m$^3$] | 2590-2670 |
| $E_e$ – equivalent elastic modulus [N/m$^2$] | stator I: 4.2; stator II: 3.6; stator III: 1.7 |
| $\Psi$ – the ratio of the contact length ($L$) and the cutting depth ($\delta$) of the impact area | 0.7÷15 |
| $\mu$ – friction coefficient between the particle and the material | 0.1÷1 |

Analysis of wear rate variation was achived for the following cases:

**1.** $E_F = f(\alpha)$ - the variation depending on the incidence angle, for the three analyzed stators (the surface hardness $H_s$ is: for stator I: 80 N/m$^2$; stator II: 75 N/m$^2$; stator III: 70 N/m$^2$), for three values of the friction coefficient ($\mu = 0.1$; $0.5$; $1$). The results are shown in figure 1.

First Finnie model Second Finnie model

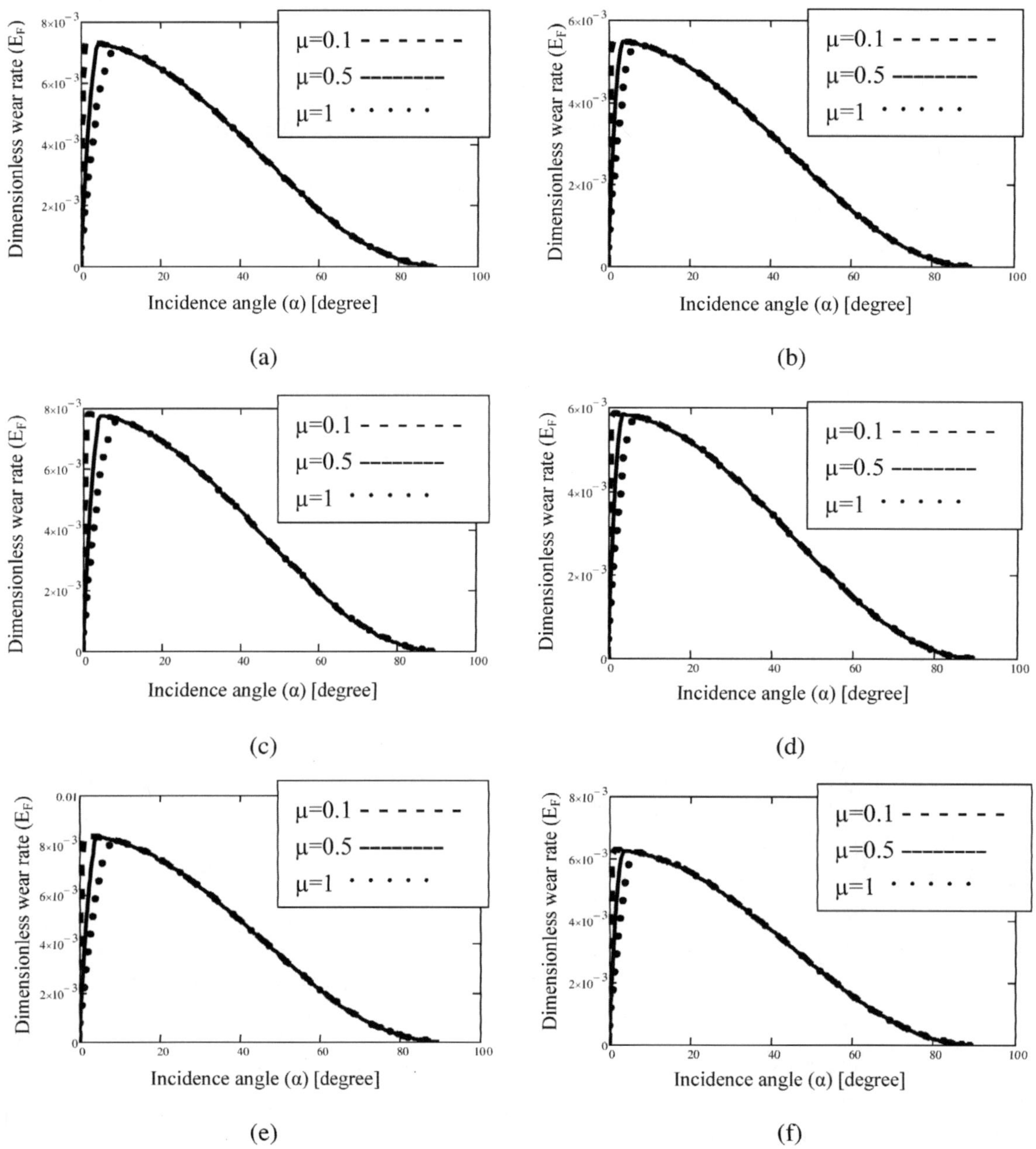


**Figure 1.** Wear rate variation depending on the angle of incidence for: (a) and (b) stator I; (c) and (d) stator II; (e) and (f) stator III.

**2.** $E_F = f(\psi)$ - the variation depending on the ratio of the contact lenght and the cutting depth of the impact area, for different values of the incidence angle ($\alpha_0$ =15°; 30°; 60°), for the three analyzed stators (the surface hardness $H_s$ is: for stator I: 80 N/m$^2$; stator II: 75 N/m$^2$; stator III: 70 N/m$^2$), for the friction coefficient of $\mu = 0.5$. The results are shown in figure 2.

First Finnie model

Second Finnie model

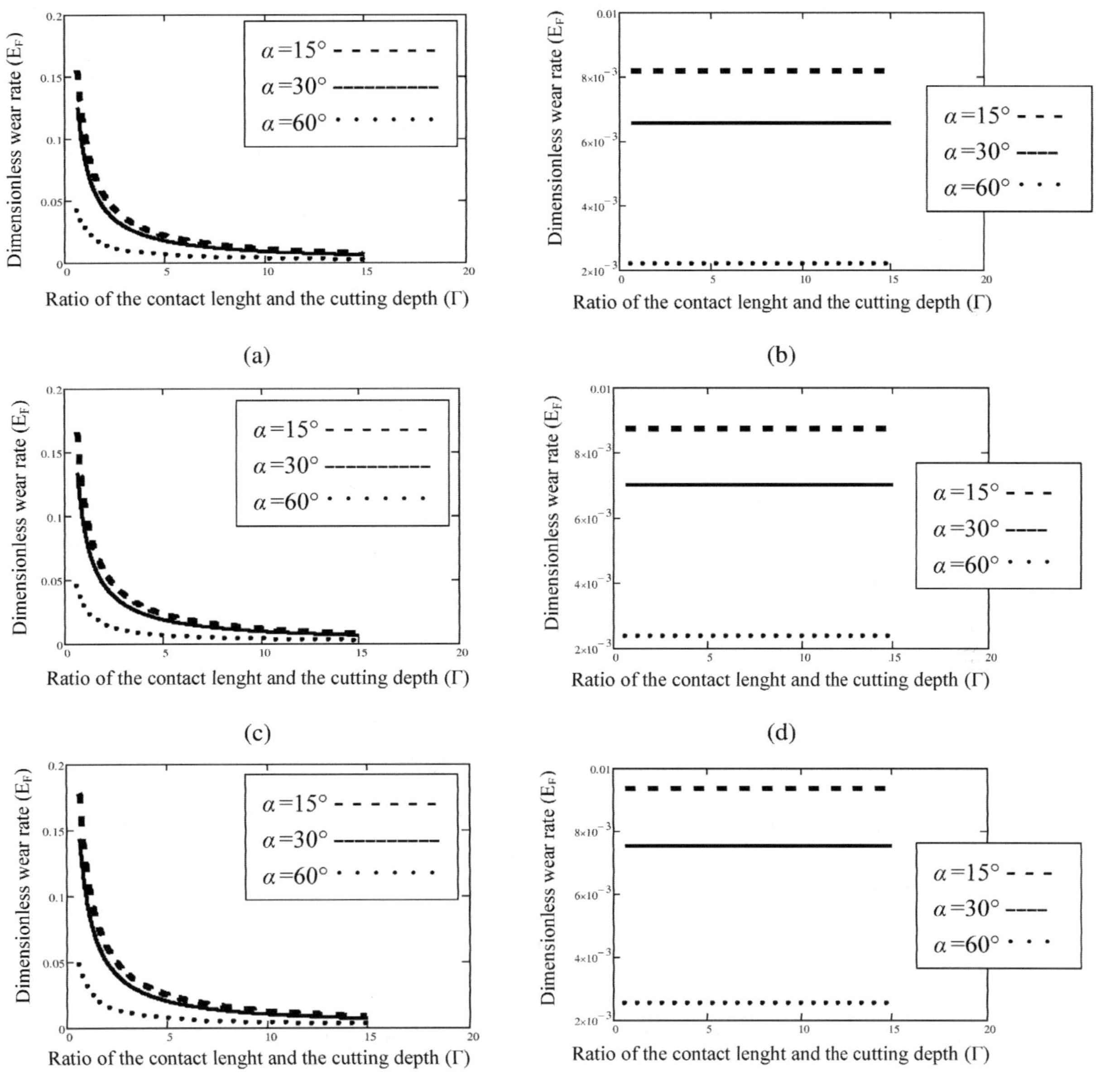


**Figure 2.** Wear rate variation depending on the ratio of the contact lenght and the cutting depth of the impact area for: (a) and (b) stator I; (c) and (d) stator II; (e) and (f) stator III.

**3.** $E_F = f(v)$ - the variation depending on the impact speed, for the three analyzed stators (the surface hardness $H_s$ is: for stator I: 80 N/m$^2$; stator II: 75 N/m$^2$; stator III: 70 N/m$^2$), for the friction coefficient of $\mu = 0.5$.The results are shown in figure 3.

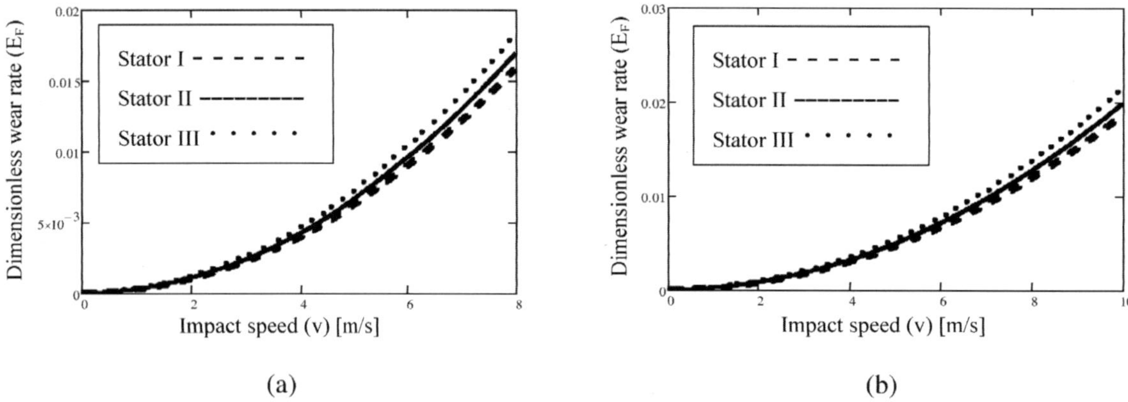


**Figure 3.** Wear rate variation depending on the impact speed for: (a) First Finnie model; (b) Second Finnie model.

In table 2 are shown the calculated values of the wear rate with Finnie models, for the angle of incidence α = 20° and the speed of v = 4.71 [m/s], and the values experimentally determined for the three analyzed stators.

**Table 2**. Wear rate calculated for the working conditions.

| Stator | The calculation model of the rate of wear | Rate of wear | |
|---|---|---|---|
| | | Calculated value | Experimentally determined value [18] |
| I | Finnie I | $6.454 \cdot 10^{-3}$ | $5.356 \cdot 10^{-3}$ |
| | Finnie II | $4.841 \cdot 10^{-3}$ | |
| II | Finnie I | $6.885 \cdot 10^{-3}$ | $5.783 \cdot 10^{-3}$ |
| | Finnie II | $5.163 \cdot 10^{-3}$ | |
| III | Finnie I | $7.376 \cdot 10^{-3}$ | $6.175 \cdot 10^{-3}$ |
| | Finnie II | $5.532 \cdot 10^{-3}$ | |

The analysis of the obtained results for the two Finnie models leads to the following findings:

- the shape of wear rate variation curves with the angle of incidence is the same for different values of the friction coefficient and the same value of the impact speed;
- is distinguished an incidence angle at which the wear rate has a maximum value; this is considered the critical angle of incidence in relation to the erosive effects it can produces;
- wear rate variation depending on the angle of incidence is different in relation to the critical angle of incidence: from zero to the critical angle the wear rate increases rapidly, then decreases slowly becoming zero at incidence angle of 90°;
- the maximum value rate of the wear rate and the angle of incidence at which is recorded this value it is changing depending on the friction coefficient: the increase of the friction coefficient leads to decreasing of the wear rate and to increasing of the critical angle of incidence;
- for angles of incidence higher than the critical value, erosive wear rate is not influenced by the friction coefficient;
- the wear rate decreases exponentially with the ψ coefficient (the ratio between the contact length and the depth of the cut) for angles of incidence smaller than 45° - for the first model, and is maintained constant for any value of the ψ coefficient – for the second model;

- erosive wear rate increases exponentially with speed impact at the same angle of incidence (for the second analyzed model); the increase is not influenced by the friction coefficient;
- for the erosion developed at the same speed and angle of incidence, the wear rate is not influenced by the friction coefficient between the erodent particle and the target area.

## 4. Conclusion

- in all analyzed cases, the estimated erosive wear rate with second Finnie model is less than the one assessed with the first Finnie model, which is explained by the fact that the percentage of abrasive particles with microcutting effects is 10% for the second model, versus 50% for the first model (the basic assumptions for the two models);
- the presented graphical results highlights the influence on the eroded surface hardness on the erosion wear: the increas hardness leads to a lower rate of erosion wear;
- the results acchived with the second Finnie model are the closest to the experimentally determined values of wear rate. It is therefore reccomanded to use the second model for the evaluation of erosive process of progression cavitation pumps.

## References

[1] Johnson K L 1985 *Contact Mechanics* (Cambridge: Cambridge University Press)
[2] Tudor A 2002 *Frecarea şi uzarea materialelor* (Bucharest: Editura Bren)
[3] Ashrafizadeh H and Ashrafizadeh F 2012 A numerical 3D simulation for prediction of wear caused by solid particle impact *Wear* **276-277** 75-84
[4] Arabnejad H, Mansouri A, Shirazi S A and McLaury B S 2015 Development of mechanistic erosion equation for solid particles *Wear* **332-333** 1044-1050
[5] Mansouri A, Arabnejad H, Shirazi S A and McLaury B S 2015 A combined CFD/experimental methodology for erosion prediction *Wear* **332-333** 1090-1097
[6] Hadavi V, Moreno C E and Papini M 2016 Numerical and experimental analysis of particle fracture during solid particle erosion, Part II: Effect of incident angle, velocity and abrasive size *Wear* **356-357** 146-157
[7] Zhu H, Lin Y, Zeng D, Zhou Y, Xie J and Wu Y 2012 Numerical analysis of flow erosion on drill pipe in gas drilling *Eng. Fail. Anal.* **22** 83-91
[8] Zhu X, Liu S, Tong H, Huang X and Li J 2012 Experimental and numerical study of drill pipe erosion wear in gas drilling *Eng. Fail. Anal.* **26** 370-380
[9] Meng H C and Ludema K C 1995 Wear models and predictive equations: their form and content *Wear* **181-183** 443-457
[10] Wang Y-F, Yang Z-G 2008 Finite element model of erosion wear on ductile and brittle materials *Wear* **165** 871-878
[11] Popovici A I 2012 *Contribuţii la studiul efectelor uzării asupra performanţelor sistemelor hidraulice de reglare automată* (Bucharest: PhD Thessis, Technical University of Civil Engineering Bucharest)
[12] Finnie I 1960 Erosion of surface by solid particles *Wear* **3** (2) 87-103
[13] Finnie I 1972 Some observations on the erosion of ductile metals *Wear* **19** (1) 81-90
[14a] Bitter J G A 1963 A study of erosion phenomena: Part I *Wear* **6** (1) 5-21
[14b] Bitter J G A 1963 A study of erosion phenomena: Part I and Part II *Wear* **6** (1) 5-21, 169-190
[15] Hutchings I M 1981 A model for the erosion of metals by spherical particles at normal incidence *Wear* **70** 269-281
[16] Johnson G R and Cook W H 1985 *Eng. Fract. Mech.* **21** 31-48
[17] Sundararajan G and Shewmon P G 1983 *Wear* **84** 237-258
[18] Popescu M 2016 *Contribuţii la realizarea constructivă şi tehnologică a pompelor cu şurub pentru industria petrolieră* PhD Thesis, Petroleum-Gas University of Ploieşti

IOP Conf. Series: Materials Science and Engineering **174** (2017) 012021 doi:10.1088/1757-899X/174/1/012021

# The effect of friction in coulombian damper

**H S Wahad[1], A Tudor[1], M Vlase[2], N Cerbu[1] and K A Subhi[1,3]**

[1] University Politehnica of Bucharest, Spl. Independentei 313, 060042, Bucharest, Romania
[2] Technical University of Civil Engineering of Bucharest, Romania
[3] AL-Furat AL-Awsat Technical University of Iraq

E-mail: haider.shahad@gmail.com

**Abstract:** The study aimed to analyze the damping phenomenon in a system with variable friction, Stribeck type. Shock absorbers with limit and dry friction, is called coulombian shock-absorbers. The physical damping vibration phenomenon, in equipment, is based on friction between the cushioning gasket and the output regulator of the shock-absorber. Friction between them can be dry, limit, mixture or fluid. The friction is depending on the contact pressure and lubricant presence. It is defined dimensionless form for the Striebeck curve ($\mu$ friction coefficient - sliding speed v). The friction may damp a vibratory movement or can maintain it (self-vibration), depending on the $\mu$ with v (it can increase / decrease or it can be relative constant). The solutions of differential equation of movement are obtained for some work condition of one damper for automatic washing machine. The friction force can transfer partial or total energy or generates excitation energy in damper. The damping efficiency is defined and is determined analytical for the constant friction coefficient and for the parabolic friction coefficient.

## 1. Introduction

In the shock-absorbers with dry and boundary friction regimes, also named coulombian shock-absorbers, the functional key element is the cushioning gasket (ring). This cushioning gasket is made of rubber, soaked with lubricant till saturation (oil or grease). These shock-absorbers are frequently used in machine building, with applications in construction equipments for different mixing materials, washing machines, and so on [1- 3].
Depending on the method of gasket radial fixing, the shock-absorbers are classified in direct absorbers (fixed in the outer shell), or reverse absorbers (attached to the rod).

The physical phenomenon of damping vibration, in equipment, is based on friction between the cushioning gasket and the output regulator of the shock-absorber. The output regulator is the steel rod of the direct shock-absorber or the shock-absorber housing for the "reverse" shock-absorber. In figure 1it is represented a shock-absorber with friction, (coulombian shock-absorber), used in automatic washing machines construction [4].

It is known that, in a friction coupling, when the sliding speed is variable (the case of starting, stopping or operating conditions with variable speed of rotation), the friction can be dry, limit, mixed or fluid, also depending on the contact pressure and lubricant presence. In these conditions, the friction is characterized by a Striebeck curve (friction coefficient $\mu$ - sliding speed v) [4], [5]. And also it is known that friction may damp a vibratory movement or can maintain it (self-vibration),depending on the friction coefficient variation with speed (increase, decrease or it can be relative constant)[6-11].

In all cases, the friction force is a disturbing force which is, partial or total, transferred through the shock-absorber or it can maintain the vibratory movement.

The paper aim is to analyze the damping phenomenon in a system with variable friction, Striebeck type, and to determine the work conditions of the direct coulombian shock-absorber when the effect of self - vibrations appears. It is defined and determined the damping efficiency of shock-absorber.

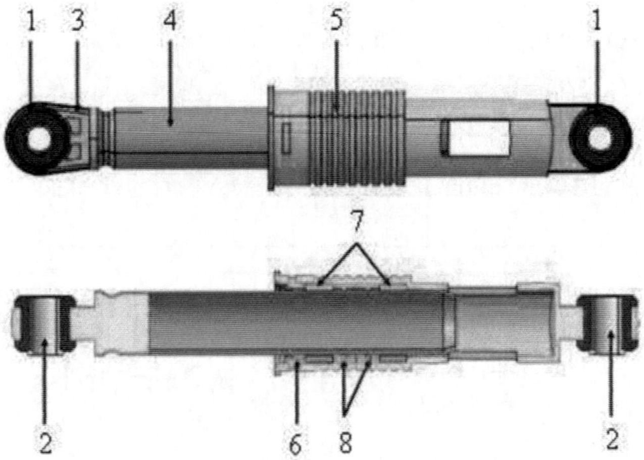


**Figure 1a.** Shock-absorber with friction [4]. 1 - Sealing member made of rubber; 2 - Plastic inner lining; 3 - Clamping sleeve of the shock-absorber; 4 - Steel rod; 5 - Body shock-absorber equipped with clamping sleeve; 6 - Limiting sleeve; 7 - Shock-absorber rings; 8 – Spacers.

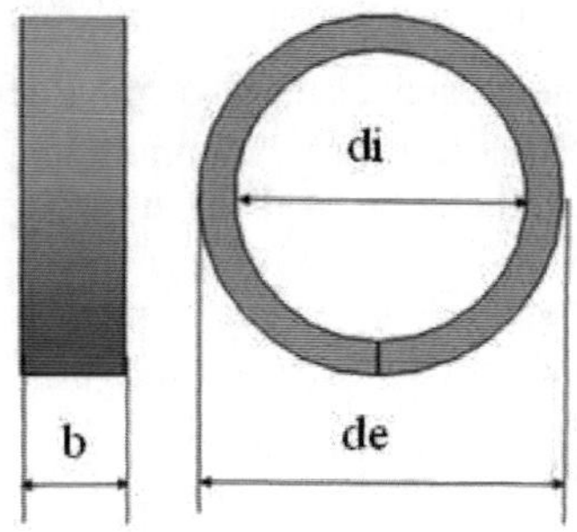


**Figure 1b.** The geomet shock-absorber ring [4].

## 2. Model for Striebeck friction analysis

When a friction coupling is in a state of film lubrication, the friction coefficient has a linear dependence with velocity for fluids with threshold flow

$$\mu_n = c_h v + \mu_o \tag{1}$$

where: $c_h$ parameter depends on lubricant viscosity, the medium contact pressure and the couple geometry; $\mu_0$ – coefficient, dependent on the lubricant rheology (threshold flow).

In the area of low and very low sliding speed, even if lubricant exists, it can`t be formed a continuous film bearing of lubricant. In this case, at zero velocity, there is static friction characterized by $\mu_S$ coefficient, which is dependent on the couple material, real contact pressure, and so on. It is accepted that in this zone, the friction coefficient is maintained constant until a certain velocity $v_0$:

$$\mu_0 = \mu_S \quad \text{for} \quad 0 \le v \le v_0 . \tag{2}$$

When velocity increases over $v_0$ limit, in contact, the normal force is transmitted through the real area which increases by velocity and friction coefficient decreases curvilinear by velocity until it reaches a limit value $\mu_m$. Analyzing the results for different Striebeck curves, [4], [5], it is accepted in the zone of dry, limit and mixed friction a parabolic variation as:

$$\mu_u = av^2 + bv + c , \tag{3}$$

with a, b, c determinable constants for certain known points ($v_0$, $\mu_S$), ($v_m$, $\mu_m$) – minimum point.

For velocities higher than $v_m$ a fluid friction begins, with a friction coefficient having an analytical form as equation (1). To resolve the analytical equation of vibratory movement in the shock-absorber it is accepted the existence of a point on the friction coefficient curve (3) for which line (1) is tangent.

The coordinates of this point ($v_{cr}$, $\mu_{cr}$) define the transition from dry, limit and mixed friction to fluid friction. These coordinates satisfy the conditions of continuity and derivability of friction coefficient.

The following notations are made: $\mu_a=\mu/\mu_s$; $\mu_{ma}=\mu_m/\mu_s$; $v_{am}=v/v_m$; $v_{0m}=v_0/v_m$; $v_{0a}=v_0/v_m$; $v_{a\,cr}=v_{cr}/v_m$; $c_{ha}=c_n \cdot v_m/\mu_s$. Putting the continuity conditions everywhere and derivability in the zone of limit and mixed friction of friction coefficient function, Striebeck type, the constants *a, b, c,* $\mu_0$ can be determined.

Thus, the relative friction-sliding coefficient for static friction ($\mu_a=\mu/\mu_S$) has the expression

$$\mu_a = \begin{cases} 1 \quad if \quad 0 \le v_{am} \le v_{om} \\ a_a v_{am}^2 + b_a v_{am} + c_a \quad if \quad v_{0m} < v_{am} < v_{acr} \\ c_{ha} v_{am} + \mu_o \quad if \quad v_{am} \ge v_{acr} \end{cases} . \tag{4}$$

The Striebeck curves are three kinds, depending on lubricant viscosity ($c_{ha}$), minimum sliding velocity ($v_{0m}$) and the minimum value of the friction coefficient $\mu_{ma}$. As an example, the figure 2 shows the theoretical curve for some minimum value of the friction coefficient.

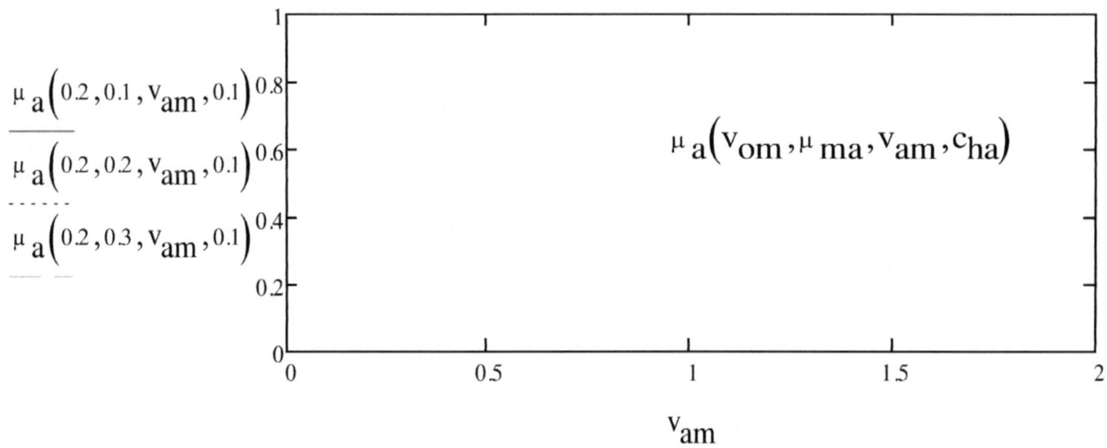


**Figure 2.** Striebeck curves for some minimum friction coefficients.

## 3. Analytical model for vibratory movement in the coulombian damper

To analyze the outer shell shock-absorber movement witch is in contact with the two damping gaskets, it is considered as known: the outer shell mass (*m*), the axial rigidity of the outer shell along with the drive support (*k*), the viscous damping in the drive support joint (*c*), the gasket geometry (inner diameter $D_g$, width $b_g$ and friction curve, Striebeck type ($\mu_a - v_{am}$) (equation 4).

As a model, it is proposed the system from figure 3, consisted of a material point having "*m*" mass, supported with Striebeck friction ($F_f$) on a body. This body moves with variable speed: $u = u_0 \cos(\omega t)$, where $u_0$ – is the drive speed amplitude; $\omega$ - pulsation; *t* – time (figure 3).

The friction force between the material point and the rigid support depends on the relative velocity $v = u - dx/dt$. If the "*u*" velocity is high enough ($u > dx/dt$), then the friction force will act in the positive direction of Ox axis, otherwise it will act in the negative direction of the axis. Thus, upon the material point will act the forces: $-c\,(dx/dt)$, $kx$ and $F_f$ (figure 3b). The friction force depends on the friction coefficient and the normal force, generated by the contact pressure, ($p_m$), between the rod and the gasket:

$$F_f = \mu F_n = \mu_S \mu_a 2\pi D_g b_g \cdot p_m = k_a \mu_a \quad , \tag{5}$$

where parameter $k_a = \mu_S 2\pi D_g b_g p_m$ depends on the static friction characteristics of the gasket material ($\mu_S$), gasket geometry ($D_g$, $b_g$) and the radial contact pressure ($p_m$), determined by the fitting assemblage and the elastic characteristics of the gasket material.

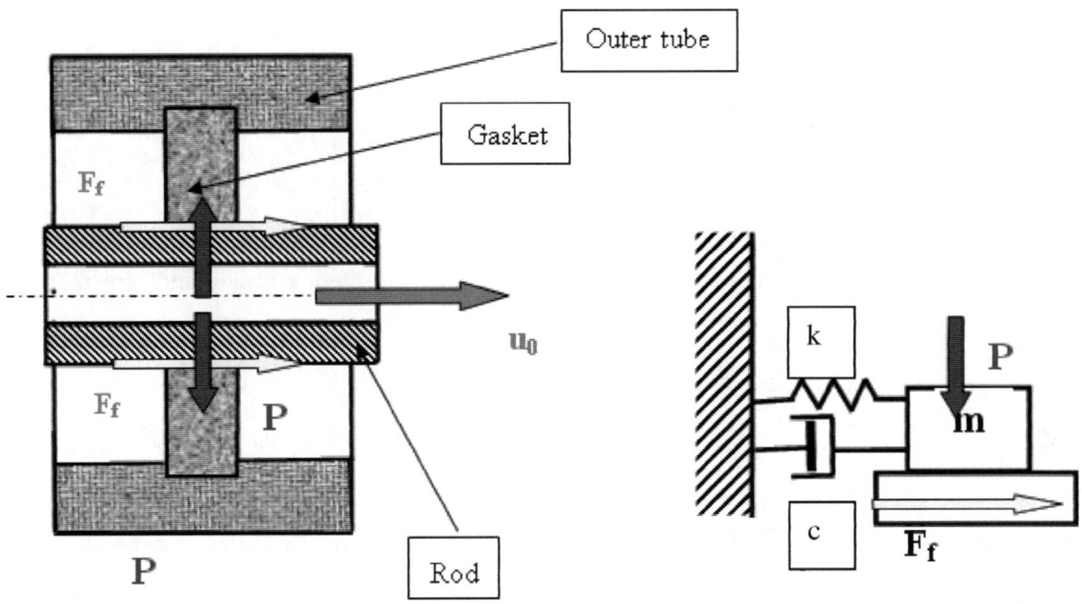


a) b)

**Figure 3**. Scheme of forces (a) and the dynamic model (b) in the coulombian shock-absorber.

The sliding friction coefficient relative to the static one, $\mu_a$, has the expression (4), in which the sliding velocity is

$$v = u - dx/dt \quad \text{or} \quad v_{am} = \frac{v}{v_m} = \omega_{mi}\cos\omega t - dx_m/dt \tag{6}$$

where $u_m = u_0/v_m$ and $dx_m/dt = (dx/dt)/v_m$ are the relative driving velocity, respectively, the relative oscillation velocity.

The differential equation of motion will be:

$$m\frac{d^2x}{dt^2} + c\frac{dx}{dt} + kx = F_f = k_a\mu_a \, . \tag{7}$$

Taking into account $\mu_a$ (equation (4)), the solution of equation (7) has different forms, depending on the driving speed, u, and Striebeck friction characteristics.

*Case 1 – constant friction coefficient (first part of Striebeck curve).*

For the first part of Striebeck curve ($v < v_0$) for which the friction coefficient is constant ($\mu_a=\mu_S$), equation (7) can be written in the classic mode [3], [12]:

$$\frac{d^2x}{dt^2} + 2\alpha\frac{dx}{dt} + p^2x = q \, , \tag{8}$$

where: $2\alpha = c/m$; $p^2 = k/m$; $q = k_a\mu_S/m$.

The values of α and p parameters lead to different forms of solutions for the equation (8). Thus, noting $c_0 = 2\sqrt{km}$ the critical coefficient of damping, which corresponds to the case $\alpha = p$, $\xi = c/c_0$ – the damping factor, there are three different situations: $c < c_0$, $c > c_0$ şi $c = c_0$.

*a) Sub case* $c < c_0$ $(\alpha < p)$ (small damping). In this case, the homogenous equation roots attached to it (8) $(r^2+2\alpha r+p^2=0)$ are complex and the solution of the (8) equation is

$$x_1 = ae^{-\alpha t} \cdot \cos(\beta t - \varphi) \tag{9}$$

$\beta^2 = p^2 - a^2$, a and φ are the integration constants that can be determined from the initial conditions:

For the initial moment $t = 0$, $x = 0$ and $\frac{dx}{dt} = v_{x_0} = \max; \left( \frac{d^2x}{dt^2} = 0 \right)$. It results $\varphi = \frac{\pi}{2}$; $a = v_{x_0} / \beta$.

The period of movement is $\mathrm{T} = 2\pi / \beta$

The particular solution of the equation (8), ($x_2$), has the right member form and replacing it in equation (8) it results:

$$x_2 = q / p^2 \tag{10}$$

From the equations (9) and (10) it results:

$$x = x_1 + x_2 = \frac{v_{x_0}}{\beta} e^{-\alpha t} \cos\left( \beta t - \frac{\pi}{2} \right) + \frac{q}{p^2} \tag{11}$$

or $\qquad x_a = xp^2 / q$.

By differentiating with respect to time, it results the vibration speed ($v_x$) and the acceleration of the vibration ($a_x$):

$$v_x = \frac{dx}{dt} = \frac{v_{x_0}}{\beta} e^{-\alpha t} [\beta \cos \beta t - \alpha \sin \beta t]; a_x = \frac{dv_x}{dt} = \frac{v_{x_0}}{\beta} e^{-\alpha t} [(\alpha^2 - \beta^2)\sin \beta t - 2\alpha\beta \cos \beta t] \tag{12}$$

Putting the condition that for the initial moment $t=0$, $v=0$, it results $v_{x0} = u_0$.

The relative sliding velocity between the gasket and the rod of the shock-absorber, in the presence of which friction occur, is

$$v_{am} = \frac{u - v_x}{v_m} = u_{0m} \left[ \cos \omega t - \frac{1}{\beta} e^{-\alpha t} (\beta \cos \beta t - \alpha \sin \beta t) \right] \tag{13}$$

where $u_{0m}=v_0/v_m$, $v_{am}=v/v_m$ .

If it is considered the initial hypothesis $\mu_a = 1 = constant$ and this coefficient is maintained constant when the instantaneous sliding velocity $v_{am} < v_{0m}$, it can be deduced the time range from a period when the shock-absorber dissipates energy and the time range when the shock-absorber becomes vibrator, thus it generates self-vibrations.

This phenomenon is illustrated in figure 4, where the instantaneous sliding velocity ($v_{am}$) is a parameter compared with the specific velocity from the Striebeck curve ($v_{om}$). This relative velocity ($v_{a1}$) is an important work parameter for analyze efficiency of damper:

$$v_{a_1} = v_{am} - v_{om} \, . \tag{14}$$

Thus, for $\alpha = 1$, $p = 2$, $q = 5$, $u_{0m} = 2$, $v_{0m} = 0{,}1$, $\omega = 2$, it results that during a certain period $T = 2\pi/\beta = 3{,}68$ s. In the time elements $t = 0.26...0.867$ s, $t = 2.37...3.68$ s „the damper" generates vibrations, having the role of a vibrator (excitatory). Its basic function (damping) is accomplished in the time ranges $t = 0...0.26$ s and $0.867. ...2.37$ s.

The area situated under the line $v_{a1} = 0$ is an indicator of the damping, and the area situated above the line $v_{a1} = 0$ is an indicator of the self-vibrations. It is defined the *damping efficiency* as the ratio between the dissipated energy due to friction and the total energy during a certain time element

$$\eta_a = \frac{W_a}{W_t} - \frac{\frac{1}{2}m\frac{1}{T}\left[\int_0^{tcr1} v_{a_1}^2 dt + \int_{tcr2}^{tcr3} v_{a_1} dt + \ldots\right]}{\frac{1}{2}m\frac{1}{T}\int_0^T v_{a_1}^2 dt} = \frac{\sum_{i=1}^{n} I_{ai}}{I_t}, \tag{15}$$

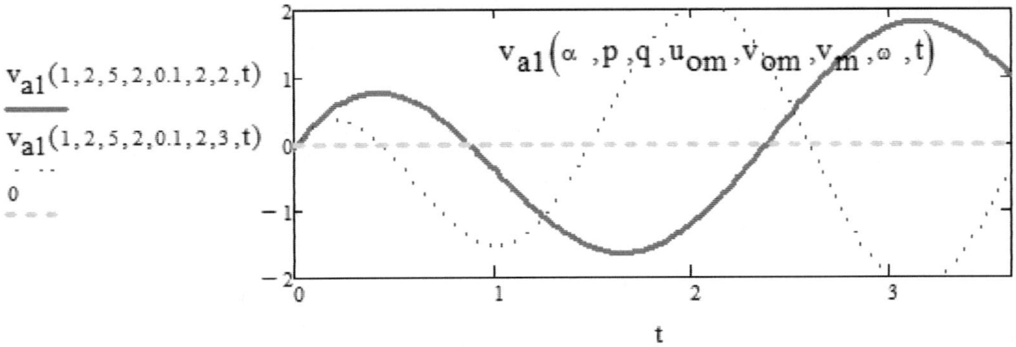


**Figure 4**. The critical speed for functioning as a damper.

where $t_{cr_1}, t_{cr_2}, t_{cr_3}, \ldots, t_{cr_m}$ are the time elements when the instantaneous velocity is null [roots of equation (20)]; $I_{ai}$ – is the integral of the specific energy of damping due to friction; $I_t$ – is the integral of the specific total energy. Thus, for the two curves in figure 4, the *damping efficiency* is $\eta_a = \eta_{a3} = 0.433$ for the curve with $\omega = 2\, rad/s$ and $\eta_a = \eta_{a4} = 0.624$ for the curve with $\omega = 3\, rad/s$.

Curve 1 ($\omega = 2\ rad/s$) intersect the axis in three points, and curve 2 ($\omega = 3\ rad/s$) in four points.

For a real shock-absorber, used in washing machines, $\alpha = 790$; $p = 880$; $q = 2$; $u_{om} = 2$; $v_{om} = 0,1$; $v_m = 2$ (SI measurement units), it results that the critical time is $t_{cr1} = 32.21\ \mu s$ and the *damping efficiency* is only $1,78 \cdot 10^{-6}$ (negligible), when the working pulsation is $\omega = 2\ rad/s$.

In the case of the shock-absorber housing that is attached to equipment frame through an aluminum socket, the housing movement (distance, velocity, acceleration, damping efficiency) is obtained by particularizing the expressions (9) – (15) for $\alpha = 0$. It results $\beta = p$,

- the deflection $x = \frac{v_{x_0}}{p}\sin pt + \frac{q}{p^2}$; velocity $v_x = v_{x_0}\cos pt$ - acceleration $a_x = -p v_{x_0} \sin pt$

-damping efficiency $\eta_{a2} = 0.51$ (for $p = 2$; $q = 5$; $u_{0m} = 0.1$; $v_{0m} = 0.1$; $v_m = 2$; $\omega = 3$ - SI measurement units); $\eta_{a4} = 0.376$ (for $p = 2$; $q_{0m} = 5,0ul$; $v_{0m} = 0.1$; $v_m = 2$; $\omega = 4$).

It can be noted that for this particular case, the damping efficiency is obtained by constant friction ($\mu = \mu_s$) for small instantaneous velocities.

*b. Sub case* $c > c_0$ $(\alpha > p)$ (high damping)

The homogenous equation roots attached to equation (7) are real and both negative. The solutions of the homogenous equation (natural vibration) (equation 11) is [12],

$$x_1 = b\, e^{-\alpha t} \sinh(\gamma t), \tag{16}$$

where $\gamma = \sqrt{\alpha^2 - p^2}$, $b$ – the determinable amplitude from the initial conditions:

- at zero moment $t = 0$, $x_1 = 0$ and $\frac{dx_1}{dt} = v_{x0} = \max\left(\frac{d^2x_1}{dt^2} = 0\right)$. It results $b = v_{x0}/\gamma$.

The particular solution of the equation (11), ($x_2$), has the same form as the right term, $x_2 = q/p^2$. Hence it results that:

-deflection (space) $x = x_1 + x_2 = \frac{v_{x_0}}{\gamma} e^{-\alpha t} \sin h(\gamma t) + \frac{q}{p^2}$; - velocity $v_x = \frac{dx}{dt}$; acceleration $a_x = \frac{dv_x}{dt}$.

Analog to case a) ($\alpha < p$), it is determined the instantaneous sliding velocity ($v_{a1}$) and it is presented in figure 5. It is noticed the existence of certain zones with normal operation as a damper ($v_{a1} < 0$) and zones with self-vibrations [vibrator, ($v_{a1} > 0$)].

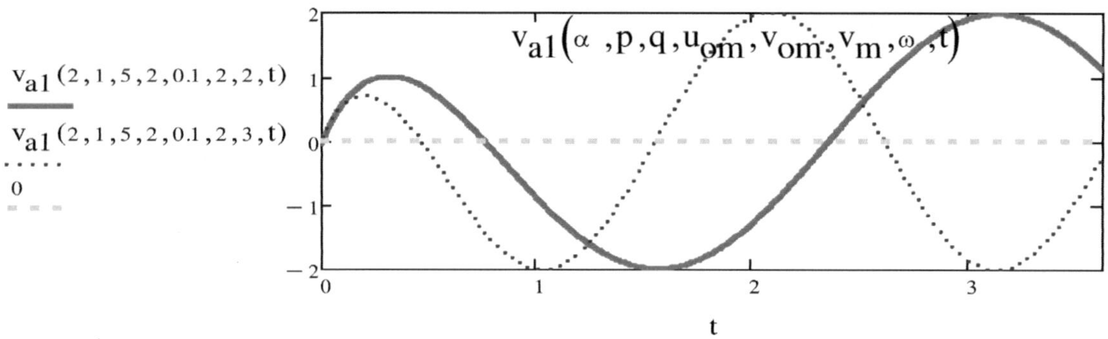


**Figure 5.** The instantaneous sliding velocity for high damping.

Thus, for $\alpha = 2$, $p = 1$, $q = 5$, $u_{0m} = 2$, $v_{0m} = 0.1$, it results that, during a certain period $T = 2\pi/\beta = 3.68$ s, in the time range $t = 0.013...0.76$ s, $t = 2.36...3.68$ s „the damper" generates vibrations, having the role of a vibrator. Its basic function (damping) is accomplished in the time ranges $t = 0...0.013$ s and $0.76....2.36$ s.

For $\alpha = 2$ and $p = 1$ (the numerical situation reversed as in case *a)* ($\alpha = 1$, $p = 2$), the damping efficiency is net superior $\eta_{a2} = 0.941$ ($\omega = 2$) and $\eta_{a4} = 0.665$ ($\omega = 3$) comparing with $\eta_{a2} = 0.443$ ($\omega = 2$) and $\eta_{a4} = 0.624$ ($\omega = 3$).

*c) Sub case* $c = c_0$ ($\alpha = p$) (critical damping)

The characteristic equation roots are real and equal to - $\alpha$. In these conditions, the movement can be characterized by

deflection $x = x_1 + x_2 = v_{x_0} t\, e^{-\alpha t} + \frac{q}{\alpha^2}$ ; - velocity $v_x = \frac{dx}{dt}$ and acceleration $a_x = \frac{dv_x}{dt}$.

The *damping efficiency* can be evaluating similarly. At example, for $\alpha = 1$, $p = 1$, $q=5$, $u_{0m} = 2$, $v_{0m} = 0.1$ is $\eta_a = 0.908$ ($\omega = 2$) and $\eta_a = 0.918$ ($\omega = 3$).

*Case 2. The friction coefficient variable by speed as a parabolic law (second part of Striebeck curve).*
In this case, the friction coefficient depends on the instantaneous sliding speed (7):
The differential equation of movement (8) becomes

$$m\frac{d^2x}{dt^2} + c\frac{dx}{dt} + kx = k_a\left(a_a v_{am}^2 + b_a v_{am} + c_a\right) \quad \text{where} \quad v_{am} = \frac{1}{v_m}\left(u_0 \cos\omega t - \frac{dx}{dt}\right). \tag{17}$$

After some algebraic calculus, it results:

$$\frac{d^2x}{dt^2} + \left(2\alpha + A_u \cos\omega t - B\frac{dx}{dt}\right)\frac{dx}{dt} + p^2 x = D\cos^2\omega t + E\cos\omega t + F_a \tag{18}$$

were $p^2 = k/m$, $2\alpha = c/m$ – the viscous damping parameter in the system itself; $A_u$, $B$, $D$, $E$, $F_a$ – the geometrical and functional parameters of the shock-absorber:

$$A_u = \frac{2u_0 a_a k_a}{v_m^2} + \frac{u_0 b_a k_a}{v_m};\quad B = \frac{a_a k_a}{v_m^2};\quad D = \frac{u_0^2 b_a k_a}{v_m^2};\quad E = \frac{u_0 b_a k_a}{v_m};\quad F_a = \frac{F_a k_a}{m}.$$

The differential equation (18) is nonlinear and thus, it is proposed the MATHCAD 2000 program for its numerical solving („Odesolve" subroutine). The curves from figure 6 are obtained for the experiment conditions that define the parameters:

$\mu_s = 1.4; \mu_m = 0.12; v_m = 0.07\ m/s; c_h = 0.08\ s/m; \omega = 2\ rad/s; v_o = 0.1 v_m; v_{cr} = 1.2 v_m.$

Thus, it results the geometrical and functional parameters of the shock-absorber:

$A_u = 1.8\cdot10^3; B = 1.612\cdot10^4; D = 58.05; E = -135.45; F_a = 2.32\cdot10^3.$

The instantaneous sliding velocities, $v_{a_1} = v_{am} - v_{om}$, $v_{a11}$ for $\alpha = 1848$ and $v_{a110}$ for $\alpha = 8758$ are presented in figure 6.

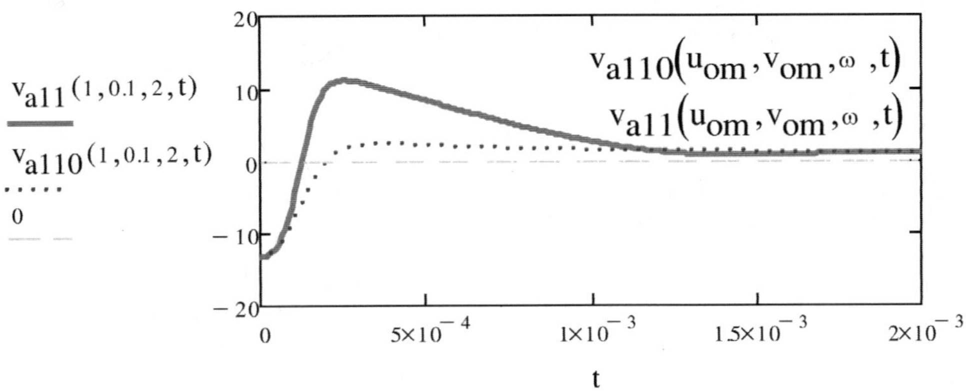


**Figure 6**. The instantaneous sliding speed in the shock-absorber that function in the parabolic zone.

For the two values of the damping ($\alpha$), taken as example, it is observed that the shock-absorber works more like a „vibrator" by friction self-vibration.

The damping efficiency is small. Thus, for $\alpha = 1848$, $\eta_a = 0.126$, and for $\alpha = 8758$, $\eta_a = 0.144$.

## 3. Conclusions

The friction can be fluid, limit and mixed or conventionally dry in the coulombian damper with rubber rings, saturated with liquid lubricant. The relative sliding speed between the rod and the ring and the working period are principal parameters of friction type.

The friction curve, on a large range of working speed of the damper can be considered as a Striebeck type curve.

The curve of the friction coefficient dependent on the relative sliding speed is simulated as a parabola for the dry, limit and mixed friction regime, and being tangent to a line for the Newtonian fluid friction.

Any kind of coulombian damper with Stribeck friction works in vibratory regime with different damping, depending on the pressure between rubber rings and the working rod and also the state of friction. The theoretical model of movement can explain and determine the damping efficiency.

The coulombian dampers are sources of vibrations when the friction coefficient decreases with the increase of the relative speed between the rings and the rod.

The damping efficiency is one important parameter to characterize coulombian damper and depends to geometry, contact pressure and material of damper rings and rod.

## References

[1] Draghici I, Ivan M, Kremer S, Lacatus V, Macarie V and Petrescu N 1979 *Suspensii si Amortizoare* (Bucuresti: Ed Tehnica)

[2] Ye Shao Chun and Williams Keith A 2006 Torsional Friction Damper Optimization *J. Sound Vib.* **294** 529–546

[3] Bratu P 1982 *Izolarea si amortizarea vibratiilor la utilajele de constructii* (Bucuresti: INCERC)

[4] Cerbu N, Tudor A and Paduraru G I Some Experimental Aspects Of The Mechanical Behaviour of Coulombian Dry Friction Damper (Part II), *U P B Sci. Bull., D Series* (to be published in 2016)

[5] Al Sayed Baraa, Chatelet E, Baguet S and Georges J K., 2011 Dissipated Energy And Boundary Condition Effects Associated to Dry Friction on the Dynamics of Vibrating Structures, *Mech. Mach. Theory* **46** 479–491

[6] Cerbu N, Tudor A and Paduraru G I Some theoretical aspects of the mechanical behavior of coulombian dry friction damper (Part I), *U.P.B. Sci. Bull. D Series* (to be published in 2016)

[7] Campbell W E 1969 Boundary lubrication Ch. 6 in *Boundary lubrication; an appraisal of world literature*, Eds: Ling F F, Klaus E E and Fein R S *ASME* pp 87-117

[8] Kalin M, Vizitin J and Velkavrh I 2009 The Stribeck curve and lubrication design for non-fully wetted surfaces *Wear* ***267*** 1232-40

[9] Braun O M and Peyard M Dependence of Kinetic Friction on Velocity: Master Equation Approach *Physical Review E* **83** (4) 046129

[10] Chengwu D, Rajendra S 2006 Dynamics of a 3dof torsional system with a dry friction controlled path *J. Sound Vib.* **289** 657–688

[11] Wu Z, Liu H, Liu L and Yuan D 2006 Identification of nonlinear viscous damping and coulomb friction from the free response data *J. Sound Vib.* **304** 407–414

[12] Voinea R, Voiculescu D and Ceausu V 1983 *Mecanica (*Bucuresti: Ed. Didactica si Pedagogica)

IOP Conf. Series: Materials Science and Engineering **174** (2017) 012022 doi:10.1088/1757-899X/174/1/012022

# Analysis of progressive cavity pumps specific wear processes using Bitter and Hutchings models

**A I Popovici[1], M Popescu[2], N N Antonescu[2] and F N Petrescu[1]**

[1]Department of Mechanical Technology, Technical University of Civil Engineering Bucharest, 59 Plevnei Way, code 010223, Bucharest, Romania
[2]Petroleum-Gas University of Ploiesti, Blvd. Bucureşti no. 39, code 100680, Ploiesti, Romania

E-mail: arys_jo@yahoo.com

**Abstract**. Progressive cavitation pumps are designed to work in aggressive environments thus their wear is inevitable. The specific tribological coupling of these pumps is composed of a helical rotor with a single outward helix and a stator with a double inside helix. These elements are in relative motion and direct contact to each other and also in direct contact with the pumped oil. Therefore the main forms of wear of rotor-stator coupling elements are the abrasive wear and the erosion wear. In this paper is presented the analysis of erosion with Bitter and Hutchings models. The results are useful for estimation of progressive cavity pumps specific abrasive-erosive wear.

## 1. Introduction

The researchers attention was focused on the study of the influence factors from the abrasive erosion models relations, to specify the share of their influence, the precision in describing the erosive processes, the generalization degree and the limits of applicability.

The main assumptions on which erosion models were developed are as following [1-3]:

a – the erosive wear particle occurs due to the mechanical action of abrasive particles on the surfaces that cause microcracking and microcutting of the incident material layer (the microcutting component of erosion);

b - the kinetic energy of the abrasive particles is generated by their impact with the target surface causing the deformation of the material, initiation and propagation of crack and detachement of wear particle when the cracks have reached a critical length (the localized plastic deformation component associated with the mechanical fatigue process).

The first case is typical to ductile material and the second one to fragile material. The assumptions made are based on Johnson, Cook and Holmquist studies [4-5] and analyzed by Wang [6].

## 2. Bitter and Hutchings analitical models of abrasive erosion wear

### *2.1. Bitter model*

In 1963 Bitter [7] analyzed the abrasive erosion regarding two components:

- the microcutting component
- the component of plastic deformation of erodent material surface.

Bitter considers that the microcutting process is attend by particle sliding on the surface.

The sliding speed is distinguished when:

- tangential component different from zero at the time of detachement of the incident surface;
- tangential component is zero at a given time of the impact.

Thus, Bitter proposed relations for calculating the adimensional rate of wear, differentiated according to the size of the speed tangential component.

The equation for the tangential speed component different from zero is:

$$E_{B,C_1} = \frac{2\rho C_1 (U_i \sin\alpha - U_{el})^2}{(U_i \sin\alpha)^{0.5}} \cdot \left( U_i \cos\alpha - \frac{\varepsilon_C C_1 (U_i \sin\alpha - U_{el})^2}{(U_i \sin\alpha)^{0.5}} \right) \tag{1}$$

where:

$$C_1 = 0.361 \frac{\sigma^{0.25}}{H_S^{1.25}} \tag{2}$$

and for the tangential speed component is zero, the equation is:

$$E_{B,C2} = \frac{\rho\left(U_i^2 \cos\alpha - C_2 (U_i \sin\alpha - U_{el})^{1.5}\right)}{2\varepsilon_C} \tag{3}$$

where:

$$C_2 = 0.547 \frac{H_S^{2.25}}{\sigma^{0.25} E_c^2} \tag{4}$$

For repeated surface deformation component, Bitter suggest the following relation for calculating the rate of wear:

$$E_{B,DW} = \frac{\rho (U_i \sin\alpha - U_{el})^2}{2\varepsilon_D} \tag{5}$$

where:

$$U_{el} = 1.25 \frac{H_S^{2.5}}{\sigma^{0.5} E_e^2} \tag{6}$$

The terms used in previous equations have the following meanings:

$E_{B,C1}$, $E_{B,C21}$ and $E_{B,DW}$ – erosion wear rate

$v_1$ - impact speed of the particle;

$v_{el}$ - the impact speed that is reached at the elastic limit of the material;

$\varepsilon_D$ - the specific energy of deformation;

$\rho_{me}$ - density of the erosive particle material;

$E_c$- modulus of elasticity on impact;

$\varepsilon_c$ - the specific wear energy for microcutting;

$C_1$ and $C_2$- constants with specified relations.

The overall erosive rate of wear is calculated as sum of the specified components, differentiated by the angle of incidence

$$E_B = E_{B,DW} + E_{B,C_1} \text{ - for } \alpha \le \alpha_0 \tag{7}$$

$$E_B = E_{B,DW} + E_{B,C_2} \text{ - for } \alpha > \alpha_0 \tag{8}$$

where $\alpha_0$ corresponds to the angle of impact at which the tangential component of speed is zero, which occurs when the particle leaves the incident surface.

### *2.2. Hutchings model*

Hutchings [1] consider two models for the erosion wear: the microcutting erosion model and the plastic deformation erosion model. For microcutting erosion Hutchings suggests two equations for calculating wear rate:

- for the normal angle of incidence $\alpha_0 = 90^o$, the relation is:

$$E_{H,n} = K_{er} \cdot \frac{\rho_m \cdot v^2}{H} \tag{9}$$

where: $K_{er}$ – erosive wear coefficient (the ratio between the volume of detached material by wear and the total volume of material affected by the microcutting process).

- for oblique impact, $\alpha_0 < 90^o$, the relation is:

$$E_{H,n0} = K_{er} \cdot \frac{\rho \cdot v^n}{H} f(\alpha) \tag{10}$$

where: $n$ – speed exponent with values in the range 2÷2.5;
$f(\alpha)$ - depending on the angle of incidence.

When the erosion is caused by plastic deformation due to spherical abrasive particles that strikes the target surface for incidence angle less than 90°, the relation is:

$$E_H = 0.033 \cdot \alpha_p \cdot \frac{\rho_m \cdot \rho_a^{0.5} \cdot v^3}{\varepsilon_{crt}^2 \cdot H_d^{1.5}} \tag{11}$$

The $\alpha$ coefficient is defined by the ratio between the volume of plastically deformed material and the indentation volume. This coefficient depends on the geometry of the indentation, impact speed and the characteristics of the target material.

Hutchings consider the Coffin-Manson equation to calculate the $\varepsilon_{crt}$ (fatigue model for normal impact):

$$\varepsilon_{crt} = \Delta\varepsilon_p \cdot N_c^{1/2} \tag{12}$$

where: $N_c$ – the number of stress cycles at which occurs the particle of wear.
$\Delta\varepsilon_p$ (plastic deformation) determined with the relations [8]:

$$\Delta\varepsilon_p = 0.2 \cdot \frac{a}{r} \tag{13}$$

$$a = r\sqrt{2 \cdot v} \cdot \sqrt[4]{2 \cdot \sigma / 3 \cdot H} \tag{14}$$

where: $r$ –particle radius; $a$ – indentation radius.

## 3. Results and analysis

### *3.1. Bitter model*

Bitter [7] describes the mechanism of abrasive erosion considering two components that occurrs simultaneously: the microcutting and repeated deformation of the target surface.

Bitter parameters involved in the model are listed in table 1.

**Table 1.** Bitter model parameters.

| parameter | calculation value | |
|---|---|---|
| | elastic deformation | microcutting |
| $\rho_m$ – density of the targe material [kg/m$^3$] | 1.19 | |
| $H_s$ – surface hardness [N/m$^2$] | stator I: 80; stator II: 75; stator III: 70 | |
| $E_m$ – equivalent elastic modulus [N/m$^2$] | stator I: 4.2; stator II: 3.6; stator III: 1.7 | |
| $\rho_a$ – density of the particle material [kg/m$^3$] | 2590÷2670 | |
| $v$ – particle impact speed [m/s] | 4.71 | |
| $c_r$ – coefficient of restitution | 0.5 | |
| $\alpha$ – the angle of incidence [°] | 30 ° | |
| $\alpha_0$ – the angle at which the speed is zero | 0° | |
| $\varepsilon_D$, $\varepsilon_C$ – the specific energy for wear [J/m$^3$] | $\varepsilon_D = 4.7\cdot10^{10}$ | $\varepsilon_C = 2.2\cdot10^{10}$ |

The analysis of wear rate was achived for two cases:

1. $E_B = f(\alpha)$ - the variation depending on the angle of incidence, for the three analyzed stators ($H_s$ – surface hardness: stator I: 80 N/m$^2$; stator II: 75 N/m$^2$; stator III: 70 N/m$^2$);
2. $E_B = f(v)$ - the variation depending on the impact speed, for the three analyzed stators ($H_s$ – surface hardness: stator I: 80 N/m$^2$; stator II: 75 N/m$^2$; stator III: 70 N/m$^2$).

The results are shown in figure 1 and figure 2.

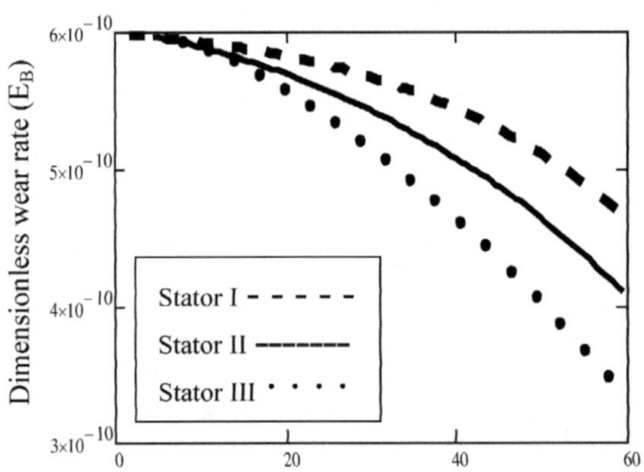


**Figure 1.** Wear rate variation depending on the angle of incidence.

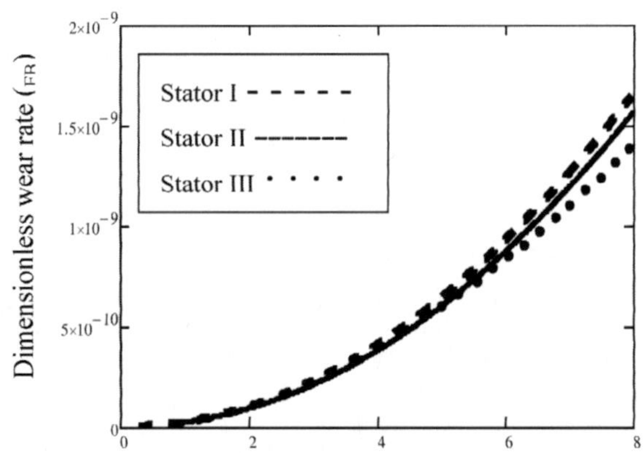


**Figure 2.** Wear rate variation depending on the impact speed.

*3.2. Hutchings model*

Hutchings [1] has developed two models for abrasive erosion wear differentiated by the erosion mechanism, as follows:

- microcutting erosion model (the first model);
- erosion by plastic deformation model (second model)

The involved parameters in the calculating equations, for both models, are listed in table 2.

IOP Conf. Series: Materials Science and Engineering **174** (2017) 012022 doi:10.1088/1757-899X/174/1/012022

**Table 2.** Hutchings model parameters.

| parameter | calculation value | |
|---|---|---|
| | microcutting | plastic deformation |
| $\rho_m$ – density of the targe material [kg/m$^3$] | 1.19 | |
| $H_s$ – surface hardness [N/m$^2$] | stator I: 80; stator II: 75; stator III: 70 | |
| $v$ – particle impact speed [m/s] | 4.71 | |
| $c_r$ – coefficient of restitution | 0.5 | |
| $\rho_a$ – density of the particle material [kg/m$^3$] | -- | 0.33 |
| $\Gamma$ – rate of removed volume depending on the particle volume and the critical stress | -- | 0.7÷13 |

The analysis of wear rate was achived for the following cases:

**1.** $E_H = f(v)$ - the variation depending on the impact speed, for the three analyzed stators ($H_s$ – surface hardness: stator I: 80 N/m$^2$; stator II: 75 N/m$^2$; stator III: 70 N/m$^2$);

The results are shown in figure 3 and figure 4.

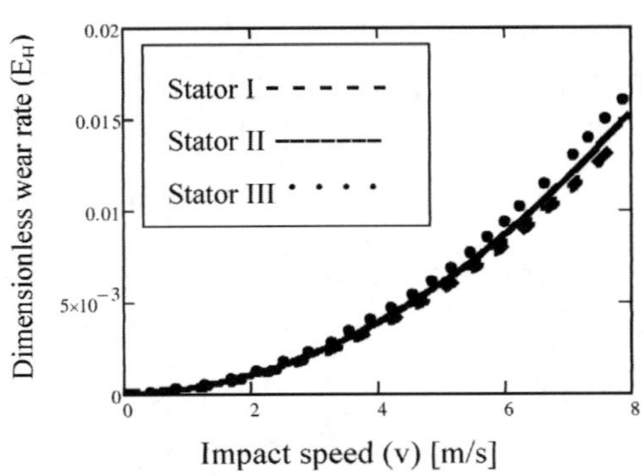


**Figure 3.** Wear rate variation depending on the impact speed for the microcutting model.

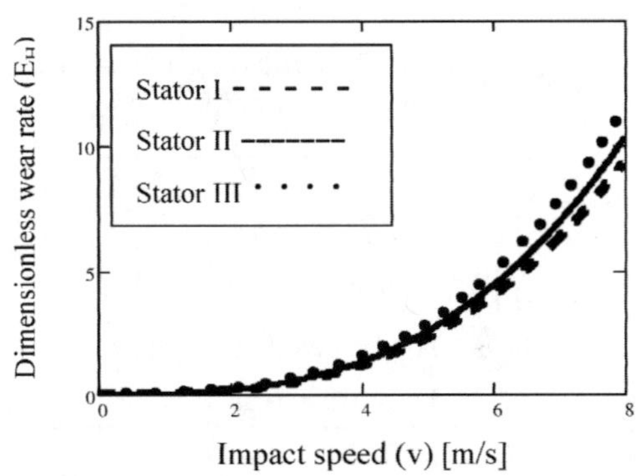


**Figure 4.** Wear rate variation depending on the impact speed for the plastic deformation model.

**2.** $E_H = f(\Gamma)$ - the variation depending on the ratio of removed volume depending on the particle volume and the critical stress, for the three analyzed stators ($H_s$ – surface hardness: stator I: 80 N/m$^2$; stator II: 75 N/m$^2$; stator III: 70 N/m$^2$).

The results are shown in figure 5.

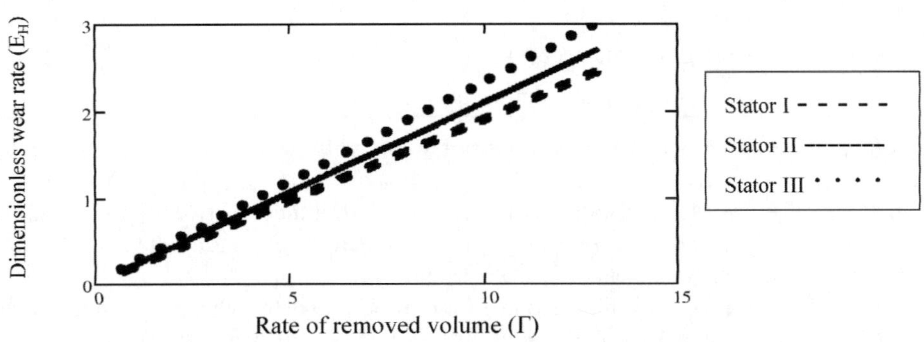


**Figure 5.** Wear rate variation depending on the rate of removed volume for the plastic deformation.

The analysis of the obtained results for Bitter and Hutchings models leads to the following findings:

- the shape of wear rate variation curves with the angle of incidence is the same for different values of the friction coefficient and for the same value of the impact speed;
- both Hutchings models consider exponentially increases of erosive wear rate with the increasing of the impact speed, indicating that the first model estimates higher values for erosive wear;
- the second Hutchings model shows a linear increasing variation of the erosive wear rate with the value of the ratio of removed material volume that depends on the particle volume (coefficient $\boldsymbol{\Gamma}$);
- Bitter model shows that the intensity of erosive wear increases linearly with the angle of incidence to its value when the speed is zero ($\alpha = \alpha_0$), then the variation of the intensity of erosion decreases exponentially;
- the impact speed also produces an exponential increase of the wear intensity up to $\alpha > \alpha_0$ and then it decreases exponentially for $\alpha < \alpha_0$;

## 4. Analysis and validation of the mathematical models

The calculated values of the wear rate, for the angle of incidence $\alpha = 20°$, for the speed $v = 4.71$ [m/s], are shown in table 3, for Finnie, Bitter and Hutchings model, compared to the values experimentally determined for the three analyzed stators.

**Table 3.** Wear rate calculated for the working conditions.

| Stator | The calculation model of the rate of wear | Rate of wear | |
|---|---|---|---|
| | | Calculated value | Experimentally determined value [9] |
| I | Finnie I | $6.454 \cdot 10^{-3}$ | $5.356 \cdot 10^{-3}$ |
| | Finnie II | $4.841 \cdot 10^{-3}$ | |
| | Bitter | $5.533 \cdot 10^{-10}$ | |
| | Hutchings (microcutting) | $4.928 \cdot 10^{-3}$ | |
| | Hurchings (plastic deformation) | $1.887 \cdot 10^{-1}$ | |
| II | Finnie I | $6.885 \cdot 10^{-3}$ | $5.783 \cdot 10^{-3}$ |
| | Finnie II | $5.163 \cdot 10^{-3}$ | |
| | Bitter | $5.523 \cdot 10^{-10}$ | |
| | Hutchings (microcutting) | $5.257 \cdot 10^{-3}$ | |
| | Hurchings (plastic deformation) | $2.079 \cdot 10^{-1}$ | |
| III | Finnie I | $7.376 \cdot 10^{-3}$ | $6.175 \cdot 10^{-3}$ |
| | Finnie II | $5.532 \cdot 10^{-3}$ | |
| | Bitter | $5.577 \cdot 10^{-10}$ | |
| | Hutchings (microcutting) | $5.633 \cdot 10^{-3}$ | |
| | Hurchings (plastic deformation) | $2.306 \cdot 10^{-1}$ | |

The comparative analysis of calculated rate of wear with the one experimentally determined, shows the following:

- first Finnie model highlights the biggest difference between the calculated rate and experimentally determined rate of wear. The equation can be applied only in the rare cases in practice, where the content of the erosive solids particles reaches 60%;
- the results achived with the second Finnie model and the Hutchings model (microcutting) are the closest to the experimentally determined values of wear rate. It is therefore reccomanded their use in the evaluation of erosive process of progression cavitation pumps. Hutchings model

(microcutting) is suitable for the cases where the microcutting is predominant, and the second Finnie model is suitable where the microcutting is accompanied by plastic deformation caused by the impact of abrasive particle on the target material;

- the values obtained for the erosive wear rate with Bitter model are practically invalide, so it can't be applyed in the present case;
- the results obtained with Hutchings model (plastic deformation) can not be used because the plastic deformation of stator materials (the base assumption for the Hutchings equations) is not specific to the dynamic of the erosive process for progressive cavitation pump.

## 5. Conclusions

The presented results of the erosion wear analysis using Finnie, Bitter and Hutchings models are useful for the followings:

- quantitative estimation of erosive wear;
- highlighting the erosion evolution;
- quantify the direct influence and in interaction of the most significant parameters of erosion.

Analysis of factors influence on abrasive erosion provides data reguarding: the share of their influence, their generalization degree and the limits of applicability.

The obtained results provides solutions necessary for the design and for constructive optimization of the progressive cavitation pump or for other mechanical systems that are functioning in a similar condition of erosive wear.

## References

[1] Hutchings I M 1981 A model for the erosion of metals by spherical particles at normal incidence *Wear* **70** 269-281
[2] Bingley M S and O'Flynn D J 2005 Examination and comparison of various erosive wear models *Wear* **258** 511-525
[3] Tudor A 2002 *Frecarea şi uzarea materialelor* (Bucharest: Editura Bren)
[4] Johnson G R and Cook W H 1985 *Eng. Fract. Mech.* **21** 31-48
[5] Johnson G R and Holmquist T J 1999 Response of boron carbide subjected to large strains, high train rates and high pressures *J. Appl. Phys.* **85** 8060-8073
[6] Wang, Yu-Fei, Yang, Zhen-Guo 2008 Finite element model of erosion wear on ductile and brittle materials *Wear* **165** 871-878
[7] Bitter J G A 1963 A study of erosion phenomena: Part I and Part II *Wear* **6** 5-21, 169-190
[8] Tabor D 1951 *The hardness of metals* (Oxford: Clarendon Press)
[9] Popescu M 2016 *Contribuţii la realizarea constructivă şi tehnologică a pompelor cu şurub pentru industria petrolieră* PhD Thesis, Petroleum-Gas University of Ploieşti

IOP Conf. Series: Materials Science and Engineering **174** (2017) 012023 doi:10.1088/1757-899X/174/1/012023

# A study regarding friction behaviour of lysine and isoleucine modified epoxy matrix

**I Bălan[1], R Bosoancă[1,2], A Căpăţină[1], I Graur[1,2], V Bria[1] and C Ungureanu[1]**

[1]"Dunărea de Jos" University of Galati, Domneasca 47, 800008, Galati, Romania
[2] Diagnose and Measurement Group, Rosiori 41, 800055, Galati, Romania

E-mail: claudia.popa@ugal.ro

**Abstract.** The aim of this study is to point out the effect of L-lysine and L-isoleucine used as modifying agents for epoxy resins. The amino acids are largely used to turn the usual polymers in bio-compatible materials but they effect also other significant proprieties of formed materials. The general study developed in Polymer Composite Laboratory is focused on analysis of 14 amino acids used as modifying agents but the two above mentioned showed a special behaviour namely they re-crystalized during the polymerization of the matrix. The coefficient of friction was obtained through the calculation of friction torque measured with a loaded cell sensor. As far as we know, there is no report on the friction proprieties of amino acids modified epoxy resins.

## 1. Introduction

Epoxy resins have long been receiving a lot of scientific and technical interests and are a family of thermosetting materials widely used as adhesives, coatings and matrices in polymer composites because of the low viscosity of the formulations [1]. Epoxy resins possess excellent thermal stability, moisture resistance, chemical stability, superior electrical and mechanical properties [2]. Diglycidylether of bisphenol A (known as DGEBA) is the most common epoxy monomer. These resins are synthesized via the addition of epichlorohydrine and bisphenol A so oligomers with a relatively narrow distribution of polymerization degrees are obtained [3].

Because of the high adhesive strength and low cost, epoxy-type resins have been primary matrix resins of the adhesives, and various types of curing agents, such as nitrogen - (amines and polyamides), oxygen- (anhydrides), and sulphur – (mercaptans) containing agents, have been used for crosslinked adhesives. However, these systems do have some environmental problems that have been especially noted in recent years.

One problem is that the systems usually generate dense smoke and toxic decomposition products during combustion. Another environmental problem of epoxy resins is that all the curing agents are toxic before curing. The curing agents, such as aromatic and aliphatic amines, which are most commonly employed for epoxy curing, are known to be toxic, and to reduce the toxicity, adduct types of amines are sometimes employed. Recently was demonstrated the feasibility of using an amino acid as a novel ecofriendly curing agent. Therefore, the development of environmentally friendly epoxy systems is a great importance for designing green and biocompatible materials in many applications. L-lysine is one of the essential amino acids used in a wide range of applications. In this study an amino acid was used for the preparation of polymeric materials.

The more important epoxy hardeners are the amines and their derivatives, because are less reactive but with higher mechanical properties than those of aliphatic amines. However, these aromatic amines can be also toxic and there is a need for biobased and nonharmful amine hardeners. Unfortunately, there are very few biobased amines; some have been prepared such as isosorbide diamine but through complex chemistries. More simple is the hydrolysis and decarboxylation of amino acids which can lead to diamines, mostly aliphatic diamines. One example is lysine, which can be an interesting platform chemical and which can be converted into a number of industrial monomers, including 1,5-diaminopentan. Li at al., [4] reported that lysine was used as a novel curing agent for a cycloaliphatic-type epoxy. Also, the reaction between L-lysine amino acid and bisphenol A epoxies was not been widely investigated yet. Yong Lv et al. [5] studied the curing kinetics of a diglycidyl ether of bisphenol A with a methanol etherified amino resin [2].

There are many studies concerning the problem of turning polymers into bio-compatible materials and many attempts were developed by using amino acids as agents to ensure the bio-compatibility. The selection of amino acids is due to their availability with high optical purity and lower cost [6, 7, 8].

The fundamental understanding of synergy in tribological performance [9] among various functional fillers is essential for successful applications [10]. The tribological behaviour of epoxy matrix can be significantly improved by addition of suitable filler materials as lysine and isoleucine. Besides the type, the shape and size of the materials added also influence the tribological properties [11].

The aim of this study is to point out the effect of L-lysine and L-isoleucine used as modifying agents for epoxy resins and to explore their role in the friction of the composite. We studied the influence of test speed and applied force values on the friction behavior of three epoxy systems with certain ratio of L-lysine and L-isoleucine as modifying agents. To highlight the effect of these aminoacids in epoxy matrix it is mandatory to perform some mechanical tests as three-point bending and compression to establish where such modified polymers might be used as.

## 2. Experimental procedure

### *2.1. Materials*

Three epoxy systems were chosen mostly because of their different bisphenol A content namely Epiphen RE4020-DE 4020 78% bisphenol A diglycidyl ether (Bostik), Epoxy Resin C 75% bisphenol A diglycidyl ether (R&G Gmbh Waldenbuch), and Epoxy Resin HT-2 50% bisphenol A diglycidyl ether (R&G Gmbh Waldenbuch). All these epoxy systems are bicomponent, namely consist of a resin and a hardener mixed in exact proportions given by producer. All the systems are slow resins with a gel time between 20 and 45 minutes time which allows the manoeuvres required by moulding.

All these systems were modified by mixing the main component (the resin) with various amounts of amino acids such as finally their weight ratios into the formed materials to be 1%, 3%, and 5%. L-lysine and L-isoleucine obtained from Sigma-Aldrich were the chosen filler.

### *2.2. Fabrication of the samples*

L-isoleucine was used as received. L-lysine is not miscible with epoxy resin, it cannot be incorporated into epoxy as such. For this reason, L-Lysine was dispersed in methanol (1%, 3% and 5% of epoxy resin). Thereafter, the epoxy resin was added. Methanol was evaporated at 50°C on a magnetic heating stirrer and curing agent for each resin was added. For each pair of substances and for each concentration first of all the mixtures were homogenised by mechanical stirring at 300 rotation/min for 15 minutes. After the mixtures were uniformed the right amount of hardener was stirred for ten minutes to ensure the homogeneity. The pre-polymer mixtures were moulded in cylindrical moulds 8 mm diameter and 200 mm height. After curing, three thermal treatments were applied: 8 hours at 60°C, 2 hours at 80°C and 1 hour at 90°C.

The codifications for the materials used in this study are shown in table 1.

**Table 1.** Materials codification.

| Epoxy resins | | |
|---|---|---|
| C | Epiphen (E) | HT2 |
| C1i-1% L-isoleucine | E1i-1% L-isoleucine in E | HT1i-1% L-isoleucine in HT |
| C3i- 3% L-isoleucine in C | E3i-3% L-isoleucine in E | HT3i-3% L-isoleucine in HT |
| C5i-5% L-isoleucine in C | E5i- 5% L-isoleucine in E | HT5i- 5% L-isoleucine in HT |
| C1l-1% L-lysine in C | E1l-1% L-lysine in E | HT1l-1% L-lysine in HT |
| C3l-3% L-lysine in C | E3l- 3% L-lysine in E | HT3l- 3% L-lysine in HT |
| C5l-5% L-lysine in C | E5l- 5% L-lysine in E | HT5l- 5% L-lysine in HT . |

*2.3 Details on the friction tests*

Coefficient of friction of lysine and isoleucine modified epoxy matrix was measured using pin-on-disc tribometer TRM 1000 (Wazau®, Germany) in the Polymer Composites Laboratory of *Dunărea de Jos* University of Galați.

All friction tests were carried out in dry condition for pin on disc geometry, the pin having a flat contact surface (figure 1).

**Figure 1**. Testing device WAZAU Tribometer TRM 1000 pin-on-disc module.

Cylindrical samples (Ø 8×2.5 mm) were manufactured from the produced materials. The counterbody consisted of a steel disc with 100 mm diameter). All tests were performed in laboratory environment (20 ± 2°C).

The friction tests were carried out under three different regimes, such as:

R1 – 1 m/s sliding speed and 10 N normal load,

R2 - 1.5 m/s sliding speed and 15 N normal load,

R3 – 2 m/s sliding speed and 20 N normal load.

The sliding distance was constant for all studied regimes, namely 1000 m.

For each material three tests were carried out for each regime and the presented results are average values of the three tests.

## 3. Results and discussion

The evolutions of the coefficient of friction during sliding for L-isoleucine and L-lysine in C, E and H epoxy resins against steel disk are shown in Figures 2, 3 and 4. In each representation three curves are shown, which correspond to the three different regimes: R1, R2 and R3.

It is well known that during sliding there are two regions: the running-in region regime (where surfaces adapt to each other and coefficient of friction changes significantly during the first period of sliding, increasing with the sliding distance) and the steady-state region (coefficient of friction remains approximately constant during the whole test).

A typical behaviour of the coefficient of friction with the sliding distance for a sliding test can be observed for all the studied cases. After a very short period (approximately 20 m sliding distance) corresponding to the running-in the coefficient of friction enters in steady-state conditions and remains approximately constant during the whole test [12, 13].

Figure 2 presents the evolution of coefficient of friction for C epoxy system with L-isoleucine and L-lysine for all concentrations with applied force and sliding speeds, respectively. For C1i material, at 1 and 1.5 m/s sliding speed, the coefficient of friction decreases compared to 2 m/s sliding speed. The best graphical representation is for C1l, C3i and C5i for all applied forces. Concerning the C3l material the best coefficient of friction value is for 10N applied force.

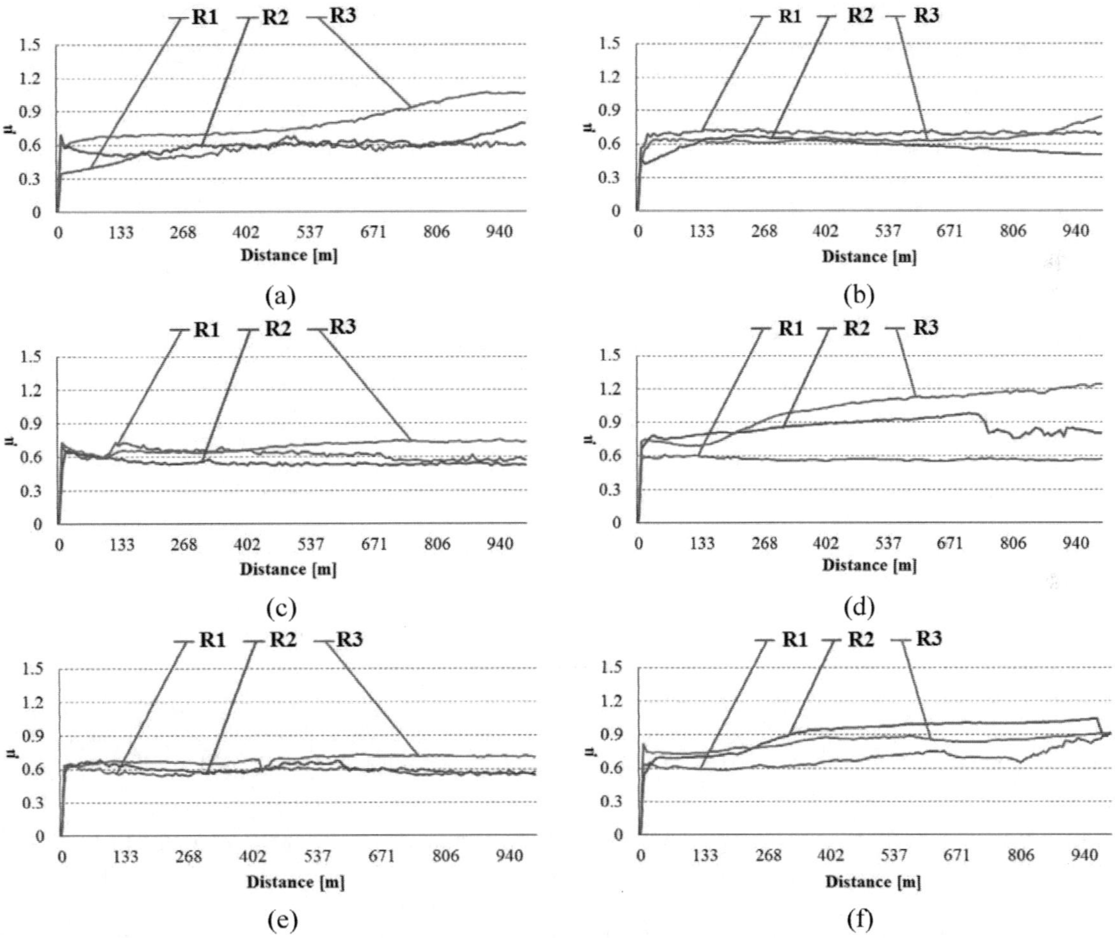


**Figure 2**. Evolution of the coefficient of friction with the sliding distance for L-isoleucine (a-C1i, c-C3i, e-C5i) and L-lysine (b-C1l, d-C3l, f-C5l) in C epoxy resin for all concentrations.

From figure 2 can be oserved that during the running-in period the values recorded for the three studied regimes are different. The lowest coefficient of friction values for the running-in period are observe for regime R1. The the coefficient of friction value increased with the increasing the normal load. The same behaviour was reported by Cui ş.a. [14] in their study about epoxy composites reinforced by carbon nanotubes for higher normal load 40-120 N. The increase of the coefficient of friction with the increasing normal load was explined based on the adhesion phenomenon (due to hugher loads the material is subjected to plastic deformation).

Some other authors obseved a decrease of the cefficient of friction with the increase of the applied normal load [15].

Figure 3 presents the variation of coefficient of friction for E epoxy system with L-isoleucine and L-lysine for all concentrations with applied force and sliding speeds, respectively. In the case of E epoxy system it is clear that the sliding speed influence is rather low excepting E3l material where 1.5 m/s sliding speed leads to a reduction of coefficient of friction value.

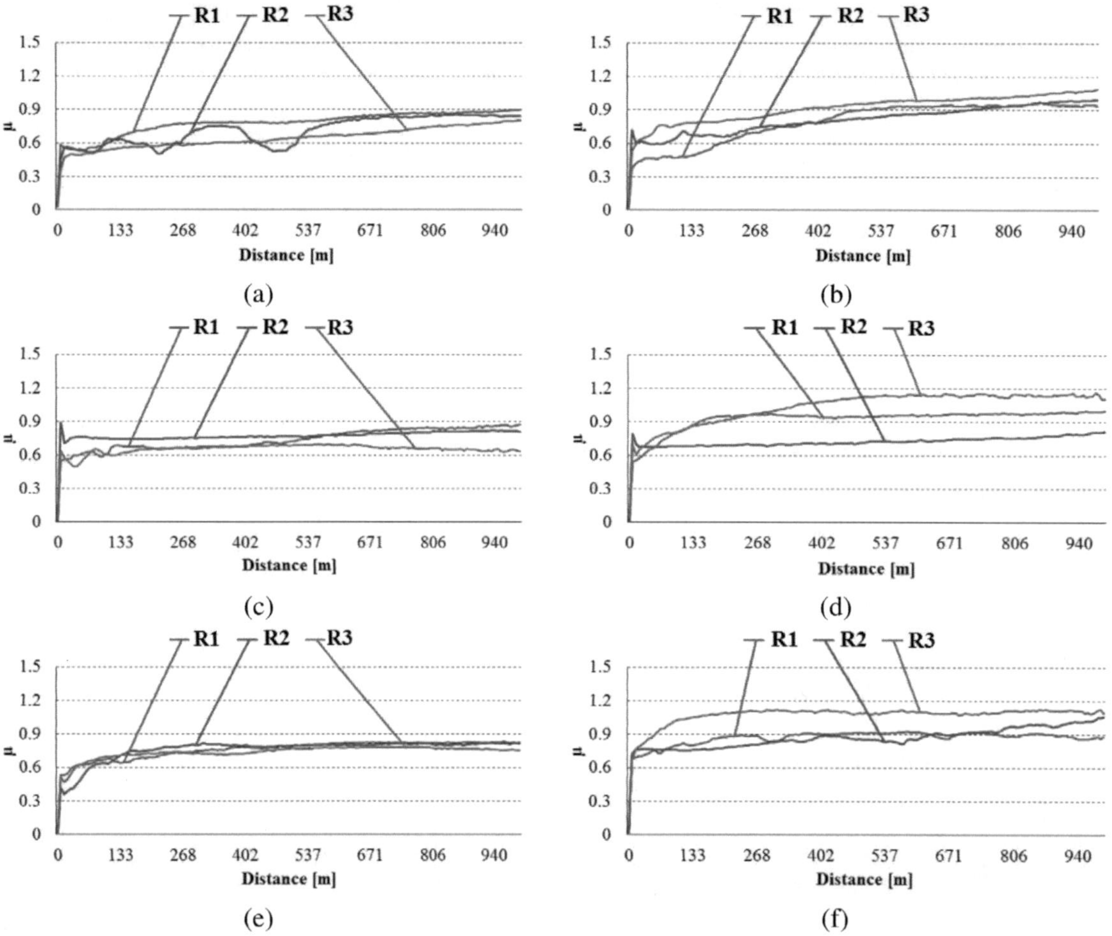


**Figure 3.** Evolution of the coefficient of friction with the sliding distance for L-isoleucine (a-E1i, c-E3i, e-E5i) and L-lysine (b-E1l, d-E3l, f-E5l) in E epoxy resin for all concentrations.

Figure 4 presents the variation of coefficient of friction for HT epoxy system with L-isoleucine and L-lysine for all concentrations with applied force and sliding speeds, respectively. In figure 4, for all composite tested and within the applied force, it is shown that the values of coefficient of friction are very much influenced by applied force.

It can be seen from figure 4 that the coefficient of friction has higher oscillation HT epoxy system as compared to the other two types of resides studied (figures 2 and 3). It can be oseved that these oscilations are slighly more pronounced in the case of R1 and R2 regimes.

It is well known that the sliding speed is another parameter that has a great influence on the evolution of the coefficient of friction. It can be seen that in almost of the studied cases the coefficient of friction is increasing with the increasing sliding speed. A higher sliding speed is leading to softening of the epoxy resin (due to the physical and/or chemical interactions) in the contact zone and these as a consequence some particles may be detaced fom the material (abrasive wear).

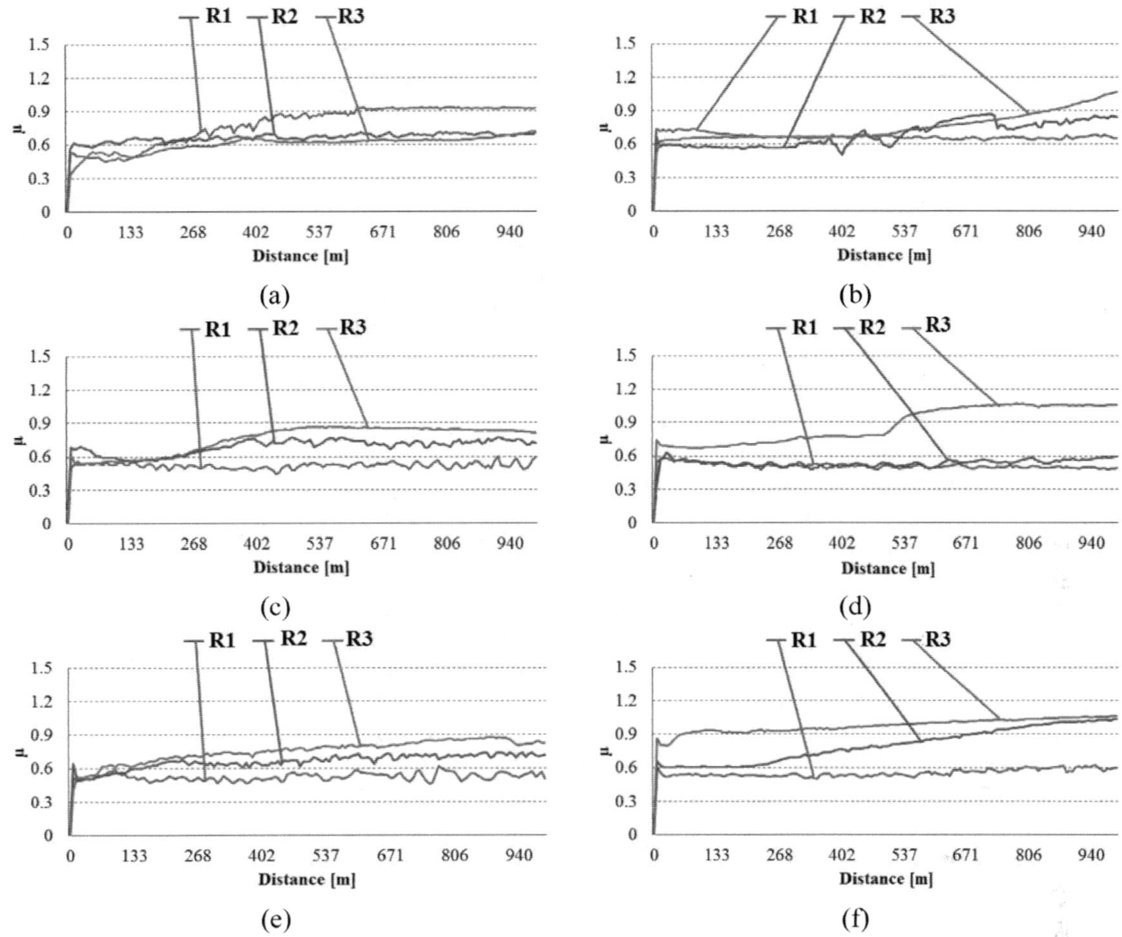


**Figure 4**. Evolution of the coefficient of friction with the sliding distance for L-isoleucine (a-HT1i, c-HT3i, e-HT5i) and L-lysine (b-HT1l, d-HT3l, f-HT5l) in HT epoxy resin for all concentrations.

## Conclusions

From the results of this research, the following conclusions can be made:

1. Coefficient of friction is very much affected by sliding distance and applied force.
2. In the range of applied force studied in this research, the coefficient of friction values for C1l and C3l materials at 10N applied force remains constant. The same remark for C5i at all applied forces.
3. The sliding speed has stronger affect the E1l and E3l materials.
4. HT3l material proved worst in terms of coefficient of friction values for 20N applied force.
5. Finally it was concluded that it is mandatory to perform some mechanical tests for a better understanding of the effect of these aminoacids in epoxy matrix.

## Acknowledgments

The authors would like to acknowledge the financial contribution of the Project 12 P01 024 21 (C11) /31.08.2012 (code SMIS 50414).

## References

[1] González M G, Baselga J and Cabanelas J C 2012 Applications of FTIR on epoxy resins-identification, monitoring the curing process, phase separation and water uptake Ed. INTECH Open Access Publisher pp 272–275

[2] Zhao S F et al. 2014 Curing kinetics, mechanism and chemorheological behavior of methanol etherified amino/novolac epoxy systems *Express Polym Lett* **8 (2)** 95–102

[3] May C A 1988 *Epoxy resins: Chemistry and Technology* Ed. May C A (New York: CRC Press) pp 624–629

[4] Li Y, Fei X, Kyoung S M and Wong C P 2006 Novel curing agent for lead-free electronics Amino acid Journal of Polymer Science Part A *Polym Chem* **44 (2)** 1020–1027

[5] Lv Y, Long Z, Luo S Y and Dai L 2012 Kinetics of the Curing Reaction of a Diglycidyl Ether of Bisphenol with a Methanol Etherified Amino Resin In Green Printing and Packaging Materials *Adv Mater Res* **380** 60–63

[6] Mallakpour S and Asadi P 2010 Novel chiral poly (ester-imide)s with different natural amino acids in the main chain as well as in the side chain: synthesis and characterization *Colloid Polym Sci* **288 (12)** 1341–1349

[7] Mallakpour S, Soltanian S and Sabzalian M R 2011 Studies on synthesis and in vitro biodegradability of novel optically active nanostructure poly(ester-imide)s containing l-phenylalanine and l-isoleucine linkages *Colloid Polym Sci* **289 (1)** 93–100

[8] Mallakpour S and Seyedjamali H 2009 Fast synthesis of optically active polyamides containing l-methionine linkages in ionic liquid via a microwave-assisted process *Colloid Polym Sci* **287 (9)** 1111–1116

[9] Brostow W, Kovacevic V, Vrsaliko D and Whitworth J 2010 Tribology of polymers and polymer-based composites *J Mater Educ* **32 (5-6)** 273–290

[10] Chang L, Zhong Z, Lin Y and Klaus F 2007 Tribological properties of epoxy nanocomposites III Characteristics of transfer films *Wear* **262 (5-6)** 699–706

[11] Tanaka K 1977 Friction and Wear of Glass and Carbon Fiber-Filled Thermoplastic Polymers *J. of Lubrication Tech* **99 (4)** 408–414

[12] Unal H, Sen U and Mimaroglu A 2004 Dry sliding wear characteristics of some industrial polymers against steel counterface *Tribol Int* **37 (9)** 727–732

[13] Wang Y Q, Li J 1999 Sliding wear behaviour and mechanism of ultra-high molecular weight polyethylene *Mater Sci Eng* **266** 155–160

[14] Li-Jun Cui, Hong-Zhang G, Wen-Yi W, Li-Ting C and Jing G 2012 Functionalization of multi-wall carbon nanotubes to reduce the coefficient of the friction and improve the wear resistance of multi-wall carbon nanotube/epoxy composites *Carbon* **54 (2)** 77–282

[15] Mudradi S, Ravikantha P, Kandavalli and Thirumaleshwara B 2014 Effect of Filler Content on the Performance of Epoxy/PTW Composites *Adv Mater Sci Eng* **6 (4)** 975 – 987

IOP Conf. Series: Materials Science and Engineering **174** (2017) 012024 doi:10.1088/1757-899X/174/1/012024

# Comparative study regarding friction coefficient for three epoxy resins

**G Mihu[1], I Mihalache[1,2], I Graur[1,2], C Ungureanu[1]and V Bria[1]**

[1]"Dunărea de Jos" University of Galati, 47 Domneasca Str., Galati, Romania
[2]Diagnose and Measurement Group, 41 Rosiori Str., 800055, Galati, Romania

E-mail: vasile.bria@ugal.ro

**Abstract**. Three commercial epoxy diglycidylether of bisphenol-A (DGEBA) were used in this study namely Epiphen RE4020-DE 4020 (Bostik), Epoxy Resin C (R&G Gmbh Waldenbuch), and Epoxy Resin HT-2 (R&G Gmbh Waldenbuch). Epoxy resins are often used for the friction purpose but their friction resistance is quite low and it is thus necessary to enhance their friction resistance. In this paper it is shown how load, sliding velocity, and distance affect friction coefficient of epoxy resins.

## 1. Introduction

The name epoxy resins has over the years become synonymous with performance; epoxy resins have established themselves as unique building blocks for high-performance coatings, adhesives, and reinforced plastics. They are a class of thermoset materials used extensively in structural and specialty composite applications because they offer a unique combination of properties that are unattainable with other thermoset resins [1], [2].

Some of their most interesting applications are found in the aerospace and recreational industries [3]. As substitutes for metal in the construction of mechanical apparatus, engineering plastics offer advantages such as adhesives, coatings, encapsulates, and molding compounds, and economy in fabrication and decorating. Replacement of metals by plastics is favored as the physical properties and operating temperature ranges of plastics improve and the cost of metals and their fabrication increases [4]. Many polymers and polymer based composites are widely used for sliding couples against metals, polymers and other materials [5].

Formaldehyde-based systems are based on reacting formaldehyde with active hydrogen containing compounds such as phenol, urea, or melamine. This family of thermosets has low flammability, high rigidity, good dimensional stabilit, and low cost. Formaldehyde-based systems are heat-activated and are processed using compression or injection molding.

The most common resin is based on bisphenol A and is referred to as the "diglycycidyl ether of bisphenol A" or DGEBA [4]. The materials must provide unique mechanical and tribological properties combined with a low specific weight and a high resistance to degradation in order to ensure safety and economic efficiency [6].

There have been numerous investigations exploring the influence of test conditions, contact geometry and environment on the friction and wear behaviour of polymers [7], [8], [9], [10]. The main advantage of polymers, from a tribological point of view, is their reasonably low rate of wear, especially corrosion type wear [11]. In this investigation, the influence of test speed and applied force values on the friction coefficient behaviour of three unfilled epoxy resins was studied. Friction tests of epoxy resins systems were carried out at dry condition on a pin-on-disc geometry.

## 2. Experimental section

The experiments were carried out with three unfilled epoxy resins: Epiphen RE4020-DE 4020 (Bostik), Epoxy Resin C (R&G Gmbh Waldenbuch), and Epoxy Resin HT-2 (R&G Gmbh Waldenbuch). Friction coefficient studies of epoxy systems against steel disc are usually carried out at different load, sliding velocity, and distance conditions.

Tribological tests were done at room temperature for 5, 10, 15, 20, 25, 30 and 35 N force, and 0.5, 1, 1.5, 2, 2.5, 3 and 3.5 m/s speed for 1000 m distance using a TRM 1000 tribometer from Wazau pin-on-disk friction and the pin has a flat contact surface (figure 1).

The cylindrical pin specimens of 8 mm diameter and 20 mm length were tested against steel discs. Although using 0.5 m/s sliding speed lead to test duration of 34 min, it was insured that the test was carried within the steady-state region.

The calculations of friction coefficient were carried out using Tribo Control Wazau software. The weight loss was measured taken after 1000 m of sliding distance.

**Figure 1**. Tribometer TRM 1000 from Wazau.

## 3. Results and discussion

In general, epoxy is not an ideal material used in sliding wear applications due to its three-dimensional network structure [12] as compared with thermoplastics. As can be seen from the figures below, frictional coefficient is quite high. For all polymer materials, the friction coefficient starts with a running-in period followed by steady-state period [13]. Figure 2 shows the friction coefficient for C epoxy resin system as a function of applied loads. As can be observed friction coefficient increased together with applied load. The lowest value of the friction coefficient is for 5N. Figure 3 presents friction coefficient values for E epoxy resin system.

The higher the sliding speed the higher is the rise in temperature [14], and as the temperature reaches the softening point of the polymer, the friction coefficient increases. This situation can be however, improved by the addition of some fillers [15].

Figure 4 presents the variation of the friction coefficient with sliding distance for HT-2 epoxy resin system. The lowest average friction coefficient of $\mu = 0.2$ was obtained for HT-2 epoxy resin system at applied force of 10N. The surface changes were analyzed by using a Olympus Microscope but there we could not identify major changes.

## 4. Conclusions

Performed studies were carried out in order to analyze the influence of test speed and applied force values on the friction coefficient behaviour of three unfilled epoxy resins. For all polymers of this study the sliding distance was the same. The friction coefficient of all studied epoxy resins increases with the applied force.

IOP Conf. Series: Materials Science and Engineering **174** (2017) 012024 doi:10.1088/1757-899X/174/1/012024

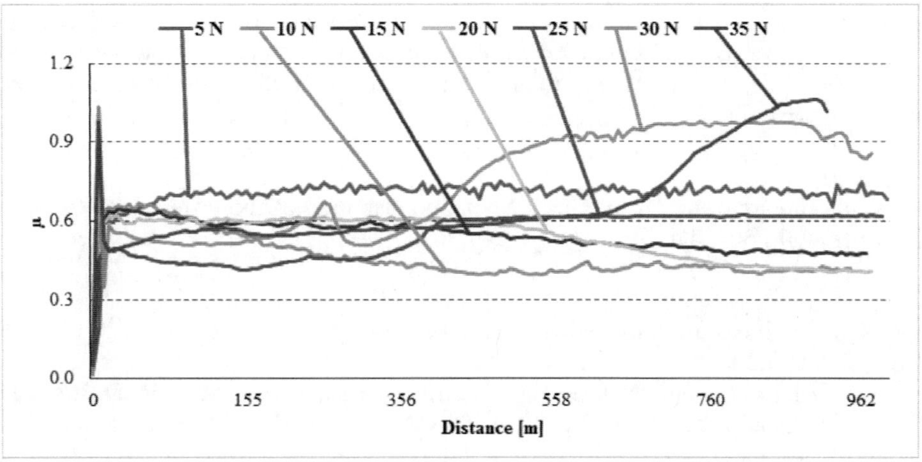


**Figure 2**. Friction coefficient for C resin system.

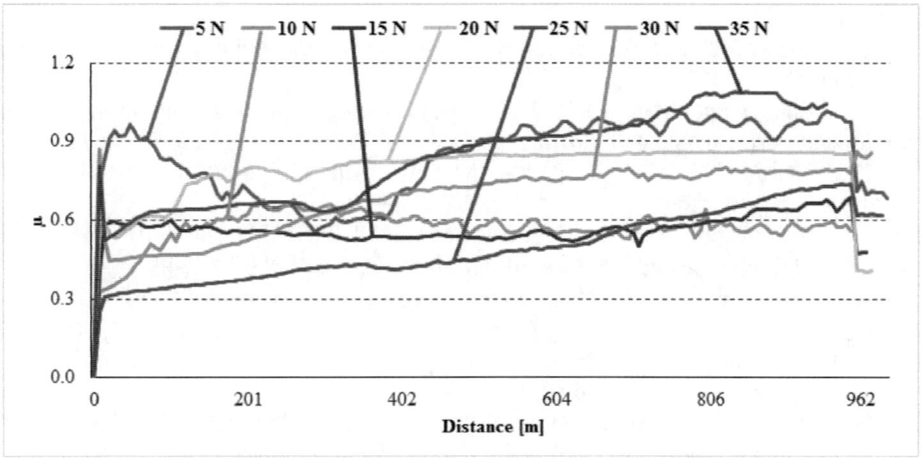


**Figure 3**. Friction coefficient for E resin system.

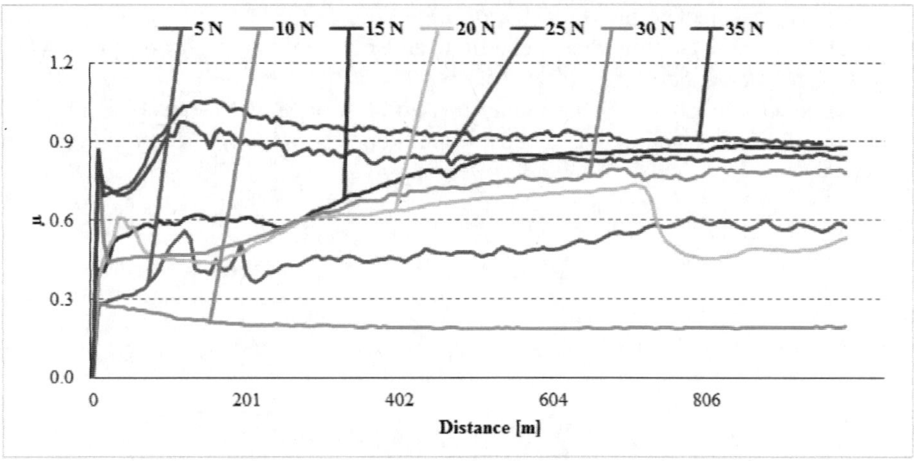


**Figure 4**. Friction coefficient for H resin system.

The sliding speed should be sufficient low in order to restrict the friction coefficient rise of the polymers used. It is observed that after 1000 m distance, μ for all materials tested varied in the range $0.2 < \mu < 1$. To conclude, epoxy resins polymers have a lower friction coefficient for every sample tested at applied force of 5N.

**Acknowledgments**
The authors would like to acknowledge the financial contribution of the Project 12 P01 024 21 (C11) /31.08.2012 (code SMIS 50414).

**References**

[1] May C A 1988 Epoxy resins Chemistry and Technology Second Edition Ed. CRC Press New York pp 624–629
[2] Boyle A M, Martin C J and Neuner J D 2001 Epoxy Resins. Eds. Miracle D.B. and Donaldson S.L. *ASM Handbook volume **21** Composites* (Materials Park: ASM International) pp 78-89
[3] Pascault J P, Williams R J J 2010 *Epoxy Polymers. New Materials and Inovations* (Hoboken: John Wiley & Sons) pp 79-108
[4] Clara D, Craver E, Charles E and Carraher Jr *Applied Polymer Science 21st Century* Ed. Elsevier pp 709-913
[5] Hutchings IM 1992 *Tribology friction and wear of engineering materials* London Edward Arnold pp 51
[6] Bernd W, Frank H and Ming Q Z 2003 Epoxy nanocomposites with high mechanical and tribological performance *Composites Science and Technology* **63** 2055–2067
[7] Hooke C J, Kukureka S N, Liao P, Rao M and Chen Y K 1996 The friction and wear of polymers in non-conformal contacts *Wear* **200** 83 –94
[8] Ludema K C and Tabor D 1996 Friction Properties of Poly(vinyl alcohol) Hydrogel Effects of Degree of Polymerization and Saponification Value *Wear* **9** 329 –348
[9] Brentnall A B and Lancaster J K *1989 Proceedings of Wear Material Congress ASME* pp 596–603
[10] Zafer D 2013 Tribological performance of polymer composites used in electrical engineering applications *Bull. Mater. Sci.* **36** pp 341–344
[11] Bekhet N E 1999 Tribological behaviour of drawn polypropylene *Wear* **236** 55–61
[12] Shi G, Zhang M Q, Ronga M Z, Wetzel B and Friedrich K 2003 Friction and wear of low nanometer Si3N4 filled epoxy composites *Wear* **254** 784–796
[13] Unal H, Sen U and Mimaroglu A 2004 Dry sliding wear characteristics of some industrial polymers against steel counterface *Tribol. Int.* **37** 727–732
[14] Wang Y Q, Li J 1999 Sliding wear behaviour and mechanism of ultra-high molecular weight polyethylene *Mater. Sci. Eng.* A **266** 155–60
[15] Graur I, Bosoanca I, Bosoanca R, Cîrciumaru A and Bodor M 2015 Effect of ultra-sonication on some properties of ionic substances doped epoxy *SICOMP Conference* (Gothenburg, Sweden)

# A study of friction wear behaviour of nano-ferite modified epoxy resins

**C Munteniță[1], C Eni[1], I Graur[1,2], C Ungureanu[1] and M Bodor[1,2]**

[1]"Dunărea de Jos" University of Galati, Domneasca 47, 800008, Galati, Romania
[2] Diagnose and Measurement Group, Rosiori 41, 800055, Galati, Romania

E-mail: cristian.muntenita@ugal.ro

**Abstract**. Ferrites are generally used to obtain soft magnets for domestic or industrial applications but their use in this direction had generated an increasing interest in modifying the polymer properties by dispersing in their volume certain amounts of these ceramic compounds. The current study focused friction wear behaviour induced by the presence of nano-sized strontium ferrite and barium ferrite in an epoxy matrix. The wear behaviour of nanoferite modified epoxy resins tests were conducted on a pin-on-disc geometry.Three types of epoxy resins had been used with 5% weight ratio of each type of ferrite and 10% weight ratio when both ferrites were used.

## 1. Introduction

In recent years one of the main classes of high performance engineering materials, next to metals and alloys, ceramics and polymeric materials are composite materials. These engineering materials are prepared from minimum two separate substances [1]. Composite materials, one of the most rapidly growing classes of materials, are being used increasingly for tribological applications [2]. However, because the polymer matrix must withstand high mechanical and tribological loads, it is usually reinforced with fillers. Nano-scale materials have attracted much attention recently due to many unusual predicted properties [3]. Magnetite and ferrite materials have been widely used as magnetic layers on polyethylene terephthalate (PET) film, paper or metallic substrate [4]. Ferrites have high dielectric constants which makes them very useful for microwave applications. They remain the best magnetic material and cannot be replaced by any other magnetic element because they are inexpensive, more stable and have a wide range of technological applications in transformer cores, high quality filters, radio wave circuits and operating devices [5]. Strontium ferrite ($SrFe_{12}O_{19}$), as an important member of hexaferrites, has higher saturation magnetization, coercivity and the Curie temperature because of high magnetocrystalline anisotropy compared to barium ferrite ($BaFe_{12}O_{19}$) [6], [7], [8].

Wear is defined [9] as the damage to a solid surface, generally involving progressive loss of material, due to relative motion between that surface and contacting substance or substances [10]. In this work, an attempt has been made to obtain the friction wear behaviour of nanoferite modified epoxy resins under various conditions such as external load and sliding distances of 1000 m. In a review of some of the literature concerning friction wear behaviour of nanoferite modified epoxy resins have not been found. Therefore, an evaluation of the tribological characteristics of nanoferite modified epoxy resins under various sliding conditions is needed.

## 2. Experimental procedure

Polymer matrix consisting of epoxy resin bisphenol A diglycidyl ether-based were used (code H representing resin epoxy - 50% bisphenol A diglycidyl ether, code C representing resin epoxy - 75% bisphenol A diglycidyl ether and code E (Epiphen DE 4020) representing resin epoxy - 78% bisphenol A diglycidyl ether). Strontium ferrite and barium ferrite obtained from Sigma-Aldrich producer and they have been used as additives. All these systems were modified by mixing the main component (the resin) with 5% weight ratio of each type of ferrite and 10% for the materials using both ferrites. Then the mixture of nano-ferrite modified epoxy resins was mechanically mixed at 200 rotations /min for 15 min, at room temperature. After the mixtures were uniformed the right amount of hardener was stirred for ten minutes to ensure the homogeneity. The pre-polymer mixtures were moulded in cylindrical moulds 8 mm diameter and 200 mm height. After 24 hours of polymerization, all the samples were extracted from moulds and the recommended thermal treatment was applied for 8 h at 60°C.

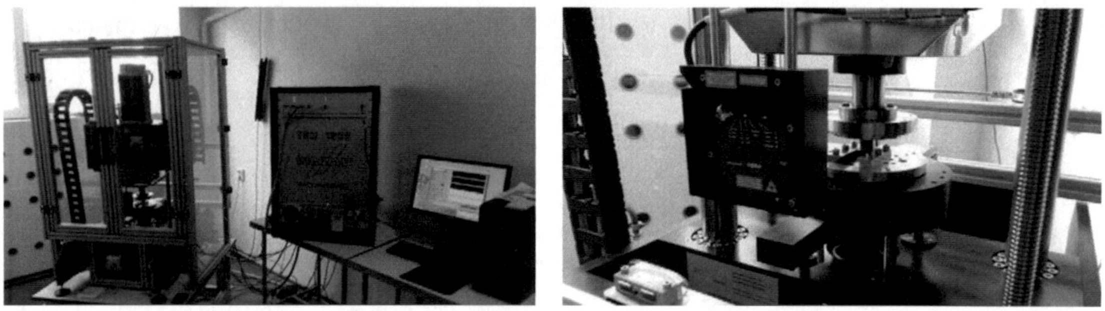

**Figure 1.** Testing device, tribometer TRM 1000 (Wazau®, Germany)

Tribological characterization was performed using TRM 1000 tribometer (Wazau®, Germany) on tribological test module dedicated to block-on-disk tests that allows setting and viewing of control and measured parameters: test rotational speed, loading force, friction and wear coefficients. The block is made of the material to be analyzed and is secured in the support of the machine, positioned perpendicularly to the disk (figure 1). Tribological tests for all types of composite were performed under dry conditions on a disk made of steel with outside diameter of 110 mm and 10 mm thickness. The length was set at 1000 meters and speed test between 0.66 [m/s] and 2 [m/s]. Testing regimes have been established as follows:

R1: speed testing of 0.66 [m / s], load of 15N, distance 1000 meters.
R2: speed testing of 1 [m / s], load of 10N, distance of 1000 meters.
R3: speed testing of 2 [m / s], load of 5N, distance of 1000 meters.

## 3. Results and discussion

The resins and the code designation for the test materials made from them are as follows: C1, E1, H1- epoxy resin with 5g barium ferrite; C2, E2, H2- epoxy resin with 5g strontium ferrite; C3, E3, H3- epoxy resin with 5g strontium and 5g barium ferrite.

Figure 2 shows friction coefficient of all samples containing C epoxy resin and the three mentioned amounts of ferrites for all three applied forces during sliding distance. The results indicate that for the same concentration of barium ferrite the friction coefficient it remains nearly constant for any tested forces. At low concentrations of strontium ferrites materials, the behavior is completely atypical to those that contain barium ferrites. At initiation stage of the process achieved friction coefficient values are different for the three regimens studied. The lowest friction coefficient for the initiation stage of the process (running-in period) are observed in the case of the R1.This value increases with decreasing the applied force because the friction coefficient is inversely proportional to normal force applied. A very unstable behavior of friction is for C2 material who present some oscillations during the tribological test. They are more pronounced in the case of R1 and available until around 500 m of sliding distance. After this distance stable friction coefficient is available until the end of the test.

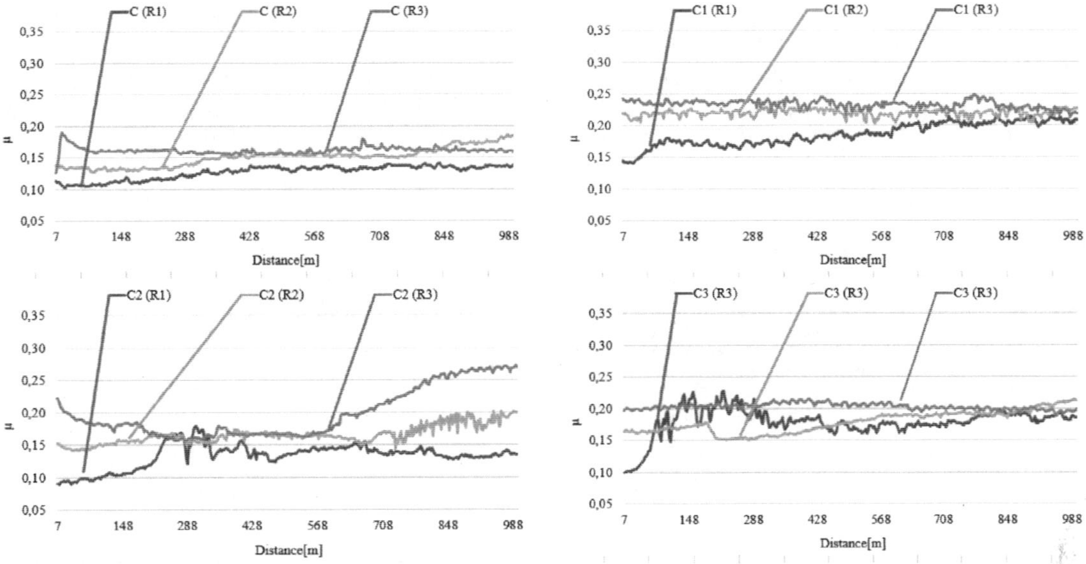


**Figure 2.** Variation with friction coefficient slide distance for C materials.

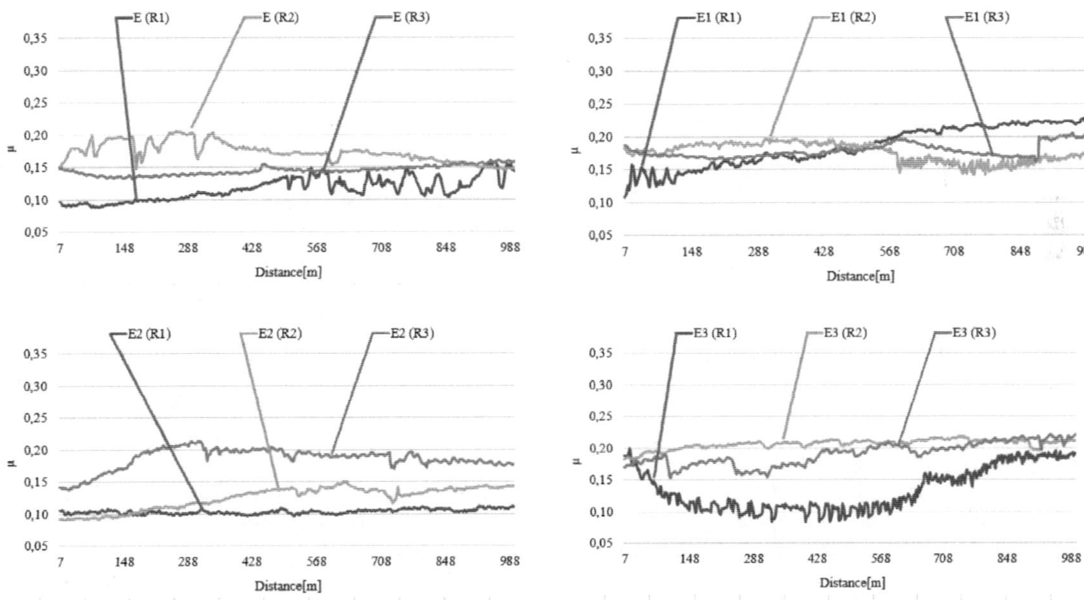


**Figure 3.** Variation with friction coefficient slide distance for E materials.

For E3 material which contain the two types of additives, can notice a significant decrease in the friction coefficient value in the first part of the test, compared to unmodified epoxy system. Small concentration of barium ferrite and strontium ferrite in E epoxy resin leads to significant increase of friction coefficient see (figure 3). The same as C type materials, the lowest values of friction coefficient for E type materials is for R1 working regime. The friction coefficient increases with decreasing the applied force. The highest value of the friction coefficient is registered for R2 regime. To some extent the evolution of friction coefficient over the entire sliding distance is relatively stable, except E3 material for R1 regime, where a noticeable decrease of friction coefficient over the entire distance is notice due to pulling phenomenon on the contact surface of an amount of material. Oscillations of friction coefficient appear at all materials but they are less obvious for materials containing small amounts of ferrite nanoparticles. These oscillations are caused by the fact that some additives show a higher hardness resulting in easy removal of nanoparticles of contact surface.

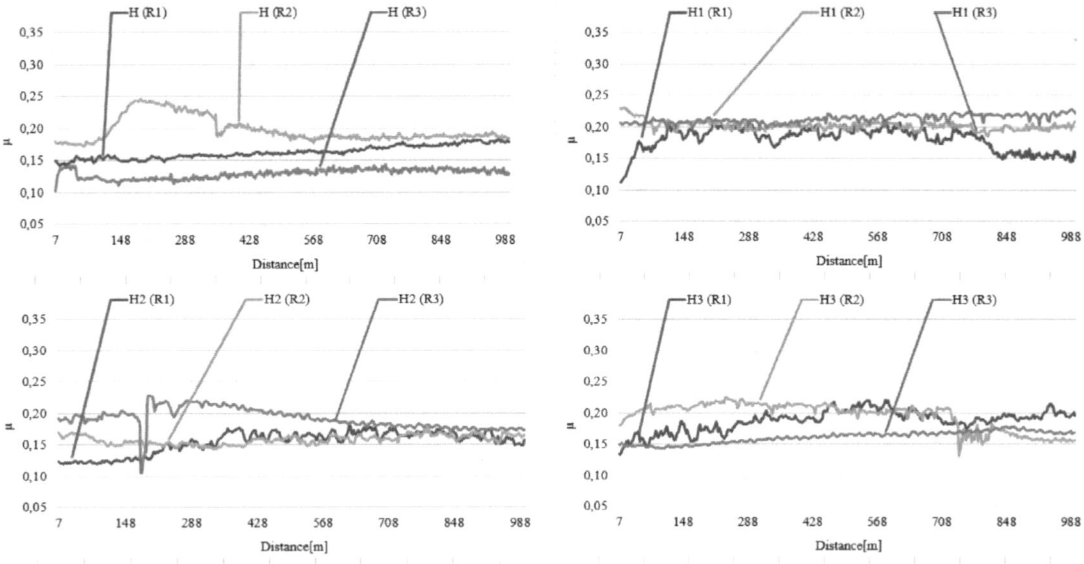


**Figure 4.** Variation with friction coefficient slide distance for H materials.

For H epoxy resin (figure 4) we can say that the friction coefficients are almost the same throughout the range 0.10 – 0.20. The modified materials behave identically, except H1 material friction coefficient values which remain almost constant throughout the test. The lowest values of the friction coefficient in the first phase are observed in R1 regime, after which they showed a slight upward trend, followed by stabilization until the end of the test.

Table 1 shows the degree of wear values calculated for each component. These values were determined using the following formula:

$$W_S = \frac{\Delta m}{F \cdot L}$$

where $\Delta m$ weight loss; F applied force; L sliding distance of 1000 m.

C-type materials testing regime had a significant influence on the wear behavior (table 1). The lowest rate of wear was obtained for C1 and C2 materials. The amount of additive has a significant influence on the tribological behavior. C3 material wear rate was similar in the case of R2 without additives and increased for R1 and R3. By adding an amount of 5% additive to improve the tribological properties of the C1 and C2 materials, while the amount of additive up to 10% resulted in degradation of the wear properties.

E-type materials testing regime had a significant influence on the wear behavior. The lowest rate of wear was obtained for E1 and E2 materials. The amount of additive has a significant influence on the wear behavior. By adding an amount of 5% additive, the tribological properties of E1 and E2 materials improves while the amount of additive up to 10% resulted in degradation of the wear properties of the E3 material.

For H-type materials the highest wear rate is for H3 material for R2 regime. For the others materials wear rate was very low, which makes the conclusion that regardless of the amount and type of used additive tribological properties are improved.

**Table 1.** Wear rate of epoxy matrix materials

| **Wear rate x $10^{-6}$ [g/N·m]** | | | | | |
|---|---|---|---|---|---|
| Materials | Values | Materials | Values | Materials | Values |
| C (R1) | 4.340 | E (R1) | 1.080 | HT (R1) | 4.200 |
| C (R2) | 1.510 | E (R2) | 0.450 | HT (R2) | 1.770 |
| C (R3) | 1.380 | E (R3) | 0.340 | HT (R3) | 1.080 |
| C1 (R1) | 0.580 | E1 (R1) | 6.800 | HT1 (R1) | 2.440 |
| C1 (R2) | 0.100 | E1 (R2) | 3.000 | HT1 (R2) | 1.300 |
| C1 (R3) | 0.140 | E1 (R3) | 2.000 | HT1 (R3) | 0.880 |
| C2 (R1) | 0,300 | E2 (R1) | 0.400 | HT2 (R1) | 5.000 |
| C2 (R2) | 0,270 | E2 (R2) | 0.200 | HT2 (R2) | 3.700 |
| C2 (R3) | 0,207 | E2 (R3) | 0.127 | HT2 (R3) | 1.270 |
| C3 (R1) | 2.520 | E3 (R1) | 5.000 | HT3 (R1) | 1.200 |
| C3 (R2) | 1.590 | E3 (R2) | 3.700 | HT3 (R2) | 0.800 |
| C3 (R3) | 0.867 | E3 (R3) | 1.270 | HT3 (R3) | 0.487 |

## 4. Conclusion

The key step of the present research and of materials formation was to choose the fillers that could be used to modify the resin basic properties. The decision regarding the use of the two above mentioned ferrites was imposed by their flexible way of use and the novelty of their usage to form epoxy composites. As a conclusion we can say that the best used filler is strontium ferrite, but this is only for C resin. This significantly influences the friction coefficient of C resin by the fact that it maintains it nearly constant for the whole distance for all tested forces.

## Acknowledgments

The authors would like to acknowledge the financial contribution of the Project 12 P01 024 21 (C11) /31.08.2012 (code SMIS 50414) and project no. POSDRU/159/1.5/S/132397, 2007-2013.

## References

[1] Stabik J Dybowska A and Chomiak M 2010 Polymer composites filled with powders as polymer graded materials *Journal of Achievements in Materials and Manufacturing Engineering* **43 (1)** 153-161
[2] Friedrich K, Lu Z, Hager A M 1995 Recent advances in polymer composites' tribology *Wear* **190 (2)** 139-144
[3] Geim A K and Novoselov K S 2007 The rise of grapheme *Nat. Mater.* **6 (3)** 183–191
[4] Tochihara S 1982 Magnetic coatings and their applications in Japan *Prog Org Coat* **10** 195–204
[5] Kulikowski J 1984 Soft magnetic ferrites - development or stagnation *J. Magn Magn Mat* **41 (1-3)** 56-62
[6] Liu J R, Honga R Y, Feng W G, Badami D and Wang Y Q 2014 Large-scale production of strontium ferrite by molten-salt-assisted coprecipitation *Powder Technol* **262** 142–149
[7] Xie T P, Xu L J, Liu C L, Ding S H and Wu W L 2013 Synthesis and adsorption properties of high specific surface area strontium ferrite from industrial strontium residue *Vacuum* **93** 71–78
[8] Pullar R C and Bhattacharya A K 2002 Crystallisation of hexagonal ferrites from a stoichiometric sol–gel precursor without formation of the α-BaFe2O4 intermediate phase *Mater. Lett.* **57** 537–542
[9] Mulchings I M 1993 Mechanisms of wear in powder technology *Powder Technol* **76** 3–13
[10] Unal H, Sen U and Mimaroglu A 2005 Abrasive wear behaviour of polymeric materials *Mater Des* **26** 705–710

IOP Conf. Series: Materials Science and Engineering **174** (2017) 012026 doi:10.1088/1757-899X/174/1/012026

# The determination of the thickness of the layers deposited on the electronic circuit boards through tribological methods

**A M Petrescu[1], A Tudor[1], G Chişiu[1], N A Stoica[1] and U Cihak Bayr[2]**

[1] University Politehnica of Bucharest, Department of Machine Elements and Tribology, Splaiul Independentei 313, Bucharest, ROMANIA
[2] Austrian Excellence Center for Tribology, Departament of Material Analysis, Street Viktor Kaplan 2, 2700 Wiener Neustadt, AUSTRIA

E-mail: am.petrescu@upb.ro

**Abstract**. The purpose of the paper is to determinate the thickness of the copper layer deposit on the electronic circuit boards, the thickness of the soldering alloy SAC 307 (96.5%Sn/3.0%Ag/0.7%Cu) deposit on the copper-PCB assembly used in electronic industry and also to determinate the sliding length of the sphere on those materials. Slurry composed of water and SiC was used to reduce the testing time. For the experiment a CSEM Calowear equipment was used and the tested materials were the layer of FR4(flame retardant 4) with copper deposit and the soldering alloy SAC 307.

## 1. Introduction

It is known that the mechanical function is one of the most important three functions (electrical, thermal, mechanical) of the electronic assembly, according to the electronic technology theory. The mechanical integrity of the assembly is provided by this mechanical function. The subject of the present paper is related to the mechanical attachment of electronic assembly [1, 2].

As a printed circuit board (PCB) the FR4 material was used, which is a woven glass fabric with epoxy resin system. The lead-free solder paste SAC 307 deposited on the copper layer from the PCB layer was also used.

For the tribological test a CSEM Calowear equipment was used, that can provide a simple method to determine the coating thickness. A steel sphere with a diameter $d$ = 36.5mm was used for the experiment. The experiment consists in two tests with different time lengths performed in the presence of a slurry composed of water (250 ml) and silicon carbide (SiC- 5 g) that falls onto the contact region between the sphere and the samples and one test without the slurry composed of water and SiC.

The Calowear test (micro-abrasive test) is used for the quality evaluation of thick coatings.

It is important to know the thickness of the copper layer deposited on the PCB and also the thickness of the solder alloy paste deposited on the PCB that is covered with copper. It is also wanted to determine the sliding length of the sphere over the samples.

The real thickness of the copper layer deposited on the PCB, the real thickness of the solder alloy paste deposited and also the surface characteristics of the layer (at the interface between the rigid support and the free surface where the electronic components can be mounted) represent a quality element for the PCB [3].

IOP Conf. Series: Materials Science and Engineering **174** (2017) 012026 doi:10.1088/1757-899X/174/1/012026

The thickness of the copper layer on the PCB affects the behavior of the entire electrical circuit and it also affects the process of the layer adhesion. It is important to know the thickness of the copper layer in order to know the adhesion to the substrate and thus it's mechanical behavior (shear resistance, behavior to vibrations, behavior in a variable thermal regime).

In the manufacturing technology for electric and electronic components a lot of measurements for the thickness of the copper layer are found. The values measured for commonly used thicknesses are: 17.5 μm, 25.5 μm, 35 μm, 71 μm, 105 μm. But in real situations we find that the real thickness of the copper layer has none of these values. From the analysis of the copper layer thickness through tribological methods (controlled wear), with or without abrasive suspension, it can be seen that the actual layer thickness sometimes differ significantly from those prescribed in the technological standards.

Any experimentally determined nominal adhesion parameter is a function of the elastic or plastic properties of the adhering material pair, interface conditions and loading conditions. The interface conditions include material structures, geometrical dimensions (thickness), solubility and amount of plastic deformation of each material in the adhering system, surface roughness, while the loading conditions include normal, shear or mixed loading, temperature, loading time, etc. [4, 5, 6, 7].

The present paper aims to determine through tribological methods the thickness of the layers used in electronics and to reduce the testing time through the controlled wear with slurry composed of water and SiC.

## 2. Experiment description

The CSEM Calowear Equipment used for the experiments is shown in the figure 1a. The configuration of the tests with a side view of the inclination of the plate is shown in figure 1b.

The measurement principle of the equipment is based on a rotating sphere made of bearing steel, with a diameter $d$ = 36.5 mm, that presses the coating surface with a certain load. The values for the drive shaft number of revolutions are presented in table 1. The drive shaft radius is $R$= 6.5 mm and the length of the contact between the sphere and the drive shaft is $p$ = 10 mm.

For the FR4 sample with a copper layer deposit the wear of the copper layer is very low (transfer phenomena on the ball) and the testing time for the determination of the thickness is high. In this case, the use of a slurry composed of water and SiC is proposed, thus the wear of the layer will be abrasive with rough particles in suspension (SiC in water).

During the tests conducted in a conventional dry regime for the FR4 sample with a copper layer deposit the controlled wear speed was very low and also a transfer of copper on the steel sphere could be observed (selective transfer). In this case the testing time was high and the wear phenomenon are influenced by selective transfer of the copper. Thus, for the layers containing copper it is recommended to adopt the solution of the controlled wear with slurry composed of water and SiC.

A consistent statistical result is considered to be composed of three identical tests. The statistic coefficient of variation for the wear of electronic layers in the CSEM testing equipment is 0.08. This coefficient is defined as the ratio between the standard deviation and the arithmetic mean and was determined by the previously experiments (ten identical experiments). For the mentioned statistic coefficient of variation, a consistent statistic result is comprised of the mean value of three identical tests.

Previous preliminary tests have shown that the friction force between the sphere and the layer varies significantly when the substrate is reached. Thus the determination of the thickness layer can be done, even after the complete wear, through optical measurements with a special magnifying glass without the removal of the sample.

The determination of the number of cycles and thus the sliding length at the damage of the layer appear was done first through a cycle of successive short tests (approximately 50 cycles). In the final part of the experiments was observed that it is need more than 1500 cycles to worn the layers. The decision concerning the number of cycles at which the layer was worn was taken through the calculation of the arithmetic mean between the number of cycles at which the damage was observed and the number of from the last cycles where there was no damage.

During the tests, an abrasive slurry composed of water and SiC falls over the contact zone. When the sphere rotates against the sample, a wear spot appears. The dimensions of the wear spot was determined with the help of an Leica DCM 3D Dual Core optical equipment.

a. b.

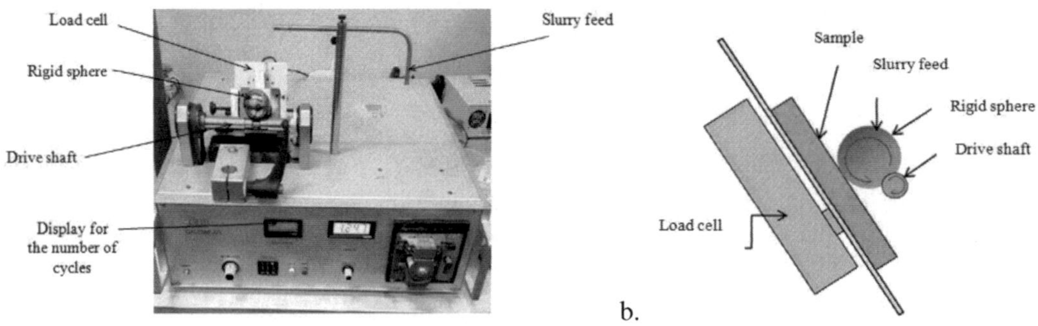


**Figure 1** a. CSEM Calowear Equipment and b. The configuration of the tests

From the optical inspection of the wear spot we can measure the geometrical parameters *a, b, x* and *y* that are shown in figure 2 and knowing the sphere diameter *d*, with a simple equation 1 we can determine the thickness of the coating

$$h = \sqrt{r^2 - \frac{b^2}{4}} - \sqrt{r^2 - \frac{a^2}{4}} \approx \frac{x \cdot y}{2r} \tag{1}$$

where *x, y, a* and *b* are the geometrical parameters optically determined, *r* is the radius of the sphere.

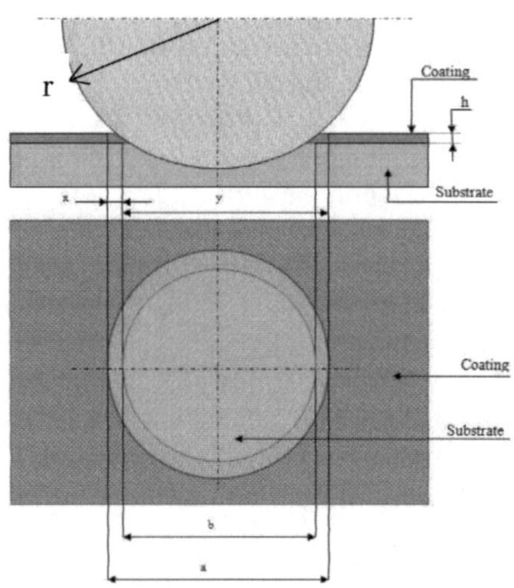


**Figure 2.** Calowear principle for the thickness determination of the covered material.

For the samples different time lengths in which the sphere slides on them were used. The first test (test 1) with the FR4 sample which has a copper layer deposited was done in the presence of the slurry composed of water and SiC with a working time of 1500 sec. The fixture of the sample is shown in figure 3a. Test number 2, done with a FR4 sample that has a copper layer and a soldering alloy SAC 307 deposited, was also done in the presence of the slurry composed of water and SiC. This test had a working time of 300 sec and the fixture is shown in figure 3b. Test number 3 was done for the same

FR4 sample that has a copper layer and a soldering alloy SAC 307 deposited, but without the presence of the slurry composed of water and SiC and with a working time of 1200 sec. Previous experiments were made on the same samples used in this experiment at different increasing time lengths.

a.  b. 

**Figure 3.** a. The sample FR4 with the copper layer deposited and b. the sample FR4 with the soldering alloy deposited on the copper layer.

The sliding length of the sphere is influenced by the diameter of the sphere and by the values of the elements from the contact between the sphere and the drive shaft shown in figure 4 [3].

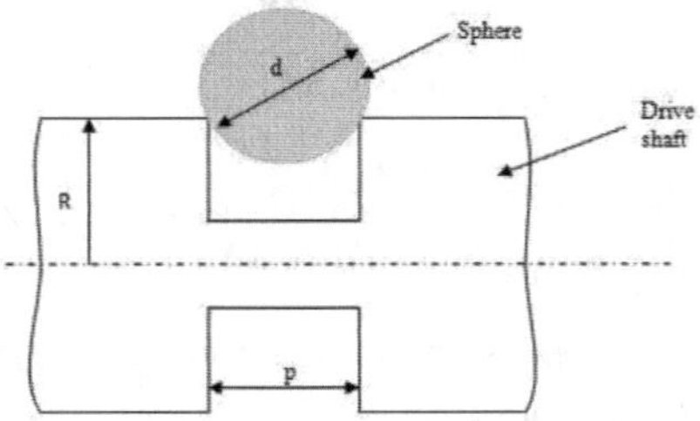


**Figure 4.** The contact between the sphere and the drive shaft.

The sliding length can be calculated using the equation 2 [3].

$$L = \frac{\pi \cdot R \cdot n \cdot d}{\sqrt{\left(\frac{d^2}{4} - \frac{p^2}{4}\right)}} \quad (2)$$

where $d$- sphere diameter; $L$- sliding length; $n$- drive shaft number of revolutions; $R$- drive shaft radius; $p$- length of the contact between the sphere and the drive shaft

## 3. Results

The present experimental results were processed in order to determine the sliding length using equation 2 and the mean values (three experiments) obtained are presented in table 1.

**Table 1.** Values of the sliding length

| Test number | Sample | Time [s] | Number of cycles [rot] | Sliding length [mm] |
|---|---|---|---|---|
| 1 | FR4+Cu (SiC) | 1500 | 8051 | 340960 |
| 2 | FR4+Cu+SAC 307(SiC) | 300 | 1614 | 68353 |
| 3 | FR4+Cu+SAC 307 (without SiC) | 1200 | 6436 | 272649 |

The images of the wear spot on the samples can be seen in figure 5. First the diameters $a$ and $b$ of the wear spot were measured and then with equation 1 the thickness of the copper layer and the thickness of the soldering alloy SAC 307 deposited on the other sample were calculated. The results are presented in table 2.

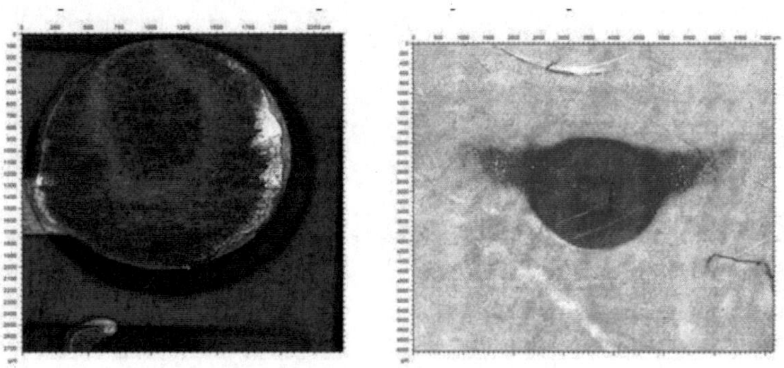

**Figure 5.** The optical inspection of the samples.

**Table 2.** Values of the thickness and wear parameter for the copper layer and for the soldering alloy SAC 307 layer

| Sample | Layer | Thickness | Wear parameter |
|---|---|---|---|
| FR4+Cu | Copper | 31 μm | $9.09 \times 10^{-11}$ |
| FR4+Cu+SAC 307 | Soldering alloy SAC 307 | 2.6 μm | $3.82 \times 10^{-11}$ |

The dimensionless wear parameter of layers (Ih) is considered that report of the thickness (h) to the sliding friction length (L) (Ih = h /L). This wear parameter is reported in table 2.

## 4. Conclusions

The real thickness of the layers is determined through controlled wear.

It is difficult to determine the thickness of the copper layer through the controlled adhesive wear in a conventional dry regime as a result of the selective transfer phenomenon of the copper on the steel sphere.

The thickness of other layers (without copper) can be determined even in a conventional dry friction regime.

The friction behavior of the layer, especially at the separation area with the rigid support, emphasizes the adhesion of the layer. This aspect will be detailed in a future paper.

The testing time is reduced significantly (approximately four times) when abrasive particles in suspension (SiC in water) are used.

**Acknowledgments**

This work was funded by the Austrian COMET-Program (Project K2 XTribology, Grant No. 849109) and has been carried out within the Excellence Centre of Tribology, Vienna, Austria. The authors would like to thank Professor Friedrich Franek for his support and also want to thank to Tetyana Khmelevska.

## References

[1] Dumitru G I, Tudor A, Chişiu G and Plotog I 2014 Mechanical characteristics of electronic printed circuit obtained by the vapour phase soldering process *Applied Mechanics and Materials* **658** 71-76

[2] Cucu T C, Plotog I and Branzei M 2014 Mechanical tests regarding low-temperature lead-free solder pastes application in automotive electronics *IEEE 20th International Symposium for Design and Technology in Electronic Packaging* pp 63-68

[3] Dumitru G I 2014 *PhD thesis of Contributions related to the mechanical characterisics of electronic connections developed with vapor phase tehnology* PhD thesis (Bucharest: University Politehnica of Bucharest) pp 118-119

[4] Lei W S and Kumar A 2014 Adhesion Measurement of Thin Films and Coatings: Relevance to Microelectronics *Adhesion in Microelectrinics* (Beverly, USA: Scrivener Publishing LLC) pp 33-64

[5] Păduraru G I, Petrescu A M, Stoica N A and Tudor A 2015 Analytical and Experimental Model for Tribological Behaviour of Thin Copper Layers Subjected to Punctual Stress *Tribological Journal Bultrib* **5** 125-132

[6] Mikicic D, Kunosic A and Zlatanovic M 2005 Contact Force Determination in Abrasive Wear Test *Tribology in industry* **27** 34-37

[7] Hutchings I M 2000 Applications of the micro-scale abrasion test for coatings and bulk materials *Tribology Series* **38** 37-42

IOP Conf. Series: Materials Science and Engineering **174** (2017) 012027 doi:10.1088/1757-899X/174/1/012027

# Aspects of the tribological behaviour of powders recycled from rapid steel treated sub-zero*

**S Radu and M Ciobanu**

Department of Engineering and Management of Technological Systems, Faculty of Mechanics, University of Craiova, Str. Al. I. Cuza 13, 200585 Craiova, Romania

e-mail: stefan.radu@yahoo.com

**Abstract**. The recycling of high-alloyed steels represents a significant opportunity in Powder Metallurgy as it permits the use of raw materials with relatively low prices compared to the conventional methods. Recycling can be achieved by two methods: from spraying debris resulted from worn cutting tools and processes obtained from processing chip drilling and re-sharpening of tools. The research aims to confirm that wastes from rapid steels can become, by the successive processing, metal powders that can thereafter be used for cutting tools of lathe type removable plate. After pressing and sintering the recycling powder, cylindrical samples were obtained that were subsequently applied a subcritical annealing. Wear tests conducted on a tribometer type TRB-01-02541 confirmed that their wear resistance is superior to the same samples that were sintered, hardened and tempered in oil.

## 1. Introduction

The paper presents studies conducted by a research team from the Faculty of Mechanical Engineering of the University of Craiova on the mechanical characteristics of pieces obtained from recycled metal powders that were sintered and treated sub-zero.

Recent investigations undertaken worldwide in connection with the changes that take place in steel when subjected to very low temperature following the usual heat treatment [1, 2, 3, 4] open up an entirely new field in the treatment of cutting tools and alloyed steels in general.

Recycling of rapid steels is a significant opportunity in Powder Metallurgy because it allows the obtaining of raw materials with relatively low prices as compared to the classical powders.

In the research was used only rapid steel type HS 18-0-1 EN ISO 4957:2002, obtained by recycling steel cutting tools from same material type.

Two possibilities for achieving the recycling were identified: spraying molten metal from the remains of cutting tools and machining waste chips resulting from sharpening and re-sharpening processes of cutting tools.

Research processes aim to confirm that waste from rapid steels can be used after the successive processing by pressing, sintering and subcritical annealing of the metal powders as removable plates to knives cutting lathe. Therefore, the project benefits the environment as well as in the Powder Metallurgy industry.

* This paper was accepted for publication in Proceedings after double peer reviewing process but was not presented at the Conference ROTRIB'16.

IOP Conf. Series: Materials Science and Engineering **174** (2017) 012027 doi:10.1088/1757-899X/174/1/012027

## 2. Experimental research

The chemical composition of the initial chips and distribution of alloying elements was determined with an electronic microscope SEM equipped with EDS spectrometer and analysis software INCA 200. The chemical composition of rapid steel chips used in research is shown in figure 1.

EDAX ZAF Quantification (Standardless)
Element Normalized
SEC Table : Default

| Elem | Wt % | At % | K-Ratio | Z | A | F |
|---|---|---|---|---|---|---|
| C K | 2.28 | 9.52 | 0.0059 | 1.2179 | 0.2131 | 1.0003 |
| O K | 6.20 | 19.41 | 0.0300 | 1.1938 | 0.4044 | 1.0024 |
| V K | 1.15 | 1.13 | 0.0120 | 0.9950 | 0.9698 | 1.0788 |
| CrK | 3.88 | 3.73 | 0.0432 | 1.0142 | 0.9792 | 1.1224 |
| MnK | 0.99 | 0.90 | 0.0097 | 0.9969 | 0.9849 | 1.0046 |
| FeK | 67.30 | 60.35 | 0.6796 | 1.0167 | 0.9865 | 1.0068 |
| W L | 18.20 | 4.96 | 0.1339 | 0.7430 | 0.9905 | 1.0000 |
| Total | 100.00 | 100.00 | | | | |

| Element | Net Inte. | Backgrd | Inte. Error | P/B |
|---|---|---|---|---|
| C K | 4.72 | 1.82 | 4.33 | 2.60 |
| O K | 53.48 | 2.44 | 1.01 | 21.87 |
| V K | 7.51 | 7.45 | 4.46 | 1.01 |
| CrK | 22.69 | 6.98 | 1.89 | 3.25 |
| MnK | 4.22 | 6.05 | 6.77 | 0.70 |
| FeK | 239.10 | 5.55 | 0.47 | 43.08 |
| W L | 6.82 | 3.59 | 3.88 | 1.90 |

\\Inspect_d8368\ovidiu\03.07.08\Nes\Rp.spc
Label :Rp
Acquisition Time : 16:22:30 Date : 3-Jul-2008

kV: 15.01 Tilt: 0.00 Take-off: 35.80 AmpT: 102.4
Det Type:SUTW, Sapphire Res: 130.42 Lsec: 200

**Figure 1.** Parameters and results of EDAX analysis.

Original chips are generally rough and irregular shaped needle containing abrasive grains on the surface, therefore magnetic separation is required. After 10 hours of grinding the grain powder is in the range [0.1 to 150] μm and after 20 hours of grinding grain powder is in the range [0.04 to 25] μm.

Both the chemical analysis of the chips used in research as well as the working procedure used for obtaining recycled powder was described in detail in the paper [5].

Recycled powder from rapid steel was compacted by into an electro-hydraulic press and then unilaterally pressed into a mold of Ø12 mm at a pressure of 800 MPa, as shown in figure 2. Subsequently, the sintering process of green samples was carried in a sealed box in a neutral atmosphere (Argon) at a rate of 2.5 l/h at three different temperatures: 1150°C, 1200°C and 1250°C.

The research conducted used a constant rate of heating in order to ensure a uniform distribution of the heat flow in the total mass of parts, thereby reducing the possibility of cracks or internal defects occurring. The best density values for the sintered parts were obtained when two combined conditions were used: maximum compacting pressure and the grain aggregate as small as possible [6]. One of the main objectives of the study was to obtain the compacts with good wear resistance, which can then be used as removable plates in the cutting process.

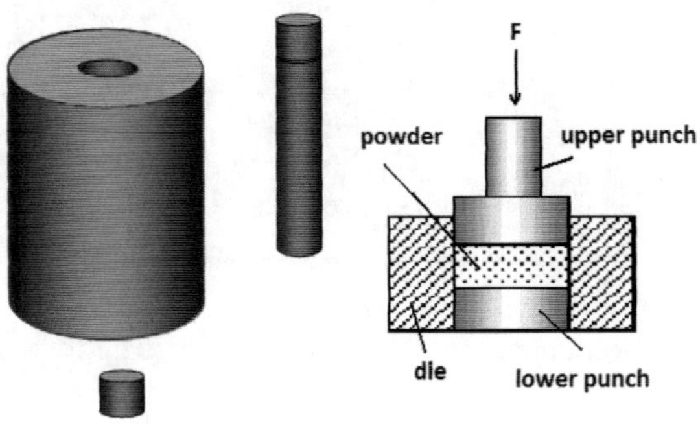


**Figure 2.** Unidirectional cylindrical mold.

According to the measurements conducted in the study, the density of samples sintered at 1150 °C samples ranged between [4.56 to 6.14] g/cm³, at the samples sintered at 1200 °C ranged between [4.89 to 6.38] g/cm ³, and at the samples sintered at 1250°C, the density of compacted samples was in the range [5.21to 6.75] g/cm³. The porosity of the compacts is complementary to the density, the lowest porosity of 13.46% was recorded for sintered samples at 1250°C. Also after sintering, the samples were cooled in the furnace under with protective atmosphere of argon throughout process.

In order to ensure a high wear resistance of the samples of rapid steels recovered powder, compacted at a pressure of 800 MPa and sintered under the conditions mentioned above, the samples were thereafter subjected to two types of heat treatment:

- hardening heat treatment and oil annealing, as shown in diagram of figure 3a
- hardening treatment sub-zero, according to the diagram in figure 3b

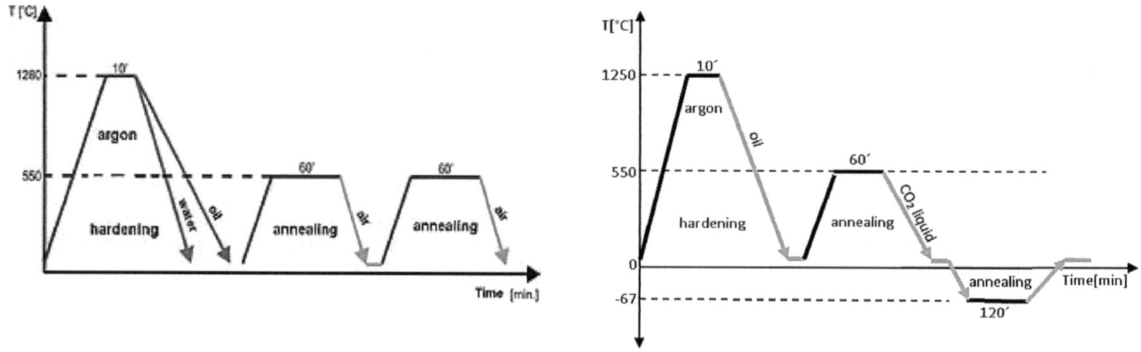


(a) (b)

**Figure 3.** Diagram of hardening and annealing: oil (a) and the diagram of hardening and annealing sub-zero (b).

Before adding samples in dry ice, they were treated by oil annealing. The procedure applied for the sub-zero treatment was as following: [7, 8].

- sub-zero thermal treatment was carried out in a mixture of dry ice (solidified $CO_2$) and technical alcohol, according figure 3.
- in order to maintain the proposed 206 K temperature, dry ice was added during the long sub-zero treatment (120 min);
- treatment dry ice was carried out in a thermally insulated box.

(a) (b)

**Figure 4.** Details on how to achieve the sub-zero treatment: (a) box insulation used for cooling samples (b) the thermal treatment with dry ice.

Dry ice thermal treatment parameters are shown in table 1.

**Table 1.** Parameters of treatment dry ice.

| Treatment | The procedure | Parameters |
|---|---|---|
| Treatment with dry ice | The temperature of the cooling | 206K (-67°C) |
| | Cooling medium | solution $CO_2$ (dry ice) |
| | Retention time | $\tau_l$ = 120 min |
| | The rate of cooling | 25° C/min |

IOP Conf. Series: Materials Science and Engineering **174** (2017) 012027 doi:10.1088/1757-899X/174/1/012027

## 3. Results and discussion

Heat treatments have led to the transformation of residual austenite into martensite and of partial decomposition of martensite and carbides in finely dispersed precipitation [9, 10].

The sintered parts were embedded in epoxy resin using an embedded machine brand METAPRESS-A after which they were prepared metallographic on a grinding machine type FORCIPOL with two workstations and speed platters between 50 and 500 rpm.

Two types of measurements were performed on the annealed samples. In the first stage the hardness of the samples was determined after heat treatment and oil annealing and in the second stage the hardness was determined after thermal treatment and annealing sub-zero in ice dry, followed by the measuring the friction coefficients in both situations After polishing the surface of the samples, it was attacked with alcohol and NITAL solution in concentrate 2%

Hardness measurements were performed with Vickers Micro-Hardness System CV-400DM$^{TM}$, from CV INSTRUMENTS EUROPE BV, according to figure 4.

The hardness after the application of the two variants of heat treatment is shown in figure 5.

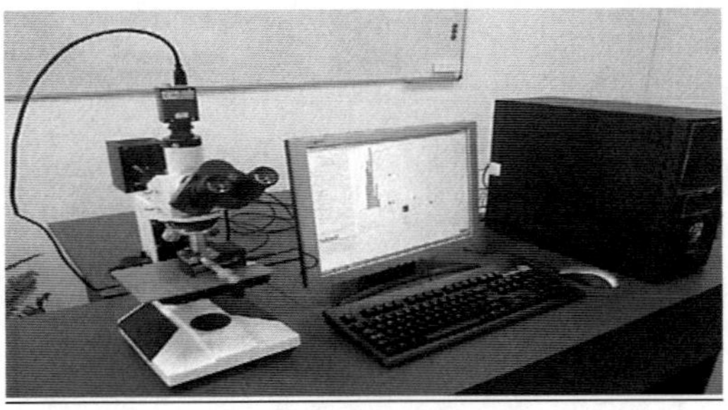
(a)

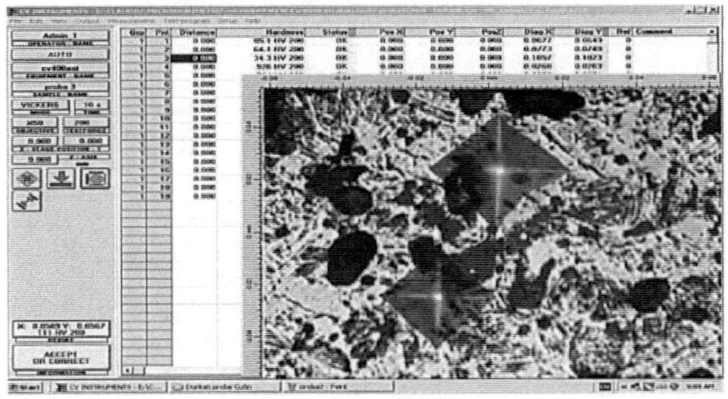
(b)

**Figure 5.** Hardness measuring equipment:
(a) Vickers Micro-Hardness System CV-400DM$^{TM}$ and
(b) hardness measurement mark.

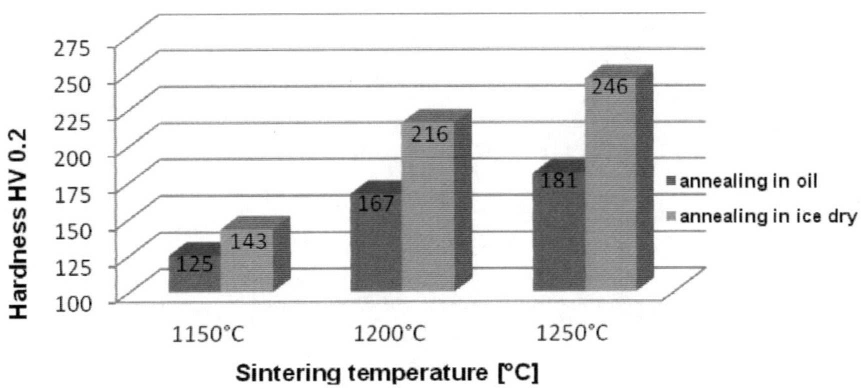


**Figure 6.** The hardness values according to applied heat treatment.

For evaluating the friction coefficient a tribometer brand CSM Instruments, type TRN 01-02541 was used, which allowed for the determination of friction under pure sliding friction torque using Class I under dry friction. The way of measuring the friction coefficient is shown in figure 7.

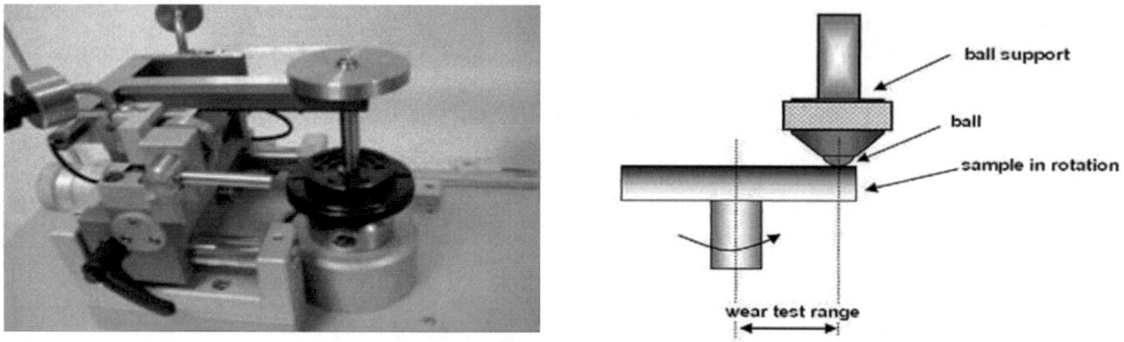


(a) (b)

**Figure 7.** Measuring of cylindrical parts wear coefficient:
(a) tribometer type TRN 01-02541 and (b) the measurement principle.

The test parameters used to determine the coefficient of friction were:

- working version: linear sliding;
- normal force: 2 N;
- radius: 3 mm;
- sliding speed: 1 cm/s;
- 100Cr6 steel ball, diameter 6 mm;
- working environment: air;
- air temperature: 27.3 ° C;
- relative humidity RH = 35.5%.

The tribological behaviour of sintered samples was tested in two different situations, and the average values of friction coefficients it is shown in figure 7:

- after sintering treatment followed by oil annealing;
- after sintering, treatment followed by annealing sub-zero.

The data on the tribological behaviour of sintered compacts from powder rapid steel recovered, results in the experimental research, revealed that the friction coefficient depends on the compaction pressure, the parameters of the process of sintering (temperature and dwell time) and the applied thermal treatment [11].

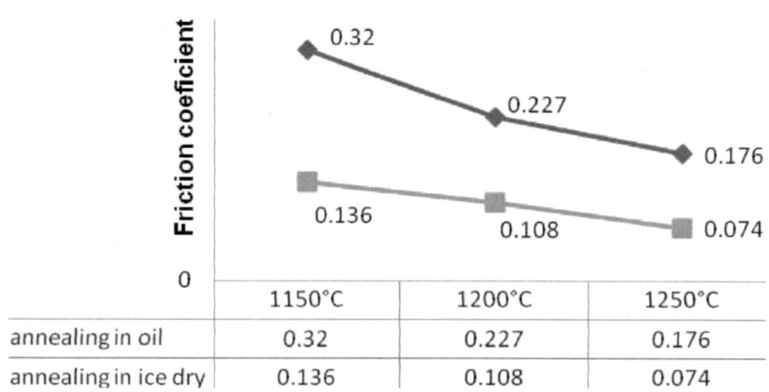


**Figure 8.** The evolution of friction coefficient depending on thermal treatment.

## 4. Conclusions

Research conducted on samples of rapid steel sintered and hardened in advance and then annealed at sub-zero temperatures shows structural improvements compared to sintered steels hardened in oil, so in conclusion it was observed that they had a high resistance to wear which means an increase in life cycle.

The role of sub-zero treatment is to make it possible to reduce residual austenite in the steel and martensite transformation to obtaining an improved hardness, good wear resistance and dimensional stability.

After applying thermal treatment sub-zero, it was observed an increase by approximately 30% in hardness, comparative to conventional thermal treatment applied of rapid steel and low friction coefficients and thus a greater durability that can justify the application in practice of this thermal treatment even if the result will increase final costs.

## References

[1] Kalsi N S and Sehgal R 2010 Cryogenic treatment of tool materials *Mater Manuf Process*, **25** 1077-1100
[2] Liang Y and Chen D 2011 Effect of cryogenic treatment on WC-Co cemented carbides *Mater Sci Eng* **528** 1735-1739
[3] Barron R F 1980 Dry cryogenic treatment increases machine tool life, *Boxboard Containers International PRIMEDIA Business Magazines & Media Inc* **(May)** 27
[4] Stojko A 2001 *Sub-zero treatment of tool steels* Diss. Technical University of Denmark DTU
[5] Radu S and M Demian 2013 Researches about wear comportment of sintered parts elaborated from recovered powders *J Optoelectron Adv M* **15** (7-8) 841-846
[6] Šalak A 1995 *Ferrous powder metallurgy* (Cambridge: Cambridge Intern. Science Publishing)
[7] Dubovan S 2014 Effects of sub-zero treatment applied to sintered steels *UPB Sci. Bull. B* **76 (1)** 197-208
[8] The Linde Group 2010 Sub-zero treatment *Technology processes and equipment* 1-23
[9] Hering D H 2005 Cold and cryogenic treatments *Ind Heat* **72** (5) 16-18
[10] *Holm* T 1997 *Cryotreatment – State of the art – an update* AGA Internal Report
[11] Podgornik V Leskovsek and J Vizintin 2009 Influence of deep-cryogenic treatment on tribological properties of p/m high-speed steel *Mater Manuf Process* **24** (7-8) 734-738

IOP Conf. Series: Materials Science and Engineering **174** (2017) 012028 doi:10.1088/1757-899X/174/1/012028

# Investigation on the contact behaviour of MEMS micromembrane with serpentine hinges

**M Pustan, C Birleanu, C Dudescu and F Rusu**

Micro and Nano Systems Laboratory, Technical University of Cluj-Napoca, 400641 Cluj-Napoca, Romania.

E-mail: Marius.Pustan@omt.utcluj.ro

**Abstract.** This paper presents the study of micromembranes supported by serpentine hinges. Widely used in microelectromechanical systems as switches or optical micromirrors, these micromembranes are deflected to the substrate in order to close a circuit or to process a signal. The investigated micromembranes are electroplated from gold in different geometrical dimensions. Furthermore, the central plate of micromembrane is suspended by two or four serpentine hinges. The stiffness of micromembranes is given by the geometry of hinges. One of the failure causes of micromembranes, which are directly deflected to substrate, is the adhesion effect between the flexible plate and the substrate. The adhesive force depends on the mechanical restoring force given by the hinges stiffness. In the case of micromembranes for optical applications, one additional stress is provided by temperature. A temperature gradient applied on micromembranes changes the stiffness with influence on the adhesion force. The study of temperature effect on stiffness and adhesion force is performed using an atomic force microscope and a thermal controlled stage. Experimental results of stiffness as a function of temperature are compared to numerical data.

## 1. Introduction

Microelectromechanical systems (MEMS) find their application in many domains where a miniaturization of products is necessary, such as: automotive and aerospace industry, chemical and bio-medicine, optical applications including displays and wireless communications. MEMS devices include sensors and actuators, switches, microrobots, optical scanners, micromotors, micropumps etc. Numerous bibliographic sources provided by the literature highlights the growing interest of researchers on MEMS generally, respectively of MEMS-based optical applications in particular [1-10]. Over the past decade, the telecommunications have become a strong market for optical MEMS devices. This is due to the explosion of internet network systems through fiber optics, digital scanning systems based on optical mirrors and of switches and transmission systems for the optical signal. In medical devices, micro optical scanners have resulted in three-dimensional scanning endoscopic systems. There are several applications, which include micromembranes as a mechanical flexible component. In pressure sensors, a capacitive micromembrane deflects when the pressure is applied, changing the distance between electrodes and the capacitance. In optical MEMS, micromembranes are used as micromirrors supported by hinges (springs) with high mobility. In an interferometer, the deformation of a micromembrane changes the optical light path and the propagation speed, the result being a phase shift. Micromembrane surface stress sensors in chemical and biological applications are fabricated from thin gold layer. The molecular interaction between probe molecules and target molecules generates a surface stress on the micromembrane. This surface stress causes the structural deflection of micromembrane, which generates the capacitance change in electrical sensing [1, 2].

In optical applications, high mobility micromembrane with optical control ability in different planes is needed. Figure 1 shows two micromembranes used in optical applications with the possibility of orientation of the signal by rotating the micromembrane with 10° (Figure 1a) [11].

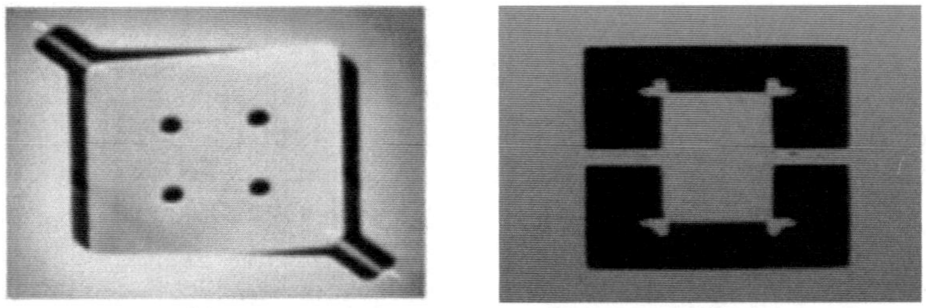

(a) (b)

**Figure 1**. Micromembranes from optical applications [11].

These micromembranes are used for optical signal processing by turning the micromembrane and/ or its linear movement. In a situation when an optical signal is desired to be orientated in different direction, these micromembranes must be replaced by other micromembrane with different systems of joints which may serve as a disadvantage of the optical quality as well as an increase in costs. Such micromembrane with multiple degrees of freedom can be represented by a micromembrane with serpentine hinges [12] where the hinges are deformed in bending and torsion depending on the applied force. When analysing the stress behaviour of a hinge, the critical issue is to understand how this deflects under the applied loads. Finite element analysis is a useful method to simulate the deflection of micromembrane and to compute the stress in hinges [13].

In RF switches (ohmic switch) a micromembrane is electrostatically deflected down to rest on a thin dielectric metal conductor. Stiction is one of the most important and unavoidable failure problems of micromembranes which deflect to substrate. Stiction is the adhesion of the contacting surfaces due to surface forces (van der Waals, capillary forces, Casmir forces, hydrogen bridging, and electrostatic forces). The restoring force acts against the adhesion force and depends on the micromembrane stiffness. In a stiction failure case, the restoring force of the micromembrane is not enough to remove the flexible plate to its original shape.

The atomic force microscopy (AFM) technique due to force and displacement sensing capabilities has been extensively used for the measurement of mechanical behaviour of flexible microcomponents. The AFM spectroscopy in point is used to deflect the central plate directly to substrate in order to measure the stiffness and the adhesion force.

In some MEMS applications the operating temperature of micromembranes can be significantly higher than the ambient one, for example in MEMS carrying several hundred of mW of power the membrane can reach 200ºC and in a video projector near the incandescent light source can exceed 100ºC. The temperature introduces residual stress and stress gradients that deform the micromembrane but it also decreases the modulus of elasticity of the material with influence on the mechanical response of the system. For reliability design, it is recommended to investigate the structure behaviour under a thermal field including a thermo-mechanical analysis. A thermal phenomenon introduces softening due to Young's modulus-temperature relation and a thermal relaxation which affects the rigidity of material - less force is needed to deflect the micromembrane if temperature increases, in order to produce the same displacement as at the initial temperature. In a case of a heated micromembrane under bending, the relaxation of Young's modulus has to be considered. Experimentally, the dependence between the force $F_z$ and the displacement in z- direction of micromembranes as a function of temperature is determined.

This paper presents experimental investigations and numerical analysis performed on micromembranes with serpentine hinges fabricated from gold as described in section 2. The interest is to determine the temperature influence on the mechanical and tribological behaviour of micromembranes. Stiffness and adhesion force are experimentally determined as a function of temperature, work included in section 3 of this paper. In section 4 of the paper, numerical analysis on stiffness is presented and the comparative results are included in section 5. The paper is ended with conclusions.

## 2. Sample description

The samples for experimental tests are micromembranes gold electroplated with rectangular hinges and different geometrical dimensions as presented in (Figure 2). The samples were manufactured in the Laboratory for Analysis and Architecture of System in Toulouse (France). The material used to fabricate the micromembranes is gold and the structure was fabricated in 10 lithography and deposition steps. The gold material has high thermal efficiency and short thermal time constant of relatively low temperature (< 200°C). The serpentine hinges used in our model to connect the mobile plate of micromembrane to anchors are formed of seven series-connected units. The in-plane displacement of the mobile plate gives extension and compression of each hinges. Out of plane movement of the proof mass performs the bending as well as the torsion of hinges.

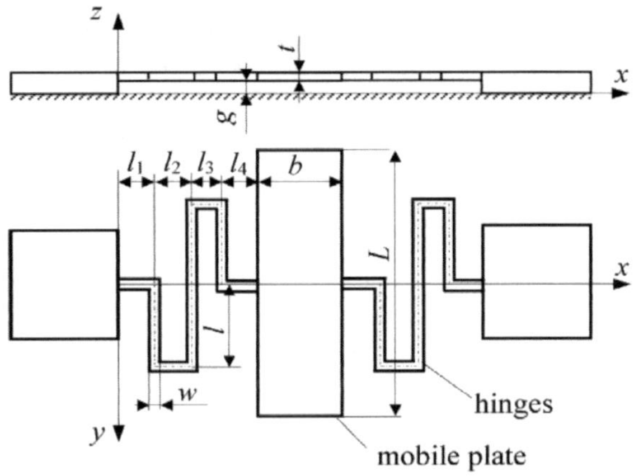


**Figure 2.** Geometry of micromembranes supported by serpentine hinges [12].

The micromembrane supported by two serpentine hinges (Figure 2) is sensitive to in-plane and out of the plane moments. While the central mass translates at least on one of the in-plane directions, the hinge legs that are directed perpendicularly to the motion direction will be bent, whereas the other legs will be subjected to axial extension and compression in addition to bending. Out of the plane displacement gives bending and torsion of the hinge legs. A simplified analytical formula for the bending stiffness of a micromembrane supported by serpentine hinges when a force is applying on $z$-direction in the mid-position of the mobile plate (Figure 2), is given in [14]. The model lead to approximate results due to the fact that lengths $l_1$ and $l_4$ are neglected and the central plate was considered rigid.

In our paper only numerical and experimental analyses were performed. Micromembranes supported by two and four serpentine hinges are investigated in order to estimate the samples' deflections under a mechanical force as a function of temperature. The thickness of micromembrane is 3μm and the flexible part is suspended at 2μm above a silicon substrate. The gold material is the most used in optical and switching applications. The width of mobile plate is 38μm and the length is 118μm (which define the contact area with the substrate). The mobile plate is supported by two and four serpentine hinges. The hinges have the following geometrical dimensions (Figure 2): width $w$ = 6μm, $l_1$=$l_4$=13μm, $l_2$=$l_3$=16μm. Hinges with different length $l$ are used to suspend the mobile plate. The images with the investigated micromembranes are presented in Figure 3. The *micromembrane-1* presented in Figure 3a is supported by 2 hinges as the *micromembrane-2* (Figure 3b). The difference between them consists of different dimensions $l$ of the hinges (Figure 2). *Micromembrane-1* has the length $l$=14μm and the second micromembrane is characterized by $l$=39μm. The *micromembrane-3* and 4 are fabricated with 4 hinges. The length $l$ of *micromembrane-3* (Figure 3c) is 14μm comparatively with *micromembrane-4* (Figure 3d) for which $l$=39μm.

IOP Conf. Series: Materials Science and Engineering **174** (2017) 012028 doi:10.1088/1757-899X/174/1/012028

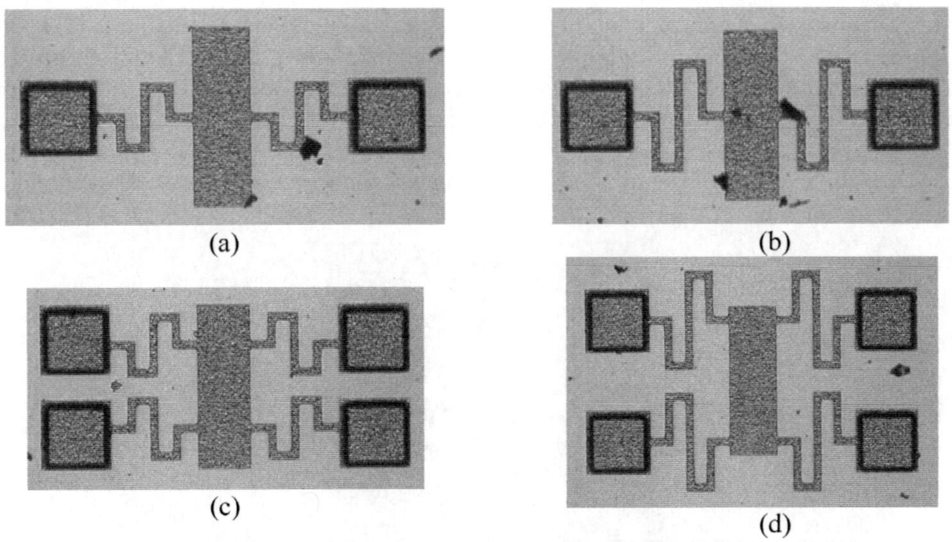


**Figure 3.** Micromembranes with different hinges: (a) Micromembrane-1; (b) Micromembrane-2; (c) Micromembrane-3; (d) Micromembrane-4.

## 3. Experimental stiffness and adhesion force

The scopes of this analysis are the following: i) to determine the variation of the micromembranes stiffness as a function of temperature; ii) to analyse the temperature influence on the adhesion force between micromembranes and substrate. The experimental tests were done at Technical University of Cluj-Napoca in the Micro and Nano System Laboratory using an AFM XE 70 produced by Park System Co and a temperature controlled heating unit.

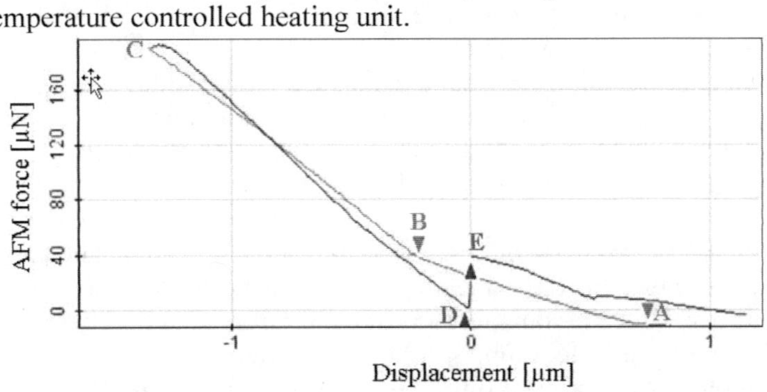


**Figure 4.** AFM experimental curve of micromembrane-1.

An AFM probe with known stiffness is used to bend the flexible plate directly to the substrate. Figure 4 shows the AFM experimental curve of the micromembrane 1 from Figure 3a. During testing the vertical approach of the scanning head and AFM probe toward to the substrate is controlled and the bending deflection of AFM probe is optically monitored. The experimental curve gives the dependence between the displacement of the scanning head and the deflection of AFM probe. The first part (A-B) of the loading curve corresponds to the bending of AFM probe and sample. In the position B the micromembrane comes in contact with the substrate. On the part (B-C) there is bending only of AFM probe. The unloading process starts from position C when the AFM probe is coming in its initial position. As it can be observed on unloading AFM curve, there is a jump (D-E) that is given by the adhesion effect between micromembrane and substrate. The detachment of micromembrane from substrate is delayed due to the adhesion effect. The jump size is proportional to the strength of adhesion between the contacted surfaces. After the contact between micromembrane and substrate is broken in position D the sample and the AFM probe are coming to the initial position. The first part

(A-B) from experimental curve is used to compute the micromembrane stiffness. The detected deflection of the AFM probe multiplies by its stiffness gives the acting force. The difference between the vertical displacement of AFM scanning head and the detected deflection of AFM probe taken from the first part (A-B) of the experimental curve is the micromembrane bending deflection ($Z_{sample}$). The applied force divided by $Z_{sample}$ gives the micromembrane stiffness. During experimental tests, the force is applied in the mid-position of mobile plate as presented in Figure 5 for micromembrane 4 and deflects it directly to substrate.

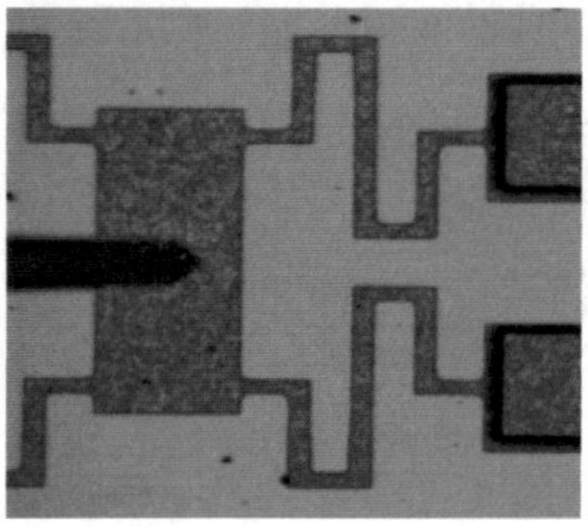

**Figure 5.** AFM probe in contact with the micromembrane-4.

Using this AFM tests, a stiffness of 64.4N/m is determined for the micromembrane 1, if the force is applied in the mid-position of the mobile plate. The adhesion force between micromembrane and substrate given by the distance between D and E (Figure 4) is 36μN. The same experiment is used to estimate the stiffness of the micromembrane 2 (Figure 3b) with a characteristic length of 39μm of hinges. An out of the plane stiffness of 14.9N/m was experimentally determined if the force is applied in the mid-position of the mobile plate and the adhesion force between flexible plate and substrate is equal to 124μN. For the same contact area from mobile plate and substrate the adhesion force increases if the stiffness decreases, respectively.

The same AFM test is also used to measure the stiffness of micromembranes 3 and 4 supported by four serpentine hinges for a force applied in the mid-position of mobile plate, as presented in Figure 5. The geometrical dimensions of hinges are the same as those previously described for the micromembranes supported by two hinges. Experimentally, a bending stiffness of 127.5N/m is determined for the micromembrane 3 and it is 26N/m for the micromembrane 4. The adhesion force between mobile plate and substrate of micromembrane 3 is 1.27μN and for the micromembrane 4, with smaller value of stiffness, for the same contact area, the adhesion force increases to 78.2 μN. These preliminary tests were done at room temperature (20°C) and a humidity equal to 40%RH.

Moreover, the temperature effect on stiffness and adhesion force is investigated. During experimental tests, the following temperatures 20 °C, 40 °C, 60 °C, 80 °C and 100 °C are applied directly on substrate with the help of a temperature controlled heating unit. In order, to limit the thermal effect on the AFM tip ($Si_3N_4$), after each measurement, the AFM scanning head is moved to the zero position that is the initial starting location of measurements. The AFM probe used in our experiment has the tip height of 25μm. So, the cantilever of AFM probe that is optically monitored during scanning is suspended at 25μm above the sample that reduces the influence of temperature on the AFM probe. The experimental stiffness of the micromembrane 1 decreases with 4.1% if the micromembrane operates at 100°C, instead of 20 °C. In the same way, the adhesion force increases but not significantly. For the micromembrane 2, the stiffness increases with 3.3% if temperature increases from 20 °C to 100 °C. The small influence of temperature is determined for the micromembrane 3 as its stiffness changes only with 0.05%. For the micromembrane 4 the stiffness is modified with 4.1%. Because the stiffness of the structure is not strongly influenced by temperature, the adhesion forces is not significantly changed. Based on this investigations, the temperature has a small influence on stiffness and adhesion force that making these micromembranes with serpentine hinges suitable for MEMS applications where a temperature gradient occurs.

## 4. Numerical simulations of stiffness for different temperatures

A numerical analysis of micromembranes stiffness was performed by Finite Element (FE) method using the thermo-mechanical module in ANSYS Workbench 13 software. During the virtual tests, a thermal gradient is applied on the micromembrane and a mechanical force is used to bend the mobile plate toward the substrate. Using the same geometrical dimensions as in the experiments, the samples were modelled and their response (out of plane deflection) for a given temperature and a central force was computed. The own weight of the mobile plate upon the hinge deformation is very small and has been neglected. The simulation consists of two steps: first, only the temperature was increased with no mechanical load; in the second step, the temperature was kept constant and a mechanical load was applied in the central position of the mobile plate. The stiffness calculation took into account only the out of plane deflection due to mechanical force, thus, from the total deflection, those due to temperature increase were extracted. The Young's modulus of the investigated samples used in the FEA, according to the literature (www.totalmateria.com; http://silicon.mhopeng.ml1.net/Silicon/), are 79GPa for gold and 169GP for silicon. The thermal expansion coefficients (CTE) considered in these simulations were: $14\times10^{-6}$ [1/°C] for gold and $2.1\times10^{-6}$ [1/°C] for silicon. The CTE values were considered constant in the temperature range from 20 °C to 100 °C. In Figure 6, the total deflections (thermal & mechanical) of the investigated micromembranes for a temperature change of 20°C (from 20°C to 40°C) and a 1μN central force are shown.

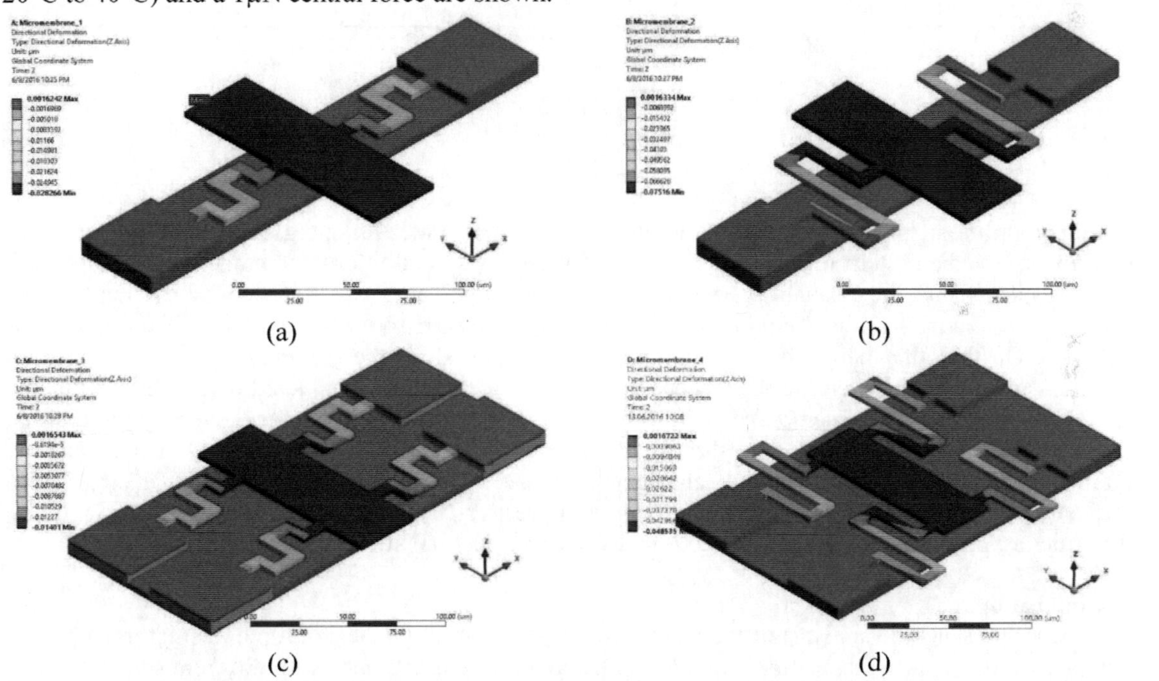


**Figure 6.** Out of plane deflection of micromembranes due to change in temperature from 20°C to 40°C and applied central concentrated force (1μN): (a) Micromembrane- 1; (b) Micromembrane- 2; (c) Micromembrane- 3; (d) Micromembrane- 4.

## 5. Results and discussions

Comparative results (numerical vs. experimental) for out of the plane stiffness of the investigated micromembranes as a function of temperature are included in table 1. Table 2 presents the experimental results of the adhesion force between flexible plate and substrate for the investigated micromembranes for different temperatures. The tests were done with a mechanical force applied in the mid-position of mobile plate. In order to measure the adhesion force, the micromembrane is directly deflected to substrate with a controlled AFM force. The materials in contact are the

IOP Conf. Series: Materials Science and Engineering **174** (2017) 012028 doi:10.1088/1757-899X/174/1/012028

micromembrane material (gold) and its substrate (silicon). In order to determine the temperature effect on stiffness and adhesion, a thermal controlled stage was used. During tests, the temperature increases until 100°C.

**Table 1.** Out of the plane stiffness [N/m] as a function of temperature.

| Temperature [°C] | Membrane 1 | | Membrane 2 | | Membrane 3 | | Membrane 4 | |
|---|---|---|---|---|---|---|---|---|
| | N[a] | E[b] | N[a] | E[b] | N[a] | E[b] | N[a] | E[b] |
| 20 | 63.90 | 64.40 | 15.35 | 14.90 | 130.82 | 127.50 | 28.80 | 26.00 |
| 40 | 63.42 | 63.71 | 15.40 | 15.12 | 130.84 | 127.42 | 30.59 | 26.20 |
| 60 | 62.89 | 62.92 | 15.45 | 15.20 | 130.89 | 127.66 | 30.72 | 25.89 |
| 80 | 62.17 | 62.24 | 15.50 | 15.38 | 130.89 | 127.81 | 30.10 | 26.38 |
| 100 | 62.09 | 61.75 | 15.55 | 15.42 | 131.20 | 128.06 | 30.65 | 27.13 |

[a]N- numerical results / [b]E- experimental results

**Table 2.** Experimental results of adhesion force [μN].

| Temperature [°C] | Membrane 1 | Membrane 2 | Membrane 3 | Membrane 4 |
|---|---|---|---|---|
| 20 | 36.0 | 124.0 | 1.27 | 78.20 |
| 40 | 38.2 | 122.8 | 1.27 | 76.80 |
| 60 | 40.0 | 122.2 | 1.30 | 76.20 |
| 80 | 40.5 | 121.6 | 1.32 | 77.10 |
| 100 | 40.2 | 121.1 | 1.46 | 76.72 |

The coupling of the strain field to a temperature field provides an energy dissipation mechanism that allows for the system to relax. In the case of the investigated micromembranes under out-of-the plane displacement, the relaxation strength/stress to be considered is influenced by the modulus of elasticity. A thermal gradient applied on the structure decreases both the modulus of elasticity and stiffness. On the other hand, the temperature gradient applied on the structure introduces a thermal prestress in hinges that increase the stiffness. These two aspects are considered of in the investigated samples. As for the micromembranes with higher stiffness the thermal prestress given by the thermal dilatation of hinges is smaller than that of the micromembrane with less stiffness. In this case the thermal relaxation of the modulus of elasticity is consider as the reason for stiffness decrease. For micromembranes with smaller stiffness, the thermal prestress is significant and the stiffness given by the structure geometry is increased based on an additional thermal stiffness.

## 6. Conclusions

This paper presents the analysis of thermal effect on stiffness and adhesion force of micromembranes with serpentine hinges. Depending on their application, micromembranes with different sensitivity can be obtained by changing the number of hinges and their geometrical configurations. Micromembranes with two and four hinges were investigated in order to evaluate their static response under a mechanical load and a thermal gradient. The AFM spectroscopy in point was used to measure the stiffness and adhesion force between mobile plate and substrate. The force was applied in the mid position of mobile plate and it deflects it directly to substrate. The same force was used for all investigated micromembranes. Because the dimensions of mobile plate are the same for all micromembranes, the contact areas with substrate are equal and the roughness of micromembranes should be in the same range. The adhesion force is mainly influenced by the restoring force of micromembranes from substrate, this force depending of the membrane stiffness. Micromembranes with higher stiffness are accompanied by smaller values of adhesion forces comparatively with micromembranes with smaller stiffness. Increasing the temperature, the stiffness as well as the adhesion forces are modified. Experimentally, the dependence between force and displacement of micromembranes can be determined as a function of temperature and the Young's modulus – temperature dependence can be estimated. Temperature introduces an additional prestress of structure given by the thermal dilatation of hinges that increases the stiffness especially for micromembranes with small initial stiffness. The experimental results of stiffness as a function of temperature are compared to the numerical values and these are in good agreement. Comparing to with micromembranes supported by the other geometry of hinges, the micromembranes with serpentines hinges provide a compensation of thermal effect with small influence of stiffness and on adhesion, thus, these micromembranes being suitable for applications where a thermal gradient occurs.

**Acknowledgments**
This work was supported by a grant of the Romanian National Authority for Scientific Research and Innovation, CNCS-UEFISCDI, project number PN-II-RU-TE-2014-4-1271.

**References**
[1] Syms R R A and Moore D F 2002 Optical MEMS for telecoms, *Mater Today* **5** 26-35
[2] Gorelick S 2012 XMEMS: dynamic diffraction gratings by MEMS technology for X-ray imaging applications *Procedia Eng* **47** 277-280
[3] Hashimoto T, Maeya J, Fujita T and Maenaka K 2009 Concept of MEMS ring laser gyroscope with movable optical parts *Procedia Chem* **1** 564-567
[4] Kenda A, Frank A, Kraft M, Tortschanoff A, Sandner T, Schenk H and Scherf W 2009 Compact high-speed spectrometers based on MEMS devices with large amplitude in-plane actuators *Procedia Chem* **1** 556-559
[5] Yan J, Watanabe K and Nakagawa Y 2012 Fabrication of thin-film Fresnel optics by combining diamond turning and MEMS techniques *Procedia CIRP* **4** 146-150
[6] Bourim E M, Kim H-Y, Yang J-S, Yang J-W, Woo K-S, Song J-H and Yun S- K 2009 Creep behavior of undoped and La-Nb codoped PZT based micro-piezoactuators for micro-optical modulator applications *Sensors Actuat A Phys* **155** 290-298
[7] Emadi A, Wu H, Grabarnik S, de Graaf G, Hedsten K, Enoksson P, Correia J H and Wolffenbuttel R F 2010 Fabrication and characterization of IC-compatible linear variable optical filters with application in a micro-spectrometer *Sensors Actuat A Phys* **162** 400-405
[8] Yang D, Tian H and Ji Y 2011 The properties of lattice-shifted microcavity in photonic crystal slab and its applications for electro-optical sensor *Sensors Actuat* **171** 146-151
[9] Pang C, Yu M, Zhang X M, Gupta A K and Bryden K M 2012 Multifunctional optical MEMS sensor platform with heterogeneous fiber optic Fabry-Pérot sensors for wireless sensor networks, *Sensors Actuat A Phys* **188** 471-480
[10] Linslal C L, Syam Mohan P M, Halder A and Gangopadhyay T K 2013 Analysis and modeling of an optical fiber loop resonator and an evanescent field absorption sensor for the application for chemical detection *Sensors Actuat A Phys* **194** 160-168
[11] Wiswell N 2012 *Design and fabrication of serpentine – hinged silicon micro-mirror devices-Research Report* (San Luis Obispo CA: Material Engineering Department, California Polytechnic State University)
[12] Pustan M, Birleanu C, Dudescu C and Belcin O 2012 *Proc. Int. Conf. on Thermal, Mechanical and Multi-Physics Simulation and Experiments in Microelectronics and Mycrosystems (Cascais)* (IEEE Catalog Number: CFP12566-CDR, DOI: 10.1109/ESimE.2012.6191802)
[13] Strohm K M, Schauwecker B., Pilz D., Simon W and Luy J F 2001 *IEEE Silicon Monolithic Integrated Circuits in RF Systems. Digest of papers* (Mi, USA) 42-46
[14] Lobontiu N and Garcia E 2004 *Mechanics of microelectromechanical systems* (Boston: Springer Science & Business Media).

IOP Conf. Series: Materials Science and Engineering **174** (2017) 012029 doi:10.1088/1757-899X/174/1/012029

# Analysis of the bending stiffness and adhesion effect in RF-MEMS structures

**C Birleanu[1], M Pustan[1], C Dudescu[2], V Merie[3] and I Pintea[1]**

[1]Mechanical Systems Engineering Department, Technical University of Cluj-Napoca, Blv. Muncii no. 103-105, 400641, Cluj-Napoca, Romania
[2]Mechanical Engineering Department, Technical University of Cluj-Napoca, Blv. Muncii no. 103-105, 400641, Cluj-Napoca, Romania
[3]Materials Science and Engineering Department, Technical University of Cluj-Napoca, Blv. Muncii no. 103-105, 400641, Cluj-Napoca, Romania

E-mail: Corina.Barleanu@omt.utcluj.ro

**Abstract.** Microelectromechanical system (MEMS) is a special branch with a wide range of applications in sensing, switching and actuating devices. Designing the reliable MEMS for thin free-standing structures like as bridges and cantilevers requires understanding of the tribomechanical properties of the materials and structures.
The effect of geometrical dimensions (cross-section dimensions and length) on mechanical and tribological behavior of free-standing MEMS structures made of electroplated gold was analyzed in this paper. Special attention was given to the dependences between stiffness and cantilever length and the dependences between bending stress and variable travel range of actuated load. The force position was moved from the beams free-end toward to the anchor. The tests were performed at room temperature (22°C) and relative humidity RH of 40% with a noise- and vibration-isolated and environment-controlled XE-70 AFM from Park Systems using the contact mode. Each measurement was repeated many times in order to improve the accuracy of the experimental results. The stiffness of a microcantilever varies if the position of the acting force is changed. The experimental results obtained were in good correlation with those obtained analytically.

## 1. Introduction

Microelectromechanical systems (MEMS) are integrated electromechanical system where the components size and the actuating range are within the limits to micro-scale. Nowadays, due to increasingly mature process technology, several micro structural and functional sophisticated modules are available. Among the current MEMS devices, multiple MEMS devices are driven electrostatic such as capacitive pressure sensors, micro pumps, radio frequency (RF) switches, vacuum resonators etc. Electrostatic-driven MEMS devices have advantages of quick response, lower power consumption, and standard integrated circuit process compatibility [1]. One of the fundamental issues in building MEMS is scale effect [2]. In MEMS devices where size effects become prevalent determine different properties which deviate from their bulk characteristics. This leads to a grey area regarding the application of specific characterization technologies.

For the microcomponents from thin films is evident that stress-strain relationship is affected by the relatively high area to volume ratio. As a result, typical properties used for description of bulk material strength and deformation do not apply for thin films and micro devices. In particular, this size effect relationship is not yet well understood when structural dimensions decrease from millimeters to micro and nanometers [1].

RF-sensors have been a subject of interest in recent years [3]. These sensors with high sensitivity to applied force, high resolution, low power consumption and digital output are an attractive alternative to conventional piezoresistive sensors.

The effect of geometrical dimensions on stiffness, stress and adhesion force of clamp-free MEMS structures (microcantilevers) is analysed in this work. Experimental investigations are performed on samples fabricated from gold with different geometrical dimensions using Atomic Force Microscope. For soft cantilevers, such as those employed in atomic force microscopy, the bending sensitivity can approach a few tenths of a newton per meter [4].

Microcantilever deflection is most commonly measured by reflecting a laser from the free end of the cantilever. An AFM cantilever experiences a point load at its free end (where the probe tip is located), whereas a cantilever with a functional coating for sensing applications experiences a distributed surface stress. Stress distribution in MEMS deformable component is changed depending on the contact areas between the flexible plate and the substrate. In this paper it is worth to discuss the overall influence of length scale on the performance of mechanical devices.

## 2. Theoretical considerations

### *2.1. Stiffness and stress*

Electrostatic actuation and sensing are largely utilized in MEMS transducers due to advantages such as sensitivity, fast response, precision, relatively easy fabrication. It is well known that the flexible plate displacement of an electrostatic actuated MEMS resonator depends on the applied load.

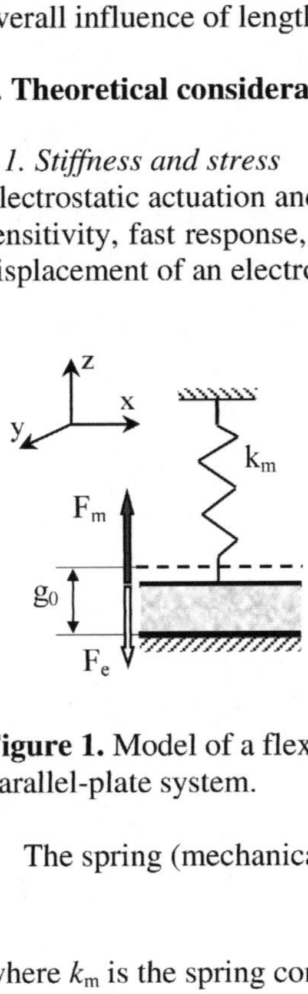


**Figure 1.** Model of a flexible parallel-plate system.

The maximum gap $g_0$ between the fixed and the mobile plates, initially occurs for $u_z = 0$ (figure 1). The static equilibrium sets in situation when the two opposing forces, the electrostatic force $F_e$ and the spring force $F_m$, are equal:

$$F_e = F_m \tag{1}$$

The force produced through transverse electrostatic actuation is given by [2]:

$$F_e = \frac{\varepsilon \cdot A \cdot U^2}{2(g_0 - u_z)^2} \tag{2}$$

where: $\varepsilon$ is the electric permittivity; $A$ - overlapped area; $U$ - the actuation voltage); $g_0$ - the initial gap in the z direction; $u_z$ - the displacement produced through attraction electrostatic forces.

The spring (mechanical) force $F_m$ is:

$$F_m = k_m \cdot u_z \tag{3}$$

where $k_m$ is the spring constant.

When the electrostatic force $F_e$ is larger than the mechanical force $F_m$ given by the spring constant, the system becomes unstable. In this way for displacements larger than one-third of the initial gap (pull-in position), the mobile plate collapses against the fixed one [5].

The main failure cause in MEMS is stiction when the flexible plate collapses into nearby surfaces, resulting in their permanent adhesion. This situation can arise either during fabrication or operation. Under stiction the contact area between flexible plate and substrate increases if the applied load increases, respectively. The cantilevers with rectangular cross-sections under investigation are subjected to a mechanical force applied in the position 1 on samples as presented in figure 2. In the first case, the force is applied at the cantilever free- end. Then, the force is moved toward to the beam

IOP Conf. Series: Materials Science and Engineering **174** (2017) 012029 doi:10.1088/1757-899X/174/1/012029

anchor. In both situations the investigated cantilevers are bending directly to substrate. The cantilevers of rectangular cross-sections under our investigation have the configuration (*w* - width, *t* - thickness and *l* - length) as presented in figure 3 and are subjected to a force *F*.

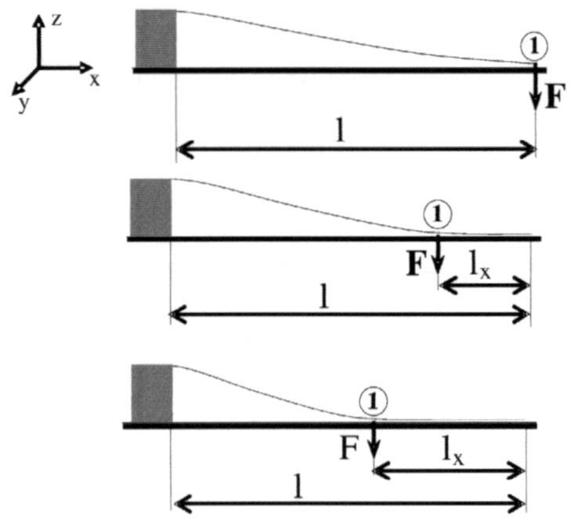


**Figure 2**. Bending of a microcantilever to substrate for different force positions [6, 8].

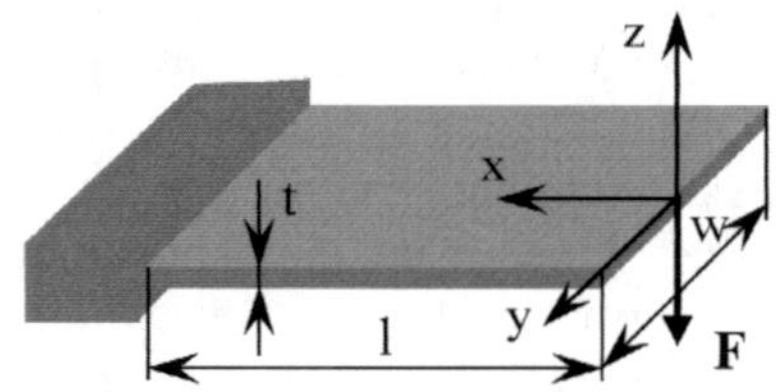


**Figure 3.** Microcantilever subjected to a mechanical force.

The stiffness and stress in beam is changed when the position of the applied force which bends the flexible plate to substrate, is moved towards to the anchor. As a consequence, two situations should be considered in our study. First, for long beams when the distance $(l-l_x)$ is at least 5 times larger than the largest cross-sectional dimension when the cross-sections remain perpendicular to the neutral axis during displacement and the Euler-Bernoulli beam model is used [5].

Secondly, for relatively-short distance $(l-l_x)<5w$, shearing effects become important, and the regular bending deformations are augmented by the addition of shearing deformations, according to the Timoshenko's beam model. In this case, the cross-sections are no longer perpendicular to the neutral axis in the deformed state [5]. Corresponding to this situation, when the shearing effects become important, normal and tangential stresses are produced simultaneously in flexible plates.

If $(l-l_x)\geq 5w$, the deflection at the point 1 of the microcantilever (figure 2) can be determined as:

$$u_{1z}=F\cdot\frac{(l-l_x)^3}{3E\cdot I_y} \tag{4}$$

where $I_y$ is the axial moment of inertia and $E$ is the longitudinal modulus of elasticity.

The z-direction stiffness of the microcantilever is:

$$k=\frac{3E\cdot I_y}{(l-l_x)^3} \tag{5}$$

The bending stress appearing in cantilever can be analytically computed as:

$$\sigma_b=\frac{6F\cdot(l-l_x)}{w\cdot t^2} \tag{6}$$

If the force is applied at a distance $(l-l_x)<5w$, the stiction area is large and shearing effects became important. The deflection and force dependence is given by the relation:

$$u_{1z}=F\cdot(l-l_x)\cdot\left[\frac{(l-l_x)^2}{3EI_y}+\frac{\kappa}{GA}\right] \tag{7}$$

where $A$ is the cross-sectional area, $G$ is the shear modulus and $\kappa=6/5$ for rectangular cross-sections from Young and Budynas [7].

Based on shearing effect the stiffness of cantilever $k_{(sh)}$ could be expressed by [5]:

$$k_{(sh)} = \frac{3E \cdot G \cdot I_y \cdot A}{(l-l_x)\left[G \cdot A \cdot (l-l_x)^2 + 3\kappa \cdot EI_y\right]} \tag{8}$$

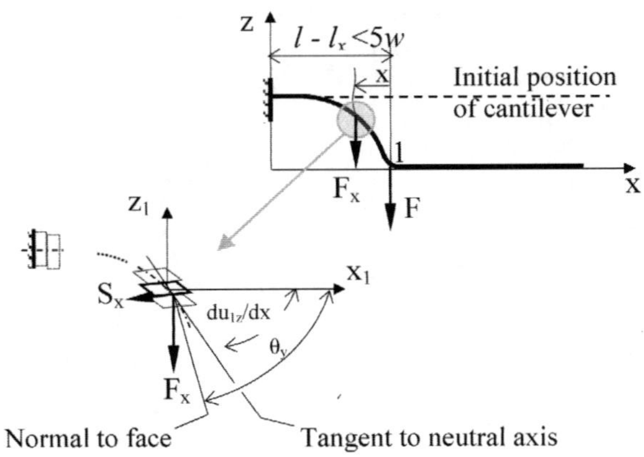


**Figure 4.** Microcantilever under bending and shear deformations [6, 8].

In this situation, the cross-section is no longer perpendicular to the neutral axis in the deformed state (figure 4) shearing effects produces an additional angular deformation $\theta_y(x) - \frac{\mathrm{d}u_{1z}(x)}{\mathrm{d}x}$.

The angle $\theta_y$ is determined based on:

$$\theta_y(x) = \frac{F \cdot \left[(l-l_x)^2 - x^2\right]}{2EI_y} \tag{9}$$

The equivalent stress for a cross-sectional element located to the $x$ distance from the point 1 where the force $F_x$ is applied can be determined using:

$$\sigma_{ech} = \sqrt{\sigma_b(x)^2 + 3\tau(x)^2} \tag{10}$$

where:

$$\tau(x) = \kappa \frac{S_x}{A} = \kappa \frac{F_x \cdot \cos\left[\theta_y(x) - \frac{\mathrm{d}u_{1z}(\mathrm{x})}{\mathrm{d}x}\right]}{w \cdot t} \tag{11}$$

$$\sigma_b(x) = \frac{6F_x \cdot (l - l_x - x)}{w \cdot t^2} \tag{12}$$

MEMS component failure as a function of the force position (figure 2), can occur in various situations: (a) fracture - where a high value force is applied close to the anchor and the shear stress becomes significant in comparison to the bending moment, and (b) excessive deformations, both elastic and plastic - when flexible plate not recover its shape after loading [6, 8].

### *2.2. Stiction and adhesion considerations*

*2.2.1. Peel number.* The surface effects produce strong adhesion, friction and wear and are major problems which limit both manufacturing efficiency and lifetime of several MEMS devices [9]. It is known that adhesion is caused by capillary, electrostatic, van der Waals forces, and can also be caused by inertia (shock, rapid air flow) forces. On the other hand, stiction phenomenon is an unintentional adhesion of microstructure on the substrate when recovery forces are unable to overcome the interfacial forces. The problem of MEMS stiction can be divided into two categories: release-related stiction and in-use stiction [9]. Consequently adhesion and stiction can be a catastrophic failure that deserves attention.

We have focused on MEMS devices exposure to environment conditions when stiction phenomenon appears. Theories used in this paper are environmental theories of continuous elastic medium and therefore assume that no plastic deformation occurs.

To study the phenomenon of adhesion of mobile MEMS microstructures brought to the substrate, it was proposed by Mastrangelo and Hsu in [10], a dimensionless number, called peel number. This peel number, $N_P$, is the ratio of the elastic deformation energy stored in the deformed microstructure to the work of adhesion between the microstructure and the substrate. The microstructure will not join to the substrate if $N_P > 1$, more exactly the restored elastic strain energy is greater than the work of adhesion. Furthermore when $N_P \leq 1$, the deformed microstructure does not have enough energy to overcome the adhesion between the beam and the substrate.

For a long cantilever of thickness $t$ and elastic modulus $E$ suspended at a distance $g_0$ from the substrate, illustrated in figure 2, the peel number is [9, 10]:

$$N_p = \frac{3E \cdot t^3 \cdot g_0^2}{2(l - l_x)^4 \cdot \omega} \tag{13}$$

where $\omega$ is the work of adhesion between the cantilever and the substrate.

For a short cantilever beam with just its free-end adhered to the substrate, the corresponding peel number is [9, 10]:

$$N_p = \frac{3E \cdot t^3 \cdot g_0^2}{8l^4 \cdot \omega} \tag{14}$$

In this case, provided that the stored elastic strain energy is equal to the energy of adhesion, the peel number $N_P = 1$, from equation (14) could be determined the maximum dimensions of the microstructures (length of cantilever) that will not stick to the substrate:

$$l_{max} = \left( \frac{3E \cdot t^3 \cdot g_0^2}{8 \cdot \omega} \right)^{1/4} \tag{15}$$

*2.2.2. The influence of surface roughness on adhesion.* The energy required of detachment the cantilever from the substrate taking into account the influence of surface roughness on adhesion could be expressed by [9]:

$$U_s = w \cdot l_x \cdot \omega' = w \cdot l_x \cdot f(\delta) \cdot \omega \tag{16}$$

Where the effective work of adhesion $\omega'$, taking into consideration the surface roughness is given by [9]: $\omega' = f(\delta) \cdot \omega$ and $f(\delta)$ is a dimensionless roughness function which reflects the influence of surface roughness on adhesion.

Elastic strain energy is [9]:

$$U_E = \frac{6E \cdot I_y \cdot g_0^2}{(l - l_x)^3} \tag{17}$$

At equilibrium $\frac{d(U_S + U_E)}{d(l - l_x)} = 0$ results the corresponding peel number for cantilever beam adhesion to a rough surface $\overline{N}_p = \frac{N_p}{f(\delta)}$. This express indicates that the adhesion of a cantilever beam with a rough substrate is reduced with increasing adhesion parameter $\delta$ [9].

## 3. Experimental program

### *3.1. Investigated sample*

The investigated samples (figure 5) are microcantilevers with various geometrical dimensions made from gold (electroplated + about 40nm evaporated Au) with a Young's modulus of $E$ = 86GPa. The structures were fabricated by LAAS laboratory in Toulouse (France) in 10 lithography and deposition steps using a wafer substrate of silicon, with a gap between flexible part and substrate of about 3μm. The geometrical dimensions of the microcantilevers are: $l$ = 300, 350, 400, 450, 500, 550, 600, 650, 700, 750, 800μm; $w$ = 35μm; $t$ = 2μm.

*3.2. Experimental procedure*

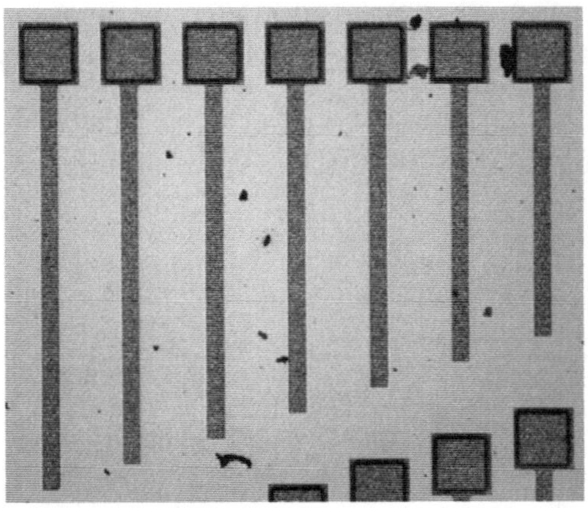

**Figure 5.** Samples for experimental tests.

Stiffness of investigated cantilevers is experimentally determined based on force versus deflection of the samples. Experimental-analytical evaluation of stress implies to find the experimental force that bends the flexible plate to substrate and to compute the stress by using relation (6) for long beam and relation (10) for shortest one.

Investigations were performed with an acoustic isolated and environment controlled atomic force microscope (AFM) XE 70 using the contact mode. The introduction in 1985 of the AFM based technique, due to its high spatial resolution and also force and displacement sensing capabilities is provided as a method for measuring ultra-small forces between a probe tip and an engineering surface and also has been used for morphological and surface roughness on the nanoscale, as well as for friction and adhesion measurements.

Stiction problems are typical to the electrostatic actuated MEMS cantilevers. In our experiments the electrostatic force is changed with a mechanical force performed by the bending deflection of an AFM probe with known stiffness, which deflects the samples directly to substrate. For a given deflection of investigated cantilevers, the correlation between the electrostatic force and the used mechanical one, based on equations (2) and (3), can be written as:

$$k_{AFM} \cdot z_{AFM} = \frac{\varepsilon \cdot w(l - l_x) \cdot U^2}{2(g_0 - u_z)^2} \tag{18}$$

where $k_{AFM}$ is the stiffness of AFM probe and $z_{AFM}$ is its bending deflection.

During tests, the position of the acting force (position of AFM probe) which bends the flexible plate to substrate is applied at the beams free-end. After, it is moved from the free-end towards the anchor. In this situation, the contact area between flexible plate and substrate increases.

The tests were performed at room temperature (22°C) and at relative humidity RH of 40%. Each measurement was repeated many times in order to improve the accuracy of the experimental results.

The cantilever used for tribological investigation in contact mode was HQ-NSC35/Hard/Al BS from $Si_3N_4$ with 30nm coating aluminum on the backside, with a nominal value of the spring constant $k$ = 16Nm$^{-1}$ and the following geometrical parameters: length 130μm, width 35μm, thickness 2μm.

During AFM tests the vertical piezoelectric displacement of the scanning head $z_{piezo}$ is controlled and the bending deflection of AFM probe is optically monitored. The bending deflection of cantilevers can be determined as $z_{sample} = z_{piezo} - z_{AFM}$. The normal load $F$ applied on investigated samples is the same as the force on the AFM probe that is obtained based on the bending deflection $z_{AFM}$ and the stiffness of AFM probe as included in equation (18). Slope of the load vs. displacement experimental curve provides the stiffness of the microcantilevers. The maximum AFM force that bends the samples to substrate is used to determine the bending stress. The spectroscopy in point is used to determine the adhesion between flexible structure and their substrate.

### *3.3. Stiffness and stress evaluation*

*3.3.1. Stiffness and stress for a force applied on the cantilevers free-end.* Samples with the length between 300µm and 800µm were investigated in order to determine the influence of length on the cantilever stiffness. A mechanical AFM force was applied at the free end of each cantilever until it reached the substrate, i.e. a distance of 3µm. Based on the AFM experimental curve and the methodology described above the stiffness was obtained between between 0.28N/m for sample of 300µm length and 0.012N/m for sample of 800µm length. In figure 6 are plotted the dependence between applied force and cantilever deflection for two samples with the length of 300µm and 800µm. It is observed that as the length of samples increases, the stiffness decreases, respectively. With other words, increasing the length of the sample the force required to bend the sample to the substrate decreases.

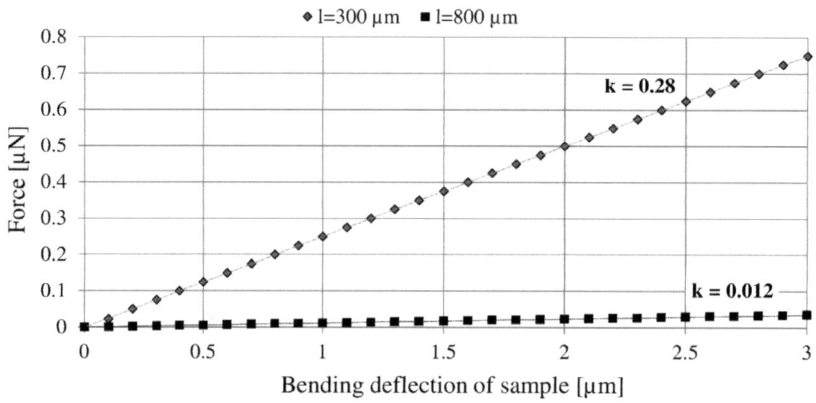


**Figure 6.** Experimental bending force versus deflection of samples.

The investigated cantilevers were also numerically analyzed by Finite Element (FE) method using ANSYS Workbench 13.0 software. The static structural analysis included large deflection option and uses a FE model consisting of hexahedral elements with an element size of 1.5µm. For the shortest cantilever the generated mesh has about 197000 elements and 245000 elements for the longest. Deflection under a unit load applied as in the experiment at the free end of the shortest cantilever ($l$ = 300µm) is presented in figure 7a. Under the applied force, a bending stiffness equal to 0.2261 was numerically determined for the cantilever with a length of 300µm.

The same experimental and numerical analysis was applied for all samples and the comparative results of stiffness are presented in figure 7b. It can be observed that the stiffness decreases as the length increases. A good correlation between the stiffness obtained based on numerical analysis and experimental test is observed.

Using the experimental applied forces and the modulus of elasticity, the bending stress for different deflections positions are computed based on equation (6). Figure 8 presents the stress variation as a function of the bending deflection of the shortest and longest samples. The force was applied at the beam free-end and a good correlation between the bending stresses obtained based on numerical analysis and experimental test is observed.

IOP Conf. Series: Materials Science and Engineering **174** (2017) 012029 doi:10.1088/1757-899X/174/1/012029

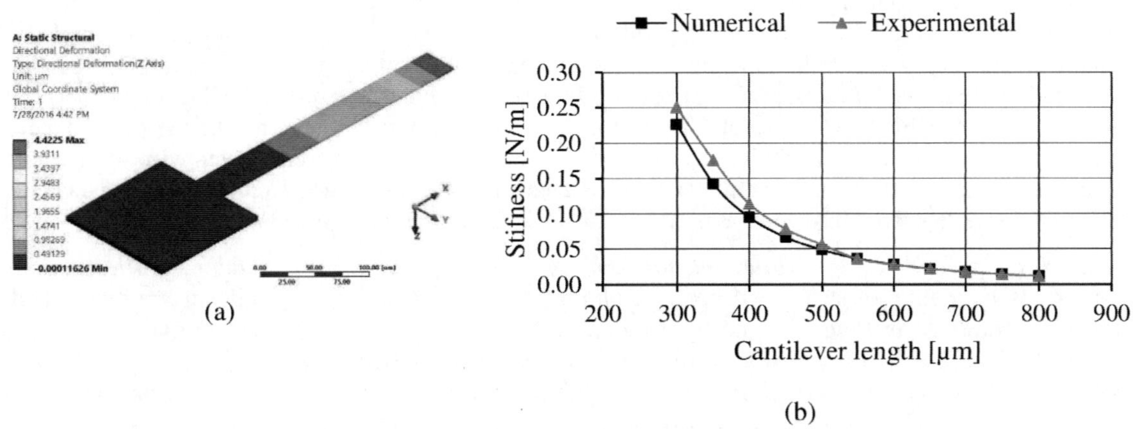


**Figure 7.** (a) FE analysis of the cantilever deflection with a length of 300µm, (b) Experimental and numerical bending stiffness versus the length of samples.

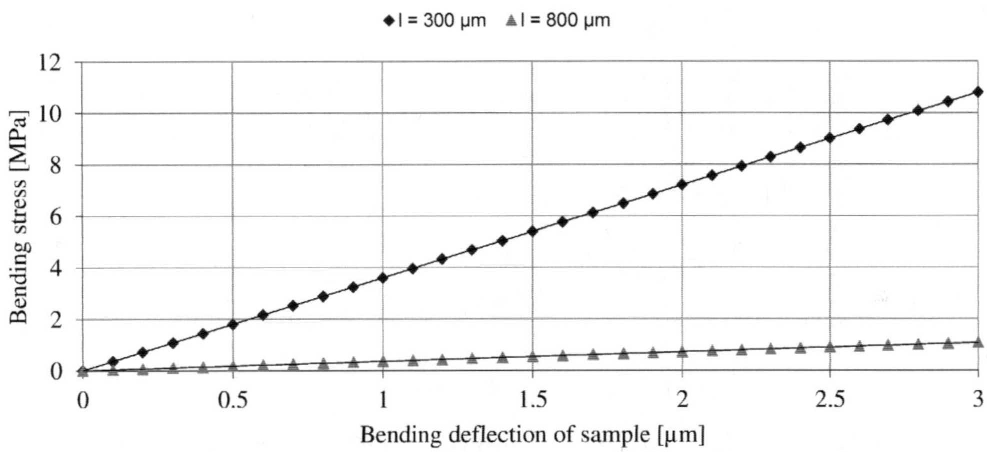


**Figure 8.** Bending stress versus bending deflection of samples with the lengths of 300µm and 800µm (the force acts at the free-end of cantilevers, gap = 3µm).

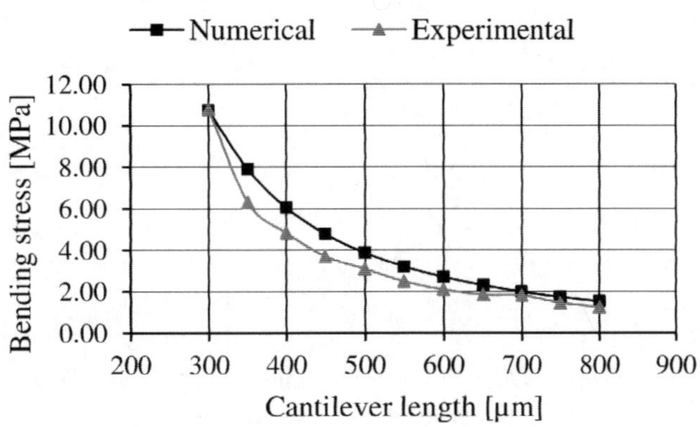


**Figure 9.** Experimental and numerical maximum bending stress as a function of the samples length.

The maximum bending stress at the fixed end of the microcantilever is determined for all microcantilevers with the length between 300μm and 800μm (figure 9). The same experimental and numerical analysis was applied for all samples and the comparative results of bending stress are presented in figure 9. It is observed that the bending stress decreases from 10.8MPa for the shortest sample to 1.2MPa for the sample of 800μm for the same bending deflection. On the other hand a bending stress equal by 10.742MPa was numerically determined for the cantilever with a length of 300μm and 1.512MPa for the cantilever with a length of 800μm. A good correlation between the bending stresses obtained based on numerical analysis and experimental test is observed.

*3.3.2. Stiffness and stress evaluation for different force positions on a microcantilever.* The experimental variation of stiffness depending on the force positions for the cantilever with the length 800μm is presented in figure 10 and the bending stresses variation of the same sample is shown in figure 11. During tests, the position of the acting force, which deflects the samples to substrate, was changed in the case of $(l - l_x) \geq 5w$. In this situation the Euler-Bernoulli beam model was considered (cross-sections remain perpendicular to the neutral axis). If we go too close to the anchor the shearing effects become important, and the regular bending deformations are augmented by the addition of shearing deformations, according to the Timoshenko's beam model.

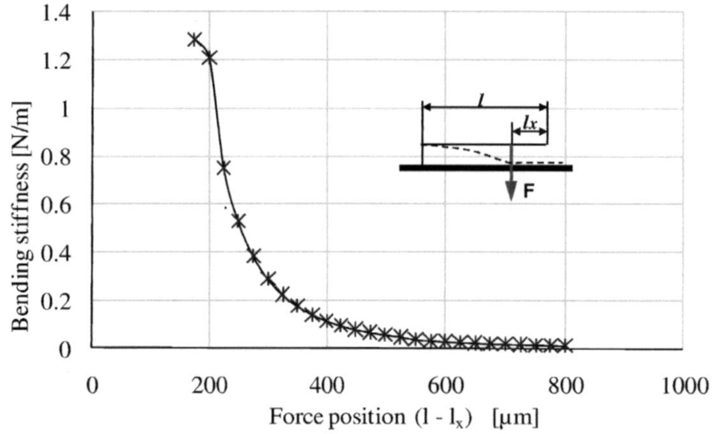


**Figure 10.** Experimental variations of stiffness as a function of the force positions *$(l - l_x)$* on 800μm length cantilever.

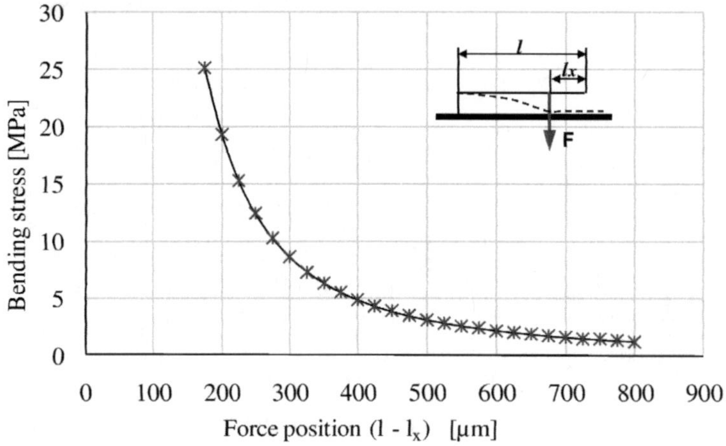


**Figure 11.** Experimental variations of bending stress as a function of the force positions *$(l - l_x)$* of the cantilever with 800μm length.

As expected, an increase of the sample stiffness is observed as we apply a force close to the anchor.

Moreover, we can observe that, the bending stress is increasing if the force is applied close to anchor and produces sample deformation until substrate. Indeed, the stress is mainly influenced by the width $w$ and thickness $t$ of samples. Based on these graphs at any time we can determine the stiffness of the sample as a function of the acting electrode position as well as the maximum bending stress from structure.

### *3.4. Adhesion evaluation*

Using AFM experimental curves, adhesion between microcantilevers and substrate can be determined. An experimental AFM curve of flexible structures has two different slopes as shown in figure 12 and figure 13. The unloading part of curve (from point 1 to 3) corresponds to the retracting of the AFM probe to its initial position. This part is used to determine the adhesion between microcantilever and substrate, phenomenon that occurs in the position 2 where the cantilever should be detached from substrate and moved to its initial position. A peak in the unloading curve can be observed in this point, and the peak height is directly proportional with the adhesion force between cantilever and substrate. The force is applied at the beam free-end. The adhesion force increases as the stiffness of samples decreases which can be seen in figure 12 for the cantilever with the length of 450μm comparatively with the sample with the length of 650μm (figure 13).

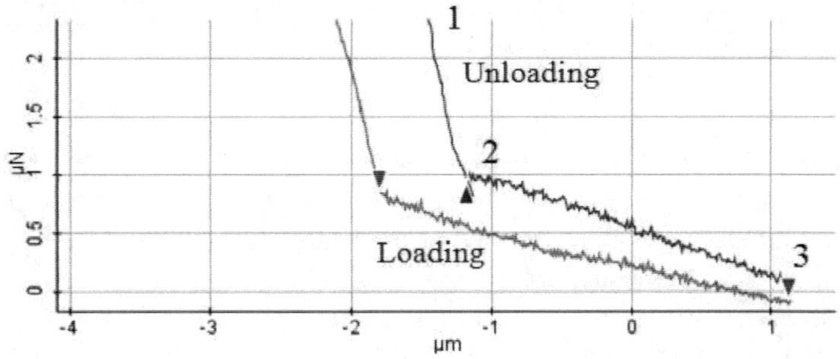


Cursors

| Cursor | ΔX(μm) | ΔY(nN) | Left X(μm) | Left Y(nN) | Right X(μm) | Right Y(nN) |
|---|---|---|---|---|---|---|
| Trace | 2.930 | -916.272 | -1.796 | 836.278 | 1.134 | -79.994 |
| Retrace | 0.018 | 54.090 | -1.179 | 945.203 | -1.161 | 999.294 |

**Figure 12.** AFM experimental curve of a cantilever of 450μm length.

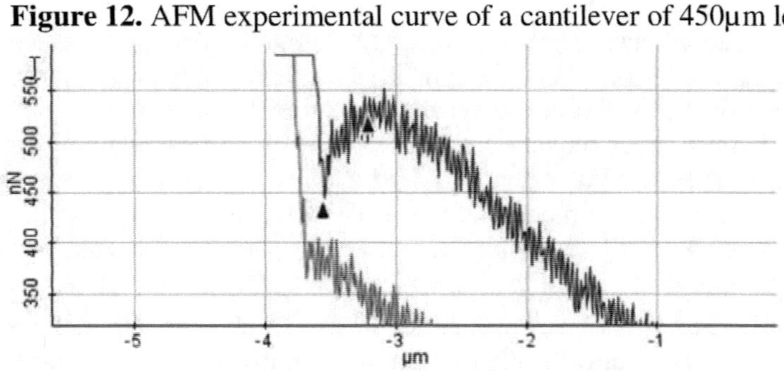


Cursors

| Cursor | ΔX(nm) | ΔY(nN) | Left X(μm) | Left Y(nN) | Right X(μm) | Right Y(nN) |
|---|---|---|---|---|---|---|
| Trace | 0.000 | 0.000 | 0.791 | -46.043 | 0.791 | -46.043 |
| Retrace | 338.982 | 84.147 | -3.555 | 445.583 | -3.216 | 529.730 |

**Figure 13.** Experimental adhesion of a cantilever of 650μm length.

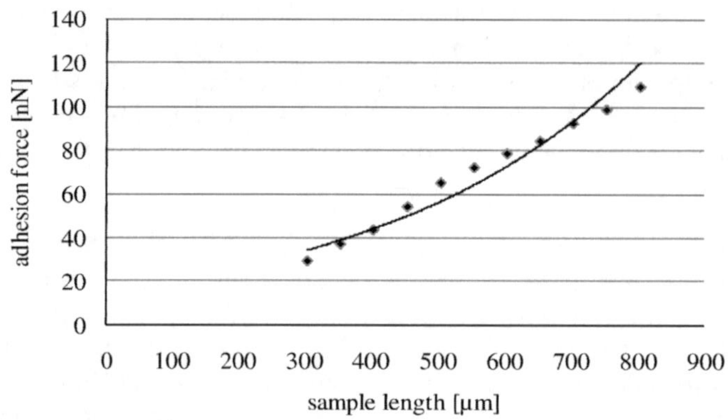


**Figure 14.** Experimental adhesion force versus sample length.

The same tests were performed for all investigated samples when a force is applied at the beam free-end. The variation of adhesion force obtained for the whole set of experimentally cantilevers is represented in figure 14. It is observed that the adhesion force between investigated cantilevers and substrate decreases with decreasing the length of samples. From the chart we can determine the value of adhesion force for the membrane that interests us.

## 4. Conclusions

The effect of geometrical dimensions (length) on mechanical and tribological behavior of MEMS cantilevers made of electroplated gold was analyzed in this paper. The force position (corresponding to the position of the acting electrode) was applied at the beams free-end and after it was moved from the cantilevers free-end towards to the anchor.

The dependences between stiffness and stress as a function of the cantilever length as well as their dependences on the variable travel range of load have been investigated. It has been experimentally determined that the cantilever stiffness decreases as the length increases, more exactly from 0.28N/m for sample of 300µm length to 0.012N/m for sample of 800µm length. Using numerical simulation a bending stiffness equal to 0.2261 was determined for the cantilever with a length of 300µm and 0.01175 for the cantilever with a length of 800µm. The numerical results are in good correlation with experimental values.

Experimental investigations have shown that deflections and stresses generated by movement of applied force towards the anchor should be analytically described using two hypotheses. For positions far from the anchor (length at least 5 times larger than the largest cross-sectional dimension) the Euler-Bernoulli beam model must be employed. For closer force positions relatively to anchor shearing effect become important and the regular bending deformations has to be corrected by the addition of shearing deformations, according to a model known as Timoshenko's beam model.

In our experiments, in order to avoid plastic deformation of cantilevers, we stopped the force position at the situation when only the Euler-Bernoulli beam model (cross-sections remain perpendicular to the neutral axis) have to be considered ($l - l_x \geq 5w$). Experimental and numerical maximum bending stresses as a function of the samples length show a very good correlation of the results.

The differences in terms of stiffness and bending stresses between the situations of separate analysis of each cantilever (the force applied at free-end) or by moving the force towards the anchor, in the same loading position, in case of the longest cantilever are small. Due to the restricted movement of the free end (friction between the cantilever and substrate) when the cantilever was deflected to the substrate a low stiffness increase can be noticed in the second analysis.

A theoretical consideration about stiction and adhesion has been presented in the paper followed by experimental test. Using the AFM experimental force versus deflection curve variation of adhesion force for the whole set of cantilevers was obtained for a force applied to the free end of the samples. It can be stated that the adhesion force increase when the cantilever length increase. The phenomenon is mainly due to the restoring forces that depend on the stiffness value. A smaller stiffness will produce a lower restoring force. Considering the experimental results presented in the paper concerning to the adhesion effect we can determine the value of adhesion force for the cantilever that interests us. Based on experimental results obtained it is possible to appreciate the tribomechanical behavior of thin gold cantilevers when a moving force acts over the cantilever.

**Acknowledgments**
This work was supported by the Romanian Space Agency – STAR Project no. 97 - 2013 grant from the Research Program for Space Technology Development and Innovation and Advanced Research (STAR).

**References**

[1] Chuang W C Lee, H L Cang P Z and Hu Y C 2010 *Sensors* **10** 6149-6171
[2] Spearing S M 2000 *Acta Mater* **48** 179-196
[3] Abdel-Rahman E M, Younis M I and Nayfeh A H 2002 *J Micromech Microeng* **12** 759
[4] Loui A, Goericke F T, Ratto T V, Lee J, Hart B R and King WP 2008 *Sensor Actuator A- Phys* **147** 516-521
[5] Lobontiu N and Garcia E 2004 *Mechanics of microelectromechanical systems* (Boston: Springer Science & Business Media).
[6] Pustan M, Rochus V and Golinval J-C 2010 Effects of the geometrical dimensions on stress and strain of electrostatically actuated MEMS resonators at pull-in and stiction positions *Proc. IEEE Conference on Thermal, Mechanical & Multi-Physics Simulation, and Experiments in Microelectronics and Microsystems (EuroSimE)* pp. 1-5
[7] Young W C and Budynas R G 2002 *Roark's formulas for stress and strain* vol 7 (New York: McGraw-Hill)
[8] Pustan M S 2015 *Mechanical and tribological characterization of MEMS* (Hab. Thesis, Cluj Napoca) available at http://www.utcluj.ro/media/documents/2015/Teza_de_Abilitare_Pustan_Marius_1.pdf (15.06.2016)
[9] Zhao Y P, Wang L and Yu T 2003 *J Adhes Sci Technol* **17** 519-546
[10] Mastrangelo C H and Hsu C 1993 *J Microelectromech S* **2** 44-55

# Two-way coupled Reynolds, Rayleigh-Plesset-Scriven and energy equations for fully transient cavitation and heat transfer modeling

**M Braun, K Pierson and T Snyder**

Department of Mechanical Engineering, The University of Akron, Akron, OH, USA

E-mail: mjbraun@uakron.edu

**Abstract.** The Rayleigh-Plesset-Scriven (RPS) equation representing the source of void production is coupled with the Reynolds (RE) and energy equations in a fully transient solution with feedback between these equations. Temperature effects are accounted for by using an energy equation integrated across the bearing film. Dynamic enlargement of the surface of each of the discrete bubbles forming the pseudo-cavitation zone is accounted for by introducing the surface dilatational viscosity term in the RPS equation; this represents a continuation of previous work by Snyder et al [1] that established parametrically the significance of the surface dilatational viscosity ($k^s$=3.75*$10^{-3}$ N.s/m) both in the development of the cavitation bubble and the formation and sustenance of the subcavitation tensile forces. The study presents the interlaced effects of residual fluid internal energy, eccentricity, angular velocity and heat transfer coefficient on the pressure and pseudo-cavitation development.

## Nomenclature

| | |
|---|---|
| $c$ | Specific heat of oil |
| $h, h_0, h_h$ | Heat transfer coefficient |
| $H$ | Thickness of thin film |
| $k$ | Thermal conductivity of oil |
| $P, P_B, P_G, P_v$ | Pressure, total pressure inside the bubble, pressure of the gas, vapor pressure |
| $P_R, P_\infty$ | Pressure directly outside of bubble wall, and pressure in the liquid far from the bubble |
| $Q_{gen}$ | Heat generation due to viscous dissipation |
| $Q_{res}$ | $Q_{gen}$ less the amount of heat lost from convection to bearing components |
| $R_B$ | Cavitating bubble radius |
| $t$ | Time |
| $T$ | Temperature of mixture |
| $\bar{u}, \bar{w}$ | Average velocity of mixture in $x-$ and $y-$ direction respectively |
| $V_{cell}$ | Volume of computational cell |
| $x, y$ | Spatial coordinates: along the unwrapped surface, axial |
| $\alpha$ | Void fraction |
| $\gamma$ | Surface tension of bubble |
| $\varepsilon$ | Bearing eccentricity ratio |

| | |
|---|---|
| $\kappa$, $\kappa^s$ | Bulk dilatational viscosity, Surface Dilatational Viscosity |
| $\mu, \mu_l, \mu_B$ | Dynamic viscosity of mixture, liquid (oil) and bubble. |
| $\nu_L$ | Kinematic viscosity of liquid (oil) |
| $\rho, \rho_l, \rho_B$ | Density of mixture, liquid and gas bubble respectively |

## 1. Introduction

There is a plethora of literature treating pseudo-, vaporous and gaseous cavitation, a difficult and complex problem that requires an interdisciplinary approach to solution. There has been steady progress in the understanding of this problem, starting with the initial treatise of Osborne Reynolds [2] in 1886 and Sommerfeld [3] in 1904, to today's state of the art which offers a multi physics model and numerical implementation (Someya [4], Natsumeda and Someya [5], Gehanin et al. [6, 7], Snyder et al. [1]). Early, nonconservative cavitation models were followed by mass conserving models, which assumed striated flows formed out of gas fingers interspersed with liquid regions extending across the clearance. In the last 75 years, numerous innovative analytical and numerical solutions, as well as seminal experimental findings were offered. In this context, the reader is further referred to the review publications of Dowson et al. [8], Brewe et al. [9], Braun and Hannon [10] and Snyder et al [1]. Zuber and Dougherty [11] in 1982 proposed a Reynolds equation (RE) that introduced a homogeneous two-phase lubricating film mixture as the working fluid. Natsumeda and Someya [5] in 1987 also used the concept of a homogeneous two-phase flow RE coupled with the Rayleigh-Plesset-Scriven (RPS) equation as the void generation source for the homogeneous two-phase film. Notably, the RPS equation differs from the classical Rayleigh-Plesset equation through the inclusion of viscous surface dilatation effects ($\frac{4\kappa^s}{\rho_l R_B^2}\dot{R}_B$). These effects have proven to have a major role in both the physical sustenance of the subcavity tensile forces (measured to be as high as 1.2 MPa), and the numerical stability of the RPS equation. The importance for the stability of the numerical computations has been clearly recognized by both work of Gehannin et al. [7, 12] and Natsumeda [5] and Someya [4]. The RPS equation solves for the bubble radius growth $R(t)$, independent of inertia effects, thus providing the mechanism for void fraction, $\alpha$, generation and its change with respect to time, implicitly contributing to the calculation of the two-phase homogeneous $\rho$ and $\mu$ used in the RE.

## 2. Scope of work

By adding the energy equation, the present work extends the work of Snyder et al. [1] regarding the development of pseudo-cavitation (no mass transfer by diffusion, or evaporation). In the present development, the collection of kernel bubbles grow both due to depressurization and temperature change, as the RE and RPS equations are coupled in a feedback loop with a two-dimensional energy equation lumped across the film thickness. Thus, the void fraction $\alpha$, the liquid and bubble densities ($\rho_L$and $\rho_B$), as well as their viscosities ($\mu_L$and $\mu_B$), vary both in the circumferential and axial directions ($x$, $z$) with both pressure and temperature. The two dimensional energy equation ($x$, $z$) incorporates the effects of convection-conduction heat transfer at the fluid boundaries with the shaft and bushing. The calculation of the void fraction, $\alpha$, as a function of ($x$, $z$, $t$, and $T$) at every grid point allows transient calculation of local homogeneous two-phase film density and viscosity in conjunction with the transient form of the RPS, RE and energy equations. The results account for cavitation, pressure, temperature and void fraction development when the fluid heat transfer coefficient, $h_0=h_h=h$, is varied parametrically, while eccentricity, angular velocity, and surface dilatational viscosity are kept constant.

## 3. On the surface dilatational viscosity

In a fluid at rest, the hydrostatic stress ($-\bar{P}$) is the same as the thermodynamic pressure. However, for a fluid in motion, the total normal stress, $\sigma_{ii}$, deviates from the thermodynamic pressure. One can write a relationship between the symmetric components of the deviatoric stress tensor and the thermodynamic pressure and relate to the average pressure, $\bar{P}$, equation 1

$$-\bar{P} = \bar{\sigma} = \frac{\sigma_{ii}}{3} = \left(\frac{2\mu}{3} + \lambda\right)\frac{\partial u_i}{\partial x_i} - p \tag{1}$$

where $(\frac{2\mu}{3} + \lambda) = \kappa$ is defined as bulk (dilatational) viscosity. While $\mu$ represents the dynamic shear viscosity of the fluid, the physical meaning of $\lambda$ is associated with viscous dissipation as it relates to volumetric change, Schlichting [13]. In three-dimensional fluids, generally Stokes hypothesis is applied, wherein the thermodynamic and mechanical pressures are taken to be equal, i.e. $\lambda = -2\mu/3$ and $\kappa = 0$. However, unlike $\kappa$, its two-dimensional analog $\kappa^s$ (surface dilatational viscosity) does not vanish when it is present in two-phase fluids and can greatly affect the dynamic interfacial balance of forces for a bubble growing through pseudo-cavitation, evaporation or gas diffusion. A stress balance calculated at the bubble's wall, yields

$$\Delta\sigma_{fluid-bubble} = \frac{2\gamma}{R_B} + \frac{4\kappa^s}{R_B^2}\dot{R}_B \tag{2}$$

where, in an active expanding/contracting environment the dynamic tension at the interface exceeds the equilibrium tension, $\frac{2\gamma}{R_B}$, by the quantity $\frac{4\kappa^s}{R_B^2}\dot{R}_B$. The latter through its derivative $\dot{R}_B$, always exerts an action opposing the motion of the bubble wall resulting in an interfacial stress jump that causes the deviation from the static case.

## 4. Bubble dynamics

The Rayleigh [14] bubble dynamics equation describes the growth and collapse of an isolated, spherical bubble driven only by a varying external pressure field. The inclusion of interfacial rheological stresses is necessary (Scriven [15]) when the ratio of the interfacial area-to-volume is large, which is the case of the journal bearing or squeeze film dampers. If viscous effects within the gas bubble are neglected and the liquid surrounding the bubble is considered to be Newtonian and incompressible ($\kappa = 0$, Stokes hypothesis), the dynamic surface stress balance yields to a modified pressures in the liquid adjacent to the bubble wall, $P_R$, of the form (Snyder et al. [1])

$$P_R = P_B - 4\frac{\mu_L \dot{R}}{R} - \frac{2\gamma}{R} - \frac{4\kappa^s \dot{R}}{R^2} \tag{3}$$

Further, replacing $P_R$ in Rayleigh's equation with its expression given by equation 3 results in a bubble dynamics equation which accounts for all of the interfacial properties

$$(R\ddot{R} + \frac{3}{2}\dot{R}^2) + 4\frac{\nu_L \dot{R}}{R} + \frac{2\gamma}{\rho_L R} + \frac{4\kappa^s \dot{R}}{\rho_L R^2} = \frac{P_B - P_\infty}{\rho_L} \overset{Daltonlaw}{\cong} \frac{P_G + P_v - P_\infty}{\rho_L} \tag{4}$$

In equation 4, the bubble contents are considered to be a mixture of both vapor and non-condensible gas(es) where by virtue of Dalton law of partial pressures one can write that $P_B = P_G + P_v$. For the $P_G$ variation one may consider a polytropic transformation, Snyder et al. [1], ($P_G = \frac{P_{G0}}{\rho_L}\left(\frac{R_0}{R}\right)^{3n}$) or the Combined Gas Law (CGL) ($P_G = \frac{P_{G0}}{\rho_L}\left(\frac{R_0}{R}\right)^3 \frac{T_G}{T_{G0}}$). In the latter case the variation of temperatures require that the energy equation be coupled with the RE and the RPS equations.

## 5. The physical model

The proposed pseudo-cavitation model involves the feedback coupling between the RE, RPS and the energy equations. This occurs through the exchange of values of the film pressure (RE) and variable bulk properties $\mu = \mu[\alpha\,(x,z,T)]$ and $\rho = \rho[\alpha(x,z,T)]$ of the two phase mixture, and adjustment of the void fraction by means of the combination of equation 4 with the Combined Gas Law and energy equation. The local film pressure $P$ is computed by the RE, and by entering the pressure driving term into the RPS equation, these two equations are coupled. The bubble is assumed to contain a single non

-condensable gas kernel undergoing a transformation consistent with equation 4 and the CGL, but neglecting the inertia terms.

$$4\frac{\nu_L \dot{R}}{R}+\frac{2\gamma}{\rho_L R}+\frac{4\kappa^s \dot{R}}{\rho_L R^2}=\frac{P_{G0}}{\rho_L}\left(\frac{R_0}{R}\right)^3\frac{T_G}{T_{G0}}-\frac{P}{\rho_L} \tag{5}$$

The change in bubble radii is used to compute the local cell void fraction through the ratio of a spherical bubble volume to the local computational cell volume as

$$\alpha=\frac{4/3\pi R^3}{V_{cell}} \tag{6}$$

The particular shape of the bubble is not considered to be of consequence insomuch as limiting the validity of the bubble dynamics equation, or its usefulness for the present application. With no simple or practical alternative approach for non-spherical bubble dynamics, the spherical bubble symmetry is assumed to be maintained.

If the bulk density and dynamic viscosity of the fluid film are considered to be continuous functions of a homogeneous two-phase fluid similar to the one used by Zuber and Dougherty [11], they may be expressed through the component properties and the void fraction as

$$\rho=\alpha\rho_G+(1-\alpha)\rho_L;\quad \mu=\alpha\mu_G+(1-\alpha)\mu_L \tag{7a,b}$$

In equations 7a and 7b, $\rho_L$ and $\mu_L$ are functions of the temperature [16] yielded by the energy equation presented below, equation 9. In regard to the fluid film, the laminar, transient, variable property form of the Reynolds equation can be written as

$$\frac{\partial}{\partial x}\left(\frac{\rho(\alpha)h^3}{12\mu(\alpha)}\frac{\partial P_R}{\partial x}\right)+\frac{\partial}{\partial y}\left(\frac{\rho(\alpha)h^3}{12\mu(\alpha)}\frac{\partial P_R}{\partial y}\right)=\frac{\partial}{\partial x}\left(\frac{\rho(\alpha)hU}{2}\right)+\frac{\partial(\rho(\alpha)h)}{\partial t} \tag{8}$$

The energy equation in its transient form, lumped across the fluid film takes the form

$$\frac{\partial(cT)}{\partial t}+\bar{u}\frac{\partial(cT)}{\partial x}+\bar{w}\frac{\partial(cT)}{\partial y}=\frac{1}{\rho}\frac{\partial}{\partial y}\left(k\frac{\partial T}{\partial y}\right)-\frac{1}{\rho H}\left[h_h(T-T_{\infty 1})+h_0(T-T_{\infty 2})-Q_{gen}\right] \tag{9}$$

The last three terms of the RHS of equation 9 represent the heat lost by convection to the shaft and bushing (where $h_0=h_h=h$), and the heat generated due to frictional dissipation, respectively. The field equations written in Cartesian coordinates are applied to an unwrapped bearing fluid film, figure 1.

## 6. Results and discussion of the parametric variation the heat transfer coefficients.

All results presented herein are obtained for a journal bearing with an eccentricity of $\varepsilon=0.8$ running at an angular velocity of 5 krpm. Figure 2 presents a composite of 2-D line graphs at the axial center line for pressure, void, mixture viscosity, residual heat carried by the fluid and temperature, as the convective heat transfer coefficient (HTC) varies from 10 to isothermal conditions (HTC>>1000 $W/m^2$°C). This combined presentation is intended to facilitate the understanding of the interaction between these parameters. In figure 2a, the pressure decreases, as the HTC decreases from 1000 to 10 $W/m^2$°C. It is noteworthy that for the same trend of HTC (decrease by two orders of magnitude), the temperature raises by a factor of 2, figure 2d, while viscosity decreases by a factor of approximately 4 in the region confined between 90 and 200, figure 2e. The decrease in viscosity at the same temperature is driven mainly by the change in the void fraction, while its decrease at the different HTCs different is driven by the increase in temperature as HTC goes down. A comparative look at figures 2b and 2e, shows the most dramatic decrease in the mixture viscosity in the regions between 90° and 200° even though, between the same angle confines, the void fraction is essentially zero, figure 2b. This indicates that $\mu_L$ is the dominant quantity in the mixture viscosity with void playing almost no role. The driver of the entire development described above is $Q_{res}$, representing the amount of residual enthalpy left in the fluid when parts of the $Q_{gen}$ are lost due to convection to the shaft and bushing. Most generally, as HTC decreases, less heat is lost, resulting in highest temperatures at $h = 10\ \mathrm{W/m^2 °C}$. As the HTCs are increased by a factor of 10 and 100 respectively, the temperatures trend lower (by almost a factor of 2). This is justified by an increased heat loss as the increase in the HTC outweighs the decrease in temperatures.

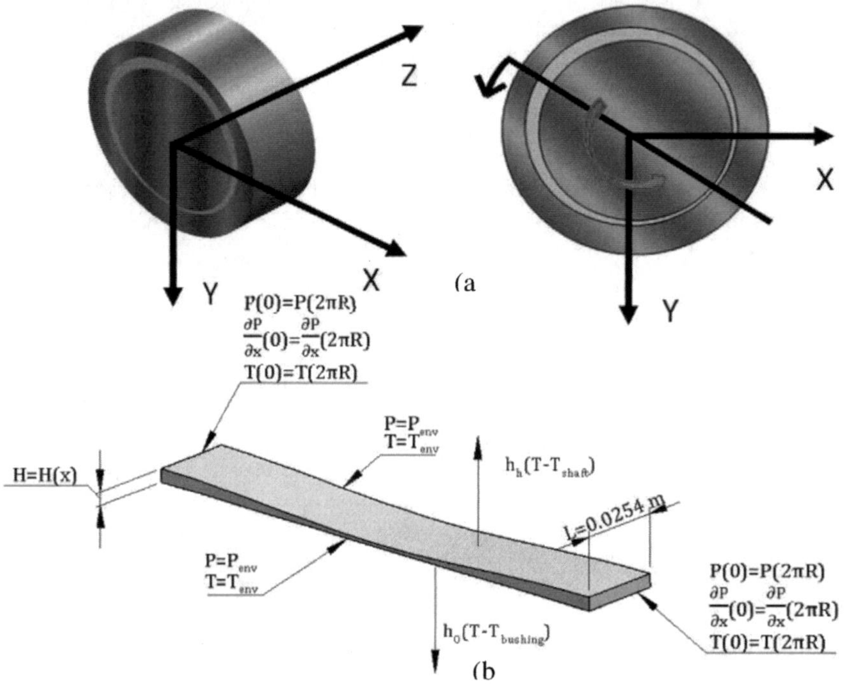


**Figure 1.** Geometry of the unwrapped bearing: (a) lateral and frontal view; (b) unwrapped film with Reynolds and energy equations boundary conditions. $T_{env} = T_{shaft} = T_{bushing} = 80°\mathrm{C}$; $P_{env} = 1\ \mathrm{atm}$; $h_h = h_0 = h$.

Thus, counter-intuitively more heat is lost at lower temperatures than at the higher ones. The HTC, and not the temperature magnitude is responsible for this phenomenon. Figure 3 presents the temperature development in the fluid over the entire surface of the fluid in the unwrapped fluid film when $\varepsilon = 0.8$ and the angular velocity is 5 krpm. For direct comparison purposes all three figures are plotted at the same scale. Figure 3a presents the temperature variation when HTC is $10\mathrm{W/m^2\ °C}$. Highest temperatures are observed here since this case corresponds to the lowest HTC. It appears the highest values occur, as expected, at the center line and continue almost unabated for the entire circumference (see also figure 2), though the width of this band is rather narrow. The lower temperatures that prevail on the outer axial regions are due to the boundary conditions which provide for fluid at 80°C temperature entering axially the bearing. The effects of these lower temperatures penetrate gradually towards the center where they meet a high temperature band. The same effect is visible, though less pronounced in figure 3b. The higher pressure achieved in the case of $h = 100\frac{\mathrm{W}}{\mathrm{m^2}}°\mathrm{C}$, allows more penetration of the temperatures towards the axial ends of the bearing creating larger bands of higher temperatures throughout the bearing. Figure 3c presents the case for $h = 1000\ \mathrm{W/m^2 °C}$. Even though this case is associated with the lowest temperature developing inside the bearing, the heat generation is the highest (figure 2c) due to higher viscosity (figure 2e).

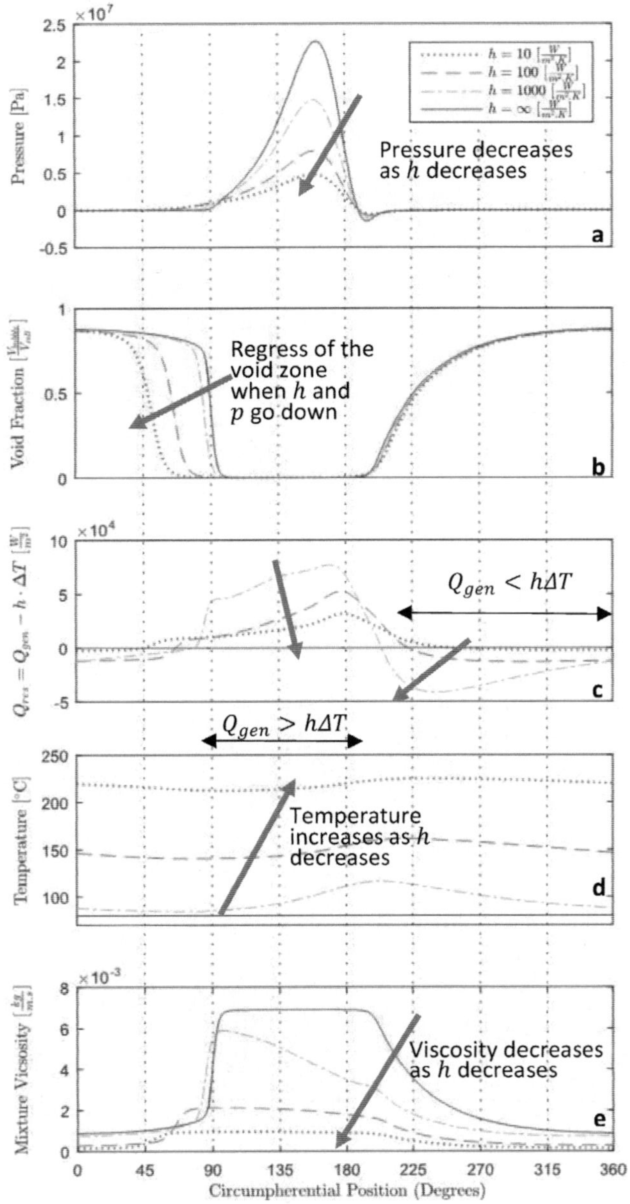


**Figure 2.**Study of the parametric effects of the heat transfer coefficient on the journal bearing behavior, $\varepsilon = 0.8$, angular velocity = 5 krpm. a) Pressure; b) void, $\alpha$; c) fluid residual heat $Q_{res}$; d) temperatures; e) mixture dynamic viscosity.

Figure 4 presents the pressure development when the heat transfer coefficient is varied. The discussion of the centerline pressures behavior has been offered in figure 2. This figure presents the 2-D contour plots covering the entire surface of the bearing and rendering the two dimensional formation both of the cavity zone and the subcavity tensile region. Figures 4a-4d present the evolution of the overall pressures as the heat transfer coefficient is increased from 10 to 1000 W/m$^2$°C. For comparison purposes the isothermal case ($h \rightarrow \infty$) is presented in figure 4d.

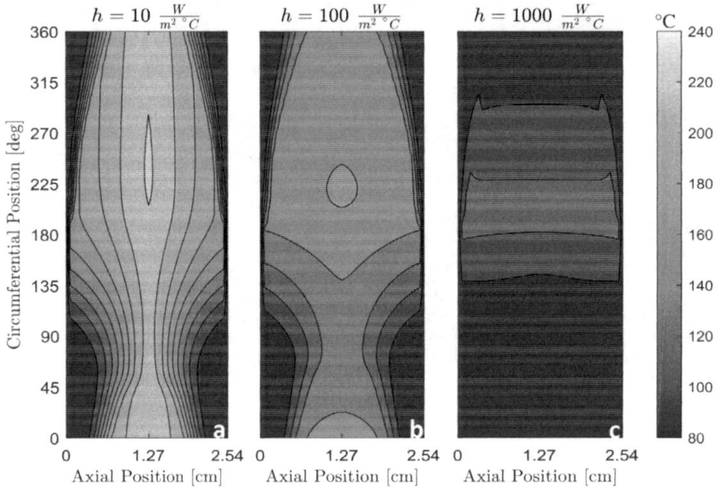


**Figure 3.**Temperature contour plots for bearing temperature for parametric variation of the heat transfer coefficient when $\varepsilon = 0.8$, angular velocity=5 krpm.

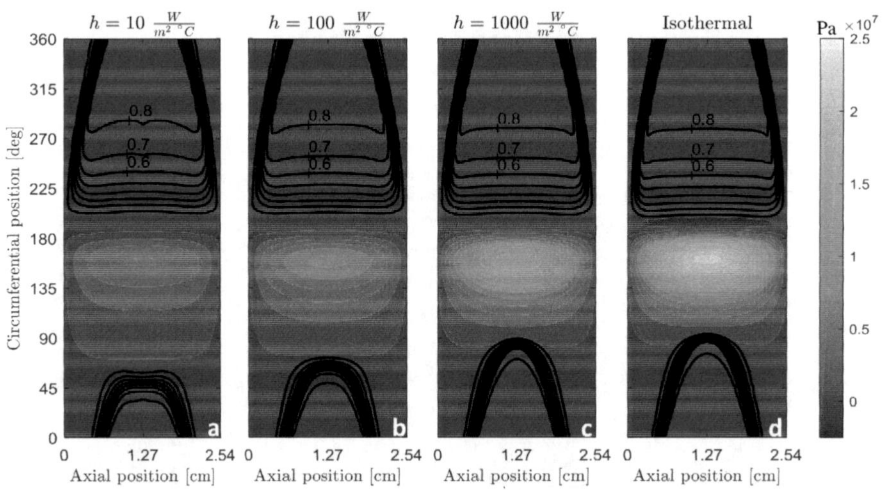


**Figure 4**. Pressure contour plots for bearing pressure and void fraction for parametric variation of the heat transfer coefficient when ε=0.8, angular velocity=5 krpm.

One can see clearly the evolution of the pressure contour curves from the lowest pressures $at\ h = 10\ W/m^2$°C, with practically no subcavity pressure present to the highest pressure for the isothermal case and the largest subcavity zone. The size of the subcavity is related to both surface dilatational viscosity [1] and the overall temperature levels. The figure also presents, in 2-D, the extent and growth of the cavity zone from its lowest expanse associated with the lowest pressure levels, to its highest, associated with the highest pressures.

## 7. Conclusions

The present study endeavors to detail the temperature effects on pressure, void, mixture viscosity and subcavity development when an axial-circumferential energy equation is fully coupled with a variable properties Reynolds, Rayleigh-Plesset-Scriven equations as well as the temperature dependent μ and ρ. It has been shown that the void development is largely dependent on the pressure variations while also being influenced by the temperature levels (figure 2b). Viscosity decreases significantly with temperature (figure 2c), while the latter is basically dependent on the $Q_{res}$ left in the fluid following heat losses from the film to the shaft and bushing. Counterintuitively, both the highest heat generation and heat losses happen for the lowest temperature, (figures 2c and 2d), resulting from the joint effects of temperature decrease with increasing HTC and decrease of mixture viscosity with temperature.

**References**

[1] Snyder T, Braun M and Pierson K 2016 Two-way coupled Reynolds and Rayleigh–Plesset equations for a fully transient, multiphysics cavitation model with pseudo-cavitation *Tribol. Int.* **93** 429-445

[2] Reynolds O 1886 On the theory of lubrication and its application to Mr. Beauchamp Tower's experiments including an experimental determination of the viscosity of olive oil *Phil. Trans. R. Soc. Lond.* **177** 157-234

[3] Sommerfeld A 1904 Zur hydrodynamische Theorie der Schmiermittelreibung *Zeit. Math. Phys.* **50** 97-155

[4] Someya T 2003 On the development of negative pressure in oil film and the characteristics of journal bearing *Meccanica* **38** 643-658

[5] Natsumeda S and Someya T 1987 negative pressures in statically and dynamically loaded 6journal bearings Paper III (ii) *Proc. of the 13th Leeds-Lyon Symp. on Tribology (Tribology Series* **11***)* ed D. Dowson et al. (Amsterdam: Elsevier) 65-72

[6] Gehannin J, Arghir M and Bonneau O 2015 A volume of fluid method for air ingestion in squeeze film dampers *Tribol T* **59** 208-218

[7] Gehannin J, Arghir M and Bonneau O 2010 Complete squeeze film damper analysis based on the bulk flow equations *Tribol T* **53** 84-95

[8] Dowson D, Godet M and Taylor CM 1974 Cavitation and related phenomena in lubrication *Proc. of the 1st Leeds-Lyon Symp. on Tribology* (London; New York: Mechanical Engineering Publications Ltd) 1-238

[9] Brewe D E 1986 Theoretical modeling of the vapor cavitation in dynamically loaded journal bearings *J. Tribology* **108** (4) 628-638

[10] Braun M and Hannon W 2010 Cavitation formation and modelling for fluid film bearings: a review *P. I. Mech. Eng. J-J Eng.* **224** (9) 839-86

[11] Zuber N and Dougherty D 1982 The field equations for two-phase Reynolds film flow with change of phase *ASLE Trans.* **25** (1) 108-116

[12] Gehannin J. Arghir M and Bonneau O 2009 Evaluation of the Rayleigh-Plesset equation based cavitation models for squeeze film dampers *J. Tribology* **131** (2) 15-25

[13] Schlichting H 1979 *Boundary layer theory* (7th ed.) (New York: McGraw Hill) 60-63

[14] Rayleigh L 1917 On the pressure developed in a liquid during the collapse of a spherical cavity *The London, Edinburgh, and Dublin Philosophical Magazine and Journal of Science* **34** (200) 94-98

[15] Scriven L 1960 Dyamics of a fluid interface. Equation of motion for Newtonian surface fluids *Chem. Eng. Sci.* **12** (2) 99-108

[16] Bruce RW 2012 *Handbook of lubrication and tribology, Volume II: Theory and design, Second Edition* (Boca Raton: CRC Press) 458-459

# Saturated porous layers squeezed between parallel disks in enclosed cells

**I C Melciu, T Cicone and M D Pascovici**

Department of Machine Elements and Tribology, University Politehnica of Bucharest, Splaiul Independenţei 313, Bucharest 060042, Romania

E-mail: melciu.catalin@yahoo.com

**Abstract.** Theoretical and experimental evidences show that high lift forces can be generated when a porous layer imbibed with a fluid is subjected to compression by a rigid and impermeable component in normal (approaching) relative motion. If the porous layer is soft enough to neglect its solid structure reaction to compression then the pressure increase can be entirely attributed to the flow resistance of the porous structure when the fluid is squeezed out. The mechanism is highly dependent on the variation of permeability with porosity at its turn variable with the rate of compression. Such a mechanism can be used for impact damping but realistic applications need to consider an enclosed system which keeps the squeezed fluid inside and allows for re-imbibition. The paper presents a simple analytical model for the effects produced in highly compressible porous layers imbibed with Newtonian liquids, during compression between two parallel rigid disks placed in enclosed cells with variable volume buffer, similar to a hydro-pneumatic accumulator.

## 1. Introduction

Porous materials and squeeze-film systems have become increasingly popular for a variety of applications in engineering (e.g. shock absorbers and squeeze dampers). Various squeezing effects have been analysed with relatively rigid, porous materials, in different geometrical configurations (spherical, cylindrical or plane contacts) and loading conditions (constant speed, constant force or constant energy). All these applications were characterized by constant permeability. A more complex squeezing mechanism in the presence of highly deformable porous media has been relatively recently put into light [1]. It was named by professor M.D. Pascovici ex-poro-hidrodynamic (XPHD) lubrication. Its characteristic is the variation of permeability during the process, due to high compression rates that modifies (decrease) material porosity. It was shown, both theoretically and experimentally its potential in generating high lift forces produced by the resistance to flow through the porous structure [2,3]. One of the possible applications is in shock absorption.

This complex process involves the squeeze of fluid from the porous layer at different levels of permeability. XPHD mechanism is strongly dependent on porosity variation and describes the lift effects produced by the flow of fluid through a porous material. As the porous structure is highly deformable (very soft) the normal forces generated by elastic compression of the fibers comprising the solid phase are negligible compared to the pressure forces.

Similar lift effects have been extensively studied by Weinbaum [4], with special emphasis on sliding motion over highly porous and soft materials filled with air. Even if, apparently, the mechanism differs from that of normal motion, both show a lift force generated under rapid compaction of the soft material.

Until now the XPHD mechanism proved to be efficient for the different configurations (sphere/plane, cylinder/cylinder, cylinder/plane, disk/plane) in open squeeze system. It means that the relative pressure on the contact boundaries is null (atmospheric pressure). Pascovici and co-workers [5] analyzed the squeeze flow at the contact between a disk and a conjugated plane surface, in case of non-zero pressure, modelling the flow into an extended porous layer (porous layer diameter is greater than that of the squeezing disk).

However for practical applications the system should be able to operate continuously, that is after squeezing out, the fluid must be re-imbibed into the porous layer, able to support a new loading. As a consequence, the fluid expelled out must be collected into a variable volume buffer, able to collect the whole volume of the dislocated fluid. The buffer can be represented either by a volume with elastic limits (toroidal membrane similar to an elastic bladder) or by a variable volume accumulator separated by a differential piston from a gas buffer chamber (solution similar to hydro-pneumatic systems used in shock dampers). This arrangement is called closed-cell as there is no fluid flow to the environment.

The model presented in this paper is focused on squeeze effects with variable pressure at the boundaries of the disks corresponding to the squeeze flow case in an encapsulated system when the fluid expelled out by the squeezing disks flows into a variable hydro-pneumatic buffer. The present work is originated from the necessity to investigate the damping capacity of the porous media imbibed with liquids, in cyclic loading conditions, a relatively new subject but which experiences an increasing interest due to its innovative approach and the variety of its possible applications.

## 2. The model

The model for the closed-cell in hydro-pneumatic-like configuration is schematically presented in figure 1. Two disks of radius, $R$, both perfectly rigid, separated by a soft porous layer, are placed into an enclosed cell. The porous layer of initial thickness, $h_0$, and initial porosity, $\varepsilon_0$, is completely imbibed with a Newtonian fluid of viscosity, $\eta$, assumed constant during the squeeze process. The upper disk can move downwardly, squeezes the porous layer and the liquid expelled out is collected in a variable volume accumulator created by a differential piston that separates it from a buffer gas chamber. The initial volume of the gas chamber is $V_0$ and its pressure is $p_0$. On the other side of the piston, the initial volume of the accumulator is assumed zero. During the squeeze, the fluid expelled out radially from the compressed porous layer is collected into the accumulator with simultaneous displacement of the piston toward the gas chamber, compressing the gas. The new volume of the gas chamber is reduced with the volume of the dislocated fluid. Neglecting the piston weight and the friction force between the piston and cylinder wall, the pressure in the accumulator equals the pressure in the gas chamber.

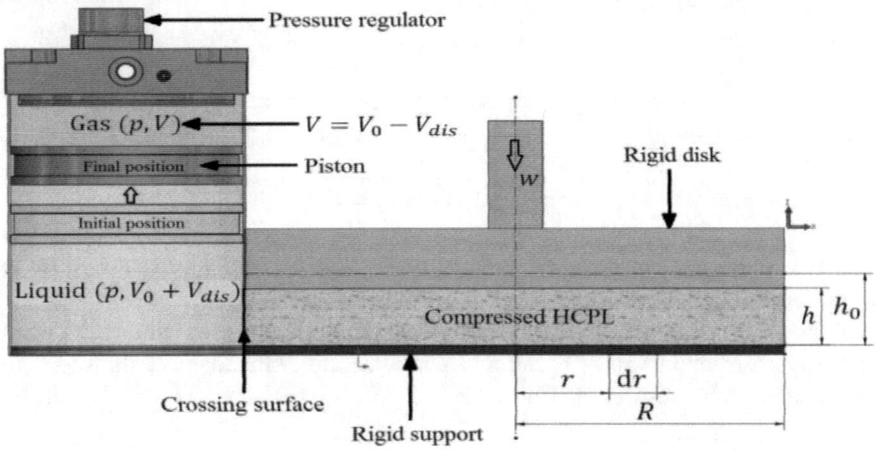


**Figure 1.** The working principle of an enclosed system.

The expelling of fluid from the porous material generates high lift forces on the moving disk and compresses the gas from the buffer. For this hydro-pneumatic system it is customary to assume that the heat generated by compression dissipates into the environment and the temperature remains constant during the process; thus isothermal change of state according to Boyle-Mariotte law can be used.

So that the pressure, $p$ can be expressed function of the initial volume of the buffer, $V_0$, and of the porous layer thickness, $h$, as follows:

$$p = \frac{p_0 V_0}{V_0 - \pi R^2 (h_0 - h)} \tag{1}$$

Introducing the relative buffer initial volume, $\bar{V}_0 = \pi R^2 h_0 \varepsilon_0 / V_0$ and the dimensionless thickness, $H = h/h_0$, relation (1) can be rewritten in the following form:

$$p = \frac{p_0 \varepsilon_0}{\varepsilon_0 - \bar{V}_0 + \bar{V}_0 H} \tag{2}$$

The XPHD lubrication modelling is based on the following simplifying hypotheses:

- The fluid is Newtonian, the flow is laminar, isothermal and isoviscous;
- The contact surfaces are rigid and impermeable;
- The highly compressible porous layer (HCPL) is homogeneous, isotropic and relatively thin; as consequence the pressure is assumed constant across the thickness of the porous layer;
- During the squeeze process the rigid plates remain parallel so that the thickness of the porous layer is constant in space and varies only in time;
- The flow through the porous layer is modelled with Darcy equation;
- The elastic forces of the porous medium are negligible compared to the flow resistance forces.

Highly porous materials are mainly characterized by porosity (voids in the structure of the solid material) and by permeability, which is the ability of the material to allow the flow of fluid through it. Throughout this paper, we will define the porous layer using compactness which is the inverse of porosity, and defined by this equation:

$$\sigma = 1 - \varepsilon \tag{3}$$

During compression, the pore geometry changes, expelling the fluid from the material. In this process, the solid volume of the structure is compressed until it reaches zero porosity. Assuming that the porous layer does not change its circular area (does not expand during compression), the solid fraction conservation yields to the following equation:

$$\sigma h = \sigma_0 h_0 \tag{4}$$

The permeability of the porous layer is associated to the porosity/compactness according with Kozeny–Carman law [6], regularly used in previous studies ([1]-[3], [5], [8]-[12]):

$$\phi = \frac{D(1-\sigma)^3}{\sigma^2} \tag{5}$$

where $D = d^2/16k$ is porous structure parameter, dependent on a characteristic dimension of the porous structure, $d$. Originally proposed for fluid flow through porous media consisting of spherical particles with relatively constant diameters, Kozeny–Carman equation proposes for the correction factor, $k$, correction values between 5 and 10. Later, Ghaddar [7] showed that equation 5 can be used with satisfactory results for fibrous-based porous structures, where $d$ is the average fiber diameter.

## 3. Analytical solution

The model described in previous section will be analysed in two loading conditions: constant velocity ($w = ct$) and constant force ($F = ct$), respectively. Each of them will be presented separately in the following chapters.

In both cases, as the squeezing surfaces remain parallel, the axisymmetry prevails and, assuming constant pressure across the thickness of the porous layer, the analysis becomes 1-D [13].

### 3.1. *Constant velocity squeeze* ($\mathbf{w = ct}$)

The flow rate conservation equation at any radius, $r$ for a given normal velocity, $w$, can be written by equating the fluid dislocated during compression with the fluid flowing radially across the circular boundary, according to Darcy law:

$$\pi r^2 w = -2\pi r \frac{\phi h}{\eta}\frac{\mathrm{d}p}{\mathrm{d}r} \tag{6}$$

Combining equations (4), (5) and (6) and rearranging the terms, the pressure gradient can be expressed as:

$$\frac{\mathrm{d}p}{\mathrm{d}r} = -\frac{\eta w {\sigma_0}^2 r}{2Dh_0(H-\sigma_0)^3} \tag{7}$$

Integrating equation (7) for variable pressure on the outer edge of the porous layer $p = p(H)$ – equation (2) at $r = R$, we get the parabolic pressure distribution on the disk:

$$p = \frac{p_0(1-\sigma_0)}{\bar{V}_0 H - \bar{V}_0 - \sigma_0 + 1} + \frac{\eta w {\sigma_0}^2}{4Dh_0(H-\sigma_0)^3}(R^2 - r^2) \tag{8}$$

Integrating the pressure distribution on the surface of contact, we obtain the lift force at a given compression, expressed in terms of dimensionless layer thickness, $H$:

$$F = 2\pi \int_0^R rp\mathrm{d}r = \frac{\pi R^2 p_0(1-\sigma_0)}{\bar{V}_0 H - \bar{V}_0 - \sigma_0 + 1} + \frac{\pi \eta w {\sigma_0}^2 R^4}{8Dh_0(H-\sigma_0)^3} \tag{9}$$

Finally, using a complex parameter, $P = \eta w R^2 / D h_0 p_0$ and dimensionless lift force, $\bar{F} = F/\pi R^2 p_0$, equation (9) takes the form:

$$\bar{F} = \frac{1-\sigma_0}{\bar{V}_0 H - \bar{V}_0 - \sigma_0 + 1} + \frac{P{\sigma_0}^2}{8(H-\sigma_0)^3} \tag{10}$$

From equation (10) one can observe that the dimensionless lift force generated at a certain compression level, $H$, depends on three parameters: the initial compactness, $\sigma_0$, the relative buffer initial volume, $\bar{V}_0$ and the complex parameter, $P$.

For a parametric analysis we selected typical values used in preliminary tests where a disk of radius $R = 25\,mm$ moves downwardly compressing with constant velocity ($w = 5\,mm/s$) onto a $4\,mm$ thickness fiber-based textile layer imbibed with water ($\eta = 10^{-3}\,Pa \cdot s$) placed in an enclosed cell whose internal pressure is $1\,bar$. According to microscope image measurements, the fibers diameters range in between 20 and $50\,\mu m$ and thus, the complex permeability parameter ranges between $3 \cdot 10^{-12}$ and $3 \cdot 10^{-10}\,m^2$. Correspondingly, the complex parameter is close to $P = 1$ and two other possible values have been also selected for the numerical applications $P = 0.1$ and $P = 10$.

Figure 2 plots the evolution of the lift force in function of thickness, $H$ for a value of initial compactness, $\sigma_0 = 0.2$ in three cases: $P = 10$, $P = 1$ and $P = 0.1$. The force variation is different for different values of the initial gas volume, noting that the lift force increases with increasing of the pressure, $p_0$ and thus with decreasing of the volume, $V_0$. The case of $\bar{V}_0 = 0$ corresponds to the hypothetical case of an infinite volume buffer.

The parametric analysis of the lift force shows that the higher is the initial compactness, $\sigma_0$, the higher is the force generated by solid structure of the porous material (figure 2.c).

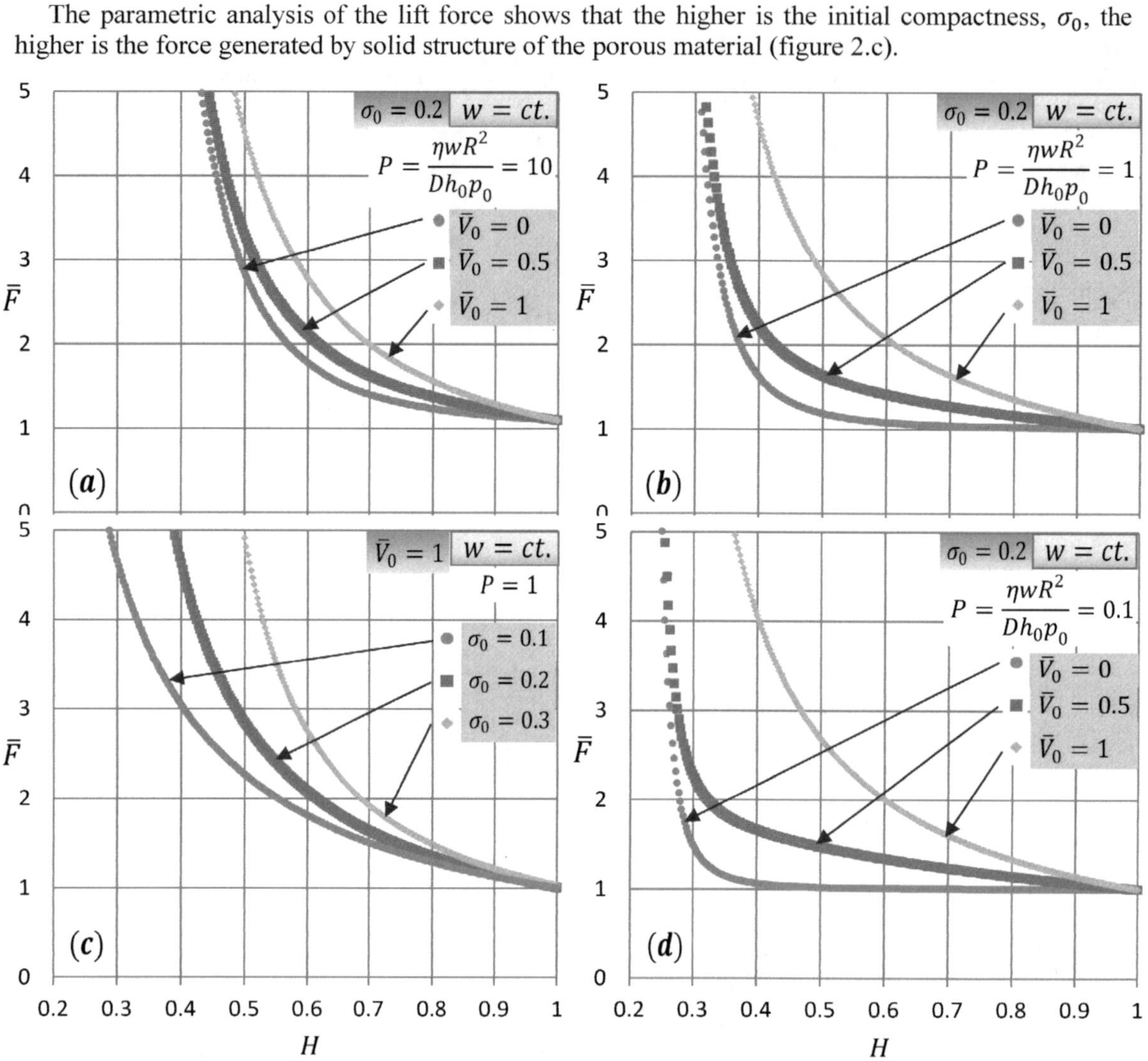


**Figure 2.** Variation of the dimensionless lift force, $\bar{F}$ versus dimensionless HCPL thickness, $H$ for different values of complex parameter, $P = 10$ (a), $P = 1$ (b, c), respectively $P = 0.1$ (d).

It can be seen that the lift force increases with the compression of the material, due to the porosity that drops down to zero, a behaviour also observed in the case of null pressure at the boundaries.

### 3.2. *Constant force squeeze* ($\boldsymbol{F = ct}$)

The solution of the model for the constant force loading is based on the fact that the upper disk velocity is variable:

$$w = -\frac{dh}{dt} \tag{11}$$

Combining equations (9) and (11) and using a new variable, $X = H - \sigma_0$, the following differential equation with separable variables is obtained:

$$\frac{\psi(\bar{V}_0X+\bar{V}_0\sigma_0-\bar{V}_0-\sigma_0+1)}{X^3[\psi(\bar{V}_0X+\bar{V}_0\sigma_0-\bar{V}_0-\sigma_0+1)-1+\sigma_0]}\mathrm{d}X=-\frac{8}{\pi{\sigma_0}^2}\mathrm{d}\tau \tag{12}$$

where $\tau=\frac{FDt}{\eta R^4}$ is the dimensionless time and $\psi=\frac{F}{\pi R^2 p_0}$ is the dimensionless constant load.

Noting that $\alpha=(1-\bar{V}_0)(1-\sigma_0)$ and $\beta=(1-\sigma_0)/\psi$, the above equation becomes:

$$\frac{\bar{V}_0X+\alpha}{X^3(\bar{V}_0X+\alpha-\beta)}\mathrm{d}X=-\frac{8}{\pi{\sigma_0}^2}\mathrm{d}\tau \tag{13}$$

The variation in time of the dimensionless thickness of porous layer, $H$, by squeezing with a constant load, results from the differential equation (13) solved with the initial condition $h=h_0$ ($H=1$) at $t=0$:

$$\tau=\frac{\pi{\sigma_0}^2}{8(\alpha-\beta)}\left\{\frac{\beta{\bar{V}_0}^2}{(\alpha-\beta)^2}\ln\frac{[1-\beta-\bar{V}_0(1-H)](1-\sigma_0)}{(H-\sigma_0)(1-\beta)}+\right.$$

$$\left.+\frac{1-H}{(1-\sigma_0)(H-\sigma_0)}\left[\frac{\alpha(1-2\sigma_0+H)}{2(1-\sigma_0)(H-\sigma_0)}-\frac{\beta\bar{V}_0}{\alpha-\beta}\right]\right\} \tag{14}$$

Figure 3 plots the evolution of the squeeze duration function of the thickness, $H$ for initial compactness, $\sigma_0=0.2$ in the case of dimensionless constant load, $\psi=0.1$. As can be seen, the longer time for the impacting disk to reach zero porosity corresponds to the lower buffer volume ($\bar{V}_0=1$).

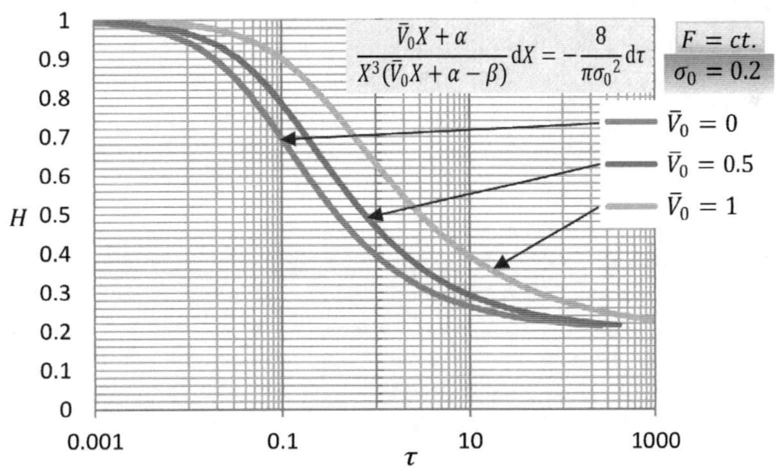


**Figure 3.** Variation of the dimensionless HCPL thickness, $H$ in time for the case $\psi=0.1$.

## 4. Concluding remarks

Analytical solutions for squeeze flow at the contact between a disk and a conjugated porous surface were developed in the case that pressure on the outer edge of the porous layer is *variable* (enclosed systems). Throughout this paper, the squeeze process of HCPL imbibed with fluids was analyzed for two loading cases: constant speed squeeze ($w=ct$) and constant force squeeze ($F=ct$).

The results show that for porosities greater than 0.6 the enclosed cell generates significant greater forces than the classical open squeeze system. Noted that the lift force increases with increasing of the buffer pressure. In addition, the lower is the buffer volume the higher is the lifting force generated by solid structure of the porous material during squeeze process.

Same as in the case of null pressure at the boundaries, it can be observed that the initial value of the lift force is not zero due to a reaction of the porous material.

These theoretical studies prove the lift effect generation and therefore the undeniable applicability in impact damping applications. For the future, parametric squeeze models that were studied theoretically for the variable pressure case can be used for model validation.

**Acknowledgement**
The work has been supported by Partnerships in Priority Areas Program – PNII implemented with the support of Romanian Ministry of Education – UEFISCDI, project number 287/2014.

**Nomenclature**

*Latin letters*

$d$ – characteristic dimension of the porous structure [m];
$D$ – porous structural parameter [m$^2$];
$F$ – lift force [N];
$\bar{F}$ – dimensionless lift force, $\bar{F} = F/\pi R^2 p_0$;
$h_0$ – thickness of the porous material [m];
$H$ – dimensionless thickness, $H = h/h_0$;
$k$ – constant in Kozeny–Carman equation;
$p$ – pressure [Pa];
$p_0$ – initial pressure in the buffer [Pa];
$P$ – complex parameter, $P = \eta w R^2/D h_0 p_0$;
$r$ – radial coordinate [m];
$R$ – disk radius [m];
$t$ – time [s];
$V_0$ – initial volume of the buffer [m$^3$];
$\bar{V}_0$ – relative buffer initial volume, $\bar{V}_0 = \pi R^2 h_0 \varepsilon_0/V_0$;
$w$ – squeeze velocity [m/s];

*Greek letters*

$\varepsilon$ – porosity [-];
$\eta$ – viscosity of the liquid [Pa·s];
$\sigma$ – compactness [-];
$\tau$ – dimensionless time (for constant force squeeze), $\tau = FDt/\eta R^4$;
$\phi$ – permeability of the porous layer [m$^2$];
$\psi$ – dimensionless constant load, $\psi = F/\pi R^2 p_0$.

**References**

[1] Pascovici M D 1994 *Procedure and device for pumping by fluid dislocation* Romanian patent no. 109469
[2] Pascovici M D 2001 Lubrication by dislocation: A new mechanism for load carrying capacity *Proc. of 2nd World Tribology Congress Vienna* p. 41
[3] Pascovici M D and Cicone T 2003 Squeeze-film of unconformal, compliant and layered contacts *Tribol. Int.* **36** (11) 791-799
[4] Feng J and Weinbaum S 2000 Lubrication theory in highly compressible porous media: the mechanics of skiing from red cells to humans *J. of Fluid Mech.* **422** 281-317
[5] Pascovici M D, Cicone T and Marian V 2009 Squeeze process under impact, in highly compressible porous layers, imbibed with liquids *Tribol. Int.* **42** 1433-1438
[6] Sheidegger A E 1974 *The physics of flow through porous media* University of Toronto Press
[7] Ghaddar C K 1995 On the Permeability of Unidirectional Fibrous Media. A Parallel Computational Approach *Phys. Fluids* **7** (11) 2563-2586

[8] Pascovici M D 2007 Lubrication processes in highly compressible porous layers, Lubrification et tribologie des revetements minces *Actes des journees internationales francophones de tribologie, Poitiers University* Presses Polytechniques et Universitaires Romandes **1** 3-12
[9] Pascovici M D, Popescu C S and Ilie M B 2009 Squeeze process in highly compressible porous layer imbibed with liquid *Proc. of 36th Leeds - Lyon Symposium on Tribology, Lyon*
[10] Ilie M B, Pascovici M D and Marian V 2011 Squeeze processes in narrow circular damper with highly compressible porous layer imbibed with liquid *P I Mech Eng J-J Eng* 2011 **225** (6: 539-549
[11] Radu M and Cicone T 2012 Constant speed squeeze process in XPHD conditions - cylinder on plane configuration *Proc. of the 16th Intern. Conf. on EHD Lubrication and Traction, Suceava* pp 86-90
[12] Radu M 2015 A modified approach of squeeze effects in porous media imbibed with newtonian fluids *6th Intern. Conf. Computational Mechanics and Virtual Engineering, Brașov, Romania*
[13] Nabhani M, El Khlifi M and Bou-Saïd B 2010 A general model for porous medium flow in squeezing film situations *Lubrication Science* **22** (2) 37-52

IOP Conf. Series: Materials Science and Engineering **174** (2017) 012032 doi:10.1088/1757-899X/174/1/012032

# Performance evolution of fully and partially textured hydrodynamic journal bearings lubricated with two lubricants

**N Tala-ighil[1] and M Fillon[2]**

[1]Research Center in Industrial Technologies CRTI, P.O.Box 64, Cheraga 16014, Algiers, Algeria
[2]Institut Pprime, CNRS, Université de Poitiers, ISAE-ENSMA, F-86962 Futuroscope Chasseneuil, France

E-mail: ntalaighil@gmail.com

**Abstract.** This study investigates the evolution of the main bearing performance of partially and fully textured hydrodynamic journal bearing. The viscosity effect is also analysed by the mean of numerical simulations for two types of oil: the oil 1 (ISO VG 32, 31.3 cSt at 40 °C) has a lower viscosity than oil 2 (ISO VG 100, 93 cSt at 40 °C). Reynolds equation is solved by finite difference and Gauss-Seidel methods with over-relaxation for various operating conditions. It is shown that, under hydrodynamic lubrication regime, the improvement of the most important characteristics (the friction coefficient and minimum film thickness) of a textured journal bearing depend strongly on the lubricant viscosity and the journal rotational speed. The fully textured journal bearing is highly favorable at very low speeds while the partially textured journal bearing is more suitable for slightly higher speeds. The gain in bearing performance due to the texturing of the bushing disappears at a critical speed of the journal and then, for higher rotational speeds, the presence of textures becomes detrimental.

## 1. Introduction

The modern theory of lubrication was introduced by the great work of Osborne Reynolds (1842-1912). The excellent hydrodynamic lubrication method permits to attain a low friction coefficients close to 0.001 [1]. Under ideal conditions, Reynold showed that the lubricant pressure was sufficient to separate the two bodies in motion and the friction in the contact is due to the viscous resistance of the lubricant. The distance between the contact surfaces reduces with higher carrying loads, small speeds, and less viscous lubricants. In the past, the choice of lubricant was essentially based on experience and knowledge [2]. Today, because of the extreme operating conditions and increasingly stringent of machines (run faster, longer and hotter), this approach is no longer reliable.

Today's lubricants must adapt to extreme operating conditions of modern machines. The minimum film thickness of the lubricant increases with the velocities increase of the contact surfaces and decreases as the load increases [3]. The minimum film thickness is normally greater than 1 μm. However it was only recently that been such textures engineered in order to improve the machine elements tribological performance [4]. Microtextures act as micro- hydrodynamic bearings enhance load carrying capacity and increase film thickness, which also leads to lower friction compared to untextured surfaces.

Recently, experimental results concerning the dimples effect on the Stribeck curve have been presented by Lu and Khonsari [5]. Load, oil type, dimple size, depth and shape were varied to explore their influence on the friction characteristics. It is shown that with proper dimensions of dimples, the friction performance of journal bearings can be improved.

## 2. Theory

In a hydrodynamic lubrication problem, the governing equations for a full hydrodynamic lubrication region can be described by the known Reynolds' equation [6].

$$\frac{\partial}{\partial\theta}\left(h^3\frac{\partial P}{\partial\theta}\right)+\left(\frac{R}{L}\right)^2\frac{\partial}{\partial Z}\left(h^3\frac{\partial P}{\partial Z}\right)=6\mu R^2\left[\left(\omega_2-\omega_1\right)\frac{\partial h}{\partial\theta}\right] \tag{1}$$

$R$ is the bearing radius, $L$ the bearing length. $\omega_1$ and $\omega_2$ are respectively, the rotational speeds of the journal and the bearing (figure 1a). The film thickness $h$ is:

$$h=C\left(1+\varepsilon\cos\theta\right)+\Delta h(\theta,Z) \tag{2}$$

$\Delta h(\theta,Z)$ is the film thickness variation due to the dimple surface, $\varepsilon$ the relative eccentricity of the journal and $C$ the radial clearance bearing.

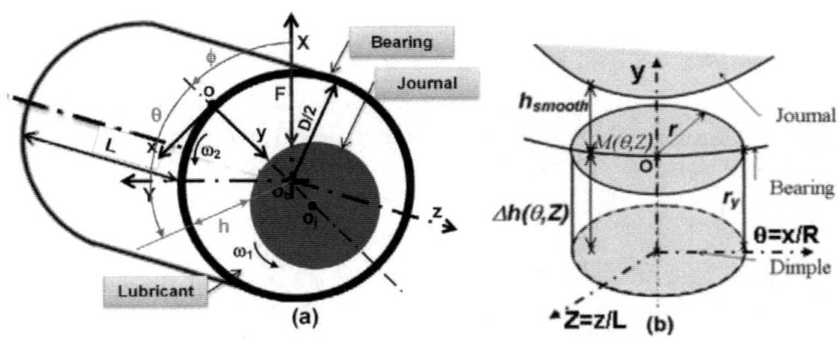


**Figure 1.** (a) Hydrodynamic journal bearing geometry and (b) Cylindrical dimple shape.

In the case of cylindrical dimple geometry (figure 1b) with radius $r$, the centre of the dimple o is located on the bearing surface (making $y_0=0$) and its coordinates are noted $(x_0,z_0)$. The depth at point $M(x,z)$ on the textured surface is defined by $\Delta h=r_y$. The equation of geometry is defined by,

$$(x-x_o)^2+(z-z_o)^2=r^2 \tag{3}$$

The film thickness variation in the dimple, can be written: $\Delta h(\theta,Z)=r_y$

The applied Reynolds boundary conditions on pressure are used to determine the film rupture zone.

They consist in ensuring that $\partial P/\partial\theta=\partial P/\partial Z=0$ and $P=0$ at the rupture limits for the film lubricant defined by the rupture angle $\theta_e$ (vary from 180 to 210°).

The Reynolds equation (1) after applying the Finite Difference Method can be written:

$$P_{i,j}=\left(1-\Omega\right)P_{i,j}+\Omega\,[A_1\,P_{i+1,j}+A_2\,P_{i-1,j}+A_3(P_{i,j+1}+P_{i,j-1})+A'_3(P_{i,j+1}-P_{i,j-1})+A_4] \tag{4}$$

$P_{i,j}$ is the pressure value at the mesh node $(i,j)$. $A_1$, $A_2$, $A_3$, $A'_3$ and $A_4$ are coefficients. $\Omega$ is an over-relaxation parameter (in lubrication problems, the value of this parameter is generally chosen between 1.5 and 1.85). The best resolution method used to calculate the pressure field is that of Christopherson [7]. The Gauss-Seidel iterative method is used to solve the linear systems (4) obtained after discretization as a consequence of the Reynolds boundary conditions. The used global computational procedure contains two computational processes linked together. The first one deals with pressure $P$ computation until convergence ($|\Delta P_i|/|P_i|\leq\varepsilon_p$), while the second one concerns the relative eccentricity $\varepsilon$ computation until convergence on the load (the applied load $F$ and the calculated ones $W$ are compared

IOP Conf. Series: Materials Science and Engineering **174** (2017) 012032 doi:10.1088/1757-899X/174/1/012032

$|F\text{-}W|/|F|\leq\varepsilon_W$). After calculating the pressure field, the static characteristics are calculated. The global process stop after the load convergence condition $|F\text{-}W|/|F|\leq\varepsilon_W$ is satisfied.
The used precisions for the calculation are: $\varepsilon_P = 10^{-7}$ and $\varepsilon_W = 10^{-5}$. The optimized mesh size is $N_\theta$ = 929 and $N_Z$ = 153 (only one-half of the bearing is meshed), respectively along the circumferential and the axial directions.

## 3. Results and discussion

The bearing surface is stationary ($\omega_2$=0) and the journal is moving. Only one-half of the journal bearing system is studied due to the bearing symmetry and the used refined uniform meshes. The Geometrical parameters for the studied journal-bearing are: the journal diameter is 24.625 mm, the bearing length is 25.400 mm and the radial clearance is 85 μm. Two oils are applied. The oil 1 (ISO VG 32, 31.3 cSt at 40 °C) has a much smaller viscosity than oil 2 (ISO VG 100, 93 cSt at 40 °C). Their properties are shown in table 1.

**Table 1**. Oil properties.

| | Viscosity (cSt) | | Specific Gravity at 15 °C |
|---|---|---|---|
| | 40 °C | 100 °C | |
| **Oil 1 (ISO VG 32)** | 31.30 | 5.25 | 0.877 |
| **Oil 2 (ISO VG 100)** | 93.00 | 10.80 | 0.890 |

Three configuration cases are studied: one smooth bearing without texture (Conventional bearing) and two cases of textured bearing (the bearing is fully textured from 0-360° or just the first half of the bearing surface is textured 0-180°). The considered texture geometry in the study is cylindrical with diameter of 4 mm and depth of 0.130 mm.

As shown in figure 2 and for a constant applied load of 667 N, the friction coefficient (figures 2a1 and 2a2) and the minimum film thickness (figures 2b1 and 2b2) are increasing with the journal speed increase. The friction coefficient and the minimum film thickness in the case of journal bearing (conventional and textured bearing) lubricated with oil 2 are higher than those with the lighter oil 1. One can observed that, for a fully textured journal bearing and for a partially (only the first one half of the bearing surface is textured) journal bearing, the positive effect of surface texturing on the carrying capacity (improvement of the minimum film thickness) and on the friction (reduction of the friction coefficient) is effective for the low shaft rotational speeds up to a critical speed which depends on the textured area.

As seen on figures 2b1 and 2b2, the full texturing is very favorable at very low speeds below 152 rpm for oil 2 and 454 rpm for oil 1, while partial texturing is more suitable for speeds between 152 and 612 rpm for the oil 2 and from 454 to 1750 rpm for the oil 1. For the configurations of fully textured bushing and above the critical speed of 298 rpm for oil 2 and 907 rpm for oil 1, the texturation effect on the journal bearing performance becomes negative. For the partially textured bushing configurations, a similar phenomenon is observed for critical speeds of 612 rpm for the oil 2 and of 1750 rpm for oil 1. For both studied oils, the transition speeds for each type of texturing correspond to constant values of the friction coefficient and minimum film thicknesses. As observed on figure 2b1, for $h_{min}$ = 2.6 μm, the positive effect of partial texturing become greater than that of fully texturing at speed greater than 152 rpm for oil 2 and 454 rpm for oil 1. For $h_{min}$ = 4.1 μm, the positive effect of fully texturing disappears at values of 298 rpm for oil 2 and of 907 rpm for oil 1. For $h_{min}$ = 7.6 μm, the texturation effect becomes unfavorable from the speeds of 612 rpm for oil 2 and of 1750 rpm for oil 1. The highest gains in minimum film thickness are observed for lowest rotational speed (80 rpm) and for the fully textured bearing surface (0 to 360°): + 35.5% and +28.7% for oil 1 and oil 2, respectively (figure 2b2). For the same configurations and operating conditions, the friction reduction is –2.50% for oil 2 (figure 2a2).

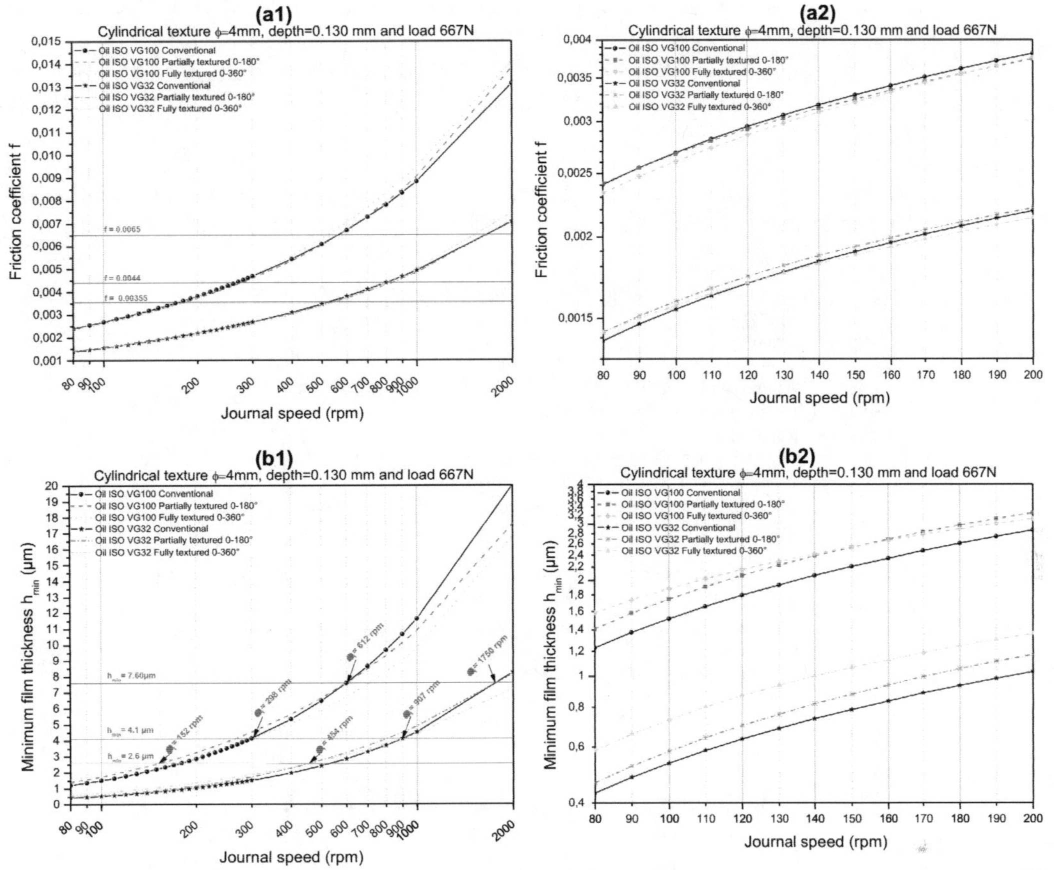


**Figure 2.** Friction coefficient and minimum film thickness versus the rotational speed of the journal for two lubricants.

As shown in figure 3 and for a fixed journal speed (80, 600 and 2000 rpm), the minimum film thickness is decreasing with the carrying load increase.

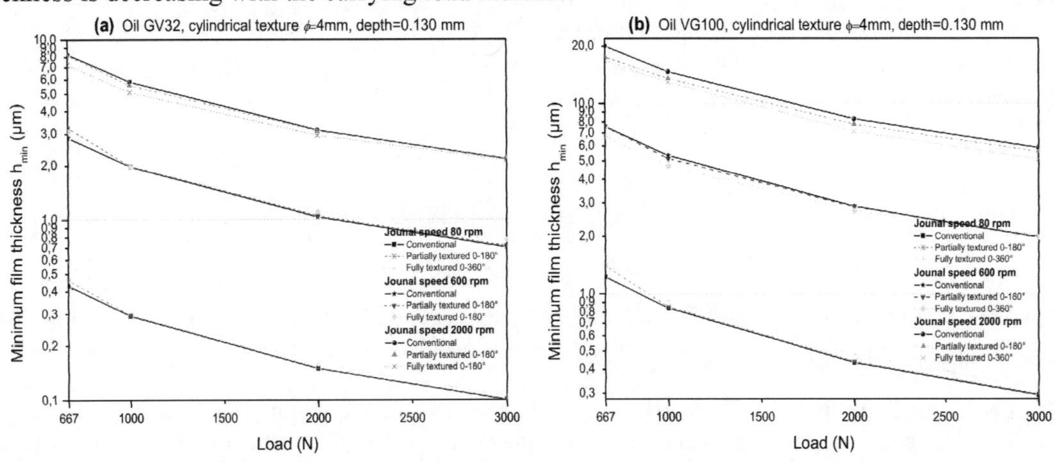


**Figure 3.** Minimum film thickness versus applied load for three speeds (80, 600 and 2000 rpm).

The minimum film thickness in the case of journal bearing (conventional and textured bearing) lubricated with oil 2 (ISO VG100) are higher than those with the lighter oil 1 (ISO VG32) for each value of the journal speed. One can observed that, for a textured journal bearing, the positive effect of surface texturing especially on the carrying capacity is effective for the low shaft rotational speeds regardless of the supported load variation.

As observed in figure 4, the main reason of the gain in bearing performance is that the presence of the dimples on the bearing surface leads to a better repartition of the pressure field for the lowest rotational speed (80 rpm), an applied load of 667 N and lubricated with less viscous oil (oil 1, ISO VG 32).

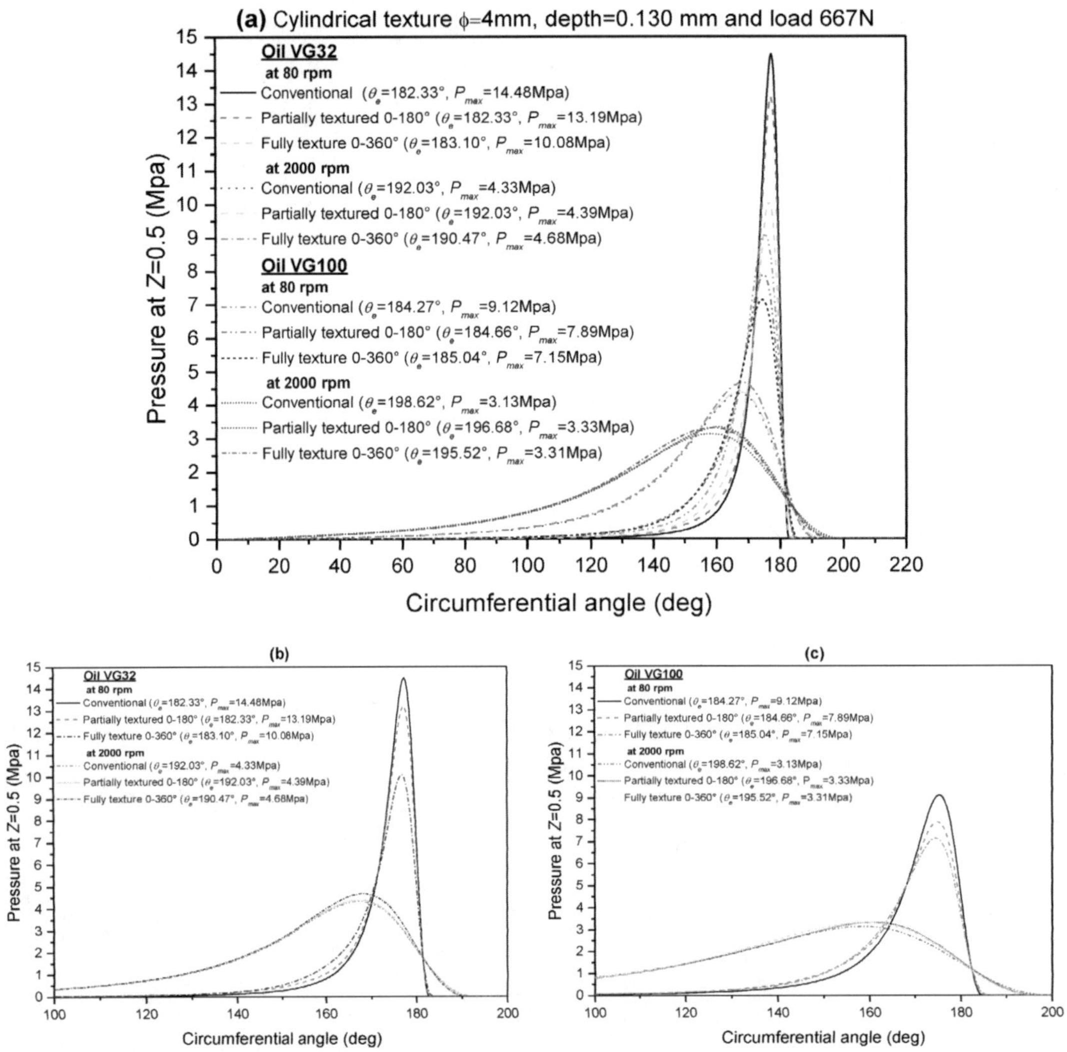


**Figure 4.** Pressure versus circumferential angle at the middle of the bearing ($Z$=1/2) for two speeds (80 and 2000 rpm) (a) for the two oils (b) for oil1 VG32 and (c) for oil2 VG100.

A higher significant hydrodynamic effect can be observed in the angular zone between 120 to 170° which also induces a significant decrease of the maximum pressure, from 14.48 MPa for the un-textured bearing configuration to 10.08 MPa for the fully texturing configuration, i.e. a reduction of 30%. For the oil 2 (ISO VG 100), the maximum pressure decreases from 9.12 MPa for the un-textured bearing configuration to 7.15 MPa for the fully texturing configuration, i.e. a reduction of 21.6%. In fact, the gain is as large as the journal eccentricity is high, i.e. the minimum film thickness is low. The partially textured bearing has a less effect and so, the reduction on the maximum pressure is 8.9% for oil 1 and 13.5% for oil 2, only.

## 4. Conclusion

This study shows that under a hydrodynamic lubrication regime, the improvement of the most important characteristics (friction coefficient and minimum film thickness) of a textured (dimpled) journal bearing depends strongly on lubricant viscosity and the shaft rotational speed. The type of texture is also very important.

The fully textured bearing has the best performance for the lowest rotational speed and the less viscous lubricant. The significant increase in minimum film thickness leads to the decrease in maximum pressure and to the reduction in friction, for the largest shaft eccentricities. This improvement could be also greatly influenced by the thermal effects (mainly because of a decrease of the lubricant viscosity and so, an increase in journal eccentricity) which will be taken into account in the future studies.

## References

[1] Reynolds O 1886 On the theory of lubrication and its application of Mr. Beauchamp Tower's Experiments *Phil. Tran. of the Roy. Soc.* **177** (1) 157-234

[2] Lauer D 2008 Tribology: the Key to Proper Lubricant Selection *Machinery Lubrication, Klüber Lubrication* **3**

[3] Grote K H and Antonsson E K 2009 *Springer handbook of mechanical engineering* (Dordrecht: Springer Science+Business Media)

[4] Etsion I, Kligerman Y and Halerin G 1999 Analytical and experimental investigation of laser-textured mechanical seal faces *Tribology Transactions* **42** (3) 511-516

[5] Lu X and Khonsari M M 2007 An experimental investigation of dimple effect on the stribeck curve of journal bearings *Tribology Lett* **27** 169-176

[6] Frêne J, Nicolas D, Degueurce B, Berthe D and Godet M 1997 Hydrodynamic lubrication-bearings and thrust bearings *In Tribology series* **33** *(Ed. D. Dowson, Elsevier)*

[7] Christopherson D G 1941 A new mathematical method for the solution of film lubrication problems *Inst. Mech. Engrs. J. Proc.* **146** 126-135

# The effect of lubricant supply and frequency upon the behaviour of EHD films subjected to vibrations

**R Glovnea[1], X Zhang[1] and J Sugimura[2]**

[1] Department of Engineering and Design, University of Sussex, Brighton, BN1 9QT United Kingdom
[2] Department of Mechanical Engineering, Kyushu University, 744 Motooka, Nishi-ku, Fukuoka, Japan, 819-0395

E-mail: R.P.Glovnea@sussex.ac.uk, sugimura.joichi.666@m.kyushu-u.ac.jp

**Abstract**. Machine elements such as rolling element bearings or gears often experience vibrations due to for example geometrical inaccuracies, shock loading, rotating unbalanced masses, and others. These machine elements rely on a very thin lubricant film to protect the metallic surfaces from direct contact and eventual damage. During rapid variation of load the elastohydrodynamic contact is influenced by the so-called squeeze film effect, however, when both entrainment and squeeze are present, the conditions of film formation are more complex. It is expected that the lubricant film thickness is influenced by the amplitude and frequency of the vibrations. At the same time, as it is known that the film thickness is established in the inlet of the contact, it is equally important to evaluate what is the role played by the supply of lubricant to the contact under oscillatory conditions. To date there are not many studies on the effect of the oscillatory motion parameters upon the behaviour of the lubricant film. In this study the focus is on the effect of the frequency of vibrations and the supply of lubricant upon the film thickness.

## 1. Introduction

### *1.1. Starved elastohydrodynamic contacts*

When two non – conformal surfaces are in relative motion in the presence of a lubricant, the later is dragged within the convergent conjunction generating the so – called hydrodynamic action which permits the complete separation of the surfaces by a thin lubricant film. It has been established that the thickness of this film depends of the conditions in front of the contact, and once the lubricant enters the Hertzian contact area it travels almost unchanged until near the exit of the contact, where a constriction is formed, to accommodate the rapid drop in pressure [1]. When there is plenty of lubricant in the inlet zone the film thickness can be predicted with a good level of confidence if the properties of the lubricant are known. Very often though, lubricated contacts, including elastohydrodynamic contacts, are forced to work in starved conditions. This means that there is not enough lubricant in the inlet zone to feed properly the contact and allow a full film to be formed. Starved lubrication occurs routinely in rolling element bearings, especially those operating at large rotational speed. This is because the passage of one rolling element pushes the lubricant aside and the next element finds its inlet almost depleted of lubricant. This phenomenon is also frequently encountered in grease lubricated bearings, where the liquid lubricant needs some time to bleed from the structure of the thickener. In starved lubrication conditions the oil film thickness cannot be accurately predicted by full film formulas.

Important research effort has been done in the past for analysing the conditions in which EHD contacts work in starved conditions and evaluating how these conditions affect the lubricant film thickness. Wedeven and co - workers [2] have studied starvation in EHD contact by measuring the central film thickness for various distances in front of the contact where the inlet meniscus is formed. They have found that when the meniscus is located at a distance corresponding to a separation between surfaces $h_b$ larger than nine times the central film thickness, the contact is in fully flooded condition. On this basis they suggest that the degree of starvation can be evaluated by the inlet distance $S_f$ which corresponds to the fully flooded condition.

$$S_f = \frac{3.52(Rh_0)^{2/3}}{a^{1/3}}, \tag{1}$$

where $R$ is the reduced radius of curvature of the surfaces, $h_0$ is the central film thickness, and $a$ is the radius of the contact. This is actually the distance where most of the pressure build – up in the inlet takes place. They have found that for zero $S_f$ the film thickness is null, so for values of this distance between zero and that given by equation (1), the contact is in a greater or lesser degree of starvation.

If the inlet distance of the meniscus is known that Wedeven et al. also suggest an equation for the film thickness, which accounts for starvation effects.

$$\frac{h_0}{(h_0)_f} = \left[\frac{S}{S_f}\left(2 - \frac{S}{S_f}\right)\right]^{1/2} \tag{2}$$

As the authors notice it is difficult to get the inlet distance S, so for many practical applications this equation is not really useful, but for laboratory studies, especially those employing optical interferometry, as it is the case of this paper, this can be used to study the effect of lubricant supply upon elastohydrodynamic contacts.

Studying the transition between fully flooded and starved condition, Cann et al [3], suggest the use of a non-dimensionless parameter, called starvation degree (SD). They observe that the degree of starvation is established by the balance between the oil being removed by the action of the rolling element and replenished by surface tension.

$$SD = \frac{\eta_0 u a}{h_{oil\infty}\sigma_s}, \tag{3}$$

In this equation $\eta_0$ is the base oil viscosity, $u$ is the velocity, $a$ the radius of contact, and $\sigma_s$ is the surface tension of the metal, oil and air interfaces. As seen this equation does not contain the inlet meniscus distance, but comes with another parameter, even more difficult to obtain, that is the thickness of the oil "ridges" at the sides of the rolling element track, $h_{oil\infty}$. The theoretical analysis indicates onset of starvation for degree of starvation SD = 1 however, experimental results show that the transition between fully flooded and starvation occurs for values of SD between 1.5 and 2. Cann and co – workers suggest that once starvation has occurred (that is for SD > 1.5) the film thickness can be evaluated by:

$$\frac{h_c}{h_{eff}} = \left(\frac{1.5}{SD}\right)^{1.67} \tag{4}$$

where $h_{eff}$ represents fully flooded central film thickness. Peper [4] uses equation (4) to evaluate the lubrication in a regime which he calls ultrastarved. He conducts experiments in which friction coefficient is measured in a contact lubricated with very precisely measured quantity of lubricant of

IOP Conf. Series: Materials Science and Engineering **174** (2017) 012033 doi:10.1088/1757-899X/174/1/012033

the order of nano – litres. Somehow surprising he finds that the fully flooded coefficient of friction has the same values as that obtained in ultrastarved conditions.

Using one version of optical interferometry named colorimetric optical interferometry, Svoboda and co – workers carried out detailed experimental investigations on starved EHD lubrication [5]. They found that their results fit well with earlier theoretical predictions by Chevalier et al [6].

Jiang et al [7] have studied starvation of elastohydrodynamic contacts caused by very large speeds up to 30 m/s. They have found that centrifugal forces play an important role, resulting in an asymmetrical oil reservoir which in turns feeds asymmetrically the contact.

### *1.2. Elastohydrodynamic contacts under oscillatory motion*

There are other studies of starvation in elastohydrodynamic contacts but they will not be mentioned here because the main focus of this paper is on the effect that transient conditions may have when the contact works in starved conditions. It has to be mentioned that transient lubrication occurs frequently and practically any EHD contact experiences regularly or accidentally this condition. A comprehensive review, up to that date, of transient lubrication can be found in reference [8]. There are a number of studies on the effect of oscillatory motion upon elastohydrodynamic contacts, but only two tackle the effect of transient condition upon starvation. The study by Cann and Lubrecht [9] on the effect of loading/unloading of a ball – on – disc contact is one of those studies. They used optical interferometry to measure film thickness during cyclic unloading and loading of the ball. This did not resulted in vibrations, as each, loading and unloading took about 2 seconds, but the authors could still see the beneficial effect of cyclic loading upon track replenishment and recovery of lubricant film thickness. They concluded that cycling loading and unloading can be one of the mechanisms which in practice help the lubrication in rolling element bearings subjected to vibrations.

Nagata et al [10] on the other hand used much larger frequencies, up to 50 Hz, but studied the effect of lateral oscillations upon track replenishment in grease – lubricated starved contacts. They measured the lubricant film thickness of a severely starved contact after the onset of the oscillatory motion and found that the lubricant film recovers completely and maintains full film conditions over long periods of time. They also noticed that, in the conditions of the tests, the inlet meniscus distance is not a true indication of the degree of starvation of the contact.

In the current paper the authors present a research into the effect of vibrations, normal to the plane of the film, upon lubricant film thickness in contacts with restricted lubricant supply.

## 2. Experimental method and calibration

Optical interferometry is the most convenient method for studying elastohydrodynamic lubrication. It can provide film thickness maps of the whole contact so it is best suited for studying starvation as the film is expected to have uneven thickness across the contact. The principles of optical interferometry as it is applied to studying elastohydrodynamic lubrication have been presented in many papers [9, 10]. In the current research white light was used which resulted in coloured images of the contact, which were then converted into film thickness profiles. The EHD contact analysed was formed between a glass disc and a steel ball. The disc is driven at desired rotational speed while the ball moves due to the traction force in the contact, thus the contact is in nominal pure rolling conditions. This allowed the support of the ball to undergo oscillatory motion in a direction perpendicular to the plane of the disc, thus effectively loading and unloading the contact.

A schematic of the experimental method and setup is shown in figure 1. This is not a detailed picture, as in reality the ball shaft is supported by two ball bearings on one side and by a needle bearing on the other. It is done in this way so that the ball support is more rigid than the EHD contact under study, between the ball and the glass disc. As it is custom to the optical interferometry method applied to study of elastohydrodynamic lubrication, the disc is coated by a very thin chromium layer and a silica layer of about 130 nanometres thickness.

The light is shown onto the contact through a specially built microscope, which also collects the reflected light and directs it towards a high – speed camera. Before the beginning of the tests the calibration of the colour intensity versus lubricant film thickness was carried out. The procedure of how the calibration is carried out was explained in many other publications [e.g. 10], and will not be repeated here. To note that the calibration is carried out for each oil studied to take into account the difference in refractive index between different fluids.

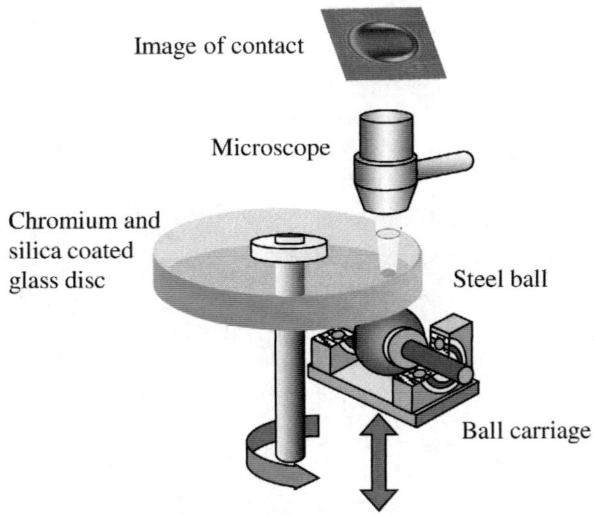


**Figure 1.** Experimental setup

The load was applied from a electrodynamic shaker through a plunger and lever. In the current experiments the load varied between zero and approximately 50 N. Figure 2 shows the load variation together with the contact diameter in static, dry contact test.

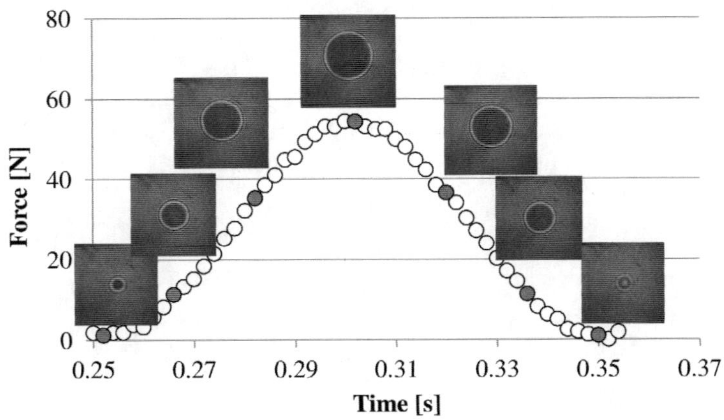


**Figure 2.** Load variation for 10Hz, dry contact.

In the present study the tests were carried out at frequencies of 10 Hz, 25 Hz and 50 Hz. The lubricant used was a synthetic lubricant, poly – alpha – olefin with viscosity 0.396 Pas at 40°C and 0.039 Pas at 100°C. The temperature of the tests was set at 40°C. The ball was about one third immersed into the oil bath and in normal conditions drags the oil into the inlet of the contact ensuring fully flooded conditions. Starved lubrication was obtained by placing wipers in front of the ball and disc. The wipers were flexible, straight – profile sheets of plastic which rubbed against the surfaces of the disc and ball respectively. This was not intended as a method to control the exact quantity of lubricant, but just to make sure that there is little lubricant supply in the inlet. The speed for all tests was 0.1 m/s.

## 3. Results and discussion

Figure 3 shows images of a steady – state contact in three different conditions: fully flooded, partially flooded and starved. In the image on the left hand side the oil fills completely the inlet region and no inlet meniscus boundary can be seen. In the middle image the inlet meniscus boundary is marked by a yellow line for better visibility. This result was obtained with wiper only on the ball. As it can be seen the meniscus is continuous around the contact however, the distance from this boundary to the contact edge is about 59 percent of the contact radius at 109 micrometres. This continuous meniscus ensures a uniform thickness film throughout the contact however the film thickness is smaller than that corresponding to fully flooded conditions. According to equation (2) the film thickness in this case should be approximately 94 percent of the fully flooded condition. The measurement shows values of 214 nanometres for fully flooded and 130 nanometres for limited supply case.

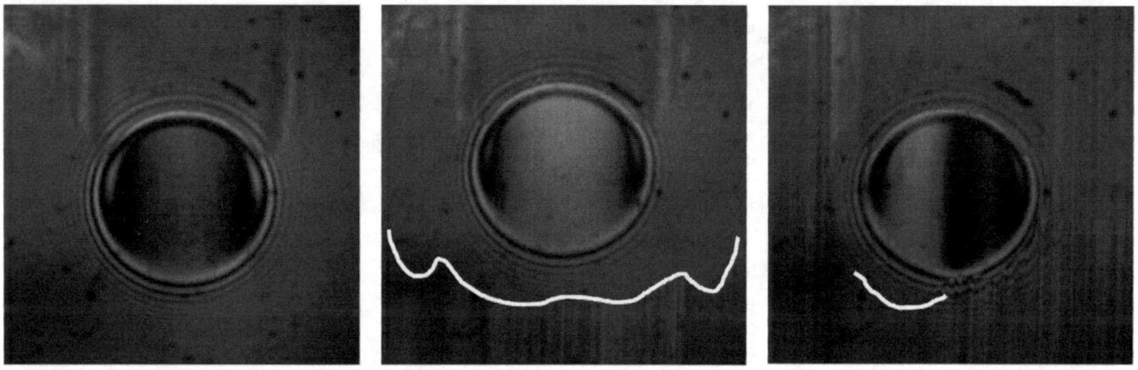

**Figure 3**. Images of the contact showing three different inlet supply conditions.

Finally the image on the right hand side shows a contact with even less lubricant supply, obtained from tests with wipers on both the ball and the disc. The inlet meniscus only covers a small fraction of contact diameter and it is close to the inlet edge of the contact. As seen on the left hand side of the contact inlet the meniscus is located at about 61 micrometres in front of the contact edge. Using again equation (2) the resulted film thickness is 78 percent of fully flooded value. On the right hand side the meniscus disappears almost completely; from the ripples in the Newton's rings the inlet distance can be approximated at less than 25 micrometres. Even for such short inlet meniscus distance the film thickness calculated in this case is about 50 percent of the fully flooded thickness.

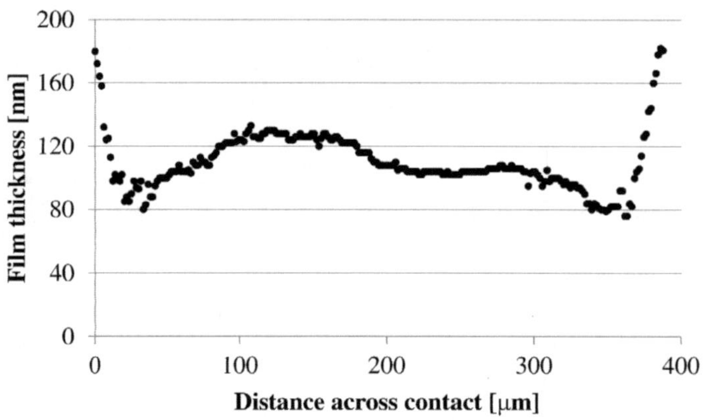


**Figure 4**. Film thickness profile for low lubricant supply (steady – state conditions).

The film thickness profile across the rolling direction, extracted from the right hand side image is shown in figure 4. As seen the film thickness in the central region of the contact varies between 100 and 128 nanometres. These values are less than those predicted by equation (2). Note that $S_f$ in equation (2) was calculated using the measured central film thickness in fully flooded condition and equation (1).

Figures 5 to 7 show the load variation for 10 Hz, 25 Hz and 50 Hz frequency of vibrations respectively. One overall view of these images shows that film thickness is more uniform over the whole contact area, unlike the steady state case. This means that vibrations of the contact have an effect of levelling somehow the quantity of lubricant in the inlet.

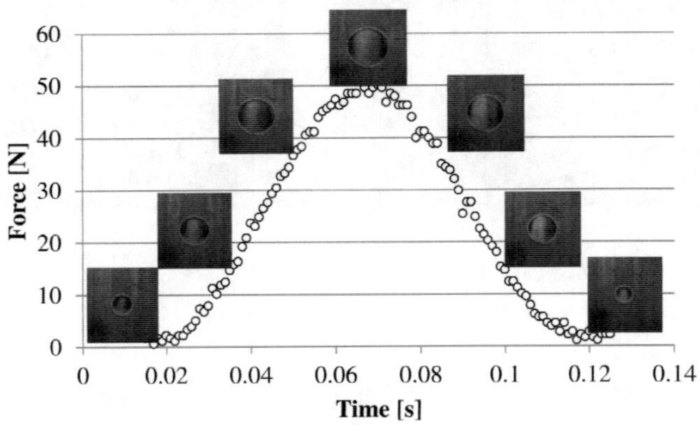


**Figure 5.** Load variation and images of contact for 10 Hz.

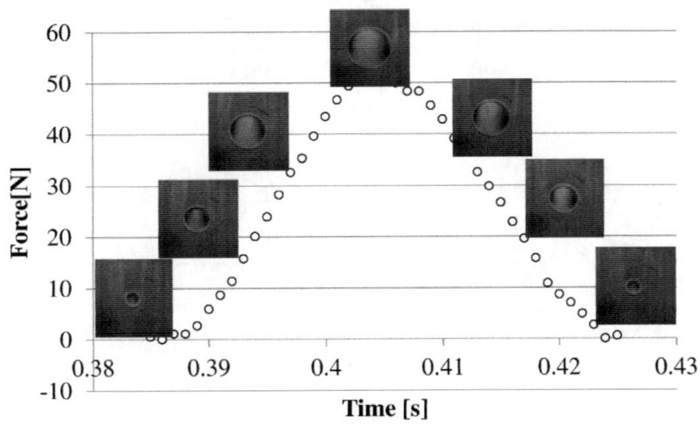


**Figure 6.** Load variation and images of contact for 25 Hz

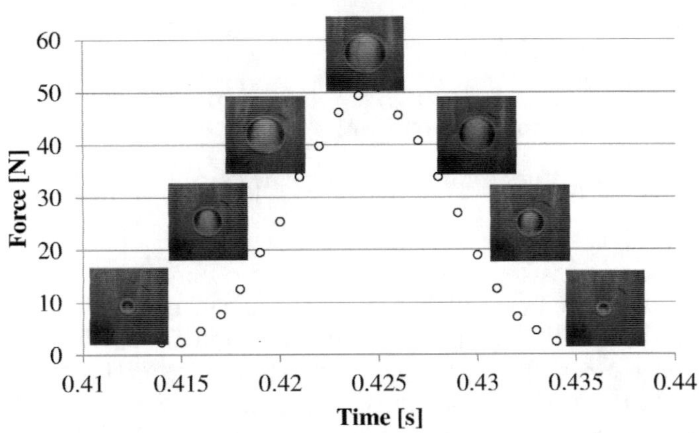


**Figure 7.** Load variation and images of contact for 50 Hz.

A closer look at the image obtained for the largest loads, for all three frequencies, indeed show that the inlet of the contact is better covered by the inlet meniscus for larger frequencies. Compared with the steady state meniscus, shown in figure 3 it is clear that vibrations have a beneficial effect upon the lubrication mechanism in limited oil supply conditions.

10 Hz 25 Hz 50 Hz

**Figure 8.** Inlet menisci for three frequencies at peak load.

Film thickness profiles extracted from the images in figure 8 are shown in figure 9. As seen the central film thickness increases with the frequency. It is thought that during the unloading of the load a suction action takes place which enlarges the meniscus of lubricant around the contact, which in turn leads to thicker film thickness. The maximum thickness is about 33 percent larger for 50 Hz than it is for the 10 Hz test.

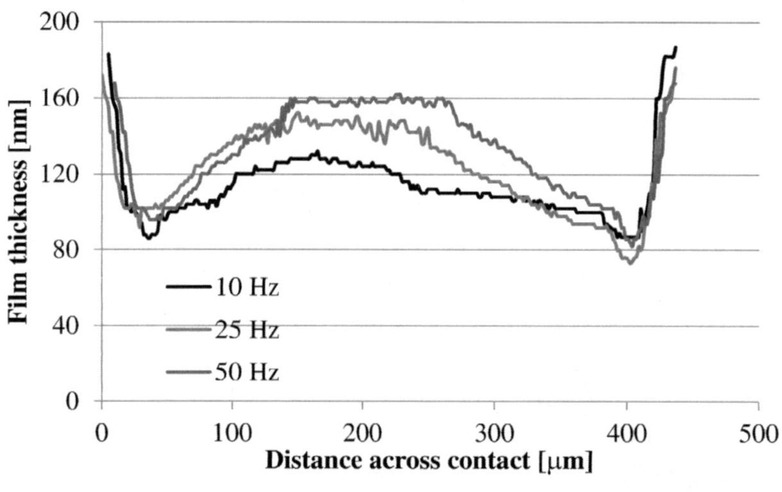


**Figure 9.** Film thickness profiles for different frequencies.

## 4. Conclusions

An experimental study on the effect of vibrations upon elastohydrodynamic films behaviour when limited supply of lubricant is available was carried out. High – speed optical interferometry allowed capturing of interferometric images of the contact during rapid variation of the force applied.

The results showed that vibrations have a beneficial role in the formation of the EHD film. It is thought that the a suction action takes place during the load decrease phase, which helps supply the contact inlet, lowers the degree of starvation and results in larger film thickness.

The supply of lubricant to the contact was restricted by placing wipers in front of the ball and disc. The intention was not to achieve severe starved contacts, but to limit the quantity of lubricant in the contact inlet. It is thought that severely starved contacts subjected to vibrations would destroy the silica layer very quickly, so these conditions were not sought in this study. The authors intend to continue this work at larger frequencies, with different lubricants and at thinner films.

## References

[1] Cameron A 1966 *The principles of lubrication* (London: Longmans)
[2] Wedeven L D, Evans D and Cameron A 1971 *J. Lubric. Technol.* **93** 349-363
[3] Cann P M, Damiens B and Lubrecht A A 2004, *Trib. Int.* **37** 859-864
[4] Pepper S V 2008 *Tribology Transactions* **51** (6) 723 – 729
[5] Svoboda P, Kostal D, Krupka I and Hartl M 2013 *Tribol. Int.* **67** 140–145
[6] Chevalier F, Lubrecht A A, Cann P M, Colin F and Dalmaz G J 1998 *J. Tribol.* **120** 126–33
[7] Liang H, Guo D and Luo J 2014 *Tribology Letters* **56** 491–500
[8] Glovnea R 2010 Transient phenomena in elastohydrodynamic lubrication (Book Section) *Recent Advances in Wear prevention, Friction and Lubrication* Ed. G. Nikas (Kerala, India: Research Signpost) pp. 227-262
[9] Cann P M and Lubrecht A A 2004 *Trans. Proc. Trib.* Elsevier, 745-750
[10] Nagata Y, Kalogiannis K and Glovnea R 2012 *Tribology Transactions* **55** (1) 91-98

IOP Conf. Series: Materials Science and Engineering **174** (2017) 012034 doi:10.1088/1757-899X/174/1/012034

# Tribological study of an aerodynamic thrust bearing in the supersonic regime

**F. Dupuy[1,2], B. Bou-Saïd[2] and M. Garcia[1]**

[1]Liebherr Aerospace Toulouse, 408 Avenue des Etats-Unis, 31016 Toulouse Cedex France

[2]Université de Lyon, CNRS, INSA Lyon, LaMCoS URM5259, F-69621, France

E-mail : benyebka.bou-said@insa-lyon.fr

**Abstract**. Nowadays, aerodynamic air thrust bearing are mainly used over a large panel of turbo-machineries. These systems become increasingly faster and up to operate in supersonic regime. They have not reached a sufficient level of research in terms of high speed. The possibility of an aerodynamic thrust bearing operating in a supersonic regime is studied. First, the air film dynamic study for high Reynolds number is based on the "modified Reynolds" equations, which take into account the inertia terms, the viscosity's variation in the film thickness, and the turbulence. It's an extension of the traditional model used in lubrication called the generalized Reynolds equation. The results show that a depression occur at the location of the change of slope of the tapper flat geometry. The hypothesis of presence of shock or rarefaction waves shows that the pressure gradient in the film thickness may be no longer negligible. The modified Reynolds equation may be not enough to describe the problem. A new system is built to study these phenomena: the Navier-Stokes equation are adapted to the film's geometry. The dynamic air film's behavior study in supersonic regime requires a shock capturing scheme called WENO scheme ("Weighted Essentially Non Oscillatory"), mainly used in shock and turbulence studies. The numerical results give the film behavior modelling of a fixed air thrust bearing pad. The evolution of the quantities shows that shock wave can occur in a thin film.

## 1. Introduction

Aerodynamic thrust bearings are mechanical tribological devices which use relative sliding to support axial loads on a rotor. They are used today on a wide scale of turbomachinery at higher and higher speeds, possibly operating in the supersonic flow regime [1]. In this paper, we seek to study this regime of behavior of a thin film of air within an aerodynamic thrust bearing. The question of a supersonic regime in thin films remains open and its study is found at the interface between two scientific domains. In effect, aerodynamic thrust bearings are studied in the realm of tribology and lubrication. By contrast, supersonic flows are confined to the fields of compressible fluid dynamics. These two disciplines are differentiated by their scale of observation, their respective geometries, and consequently the systems of associated equations. On one hand, lubrication is based on the solution of Reynolds equation [2], which does not take into account a number of phenomena. On the other hand, compressible fluid dynamics is based on the solution of the Navier-Stokes equations [3], which in the compressible case is not well adapted to the study of thin films, but does take into account diverse phenomena. It is therefore important to choose the best approach to the study of aerodynamic bearings.

Clearly, a balance is required to include the most important physical phenomena, but also, it is necessary to obtain efficient numerical resolution of the governing equations used. We present a numerical approache, solution of the Navier-Stokes equations adapted to thin film flow. The schematic drawing of figure 1 illustrates a typical configuration of a thrust bearing in a rotational geometry. The pad is fixed and faces an opposing rotating disk to resist axial forces generated within the machine. When the rotor is moving, a film of air is entrained between the two discs to establish the lubrication effect. In practice, the bearings may possess surfaces consisting of a deformable bump foil. However, in this study, for simplicity, and to focus on the fluid mechanics, the surfaces are considered rigid.

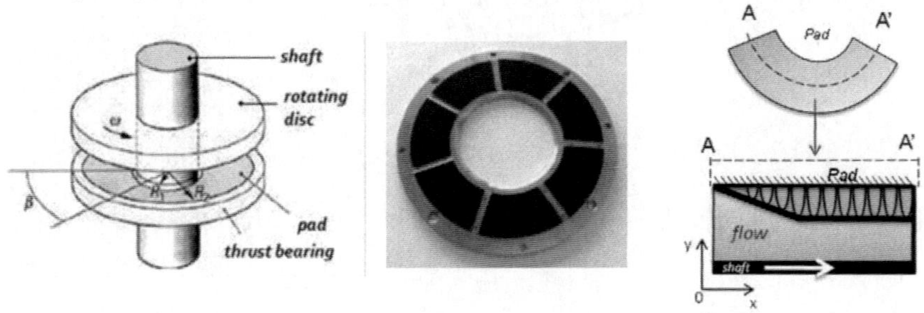


**Figure 1**. Schematic of the dual profile aerodynamic thrust bearing.

## 2. The Navier-Stokes equations adapted for thin film flow

The Navier-Stokes equations and the equation of energy are dimensionless as in [4], with the exception of the pressure. Pressure will no longer be nondimensionalized by a viscous term but rather by a dynamic inertia-based term [5]:

$$\overline{p}=\frac{p}{\rho_0 u_0^2} \quad . \tag{1}$$

This system, orders terms according to the geometric aspect ratio $\varepsilon=H/L$. All the terms are retained and the system appears in the following vector form:

$$\frac{\partial \mathbf{q}}{\partial t}=-\frac{\partial \mathbf{F}}{\partial x}-\frac{\partial \mathbf{G}}{\partial x}+\frac{\partial \mathbf{F}_{visc}}{\partial x}+\frac{\partial \mathbf{G}_{visc}}{\partial y} \quad . \tag{2}$$

The terms are separated into three classes according to their nature. The vector $\mathbf{q}$ represents the vector of variables evolving over time. Vectors $\mathbf{F}$ and $\mathbf{G}$ are the Euler vectors that contain nonlinear inertia terms. The $\mathbf{F}_{visc}$ and $\mathbf{G}_{visc}$ vectors combine all terms related to the viscosity. The various expressions are,

$$\mathbf{q}=\begin{Bmatrix}\rho\\ \rho u\\ \varepsilon^2\rho v\\ E_t\end{Bmatrix},\quad \mathbf{F}=\begin{Bmatrix}\rho u\\ \rho u^2+p\\ \varepsilon^2\rho uv\\ \left(E_t+p\right)u\end{Bmatrix},\quad \mathbf{G}=\begin{Bmatrix}\rho v\\ \rho uv\\ \varepsilon^2\rho v^2+p\\ \left(E_t+p\right)v\end{Bmatrix}, \tag{3}$$

$$\mathbf{F}_{visc}=\begin{Bmatrix}0\\ \varepsilon^2\sigma_{11}\\ \varepsilon^2\sigma_{12}\\ \varepsilon^2\left(\sigma_{11}u+\sigma_{12}v+q_1\right)\end{Bmatrix},\quad \mathbf{G}_{visc}=\begin{Bmatrix}0\\ \sigma_{12}\\ \varepsilon^2\sigma_{22}\\ \sigma_{12}u+\varepsilon^2\sigma_{22}v+q_2\end{Bmatrix}, \tag{4}$$

The deformation tensor is expressed in terms of the geometric ratio $\varepsilon$ which represents the ratio thickness on contact length:

$$\sigma_{11} = \frac{\mu}{\varepsilon Re}\left(\frac{2}{3}\frac{\partial u}{\partial x} - \frac{2}{3}\frac{\partial v}{\partial y}\right), \quad \sigma_{12} = \frac{\mu}{\varepsilon Re}\left(\frac{\partial u}{\partial y} + \varepsilon^2 \frac{\partial v}{\partial x}\right), \quad \sigma_{22} = \frac{\mu}{\varepsilon Re}\left(\frac{2}{3}\frac{\partial v}{\partial y} - \frac{2}{3}\frac{\partial u}{\partial x}\right) \tag{5}$$

The heat source and total energy terms are given by the following formulas:

$$Q_i = -\frac{1}{(\gamma-1)M_r^2}\frac{1}{\varepsilon Re Pr}\frac{\partial T}{\partial x_i}, \tag{6}$$

$$E_t = \frac{p}{\gamma-1} + \frac{1}{2}\rho\left(u^2 + \varepsilon^2 v^2\right),$$

where $Pr$ is the Prandtl number and $\gamma$ is the ratio of specific heats. The reference Mach number defined by

$$M_0^2 = \frac{u_0^2}{\gamma R T_0}, \tag{7}$$

and $R$ is the perfect gas constant. These equations do not neglect viscous terms, and since the phenomena sought are of large physical scale (shocks waves, expansion waves, etc.), a turbulence model is initially not essential. Subsequently, such a model may be integrated into the system to capture smaller scale phenomena (turbulent vortices, etc.).

This system is suitable for thin films since the nondimensionalization allows the geometrical characteristics of the problem to be taken into account. The method also allows the use of a mesh with the same degree of discretization in the length and thickness. Each term of the equations sees its contribution weighted by $\varepsilon$ and thus flows are deformed to match the actual geometry.

These equations are solved using finite differences. The time contribution is discretized using an explicit Runge Kutta 4 scheme. The nonlinear terms are discretized using a shock-capturing scheme with the acronym WENO3 [6], [7], [8] (the Weighted Essentially Non-Oscillatory scheme). The viscous terms are discretized using a central differencing method of order two.

The boundary conditions for this system are that the pressure and density are imposed at the entry, the velocity $u = 0$ at the pad and $u = 1$ on the rotor, and the pressure is imposed at the exit output only if the flow is subsonic.

## 3. Results and Discussion

The geometrical configuration studied is as follows: the pad length $L = 2.5$ cm, the maximum to minimum ratio film thickness is 1.5, and the length of the inclined ramp is half of the total length of the pad. These data are portrayed in figure 2 below.

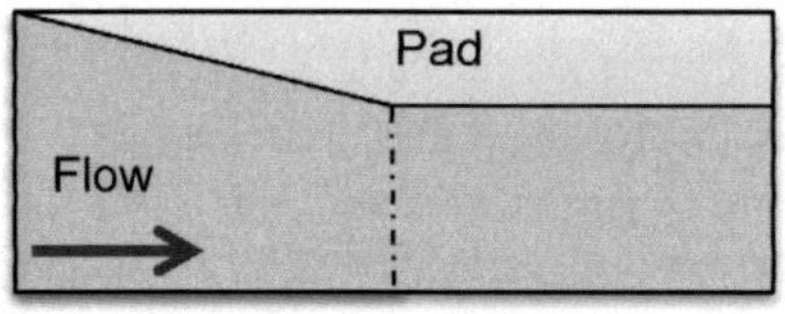


**Figure 2**. Schematic configuration of the bearing double profile studied.

Looking at the theory of supersonic flows, a well-known effect appears when a supersonic flow along a wall meets a wall change in direction (figure 3). Depending on the angle of the turn, either a shock wave or a series of expansion waves may occur as shown in the following figure. Through these

IOP Conf. Series: Materials Science and Engineering **174** (2017) 012034 doi:10.1088/1757-899X/174/1/012034

phenomena, the variables (such as pressure) sometimes evolve abruptly as portrayed in the above figures.

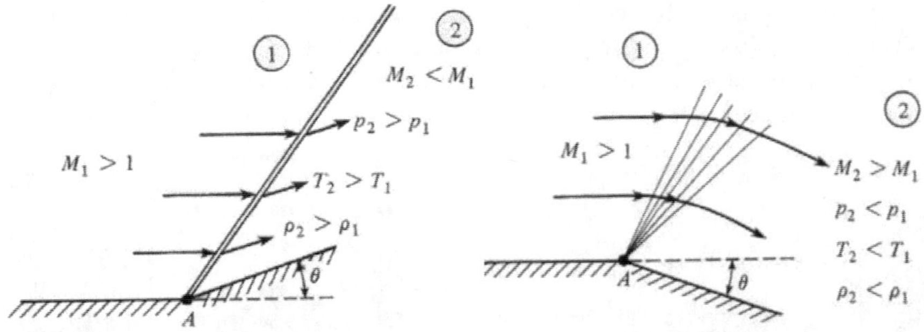


**Figure 3.** Direction change at the wall creating shock or expansion behaviour.

There appears to be no information in the tribology literature that may help, the behaviour of the fluid in the supersonic regime is quite different from the lower speed case. It is interesting to note under what conditions shock and expansion wave phenomena may occur. We conduct a study on the influence of the geometrical aspect ratio $\varepsilon$. For the same geometric shape, this factor is gradually increased, by changing the entry and output height film thickness to keep the same inclination. The slider surface is 4 cm in length and the tilt point lies in the middle, see figure. 2. The ratio of the height at the input to the output is one half. The results are given for different values of $\varepsilon$ in the following figure 4.

Several remarks can be made about these results. Three regimes are distinguished on the basis of the value of the height to length ratio $\varepsilon$.

• A first regime corresponding to a lubrication flow occurs for values of $\varepsilon$ less than 0.1, figures 4 (a) and (b). The properties of this case are a constant pressure field across the gap, a maximum pressure greater than the pressure at the inlet, and the absence of shock or expansion wave phenomena.

• A second regime corresponding to a transition phase occurs for values of $\varepsilon$ ranging from 0.1 to 0.24, Figs 4 (c) and (d). This regime is distinguished by a varying pressure field across the gap, a maximum pressure near the moving wall, and the absence of shock or expansion phenomena.

• The third regime corresponds to shock flow for values of $\varepsilon$ greater than 0.24. The properties of this regime are the following: the pressure varies across the gap, a shock is present at the entry creating a very strong compression, Figs 4 (e)-(h). Expansion waves appear at the change of inclination of the pad creating a strong decrease of the pressure. It can also be noted that the shock increasingly tilts as the value of $\varepsilon$ increases.

These findings are consistent with theories of supersonic flows. Below $\varepsilon = 0.1$ the results are close to those of classical boundary layer theory, dissipating the shock phenomena. It is interesting to observe this transition by keeping the same number of discretization points.

In addition, looking at the influence of the nonlinear terms present in the Navier-Stokes system, it would seem that the following term (in bold) is the one responsible for the disappearance of the shock with decreasing $\varepsilon$.

$$\mathbf{F} = \begin{Bmatrix} \rho u \\ \rho u^2 + p \\ \boldsymbol{\varepsilon^2 \rho u v} \\ (E_t + p)u \end{Bmatrix}, \tag{8}$$

This term should be predominant in the convergent part of the pad to allow for the appearance of the shock. However, its contribution is greatly diminished due to the geometric factor, and for this reason, there appears to be no shock in this geometry in thin film.

These results are only meant to scratch the surface in revealing the complete nature of shock behaviour in thin film high speed compressible lubrication. The film thickness is very high in the Figures 4(c) to '(h) and thus the Reynolds equation is no longer applicable for such configurations. Indeed, this study must also be conducted on geometries with a greater degree of convergence, different bearing numbers, and also for other shapes than the double profile. It is still possible that shock phenomena may appear in thin films and this will be the continuation of the further study.

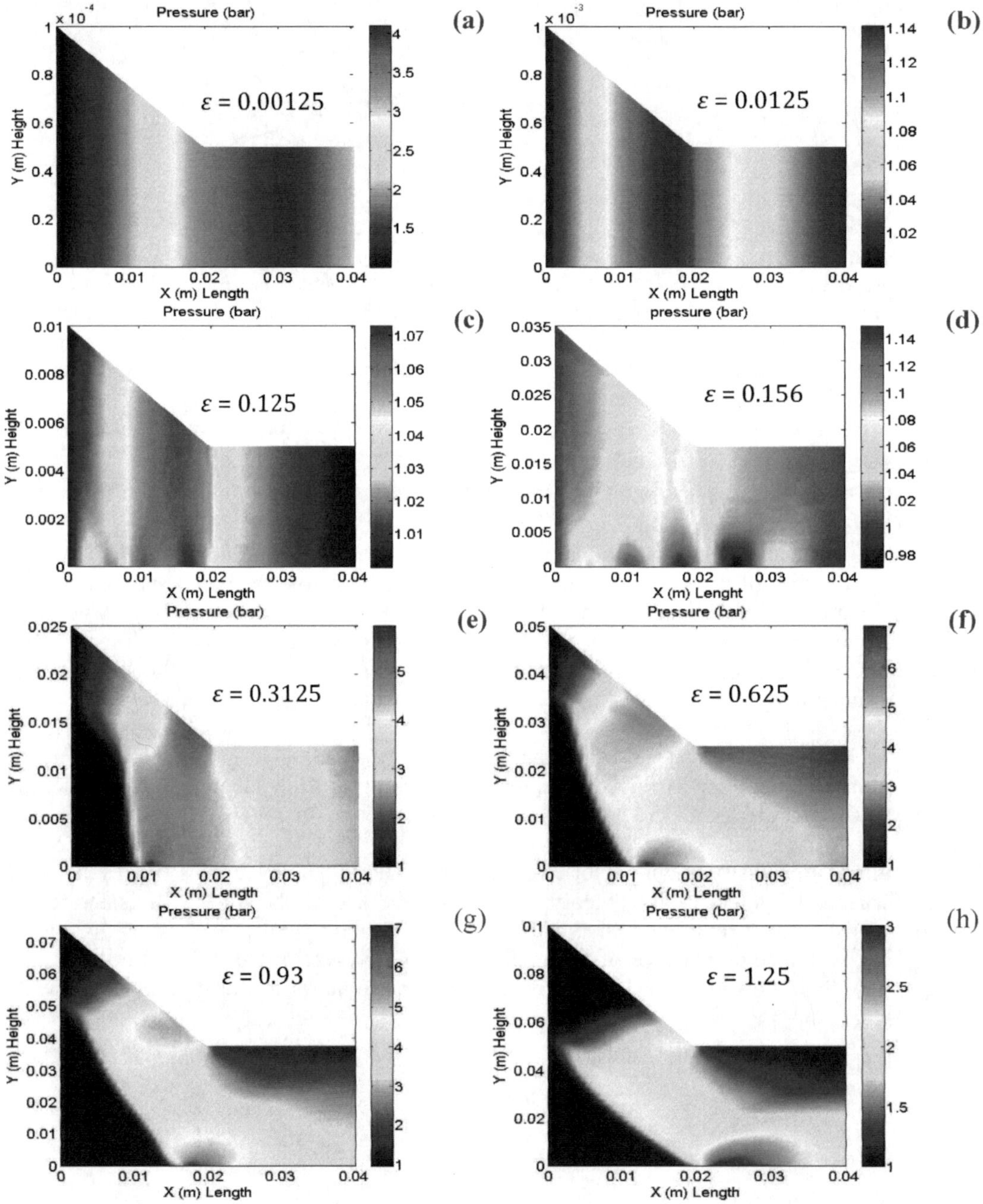


**Figure 4.** Pressure versus position near the exit. Influence of the geometric ratio $\varepsilon$ on the occurrence of shock and expansion waves.

## 4. Conclusions

The study of an aerodynamic thrust bearing in a supersonic regime poses a choice of approaches to be taken. The chosen approach is suited to addressing supersonic behaviour, being based on the fundamental equations. The advantage of this approach is that the terms are weighted by the geometric aspect ratio $\varepsilon$ factor (ratio of gap height to length), allowing study of the system in a computational domain comprising the same number of discretization points in both directions.

The results show that inertia plays a role that is predominant on the behaviour of the fluid, strongly increasing the maximum pressure. The influence of the geometric ratio parameter is able to show the transition between a lubricated and a supersonic shock regime.

It should be noted that the absence of shock or expansion waves in these results is not an absolute certainty, indeed only one type of geometry has been studied. The possibility of the presence of shock phenomena in thin films is not entirely excluded, and one should study other geometries such as more severe convergence, bearing numbers and other pad shapes (such as the Rayleigh step, for example).

## References

[1] Heshmat H and Hermel P 1993 Compliant foil bearings technology and their applications to high speed turbomachinery *Tribology Series* **25** 559-575

[2] Dowson D 1962 A Generalized Reynolds Equation for Fluid-Film Lubrication *Int. J. Mech. Sci.* **4** 159-170

[3] Hirsch C 2017 *Numerical computation of internal and external flows* (Oxford: Butterworth-Heinemann)

[4] Brunetiere N and Tournerie B 2007 Finite Element Solution of Inertia Influenced Flow in Thin Fluid Films *J. Tribol.-T. ASME* **129** 876-886

[5] Brunetiere N 2010 *Les garnitures mécaniques. Etude théorique et expérimentale. Habilitation à diriger des recherches* Université de Poitiers - CNRS

[6] Courant R and Friedrichs K O 1976 *Supersonic flow and shock waves* (New York: Springer-Verlag)

[7] Liu X, Osher S and Chan T 1994 Weighted Essentially Non-Oscillatory Schemes *J. Comput. Phys.* **115** 200-212

[8] Harten A 1997 High Resolution Schemes for Hyperbolic Conservation Laws *J. Comput. Phys.* **135** 260-278

# Down to the roughness scale assessment of piston-ring/liner contacts

**H M Checo[1], A Jaramillo[1], R F Ausas[1], M Jai[2] and G C Buscaglia[1]**

[1]Instituto de Ciências Matemáticas e de Computação, Univ. São Paulo, 13560-970 São Carlos, Brazil
[2]ICJ, INSA de Lyon (Pôle de Mathématiques), 69621 Villeurbanne, France

Email: hugocheco@gmail.com

**Abstract**. The effects of surface roughness in hydrodynamic bearings been accounted for through several approaches, the most widely used being averaging or stochastic techniques. With these the surface is not treated "as it is", but by means of an assumed probability distribution for the roughness. The so called *direct*, *deterministic* or *measured-surface* simulation) solve the lubrication problem with realistic surfaces down to the roughness scale. This leads to expensive computational problems. Most researchers have tackled this problem considering non-moving surfaces and neglecting the ring dynamics to reduce the computational burden. What is proposed here is to solve the fully-deterministic simulation both in space and in time, so that the actual movement of the surfaces and the rings dynamics are taken into account. This simulation is much more complex than previous ones, as it is intrinsically transient. The feasibility of these fully-deterministic simulations is illustrated two cases: fully deterministic simulation of liner surfaces with diverse finishings (honed and coated bores) with constant piston velocity and load on the ring and also in real engine conditions.

## 1. Introduction

Fluid-dynamic bearings are present in many key high-load, high-speed precision applications where other mechanisms like ball bearings would not perform as well due to noise and vibrations, or because it is even impossible to apply another solution. In car engines they are present in the piston assembly, crankshaft bearings, valves, valve seats, camshaft contacts and seals in general. The piston assembly accounts for 50% of the engine friction losses, which is approximately 10% of the total fuel consumption. Half of that takes place in the piston ring/liner contact [1], thus being one of the major contributors to the mechanical power losses. Thus, it is important to understand the key factors influencing film formation and friction in fluid film bearings.

All the mentioned engine components work under severe loading conditions. A complete separation of both bearing surfaces (*hydrodynamic* lubrication regime) is not assured, and partial solid-solid lubrication (*mixed* lubrication regime) or mainly solid-solid contact (*boundary* lubrication regime) takes place. Under these conditions the real topography of the surfaces is a determining factor in the bearing performance.

Honing is the most common machining process used by the industry in the finishing of engine liners. These micropatterned surfaces consist of grooves whose width and depth is in the tens of microns, and a plateau with even smaller features (the surface roughness), in the order of the micrometer or even lower. With a domain in the centimeters (size of the ring pack, journal bearing, seal, etc.), this enforces the use of very fine meshes which is traduced in a greater computational cost. In this work an efficient numerical method to solve thin film (lubrication) problems down to the roughness scale is presented. As an application, these methods will help with the assessment of surfaces actually being produced by the industry, to analyze their performance in the piston ring/liner contact.

This paper is organized as follows: a general model for fluid lubricated bearings is presented in Section 2, along with a brief overview of the numerical methods employed. Results of numerical simulations with several measured rough surfaces are given in Sections 3 and 4. Finally, conclusions are drawn in Section 5.

## 2. Modeling and numerical method

The arrangement of Figure 1 shows a ring sliding over a textured surface (the liner) with velocity $-u(t)$. This corresponds to a section in the piston's movement direction, however, the mathematical formulation of the problem remains general enough to represent other lubricated contacts (e.g., journal-bearing-like devices). The domain is a strip $\Omega = (x_l, x_r) \times (0, w)$. Notice that the domain is two-dimensional, in the directions $x_1$ and $x_2$. Figure 1 shows just a section along $x_1$. The geometry is determined by the functions $h_U$ and $h_L$ describing the upper (fixed) surface and lower (moving) surfaces respectively, and $Z(t)$, which parametrizes the ring movement.

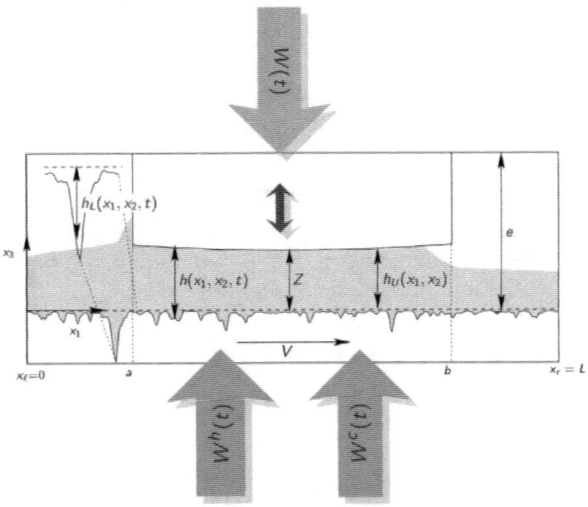


**Figure 1.** Section of the domain along the ring's direction of motion, along with the forces acting on it.

The ring dynamics is governed by Newton's Second Law. The forces $W(t)$ acting on it are the applied ones (force of the gases of the combustion chamber $W^{gp}(t)$, elastic forces $W^{ps}$) and the reactive forces. The latter ones come from the interaction between the ring and the lubricating oil (the hydrodynamic forces) and between the ring and the liner (contact forces), here named $W^h(t)$ and $W^c(t)$ respectively. $W^c(t)$, usually given by the Greenwood-Tripp model [14], is a function of the separation $Z(t)$ between the surfaces. To be able to compute $W^h$, the hydrodynamics must be solved first. Due to the geometry of the contacts and the high velocities involved *cavitation* takes place. It has been shown by Ausas et al [3] that a mass-conserving model is of foremost importance, which is confirmed by several authors, among them [4] who performed a careful comparison with experiments. Here we employ the Elrod- Adams model [5], which includes into the Reynolds equation conservative boundary conditions for cavitation.

The complete mathematical problem [6] to be solved reads:

"Find trajectory $Z(t)$, and fields $p(t)$, $\theta(t)$, defined on $\Omega \times [0, T] = \{(x_{1l}, x_{1r}) \times (0, w)\} \times [0, T]$ and periodic in $x_2$, satisfying

$$\begin{cases} Z(0) = Z_0, Z'(0) = V_0, \\ \theta(x_{1l}, x_2, t) = \dfrac{d_{oil}}{h(x_{1l}, x_2, t)} \forall x_2 \in (0, w), \forall t \\ \theta(x_{1r}, x_2, t) = \dfrac{d_{oil}}{h(x_{1r}, x_2, t)} \forall x_2 \in (0, w), \forall t \\ p > 0 \Rightarrow \theta = 1, \theta < 1 \Rightarrow p = 0, 0 \leq \theta \leq 1 \end{cases} \tag{1}$$

and

$$m\frac{d^2 Z}{dt^2} = W(t) + W^h(t) + W^c(t), \;\; W(t) = W^{gp}(t) + W^{ps}, \tag{2}$$

$$\nabla \cdot \left( \frac{h^3}{12\mu} \nabla p \right) = \frac{u(t)}{2} \frac{\partial h\theta}{\partial x_1} + \frac{\partial h\theta}{\partial t} \tag{3}$$

where

$$h(x_1, x_2, t) = \begin{cases} h_L(x_1 - ut, x_2) + h_U(x_1, x_2) + Z(t) & if a \leq x_1 \leq b \\ h_L(x_1 - ut, x_2) + e & otherwise \end{cases}, \tag{4}$$

$$W^h(t) = \frac{1}{w} \int_a^b \int_0^w p(x_1, x_2, t) dx_1 dx_2, W^c(t) = \frac{1}{w} \int_a^b \int_0^w p_{con}(x_1, x_2, t) dx_1 dx_2, \tag{5}$$

and the functions $h_L$, $h_U$, $W(t)$, and $u(t)$ are known explicitly, $d_{oil}$ and $e$ are positive constants, $m$ is the ring's linear mass and the contact pressure $p_{con}$ is given by the Greenwood-Tripp model".

The honed surface is set on the moving surface of the lubricated pair. Moving textures are more challenging from the numerical point of view since the problem becomes intrinsically transient. Then, the dynamics of the ring must be solved coupled with the hydrodynamics and the other forces involved.

The tribological quantities computed are the *clearance* and the *hydrodynamic lift*

$$C(t) = \min_{(x_1, x_2) \in \Omega} h(x_1, x_2, t), \;\; L(t) = \frac{1}{w} \int_a^b \int_0^w p(x_1, x_2, t) dx_1 dx_2, \tag{6}$$

and the total *friction force*

$$F(t) = \frac{-1}{w} \int_a^b \int_0^w \left( \mu \frac{u g(\theta)}{h} + \frac{h}{2} \frac{\partial p}{\partial x_1} - p \frac{\partial h_L}{\partial x_1} + \mu_s p_{con} \right) dx_1 dx_2 \tag{7}$$

It can be noticed that the last term of equation (7) is the contact friction force, and the rest of the terms represent the hydrodynamic friction force. The instantaneous power loss is also computed

$$P(t) = F(t) u(t). \tag{8}$$

Function $g$ in equation (7) is defined by

$$g(\theta) = \begin{cases} 1 & if \theta > \theta_s \\ 0 & otherwise \end{cases} \tag{9}$$

The parameter $\theta_s$ is a threshold for the onset of friction, which can be interpreted as the minimum oil fraction for shear forces to be transmitted from one surface to the other. Here is taken as $\theta_s$=0.95 for all cases.

The conservative finite volume method described in [7] is used to discretize the Elrod-Adams equation (eq. (3)), coupled with the dynamics through a Newmark scheme. Results are accelerated through a multigrid implementation [8] and shared-memory parallelization.

## 3. Measured surface simulations

The usual treatment given to surface roughness by means of stochastic models [9, 10], averaged models [11, 12] or homogenization methods [13, 14, 15, 16] is avoided. The measured topography of the surface is employed, and its impact on tribological quantities of interest is studied.

To describe each of these surfaces, we resort to their Abbott-Firestone parameters [17]. Its related roughness parameters $R_k$ (plateau height), $R_{pk}$ (reduced peak height), $R_{vk}$ (reduced valley height), $M_{r1}$ (percentage of $R_{pk}$ peaks) and $M_{r2}$ (1-$M_{r2}$ can be understood as the percentage of $R_{vk}$ valleys) are shown in table 3.

To compare performance in terms of friction and clearance, a fixed load $W(t) = W = -0.6N$ was imposed on the ring. This load was chosen large enough so that the small clearance enforces a deterministic simulation, but not as high as to originate contact between the ring and the liner surface. For the same reason, surface $A$ was omitted from this group of simulations. It is a surface with large peaks as its parameter $R_{pk}$=2.64 $\mu m$ shows, and consequently contact takes place for the chosen load. Surface data pre-processing (peak removal and surface smoothing) will be treated in the next section.

The ring's linear mass was set to $m$ = 0.01 kg/m, and its initial position for all cases is $Z_0$=1.5 $\mu m$. Enough oil is provided at the entrance ($d_{oil}$ = 10 $\mu m$) so as to achieve fully flooded conditions, and variable is set to the large value $e$ = 500 $\mu m$. No contact takes place in the simulations shown in this section, so $p_{con}$ = 0 and 6000 time steps are solved with
$\Delta t$= 0.1 $\mu s$. For the simulations here performed the lowermost strip of width $w$ = 0.2 mm ( $(x_1, x_2) \in [0, L_s] \times [0,0.2]$ ) is chosen from each of these surfaces. This choice of domain size is a good trade-off between computational time and representativity of the whole sample [8]. The domain is discretized in 2000×400 finite volumes. Details of these simulations can be seen in [8].

Minimum clearance $C(t)$ and friction force $F(t)$ are shown in Figure 2 for a periodic solution in each measured surface. It seems proper to recall at this point that each measured-surface is extended periodically so as to fulfill the complete distance travelled by the ring.

Results were obtained for rings with circular profiles with curvature radius$R$= 8 and 128 mm. The surfaces with lowest minimum clearance, listed from smaller to larger values are:$D$, $B$, $C$, $F$and$E$. This can be understood taking a look at its respective $R_{pk}$ values: 0.85, 0.84, 0.71, 0.14 and 0.18$\mu m$. Surfaces with higher asperities have a lower minimum clearance $C$.

$C$ and the averaged friction force $\acute{F}(t)$ over each period $2L_s$ ($L_s \times L_s$ is the sample size) are given in Table 2. In spite of the minimum clearance being closely related to wear, it might not always correlate well with hydrodynamic friction. Surface $D$ presents the largest friction force for $R$=8 but the second largest for $R$=128. This makes surface $F$ particularly interesting, given that it shows a higher $C$ with higher friction. This can be explained by looking at $Z(t)$ for each surface, as in Figure 3(b). It is much lower for $F$, leading to a lower average $h$ which means higher shear stresses. The high peaks of surface $D$ ($R_{pk}$=0.85) induce its small minimum clearance.

**Table 1.** Dimensions and resolution of the surfaces of measured topography.

| Surface | $L_s$ (mm) | Resolution ($\mu m$) |
|---|---|---|
| $A$ | 1.241 | 0.982 |
| $B$ | 1.241 | 0.982 |
| $C$ | 0.84 | 0.822 |
| $D$ | 0.84 | 0.822 |
| $E$ | 1.8 | 2.0 |
| $F$ | 1.0 | 1.0 |

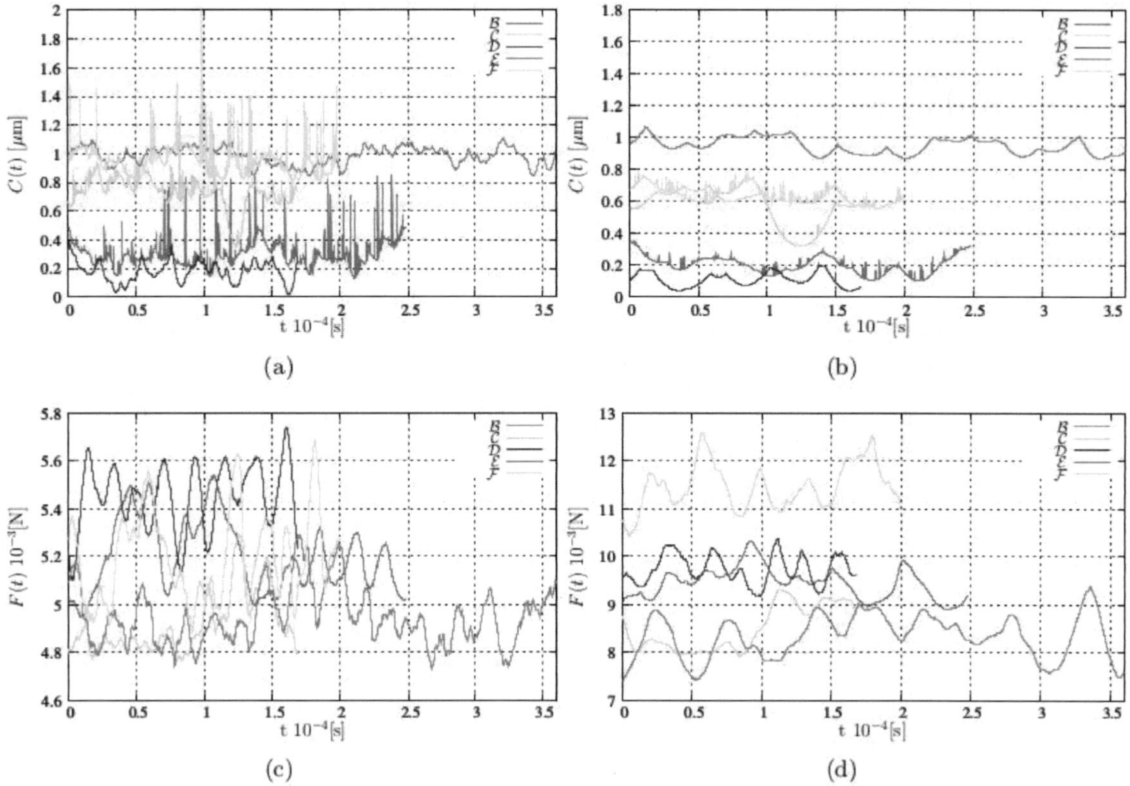


**Figure 2.** Minimum clearance $C(t)$ and friction force $F(t)$ for five of the six measured surfaces of Table 3. Only the last period of the periodic solution is shown for each surface. Figures (a) and (b) display results for $R$ = 8 mm, while Figures (b) and (d) do the same for $R$=128 mm.

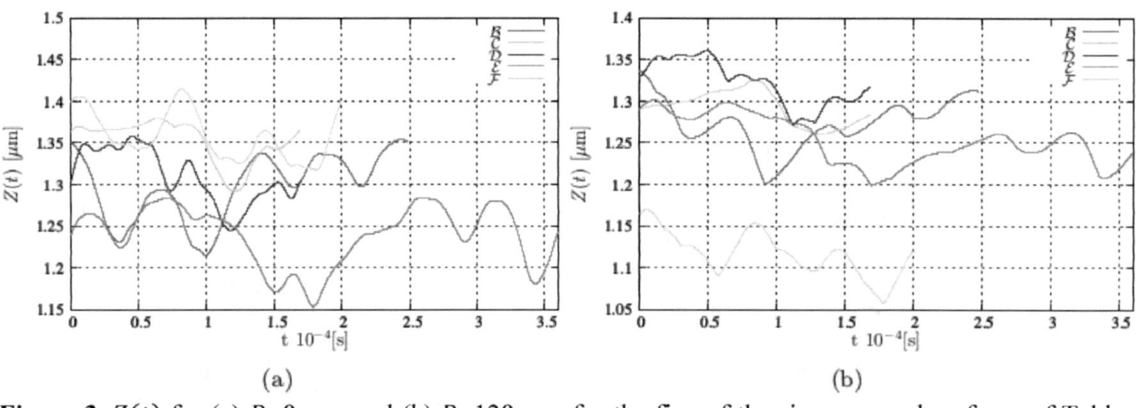


**Figure 3.** $Z(t)$ for (a) $R$=8 mm and (b) $R$=128 mm for the five of the six measured surfaces of Table 3.

## 4. Engine cycle simulations

Let us inspect for a moment the surface $E$. Each measurement point corresponds to a $2\mu m \times 2\mu m$ cell. This gives for the measured surface a total of 810.000 measurement points. Computational domains are usually larger than the sample size, so the total number of cells required to solve the problem can reach several million. Convergence studies suggest a mesh spacing of one fourth of the measurement with a Courant number ( $u_{max}\frac{\Delta t}{\Delta x}$) of 1.0 [18]. This further increases the computational burden.

**Table 2.** Friction and minimum clearance for the measured surfaces of table 3.

| Surface | $R = 8$ mm | | $R = 128$ mm | |
|---|---|---|---|---|
| | $F(t)\ (10^{-3}N)$ | $C\ (\mu m)$ | $F(t)\ (10^{-3}N)$ | $C\ (\mu m)$ |
| $B$ | 5.21 | 0.13 | 9.45 | 0.11 |
| $C$ | 4.92 | 0.35 | 8.50 | 0.32 |
| $D$ | 5.45 | 0.02 | 9.74 | 0.04 |
| $E$ | 4.95 | 0.84 | 8.34 | 0.87 |
| $F$ | 5.2 | 0.81 | 11.47 | 0.56 |

Deterministic simulations of actual liner surfaces face a wide variety of challenges. Ring, piston and fluid dynamics, contact forces among others must be modeled. Once this has been dealt with, there remain difficulties associated with the numerics. Supposing a domain large enough to hold all three rings of a pack, of 15 mm in the movement direction and 1.8 mm in the radial direction (the complete width of the measured surface of figure 2), a structured mesh large enough to capture the level of detail of the surface would have approximately *32 million finite volumes*. If a Courant number of 1.0 is to be kept with a constant time step $\Delta t$, more than *a million* iterations in time will be necessary to simulate just one engine cycle. This, in a single desktop machine will take **months** to compute. However, some pre-processing and approximations with acceptable error can be done to decrease the computational time.

## 5. Engine cycle simulations with a single compression ring

Surfaces $A$ and $B$ are used in the simulations in this section. As the samples sizes are 1.241mm in $x_1$ and 0.913 in the $x_2$ direction it must be repeated somehow all along the stroke, which for the provided engine data is 156mm. The election here was the most straightforward: they were repeated contiguously until the 156mm of the stroke were fulfilled.

The ring geometry employed is the one of a compression ring used, which is shown in Figure 4, while the engine data is given in Figure 5, corresponding to a rotational speed of 1900 rpm. The stroke's length is 156 mm and the bore diameter 132 mm. This constitutes a challenging simulation, as the piston velocity and pressure coming from the combustion chamber are really large.

Different liner configurations were considered. Each configuration is made of three regions as illustrated in Figure 8, for which the $A$ and the $B$ textures are alternatively used. The three cases actually considered are given in the table below.

**Table 3.** Textures on each region of the liner.

| Liner | Region I | Region II | Region III |
|---|---|---|---|
| 1 | $A$ | $B$ | $A$ |
| 2 | $B$ | $A$ | $A$ |
| 3 | $B$ | $A$ | $B$ |

The sizes of the different regions are also indicated in the figure as well as the position of the top dead center (TDC) and the bottom dead center (BDC). In behalf of this and Figure 7, the ring's transition between regions during the power stroke are at the 38.2 (region III to II) and 140.0 (region II to I) crank angle degrees. Details of the physical and numerical parameters adopted in the simulations are given in [8].

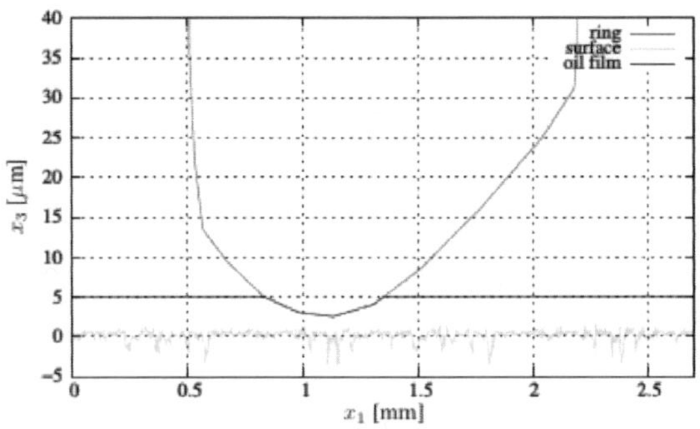


**Figure 4.** Section of the domain showing the ring and liner arrangement used in every simulation in this section.

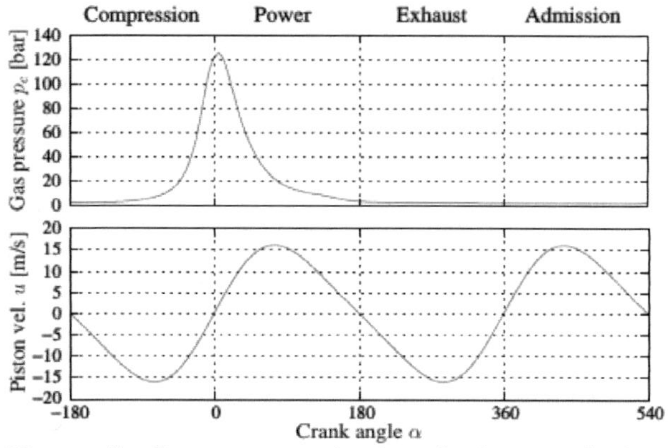


**Figure 5.** Gas pressure (top) and piston velocity (bottom) as functions of the crank angle.

## 6. Comparison of results for the three liners

Results for the three liners are shown in Figure 6. These simulations have taken approximately 20 hours on a desktop computer with a six-cored Intel i7-4930K processor at 3.4GHz. The plotted variables are the minimum clearance, the contact friction force, the total friction force and the total friction power as functions of the crank angle. The region over which the ring is sliding at each angle is marked on the horizontal axis of the top figure. Jumps in the solutions are due to the differences in roughness between surfaces $A$ and$B$.

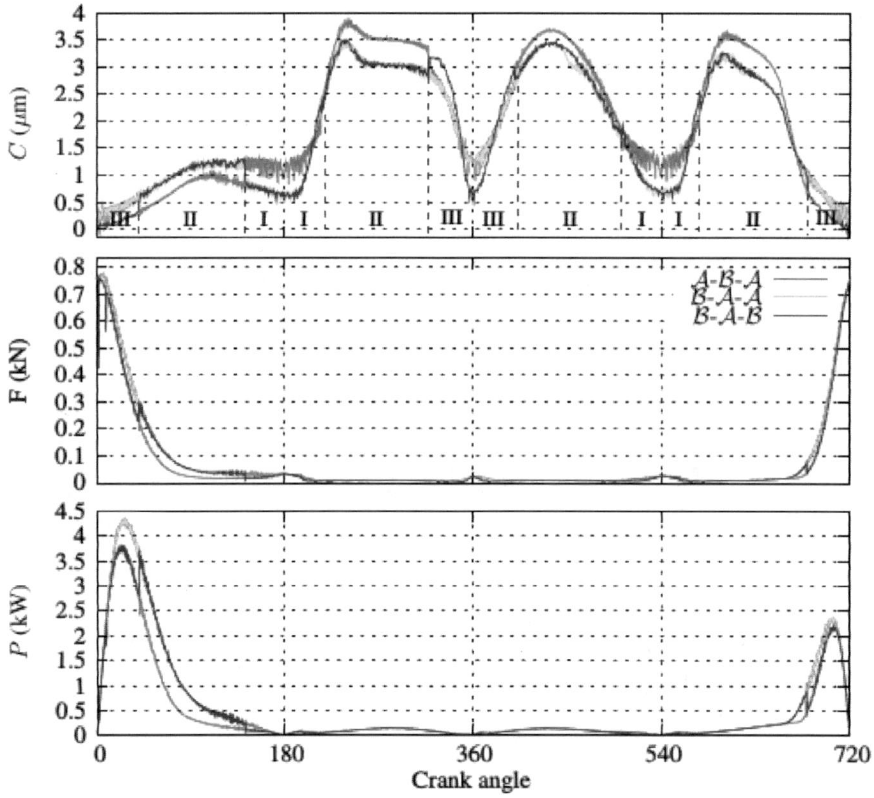


**Figure 6.** Minimum clearance, total friction force and friction power loss as functions of the crank angle for the three different liner configurations.

The maxima in clearance are reached close to the mid-strokes, simultaneously with the highest piston velocities. Clearance reaches its minimum points at the TDC and BDC, being its global minimum at the TDC during the power stroke. Notice that some surface penetration occurs. This is due to the highest loads on the ring taking place during the explosion in the combustion chamber. As piston velocity is null close to the dead centers, friction force is mostly due to contact.

The behavior of the three liner arrangements is similar. In general the $B$ texture exhibits less friction at each location. Averaging the friction power over the four-stroke cycle we obtained:

**Table 4.** Average power loss per cylinder for each configuration.

| Liner | Average power loss |
|---|---|
| $1{:}A - B - A$ | 442 W/cylinder |
| $2{:}B - A - A$ | 519 W/cylinder |
| $3{:}B - A - B$ | 479 W/cylinder |

The least friction losses correspond to Liner 1, which has texture $B$ at Region II (middle) and $A$ at Region III (top). The intermediate friction losses correspond to Liner 3, which has texture $A$ at Region II and $B$ at Region III. In the current configuration the region in which the smoother texture $B$ is more beneficial is Region II (middle).

## 7. Conclusions

The modeling, numerical treatment and numerical simulations of surfaces with measured topography in hydrodynamic and mixed lubrication regime have been summarized in this work. These surfaces with measured topography move with respect to the ring surface (in a piston ring/liner contact) and thus are intrinsically transient. Cavitation effects are taken into account and also the ring dynamics, which are governed by the hydrodynamic forces and the solid-solid forces arising from the contact of both the ring and the liner.

Dynamic simulations are performed on several samples of measured surfaces, corresponding mostly to honed cylinder liners of car engines. The effects of surface roughness are discussed. Some strategies to decrease the computational time in highly demanding complete engine cycle simulations while keeping errors at a low level are presented, and the results of these simulations compared. These studies were made possible due a multigrid implementation and a shared memory parallelization.

**Acknowledgements**

The authors thank the financial support of this work provided by CAPES (grant PROEX-8434433/D), FAPESP (grants 2011/24147-5 and 2014/19249-1) and CNPq (grant 447904/2014-0).

**References**

[1] Holmberg K, Andersson P and Erdemir A 2012 Global energy consumption due to friction in passenger cars *Tribol Int* **47** 221-234

[2] Greenwood J and Williamson J 1966 Contact of nominally flat furfaces *Proceedings of the Royal Society of London* Series A *Mathematical and Physical Sciences* **295** 300-319

[3] Ausas R, Ragot P, Leiva J, Jai M, Bayada G and Buscaglia G 2009 The impact of the cavitation model in the cnalysis of micro-textured lubricated journal bearings *J Tribol-T ASME* **129** 868-875

[4] Zhang J and Meng Y 2012 Direct observation of cavitation phenomenon and hydrodynamic lubrication analysis of textured surfaces *Tribol Lett* **46** 147-158

[5] Elrod H G and Adams M 1974 A computer program for cavitation. Technical report 190 *1st LEEDS LYON Symposium on Cavitation and Related Phenomena in Lubrication, I.M.E.* **103** 354

[6] Checo H, Ausas R, Jai M, Cadalen J P, Choukroun F and Buscaglia G 2014 Moving textures:

simulation of a ring sliding on a textured liner. *Tribol Int* **72** 131-142
[7] Ausas R, Jai M, and Buscaglia G 2009 A mass-conserving algorithm for dynamical lubrication problems with cavitation *J Tribol-T ASME* **131**, 031702
[8] Checo H 2016. Models and methods for the direct simulation of rough and micropatterned surfaces Ph. D. thesis Instituto de Ciências Matemáticas e de Computação Universidade de São Paulo
[9] Tzeng S T and Edward S 1967 Surface roughness effect on slider bearing lubricatio. *Tribol T* **10** 334-338
[10] Christensen H 1969 Stochastic Models of hydrodynamic lubrication of rough surfaces *Proc. Int. Mech. Eng* **55** 1013
[11] Patir N and Cheng H 1978 An average flow model for determining effects of three-dimensional roughness on partial hydrodynamic lubrication *Transactions of the ASME* **100** 12-17
[12] Patir N and Cheng H 1979 Application of average flow model to lubrication between rough sliding surfaces *Transactions of the ASME* **101** 220-229
[13] Bayada G, Chambat M and Faure J 1989 A double scale analysis approach of the Reynolds roughness. Comments and application to the journal bearing *J Tribol-T ASME* **111** 323-330
[14] Jai M 1995 Homogenization and two-scale convergence of the compressible Reynolds lubrication equation modeling the flying characteristics of a rough magnetic head over a rough rigid-disk surface *ESAIM-Math Model Num* **29** 199-233
[15] Buscaglia G, Ciuperca I and Jai M 2002 Homogenization of the transient Reynolds equation Asymptotic Anal **32** 131-152
[16] Buscaglia G and Jai M 2004 Homogenization of the generalized Reynolds equation for ultra-thin gas films and its resolution by FEM *J Tribol-T ASME* **126** 547
[17] Abbott E and Firestone F 1933 Specifying Surface Quality - A method based on accurate measurements and comparison. *Journal of Mechanical Engineering* **55** 569-572
[18] Checo H, Jaramillo A, Jai M and Buscaglia G 2016 The lubrication approximation of the friction force for the simulation of measured surfaces *Tribol Int* **97** 390-399

| Texture | $x_1[mm] \times x_2[mm] \times h_L[\mu m]$ | Abbott-Firestone parameters | Texture | $x_1[mm] \times x_2[mm] \times h_L[\mu m]$ | Abbott-Firestone parameters |
|---|---|---|---|---|---|
| $A$ | 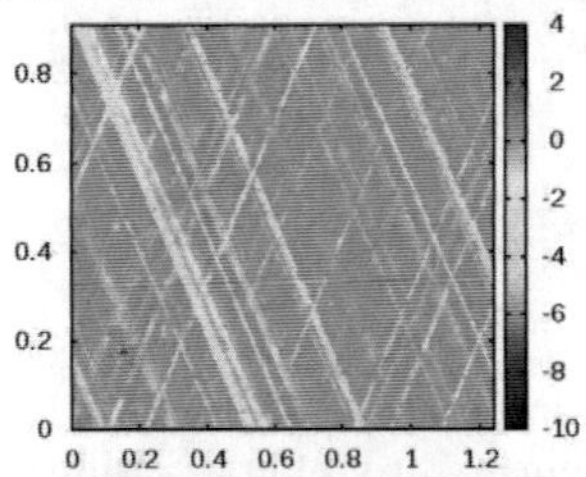 | $R_{pk} = 2.64\ \mu m$<br>$R_k = 0.91\ \mu m$<br>$R_{vk} = 7.97\ \mu m$<br>$M_{r1} = 5\%$<br>$M_{r2} = 74\%$ | $D$ | 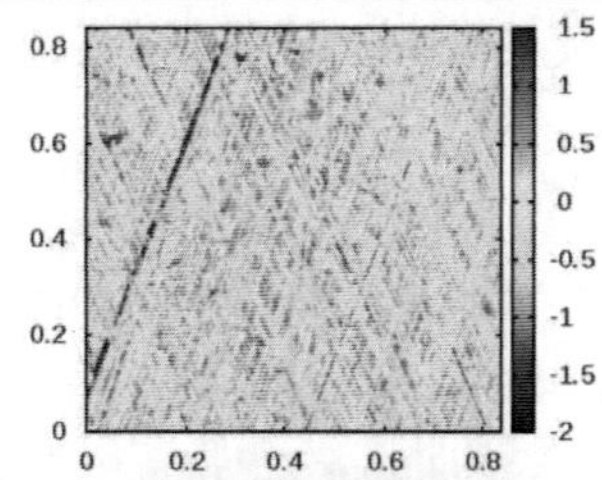 | $R_{pk} = 0.85\ \mu m$<br>$R_k = 0.86\ \mu m$<br>$R_{vk} = 1.57\ \mu m$<br>$M_{r1} = 9\%$<br>$M_{r2} = 86\%$ |
| $B$ | 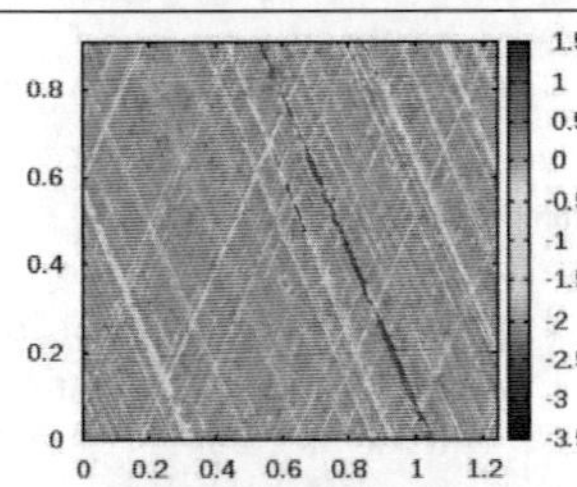 | $R_{pk} = 0.84\ \mu m$<br>$R_k = 0.62\ \mu m$<br>$R_{vk} = 2.94\ \mu m$<br>$M_{r1} = 7\%$<br>$M_{r2} = 79\%$ | $E$ | 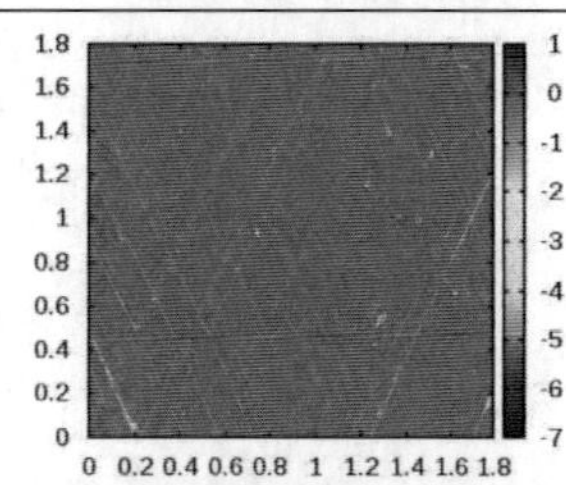 | $R_{pk} = 0.18\ \mu m$<br>$R_k = 0.21\ \mu m$<br>$R_{vk} = 6.46\ \mu m$<br>$M_{r1} = 3\%$<br>$M_{r2} = 77\%$ |
| $C$ | 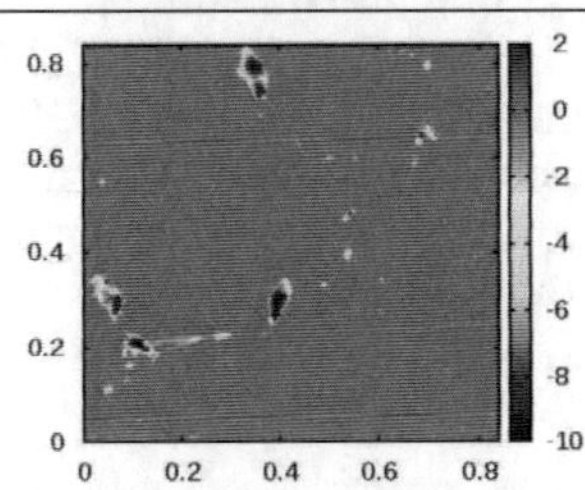 | $R_{pk} = 0.71\ \mu m$<br>$R_k = 0.27\ \mu m$<br>$R_{vk} = 9.94\ \mu m$<br>$M_{r1} = 8\%$<br>$M_{r2} = 85\%$ | $F$ | 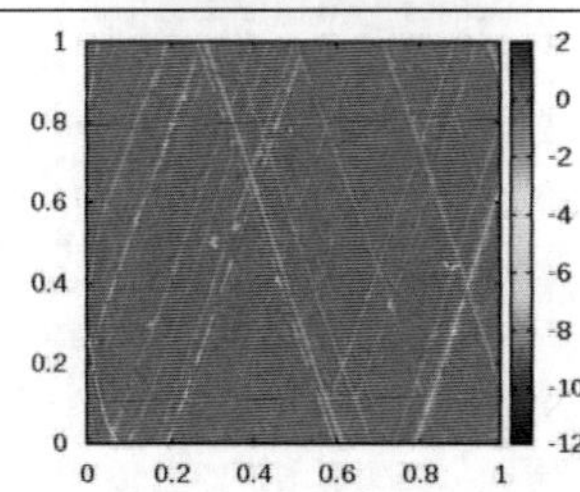 | $R_{pk} = 0.14\ \mu m$<br>$R_k = 0.21\ \mu m$<br>$R_{vk} = 10.6\ \mu m$<br>$M_{r1} = 5\%$<br>$M_{r2} = 71\%$ |

IOP Conf. Series: Materials Science and Engineering **174** (2017) 012036 doi:10.1088/1757-899X/174/1/012036

# The behaviour of lubricated EHD contacts subjected to vibrations

**X Zhang and R P Glovnea**

Department of Engineering and Design, University of Sussex, Brighton, BN1 9QT United Kingdom

E-mail: X.Zhang@sussex.ac.uk, R.P.Glovnea@sussex.ac.uk

**Abstract**. Machine components containing contacts working in elastohydrodynamic (EHD) conditions are often subjected to vibrations. These may be originated from the mechanism or machine the contact is part of, the surrounding environment and within the contact itself. The influence of vibrations upon the behaviour of elastohydrodynamic films has been studied experimentally in a number of papers, but a comprehensive study of the effect of the parameters of the oscillatory motion upon the film thickness has not been carried out yet. In this study the authors evaluate the effect of the frequency of the oscillatory motion upon the EHD film thickness. Optical interferometry is used to measure lubricant film thickness in a ball-on-flat disc arrangement. A high – speed camera records the interferometric images for later analysis and conversion into film thickness maps. The disc runs at a constant angular velocity while the ball is driven by the traction forces developed in the EHD film. In steady state conditions, this would ensure pure rolling conditions, however in the present investigation the ball is subjected to harmonic vibrations in a direction perpendicular to the plane of the film. The contact under study is lubricated by basic oils and the temperature is kept at a constant value of 60°C. The aim of this paper is to understand how vibrations influence the lubricant film formation.

## 1. Introduction

The elastohydrodynamic (EHD) regime occurs in lubricated non – conformal contacting surfaces, where, apart from the hydrodynamic action of convergent geometry, the elastic deformation of the surfaces and the increase of lubricant's viscosity with pressure govern the formation of the film. Many experimental [1 – 7] and numerical [8 – 11] studies have been carried out during the past six decades or so, which allowed the evaluation of the contribution of various parameters upon the behaviour of the lubricant film. This is valid for steady state conditions where the geometry of surfaces and various parameters are constant during the operation of the machine component however full steady state conditions are almost never encountered in practice. Classic examples of non – steady state or transient EHD conditions are cams and gears mechanisms, where rapid changes of geometry and load take place. Even rolling element bearings, normally associated with smooth, steady running experience transient conditions at the start up and shut down of the motion and in devices which inherently required variation of speed (for example the bearings of stepper motors).

Studies of the effect of transient conditions upon EHD films have received more attention during the past two decades. Many aspects of transient EHD have been approached such as elastic rebound and cyclic squeeze of a sphere on a flat surface [12 – 14], variation of geometry including artificially induced features [15 – 18], rapid change of speed, including sudden halting or start of motion [19 – 23]. These studies include both experimental and numerical approaches.

Elastohydrodynamic contacts subjected to vibrations can also be described as working in non – steady sate conditions of lubrication. The way the EHD film responds is very important in terms of the control of vibrations and noise of the whole system. Vibrations can be originated inside the contact itself, for example due to irregularities of surfaces. This situation was studied by Dareing and Johnson [24] who used a two –disc machine with the surface of one disc corrugated. Studying the effect of lubricant film on the damping of the system they found that the film has a significant contribution to the total dumping, especially at lower loads. They concluded that the fluid film contributed to damping via a squeeze film mechanism.

Another type of vibrations elastohydrodynamic contacts can be subjected to, are structural vibrations generated by shocks in loading [25, 26], or transmitted from the machinery the contact is part of [27]. It was found that a combination of lubricant entrainment and squeeze generates fluctuations in the film thickness. Parallel experiments and dynamic simulation of the whole system, including the contact has shown that the contribution of the dynamic response of the EHD contact itself is dominant. It is worth mentioning here that dampened vibrations similar to those obtained by a sudden variation of the applied load are obtained when the speed changes rapidly [28, 29]. This proves that the contacting surfaces and the lubricant can be seen as a mass, spring, damper system which responds when suddenly pushed out of a steady state condition.

Finally lubricated contacts can be subjected to forced vibration caused by cyclic, variable loading. Sakamoto and co – workers applied a pulsating load to a ball – on – flat lubricated contact and observed the variation of film thickness using the optical interferometry technique [30]. The load increased and decreased rapidly, in less than 50 milliseconds and the whole cycle was about 0.3 seconds. They found that during both, application and removal of load, oil entrapments form and travel like ripples through the contact. At larger entrainment speeds the formation of oil entrapment is suppressed.

Somehow surprising but there are not many experimental studies reporting on the behaviour of elastohydrodynamic contacts subjected forced vibrations and none, as far as the authors are aware to forced harmonic vibrations. The authors try to fill this gap in the present research where they apply forced, quasi harmonic vibrations to an elastohydrodynamic contact formed between a ball and a glass disc and evaluate the behaviour of the lubricant film.

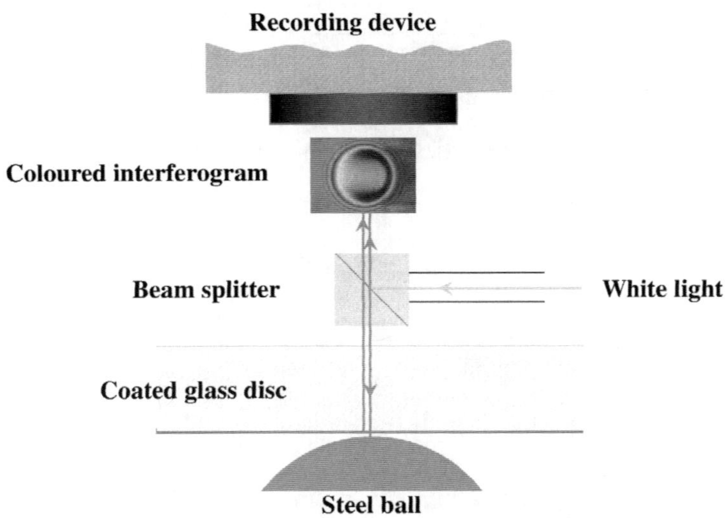


**Figure 1**. Principle of optical interferometry.

## 2. Experimental method and calibration

The experimental method widely used for quantitative, albeit modelled elastohydrodynamic contacts is optical interferometry. This technique was first used by Cameron and Gohar [1] and then employed extensively by other researchers in the field so it will only be briefly explained here. In the current arrangement the contact under study is formed between the flat surface of a glass disc and a steel ball, as seen in figure 1. The disc is coated on the contacting surface with a chromium and silica double layer. White light is used which results in film thickness maps in the form of coloured interferograms, corresponding to the wavelength of light which underwent constructive interference at a certain location on the contacting surface. In figure 1 an image of the EHD contact under study, showing the distinctive horse – shoe shape, can also be seen. The only remaining problem is to relate the colours in the image with a film thickness. This can be done in more than one way, but in this research the calibration of colour content versus film thickness is done following the procedure of the method called SLIM (spacer layer imaging method) [31].

SLIM takes advantage of the high precision of another method of measuring film thickness named Ultra – Thin – Film Interferometry (ULTRA) [2]. In ULTRA the central film thickness from a narrow band in the middle of the contact is measured. Using a test rig equipped with ULTRA, the film thickness for the test lubricant is measured in pure rolling conditions, at increasing entrainment speed. A correlation between film thickness and speed is thus obtained. In the second step the colour digital camera is attached to the rig. The spacer layer disc used in ULTRA is swapped with a glass disc coated with a thin (5 – 10 nanometres) thick chromium layer and making sure that the temperature is exactly the same as in the first test, the contact is run again through the same speeds, while coloured images are recorded. The variations of R, G, and B content with film thickness is extracted and are later used to convert the images of the contact under vibration into film thickness.

## 3. Test rig and lubricants

Optical interferometry technique, seen in figure 1 was incorporated within a test rig capable of producing forced vibrations of the ball. A schematic of the test rig is shown in figure 2.

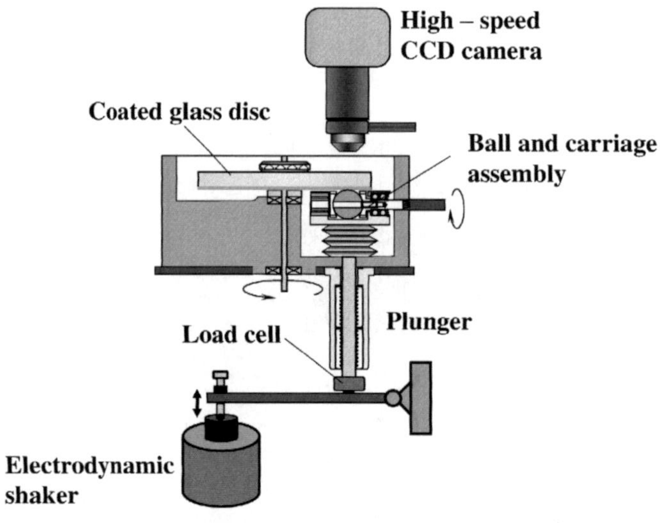


**Figure 2**. Experimental setup.

The glass disc shown is coated with a thin chromium layer (between 5 and 10 nanometres thick) and a silica layer (about 130 nanometres thickness) on the contacting surface. The disc is attached to a shaft which rotates at desired speed. The ball is also attached to a small shaft, but is free to rotate under the action of the friction force generated at the contact between the two. The shaft of the ball is supported by needle bearings on one side and a pair of angular contact bearings of the other, such that the total stiffness of these contacts is larger than that of the contact between the ball and glass disc. The ball carriage is attached to a bellow and actuated by a plunger guided by a pair of linear bearings fitted to the frame of the rig. The plunger is set in a harmonic motion by a shaker via a lever arm and a load cell. By connecting the load cell between the plunger and the lever, the force acting normal to the contact can be precisely measured, given the fact that the plunger is very rigid in compression.

The load was varied sinusoidally from zero (or close to zero) to about 50 N. For the pair of materials used this means a Hertzian pressure variation between zero and around 0.7 GPa. The frequency of the oscillatory motion of the ball was 10 Hz, 25 Hz, 50 Hz, and 100 Hz for these tests.

More than one lubricant was tested in this research, but for limitations of space only the results for a synthetic base oil, poly – alpha – olefin (PAO40) are shown. Poly – alpha – olefins are hydrocarbon synthetic oils widely used in a wide variety of applications because they have well characterized chemical composition and many steady state experimental data are available in literature. The dynamic viscosity of this oil is 0.396 Pas at 40°C and 0.039 Pas at 100°C. The temperature of the lubricant was kept at 60°C. This was chosen in order to get steady state film thickness thick enough for the transient effects to be quantified. In order to get a good number of images for each cycle of oscillation, even at larger frequencies, a speed of 2000 frames per second was set on the camera.

## 4. Results and discussion

The tests were carried out at entrainment speeds with values between 0.1 m/s to 0.5 m/s however it was found that because of the too large film thickness obtained at the larger speeds, data obtained at those conditions was not usable. For this reason only results for the lower speed will be shown in this paper. Figures 3 to 6 show the variation of the load and images of the elastohydrodynamic contact at certain points during the cycle.

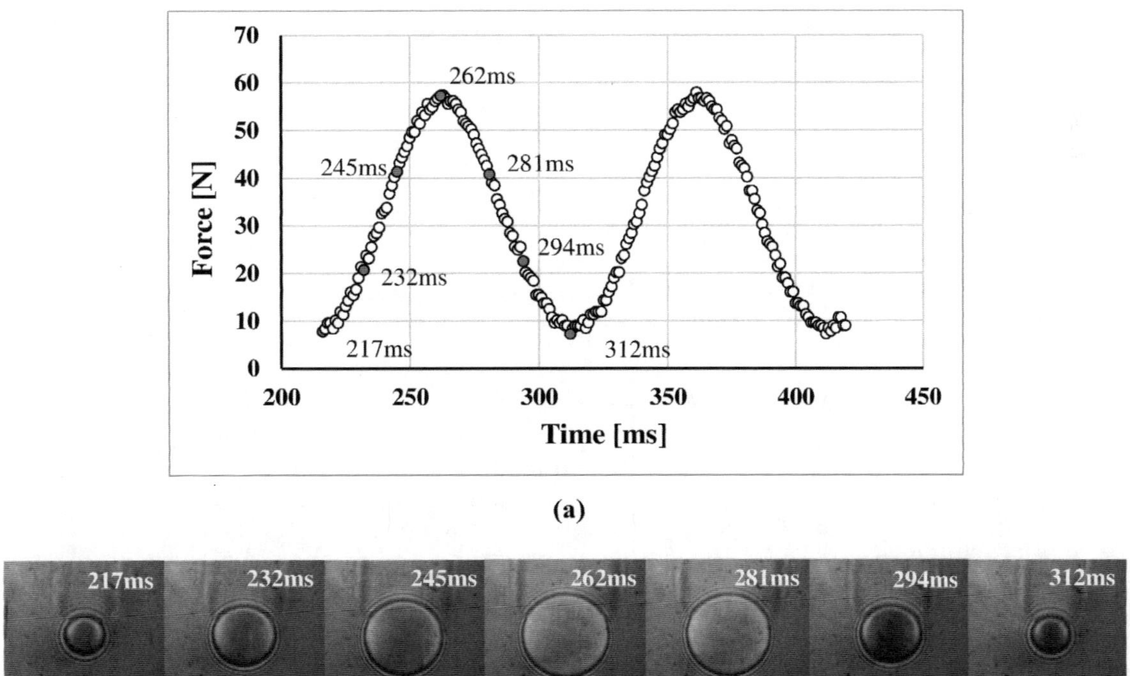


**Figure 3**. Load variation and interferograms of the contact for 10 Hz, 0.1 m/s entrainment speed.

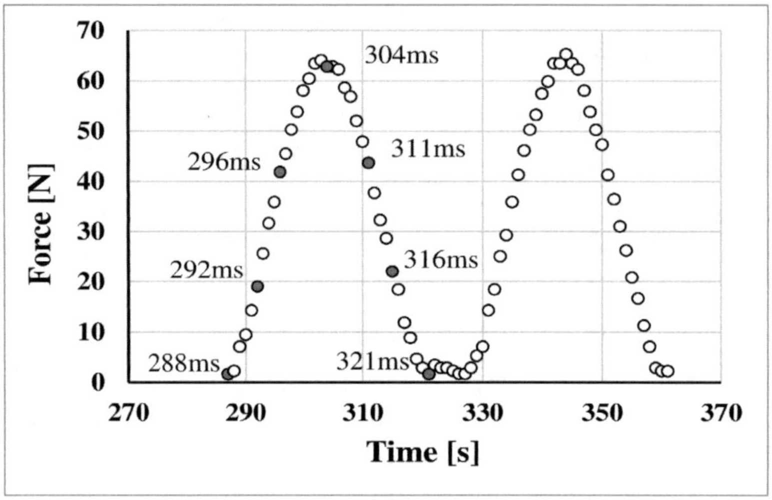


(a)

(b)

**Figure 4**. Load variation and interferograms of the contact for 25 Hz, 0.1 m/s entrainment speed.

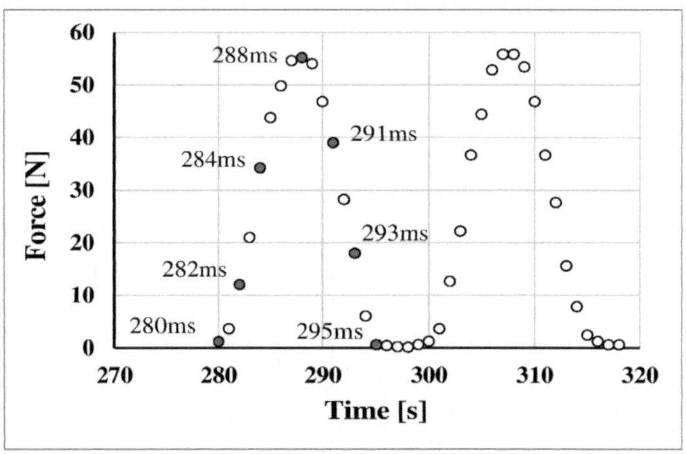


(a)

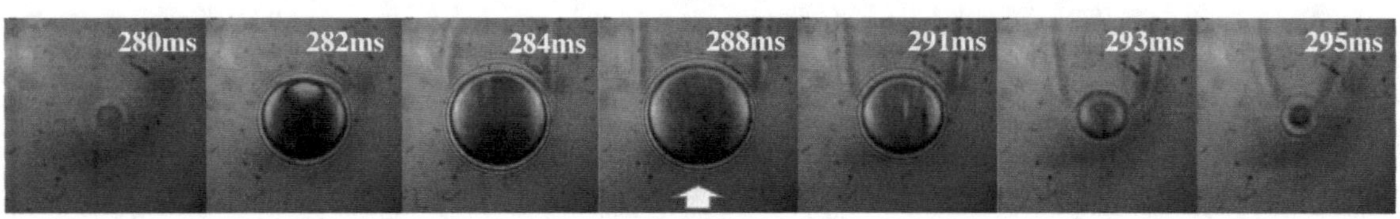


(b)

**Figure 5**. Load variation and interferograms of the contact for 50 Hz, 0.1 m/s entrainment speed.

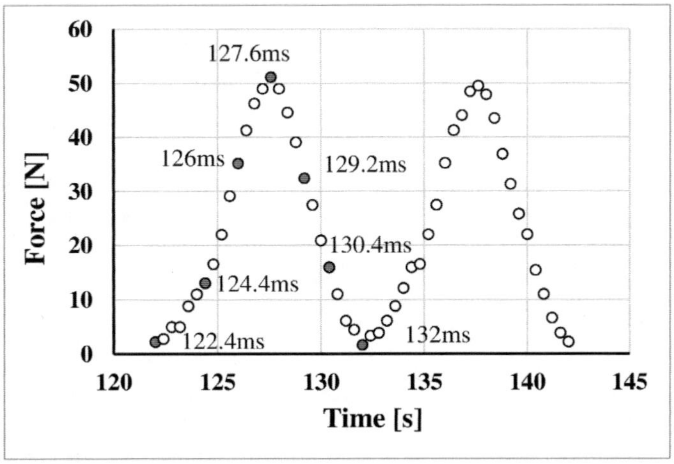


(a)

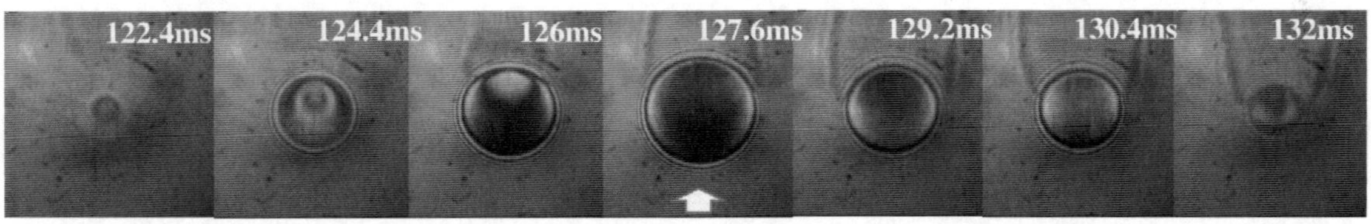


(b)

**Figure 6**. Load variation and interferograms of the contact for 100 Hz, 0.1 m/s entrainment speed.

As seen in these images at lower frequencies of vibration, which imply relatively small rate of variation of load, the film thickness does not show any special features, which leads to the conclusion that in these conditions steady state film treatment of the lubrication is satisfactory. At 50 Hz and especially at 100 Hz however, pronounced deviations from the classic flat film thickness region can be observed. These deviations are larger during the load increase phase. As it is seen in the first two images, the contact diameter increases nearly four times over a two – millisecond interval. This sudden increase of contact area implies a sudden expansion of the convergent zone in the inlet, which, in turn means that thicker film is forced into the contact. The thicker lubricant intake forces the surfaces to further deform elastically thus it is accompanied by an increase of pressure. Although calculations have not been carried out this additional elastic deformation of the surfaces is accompanied by local increase of the contact pressure. In real – life applications this may translate into reduced fatigue durability. In these tests, during the 100 Hz tests, it was noticed that the lowest durability of the silica coating was obtained at the thickest films, which may be explained by the pressure spikes due to fluid entrapment. Waves of the film thickness can be observed during the load decrease phase, but they are crescent shaped and very shallow.

The variations over half cycle of central and minimum film thickness are shown in figures 7 and 8. For the lowest frequency, at 10 Hz, the central film thickness is roughly equal to the central film thickness on a steady – state test that is about 130 nm. In the 100 Hz test the values of the film thickness during the first 2 – 4 milliseconds after the beginning of the load phase, the film thickness is much larger than the steady – state values. At 220 nanometres the central film thickness is about sixty percent larger than the steady state film.

IOP Conf. Series: Materials Science and Engineering **174** (2017) 012036 doi:10.1088/1757-899X/174/1/012036

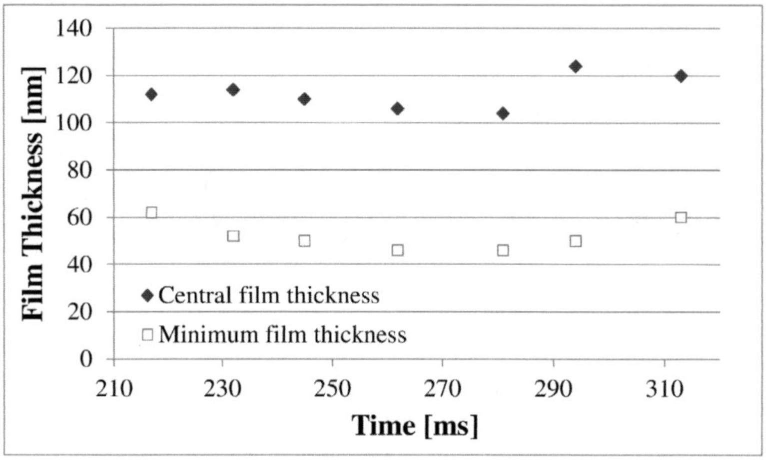


**Figure 7**. Central and minimum film thickness, 10 Hz.

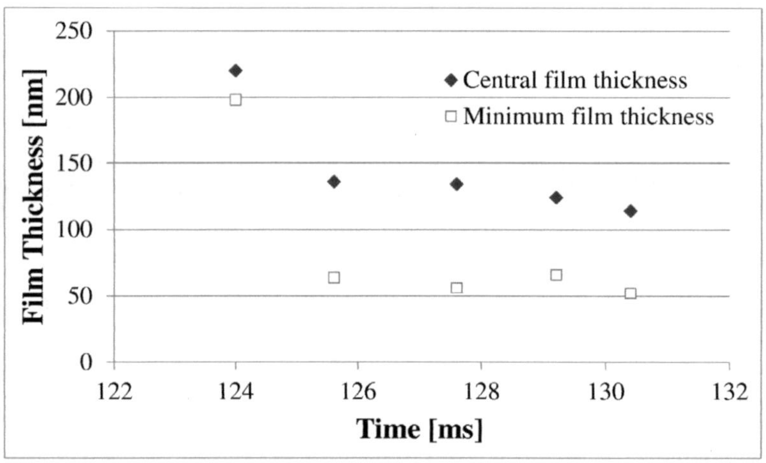


**Figure 8**. Central and minimum film thickness, 100 Hz.

## 5. Conclusions

This paper shows the results of an experimental investigation into the behaviour of elastohydrodynamic films subjected to harmonic, forced vibrations.

The elastohydrodynamic contact under study was a circular contact, formed between a glass disc and a steel ball in the presence of a synthetic base oil lubricant. The film thickness was measured by the optical interferometry. The use of white light resulted in coloured images of the EHD contact which were recorded by a high – speed camera and later analysed for film thickness at various locations on the contact area. The contact was subjected to frequencies from 10 Hz to 100 Hz while the entrainment speed was kept at 0.1 m/s.

It was found that at lower frequencies the film thickness shows very little change from the steady – state values. At 50 Hz and 100 Hz, however a very thick film formed few milliseconds after the beginning of the load increasing phase. As the load continues to increase the film thickness stabilises at around steady – state values and shows very shallow waves. The local increase of film thickness means increased elastic deformations of the contact and thus local variation of the pressure. It is expected that this local, large variations of pressure to be detrimental to fatigue life of the contact, in real – life applications.

**References**

[1] Cameron A and Gohar R 1966 *Proc. Roy. Soc. Lond.* **A291** 520-536
[2] Johnston G J, Wayte R C and Spikes H A 1991 *Trib. Trans.* **34** 187-194
[3] Gustafsson L, Höglund E and Marklund O 1994 *Proc. I. Mech. E.* **J208** 199-205
[4] Luo J, Wen S and Huang P 1996 *Wear* **194** 107-115
[5] Guangteng G and Spikes H A 1996 *Trib. Trans.* **39** 448-454
[6] Guo F and Wong P L 2002 *Proc. I. Mech. E.* **J216** 281-291
[7] Glovnea R P, Forrest A K, Olver A V and Spikes H A 2003 *Trib. Lett.***15** (3) 217-230
[8] Ranger A P, Ettles C M and Cameron A 1975 *Proc. R. Soc. Lond.* **A346** 229-244
[9] Dowson D and Higginson GR 1977 *Elastohydrodynamic Lubrication.* Pergamon, London
[10] Venner C H and Lubrecht A A 2000 *Multilevel methods in lubrication. Tribology Series 37* Ed. D Dowson. Publ. Elsevier
[11] Hughes T G, Elcoate C D and Evans H P 2000 *Proc. I.Mech.E.* **C214** 585-598
[12] Christensen H 1970 *Trans. ASME J. Lub. Tech.* **92** 145-154
[13] Dowson D and Wang D 1994 *Wear***179** 29-37
[14] Herrebrugh K 1970 *Trans. ASME J. Lubr. Tech.* **92** 292-302
[15] Guanteng G, Cann P M, Spikes H A and Olver A V 1999 *Proc. of the 25th Leeds-Lyon Symp. on Tribology. Tribology Series 36* Ed. D Dowson Publ. Elsevier pp 175-183
[16] Kaneta M and Nishikawa H 1999 *Proc. I.Mech.E. J. Eng. Trib.* **213** 371-381
[17] Venner C H, Kaneta M, and Lubrecht A A 2000 *Thinning Films and Tribological Interfaces. Proc. of the 26th Leeds-Lyon Symp. on Tribology. Tribology Series 38* Ed. D Dowson et al. Publ. Elsevier pp 25-36
[18] Krupka I and Hartl M 2006 *Trib. Int.* **40** 1100-1110
[19] Sugimura J, Okumura T, Yamamoto Y, and Spikes H A 1999 *Trib. Intern.* **32** 117-123
[20] Glovnea R P and Spikes H A 2000 *Trib. Trans.* **43** (4) 731-739
[21] Zhao J and Sadeghi F 2003 *ASME Trans. J. Trib.* **125** (1) 76-90
[22] Holmes M J A, Evans H P and Snidle R W 2003 *Tribological Research and Design for Engineering Systems, Proc. of the 29th Leeds-Lyon Symp. on Tribology. Tribology Series 41* Ed. D Dowson et al. Publ. Elsevier pp 79-89
[23] Ohno N and Yamada S 2007 *Proc. I.Mech.E. J. Eng. Trib.* **222** 279-285
[24] Dareing D W and Johnson K L 1975 *J. Mech. Eng. Sci.* **17** (4) 214 – 218
[25] Wijnant Y H, Venner C H, Larsson R and Eriksson P 1999 *J. Trib.* **121** 259 – 264
[26] El Kilali T, Perret-Liaudet J and Mazuyer D 2004 *Transient Processes in Tribology. Proc. of the 30th Leeds-Lyon Symp. on Tribology. Tribology Series 43* Ed. G. Dalmaz et al. Publ. Elsevier pp 409-418
[27] Ciulli E and Bassani R 2006 *Proc. I.Mech.E. J. Eng. Trib.* **220** 319-331
[28] Glovnea R P and Spikes H A 2001 *Proc. I. Mech. Eng J. Eng. Trib.* **215** 125-138
[29] Glovnea R P and Spikes H A 2003 *Lubrication Science* **15-4** 311-320
[30] Sakamoto M, Nishikawa H and Kaneta M 2004 *Transient Processes in Tribology. Proc. of the 30th Leeds-Lyon Symp. on Tribology. Tribology Series 43* Ed. G. Dalmaz et al. Publ. Elsevier pp 391-399
[31] Spikes H A and Cann P M 2001 *Proc. I. Mech. E. J* **215** 261-277

IOP Conf. Series: Materials Science and Engineering **174** (2017) 012037 doi:10.1088/1757-899X/174/1/012037

# A finite element solution of the Reynolds equation of lubrication with film discontinuities: application to helical groove seals

**M Jarray, D Souchet, Y Henry and A Fatu**

Institut Pprime, Université de Poitiers CNRS– ISAE-ENSMA - UPR 3346 SP2MI - Téléport 2 - 11 Boulevard Marie et Pierre Curie, BP 30179 F86962 Futuroscope Chasseneuil Cedex, Poitiers, France

E-mail: mohamed.jarray02@univ-poitiers.fr

**Abstract.** Helical groove seal, is one of the very few non-contact seals that have the capability to effectively seal a liquid. It finds use mainly in turbines and compressors. Although its reliability, this type of seals has not been investigated thoroughly because of its complex characteristics. This work presents a numerical analysis of a helical groove seal operating in laminar regime by means of solving the Reynolds equations for incompressible fluid film in steady state. Equations governing the fluid flow were solved by the finite element method.
Although the simplifying assumptions of Reynolds model help to keep the computational time at an acceptable level, the inertia effects are neglected which may lead to unreliable results especially where the film thickness is discontinuous. The present approach, inspired by Arghir *et al.* [1] is able to take into account concentrated inertia effects, as described by a generalized Bernoulli equation. Comparisons made with the classical Reynolds model show that the film discontinuities should be taken into account when dealing with helically grooved seals. In addition, the leakage of fluid towards the air side was investigated for different parameters such as the groove angle and depth.

## 1. Introduction

Although the limited studies on helically grooved seals, also known as viscoseals, that have been carried out during the past decades, some theories related to the flow in the helical seal channel have been developed, however, applying these theories to the helical grooves design has not been entirely accurate. The main reason is that they rely on a number of simplifying assumptions that lead to the celebrated Reynolds equation, which is not accurate enough for some cases, such high Reynolds number flow or a sudden change in geometry at the edge of the grooves causing film discontinuity, where the inertia effects and the energy loss can't be neglected [2].

The first study to mention a net pressure gain resulting from inertia effect was presented in 2003 by Argir *et al.* [3] By means of a computational fluid dynamics (CFD) analysis of a single macro-roughness, the authors concluded that the inertia effect is present for Re numbers larger than 1. Later in 2005, the suitability of a Reynolds-based method in analyzing discontinuous film thickness domains started to be discussed, Dobrica and Fillon [4] presented a study comparing a Reynolds-based numerical model and the Navier-Stokes model. The models were solved for a 2D geometry of a Rayleigh step. It was shown that, as long as the film thickness remains small, below 120 µm, the two methods give close results, with less than 3% difference. Subsequently, Sahlin *et al.* [5] Investigated the inertia effects in single texture configurations by mean of a 2D CFD analysis on parallel sliding contacts with circular and splined grooves. They reported that the effects of inertia were the dominant mechanism for pressure build-up and load carrying capacity. Their results indicated that load support increases with increase in Reynolds number and groove width. However, four years later, completely contrary results were presented by Dobrica and Fillon [2]. They investigated the domain limit where the Reynolds equation is valid by mean of 2D CFD study on an elementary texture cell, where the texture density and dimple depth to land film thickness are kept constant, they found that inertia in general has a negative effect on the load carrying capacity. Furthermore, it was shown that Reynolds equation is applicable in many practical configurations and that the texture aspect ratio $\lambda = l_d/h_d$ (ratio between the length of the dimple and its depth) is equally important in determining the validity of Reynolds equation, they noted also that, for small λ ratios (less than ~10), the Re model is inapplicable whatever the Reynolds number. It is noteworthy that Dobrica and Fillon stated in [2] that the inertia effects introduced by the dimple are quite similar to those observed in Rayleigh step bearings, and hence such inertia effects can be partially corrected in the region of discontinuous fluid film thickness, by means of the method used by Arghir *et al.* [1] based on equation (2), and therefore if added to the Reynolds equation, it extends the validity domain of a Reynolds-based models.

Another key issue that was the focus of a great deal of research efforts on lubrication was the treatment of cavitation phenomena on textured surfaces, as it is responsible of rupture of the lubricated film and may occur multiple times. A mass-conservative treatment of cavitation has been shown to be crucial for accurate performance predictions. Many cavitation algorithms based on Finite element method (FEM) were developed during the last decades [6] [7], and in the early 2000s, Hajjam and Bonneau [8] [9] used a modified version of the Reynolds equation, and introduced a single variable from which the fields of the pressure and a complementary variable are reconstructed. Their model was recently applied for the study of textured parallel sliders under steady and transient conditions by Gherca *et al.* [10]. Hajjam and Bonneau formulation was later used by Fatu *et al.* [11] [12] [13] for the study of dynamically loaded bearings and by Targaoui *et al.* [14], who investigated viscoseal's optimal geometrical and operating conditions. There are few other studies that aimed to analyze the flow characteristics and operating conditions of helical groove seals, for instance, Kanki and Kawakami [15] studied the leakage characteristics of helically grooved seals and reported that the leakage flow of the viscoseal was less than that of the plain seal. Ludwig et al. [16] analyzed the effect of helical profile of the seal on the pressure distribution, and determined that each groove-land pair repeats the same pressure pattern and provides a saw-tooth pressure profile around the circumference of the seal. In this paper, considering Ludwig et al. results, a periodic boundary condition will be applied at the boundary of the groove-land domain, additionally, Hajjam and Bonneau's cavitation treatment method is considered.

The aim of this work is to explore the possibility of predicting, numerically, the flow behavior in a helically grooved seals using a developed Finite Element code adopting the seal unwrapping approach, and considering, both, the modified Reynolds equation of cavitation model by Hajjam and Bonneau and the inertia correction based on the method used by Argir *et al.* [1]. The results are compared with the Reynolds only model and Navier-Stokes model using OpenFoam CFD Code. The leakage of fluid towards the air side is investigated for different helical groove parameters, such as groove angle and depth.

## 2. Problem definition and modeling approach

The problem to be considered is that of a helically grooved seals with a groove angle $\beta$, clearance $h_f$, and groove length to depth aspect ratio $\lambda$. A common approach for modelling flow in the helical channel is to unwrap the channel from the seal [14]. However the unwraping in our case is inclined in such a way that the flow domain will be viewed as a flat grooved plate inclined by an angle $\beta$, the helix angle relative to the normal to the flow entrance, as depicted in Figure 1. Helicity and curvature effects are neglected by the unwrapping procedure. In fact, this method is only valid for helical grooves channels with clearance relatively very small compared with the seal radius, as demonstrated in the work of Dai *et al.* [17] who numerically evaluated the influence of the film curvature on the pressure, and noted a slight difference, between the original polar coordinates system representation and the simplified flatted Cartesian system representation, that tend to disappear when the clearance to radius ratio $h_f/R$ approaches zero.

### 2.1. Governing Equations

The analytical formulation for cavitation treatment adopted in this paper is detailed in the work of Hajjam and Bonneau [8] and consists of the use of a modified version of the Reynolds equation, which can be applied throughout the entire domain, both in the cavitated and the non-cavitated regions. It can be written as follows

$$F\left\{\frac{\partial}{\partial x}\left(\frac{h^3}{12\mu}\frac{\partial D}{\partial x}\right)+\frac{\partial}{\partial z}\left(\frac{h^3}{12\mu}\frac{\partial D}{\partial z}\right)\right\}=\frac{1}{2}U\frac{\partial h}{\partial x}+\left\{\frac{1}{2}U\frac{\partial D}{\partial x}\right\}(1-\mathrm{F}) \tag{1}$$

where, in the active region

$$\begin{cases} D = p - p_{cav} \\ F = 1 \end{cases}$$

and in the inactive region

$$\begin{cases} D = r - h \\ F = 0 \end{cases} \text{ where } r = \frac{\rho}{\rho_0} h$$

$p_{cav}$ is the cavitation pressure, considered constant in the cavitation zone and h the film thickness.

The correction for inertia effects was introduced in the Modified Re model. The procedure leading to the Modified Re model with inertia correction (Re+I) is detailed in reference [18], and it is based on equation (2), written here for a discontinuity in the film thickness in the x-coordinates

$$p(x^-)+\frac{\rho\left(u(x^-)\right)^2}{2}=p(x^+)+\frac{\rho\left(u(x^+)\right)^2}{2}+\xi\frac{\rho\left[MAX\left(u(x^-),u(x^+)\right)\right]^2}{2} \tag{2}$$

where the coordinates *x*+ and *x*− denote the locations found immediately upstream and downstream of the discontinuity, $\xi$ pressure drop coefficient and *u* denotes the fluid velocity.

### 2.2. Modeling approach

For the finite element approach, a viscoseal of four grooves is considered, with clearance to radius ratio of 0.00625 and film thickness h accorded along the y-axis. The computational domain is assumed to be a two dimensional rectangular surface unwrapped from the cylindrical seal form by cutting it in the same direction as the grooves edge, and in which an uniform Cartesian mesh of $N_x\mathrm{x}N_z$ nodes is defined, with space of $x_i = l/N_x$ and $z_i = L/N_z$ where $N_1 = 32$ and $N_2 = 320$. A finite element algorithm is adopted to solve the governing equations, it have the advantage to easily deal with domains of complex geometry where grooves or discontinuities are located on seal faces, so the difficulties due to film thickness discontinuities can be also avoided.

The following boundary conditions and assumptions are made in the present work.

- A fixed pressure is set at the inlet and the outlet, one side maintained at atmospheric pressure and the other side at higher pressure of 0.3 MPa.
- The inclined sides are set as periodic boundary conditions.
- Steady state operation is assumed.
- The flow is laminar and the sealed fluid is ISO VG 46 with a constant dynamic viscosity of $\mu$ =0.045 Kg/m.s and density $\rho = 900\ kg/m^3$.

A grid-independence analysis was carried out before each simulation to achieve the independence of the mesh from pressure distribution and hence ensure the accuracy of the numerical results.

For Navier-Stokes (NS) calculations, a 3D geometry of the flow domain of the seal texture is created (figure 1) where L is the seal length and $l_d$ the groove length, the domain is meshed with structured hexahedron cells totaling 1838585 nodes, with 13 nodes at the clearance zone in y-direction, 241 in x-direction, and 401 nodes along the z-direction (figure 2). The simpleFoam solver of the OpenFOAM packages was used, which is a steady-state solver for incompressible fluids employing the finite volume method. The same fluid properties and boundary conditions as in the Re+I case were used in OpenFoam.

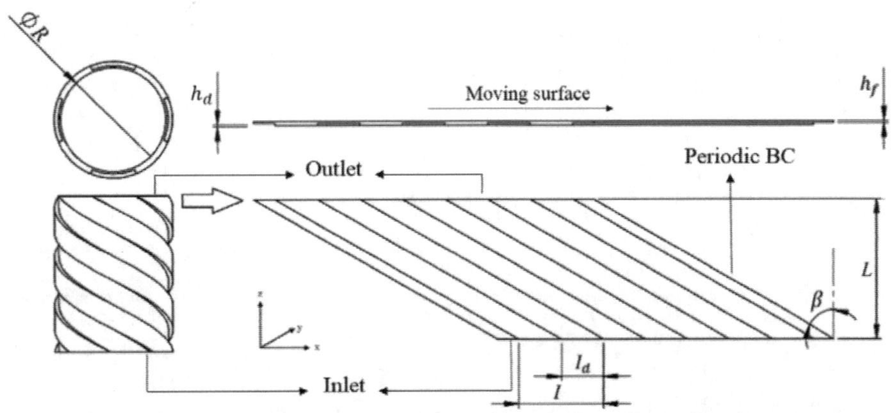


**Figure 1**. Flatted hellically grooved seal with 4 grooves.

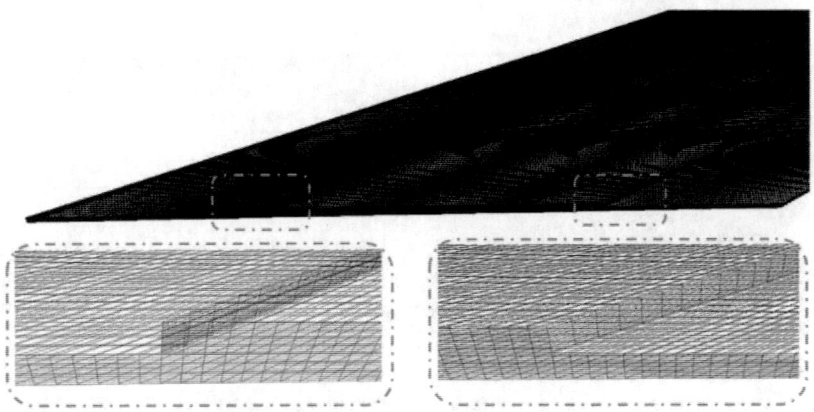

**Figure 2**. Close-up views of the 3D mesh.

## 3. Results and discussions

A significant number of numerical simulations were conducted, using either, the Modified Re+I code or the CFD simulation software (OpenFoam), for all ranges of Reynolds number.

A comparison between flatted and non-flatted geometry using NS model by mean of CFD code OpenFoam is presented in figure 3, where the pressure along x-axis at distance z/L=0.5 from the seal inlet, is shown. It can be seen that the pressure distribution in both, the grooved (a) and the ungrooved (b) faces, for flatted and cylindrical geometry, have almost the same profile.

An example of test configuration for flatted and non-flatted geometry obtained using OpenFoam is provided for illustrative purpose in figure 4.

The accuracy of the Re + I code was evaluated by comparing it with the NS results obtained using OpenFoam. Figure 5 shows the linear dimensionless pressure distributions along x-axis at z/L=0.5 for different Reynolds number ranging from 10 to 40, and based on hydraulic diameter of the seal inlet (equation 3 and 4).

$$Re = \frac{\rho v D_h}{\mu} \tag{3}$$

where $v$ is the linear velocity of the outside surface and $D_h$ the hydraulic diameter at the inlet, which is equivalent to the ratio of 4 times the inlet area to the wetted perimeter:

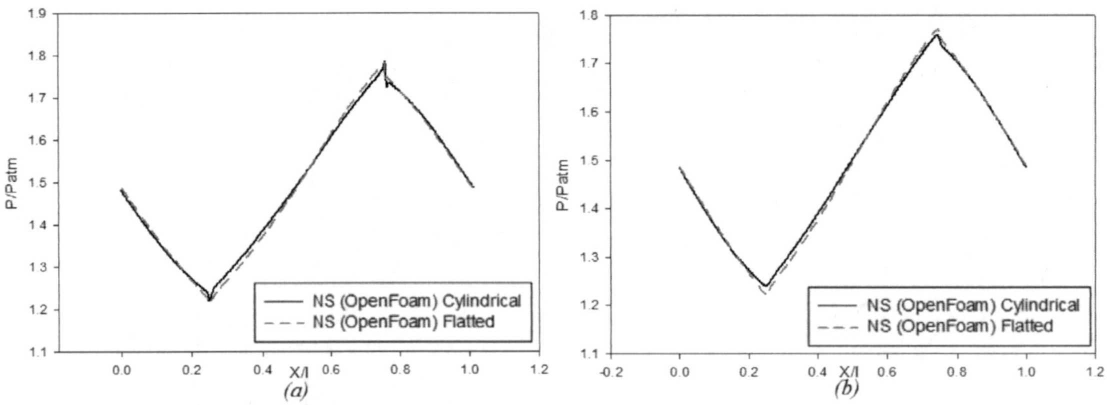


**Figure 3**. Pressure along x axis at z/L=0.5 for grooved face (a) and ungrooved face (b).

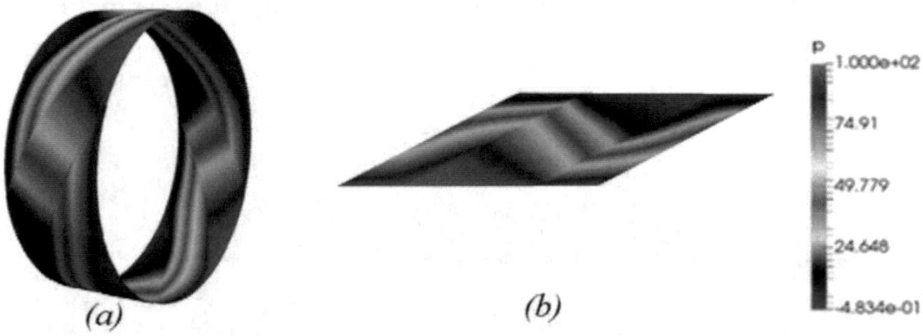


**Figure 4**. Pressure distribution obtained using OpenFoam:
non-flatted seal (a) and flatted seal's cell (b).

$$D_h = 4\frac{l.h_f + l_d.h_d}{2(l + h_d)} \tag{4}$$

where $h_d$ is the groove depth as depicted in figure 1. The results show that the seal geometry significantly affects the pressure distribution, as the developed pressure starts to increase at the start of the groove and decreases sharply at its end. A close-up view at the film discontinuity zone is presented for each case (figure 5 (a) and (b)). It is clear that Re+I and NS have nearly the same profile with a maximum relative error located around the film discontinuity zone, which does not exceed 0.01 for all cases. The relative error is the ratio of the absolute difference between the NS and Re+I model pressure values at x-coordinate, and the value obtained with NS model.

$$e_r = \frac{|P_{NS}(x) - P_{Re+I}(x)|}{|P_{NS}(x)|}. \tag{5}$$

However for Reynolds only model, it is clear that the relative error is slightly more important which can be explained by the fact that Reynolds equation does not take into account the inertia effect. It is is also noticeable that, for bigger Reynolds number, Re+I tends to overestimate the pressure loss at the discontinuity zone (a) where the groove ends, and where the relative error $e_r \approx 0.01$.

The effect of groove length to depth aspect ratio $\lambda$ on the leakage, obtained using the Modified Re+I model for different rotational speed ranged from 500 rpm to 2000 rpm is shown in Figure 6a. The leakage distance z/L is the wetted distance, starting from the inlet, divided by seal length. It can be seen that for all rotational speeds, the leakage increases with the increase of $\lambda$; furthermore it is clear that the leakage becomes more sensitive to rotational speed, as the ratio $\lambda$ increases.

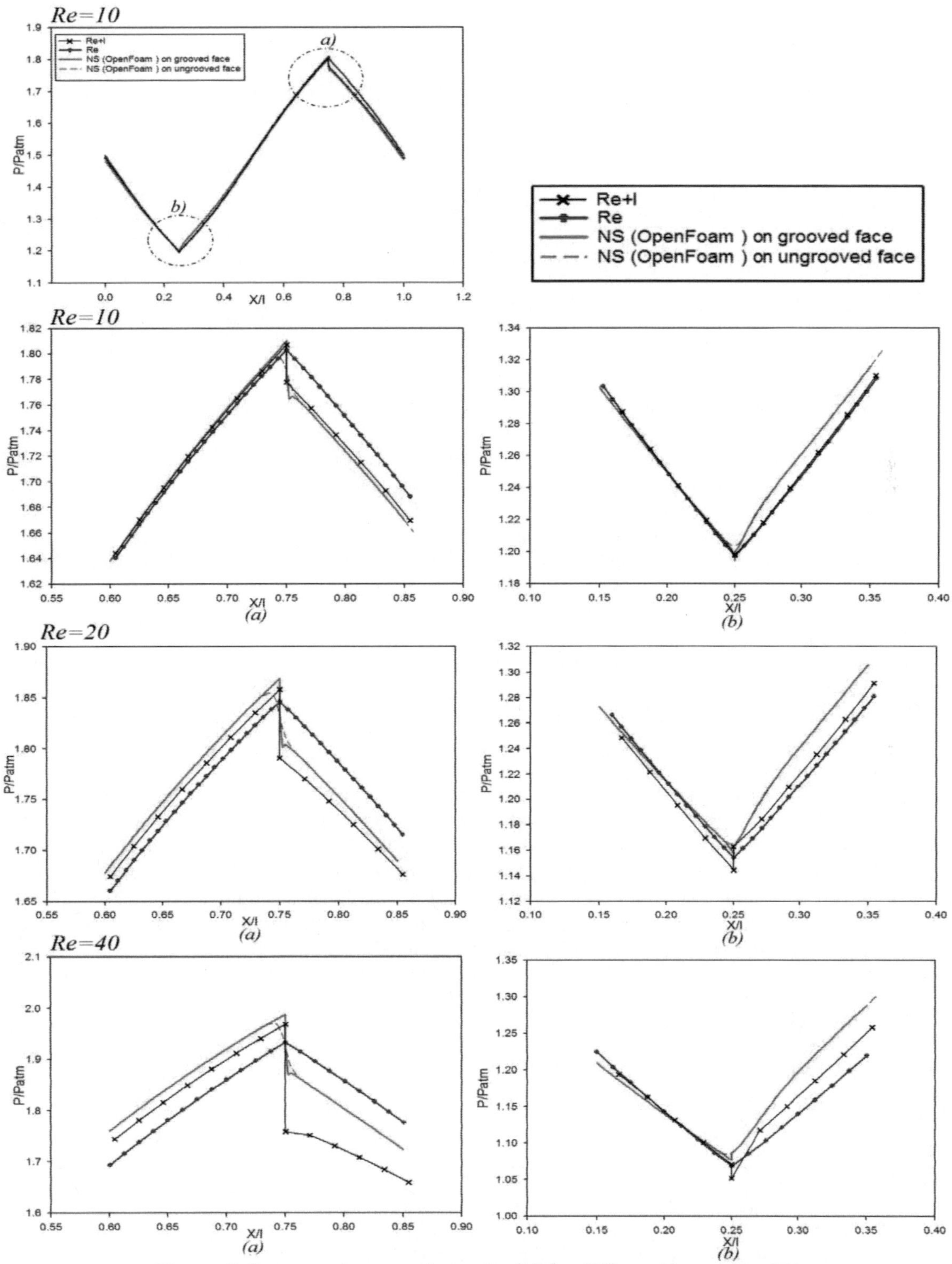


**Figure 5**. Pressure along x-axis at z/L=0.5 for different Re number, NS vs modified Re+I and Re model.

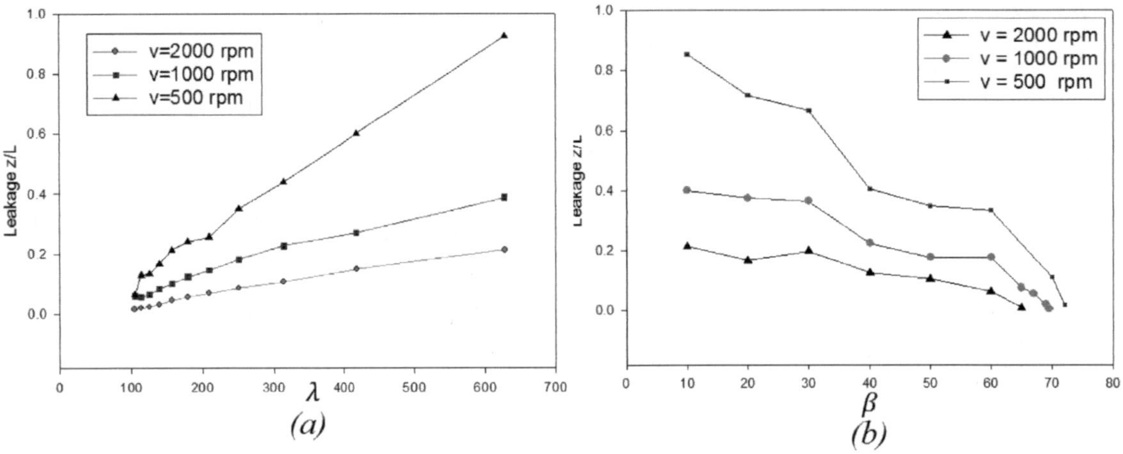


**Figure 6**. Leakage as function; of groove length to depth aspect ratio $\lambda$ (a) and of helix angle (b).

Figure 6b shows the leakage distance for different helix angles. As it can be seen, the helically grooved seal, seals better as the helix angle $\beta$ increases until reaching an optimal sealing angle situated between 65° and 75° for which the leakage distance is minimum, because the seal with a bigger helix angle has more axial component flow and longer groove in favor of energy dissipation. Furthermore fluid rotation opposes the helix angle, and it yields to a reduction of the average circumferential fluid velocity. This is in line with what has been found in previous works carried out for Re model [14].

## 4. Conclusion

The aim of this article is to summarize a methodology that numerically solve hydrodynamic lubrication problems on helically grooved seal modeled with a mass-conserving formulation, and extend the validity of the Reynolds model to account for inertia effect produced at the film discontinuity.

A significant number of numerical simulations were conducted, using either Reynolds+I code and CFD simulation software (OpenFoam), for different Reynolds number. The results showed a good agreement between the Re+I model and the NS model with a relative error that does not exceed 0.01 for all range of Reynolds number. Results also showed that the helix angle has a significant impact on the leakage and the same goes for the groove length to depth aspect ratio $\lambda$ which also have a dramatic effect on the leakage that tends to be more prominent as the rotational speed increases.

A future work will further focus on the domain of validity of the modified Re+I model for helically grooved seal for different geometry configuration by comparing it to a 3D Navier Stokes model, and modified Re+I domain limit will be a subject of further study.

## List of notations

| | |
|---|---|
| $D_h$ | Hydraulic diameter. |
| $e_r$ | Relative error. |
| F | Switch function. |
| h | Film thickness. |
| $h_d$ | Groove depth. |
| $h_f$ | Clearance. |
| $l_d$ | Groove width. |
| L | Viscoseal length. |
| $p_{cav}$ | Cavitation pressure. |
| $P_{NS}$ | Pressure obtained using NS model. |
| $P_{Re+I}$ | Pressure obtained using Re+I model. |

| | |
|---|---|
| R | Viscoseal radius. |
| $Re$ | Reynolds number. |
| $v$ | Velocity of the outside surface. |
| $\beta$ | Groove angle. |
| $\lambda$ | Groove length to depth aspect ratio. |
| $\mu$ | Dynamic viscosity. |
| $\xi$ | Pressure drop coefficient. |
| $\rho$ | Density. |

**References**

[1] Arghir M, Alsayed A and Nicolas D 2002 The finite volume solution of the Reynolds equation of lubrication with film discontinuities *Int. J. Mech. Sci* **44** 2119–2132

[2] Dobrica M B and Fillon M 2009 About the validity of Reynolds equation and inertia effects in textured sliders of infinite *width Proc Inst Mech Eng Part J: J. Eng. Tribol* **223** 69–78

[3] Arghir M, Roucou M, Helene M and Frene J 2003 Theoretical analysis of the incompressible laminar flow in a macro-roughness cell *ASME J. Tribol* **125** 309–318

[4] Dobrica M B and Fillon M 2005 Reynolds Model Suitability in Simulating Rayleigh Step Bearing THD Problems *Tribol. Trans* **48** 522–530

[5] Sahlin F, Glavatskih F, Almqvist T R and Larsson R 2005 Two-dimensional CFD-analysis of micro-patterned surfaces in hydrodynamic lubrication *J. Tribol* **127** 96–102

[6] Kumar A and Booker J F 1991 A finite element cavitation algorithm *ASME J. Tribol* **113** 276–284

[7] Shi F and Paranjpe R 2002 An implicit finite element cavitation algorithm *Comput. Model. Eng. Sci* **3** 507–15

[8] Bonneau D and Hajjam M 2001 Modélisation de la rupture et de la formation des films lubrifiants dans les contacts élastohydrodynamiques *Revue Européenne des Eléments Finis* **10** 679–704

[9] Hajjam M and Bonneau D 2007 A transient finite element cavitation algorithm with application to radial lip seals *Tribol. Int* **40** 1258–69

[10] Gherca A, Fatu A, Hajjam M and Maspeyrot P 2014 Effects of surface texturing in steady-state and transient flow conditions: Two-dimensional numerical simulation using a mass-conserving cavitation model *Proc. Inst. Mech. Eng. Part J: J. Eng. Tribol* **229** 505–22

[11] Fatu A, Hajjam M and Bonneau D 2005 Analysis of non Newtonian and piezoviscous effects in dynamically loaded connecting-rod bearings *Proc. Inst. Mech. Eng. Part J: J. Eng* **219** 209–224

[12] Fatu A, Hajjam M and Bonneau D 2005 An EHD model to predict the interdependent behaviour of two dynamically loaded hybrid journal bearings *ASME J. Tribol* **127** 416–424

[13] Fatu A, Hajjam M and Bonneau D 2006 A new model of thermoelastohydrodynamic lubrication in dynamically loaded journal bearings *ASME J Tribol* **128** 85–95

[14] Targaoui M, Souchet D and Bouyahia F 2015 On the hydrodynamic sealing in lubricated viscoseal *Tribology Transactions* **58** 527–536

[15] Kanki K and Kawakami T 1987 Experimental study on the static and dynamic characteristics of screw Grooved seals *Rotating Machinery Dynamics* **1** 273–278

[16] Ludwig L P, Strom T N, and Allen G P 1965 Experimental study of end effect and pressure patterns in helical groove fluid film seal (Viscoseal), *NASA Technical Note,* TN D-3096 Lewis Research Center, Cleveland, OH

[17] Dai R X, Dong Q and Szeri A Z 1992 Approximations in hydrodynamic lubrication *ASME J. Tribol* **114** (1) 14–25

[18] Dobrica M B and Fillon M 2006 Thermohydrodynamic behavior of a slider pocket bearing *ASME J. Tribol* **128** 312–318

IOP Conf. Series: Materials Science and Engineering **174** (2017) 012038 doi:10.1088/1757-899X/174/1/012038

# Analysis of squeeze film process between non-parallel circular surfaces

**A V Radulescu and I Radulescu**

Department Machine Elements and Tribology, POLITEHNICA University of Bucharest, Spl. Independentei 313, Bucharest, Romania

E-mail: varrav2000@yahoo.com

**Abstract**. The purpose of the paper is to determine the performance of a Newtonian fluid, in the case of squeezing process between non-parallel circular surfaces, from theoretical and experimental point of view. The theoretical analysis is based on the two-dimensional Reynolds equation, which is solved assuming the simplifying hypothesis of the "narrow bearing theory". With this approximation, it is possible to neglect certain terms in the Reynolds equation, and an analytical expression for the pressure distribution on the superior surface can be written. The theoretical results have been compared with the experimental determination of the pressure distribution, obtained on a modified Weissenberg rheogoniometer. For the squeezing experiment, two oils specific for internal combustion engines have been used. Their viscosity was measured with a con and plate Brookfield viscometer. The stand has the possibility to measure the pressure variation with the film thickness, in three points, for different squeezing velocity and for an imposed geometry of the circular plates. In conclusion it can observe a good correlation between theory and experiment, in the case of thick lubricant films. At low values of the thickness of lubricant film, the theoretical model has to be improved, using finite theory method for flow modelling.

## 1. Introduction

Squeezing phenomenon describes the large deformation (or a flow) of a soft material (or a viscous fluid) between two solid surfaces approaching one to each other. During squeezing the gap between surfaces is changing in time and the sample is ejected from the gap. In many applications the gap is small in comparison to the other dimensions of the surfaces, in particular discs diameters, so squeezing flow is mainly associated to thin film hydrodynamics, lubrication and rheology of complex fluids, [1]. The interest shown to this complex flow is due to various applications: the presence of this motion in mechanical engineering [2], rheology [3], constructions [4], industrial processes [5], biological processes such as mastication [6] etc.

The purpose of the present paper is to use the squeezing process for an engineering application, in particular for the analytical approximate formulation of the squeeze film process between non-parallel circular surfaces. The theoretical results are validated on an experimental stand, based on a modified Weissenberg rheogoniometer.

## 2. Theoretical model

The theoretical analysis is based on a generalized model for any incompressible flow, the Newtonian model, for squeezing the film between two non-parallel circular surfaces (figure 1):

IOP Conf. Series: Materials Science and Engineering **174** (2017) 012038 doi:10.1088/1757-899X/174/1/012038

$$\tau = \eta \frac{du}{dy} \tag{1}$$

where: $\tau$ - shear stress;
$\eta$ - viscosity;
$u$ - velocity;
$x, y, z$ - Cartesian coordinates.

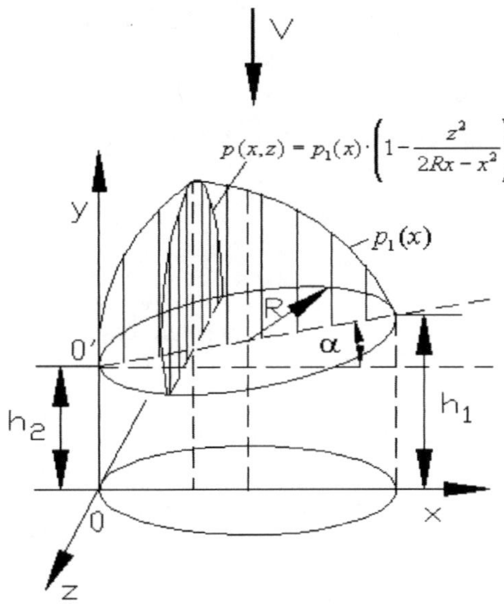


**Figure 1.** Geometry of the interstice.

The geometry of the interstice is characterized by the equation:

$$h(x) = h_2 + x \cdot \mathrm{tg}\alpha\,,\ x \in [0,\, 2R] \tag{2}$$

where: $h$ - film thickness;
$h_2$ - minimum film thickness;
$\alpha$ - tilted angle;
$R$ - radius.

The main hypotheses of the flow are:

- the flow is laminar, incompressible and stationary;
- the oil viscosity is constant on the film thickness;
- the tilted angle $\alpha$ is very small ($\alpha \leq 1^0$).

In order to determine the pressure distribution, the two-dimensional Reynolds equation for non-parallel surfaces is considered:

$$\frac{d}{dx}\left(h^3 \frac{dp}{dx}\right) + \frac{d}{dz}\left(h^3 \frac{dp}{dz}\right) = -12\eta V \tag{3}$$

where: $p$ - pressure; $V$ - squeezing velocity.

The analytical solution for equation (3) is obtained assuming the simplifying hypothesis of the "narrow bearing theory" [7 ], [8], [9]. According to this theory, the pressure variation across the Oz axis is much greater than that along Ox axis. With this approximation, it is possible to neglect certain terms in the Reynolds equation, and an analytical expression for the pressure variation on the superior surface can be written down.

IOP Conf. Series: Materials Science and Engineering **174** (2017) 012038 doi:10.1088/1757-899X/174/1/012038

By integrating equation (3) and imposing the limit conditions ($p_1(x) = 0$ for $x = 0$ and $2R$), the expression of the pressure distribution corresponding to Ox axis can be obtained:

$$p_1(x) = \frac{6\eta V}{tg^2\alpha}\left[\frac{1}{h(x)} - \frac{h_1 h_2}{h(x)^2 (h_1 + h_2)} - \frac{1}{h_1 + h_2}\right] \tag{4}$$

where: $h_1$ - maximum film thickness.

Regarding this equation for the pressure distribution corresponding to Ox axis (equation 4), an important observation is imposed: due to the fact that the tilted angle $\alpha$ appears at the denominator, when $\alpha = 0$, apparently the pressure $p_1(x)$ leads to infinite. Actually, taking into account the expression of definition for the film thickness (equation 2), which also depends on tilted angle $\alpha$, and calculating the limit value for the pressure $p_1(x)$, the following calculus expression is obtained:

$$\lim_{\alpha \to 0} p_1(x) = \frac{3\eta V (2Rx - x^2)}{h_2^3} \tag{5}$$

which is independent of the tilted angle $\alpha$.

For the Oz axis, considering the simplifying theory of the narrow bearing, the pressure distribution becomes:

$$p_2(z) = p_1(x)\left(1 - \frac{z^2}{2Rx - x^2}\right) \tag{6}$$

Finally, from the equations (4) and (6), results the two-dimensional pressure distribution:

$$p(x,z) = \frac{6\eta V}{tg^2\alpha}\left[\frac{1}{h(x)} - \frac{h_1 h_2}{h(x)^2 (h_1 + h_2)} - \frac{1}{h_1 + h_2}\right]\left(1 - \frac{z^2}{2Rx - x^2}\right) \tag{7}$$

Of course, the same observation done for the equation (4), regarding the tilted angle $\alpha$ when $\alpha = 0$, is also available for equation (7).

Figures 2 and 3 show two cases of the spatial pressure distribution in the fluid film, corresponding to two cases of the tilted angle: parallel surfaces ($\alpha = 0^0$) and non-parallel surfaces ($\alpha = 0.5^0$).

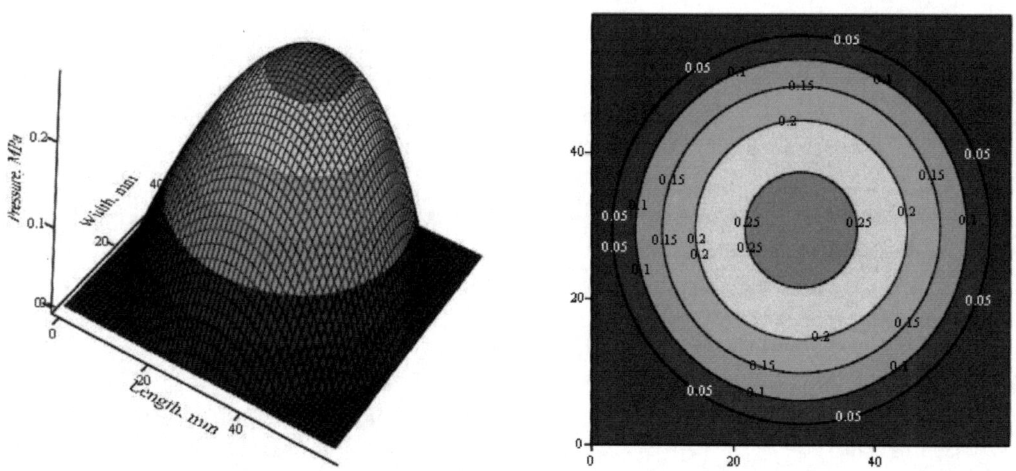


(a) (b)

**Figure 2.** Spatial pressure distribution for parallel surfaces: 3D pressure distribution (a) and isobar curves (b) ($\alpha = 0^0$, $\eta = 0.2$ Pa·s, $R = 30$ mm, $V = 0.5$ mm/s, $h_2 = 0.1$ mm)

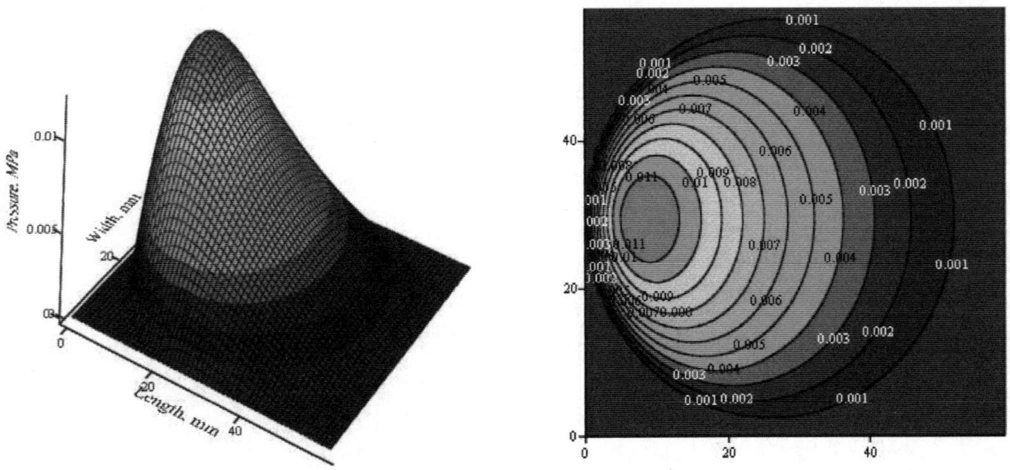


(a) (b)

**Figure 3.** Spatial pressure distribution for non-parallel surfaces: 3D pressure distribution (a) and isobar curves (b) ($\alpha = 0.5^0$, $\eta = 0.2$ Pa·s, $R = 30$ mm, $V = 0.5$ mm/s, $h_2 = 0.1$ mm)

Figure 4 shows the pressure distribution along the Ox axis, for different values of the tilted angle, in order to put in evidence the values of the maximum pressure and the corresponding axial coordinate.

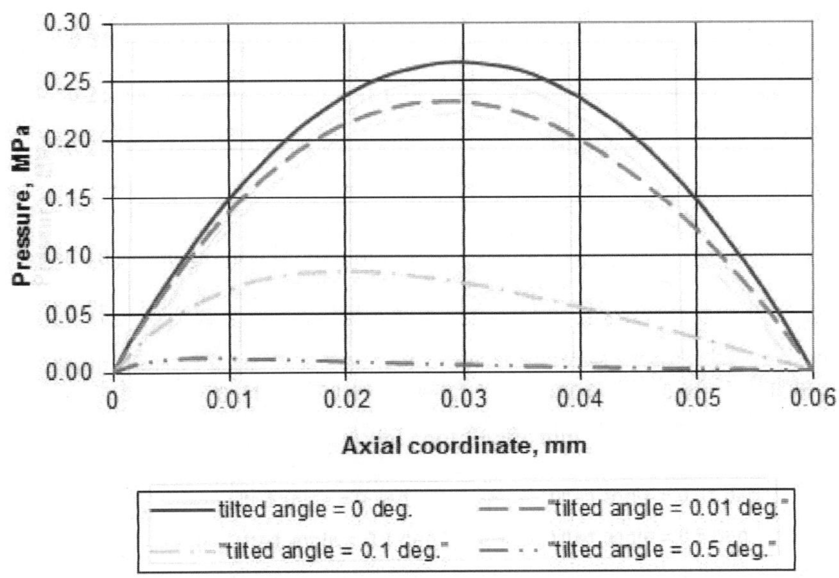


**Figure 4.** Pressure distribution along Ox axis.

It can be observed the difference between the shapes of the pressure distribution in the two cases, and also between the values of the maximum pressure in the film (0.25 MPa for parallel surfaces and 0.011 MPa for non-parallel surfaces).

According to this observation, the tilted angle between the surfaces has an important influence on the pressure distribution. Any deviation from parallelism of the surfaces has a major importance concerning the flow of the lubricant between the surfaces and the maximum pressure that occurs.

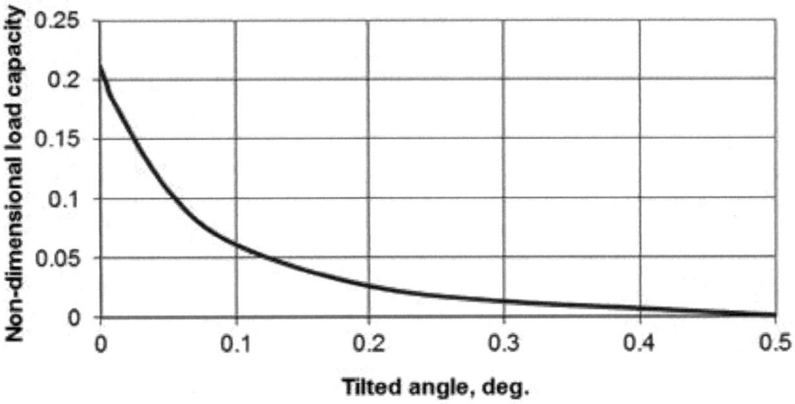


**Figure 5.** Non-dimensional load capacity versus tilted angle.

The load capacity $F$ is obtained by numerical integration of the pressure distribution, following the relation:

$$F = \iint_S p(x, z)dzdx \tag{8}$$

where $S$ is the integrating surface given by:

$$S: \ (x - R)^2 + z^2 \leq R^2 \tag{9}$$

The variation of the non-dimensional load capacity $\overline{F} = \dfrac{Fh_2^3}{6\pi\eta VR^4}$ as function of the tilted angle $\alpha$ is presented in figure 5.

The same influence of the tilted angle on the load capacity can be observed, same as for the pressure distribution.

## 3. Experimental stand

The experimental stand used for the carrying out of the determinations was a modified rheogoniometer Weissenberg (figure 6), built up from the main structural elements:

- Central working unit;
- Testing couple, which consists on two circular paralel surfaces;
- Driving system of the superior disc, corresponding to the friction couple;
- Electric and comand system for the servomotor;
- Pressure transducers;
- Displacement transducer;
- Data acquisition system.

For the acquisition and the numerical treatment of the experimental data it has been used the LabVIEW software. The measurement of the signal provided by the three pressure transducers and the displacement transducer was realized using an acquisition board NI USB-6008. The pressure transducers have been calibratated with a 200 g load, finally obtaining conversion relations between electrical signal and pressure, respectively displacement. Figure 7 presents a detail regarding the construction and the location of the pressure transducers.

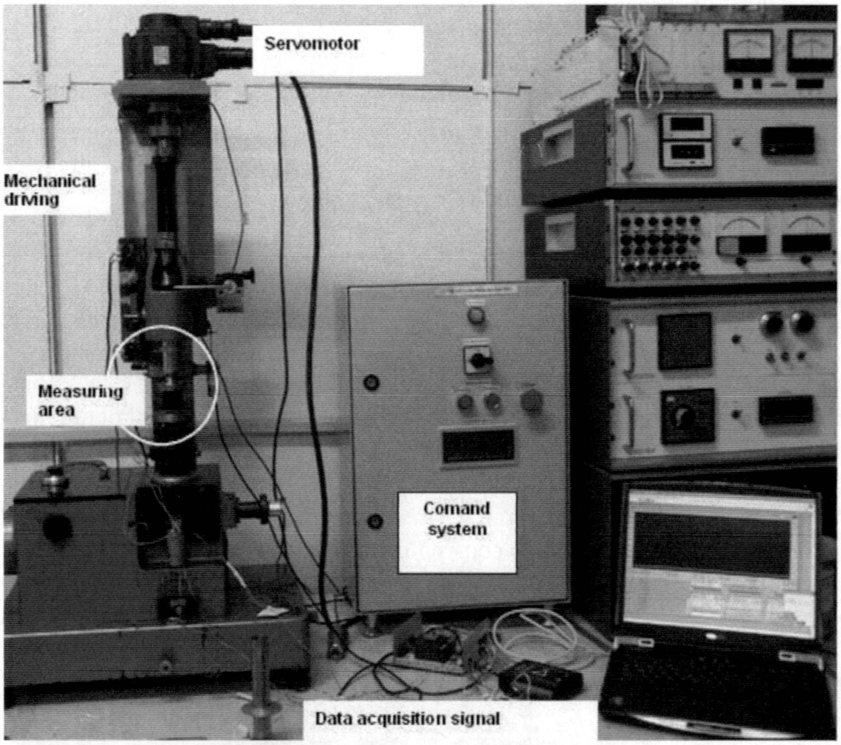


**Figure 6.** General view of the experimental stand.

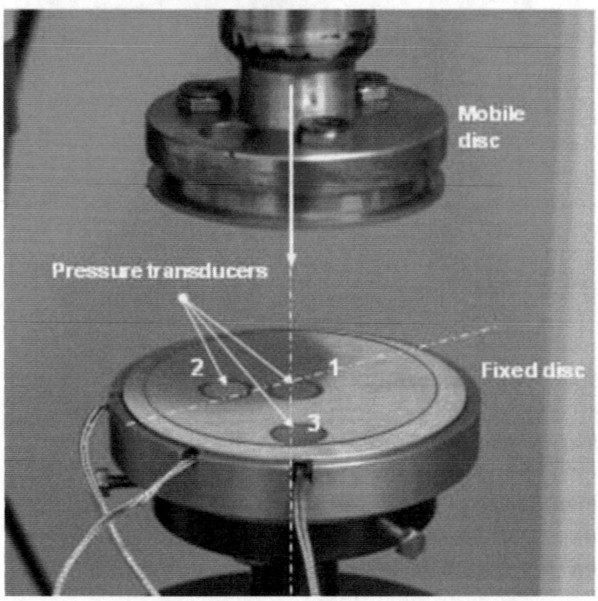


**Figure 7.** Location of the pressure transducers.

For the squeezing experiment, two oils have been used: 10W40 oil, with the dynamic viscosity $\eta$ = 0.2 Pa·s, and M30 oil, characterized by the dynamic viscosity $\eta$ = 0.5 Pa·s. The viscosity was measured at $20^0$C, with a con and plate Brookfield viscometer.

The geometry of the surfaces is characterized by the radius of the circular plate ($R$ = 30 mm) and the positions of the pressure transducers:

- Central transducer, with the coordinates $x = R$ and $z = 0$;

- Median transducer, with the coordinates $x = 5/6 \cdot R$ and $z = \sqrt{3}/6 \cdot R$;
- Lateral transducer, with the coordinates $x = 1/3 \cdot R$ and $z = 0$;

For the testing protocol, four velocities of the superior disc have been imposed (0.25, 0.50, 0.75 and 1 mm/s), and the film thickness varied from 0.6 to 0.1 mm. The stand has the possibility to measure the pressure variation with the film thickness, for different squeezing velocity and for an imposed geometry of the circular plates.

## 4. Results and discussions

In the first stage, the data obtained from the acquisition system were plotted in a pressure – film thickness diagram, for each three transducers, and they were correlated with the theoretical values obtained from equation (6), for $\alpha = 0^0$ (case of the parallel surfaces).

Analysing the experimental results, it has been observed a significant difference between theory and experiment. The explanation of this phenomenon consists in the existence of a deviation from parallelism between the superior and the inferior surfaces (tilted angle $\alpha > 0^0$).

Figure 8 shows the position of the pressure transducers on the inferior surface and the correspondence with the theoretical pressure distribution, for two values of the tilted angle: $\alpha = 0^0$ and $\alpha = 0.1^0$. The difference between the two analysed cases consists in the fact that in the second case, the transducer 1 (central) and transducer 2 (median) show the same pressure.

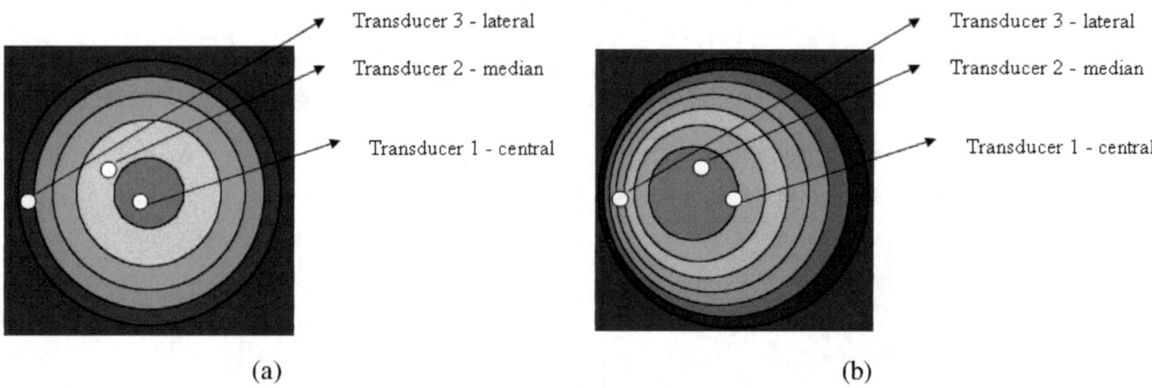


**Figure 8.** Correspondence between pressure transducers and pressure distribution for $\alpha = 0^0$ (a) and $\alpha = 0.1^0$ (b)

In the second stage, in order to correlate the experimental results with the theory, the measured values of the pressure and film thickness were plotted together with the theoretical curves for different tilted angles. These results, obtained for the two tested oils, are presented in figures 9 and 10.

## 5. Conclusions

1. A theoretical model for the squeeze film process between non-parallel circular surfaces has been developed and the theoretical results have been validated using an experimental stand;
2. The pressure distribution for the non-parallel surfaces is strong influence by the tilted angle;
3. The proposed model is in a good agreement with the experiment for mean values of the film thickness (0.1 … 0.5 mm);
4. For lower values of the film thickness, the theoretical model has to be improved using the finite element methods for the modelling of the flow;
5. The geometry of the friction couple is not stable and the tilted angle between surfaces vary during the experiments.

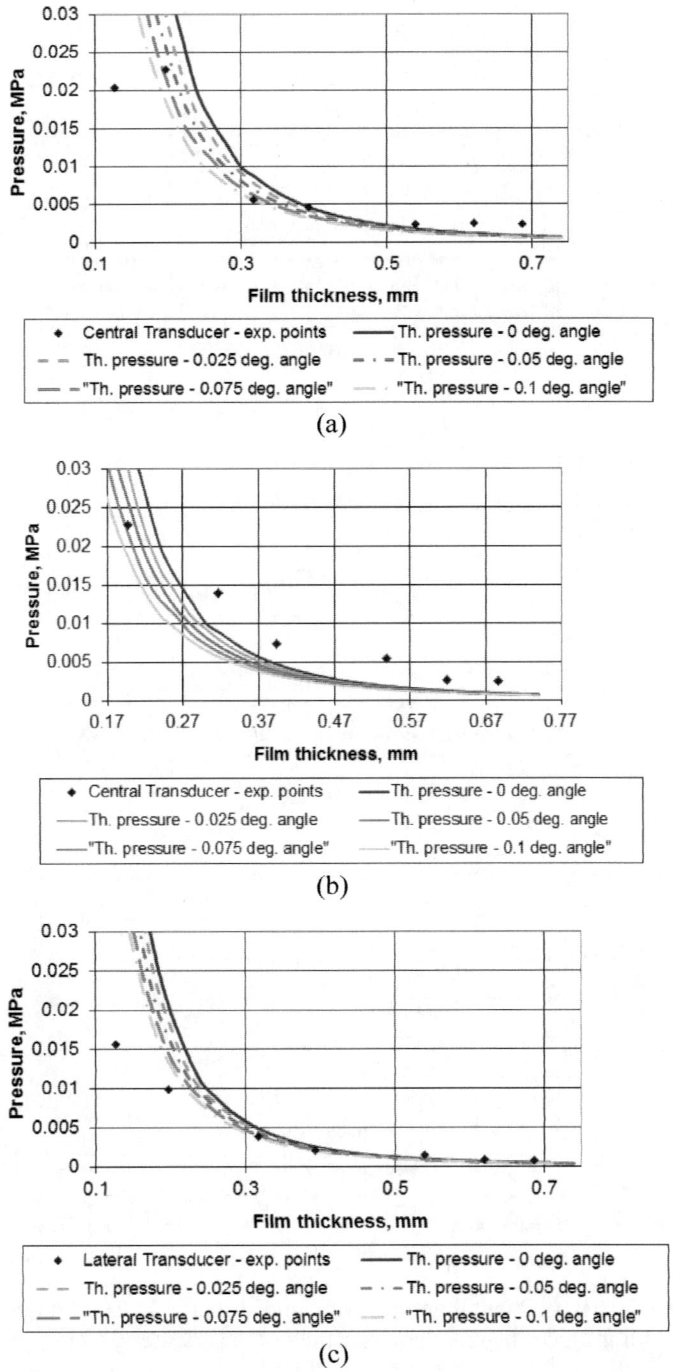


**Figure 9.** Experimental and theoretical results for 10W40 oil, in case of central transducer (a), median transducer (b) and lateral transducer (c)

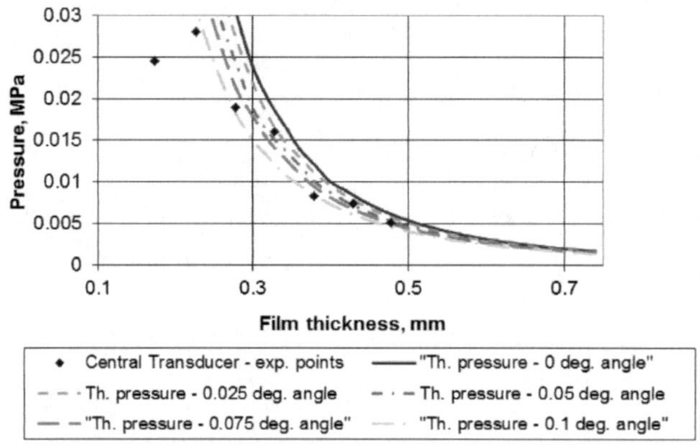


(a)

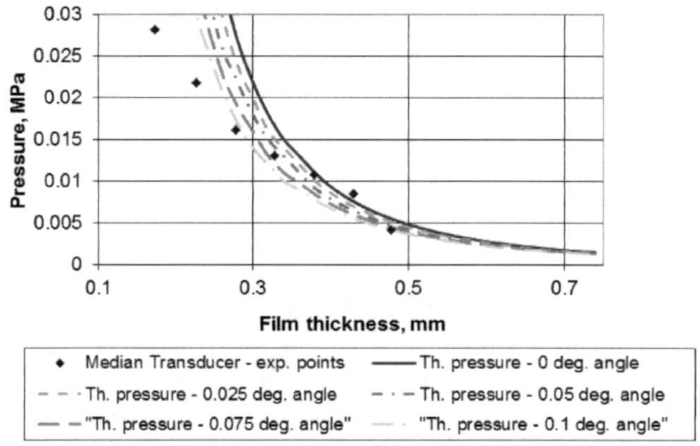


(b)

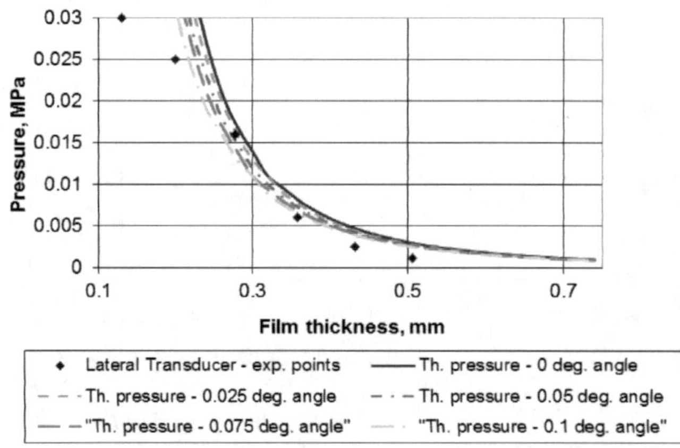


(c)

**Figure 10.** Experimental and theoretical results for M30 oil, in case of central transducer (a), median transducer (b) and lateral transducer (c)

## References

[1] Engmann J, Servais C and Burbidge A 2005 *J. Non-Newtonian Fluid Mech.* **132** 1
[2] Debbaut B and Thomas K 2004 *J. Non-Newtonian Fluid Mech.* **124** 77
[3] Kolenda F, Retana P, Racineux G and Poitou A 2003 *Powder Technol.* **130** 56

[4] Toutoua Y, Roussela N and Lanos C 2005 *Cem. Concr. Res.* **35** 1891
[5] Kompani M and Venerus D C 2000 *Rheol. Acta* **39** 441
[6] Jiang P, See H, Swain M V and Phan-Thien N 2003 *Rheol. Acta* **42** 118
[7] Goenka P K and Oh K P 1986 *J. Tribol.* **108** 294
[8] Vignolo G G, Barilà D O and Quinzani L M 2011 *Tribol. Int.* **44** 1089
[9] Bezvani M A and Hahn E J 1993 *J. Tribol.* **115** 545

IOP Conf. Series: Materials Science and Engineering **174** (2017) 012039 doi:10.1088/1757-899X/174/1/012039

# Effects of fluid inertia and bearing flexibility on the performance of finite length journal bearing

**Juliana Javorova[1] and Vassil Alexandrov[2]**

[1]Department of Applied Mechanics, University of Chemical Technology and Metallurgy, 8 Kliment Ohridski Boulevard, 1756 Sofia, Bulgaria
[2]Department of Mechanics, University of Transport, 158 Geo Milev Street, 1574 Sofia, Bulgaria

E-mail: july@uctm.edu

**Abstract**. The paper describes the theoretical study concerning the effect of lubricant inertia forces and deformability of the bearing elastic layer on the performance of a plane journal bearing. The problem is investigated for a Newtonian lubricant under isothermal and isoviscous conditions. The analysis considers the generalized Reynolds equation governing the flow of lubricant in the clearance space and the linear elasticity equation governing the displacement field in the bearing shell. An iterative numerical procedure with successive over relaxation is used to pressure distribution within the lubricated conjunction. Bearing performance characteristics have been presented for typically selected values of generalized Reynolds number Re* and elasticity parameters of the elastic liner. It has been observed that the combined effect of fluid inertia forces and bearing flexibility affects the performance characteristics of dynamically loaded journal bearing.

## 1. Introduction

It is well known that fluid inertia effect is of importance in modern bearing and seal design. This effect is most notable due to the prevailing trend towards light and compact turbomachinery operating at higher speeds, and the use of process liquids (low viscosity) and gases in fluid film bearings. Important examples include water and lubricant mixtures, liquid metals in the nuclear industry, and cryogenic fluids in space turbopumps.

The lubrication and performance characteristics of HD journal bearings are affected by the including of the lubricants inertia forces at the analysis. Some investigators have studied the performance characteristics of HD journal bearings taking into consideration the inertia effects. Among the few studies related to the fluid inertia effect must be mentioned the works of Reinhardt and Lund [1], Banerjee et al. [2], Chen and Chen [3], Kakoty and Majumdar [4, 5], Javorova et al. [6, 7], Bou-Said and Ehret [8], Prasad et al. [9], etc.

Along with that most of research on the design of HD journal bearings relies on the assumption that both the journal and bearing are rigid bodies. In some applications under high loading conditions of the bearings and/or when using bearings with layers on the contact surfaces, distortions of these contact surfaces are significant and cannot be ignored. There is increasing trend for using polymers and similar as bearings materials and layers in different applications, due to their good tribological properties [10].

The effect of elastic deformation of the bearing surface on the static and dynamic performance of HD journal bearings was studied by Fillon and co-authors [11, 12, 13], Osman [10], Elsharkawy [14], Javorova et al. [15], Ma [16], Attia et al. [17]. The major factors affecting the HD journal bearing are thermal and elastic deformation, which cause a reduction in pressure generated in the oil film. Their design requires careful attention because consideration has to be given to the steady load-carrying characteristics and also to the oil film dynamic performance. Furthermore, along last two decades many investigations are also focused on non-Newtonian, thermal, surface roughness effects, etc. as other important factors influenced on HD lubrication and performance characteristics of the bearings.

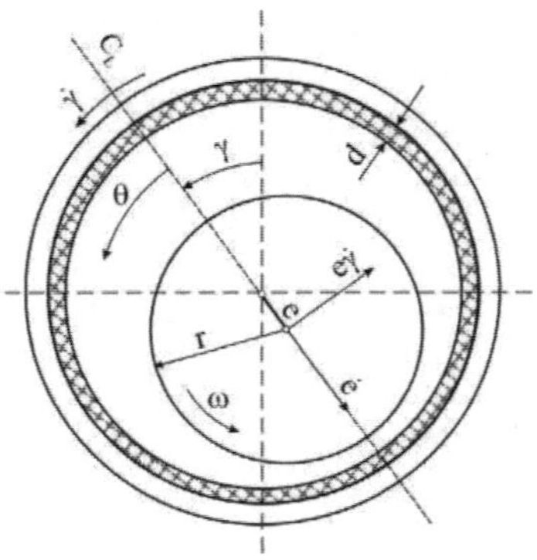


**Figure 1.** Journal bearing with a soft layer on the bush.

The present work is aimed at studying the effect of interaction of the lubricant inertia forces and elastic deformation of the bearing liner on the performance characteristics of HD journal bearings. The problem is investigated under isothermal and isoviscous conditions for a Newtonian lubricant. The bearing bush is covered with a thin resilient layer (figure 1), whose radial displacements are of the same order of magnitude as the film thickness. To take into account the inertia effects, the modified Reynolds equation for dynamic loading governing the film pressure is derived by averaged acceleration method using the Navier-Stokes equations. A finite difference scheme is implemented to solve the pressure distribution within the lubricated conjunction. The presented solution demonstrates the squeeze film effect, as is assumed vibration velocity & of the shaft centre in direction of the centers line.

## 2. Analysis

The coordinate system and the configuration of journal bearing are shown on a figure 1. It is assumed that the journal and bearing are circular and their surfaces are smooth, the load is applied in vertical direction, the groove is filled with a lubricant of a constant pressure, and the journal rotates with a constant angular velocity $\omega$ about its axis. An elastic liner with elastic properties $\mu$ and $E$ is press-fitted in a rigid housing. The liner thickness $d$ is assumed to be of the similar order of magnitude as the lubricant thickness. In the present analysis the following assumptions are also considered: the layers material is homogenous and isotropic, the variation of the pressure across the layer and the fluid film is negligible, and the thermal effects are not considered.

### *2.1. Modified Reynolds equation*

For an incompressible viscose lubricant flow under assumption that a fluid film is thin compared with the journal radius and because of which neglecting the film's curvature, the field equations (1) governing the motion of the lubricant given in Cartesian coordinates can be reduced to the following form

$$\frac{\partial u}{\partial t}+u\frac{\partial u}{\partial x}+v\frac{\partial u}{\partial y}+w\frac{\partial u}{\partial z}=-\frac{1}{\rho}\frac{\partial p}{\partial x}+\upsilon\frac{\partial^2 u}{\partial y^2} \tag{1a}$$

$$0=\frac{\partial p}{\partial y} \tag{1b}$$

$$\frac{\partial w}{\partial t}+u\frac{\partial w}{\partial x}+v\frac{\partial w}{\partial y}+w\frac{\partial w}{\partial z}=-\frac{1}{\rho}\frac{\partial p}{\partial z}+\upsilon\frac{\partial^2 w}{\partial y^2} \tag{1c}$$

$$\frac{\partial u}{\partial x}+\frac{\partial v}{\partial y}+\frac{\partial w}{\partial z}=0 \tag{1d}$$

The boundary conditions at the bearing surface are:

$$u(x,0,z,t)=u(x,0,z,t)=u(x,0,z,t)=0 \tag{2}$$

The boundary conditions at the journal-fluid film interface are

$$u(x,h,z,t)=u_h=\omega r+\dot{e}\sin\theta-e\dot{\gamma}\cos\theta \qquad v(x,h,z,t)=v_h=\omega r\frac{\partial h}{\partial x}+\dot{e}\cos\theta+e\dot{\gamma}\sin\theta$$

$$w(x,h,z,t)=w_h=0 \tag{3}$$

Integrating equations (1a) and (1c) by applying the above related boundary conditions, the velocity components can be derived as follows:

$$u(x,y,z)=\frac{1}{2}F_5(y-h)y+\frac{u_h}{h}y \qquad v(x,y,z)=\frac{v_h}{h}y \qquad w(x,y,z)=\frac{1}{2}F_6(y-h)y \tag{4}$$

Here

$$F_5(x,z,t)=\frac{1}{\eta}\frac{\partial p}{\partial x}+\frac{1}{\upsilon}a_{x,av} \qquad F_6(x,z,t)=\frac{1}{\eta}\frac{\partial p}{\partial z}+\frac{1}{\upsilon}a_{z,av} \tag{5}$$

such $a_{x,av}$ and $a_{z,av}$ are the averaged accelerations across the film on $x$ and $z$ directions, which are represented in the following manner

$$a_{x,av}=\frac{1}{h}\int_0^h\left(\frac{\partial u}{\partial t}+u\frac{\partial u}{\partial x}+v\frac{\partial u}{\partial y}+w\frac{\partial u}{\partial z}\right)dy \qquad a_{z,av}=\frac{1}{h}\int_0^h\left(\frac{\partial w}{\partial t}+u\frac{\partial w}{\partial x}+v\frac{\partial w}{\partial y}+w\frac{\partial w}{\partial z}\right)dy \tag{6}$$

Integrating the continuity equation (1d) with respect to $y$ using boundary conditions (2) and (3), a modified Reynolds equation can be derived in a form [7, 18]:

$$\frac{\partial}{\partial x}\left(\frac{h^3}{\eta}\frac{\partial p}{\partial x}\right)+\frac{\partial}{\partial z}\left(\frac{h^3}{\eta}\frac{\partial p}{\partial z}\right)=6\omega r\frac{\partial h}{\partial x}+12\frac{\partial h}{\partial t}-\frac{1}{\upsilon}\left[\frac{\partial\left(h^3a_{x,av}\right)}{\partial x}+\frac{\partial\left(h^3a_{z,av}\right)}{\partial z}\right] \tag{7}$$

At substitution of the velocity components (4) in the averaged accelerations expresses (6) and introduction of the non-dimensional variables, the dimensionless form of (7) can be obtained. From it, after serious transformations and by neglecting the higher-order terms, the modified Reynolds equation takes the final dimensionless form: / Detailed derivation of this equation is presented in [6]./

$$\left[1-\frac{\mathrm{Re}^*}{12}\left(\frac{3}{2}H\dot{H}+4\bar{u}_h\frac{\partial H}{\partial\theta}H-\bar{v}_hH+\frac{1}{2}\frac{\partial\bar{u}_h}{\partial\theta}H^2\right)\right]\frac{\partial^2\Pi}{\partial\theta^2}+\left[1-\frac{\mathrm{Re}^*}{12}\left(\frac{3}{2}H\dot{H}+2\bar{u}_h\frac{\partial H}{\partial\theta}H-\bar{v}_hH\right)\right]\alpha\frac{\partial^2\Pi}{\partial z_1{}^2}+$$

$$\left[\frac{3}{H}\frac{\partial H}{\partial\theta}-\frac{\mathrm{Re}^*}{12}\left(6\frac{\partial H}{\partial\theta}\dot{H}+\frac{3}{2}H\frac{\partial}{\partial\theta}\dot{H}+\frac{3}{2}\frac{\partial\bar{u}_h}{\partial\theta}\frac{\partial H}{\partial\theta}H+\frac{3}{2}\bar{u}_hH\frac{\partial^2H}{\partial\theta^2}+6\bar{u}_h\left(\frac{\partial H}{\partial\theta}\right)^2-\frac{\partial\bar{v}_h}{\partial\theta}H-\right.\right.$$

$$4\bar{v}_h \frac{\partial H}{\partial \theta}\Big)\Big]\frac{\partial \Pi}{\partial \theta} + \left[\frac{3}{H}\frac{\partial H}{\partial z_1} - \frac{\mathrm{Re}^*}{12}\left(6\frac{\partial H}{\partial z_1}\dot{H} + \frac{3}{2}H\frac{\partial}{\partial z_1}\dot{H} - \frac{1}{2}\frac{\partial \bar{u}_h}{\partial \theta}\frac{\partial H}{\partial z_1}H + \frac{3}{2}\bar{u}_h H\frac{\partial^2 H}{\partial \theta \partial z_1} + 6\bar{u}_h\frac{\partial H}{\partial \theta}\frac{\partial H}{\partial z_1} - \right.\right.$$

$$\left.\left.\frac{\partial \bar{v}_h}{\partial z_1}H - 4\bar{v}_h\frac{\partial H}{\partial z_1}\right)\right]\alpha\frac{\partial \Pi}{\partial z_1} - \frac{\mathrm{Re}^*}{6}\left[\bar{u}_h H\frac{\partial H}{\partial z_1}\right]\alpha\frac{\partial^2 \Pi}{\partial \theta \partial z_1} = \frac{1}{H^3}\frac{\partial H}{\partial \theta} + \frac{1}{H^3}\dot{H} + \frac{\mathrm{Re}^*}{12}\left[\frac{1}{H^2}\frac{\partial H}{\partial \theta}\dot{H} + \frac{1}{2H}\frac{\partial}{\partial \theta}\dot{H} + \right.$$

$$\left.2\bar{u}_h\frac{\partial H}{\partial \theta}\left(\frac{2}{3H}\frac{\partial \bar{u}_h}{\partial \theta} - \frac{1}{2H^2}\bar{v}_h\right) + \frac{2}{3H}\bar{u}_h\frac{\partial^2 H}{\partial \theta^2} + \frac{4}{3H^2}\bar{u}_h\left(\frac{\partial H}{\partial \theta}\right)^2 - \frac{1}{H}\left(\bar{v}_h\frac{\partial \bar{u}_h}{\partial \theta} + \bar{u}_h\frac{\partial \bar{v}_h}{\partial \theta}\right)\right] \tag{8}$$

where $H = h/c$; $\Pi = p.(c/r)^2/6\eta\omega$; $\theta = x/r$; $z_1 = z/(L/2)$; $\alpha = 2r/L$; $\beta = c/r$; $\varepsilon = e/c$; $\tau = t(\omega/2)$; $\mathrm{Re} = c\omega r/\upsilon$; $\mathrm{Re}^* = \beta\,\mathrm{Re}$; $\varepsilon_1 = \dot{e}/\omega r$; $\varepsilon_2 = e\dot{\gamma}/\omega r$; $\bar{u}_h = 1 + \varepsilon_1\sin\theta - \varepsilon_2\cos\theta$; $\bar{v}_h = (\partial H/\partial \theta) + \varepsilon_1\cos\theta + \varepsilon_2\sin\theta$; $\dot{H} = \partial H/\partial \tau$. All of the terms of above equation (8), in which the generalized Reynolds number $\mathrm{Re}^*$ participates as a coefficient, take into account the contribution of lubricants inertia forces to the hydrodynamic pressure gradient.

### 2.2. *Oil-film thickness*

The approach used in the present study aims to superimpose the deformation of the layer on the bearing bush (the other components of the bearing and the journal will be treated as rigid), caused by hydrodynamic pressure generated onto the oil film thickness. The gap thickness is then modified to account for the estimated elastic deformation as follows:

$$h = c + e\cos\theta + \delta \tag{9}$$

The last term of this equation takes into account the influence of the elastic layer deformation. In current paper the liner's surface points radial displacements are determined according to plain strain hypothesis (column or Winkler model) [13, 14, etc.]

$$\delta = \frac{(1+\mu)(1-2\mu)}{(1-\mu)}\frac{d}{E}p \tag{10}$$

Here must be remarked that the column or the Winkler model is applicable when lining is thin compared to the dimensions of the bearing [12, 13, etc.]. The applicability of this formula for mechanical deformations is also verified by comparison with the results obtained using a full deformation model.

### 2.3. *Bearing characteristics*

Once the film pressure distribution is known, the hydrodynamic film forces acting on the system can be determined by integrating the film pressure over the journal surface. Expressing in terms of dimensionless quantities give

$$\bar{W}_1 = -\int_{-1}^{1}\int_{0}^{2\pi} \Pi\cos\theta\, d\theta\, dz_1 \qquad \bar{W}_2 = \int_{-1}^{1}\int_{0}^{2\pi} \Pi\sin\theta\, d\theta\, dz_1 \tag{11}$$

where $\bar{W}_1$ and $\bar{W}_2$ are the components along and perpendicular to the line of centers, respectively.

Then the resultant dimensionless load carrying capacity $\bar{W}$ can be calculated by

$$\bar{W} = \sqrt{\bar{W}_1^2 + \bar{W}_2^2} = \frac{\beta^2}{6\eta\omega r L}W \tag{12}$$

The Sommerfeld number $S$ is defined as:

$$S = \frac{W\beta^2}{\eta\omega rL} = 6\overline{W} \qquad (13)$$

Integrating the shear stress $\tau_{xy}$ around the journal surface, the dimensionless friction force acting on the journal can be derived, subsequently the friction factor can be calculated and written as

$$\overline{F} = \int_{-1}^{1}\int_{0}^{2\pi} \left.\frac{\partial \overline{u}}{\partial y_1}\right|_{y_1=H} d\theta dz \qquad C_F = \frac{\overline{F}}{\overline{W}} \qquad (14)$$

## 3. Simulations procedure and results

### 3.1. General remarks

The solution of the problem for simultaneous influence of lubricants inertia forces and bearing's layer elastic deformations on the performance characteristics of finite journal bearings is done numerically by original program system. The system consists of two basic sections. First of them is concerned with the EHD part of the problem (elastic deformations of bearing liner) and second to the performance characteristics calculation. The pressure distribution, the film shape within the lubricant film region, and elastic deformations distribution are the outputs of the numerical solution, which is done by using a program code FULLINER. Then with program module PERFORMANCES the performance parameters such as the load carrying capacity, Sommerfeld number, attitude angle, side leakage flow and friction factor can be calculated. This program possesses convenient interface, as the output results are arranged in a suitable structure to integration in other kind of studies (for example, stability tests).

### 3.2. EHD solution – Algorithm procedure

Considering the Reynolds boundary conditions, it is clear that the hydrodynamic pressure is symmetrical in axial direction about the middle plane of the bearing. In the circumferential direction it is assumed that the positive pressure terminates where the pressure gradient angle is zero, as it represents the Reynolds boundary conditions for film pressure.

An iterative numerical procedure by applying the finite difference method is used for solving the problem. The successive over-relaxation is adopted in order to improve the convergence rate [4, 5, etc.]. Negative pressure values are immediately put to zero to be satisfying the Reynolds boundary conditions. The film shape is estimated as being due to the eccentricity of the journal with respect to the bearing, in addition to the elastic deformation of the liner on the bearing bush as result of the pressure generated in the oil film. The following steps are performed [10, 7, etc.]: a) The value of the oil film thickness all over the bearing is estimated as it is assumed an absence of deformations on the bearing layer; b) For the estimated film thickness, equation (8) can be solved numerically to obtain the generated pressure in oil film; c) For the generated pressure, the elastic deformation can be calculated in a direction normal to the bearing surfaces using the equation (10); d) The film shape, equation (9), is then modified to account for the estimated elastic deformation; e) Steps (b) to (d) should be repeated until convergence of solution is obtained.

The film domain is divided by the grid spacing. Different mesh sizes have been tried and a mesh with 72 intervals in the circumferential direction and 20 intervals across the bearing width is used. This size gives a rapid rate of convergence and agreeable computer working time. It was observed that increasing of the number of grid points almost does not affect the results. The convergence criterion adopted for pressure is $\left|\left(1-\sum \Pi_{new}/\sum \Pi_{old}\right)\right| \le 1.10^{-7}$.

### 3.3. Results and discussions

The present analysis showed that the effect of lubricants inertia forces can be presented by generalized Reynolds number Re*, whereas the effect of deformability of the bearing's layer - by parameters

$\mu$ and $E$. Then, considering the mathematical model, the governing parameters are eccentricity ratio $\varepsilon$, diameter to length ratio $\alpha$, Re* and elastic layer parameters $\mu$, $E$. The results are obtained for $\alpha$ equal to 0.5; 1.0; 2.0; $\varepsilon$ was varied from 0.1 to 0.9. To establish the validity of the solution algorithm and the computer code employed in the current study, comparison between the results of the present numerical solution and the available theoretical results from previous studies were conducted. For the purpose of the comparison the generalized Reynolds number is set to 0; 0.56; 1.4 at different elasticity parameters: $E = 2\times10^{11}$ [Pa], $\mu = 0.25$ (rigid case); $E = 7.33\times10^{7}$ [Pa], $\mu = 0.4$ (soft case).

The effect of fluid inertia and surfaces deformability on the pressure profile is presented on figures 2 and 3. Some examples are plotted: considering fluid inertia ($\text{Re}^* = 1.4$) and layers deformability (soft case), as well as neglecting acceleration terms ($\text{Re}^* = 0$) and deformability (rigid case). It has been observed that the maximum pressure values increase with the values of the generalized Reynolds

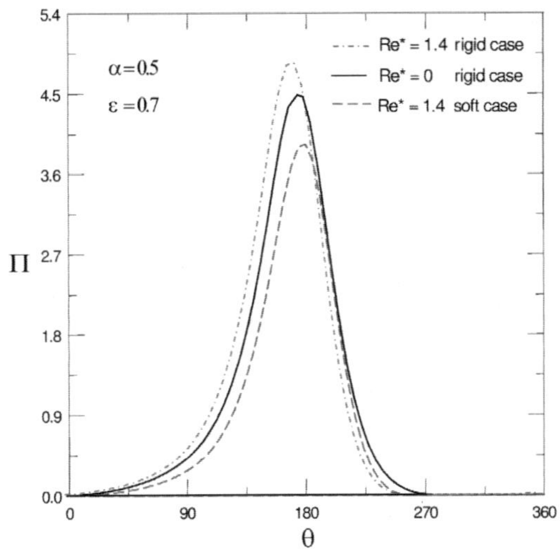


**Figure 2.** Distribution of pressure Π with/without inertia terms.

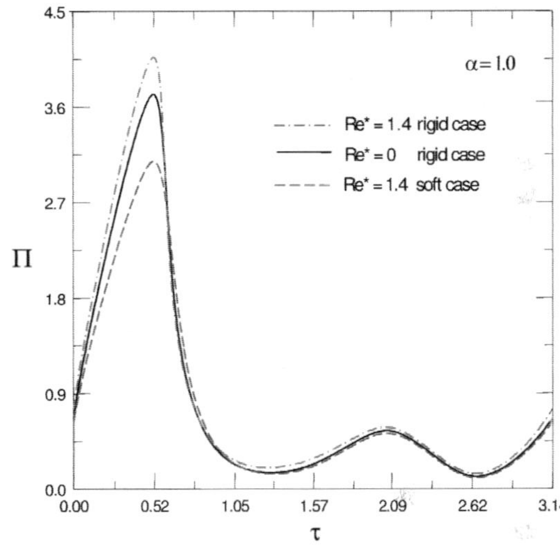


**Figure 3.** Time variation of Π with radial velocity.

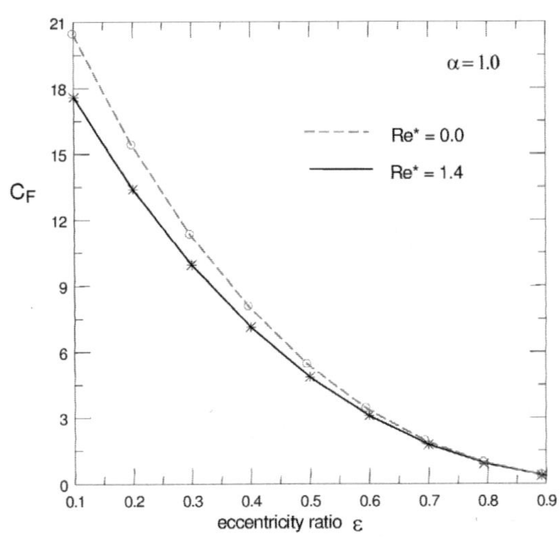


**Figure 4.** Friction factor versus eccentricity ratio.

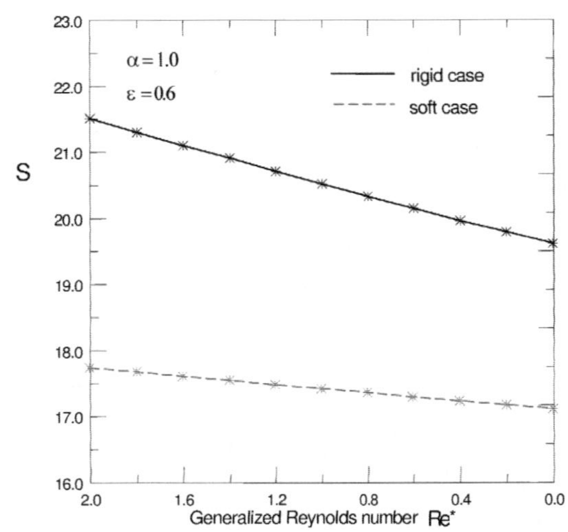


**Figure 5.** Variation of $S$ with Re*.

number but reduce at larger deformability values. When the generalized Reynolds number approaches zero, the pressure distribution represents the pure EHD lubrication case (elastic deformations of bearing liner) and if deformations are ignored the lubrication is hydrodynamically.

The influence of inertia terms is shown on a figure 4, where the friction factor is displayed as a function of eccentricity ratio $\varepsilon$ for different values of Re* (soft case). As seen in this figure, fluid inertia yields reduced values of $C_F$, such the effect of inertia terms is significantly apparent especially at low eccentricities. The dependence of Sommerfeld number on the generalized Reynolds number for rigid and soft cases is given on a figure 5. It is found that the value of load-carrying capacity coefficient decreases with reducing of Re* and the effect is more pronounced for the so called "rigid case".

A comparison between the variation of dimensionless load capacity with eccentricity ratio $\varepsilon$ at $\alpha = 1.0$, for solutions with consideration of fluid inertia and without inertia terms respectively is presented in table 1.

**Table 1.** Comparison between "inertia" and "inertia-free" solutions.

| Eccentricity ratio $\varepsilon$ | Dimensionless load-carrying capacity $\bar{W}$ | | |
|---|---|---|---|
| | Re* = 0 | Re* = 0.56 | Re* = 1.4 |
| 0.1 | 0.2980 | 0.3232 | 0.3448 |
| 0.2 | 0.3965 | 0.4268 | 0.4528 |
| 0.3 | 0.5427 | 0.5804 | 0.6128 |
| 0.4 | 0.7703 | 0.8191 | 0.8610 |
| 0.5 | 1.1498 | 1.2154 | 1.2718 |
| 0.6 | 1.8467 | 1.9390 | 2.0183 |
| 0.7 | 3.3267 | 3.4656 | 3.5850 |
| 0.8 | 7.3517 | 7.5846 | 7.7850 |
| 0.9 | 16.2333 | 16.7441 | 17.1833 |

It is found a good agreement with results, reported in [4, 3] although some theoretical treatments are different. Values which are calculated here are slightly lower, particularly at smaller eccentricities. However, the obtained results for the so called "inertia solution" (Re* = 0.56; 1.4) seem to be higher than these for "inertia-free solution" for the whole range of eccentricity ratio. Such a trend has been indicated in the literature by various investigators [3, 4, 18, etc].

## 4. Conclusions

The research numerically analyzes the performance of HD journal bearings in consideration of lubricant inertia forces and deformability of the elastic bearing layer. In order to render an account the inertia effect the modified Reynolds equation is solved simultaneously with elasticity equation, using FDM. On the base of HD pressure and film thickness values the load carrying capacity, Sommerfeld number and friction coefficient are calculated.

According to the obtained results, the conclusions can be described as follows. The fluid inertia effect is not so much significant as a whole, but the maximum values of the HD pressure increase. As the value of the generalized Reynolds number Re* approaches zero, the problem reduces to the classical EHD lubrication case. Concerning to the Sommerfeld number, a tendency to increasing is evident, but friction factor is reduced under "inertia solutions". By the other hand the pressure and load carrying capacity coefficient increase upon decreasing of deformability of the bearing surfaces. The lubricants inertia forces and elastic deformations of the bearing surfaces must be taken into consideration at the solution of similar kind of problems, because this change of HD pressure values affects the bearing performance characteristics and respectively the stability of rotor-bearing system.

**Acknowledgments**
The authors would like to thank for the support provided by Research and Development Sectors at University of Chemical Technology and Metallurgy - Sofia and University of Transport - Sofia.

**References**

[1] Reinhardt E and Lund J 1975 The influence of fluid inertia on the dynamic properties of journal bearings *J. Lubr. Technol.* **97** 159-167
[2] Benerjee M, Shandil R, Katjal S, Dube G, Pal T and Benerjee K 1986 A nonlinear theory of hydrodynamic lubrication *J. Math. Analysis Applic.* **117** 48-56
[3] Chen C H and Chen C K 1989 The influence of fluid inertia on the operating characteristics of finite journal bearings *Wear* **131** 229-240
[4] Kakoty S K and Majumdar B C 1997 The influence of fluid inertia on the steady-state characteristics and stability of journal bearings *Proc. 9th Nat. Conf. on Machines and Mechanism*, India B15-B26
[5] Kakoty S K and Majumdar B C 2000 Effect of fluid inertia on the dynamic coefficients and stability of journal bearings *Proc. Inst. Mech. Engrs.* **214**-J 229-242
[6] Javorova J G 2005 EHD lubrication of journal bearings with consideration of fluid inertia *Proc. 10th Congress on Th. and Appl. Mechanics* Bulgaria **1** 51-57
[7] Javorova J G, Alexandrov V A, Stanulov K G and Tzvetkov T 2005 Journal bearings stability with consideration of fluid inertia *Proc. World Tribology Congress III* Washington D.C.
[8] Bou-Said B and Ehret P 1994 Inertia and shear-thinning effects on bearing behavior with impulsive loads *ASME J. of Trib.* **116** 535-540
[9] Prasad E S, Nagaraju T and Sagar J P 2012 Thermohydrodynamic performance of a journal bearing with 3D-surface roughness and fluid inertia effects *Int. J. of Appl. Research in Mech. Eng. (IJARME)* **2** 1 18-24
[10] Osman T A 2004 Effect of lubricant non-Newtonian behaviour and elastic deformation on the dynamic performance of finite journal plastic bearings *Tribol. Lett.* **17** 1 31-40
[11] Monmousseau P and Fillon M 2000 Transient thermoelastohydrodynamic analysis for safe operating conditions of tilting-pad journal bearing during start-up *Tribol. Intern.* **33** 3 225-231
[12] Glavatskih S and Fillon M 2006 TEHD analysis of thrust bearings with PTFE-faced pads *ASME J. of Tribol* **128** 49–58
[13] Kuznetsov E, Glavatskih S and Fillon M 2011 THD analysis of compliant journal bearings considering liner deformation *Tribol. Intern.* **44** 1629-41
[14] Elsharkawy A 2005 Effects of lubricant additives on the performance of hydrodynamically lubricated journal bearings *Tribol.Lett.* **18** 1 63-73
[15] Javorova J G, Alexandrov V A and Stanulov K G 2009 Static and dynamic performance of EHD journal bearings in turbulent flow *Proc. Int. Conf. "Power Transmissions'09"* Greece 347-350
[16] Ma Y 2008 Performance of dynamically loaded journal bearings lubricated with couple stress fluids considering the elasticity of the liner *J. Zhejiang Univ. Sci A* **9** 7 916-921
[17] Attia H, Bouaziz S, Maatar M, Fakhfakh T and Haddar M 2010 Hydrodynamic and elasto-hydrodynamic studies of a cylindrical journal bearing *J. of Hydrodynamics* **22** 2 155-163
[18] Pincus O and Sternlicht L 1961*Theory of Hydrodynamic Lubrication* (New York: Mc Grow-Hill)

IOP Conf. Series: Materials Science and Engineering **174** (2017) 012040 doi:10.1088/1757-899X/174/1/012040

# Some considerations concerning four-ball machine testing of the polyacrylamide solutions

**A C Drumeanu**

Petroleum-Gas University of Ploiesti, Bd. Bucuresti, no. 39, 100690, Ploiesti, Romania

E-mail: drumeanu@upg-ploiesti.ro

**Abstract**. Polyacrylamide (PAM) is one of the most widely used and technically important water-soluble polymers. Polyacrylamide (PAM) is usually obtained by free radical polymerization of acrylamide (AM) and in its partially hydrolysed form is a synthetic straight-chain polymer of acrylamide monomers, some of which have been hydrolysed. The structure of HPAM molecule is a flexible chain. This kind of structure is known as a random coil in polymer chemistry. Due to the hydrolysed groups contained in its molecule, HPAM has multiple charges distributed along the chain that make it a polyelectrolyte. The paper presents the experimental results concerning the lubricant solutions based on polyacrylamide behaviour when were tested on the four ball machine. It has to be mentioned that this kind of polymer was not used until now in lubrication and the studies concerning its tribological behaviour are at the beginning.

## 1. Introduction

The synthetic lubricants utilization, especially of those liquids, was developed in the last two decades because of the new fields development such as aerospace technique, nanotechnology, cryogenic technologies, medical technique etc. In the same time the liquid lubricants based on the mineral oils showed theirs performance limits. This fact required new researches for theirs substitution.

Generally the synthetic lubricants have a good oxidation stability, a wide service temperature range, a high viscosity index and a good plastic compatibility, so that this type of products are possible alternatives for lubrication of machine elements.

Polyacrylamide (PAM) is one of the most widely used and technically important water soluble polymers. Polyacrylamide (PAM) is usually obtained by free radical polymerization of acrylamide (AM) and in its partially hydrolysed form is a synthetic straight-chain polymer of acrylamide monomers, some of which have been hydrolysed (figure 1) [1,2,3,4].

The normally available PAM often contains about 1 to 3 mole % carboxylate groups as an impurity. Because of this fact, for the researches presented in this paper it was considered HPAM, with different hydrolysis levels [2,5,6].

Some physical properties of HPAM such as polymer adsorption, shear stability and thermal stability are influenced by the hydrolysis degree (HD). However, although commercial polymers are supplied with a stated degree of hydrolysis, it is well known that at elevated temperatures the hydrolysis of the amide will continue [5,7].

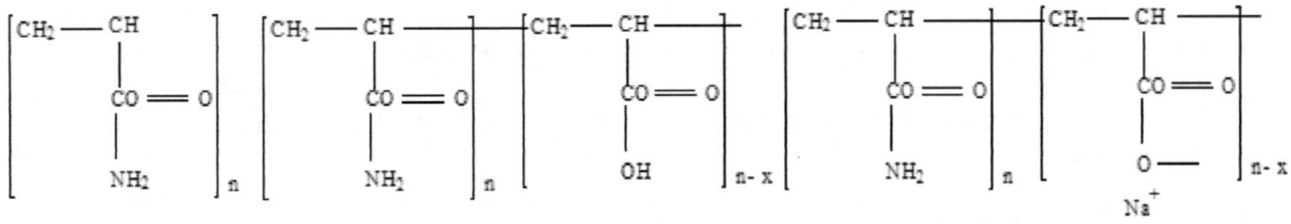

**Figure 1.** The primary chain structure of polyacrylamide (PAM) and partially hydrolysed polyacrylamide (HPAM) and its corresponding sodium salt.

The structure of HPAM molecule is a flexible chain. This kind of structure is known as a random coil in polymer chemistry. Due of the hydrolysed groups contained in its molecule, HPAM has multiple charges distributed along the chain that make it a polyelectrolyte. Generally, the polyelectrolytes are distinguished from non-ionic polymers by the large effects, which change in salt concentration, and pH has on their viscosities. These changes in viscosity, respectively its decrease with salt concentration increase, appear due to the interactions between the fixed charges along the chain and the mobile ions in solution. Because the necessity of lubrication can occur in different conditions, we considered opportune to take into account this behaviour. Also, the polyelectrolyte character of HPAM makes it to interact quite strongly with the metallic surfaces [6,7,8].

The paper presents the experimental results concerning the lubricant solutions based on polyacrylamide behavior when were tested on the four ball machine. It has to be mentioned that this kind of polymer was not used until now in lubrication and the studies concerning its tribological behavior are at the beginning [8,9,10].

## 2. Experimental conditions

There were tested six lubricant solutions based on polyacrylamide. The lubricant solutions were made using four polymers which had the main characteristics presented in table 1.

**Table 1.** The characteristics of the studied polymers.

| Polymer type | Characteristics | |
|---|---|---|
| | Molecular weight, $MW_v$ [g/mol] | Hydrolysis degree, HD [%] |
| P1 | $3 \cdot 10^6$ | 8.3 |
| P2 | $4 \cdot 10^6$ | 8.6 |
| P3 | $7 \cdot 10^6$ | 8.7 |
| P4 | $10 \cdot 10^6$ | 8.6 |

The lubricant solutions were made using different solvents based on tap water and NaCl salt. The dynamic viscosity of the lubricant solutions was determined with the Brookfield viscometer. The characteristics of used lubricant solutions are presented in table 2, and the chemical composition of the used tap water are showed in table 3.

The tests have been carried-out on a four ball machine with a classical construction configuration. The tests were performed using balls made from rolling contact bearing steel RUL1-V (AISI 52100) type. The main characteristics of these balls are: diameter – 12.7 mm; surface roughness $R_a$ – 0.02 μm; hardness HRC - 58.

The test balls are so arranged that their centers form a pyramid with a nominally equilateral base. Each of the lower balls forms three conjunctions – one with the top ball and two with the ball pot. The top ball forms three conjunctions also, one with each of the three lower balls. The upper ball was

IOP Conf. Series: Materials Science and Engineering **174** (2017) 012040 doi:10.1088/1757-899X/174/1/012040

rotated with a constant speed of 1500 rpm (0.998 $m \cdot s^{-1}$ sliding speed). For each test, the four-ball machine ran 60 seconds, and was used new balls and new lubricant sample.

**Table 2.** The characteristics of the lubricant solutions.

| Solution type | Polymer type | Solution characteristics | | |
|---|---|---|---|---|
| | | Polymer concentration [wt.%] | Dynamic viscosity [$Ns/m^2$ x$10^{-3}$] | Solvent |
| S1 | P1 | 1.0 | 240 | tap water |
| S2 | P2 | 1.0 | 320 | tap water |
| S3 | P3 | 1.0 | 530 | tap water |
| S4 | P4 | 1.0 | 1450 | tap water |
| S5 | P3 | 1.0 | 250 | tap water + 2% NaCl |
| S6 | P3 | 1.0 | 190 | tap water + 5% NaCl |

**Table 3.** Chemical composition of the tap water used as solvent.

| Chemical element | Nitrate $NO_2^-$ | Sulfate $SO_4^{2-}$ | Chloride $Cl^-$ | Potassium K | Calcium Ca | Magnesium Mg | Sodium Na |
|---|---|---|---|---|---|---|---|
| Concentration [mg/l] | 18 | 40 | 98 | 1.6 | 83 | 10 | 57 |

Before and after each test the four-ball device elements and the testing balls were cleaned with water and dried in a warm air draught. A constant quantity of 20 milliliters of lubricant solution was used for each test, so all the balls were completely covered by lubricant. The bulk temperature of the lubricant solution sample was measured with a thermocouple, mounted at the bottom side of the balls pot. The lubricant solution samples temperature during all tests was maintained at about 30…40°C, no important changes being observed in the lubricant bulk temperature, due to the short time tests. The wear spot diameter of the balls was measured using a laboratory microscope.

During each test there were measured and determined the following parameters:
- the axial load applied to the upper ball;
- the friction force that appears between the upper ball and the lower balls;
- the wear spot diameters of the lower balls determined before the welding;
- the axial load which corresponds to the balls welding;
- the friction force value before the welding.

## 3. Results and discussions

Regarding the influence of the molecular weight and the solvent on the dynamic viscosity of the lubricant solutions it can be observed in figure 2 and figure 3 that dynamic viscosity increases with the molecular weight increase and decreases with the NaCl content increase.

The behavior of the lubricant solutions during the tests is presented in figure 4 and figure 5. It can be observed that the friction force has different variations as a function of lubricant solution composition. For solutions S1, S2, S3 and S4 the friction force increases with the increase of the axial load in the range of 100…300 N. The same tendency can be remarked for solutions S1, S2, S4 and axial load range of 400…1100 N where for solution S3 the friction force decreases with axial force increase. Also, from figure 4, it can be observed that the friction force increases with the increase of polymer molecular weight, except solution S3 for which the friction force has a minimum value.

IOP Conf. Series: Materials Science and Engineering **174** (2017) 012040 doi:10.1088/1757-899X/174/1/012040

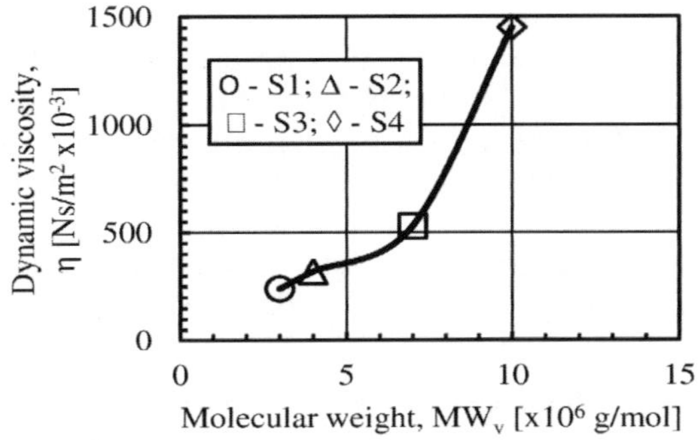


**Figure 2.** Lubricant solutions dynamic viscosity vs. molecular weight (solutions S1, S2, S3, S4).

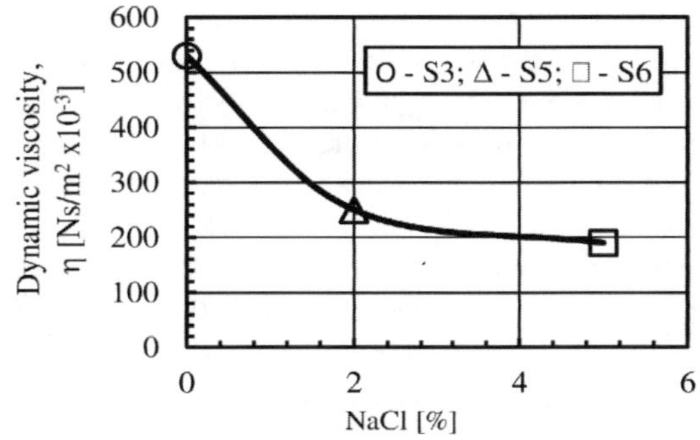


**Figure 3.** Lubricant solutions dynamic viscosity vs. NaCl content in solvent (solutions S3, S5, S6).

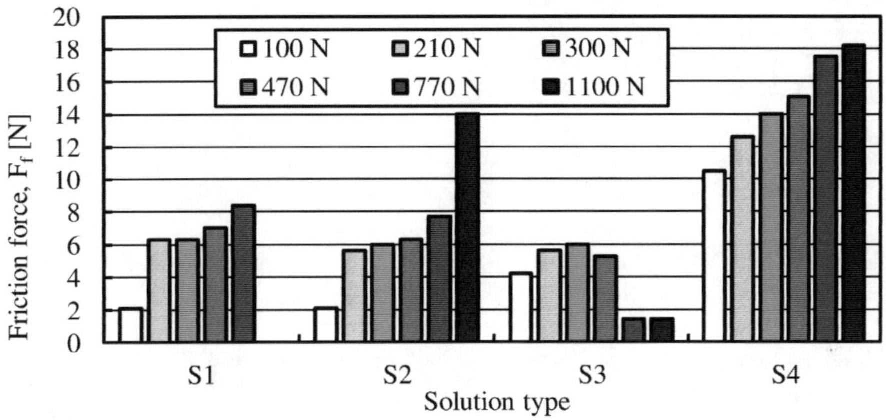


**Figure 4.** The dependence between the friction force and the axial load for the tested lubricant solutions (solvent – tap water).

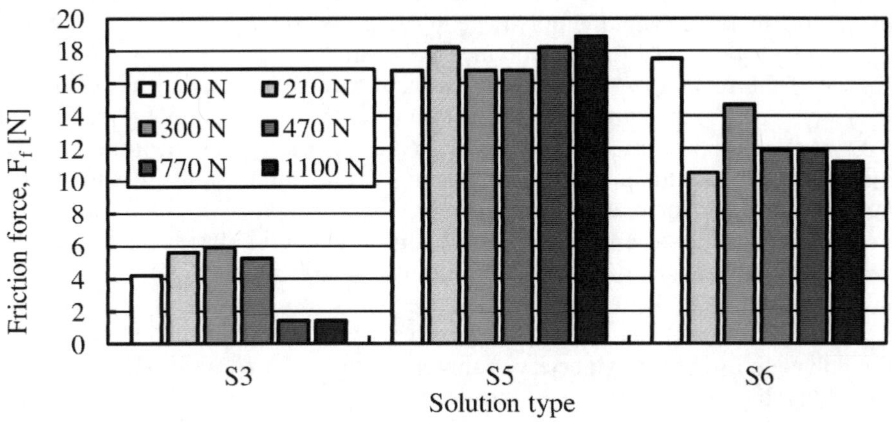


**Figure 5.** The dependence between the friction force and the axial load for the tested lubricant solutions (solvent – tap water and tap water + NaCl).

In figure 5 it is represented the influence of NaCl on the friction force, for different axial load values and for the same polymer type. For solution S3 (solvent - tap water) the lowest values of the friction force are registered for the highest values of the axial load, while for the solutions S5 and S6 the influence of NaCl content and the axial load does not influence significantly the friction force. For solution S6 (solvent - tap water + 5wt.%NaCl) the friction force values are a bit smaller compared to those registered for solution S5 (solvent - tap water + 2wt.%NaCl), and they have a decrease tendency with axial load increase. Also, it can be remarked for all the lubricant solutions that the scuffing appearance was started at about 1000 N axial load.

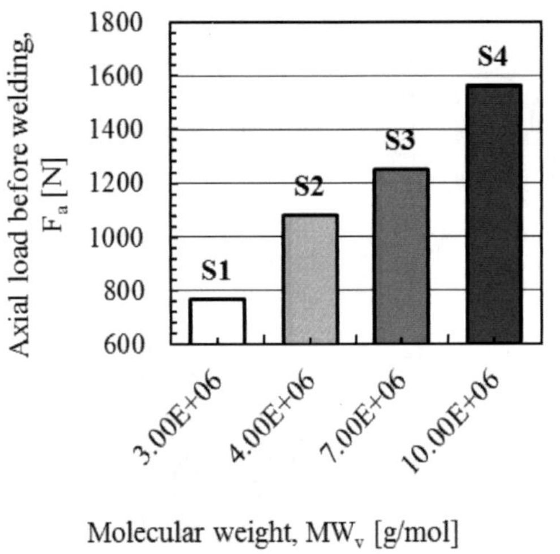


**Figure 6.** The dependence between axial load before welding and the molecular weight (solvent – tap water, polymer concentration - 1.0 wt. %).

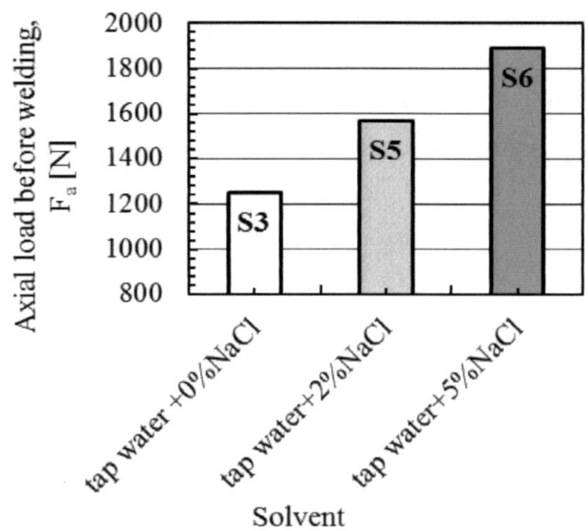


**Figure 7.** The dependence between axial load before welding and the NaCl solvent content (polymer molecular weight - $7 \cdot 10^6$ g/mol, polymer concentration - 1.0 wt. %).

For a better understanding of the influence of the solutions characteristics on their lubrication properties in figures 6 and 7 were presented the influences of molecular weight and the NaCl content on the axial load value registered before welding. Thus in figure 6 it can be observed that the axial force before welding increases with the polymer molecular weight increase for the lubricant solutions made with tap water solvent and 1.0 wt.% polymer concentration.

The influence of the solvent composition on the axial load before loading is presented in figure 7 for the same polymer (P3) and polymer concentration (1.0 wt.%). From figure 7 results that the increase of the NaCl solution content implies the axial load before welding increase. The maximum value of axial load before welding was determined for the solution S6, even if for this some of the biggest values of the friction force were registered.

To determine the optimum characteristics of the lubricant solution based on polyacrylamide it has to do the correlations between the wear and the solutions characteristics and also between the wear and the loadings. Thus, in figure 8 it is represented the dependence between the wear spot diameter (calculated like the average of the wear spot diameters of the three lower balls) and the molecular weight of the polymers used for the lubricant solutions made with tap water solvent. From figure 8 it can be observed that the minimum value of the wear spot diameter appears for the solution S3 that is made with polymer P1. In figure 9 it is showed the influence of the NaCl content in solution on the wear spot diameter. It can be remarked that the lowest value of the wear spot was registered for the solution S5 for which was determined some of the biggest friction force values (figures 4 and 5).

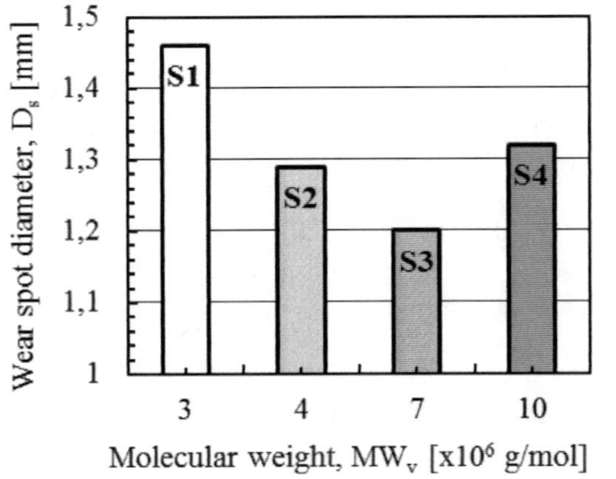


**Figure 8.** Wear spot diameter vs. molecular weight (solvent – tap water, polymer concentration - 1.0 wt. %).

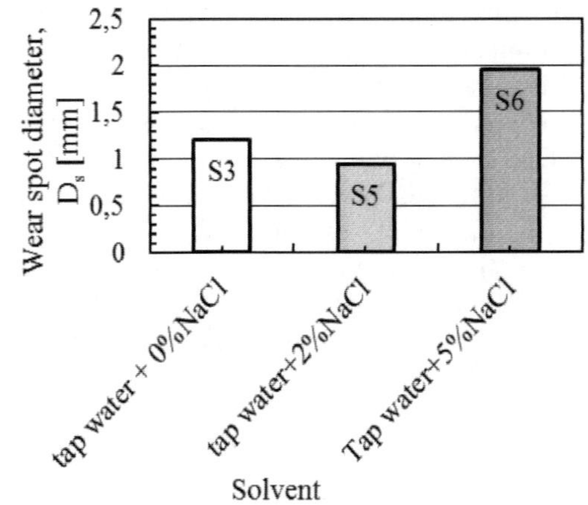


**Figure 9.** Wear spot diameter vs. NaCl solvent content (polymer molecular weight - $7{\cdot}10^6$ g/mol, polymer concentration - 1.0 wt. %).

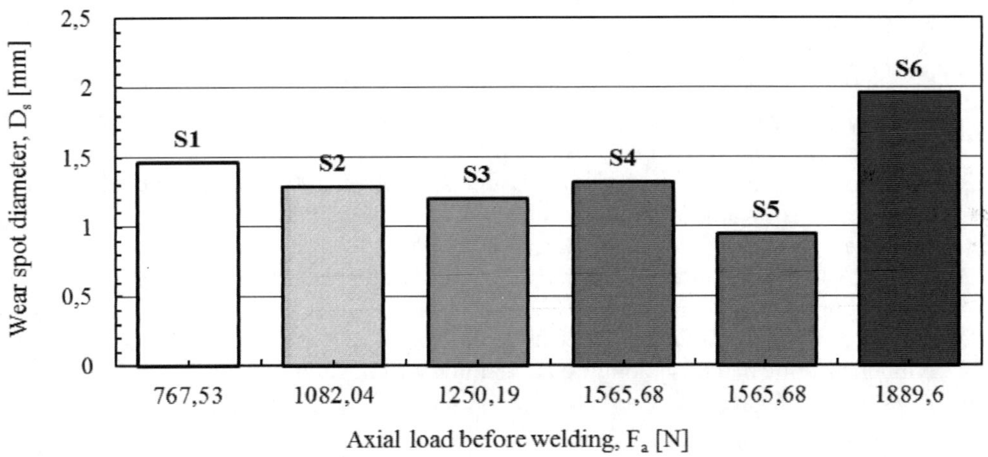


**Figure 10.** Wear spot diameter vs. axial load before welding for tested solutions.

In figure 10 there are presented the values of the wear spot diameter registered for each tested lubricant solution and the axial load before welding which corresponds of these wear values. It can be observed that the lower values for the wear spot diameter are registered for lubricant solutions S5 (solvent tap water + 2 wt. % NaCl) and S3 (solvent tap water).

## 4. Conclusions

The HPAM lubricant solutions consists a possible alternative in lubrication. Due to their properties, they can successfully remove other traditional sort of lubricants. The main characteristics (rheological and tribological) of these lubricants were determined and presented by author [7,8,9,10].

The behaviour of these lubricants in elastohydrodynamic contacts was not studied until now, and from this point of view it appears the necessity to determine the influence of their characteristics on the friction force and wear values which are registered during the tests done on four ball machine.

The main conclusions which result from this paper refer to the optimum polymer characteristics values that imply a better behaviour in elastohydrodynamic contacts:

- molecular weight in the range of $6.0 \cdot 10^6 - 10.0 \cdot 10^6$ g/mol;
- hydrolysis degree in the range of 8.0 – 9.0 %;
- solvent tap water or tap water with 2wt.% NaCl;

Due to their properties, the HPAM water solutions can be used like lubricants in fields such as:

- the industry lubrication (gears, bearings, rolling bearings, slideways, pumps etc);
- the metal working fluids;
- the drilling fluids;
- the medicine;
- the lubrication in special environments (sea water, brine etc).

## References

[1] Kulike W M, Kniewske R and Klein J 1982 Preparation, Characterization, Solution Properties and Rheological Behavior of Polyacrylamide *J Polym Sci* **8** 373-468
[2] Sorbie K S 1990 *Polymer-Improved Oil Recovery* (USA and Canada: CRC Press, Inc.)
[3] Tingting J, Haiyu W and Baohui W 2014 Chemical Structure of Surfactant-Grafted Polyacrylamide Used in Oilfields *Open Journal of Applied Sciences* **4** 451-457
[4] Wanjie L, Liqin Z, Chengcen L and Zhenhay L 2012 Preparation and Property of Poly (acrylamide-co-acrylic acid) Macromolecule Slow-releasing Fertilizer *Int J Electrochem Sci* **7** 11470-76
[5] Bird R B, Stewart W E and Lightfoot E N 1960 *Transport Phenomena* (New York: Wiley)
[6] Shokry H 2009 Corrosion protection of mild steel electrode by electrochemical polymerization of acrylamide *Chem Met Alloys* **2** 202-210
[7] Drumeanu A C, Antonescu N N and Drumeanu M R 2002 Some Considerations Concerning the Polyacrylamide Utilization in Lubrication *Proc. of 13th International Colloquium Tribology* (Esslingen: Germany/Wilfried J. Bartz) pp 2349-54
[8] Drumeanu A C, Antonescu N N and Drumeanu M R 2004 Polyacrylamide – A Possible Alternative in Synthetic Lubrication *Proc. of 14th International Colloquium Tribology* (Esslingen: Germany/Wilfried J. Bartz) pp 543-548
[9] Drumeanu A C 2005 Lubricant Composition Based on Organic Polymers *J Balk Tribol Assoc* **11** 472-478
[10] Drumeanu A C and Antonescu N N 2006 Influence of the Electrolyte on the Properties of Synthetic Lubricants Based on Polyacrylamide *J Balk Tribol Assoc* **12** 361-367

# Method and device for measurement of dynamic viscosity

**F C Ciornei, S Alaci, D Amarandei, L Irimescu, I C Romanu and L I Acsinte**

Department of Mechanics and Technologies, "Stefan cel Mare" University of Suceava, Universitatii 13, Suceava, 720229, Romania

E-mail: florina@fim.usv.ro

**Abstract**. The paper proposes a methodology and ensuing test rig for finding the viscosity of a liquid lubricant. The principle consists in obtaining a contact between two spherical surfaces, one concave and the other one convex. One of the surfaces is kept immobile and to the other, a rotation motion is imposed around the common normal in the contact point and then the law of motion for the mobile lens is found. The law of motion allows for estimation of friction torque, dependent on viscosity at its turn. Applying the method for mineral oils, values comparable to the ones presented by the producer were obtained.

## 1. Introduction

The viscosity of a fluid reflects the property of being able to transmit friction forces between particles or to exhibit resistance to shearing flows. For a quantitative characterisation of viscosity of a fluid, the laminar flow is considered [1-2] when the relative motion between two adjacent layers generates shear stresses. In the case that the magnitude of tangential stress is proportional to velocity gradient on the direction normal to flow, it is said to be a Newtonian behaviour. When the viscosity of the fluid is depending on shear rate, it is called non-Newtonian fluid. The rheological characterization of behaviour of non-Newtonian fluids is made using instruments generically named rheometers [3-8].

Next, the work refers to liquids presenting Newtonian behaviour, for which the proportionality factor between the shear stress and the magnitude of velocity gradient is the dynamic viscosity. The kinematical viscosity is obtained from dynamical viscosity, by dividing it to liquid's density. The basis of viscometers' design is the above viscosity definition, a viscometer being an instrument that allows finding the correlation between the characteristic parameters of a fluid and the shear stresses occurring in the fluid. Depending on the relative motion between the mobile surfaces between which the fluid is placed, there are met capillary viscometers [9-10], rotational viscometers when the relative motion between surfaces is a rotation about an axis, translation viscometers having as characteristic the fact that relative velocity between boundary surfaces keeps the same sign [11-13] oscillating viscometers, for which the relative motion presents periodical sense shifting. In [14-19] there are presented the main types of oscillating viscometers. In recent times there have been developed viscometers with special design, based on the principles mentioned but different from classical ones by the manner of estimating the relative motion and shear stresses from fluid, with modern data acquisition and processing systems [20-22]. In for the simplest case of Newtonian fluids, viscosity measurement is a difficult task due to numerous factors influencing it. Tipei [23] reveals among the parameters affecting decisively the viscosity of a lubricant: the type, temperature, pressure. Other aspects influencing viscosity are the nature and concentration of impurity and the presence of additives [24-25].

IOP Conf. Series: Materials Science and Engineering **174** (2017) 012041 doi:10.1088/1757-899X/174/1/012041

## 2. Viscometer proposal

The principle of the intended viscometer is based on estimation of relative velocity and friction torque occurring in an oil film placed between two spherical surfaces. The two spherical surfaces are actually two glass lenses, one concave, motionless and the other, plane-convex, mobile (figure 1). Considering an elementary volume, under the hypothesis of stationary motion, the axial symmetry of fluid motion is accepted and thus the employment of cylindrical coordinates for characterisation of relative motion is justified (figure 2).

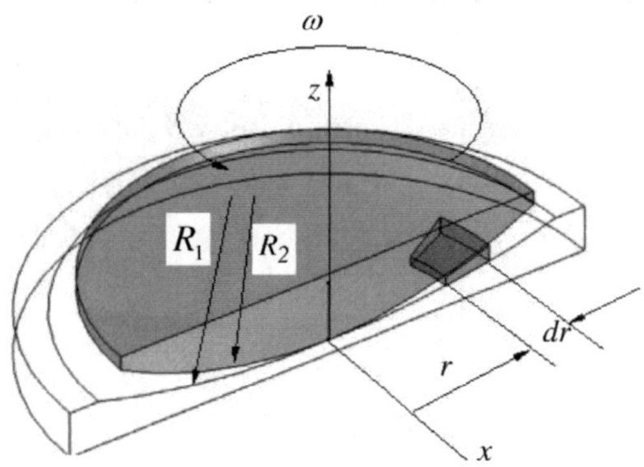


**Figure. 1.** Geometry of surfaces and elementary lubricant volume.

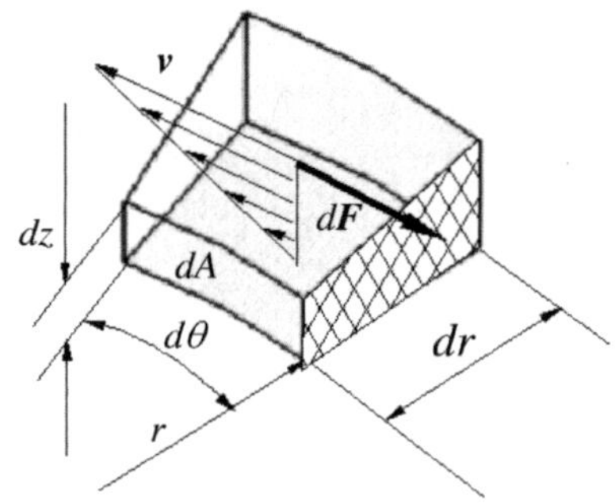


**Figure 2.** Friction force acting upon elementary contact area of lubricant elementary volume.

The gradient of velocity dimension in cylindrical coordinates has the expression [26]:

$$grad\ f = \frac{\partial f}{\partial \rho}\boldsymbol{u}_\rho + \frac{1}{\rho}\frac{\partial f}{\partial \theta}\boldsymbol{u}_\theta + \frac{\partial f}{\partial z}\boldsymbol{u}_z \tag{1}$$

Due to axial symmetry it results that $\frac{\partial f}{\partial \theta} = 0$. During experimental work, it can be noticed that the dimension of the region with fluid has constant value and thus one can consider that the velocity does not vary on radial direction. The velocity variation takes place only on vertical direction. As a result, accepting that the velocity gradient has component only on axial direction, the shear stress in current point $M(\rho,\theta)$ from contact surface is given by the relation:

$$\tau(\rho,\theta) = \eta\frac{dv}{dz} \tag{2}$$

The boundary surfaces of the bodies between which the lubricant is placed, as shown in figure 3, are two spheres of radii $R_1, R_2$ expressed by the equations:

$$x^2 + (z - R_{1,2})^2 = R_{1,2}^2 \tag{3}$$

On vertical direction, the dimension of the gap is:

$$\Delta z = R_1 - \sqrt{R_1^2 - x^2} - (R_2 - \sqrt{R_2^2 - x^2}) \tag{4}$$

Since the region where there is lubricant is situated in the vicinity of tangency point between the two surfaces, the expression (4) can be approximated by a Taylor series expansion:

$$\Delta z \cong \frac{R_2 - R_1}{R_1 R_2}\frac{x^2}{2} \tag{5}$$

The square function is used for the approximating the distance between the two surfaces and it is noticed that the opening values $\Delta z$ are much smaller than the radii of the spheres. From figure 4 it can be seen that at distance $0.2\,max(R_1,R_2)$ the gap $\Delta z$ is much smaller compared to spheres radii, $\Delta z \cong 0.5\%\,max(R_1,R_2)$.

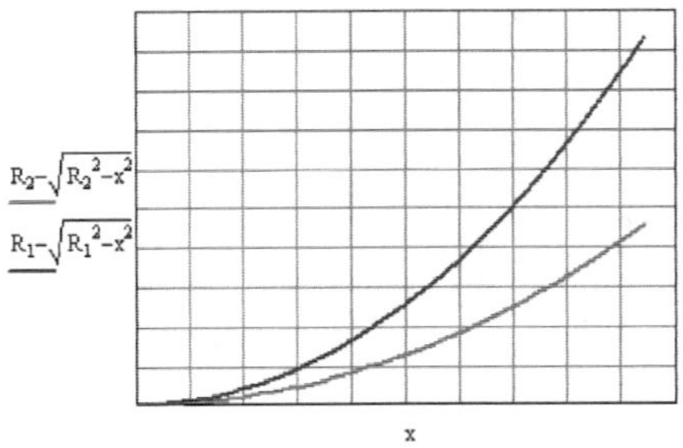


**Figure 3.** Boundary surfaces of spherical bodies.

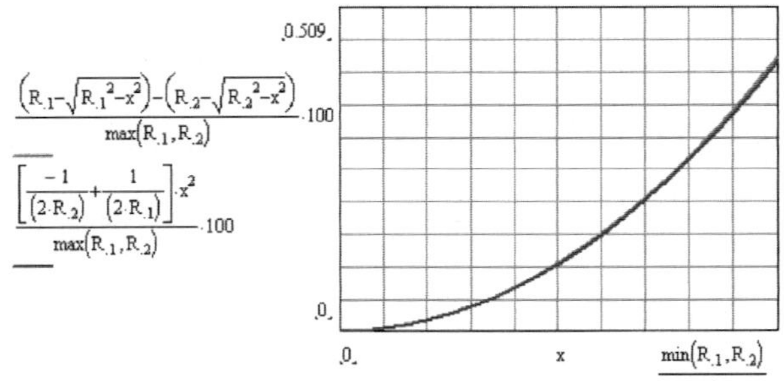


**Figure 4.** Lubricant thickness and the approximation function.

This remark allows approximation of differentials from relation (2) with incremental increases:

$$\tau(\rho,\theta) = \eta \frac{\Delta v}{\Delta z} \tag{6}$$

For estimation of velocity augmentation, one considers that the lower surface is immobile and the upper surface rotates about vertical axis with the angular velocity $\omega$:

$$\tau(\rho,\theta) = \eta \frac{\omega \rho}{\dfrac{(R_2 - R_1)}{R_1 R_2}\dfrac{\rho^2}{2}} \tag{7}$$

In relation (7) it was taken into account the fact that, due to axial symmetry, relation (5) is valid for radial direction.

The shear stress $\tau(\rho,\theta)$ will generate a friction force:

$$dF_f = \tau\, dA = \tau\, \rho\, d\rho\, d\theta \tag{8}$$

that, on its turn, will produce an elementary friction torque:

$$dM_f = \rho\, dF_f \tag{9}$$

Considering relation (7), the elementary friction torque is:

$$d\,M_f = 2\eta \frac{R_2 R_1}{R_2 - R_1}\omega\, \rho\, d\rho\, d\theta \tag{10}$$

To obtain the total friction torque, the expression (10) should be integrated on the entire area with lubricant:

$$M_f = \iint_A d\,M_f = \int_0^{2\pi}\int_0^{r_{max}} 2\eta \frac{R_2 R_1}{R_2 - R_1}\omega\, \rho\, d\rho\, d\theta = 2\pi\eta\, r_{max}^2 \frac{R_1 R_2}{R_2 - R_1}\omega \tag{11}$$

The principle of the method is based on experimental estimation of variation of mobile body's angular velocity and of the maximum radius where liquid is placed. The two theoretical surfaces were materialized by two glass lens, biconcave the immobile one and plane-convex the mobile one. From equation (11), the friction torque is proportional to the angular velocity of the lens. To broaden the applicability domain of the relations to be obtained an additional constant torque $M_0$ acting besides the friction torque is considered. The mobile lens is activated using an electrical drive, on the shaft of which a rubber disc is mounted, playing the part of clutch for torque transmission. The lens is set in motion to the required velocity and afterwards, let to move without restraint but in contact with the stationary lens. The mobile lens has five degrees of freedom, two translations and there rotations. It is observed from experimental tests that after a relatively short period of time, the translations and rotations around horizontal axes vanish and the only motion remaining is the rotation about vertical axis. This motion is described by the differential equation:

$$J_z \frac{d\omega}{dt} = -(M_0 + \kappa\omega) \tag{12}$$

where $\kappa$ notation represents:

$$\kappa = 2\eta\, r_{max}^2 \frac{R_1 R_2}{R_2 - R_1} \tag{13}$$

The general solution of equation (12) is:

$$\frac{ln(M_0 + \kappa\omega)}{\kappa} = -\frac{t}{J_z} + C \tag{14}$$

In order to find the constant of integration $C$ the initial conditions are stated:

$$t = 0,\ \omega = \omega_0 \tag{15}$$

The time $t = 0$ is the moment when it is considered that the lens has only rotation around vertical axis and $\omega_0$ is the angular velocity corresponding to this moment. The particular solution of the problem is:

$$\frac{M_0 + \kappa\omega}{M_0 + \kappa\omega_0} = e^{-\frac{\kappa t}{J_z}} \tag{16}$$

There are two unknowns to be found, $M_0$ and $\kappa$. To this end, using a non-contact tachometer the variation of angular velocity of the lens is found. The equation (16) is written for two instants $t_1$, $t_2$ to which the angular velocities $\omega_1, \omega_2$ correspond and the following system results:

$$\begin{cases} \dfrac{M_0 + \kappa\omega_1}{M_0 + \kappa\omega_0} = e^{-\frac{\kappa t_1}{J_z}} \\ \dfrac{M_0 + \kappa\omega_2}{M_0 + \kappa\omega_0} = e^{-\frac{\kappa t_2}{J_z}} \end{cases} \tag{17}$$

Elimination of $M_0$ between the equations of the system gives the following equation:

$$(\omega_0 - \omega_1)e^{-\frac{\kappa t_1}{J_z}} + (\omega_0 - \omega_2)e^{-\frac{\kappa t_2}{J_z}} = \omega_2 - \omega_1 \tag{18}$$

The equation (18) is a transcendental equation and a numerical method is required for solving it (for instance bipartition or Newton-Raphson [27]). With the assumption:

$$\frac{\kappa}{J_z} << 1 \tag{19}$$

the approximation relation is used:

$$exp(\alpha x) = 1 + \alpha x + O(x^2) \tag{20}$$

and a linear equation with $\kappa$ unknown is obtained, having the solution:

$$\kappa = \frac{\omega_1 - \omega_2}{(\omega_0 - \omega_1)t_1 + (\omega_0 - \omega_2)t_2} J_z \tag{21}$$

In the case when condition (19) is not satisfied, the value obtained by (21) can be used as "guess value" for the numerical algorithm employed in solving equation (18).

$$M_0 = -\kappa \frac{\omega_2 e^{-\kappa \frac{t_1}{J_z}} - \omega_1 e^{-\kappa \frac{t_2}{J_z}}}{e^{-\kappa \frac{t_1}{J_z}} - e^{-\kappa \frac{t_2}{J_z}}} \tag{22}$$

## 3. Experimental set-up and results

The method proposed is experimentally tested using lubricant between two glass lenses: (1) plane-convex mobile and (2) bi-concave immobile (figure 5). The mineral oil (3) was placed between the two lenses and the mobile lens was set in motion but it was noticed that the motion rapidly ceased. To increase the motion duration, to the mobile lens was attached a metallic ring (4), coaxial to the lens. Thus, the moment of inertia of the assembly lens-ring is greater and a considerable kinetical energy increase is obtained. The lens is set in motion by hand. Using a digital non contact tachometer the variation in time of lens angular velocity is found. The tachometer allows saving the data in numerical format and then graphical representation using a computer.

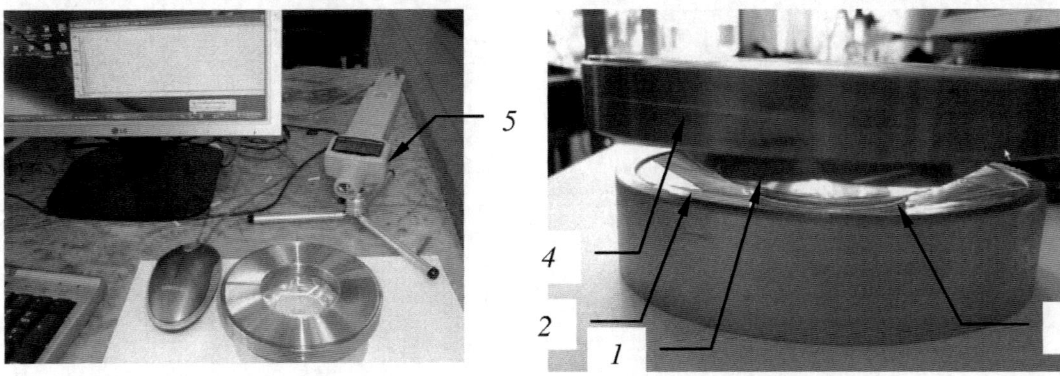


**Figure 5.** Experimental test-rig: general view (left) and detail (right).

Figure 6 presents the curves of rotation velocity variation of the lens for several successive launchings. From plots it can be noticed that two domains are visible, on any of the curves, namely a region with rapid velocity decrease followed by a second region where the rotation velocity decreases more slowly. The transition from the first domain is made at a value slightly inferior to $n = 150 rot / min$. The two domains correspond to spatial motion and rotation around a fixed axis respectively, for mobile lens. In figure 7 and figure 8 the image of the lubricated region between the two lenses is presented. In the case of spatial motion, the participating oil performs a planetary motion around the axis of immobile lens (figure 7) while after rotation motion around the axis of immobile lens is settled, the oil mark is stationary.

The disadvantage of using the tachometer for analysing the rotation motion of the lens after the stabilization, as observed from figure 6, consists in the fact that for small angular velocity values, the instrument's indications are not relevant. To eliminate this drawback, a mark was applied to the

IOP Conf. Series: Materials Science and Engineering **174** (2017) 012041 doi:10.1088/1757-899X/174/1/012041

surface of the collar-ring attached to the mobile lens. After actuating the lens, the motion was recorded using a video camera having the image acquisition performance of $120\,frames/sec$; after that, the movie was decomposed into frames using a software and the number of frames corresponding to a complete rotation was determined.

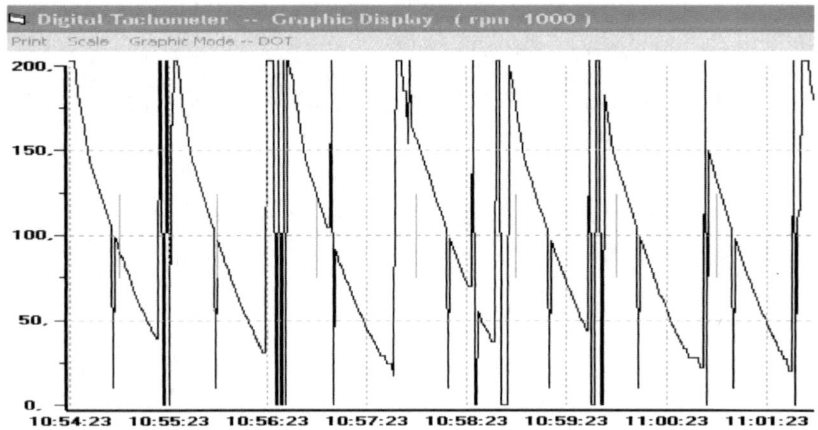


**Figure 6.** Experimental angular velocity variation.

**Figure 7.** Planetary motion of oil mark for spatial motion of mobile lens.

**Figure 8.** Oil mark settles after the stabilization of lens' motion.

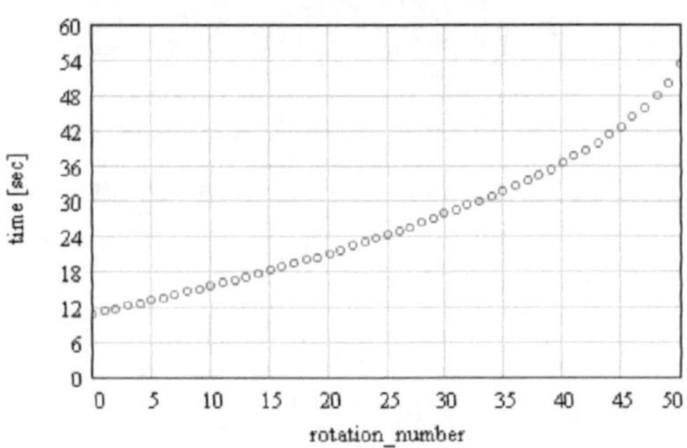


**Figure 9.** Time period versus rotation number.

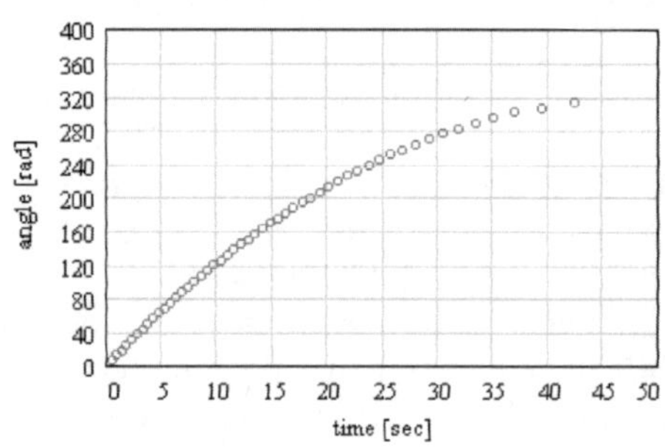


**Figure 10.** Lens's rotation angle versus time.

In figure 9 and figure 10 there are presented the values of time duration corresponding to complete rotation and the variation of rotation angle of the lens, respectively. In figure 10 it was considered that $t = 0$ instant corresponds to the moment afterwards the stationary motion of the lens was obtained and for which the rotation angle is considered $\theta = 0$.

To apply the data from figure 10 in equation (17), it was considered that $\omega_0 = 2\pi/(t_1 - t_0)$. The initial angular velocity is regarded as constant during a complete rotation. With imposed initial conditions $\theta(t=0)=0$ the equation was integrated once more and the form of the dependency is:

$$\theta(t)=\frac{J_z\omega_0}{\kappa}+\frac{J_z-\kappa t}{\kappa^2}M_0-\frac{J_z}{\kappa^2}(M_0+\kappa\omega_0)exp\left(-\frac{\kappa t}{J_z}\right) \quad (23)$$

To establish the parameters $M_0$ and $\kappa$ it is imposed the condition that the function (23) interpolates the experimental data $(t_k,\theta_k)$, plotted in figure 11, with minimum squared error. To this end, the following function is constructed:

$$\Phi(M_0,\kappa)=\sum_{k=1}^{n}[\theta(t_k)-\theta_k]^2 \quad (24)$$

and the system is proposed to be solved:

$$\begin{cases}\dfrac{\partial\Phi(M_0,\kappa)}{\partial M_0}=0\\ \dfrac{\partial\Phi(M_0,\kappa)}{\partial\kappa}=0\end{cases} \quad (25)$$

To resolve the system (25), a numerical procedure is required, and there is necessary précising the guess value for algorithm initialization. It is easily noticed that the solution of system (25) is strongly dependant on the guess values. $M_0$ and $\kappa$ found with the aid of relations (21) and (22) may be such a solution. For the present case, in order to establish a guess value the function $\theta(t)$ from relation (23) was constrained to pass through two of the experimental points from the graph given in figure 10. The experimental data (from figure 10) and the interpolation curve are presented in figure 11.

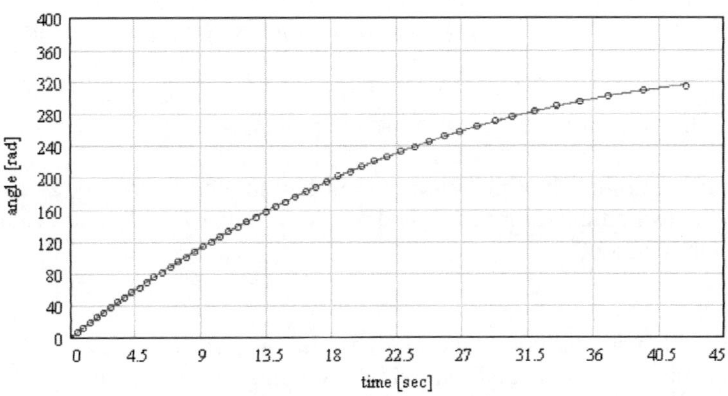


$$\begin{pmatrix}M_0\\ \kappa\end{pmatrix}=\begin{pmatrix}3.624\cdot10^{-4}\\ 3.74\cdot10^{-5}\end{pmatrix}$$

**Figure 11.** Interpolation curve for experimental data.

The characteristic parameters of the experimental set-up also used in the quantitative evaluation are: mass of the ring $M_1 = 0.894kg$; moment of inertia of the ring $J_{z1} = 2\cdot10^{-3}kg\cdot m^2$; mass of mobile lens $M_2 = 0.264kg$; moment of inertia of the lens $J_2 = 2.37\cdot10^{-4}kg\cdot m^2$; oil mark radius $r_{max} = 0.02m$; convex lens radius $R_1 = 0.07m$; concave lens radius $R_2 = 0.2m$;

$\kappa = 3.74 \cdot 10^{-5} N \cdot m \cdot s$; density of oil tested [28] $\rho_{oil} = 1030 kg/m^3$. Replacing in relation (13) the values for the parameters given above, it results a kinematical viscosity $\nu = 482$ mm$^2$/s.

## 4. Conclusions

The paper proposes a method for establishing the oil viscosity. The main goal was designing a most simple structure of the device so that only the viscous friction force should act upon the mobile body as dissipative force.

The principle of the method consists in obtaining the contact between two spherical surfaces, one concave and one convex. The concave surface is kept immobile and the convex surface is movable. Between the two surfaces, namely glass lenses- a concave one and plane-convex the other one, the lubricant to be tested is introduced. The radii of the two lenses have comparable values, thus the maximum oil thickness is less than 1mm. Under this assumption it can be accepted the hypothesis that the velocity variation along a vertical segment joining two points from the two lenses is linear, increasing from zero to a maximum value.

Another advantage of the proposed device consists in the fact that regardless of the actuation motion of the mobile lens, at a certain time, the motion of the lens becomes a rotation about an immobile vertical axis passing through the lowest point of the stationary lens.

A theoretical model is proposed, expressed by an equation in which the viscous friction is described using a damping coefficient dependant on the fluid viscosity and constant friction torque. The latter was introduced to ensure a finite time phase of mobile lens' motion.

After the stationary motion is obtained, the law of motion is experimentally found and, damping coefficient and in the end, the viscosity are found by imposing that the theoretical law of motion interpolate with minimum error the experimental data.

Concerning the experimental establishing of the law of motion, initially, the angular velocity variation was estimated using a non contact tachometer. This option was removed since at the end of the motion, the errors of the tachometer increased substantially. As ultimate solution, video recording of the lens' motion was chosen. Splitting the film into frames makes possible establishing the variation in time of the rotation angle during stationary phase. The experimental data are excellent interpolated by the theoretical curve. The tests were done using an oil of known viscosity

The viscosity obtained experimentally via the proposed methodology is comparable to the values from technical literature.

One has given up determining the angular velocity by tachometer but the method proved to be useful as it allows emphasising the transition angular velocity from spatial motion to stationary motion.

The present paper is only a starting point. The first solution in improving the methodology consists in the use of lenses with highly controlled curvature radii and of close values of curvature radii, the latter leads to an increase in velocity gradient.

## Acknowledgement

The infrastructure used for this work was partially supported by the project "Integrated Center for research, development and innovation in Advanced Materials, Nanotechnologies, and Distributed Systems for fabrication and control", Contract No. 671/09.04.2015, Sectorial Operational Program for Increase of the Economic Competitiveness co-funded from the European Regional Development Fund.

## References

[1] Gupta S V 2014 *Viscometry for liquids* (Springer) chapter 1 pp 1-17
[2] NakayamaY, Boucher R F 2000 *Introduction to Fluid Mechanics* (Butterworth-Heinemann Oxford) chapter 2 pp 9-12
[3] Macosko C W 1994 *Rheology Principles, Measurements and Applications* (Wiley-VCH NY) chapter 5 pp 181-251
[4] Barnes H A 2000 *A Handbook of Elementary Rheology* (The University of Wales Aberystwyth)

IOP Conf. Series: Materials Science and Engineering **174** (2017) 012041 doi:10.1088/1757-899X/174/1/012041

pp 51-54

[5] Barnes H A, Maia J M accessed 2016 *Rheometry, Rheology vol. I* (www.EOLSS.net) pp 1-10

[6] Barnes H A, Hutton J E, Walters 1989 *An Introduction to Rheology* (Elsevier) pp 9-10

[7] Morrison FA 2001*Understanding rheology* (Oxford University Press) chapter 10 pp 382-433

[8] Viswanath D S, Ghosh T, Prasad D H L, Dutt N V K, Rani K Y 2007 *Viscosity of Liquids: Theory, Estimation, Experiment, and Data* (Springer) chapter 2 pp 65-67

[9] ***ASTM D446-07 *Standard specifications and operating instructions for glass capillary kinematic viscometers*

[10] ***ISO 3105-1994 *Glass capillary kinematic viscometers*—Specif and operating Instructions

[11] Flowers A E 1914 Viscosity measurement and a new viscometer. *Proc Am Soc Test Mater* **14** 565–616

[12] Forgacs R L 1965 Improved Torsion viscometer for Polymer studies. *Rev Sci Instrum* **36** 307–313

[13] ***HAAKE *Instruction Manual Falling Ball Viscometer C*

[14] Kestin J, Newell G F 1957 Theory of oscillating type viscometers I: the oscillating cup. *Z Angew Math Phys* **8** 433–449

[15] Beckwith D A, Newell G F 1957 Theory of oscillating type viscometers, the oscillating cup part II. *Z Angew Math Phys* **8** 450–465

[16] Azeitia A G, Newell G F 1958 Theory of oscillating type viscometers III, a thin disc. *Z Angew Math Phys* **9a** 97–118

[17] Azeitia A G, Newell G F 1959 Theory of oscillating type viscometers IV, a thick disc. *Z Angew Math Phys* **10** 15–34

[18] Newell G F 1959 Theory of oscillating type viscometers V, disc between fixed plates II. *Z Angew Math Phys* **10** 450–465

[19] Newell G F 1959 Theory of oscillating type viscometers: the oscillating cup part II. *Z Angew Math Phys* **10** 160–174

[20] Sheen S H, Chien H T, Raptis A C 1997 *Ultrasonic methods for measuring liquid viscosity and volume percent of solids* Argonne National Laboratory (Argonne Illinois) Report ANL-97/4 chapter v2 pp 3-10

[21] Hager H E 1986 Fluid property evaluation by piezoelectric crystals operating in the thickness shear mode. *Chem Eng Common* **43** 25–38

[22] Kanazawa K K, Gordon J G II 1985 The oscillation frequency of a quartz resonator in contact with a liquid. *Anal Chem Acta* **175** 99–105

[23] Tipei N, Costantinescu, V N, Nica Al, Biţă O 1961 *Lagăre cu alunecare* (Ed Acad Bucureşti) chapter IX pp 297-302

[24] Cate W E, Deming M E 1970 Effect of impurities on density and viscosity of simulated wet-process phosphoric acid *J. Chem. Eng. Data* **15 (2)** 290–295

[25] Krantz T L,. Kahraman A 2005 *An Experimental Investigation of the Influence of the Lubricant Viscosity and Additives on Gear Wear* (Nat Aeron and Space Adm Glenn Research Center) pp 11-14

[26] Spiegel M 2009 *Schaum's Outline of Theory and Problems of Vector Analysis and an Introduction to Tensor Analysis* (McGraw Hill) chapter 7 pp 157-188

[27] Hamming R W 1987 *Numerical Methodss for Scientists and Engineers* (Dover Publications) chapter II pp 427-443

[28] REMARK Co. 2008 AGIP BLASIA S (F.T. TI 6/2008) available at www.lubrifianti.com/documente-up/fisa-tehnica_47736.pdf

IOP Conf. Series: Materials Science and Engineering **174** (2017) 012042 doi:10.1088/1757-899X/174/1/012042

# Bio and nanomaterials in tribocorrosion systems

**Lidia Benea**

Competences Center: Interfaces – Tribocorrosion and Electrochemical Systems, Faculty of Engineering, "Dunarea de Jos" University of Galati, Galati, Romania

E-mail: Lidia.Benea@ugal.ro

**Abstract**. The growing attention that the scientific community has paid in the last decades to the corrosion phenomena, including tribocorrosion is related to the huge economic, social and environmental losses (3,5 % GDP in industrialized countries as USA, UK, Japan and Germany), that result from the spread of damage of several metal constructions and devices. Tribocorrosion is defined as the chemical-electrochemical-mechanical process leading to a degradation of materials in sliding, rolling or erosion contacts immersed in a corrosive environment or even in water. That degradation results from the combined action of corrosion and wear are higher compared with addition of corrosion and wear degradation separately. This synergism between chemical, electrochemical, and mechanical processes on materials in sliding, abrasive or erosive contacts immersed in a liquid requires a multi-disciplinary approach (material science, electrochemistry, tribology, mechanics, and surface engineering). This paper presents few summary results obtained by studying the materials degradation by complex tribocorrosion processes in terms of two broad categories of applications: tribocorrosion in industrial systems with improved behaviour of nanomaterials as hybrid and nanocomposite layers and tribocorrosion in living systems with improved behaviour by surface modifications of biomaterials applying electrochemical techniques. The purpose of this paper is to provide information on the surface conditions of materials in sliding contacts and also on the kinetics of reactions that control the corrosion component in the material loss during tribocorrosion tests.

## 1. Introduction

There is a strong relation created between materials degradation (corrosion, tribocorrosion) and sustainability, with the aim to avoid excessive damage to our planet, to mitigate the effect of heavy industrialization, and this goal should be also transmitted to the future generations. Both the public and private sectors appreciate the need for engineers who have been taught corrosion engineering so that they have to take into account corrosion and tribocorrosion properties for well designing and manufacturing. The importance of materials degradation (corrosion, tribocorrosion) education in today's world continues to increase as the limits of material behaviour are stretched to improve the performance of engineered structures and devices. The basis of the academic subjects related to materials degradation is scientific, but we cannot ignore its effect on every industrial process or metallic implants and the correlation of this phenomenon with sustainability. It is necessary not only provide students with knowledge about the behaviour of materials but to raise their awareness of the importance of ecological considerations due to: metallic resources which are overexploited, the environmental impact resulting from any failure in service which is unpredictable and the necessity to point out the recycling importance. Different industries that may be affected by tribocorrosion include:

petrochemical and chemical, aerospace, automotive, marine, mining, food and biomedical. In orthopaedic field, metallic implants are often exposed to simultaneous chemical/electrochemical and mechanical stresses due to each simple or complex movement performed with the arms or legs. During the motions of the artificial joints exposed to the synovial fluid, wear debris are generated which can cause several reactions in the surrounding tissue in which they are deposited leading to implant loosening. In industry corrosion and wear are responsible for maintenance expenses and loss of productivity resulted from the shortened life of components and catastrophic failure leading to massive costs of the replacement and litigation [1,2]. Tribocorrosion is defined as the chemical-electrochemical-mechanical process leading to a degradation of materials in sliding, rolling or erosion contacts immersed in a corrosive environment or even in water. Therefore tribocorrosion is the science of surface transformation by electrochemical and mechanical synergy. That degradation results from the combined action of corrosion and wear are higher than that of adding separately the corrosion and wear. The mechanism of tribocorrosion is not yet fully understood due to the complexity of the chemical, electrochemical, physical, and mechanical processes involved. Examples of the occurrence of tribocorrosion in service are the accelerated corrosion of steel conveyors exposed to ambient air of high relative humidity, the fall out of electrical connectors in the automotive industry, the degradation of orthopedic prosthesis and dental fillers, the erosion wear of turbine blades, etc. The evidence of tribocorrosion wear in pressurized water reactors (PWR) was reported recently by Lemaire and al. [3]. Some authors have reported on a modification of the surface state of materials in sliding contacts, which results from mechanical, chemical and electrochemical processes [4-6]. That interaction is of increasing interest since it may result in a new concept of operating materials under lubrication-free sliding or rolling conditions. Indeed, such tribo-reactive surface layers may modify the corrosion process on contacting materials that in turn may modify friction and wear of materials. Recently, there has been an increase in the interest on the investigation of the combined corrosion-wear degradation of materials (tribocorrosion) by electrochemical methods [7-22].

This paper presents few summary results obtained by studying the materials degradation by complex tribocorrosion processes in terms of two broad categories of applications: tribocorrosion in industrial systems with improved behavior of nanomaterials as hybrid and nanocomposite layers and tribocorrosion in living systems with improved behaviour by surface modifications of biomaterials applying electrochemical techniques. Essential information on the surface conditions of materials in sliding and friction contacts, but also on the kinetics of reactions that control the corrosion component in the material loss during tribocorrosion tests are shown. In-situ electrochemical analysis measurements and mechanical parameters for studying tribocorrosion process were monitorized simultaneously.

## 2. Materials and methods

### *2.1. In-situ electrochemical and mechanical parameters*

Uni-directional pin-on-disk or bidirectional ball-on-flat contact geometry can be used and an electrochemical cell connected with a tribometer. Schematic diagrams of the experimental set-ups are given in figures 1 and 2.

The active surface of the working electrode is well defined by isolating with epoxy resin being the same for every experiment. The counter body (pin) usually is a hard ceramic like alumina ($Al_2O_3$) or zirconia ($ZrO_2$) cylinder (7 mm in diameter), mounted vertically on a rotating head, above the sample. The lower spherical end (radius = 100 mm) of the pin is then applied against the material surface (disc) with an adjustable normal force, correlated with the real application. When rotation is applied, the end of the pin draws a circular wear track (16 mm in diameter) on the working surface (for the uni-directional contact geometry) or a wear track with an amplitude of about 200 μm (for bi-directional contact geometry).

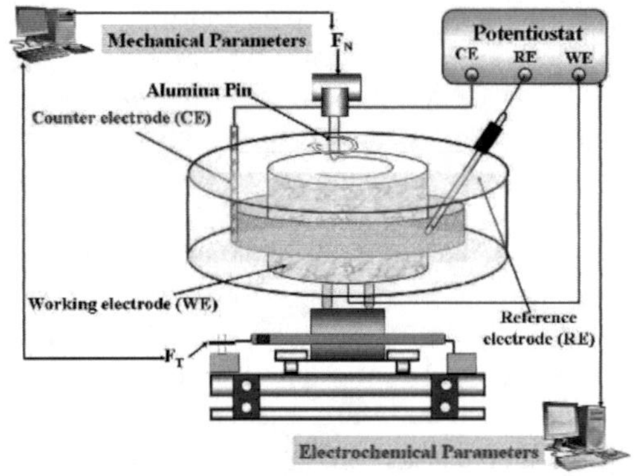


**Figure 1.** Schematically description of the pin-on-disc set-up used to study the tribocorrosion behaviour of materials and coatings in aqueous solution containing the electrochemical cell connected to potentiostat with reference electrode (RE), counter electrode (CE), working electrode (WE) which is the sample with the coating or a biomaterial on the top of a cylinder, counterbody (pin), made by corundum ($Al_2O_3$) and the wear track drawn by counterbody on the sample surface.

After the electro-codeposition of WC nanoparticles into nickel matrix the obtained Ni/ nano-WC nanocomposite layers were characterized to tribocorrosion process in pin-on-disc system, figure 1. The electrochemical protocol used for the tribocorrosion experiments in wet conditions consisted of two steps: (1) measuring the open circuit potential (OCP) of the sample for 60 min. before starting up the friction tests in order to reach a stable potential and (2) after that the $Al_2O_3$ pin was brought in contact with the tested sample and a constant normal force was applied using controlled mechanical parameters. After the end of fretting tests the normal force was released and for a period of 60 min. the OCP was monitored in order to observe its evolution without any force applied or friction motion.

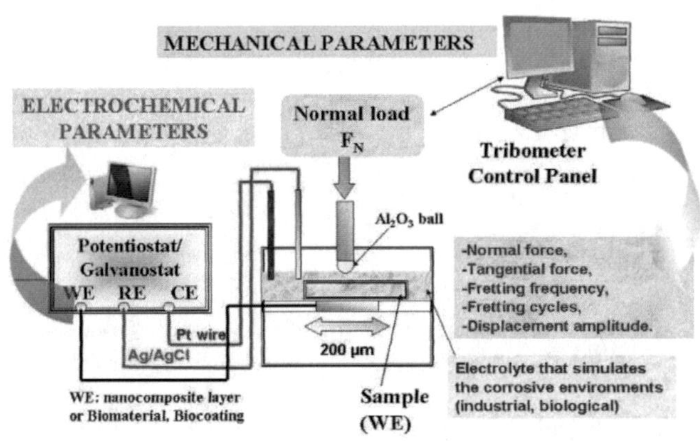


**Figure 2.** Schematically description of bi-directional fretting corrosion system with computer controlled electrochemical and mechanical parameters.

After surface modification of titanium alloy by controlled oxide film formation, the friction of untreated Ti-6Al-4V alloy and anodized formed porous $TiO_2$ layer is studied in a ball-on-flat configuration using a MUST micro-tribometer, figure 2, under reciprocating sliding condition. Alumina balls (G 10 grade - Ceratec, The Netherlands) of 10 mm diameter were used as counterbody in the fretting corrosion tests. The choice of alumina ball as the counter face material was made because of its high hardness, high wear resistance, chemical inertness and electrical insulating properties. Prior to any sliding test, the counterbodies were degreased with ethanol followed by hot air drying. After this, the alumina balls were loaded on top of the test samples at the normal forces of 400 mN, 600 mN and 800 mN, with a displacement amplitude of 500 μm and sliding frequency of 3 Hz for 1000 cycles. The mean Hertzian contact pressures which correspond in these tests were of 210 MPa at 400 mN, 240 MPa at 600 mN and 260 MPa at 800 mN. Each experiment was repeated at least twice to check for reproducibility.

All reciprocating sliding tests were performed at room temperature (23°C) in Fusayama–Mayer bio-simulated saliva. The composition of this electrolyte is 0.4 g/L NaCl, 0.4 g/L KCl, 0.8 g/L $CaCl_2$ $2H_2O$, 0.69 g/L $NaH_2PO_4$ and 1 g/L urea having a pH of 4.8. This solution was selected because this pH condition is much more aggressive than would be normal for natural saliva and its use is expected to be able to enhance the corrosion-wear process and to distinguish the effect of surface modification of titanium alloy. During the sliding test, the normal force, tangential force, coefficient of friction, number of cycles and the linear contact displacement were continuously monitored.

### *2.2. Preparation of Ni/nano-WC nanocomposite layers*

Ni/nano-WC composite layers were obtained by electro-codeposition from a standard nickel Watts plating bath to which WC nanoparticles with a concentration of 10 $gL^{-1}$ was added. The tungsten carbide (WC) nanopowder with a purity of 99,95%, grain size of 60 nm was purchased from MK Impex Corp.(Ontario, Canada).

Electro-codeposition was carried out in a typical three-electrode cell with double walls to keep constant temperature of the solution, where 304L stainless steel plates with exposed surface area of 4 $cm^2$ were used as working electrodes, a nickel plate was used as an auxiliary electrode (counter electrode) and a saturated calomel electrode (SCE - $Hg/Hg_2Cl_2$) with KCl saturated solution (E= + 244 mV vs. normal hydrogen electrode - NHE) was used as a reference electrode. Tungsten carbide as a hard phase can be included into the Ni matrix to improve the wear resistance of Ni-based coatings.

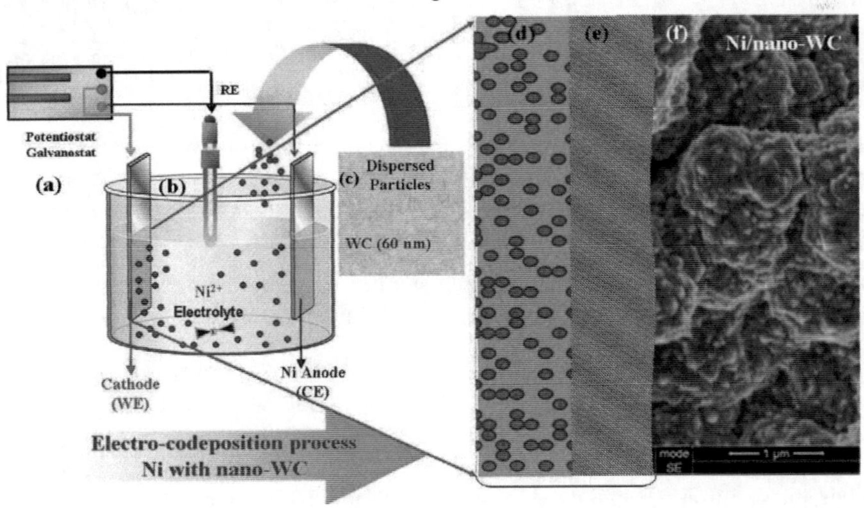


**Figure 3.** Schematic drawn of WC nanoparticles electro-codeposition into nickel matrix to obtain Ni/nano-WC nanocomposite layer: (a) the electrochemical workstation; (b) electrochemical cell with cathode (-) as the working electrode support for coatings to be electrodeposited, the anode (+) or counter electrode as nickel plate, the reference electrode, nickel plating solution and WC nanoparticles; (c) dispersed WC nanoparticles (60 nm diameter); (d) schematic cross section of Ni/nano-WC nanocomposite layer; (f) stainless steel support material; (f) SEM micrograph of Ni/nano-WC layer surface.

Schematic presentation of the electrolytic cell and electro-codeposition process is shown in figure 3, where (a) represent the electrolytic cell with cathode (-) as the working electrode support for coatings to be electrodeposited, the anode (+) or counter electrode as nickel plate in order to maintain an constant content of nickel ions into nickel plating bath and in the middle is the electrolyte or the plating solution having an homogenous dispersion of WC nanoparticles maintained by stirring. In figure 3 it is shown a schematic cross section of electrodeposited nanocomposite layer(d) as well as the SEM micrograph of nanocomposite layer surface.

*2.3. Surface modification of titanium alloy to improve the tribocorrosion resistance*
The preparation method of untreated Ti-6Al-4V alloy, nanoporous anodic oxide films were described more in detail previously [18,23]. In figure 4. there is presented a schematic drawn of controlled formation and growth of porous titanium oxide film on titanium alloy. The alloy is immersed in an anodization electrolyte connected at anode in an electrochemical cell (a). By varying the parameter as voltage, current density, temperature and time it could be obtained a porous film on titanium alloy with a given thickness, usually between 30 to 100 nm.

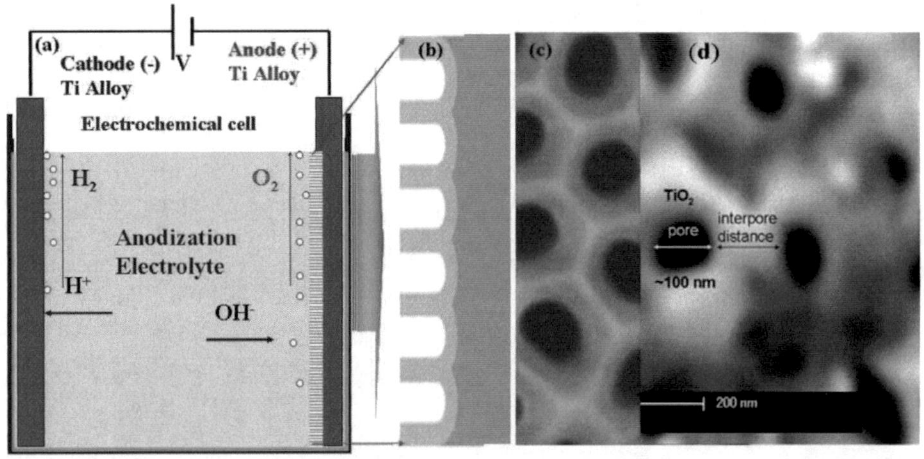


**Figure 4.** Schematic presentation of surface modification of titanium alloy by electrochemically controlled growth of $TiO_2$ nanoporous film: (a) electrochemical anodization cell; (b) cross section of $TiO_2$ film; (c) schematic surface of nanoporous $TiO_2$ film; (d) SEM micrograph of titanium oxide film

The obtained nanoporous film with estimated pore diammeter of 100 nm, figure 4 (c, d) is expected to improve the wear corrosion properties of titanium alloy in biologic solution.

## 3. Results and discussion

*3.1. Nanocomposite layers in tribocorrosion systems for energy applications*
In the nuclear power plants there is a need of wear resistant materials which work in wet conditions like primary circuit solution formed from a mixed lithium hydroxide and boric acid dissolved in purified water. Parts of nuclear pressurized water reactors (PWR), immersed in the cooling water, are submitted to both mechanical perturbation as well as corrosion. Therefore passivable alloys with high mechanical properties are used for their manufacture. The gripper latch arms (GLAs) of the control rod drive mechanisms are protected from wear and corrosion by a layer of 5 mm thickness of cobalt chromium alloy with 28 wt. % chromium which is deposited by plasma spraying.

The electrochemical analysis techniques were applied for studying specifically in-situ tribocorrosion process of cobalt chromium alloy hard coating 5 mm thick on 304 L stainless steel support [20]. It was found that this alloy is supporting a tribocorrosion mechanism consisting in a periodical destruction by friction of the $Cr_2O_3$ oxide film formed during stopping the friction (stationary time).

In this research work it was obtained a Ni / nano-WC (60 nm mean diameter) nanocomposite layer by electro-codeposition process on 304 L stainless steel support in order to be tested in the same tribocorrosion conditions as Co-Cr alloy. The nanocomposite layer with good adherence to stainless steel support, having a thickness of 10 micrometers and about 13.12 wt % of tungsten carbide included into nickel matrix was characterized to tribocorrosion process in bidirectional fretting conditions, as presented in figure 2. The Ni / nano-WC nanocomposite layers obtained by electrodeposition were comparatively characterized to nanohardness and wear corrosion by immersion in Water +Li +B, the simulated cooling water reactor.

The surface morphology and chemical composition of Ni/nano-WC composite layer is presented in figure 5. WC nanoparticles are homogeneously dispersed in the Ni matrix. This means that nickel ions from the electrolyte are adsorbed on WC nanoparticles and then are reduced on the WC surface and their deposition on the surface is totally uniformly. The SEM morphology of Ni/nano-WC composite layer also reveals a cauliflower structure [24].

The incorporation percentage of WC into Ni matrix is determined by transformation of W element wt % in molecular mass of WC, considering the standard atomic weight of W element which is 183.84.

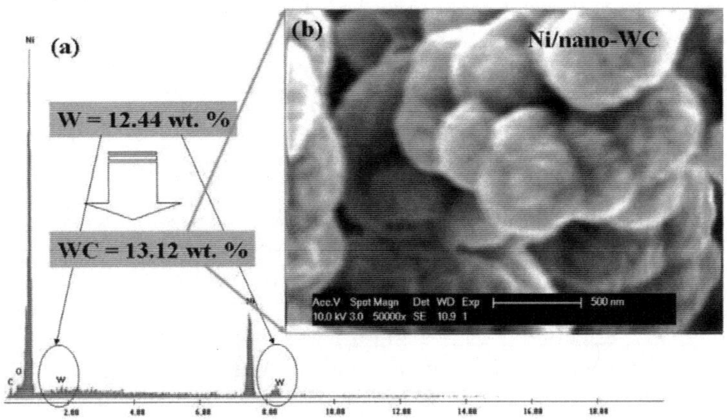


**Figure 5.** EDX-SEM analyzes and calculation of amount of reinforcing WC phase into nickel matrix (a) EDX elemental analyze of the entire surface corresponding to SEM micrograph; (b) - SEM surface morphology.

The weight percentage for W element included into Ni matrix, as shown in figure 5 was determined by EDX analyses being equal to 12.44 wt % (mean value). Therefore the wt. % of WC reinforcing the Ni/nano-WC composite layers resulted at an average value of 13.12, percentage that improved the properties of the nanocomposite layer.

The roughness of obtained layers was measured with a high resolution microtopograph STIL with optical fiber. The results of 2D and 3D surface profiles are presented in figure 6.

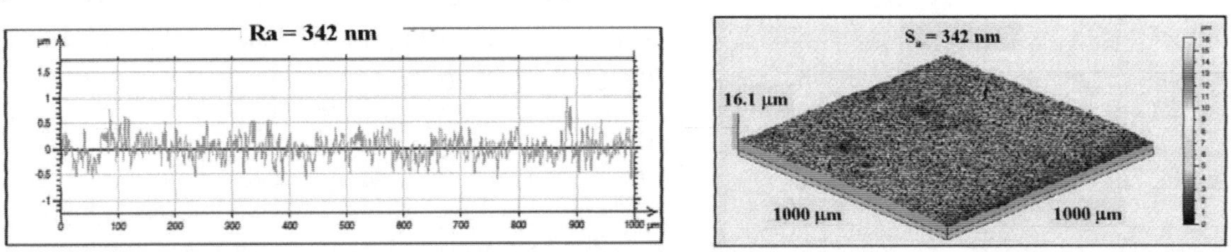


(a) (b)

**Figure 6.** Roughness of Ni/nano-WC nanocomposite layer: (a) 2D profile; (b) 3D surface profile

The nanohardness testing of the obtained layers were done in order to get the values of hardness (H) and elastic modulus (E), presented in figure 7 and Table 1.

**Table 1.** Nanoidentation results as hardness, elastic modulus and H / E ratio for Ni/nano-WC composite and pure Ni coatings

| Type of coating | Identation hardness (H) [GPa] | Vickers hardness [Hv] | Elastic modulus [GPa] | H / E ratio [GPa] |
|---|---|---|---|---|
| Pure Ni | 3.06 | 220.42 | 180.82 | 0.0169 |
| Ni/nano-WC nanocomposite layer | 3.98 | 416.84 | 194.52 | 0.0204 |

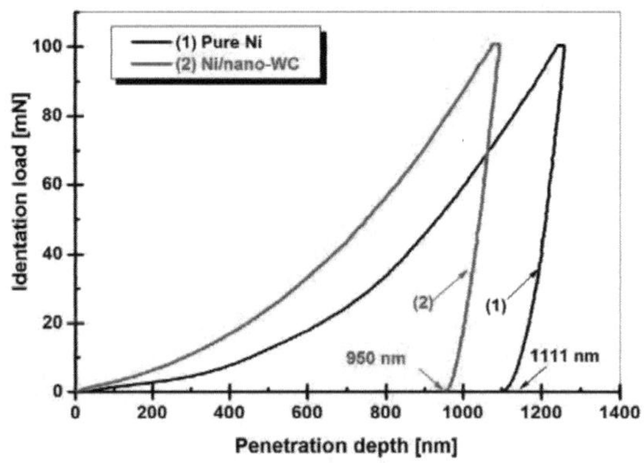


**Figure 7.** Typical cyclic nanoidentation load-displacement curves for (1) pure Ni layer and (2) Ni/nano-WC layer obtained with a Berkovich indenter.

Hardness is commonly regarded to be a significant mechanical property for evaluating wear resistance [24]. However, several authors have recognized that the ratio between the hardness and the elastic modulus is a more adequate guideline for predicting wear resistance than the hardness itself [24]. This H / E ratio is related to the elastic strain to failure [24]. Leyland et all. [25] have assumed that the most durable coatings can be obtained if the elastic strain to failure has a high value and this can be done by obtaining layers with sufficiently high hardness in order to resist at plastic deformation and simultaneously with a low elastic modulus (similar to, or slightly below, that of the substrate material).

Figure 8 provides the mean values for coefficients of friction monitored in wet conditions as a function of normal forces of 1 N, 2 N and 5 N, at the fretting frequency of 1 Hz, for a number of cycles equal to 10000 and 200 μm displacement amplitude corresponding to pure Ni coating and Ni/nano-WC nanocomposite layer.

Figure 8 shows a general behaviour of friction coefficient recorded under wet condition of Ni/nano-WC nanocomposite layers being smaller than that of pure Ni layers for all three normal forces applied.

From figure 8 it can be seen that under wet conditions, the friction coefficient of both pure Ni and Ni/nano-WC composite coatings decreases gradually with increasing the normal force. For example in the case of Ni/nano-WC nanocomposite layer the coefficient of friction has the average value of 0.27 for 1 N normal force decreasing to 0.18 at 2 N normal force and to 0.17 at 5N normal force. The coefficient of friction for pure nickel layer is 0.42 being higher at 1 N normal force and decreases to 0.32 at 2 N normal force and respectively at 0.27 at 5 N normal force. By incorporation of WC nanoparticles into nickel matrix, the friction coefficients are reduced as compared with pure Ni coatings, demonstrating an improvement of fretting wear resistance.

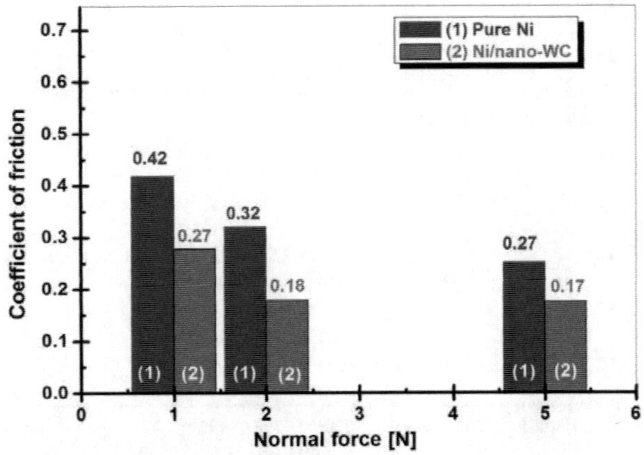


**Figure 8.** Evolution of the friction coefficient of Ni/nano-WC composite and pure Ni coatings during tribocorrosion tests performed in water +Li+B solution during 10000 cycles for the normal forces of: (a) 1 N, (b) 2 N and (c) 5 N.

The open circuit potential (OCP) data recorded on Ni/nano-WC layers and pure nickel layers before, during and after loading the forces in the same solution are shown in Figs. 9 and 10.

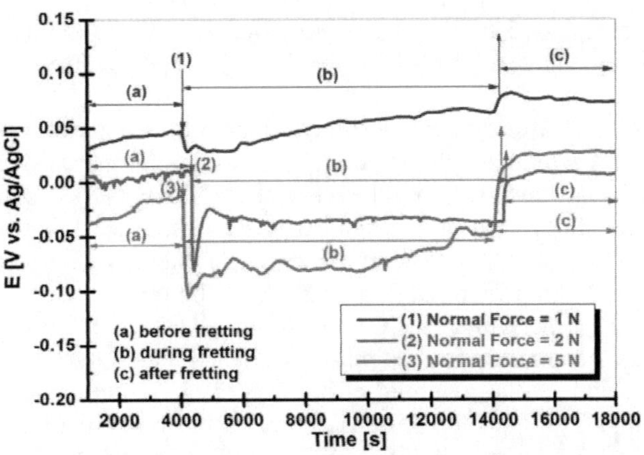


**Figure 9**. Evolution of the open circuit potential recorded (a) before, (b) during and (c) after loading the forces for Ni/nano-WC composite layers in water +Li+B solution during 10000 cycles for the normal forces of: (1) 1 N, (2) 2 N and (3) 5 N.

After immersion in the test solution, before loading the force, a large increase of the open circuit potential for both Ni/nano-WC composite layers and pure Ni layer could be observed, which confirm the stabilization of steady state of sample surface to a passive one. It must be mentioned that the potential measured before loading the force of Ni/nano-WC nanocomposite layers is more ennobled as compared to those of pure Ni layers, thus leading to a significant increase of anticorrosion effect. This means that an ennoblement of the nanocomposite surface occurs when WC nanoparticles are incorporated into the electrodeposited nickel layer reinforcing therefore the nickel matrix.

The suddenly shift in the cathodic domain at the moment when the normal force was loaded on the Ni/nano-WC nanocomposite layers surface took a slower leap of 0.018 V vs. Ag/AgCl (from 0.046 to 0.028 V vs. Ag/AgCl) for 1 N normal force (figure 9), confirming a better tribocorrosion resistance due to reinforcing nanoparticulate WC phase.

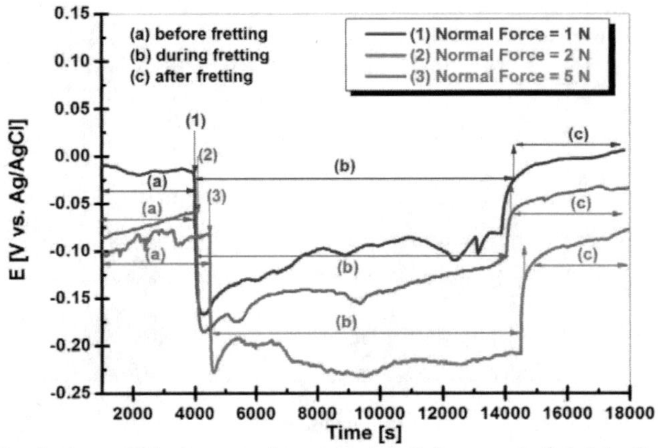


**Figure 10**. Evolution of the open circuit potential recorded (a) before, (b) during and (c) after loading the forces for pure Ni layers in water +Li+B solution during 10000 cycles for the normal forces of: (1) 1 N, (2) 2 N and (3) 5 N.

According to figure 10, at the start of fretting, the potential of the pure Ni layers drops suddenly to a more negative value, its leap being of around 0.15 V vs. Ag/AgCl (from -0.016 to -0.166 V vs. Ag/AgCl) for normal force of 1 N. Similar observations were made earlier by Ponthiaux et al. [26] for stainless steel disk immersed in 0.5 M $H_2SO_4$, by Berradja et al. [28] for stainless steel during sliding in Ringer's solution and by Mardare et al. [19] for Ti-6Al-4V alloy during sliding in bio-simulated environment.

This potential sudden shift in the negative direction favours the damage and partial or complete removal of the passive film induced by fretting, suggesting an increase in its susceptibility for corrosion [26]. By comparing the open circuit potential of Ni/nano-WC composite layers (figure 9) with that of pure Ni layers (figure 10), it can be clearly seen that the highest drop in potential for all applied forces is observed for pure nickel as compared with Ni/nano-WC composite. This observations indicate that pure Ni layers exhibit more cathodic values of the open circuit potential before and with the onset of friction force than Ni/nano-WC nanocomposite layers. This improved behaviour of nanocomposite layer could be attributed to the presence in the nickel matrix of WC nanoparticles.

At the end of the friction test, the potential of both types of studied layers exhibit an anodic shift, reaching again their original values, indicating the progressive repassivation of the worn surface. In the present study, after the fretting motion is stopped, pure Ni layers tends to reach the steady state potential before the onset of friction, whereas Ni/nano-WC protective layers go far beyond the steady state potential reached before the start of the friction, confirming again the better resistance of nanocomposite layer to tribocorrosion degradation process.

*3.2. Nanoporous thin oxide films on titanium alloy to improve the tribocorrosion resistance*

Improving corrosion and tribocorrosion resistance in physiological fluids for biomaterials by electrochemical surface modification of Ti alloy was an atteint objective of this reserach work. The friction coefficients of electrochemically formed nanoporous titanium oxide film and untreated titanium alloy, measured in saliva solution are shown in figure 11.

From Fig 11, it appears that for each separately surface the lowest coefficient of friction was recorded at the highest normal force, at constant frequency and constant number of cycles, an effect also observed by Ahmad et al. for aluminium matrix composite [28]. The reason of decrease in coefficient of friction may be due to smoothening of the surface under higher load. The average values of the coefficient of friction for both surfaces studied versus applied normal forces are shown in figure 11.

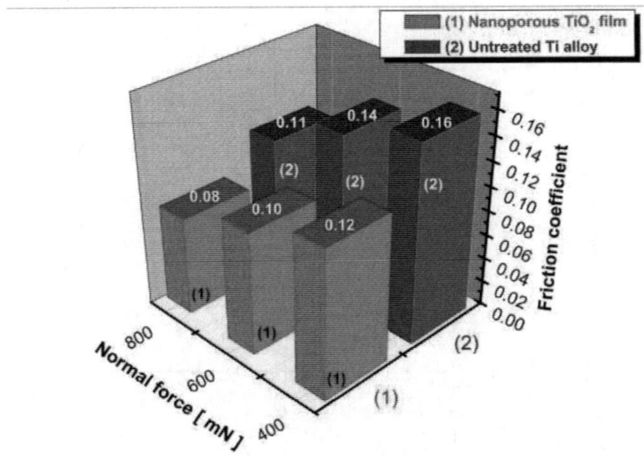


**Figure 11**. Average values of friction coefficients for nanoporous $TiO_2$ film electrochemically formed on titanium alloy (1) and for untreated titanium alloy (2) measured versus applied normal forces of 400, 600, and 800 mN at fretting frequency of 1Hz during 1000 cycles.

The coefficient of friction for untreated Ti-6Al-4V alloy decreases from about 0.16 corresponding to 400 mN to 0.11 according to 800 mN. The same trend is also maintained for anodic nanoporous $TiO_2$ layer surface but with lower values, starting from 0.12 at 400 mN normal force and slowing down to 0.08 at 800 mN normal force. It could be noticed that the lowest coefficient of friction is obtained at the highest normal force, remark applied to both types of surfaces studied. By comparing the coefficient of friction of untreated Ti-6Al-4V surface with anodized surface at the same normal force it can be seen that the lowest coefficient of friction is obtained for nanoporous oxide film formed on titanium alloy.

The open circuit potential measured before and during fretting, generated by reciprocating sliding tests at Fn=400 mN, displacement amplitude = 500 μm, 1 Hz fretting frequency and 1000 sliding cycles) performed in Fusayama Mayer saliva are shown in figure 12. By immersion in saliva solution and before fretting start, there is observed a clear ennoblement of open circuit potential of nanoporous $TiO_2$ film, which is 480 mV vs. Ag/ AgCl as compared with the untreated alloy having an open circuit potential of -150 mV vs. Ag/ AgCl. By controlled forming and growth of nanoporous titanium oxide on titanium alloy the open circuit potential of modified surface is ennobled with more than 600 mV, confirming a better behavior of nanoporous $TiO_2$ film.

By loading the fretting force of 400 mN at 1 Hz fretting frequency, the open circuit potential of untreated Ti alloy is shutting down to more cathodic active domain with about 200 mV vs. Ag/AgCl and show many oscillations during fretting period.

In the same fretting conditions the nanoporous $TiO_2$ film exhibits better behavior, the open circuit potential remaining practically constant, at the same noble value as before starting the fretting, as it is shown in figure 12 (curve 2).

By measuring the surface wear tracks generated after reciprocating sliding tests using a non-contact white light interferometry analysis Wyco NT3300 optical profilometer with Vision (version 2.210) software the better results of nanoporous $TiO_2$ film is confirmed, practically the wear track on nanoporous titanium oxide film is hardly to be observed, as it is shown on 3D surfaces of the resulted wear tracks, Figures 13 a and b.

As it can be observed in figure 13, the wear track on nanoporous titanium oxide film is not quantifiable being not drawn by the counter body because the surface is becoming highly resistant to extreme wear-corrosion conditions.

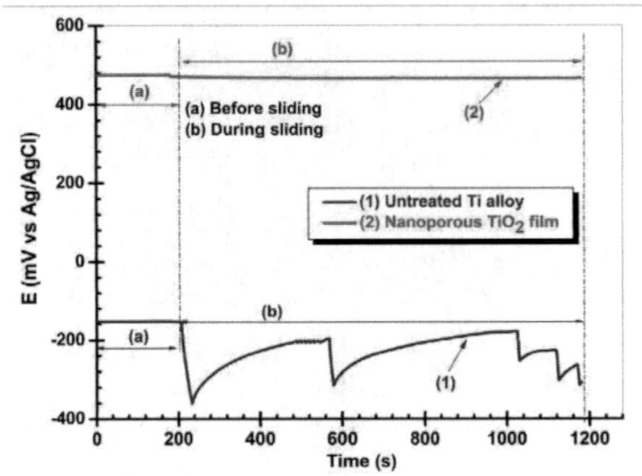


**Figure 12**. Open circuit potential of untreated Ti alloy (1) and nanoporous $TiO_2$ film (2) measured at immersion in saliva solution before fretting (a) and during fretting (b) with a normal force of 400 mN, fretting frequency of 1 Hz and 1000 cycles.

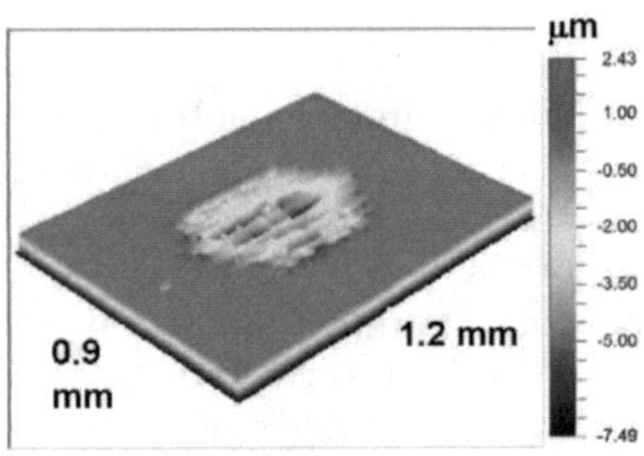


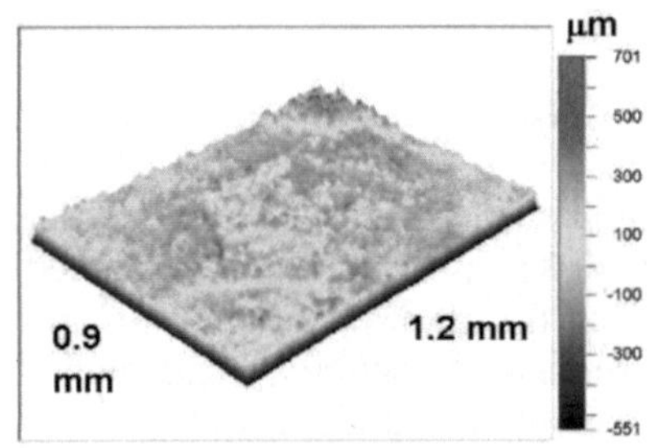


**Figure 13**. 3D surface profiles of wear tracks generated by tribocorrosion tests on untreated Ti alloy (a) and nanoporous TiO2 film (b) by fretting at 400 mN , 1Hz frequency and 1000 cycles

## 4. Conclusions

This work points out the capabilities of in - situ tribocorrosion investigations to compare the different materials and coatings in order to choose the better ones to be used under tribocorrosive conditions in sliding contacts for industrial or biomedical applications.

Electrochemical techniques as electrodeposition and controlled oxide formation and growth by anodization can lead to the a large class of nanocomposite layers as nanomaterials and nanoporous oxide films as biomaterials on different support materials and structures necessary for a future based on nanotechnology and nanomaterials to improve the resistance to aggressive environments and complex degradation processes.

Improving surface properties for corrosion and tribocorrosion resistance of materials to be used in specific environments give more valuable industrial and biomedical applications with improved life cycle.

Tribocorrosion is a complex degradation process which is present in industrial as well as in biomedical applications of materials.

The nanomaterials as nanocomposite layers having dispersed and reinforcing hard ceramic nanoparticles are an alternative possibility to increase the lifecycle of materials used in nuclear power plants in wet and moving conditions.

Functionalizing the surface of materials and biomaterials is the way to reduce or by-pass the complex degradation by corrosion and tribocorrosion processes.

IOP Conf. Series: Materials Science and Engineering **174** (2017) 012042 doi:10.1088/1757-899X/174/1/012042

**Acknowledgements**
The financial support for this research work was provided by national research project UEFISCDI-PCE–PN II, PN-II-ID-PCE-2012-4-0370, contract 10 / 2013 - 2016 – HyBioElect.

**References**
[1] Levy A V 1988 *Surf. Coat. Technol.* **36** 387
[2] Ninham A J and Levy A V 1988 *Wear* **121** 347
[3] Lemaire E and Le Calvar M 2001 *Wear* **249** 338
[4] Madsen B W and Adler T A 1994 *Wear* **171** 215
[5] Jiang X X, Li S Z, Tao D D and Yang J X 1993 *Corrosion* **49** 836
[6] Zhang T C, Jiang X X, Li S Z and Lu X C 1994 *Corros. Sci.* **36** 1953
[7] Stemp M, Mischler S and Landolt D 2003 *Wear* **255** 466
[8] Aslanyan I R, Bonino J P and Celis J P 2006 *Surf. Coat. Technol.* **200** 2909
[9] Wood R J K 2007 *J. Phys. D: Appl. Phys.* **40** 5502
[10] Fedrizzi L, Valentinelli L, Rossi S and Segna S 2007 *Corros. Sci.* **49** 2781
[11] Diomidis N, Celis J P, Ponthiaux P and Wenger F 2010 *Wear* **269** 93
[12] Smith A J, Stratmann M and Hassel A W 2006 *Electrochim. Acta* **51** 6521
[13] Henry P, Takadoum J and Berçot P 2009 *Corros. Sci.* **51** 1308
[14] Bidiville A, Favero M, Stadelmann P and Mischler S 2007 *Wear* **263** 207
[15] Yan Y, Neville A, Dowson D, Williams S and Fisher J 2009 *Wear* **267** 683
[16] Mischler S 2008 *Tribol. Int.* **41** 573
[17] Benea L, Wenger F, Ponthiaux P and Celis J P 2009 *Wear* **266** 398
[18] Benea L, Mardare-Danaila E and Celis J P 2014 *Tribol. Int.* **78** 168
[19] Mardare E, Benea L and Celis J P 2012 *Optoelectron. Adv. Mat.* **6** 474
[20] Benea L, Ponthiaux P, Wenger F, Galland J, Hertz D and Malo J Y 2004 *Wear* **256** 948
[21] García I, Drees D and Celis J P 2001 *Wear* **249** 452
[22] Benea L, Iordache V E, Wenger F, Ponthiaux P, Peybernes J and Vallory J 2005 *The Annals of Dunarea de Jos University of Galati,* Fascicle VIII Tribology **2005** 35-40
[23] Benea L, Mardare E, Mardare M and Celis J P 2014 *Corros. Sci.* **80** 331
[24] Benea L, Basa S B, Dănăilă E, Caron N, Raquet O, Ponthiaux P and Celis J P 2015 *Mater. Design* **65** 550
[25] Leyland A and Matthews A 2000 *Wear* **246** 1
[26] Ponthiaux P, Wenger F, Drees D and Celis J P 2004 *Wear* **256** 459
[27] Berradja A, Bratu F, Benea L, Willems G and Celis J P 2006 *Wear* **261** 987
[28] Ahmad F, Raza M R, Ab Rani A M and Lo S H J 2011 *J. Appl. Sci.* **11** 1673

IOP Conf. Series: Materials Science and Engineering **174** (2017) 012043 doi:10.1088/1757-899X/174/1/012043

# Design and technology parameters influence on durability for heat exchangers tube to tubesheet joints

**R G Ripeanu**

Petroleum-Gas University of Ploiesti, Mechanical Engineering Department, 39 Bucharest Blvd., 100680 Ploiesti, Romania

E-mail: rrapeanu@upg-ploiesti.ro

**Abstract.** The main failures of heat exchangers are: corrosion of tubes and jacket, tubes blockage and failures of tube to tubesheet joints also by corrosion. The most critical zone is tube to tubesheet joints. Depending on types of tube to tubesheet joints, in order to better respect conditions of tension and compression, this paper analyses the tubesheet holes shapes, smooth and with a grove, on corrosion behavior. In the case of welding tubes with tubesheet, welding parameters modify corrosion behavior. Were realized welded joints by three welding regimes and tested at corrosion in two media, tap water and industrial water. Were tested also samples made of smooth tubes, finned tubes and tubes coated with a passive product as applied by a heat exchanger manufacturer. For all samples, the roughness parameters were measured, before and after the corrosion tests. The obtained corrosion rates show that stress values and their distribution along the joint modify the corrosion behavior. The optimum welding parameters were established in order to increase the joint durability. The paper has shown that passive product used is not proper chosen and the technology of obtaining rolled thread pipes diminishes tubes' durability by increasing the corrosion rate.

## 1. Introduction

A heat exchanger is an equipment built for efficient heat transfer from one fluid to another, where the fluids are separated by a solid wall so that they never mix, or are directly contacted. They are widely used in chemical processing, heating, refrigeration, air conditioning, power production etc. Typical functions of heat exchangers are heating or cooling. Shell and tube heat exchangers are most commonly used in the process refinery industries due to a large ratio of heat transfer area to volume and weight. Shell and tube heat exchangers consists of a shell surrounding a bundle of tubes. Two fluids, of different inlet temperatures, flow through the exchanger. One fluid flows through the tubes and the other flows through the shell. Heat is transferred from one fluid to the other. The general coding guidelines [1] define heat exchangers failures as follows:

•" *State of other components*". The cause of the state of the component under consideration is due to state of another component. Examples are loss of power and loss of heating/cooling capacity.

• "*Design, manufacture or construction inadequacy*". This category contains actions and decisions taken during design, manufacture, or installation of components, both before and after the equipment is operational. Included in the design process are the equipment and system specification, material specification, and initial construction that would not be considered a maintenance function. This category also includes design modifications.

- "*Abnormal environmental stress*". Represents causes related to a harsh environment that is not within component design specifications. Specific mechanisms include chemical reactions, electromagnetic interference, fire/smoke, impact loads, moisture (sprays, floods, etc.) radiation, abnormally high or low temperature, vibration load, and severe natural events.
- "*Human actions*". Represents causes related to errors of omission or commission on the part of plant staff or contractor staff. An example is a failure to follow the correct procedure. This category includes accidental actions, and failure to follow procedures for construction, modification, operation, maintenance, calibration, and testing. This category also includes deficient training of operators.
- "*Maintenance*". All maintenance not captured by H - human actions or P - procedure inadequacy.
- "*Internal to component, piece part*". Deals with malfunctioning of parts internal to the component. Internal causes result from phenomena such as normal wear or other intrinsic failure mechanisms. It includes the influence of the environment of the component. Specific mechanisms include erosion/corrosion, internal contamination, fatigue, and wear out/end of life.
- "*Procedure inadequacy*". Refers to ambiguity, incompleteness, or error in procedures for operation and maintenance of equipment. This includes inadequacy in construction, modification, administrative, operational, maintenance, test and calibration procedures. This can also include the administrative control of procedures, such as change control.
- " *Other*". The cause of events is known, but does not fit in one of the other categories.
- "*Unknown*". This cause category is used when the cause of the component state cannot be identified.

The dominant failure causes based on ICDE codes [1] is "*Abnormal environmental stress*" accounting for 28% of the events, followed by "*Design, manufacture or construction inadequacy*" and " *Internal to component, piece part*" accounting for 26% of the events each. In the majority of the events, failures occur at equipment's working in aggressive environment and were affected heat exchanger internals as tubes, plates, chambers in multiple trains and components [2]. Observed failures have also lead to leaks and impeded flow due to corrosions (corrosion, erosion) and dirt accumulation (pitting debris, fouling) [2]. There are also direct human/operator related faults causing dependencies of heat exchanger trains, e.g. by faulty alignment of valve configuration and wrong maintenance procedures and/or practices [2].

In conclusion the main failures at heat exchangers are caused by corrosion. Not only working parameters, harsh media and materials used influence corrosion rate but also the design, the manufacture parameters and the construction influence corrosion behavior and durability of heat exchangers.

To increase the heat exchange efficiency the tubes surface could be finned or made a thread by rolling [3], but internal stresses due to plastic deformation modify electrochemical potential and material became anodic [4].

The tubes are held in place by being inserted into holes in the tubesheet and there either expanded into grooves cut into the holes or welded to the tube sheet. The most critical zone to corrosion at heat exchangers is tube to tubesheet joints.

Depending of types of tube to tubesheet joints, in order to better respect conditions of tension and compression, this paper analyses the tubesheet holes shapes, smooth and with a grove [5], above corrosion behavior. In case of welding tubes with tubesheet, welding parameters modify corrosion behavior.

There were also tested samples made of smooth tubes, finned tubes and tubes coated with a passive product as applied by a heat exchanger manufacturer. At all samples were measured the roughness parameters before and after corrosion tests.

The obtained corrosion rates show that the stress value and its distribution along the joint modify corrosion behavior.

Were established the optimum welding parameters in order to increase joints durability. It was shown that passive product used is not proper chosen and the technology of obtaining rolled thread pipes diminishes tubes durability by increasing the corrosion rate.

## 2. Heat exchanger, construction, materials and working conditions

The analysed heat exchanger is type NEN as classified by Tubular Exchanger Manufacturers Association (TEMA) [6] and it is presented in figure 1. The main working parameters are shown in table 1.

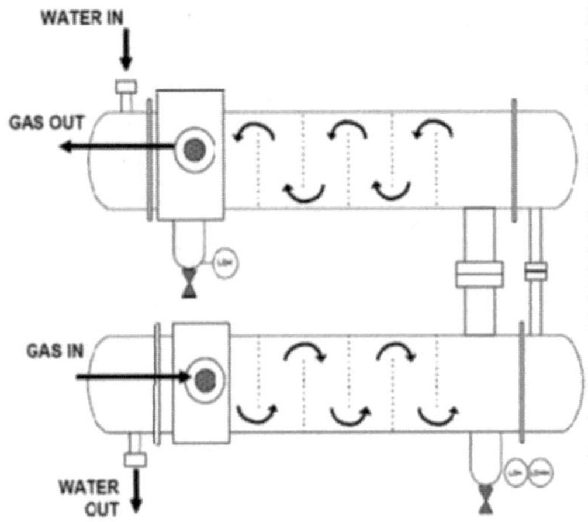

(a) (b)

**Figure 1.** Heat exchanger batery type NEN.
a) Scheme of heat exchanger; b) Photo of heat exchanger.

For the analyzed heat exchanger, steam is circulating through the jacket and through the tubes is circulating industrial water which will be heated.

**Table 1.** Main heat exchanger working parameters.

| Parameter | Shell chamber | Tubes chamber |
|---|---|---|
| Capacity ($m^3$) | 1.31 | 0.55 |
| Number of crossings | 1 | 1 |
| Fluid | Gas | Industrial water |
| Maximum working pressure (MPa) | 3.79 | 4.92 |
| IN/OUT working temperature (°C) | 199 / 185 | 21 / 49.5 |
| Material | ASTM 516 Gr.70 / EN 10028 P355 GH | ASTM 344 Gr.6 / EN 10028 P265 GH |

To design the heat exchanger, manufacturer used PV Elite software programme. PV Elite is used also for analysing heat exchangers and pressure vessels and it is compliant with ASME Section VIII Divisions 1 & 2, PD 5500, EN 13445 and API 579.

## 3. Experimental tests

### *3.1. Tests on extruded tube to tubesheets joints*

The joints between tubes and tubesheets must resist to axial stress which appear in tubes. This condition is fulfilled when tubes and tubesheets were made of steel and when the hoop stress in tubesheet is higher than in tubes [5, 7]. In order to evaluate residual stresses at tube to tubesheets expanded joints above tubes corrosion behavior were prepared samples as presented in figure 2 for tubesheets and in figure 3 for tubes. Samples were prepared at real joints dimensions and tubesheet sample of material P 355 GH and tubes of material P 265 GH EN 10028.

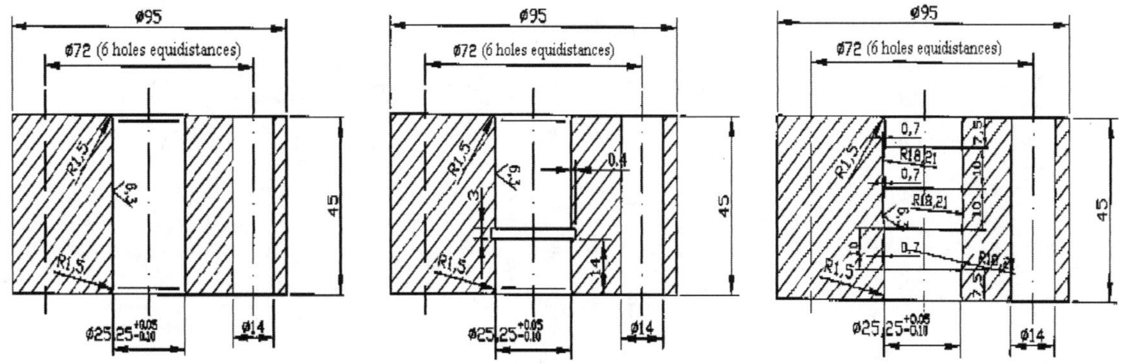


(a) (b) (c)

**Figure 2.** Types of tubesheets samples.
(a) Smooth boring surface;
(b) Rectangular grove on boring surface;
(c) Elliptic groves on boring surface.

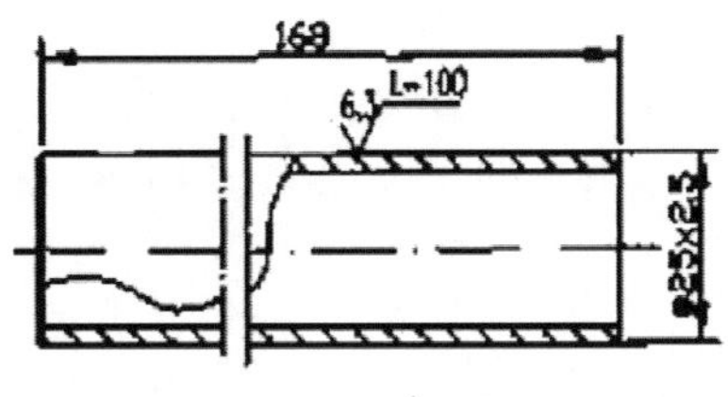


**Figure 3.** Tubes samples dimensions.

Samples were expanded to obtain joints between tubes and tubesheets. Parallelepiped shape samples for corrosion tests were extracted and machined from tubes, without affecting the material structure, as follows:

- "A" type samples extracted from tubes not used for expanded joints;
- "B" type samples extracted from tubes from expanded joint with smooth tubesheet boring surface;
- "C" type samples extracted from tubes from expanded joint with a rectangular grove on tubesheet boring surface;
- "D" type samples extracted from tubes from expanded joint with elliptic groves on tubesheet boring surface.

Working medium were industrial water with $pH$ = 7.18, *conductivity* = 1524 μS/cm, total solid deposition $TDS$ = 42 mg/l and medium temperatures were 20, 40 and 60 $^0$C [5].

For immersion corrosion tests the corrosion rate was obtained with the relation [4]:

$$v_{cor} = 8.76 \cdot \frac{m_f - m_i}{A \cdot \tau \cdot \gamma}, \text{ mm} \cdot \text{year}^{-1} \tag{1}$$

where $m_f$ is sample final mass, g;

$m_i$- initial sample mass, g;

$A$- sample area, m$^2$;

$\tau$- time, hours;

$\gamma$- specific weight, g·cm$^{-3}$.

In figure 4 it is shown the obtained corrosion rate vs. time at testing temperatures in industrial water.

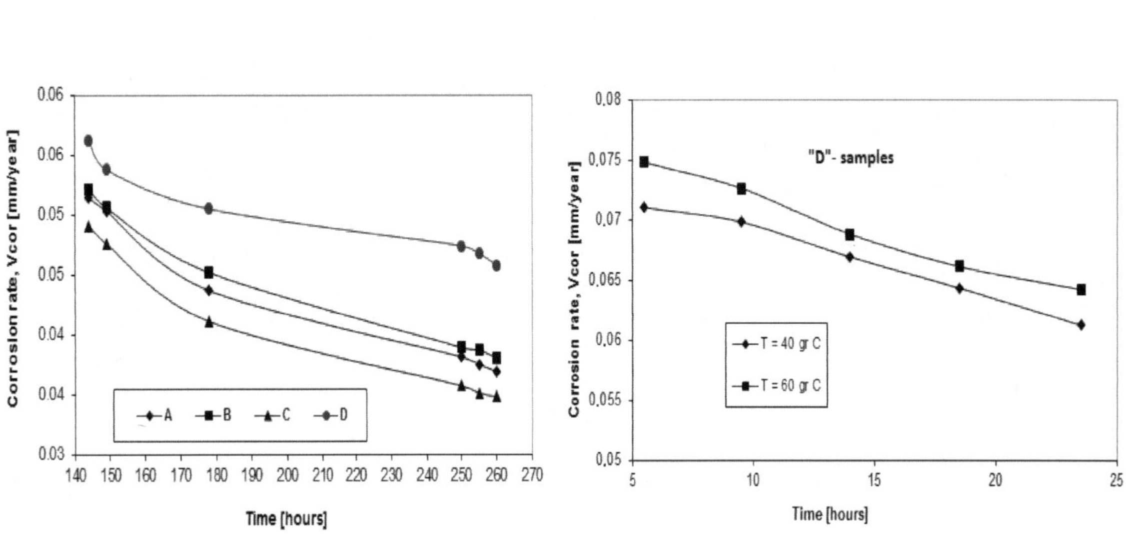


(a) (b)

**Figure 4.** Corrosion rate results for extruded tubes samples.
(a) Corrosion rate vs. time at $20^0$C in industrial water;
(b) Corrosion rate vs. time at $40^0$C and $60^0$C for samples types "D".

Similar results as presented in figure 4b for corrosion rate were obtained for samples extracted from extruded tubes types "A", "B" and "C".

### *3.2. Corrosion tests on tubes samples with different surface geometry*

In order to evaluate corrosion behavior of tubes with different geometry were prepared tubes samples as follows:
- "E" type samples with finned surfaces obtained by rolling;
- "F" type samples with smooth surface tubes with roughness parameter $R_a$ = 1.35μm as obtained by tubes manufacturer treated or not treated (passivated) with the corrosion inhibitor product type VmCI-$307^{TM}$;
- "G" type samples with smooth tubes with $R_a$ = 1.55 μm obtained by turning treated or not treated with the corrosion inhibitor product type VmCI-$307^{TM}$.

Figure 5 presents the finned tubes samples and figure 6 the smooth tubes samples.

**Figure 5.** Finned tube sample.

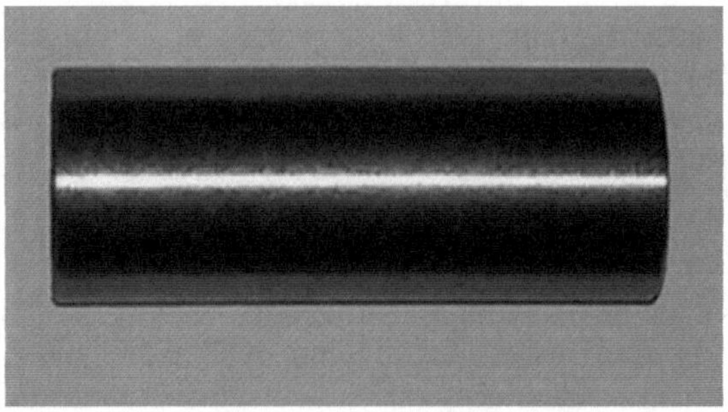

**Figure 6.** Smooth tube sample.

Testing media were tap water and industrial water with characteristics presented above in section *3.1.* Before immersion samples were measured, weighed with an analytic balance. In samples tubes holes, rubber plugs were fixed to avoid corrosion inside the tubes. The immersion time was 338 hours at ambient temperature of $20^0$C. Table 2 presents the results of corrosion rate.

**Table 2.** Corrosion rate results for different tubes surfaces.

| Sample type | Medium | Temperature, ( $^0$C ) | Time, ( hours) | Corrosion rate, $v_{cor}$ (mm·year$^{-1}$) |
|---|---|---|---|---|
| "E" | Tap water | 20 | 338 | 0.097 |
| | Industrial water | | | 0.127 |
| "F" | Tap water | | | 0.051 |
| | Industrial water | | | 0.065 |
| "F" passivated with VmCI-307$^{TM}$ | Tap water | | | 0.062 |
| | Industrial water | | | 0.069 |
| "G" | Tap water | | | 0.056 |
| | Industrial water | | | 0.066 |
| "G" passivated with VmCI-307TM | Tap water | | | 0.067 |
| | Industrial water | | | 0.071 |

*3.3. Tests on samples welded with different welding parameters*

At heat exchanger critical zones are tube to tubesheets joints as presented above. Many heat exchangers manufactures still realize welded joints between tubes and tubesheets. At heat exchanger battery type NEN presented in figure 1 were established GTAW welding parameters. In order to evaluate the influence of welding parameters on corrosion behavior, tree types of samples were prepared and were immersed for 360 hours in tap water and in industrial water, respectively, at ambient temperature of 20$^0$C:

- "H1" type samples welded with a 12% reduction of recommended welding parameters, for immersion in tap water;
- "H2" type samples welded with a 12% reduction of recommended welding parameters, for immersion in industrial water;
- "L1" type samples welded with recommended welding parameters, for immersion in tap water;
- "L2" type samples welded with recommended welding parameters, for immersion in industrial water;
- "M1" type samples welded with a rise with 12% of recommended welding parameters, for immersion in tap water;
- "M2" type samples welded with a rise with 12% of recommended welding parameters, for immersion in industrial water;

These samples were prepared by GTAW welding of real tubes to real plate as shown in figure 7.

**Figure 7.** Samples welded joints.

After welding samples were extracted by milling from plate, and microgeometry parameters were measured for tubes inside surface and plate surface in HAZ (heat affected zone) with Surtronic 3+ device. In figure 8, the measurement images are presented. Microgeometry profile was measured before and after 360 hours of immersing in testing media.

(a)

(b)

**Figure 8.** Microgeometry measurements.
(a) Measurement inside the tubes: (b) Measurement on plate surface.

Table 3 shows the sample welding parameters. Plate was of material P 355 GH and tubes of material P 265 GH EN 10028. Was used gas tungsten arc welding (GTAW) and wire with Ø1.6 mm type BÖHLER EML 5 that can be used in sour gas applications (HIC-Test acc. NACE TM-02-84).

**Table 3.** Welding parameters.

| Samples type | Welding current,(A) | Voltage (V) | Welding time (sec.) | Linear welding energy ($kJ\cdot cm^{-1}$) | Welding speed ($cm\cdot min^{-1}$) |
|---|---|---|---|---|---|
| H1 and H2 | 90 | 8 | 118 | 7.3 | 5.9 |
| L1 and L2 | 115 | 12 | 70 | 12.7 | 6.5 |
| M1 and M2 | 140 | 14 | 50 | 14.7 | 8 |

Table 4 presents the times of first observation of iron oxides on samples surfaces. Samples types "H1", "L1" and "M1"were immersed in tap water and samples types "H2", "L2" and "M2" in industrial water with the same composition as presented above in section *3.1*.

**Table 4.** Time of iron oxides first observation on samples surfaces.

| Samples type | H1 | L1 | M1 | H2 | L2 | M2 |
|---|---|---|---|---|---|---|
| Time of iron oxides first observation, (hours) | 31 | 33 | 35 | 21 | 23 | 23 |

In table 5 are presented the roughness values pointing out the sample surface modifications.

**Table 5.** Sample surfaces roughness modifications.

| Samples type | Roughness parameters (μm) | | |
|---|---|---|---|
| | $R_a$ | $R_t$ | $R_z$ |
| H1 plate | +1.203 | +8.94 | +8.42 |
| H1 tube | -0.920 | -19.70 | -10.10 |
| L1 plate | +0.673 | +9.39 | +6.87 |
| L1 tube | +2.470 | +4.80 | +5.90 |
| M1 plate | +0.897 | +7.82 | +4.60 |
| M1 tube | -3.380 | -17.80 | -14.00 |
| H2 plate | +1.875 | +15.32 | +10.80 |
| H2 tube | -2.610 | -32.40 | -20.10 |
| L2 plate | +1.671 | +11.90 | +10.96 |
| L2 tube | +3.120 | +0.80 | +8.70 |
| M2 plate | +1.584 | +9.85 | +9.37 |
| M2 tube | +1.050 | +4.00 | +4.10 |

[a] Sign "+" signifies roughness increase and "-" signifies roughness diminish.

In order to better analyze the welding parameters on HAZ were made Vickers with low load hardness measurements HV3 at samples types "H", "L" and "M". In figure 9 are shown the hardness vs. distance values for tested samples. Hardness measurements starts at 0.3 mm from inside surfaces of tubes welded in plate and continue on a distance of 10 mm on the plate surface.

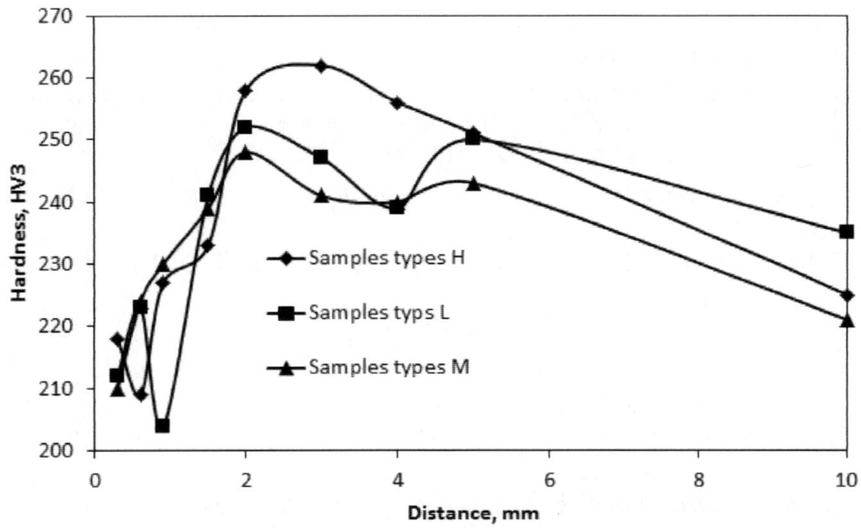


**Figure 9**. Hardness values for tested samples.

As it could be observed from figure 9, because hardness values were smaller than 230 HV3, similar to base material, was not necessary to measure hardness at greater distances on plate surfaces.

## 4. Conclusions

The obtained experimental tests results show that at expanded tube to tubesheets joints, the residual stress influences the corrosion behavior (see figure 4a). The plate boring surface with a rectangular grove shape and with smooth shape also conduct to smaller corrosion rates than those bringing surface with elliptic groves (see figure 4a).

Durability of heat exchangers depends not only on materials' corrosion resistance, but also on the joint stability under axial stresses due to tension or compression. Hence, smooth joint surfaces are not recommended because these joints do not resist to axial stresses which appear in tubes. Tubesheets with elliptic groves or rectangular groves on holes' surfaces are better.

Temperature also modifies the corrosion rates. The temperature influence in increasing corrosion rate is smaller at tubesheets with elliptic groves than those with rectangular groves or smooth holed surfaces.

Tubes shape also modify corrosion rate. Finned tubes obtained by rolling conduct at higher corrosion rates than smooth tubes (see table 2). All cutting operations, including turning (at tubes) or boring (at tubesheet), even are made with cooling fluids and at low cutting parameters modify electrochemical potential at the material surface and increase the corrosion rate.

Not all corrosion inhibitor products have a benefic effect in reducing corrosion rate and increasing equipment durability. Even at tested materials spread used for tubes and tubesheets construction, the corrosion inhibitor product type VmCI-307$^{TM}$ also spread used, conduct to a higher corrosion rate than without this product, as could be observed from table 2.

For the welded joints, welding parameters influence the corrosion rate (see table 4 and table 5), due to the microstructure transformations. The heat affected zones, where the stresses are maximum, are also critical zones for corrosion. For tested welded materials, the parameters with 12% higher values than the recommended values, conduct to better results. The times of oxide observations were longer, the surfaces of tubes and plates roughness modifications were smaller and the hardness curve presented in figure 9 shows that HAZ and the hardness gradient are smaller.

## References

[1] Nuclear Energy Agency Committee on the Safety of Nuclear Installations 2012 *International Common Cause failure data Exchange ICDE General Coding Guidelines ICDE* Organisation for Economic Co-operation and Development (OECD) (Paris, FR) pp 21-57

[2] Nuclear Energy Agency Committee on the Safety of Nuclear Installations 2015 *Collection and Analysis of Common-cause Failures of Heat Exchangers* Organisation for Economic Co-operation and Development (OECD) (Paris, FR) pp 15-27

[3] Bell K and Mueller A 2001 *Heat Transfer Databook II, vol. II* Wolverine Tube, Inc. (Decatur, Al., USA) pp31-44

[4] Tudor I and Ripeanu R G 2002 *Ingineria Coroziunii, vol. I* (Ploiesti, Ed. Universităţii Petrol-Gaze din Ploieşti) pp 45-66

[5] Iancu M, Ripeanu R G and Tudor I 2013 Heat Exchanger Tube to Tube Sheet Joints Corrosion Behavior *Tribology in Industry* **23** (1) 19-24

[6] Tubular Exchanger Manufacturers Association (TEMA) 2007 *Standards of the Tubular Exchanger Manufacturers Association, 9th Edition* Tubular Exchanger Manufacturers Association (New York, USA) pp 22-78

[7] Iancu M, Pupazescu A and Tudor I 2010 Study on the state of stresess and strain in tube to tubular plate joints *Petroleum-Gas University of Ploiesti Bulletin, Technical Series* **LXIII** (4B) 61-66

IOP Conf. Series: Materials Science and Engineering **174** (2017) 012044 doi:10.1088/1757-899X/174/1/012044

# Porous $TiO_2$-$ZrO_2$ thin film formed by electrochemical technique to improve the biocompatibility of titanium alloy in physiological environment

**L Benea[1], E Dănăilă[1], P Ponthiaux[2]**

[1] Competences Center: Interfaces – Tribocorrosion and Electrochemical Systems, Faculty of Engineering, "Dunarea de Jos" University of Galati, Galati, Romania,
[2] Laboratoire Génie des Procédés Matériaux, Ecole Centrale Paris, F–92290 Châtenay-Malabry, Paris, France.

E-mail: Lidia.Benea@ugal.ro

**Abstract.** Porous Ti and Ti alloys have received increasing research interest for bone tissue engineering, especially for dental and orthopaedic implants because they provide cell ingrowths and vascularization, improving of adhesion and osseointegration. The tribocorrosion process is encountered in orthopaedic and dentistry applications, since it is known that the implants are often exposed to simultaneous chemical/electrochemical and mechanical stresses. The purpose of this study was to carry out a systematic investigation of the tribo-electrochemical performance of porous $TiO_2$–$ZrO_2$ thin film formed by anodization of Ti–10Zr alloy surface in an artificial saliva solution and to compare the resulted performance with that of the untreated Ti–10Zr alloy surface in order to be applied for biomedical use. The in situ electrochemical technique used for investigation of tribo-electrochemical degradation was the open circuit potential (OCP) measurement performed before, during and after sliding tests. The results presented herein show that controlled anodic oxidation method can significantly improve the tribocorrosion and friction performances of Ti–10Zr alloy surface intended for biomedical applications.

## 1. Introduction

The use of commercially pure titanium (c.p. Ti) and its alloys as biomaterials, especially for orthopaedic prostheses and dental implants is increasing due to their superior biocompatibility, high mechanical strength and corrosion resistance [1-2].

Since the mechanical/tensile strength of c.p. Ti is not high enough for some implanted devices (artificial hip joint, pin, or dental screw), many titanium alloys have been developed for this purpose [3]. In the biomedical field titanium alloys have developed of to create materials with improved properties (strength and castability) compared to unalloyed Ti, for use as implants for a long period of time. Taking into account the binary alloys, Ti-Zr with different quantities of Zr have been developed and considered for applications in dental and medical fields: Ti–xZr alloys (with Zr contents ranging from 5 to 20 wt. %) [4], Ti–10Zr [5], Ti–xZr (with a zirconium content of 13–17%) [6] and Ti–50Zr [7].

The excellent corrosion resistance of Ti and Ti alloys is attributable mainly to the formation of a stable titanium oxide ($TiO_2$) layer that forms naturally on the surface of titanium metal on exposure to air or water. The thickness of this natural oxide film is a typically ranging between 4-6 nm [8-9]. Although the native oxide film considerably reduces the corrosion of Ti and its alloy, this one can be damaged easily by mechanical stress that may lead to accelerated corrosion. Moreover the native $TiO_2$ film is not bioactive enough to form a direct bonding with bone. Therefore porous structures are necessary to improve the cell adhesion and osseointegration properties [10]. Consequently, many assays were carried out to grow a thicker oxide layer on titanium or its alloys which can reduce the passivity current and hinder the ion release of titanium and its alloyed elements into physiological fluids. Anodic oxidation of titanium and its alloys might be employed for increase the thickness of the $TiO_2$ native film and can build up rough and porous surfaces suitable for osseointegration [10-11]. This electrochemical method is increasingly applied for surface treatment on titanium and its alloys and can improve the shear strength and the wear resistance properties of these materials by obtaining nanotube or nanopore structures at the surface [9,12].

The tribocorrosion process finds applications in orthopaedic and dentistry fields, since it is known that the implants are often exposed to simultaneous chemical/electrochemical and mechanical stresses. Tribocorrosion phenomena can affect living systems, as it can cause damage to restorative dentistry, prostheses and other devices, which in turn can affect life quality and overall health. Taking notice of the past studies [4,7] which have presented only the physical properties and corrosion behaviour of Ti–Zr binary alloys, the purpose of this study was to evaluate the tribocorrosion behaviour of porous $TiO_2$-$ZrO_2$ layer obtained by controlled anodic oxidation treatment compared to untreated Ti-10Zr alloy. To our knowledge there are no studies in the literature about the effect of the anodic oxidation treatment on the tribocorrosion properties of oxide films grown on Ti-10Zr alloy.

## 2. Materials and methods

### *2.1. Formation and growth of the porous oxide films*

Previously to the growth of the porous oxide films by controlled anodic oxidation method, the Ti–10Zr anode electrode (cylinder) was embedded in epoxy resin, cut and polished into disks (with exposed geometric area of about 2.5 $cm^2$) to a mirror finish using silicon carbide paper with grit from 320 to 4000 μm, diamond paste (3 and 1 μm size) and a suspension of silica (0.04 μm size of particles). Samples were ultrasonically cleaned with ethanol for 5 minutes and dried using dry, cold air. Anodic oxidation was carried out at room temperature in two-electrode electrochemical cell with Ti-10Zr samples as anode and Ti-6Al-4V alloy plate (2.4 x 2.4 cm) as cathode.

As anodizing electrolyte to grow the oxide films was used 1 M $H_2SO_4$. The growth of the oxide films was carried out using a programmable DC power supply with digital multimeter (Genesys – GEN 300-8) by applying 200 V for 2 minutes. During all anodic oxidation processes, the electrolyte was stirred using magnetic stirrer in order to speed up the escape of gas resulted in the electrochemical reaction from the surface of the Ti-10Zr alloy substrate and keep the homogeneity of the electrolyte. After anodization the specimens were washed thoroughly with distilled water in an ultrasonic bath for 10 min and finally dried with hot air.

### *2.2. Material characterization*

The surface morphology of obtained porous structures was determined by using scanning electron microscopy (SEM, FEI Quanta 200 FEG). Moreover, the energy dispersive X-ray spectrometer (EDX) equipped on the SEM system was used to detect the elemental composition of the sample surfaces.

The tribocorrosion performance of the materials was evaluated with a unidirectional pin-on-disc tribometer coupled with a Solartron 1287 electrochemical interface and a Solartron 1255 frequency response analyser. The three electrodes configuration were utilized, with the samples as work electrode (WE), Ag/AgCl (saturated KCl solution, E=200 mV vs. standard hydrogen electrode - SHE) as reference electrode (RE), and a circular Pt-Rh grid as auxiliary electrode (AE). In figure 1 is schematically presented the experimental setup for tribocorrosion experiments.

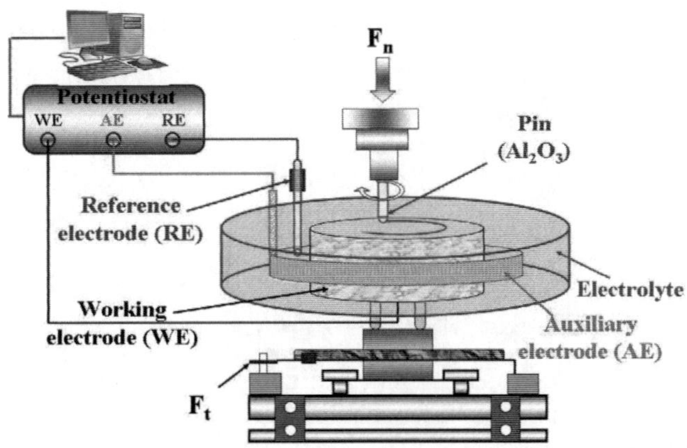


**Figure 1.** Experimental setup for tribocorrosion experiments.

The tribocorrosion experiments were performed at room temperature (23° C) in Fusayama Mayer artificial saliva solution having a pH=4.8 [10]. During tribocorrosion tests an alumina pin cylinder (7 mm diameter with 100 mm radius spherical tip) slides against the sample specimen fully immersed in electrolyte. The pins cylinder made of $Al_2O_3$ (an electrically non-conductive material with high hardness and wear resistance) were degreased with ethanol prior to sliding tests. The tribocorrosion method used consisted of three sequences: (1) stabilization of the system under open circuit potential (OCP) for 30 min (in absence of sliding) to achieve a stable passive surface; (2) sliding under OCP for 5000 cycles, and finally, (3) re-stabilization of the system under OCP (after sliding was stopped) for 30 min. All sliding tests were carried out at a constant rotation speed of 120 rpm, by applying a normal force of 5 N for 5000 cycles.

## 3. Results and discussion

### *3.1. Morphological and elemental characterization*

A simple and feasible method to construct rough and porous oxide films at ambient temperature with bioactive composition and which is able to realize a direct bonding with bone of implant surface is controlled anodic oxidation method. figure 2 (a) shows the surface morphology obtained after anodic oxidation of Ti-10Zr alloy and figure 2 (b) the corresponding elemental composition.

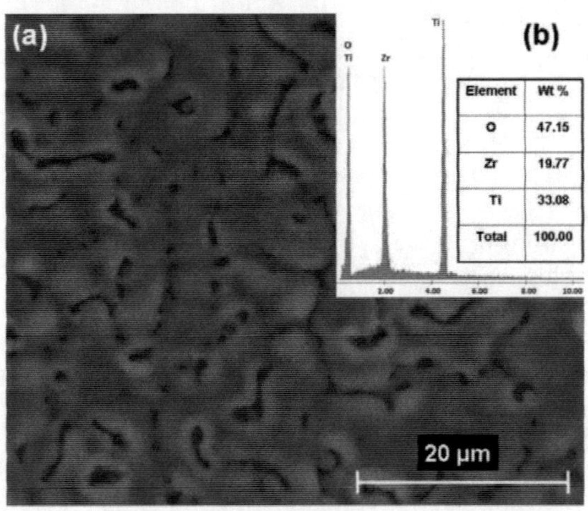


**Figure 2.** (a) SEM surface morphology and (b) EDX spectrum of anodic porous $TiO_2$-$ZrO_2$ film formed.

The main oxidation reactions at the interface of the titanium anode and electrolyte are [13]:

$$Ti \leftrightarrow Ti^{4+} + 4e^{-} \quad (1)$$

$$Ti^{4+} + 2O^{2-} \leftrightarrow TiO_2 \quad (2)$$

Regarding the formation of anodic $TiO_2$-$ZrO_2$ film, the following reactions may occur:

$$Zr \leftrightarrow Zr^{4+} + 4e^{-} \quad (3)$$

$$Zr^{4+} + 2O^{2-} \leftrightarrow ZrO_2 \quad (4)$$

$$Ti^{4+} + Zr^{4+} + 4O^{2-} \leftrightarrow TiO_2 + ZrO_2 \quad (5)$$

Zirconium element is a better reducing agent and therefore competes with titanium during anodic oxidation and will oxidize first under the same condition as Minagar et al. [13] have affirmed.

The occurrence of pores represents a positive characteristic of the materials intended for implantation, since it is well know that porous surfaces interact better with tissue, favoring the osseointegration properties.

### *3.2. Tribocorrosion behaviour tests*

The open circuit potential measurements recorded in Saliva Fusayama Meyer artificial saliva before the onset of sliding, during sliding and after the sliding motion is stopped, corresponding to untreated Ti–10Zr alloy surface and anodic porous $TiO_2$–$ZrO_2$ surface are shown in figure 3.

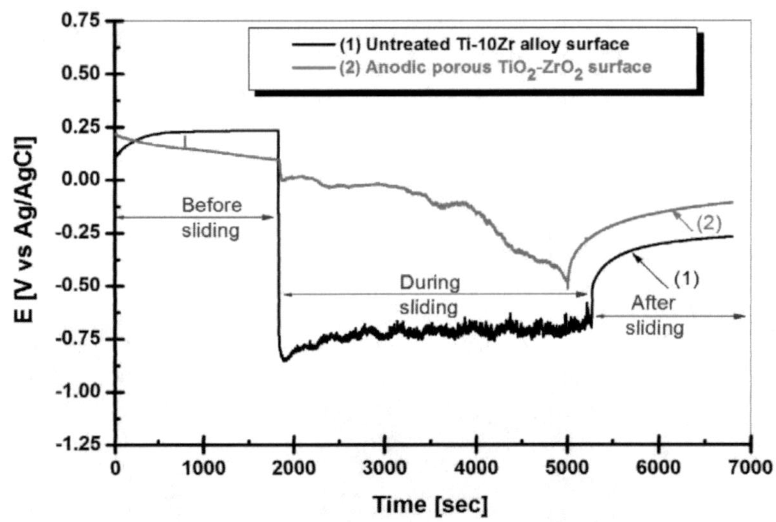


**Figure 3.** Variation of the open circuit potential of: (1) untreated Ti–10Zr alloy and (2) anodic porous $TiO_2$–$ZrO_2$ surface immersed in Fusayama–Mayer saliva before, during and after sliding tests by applying $F_n$ = 5 N, for 5000 cycles, at a constant rotation speed of 120 rpm.

Before starting the sliding tests, an increase of the OCP corresponding to untreated Ti-10Zr alloy is noticed. This behaviour indicates that a stable passive surface state is achieved due to growth of a passive film on the electrode. The extent shift of potential associate to untreated Ti-10Zr alloy in the active direction at the end of 30 min. was to +0.234 V value vs. Ag/AgCl, from the immersion value of +0.106 V vs. Ag/AgCl. The OCP of anodic porous $TiO_2$-$ZrO_2$ surface has evolved at the end of 30 min. to 0.095 V vs. Ag/AgCl from the immersion value of +0.219 V vs. Ag/AgCl.

With the onset of sliding, a sudden drop of the OCP in the cathodic domain takes place for untreated Ti-10Zr alloy. Similar observations have been made earlier for untreated Ti–6Al–4V alloy surface in phosphate buffer saline solution [14] and also for untreated Ti–6Al–4V alloy surface in Fusayama–Mayer saliva solution [10]. The sudden cathodic shift in OCP of untreated Ti–10Zr alloy indicate the removal of the passive layer with the onset of sliding therefore increase its corrosion susceptibility. In contradistinction to untreated Ti–10Zr alloy, anodic porous $TiO_2$-$ZrO_2$ surface did not reveal a sudden shift in OCP with the onset of sliding; rather it is progressive until the sliding motion is stopped. Similar observation has been made also by Kumar et al. [17] for thermally oxidized CP-Ti tested in Ringer's solution [15]. This progressive shift in OCP of anodic porous $TiO_2$-$ZrO_2$ surface indicates that the grown anodic oxide film is not totally removed with the onset of sliding, being gradually destroyed. This behavior of anodic porous $TiO_2$–$ZrO_2$ film reveals its enhanced tribocorrosion properties.

Finally on unloading of the $Al_2O_3$ counterbody from the samples, the OCP of both surfaces untreated Ti–10Zr alloy and anodic porous $TiO_2$–$ZrO_2$ film starts to increase. This behavior indicates the re-establishment of passive state (repassivation) on their surfaces in the wear track area. Both tested surfaces fails to reach the initial steady state value in the given duration of time after the sliding motion is stopped.

In figure 4 is presented the evolution of friction coefficient versus sliding cycles corresponding to untreated Ti–10Zr alloy and anodic porous $TiO_2$–$ZrO_2$ film. According to figure 4 it could be noticed that the lowest friction coefficient is obtained by anodic porous $TiO_2$–$ZrO_2$ surface with a mean value of 0.20 compared with untreated Ti–10Zr alloy surface which reveal a mean value of 0.40.

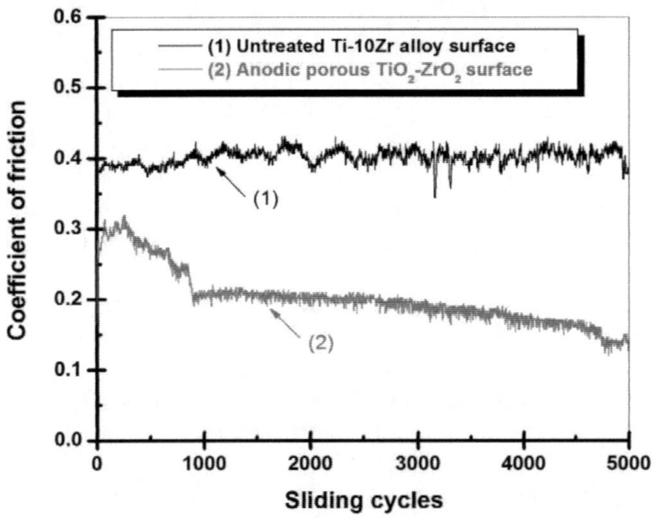


**Figure 4.** Friction coefficients versus sliding cycles obtained in Fusayama–Mayer saliva during sliding tests by applying $F_n = 5$ N at a constant rotation speed of 120 rpm: (1) untreated Ti–10Zr alloy and (2) anodic porous $TiO_2$–$ZrO_2$ surface.

The decreasing of friction coefficient on anodic porous $TiO_2$–$ZrO_2$ surface could be explained by diminishing the shear strength of the contacting interfaces. Nasab and his collaborators [16] explained that a low wear resistance or high coefficient of friction results in implant loosening. In the present work controlled anodic oxidation to form porous oxide film, reduces the friction coefficient of Ti–10Zr alloy surface. This demonstrate that anodic oxidation treatment extend the lifetime of a biomedical implant using Ti–10Zr alloy. Anodic porous $TiO_2$-$ZrO_2$ surface did not reveal a sudden shift in OCP with the onset of sliding, rather it is progressive until the sliding motion is stopped. The lowest friction coefficient is obtained by anodic porous $TiO_2$–$ZrO_2$ surface compared with untreated Ti–10Zr alloy surface.

## 4. Conclusions

The evolution of open circuit potential shown that at the moment when the sliding start, untreated Ti–10Zr alloy surface expose a sudden cathodic shift caused by the mechanical destruction and removal of the native passive oxide film induced by sliding. This behavior confirms its increase in susceptibility for corrosion. Anodic porous $TiO_2$-$ZrO_2$ surface did not reveal a sudden shift in OCP with the onset of sliding, rather it is progressive until the sliding motion is stopped. The lowest friction coefficient is obtained by anodic porous $TiO_2$–$ZrO_2$ surface compared with untreated Ti–10Zr alloy surface.

Tribocorrrosion properties in simulated physiological fluid (Fusayama Mayer saliva) revealed that in their native form, $TiO_2$ layers have poor tribological properties and are easily fractured under tribocorrosion conditions. The tribocorrosion performance of anodic formed porous $TiO_2$–$ZrO_2$ surface is higher compared with that of untreated Ti–10Zr alloy surface under similar experimental conditions.

**Acknowledgements**

UEFISCDI – Ministry of Education and Research is acknowledged for the financial support to Competences Centre Interfaces – Tribocorrosion and Electrochemical Systems (CC–ITES) – Dunarea de Jos University of Galati – Research Project: HyBioElect, contract 10 /30–08–2013 (2013 – 2016) in the frame of National Research Programme Romania – PN II PCE.

## References

[1] Chen Q and Thouas G A 2015 *Mater. Sci. Eng. R* **87** 1
[2] Marques I da S V, Barão V A R, Cristino da Cruz N, Yuan J C-C, Mesquita M F, Ricomini-Filho A P, Sukotjo C and Mathew M T 2015 *Corros. Sci.* **100** 133
[3] Ho W F, Chen W K, Wu S C and Hsu H C 2008 *J. Mater. Sci: Mater. Med.* **19** 3179
[4] Han M K, Hwang M J, Yang M S, Yang H S, Song H J and Park Y J 2014 *Mater. Sci. Eng. A* **616** 268
[5] Lin H C, Wu S C, Ho W F, Huang L H and Hsu H C 2010 *J. Alloy. Compd.* **505** 332
[6] Walter M S, Frank M J, Satué M, Monjo M, Rønold H J, Lyngstadaas S P and Haugen H J, 2014 *Dent. Mater.* **30** 200
[7] Ferreira E A, Rocha-Filho R C, Biaggio S R and Bocchi N 2010 *Corros. Sci.* **52** 4058
[8] Sittig C, Textor M, Spencer N D, Wieland M and Vallotton P H 1999 *J. Mater. Sci. Mater. Med.* **10** 35
[9] Benea L, Mardare-Danaila E and Celis J P 2014 *Tribol. Int.* **78** 168
[10] Benea L, Danaila E and Ponthiaux P 2015 *Corros. Sci.* **91** 262
[11] Sun Y S, Liu J F, Wu C P and Huang H H 2015 *J. Alloy. Compd.* **643** S124
[12] Benea L, Mardare-Danaila E, Mardare M and Celis J P 2014 *Corros. Sci.* **80** 331
[13] Minagar S, Berndt C C, Gengenbach T and Wen C 2014 *J. Mater. Chem. B* **2** 71
[14] Manhabosco T M, Tamborim S M, Santos C B dos and Muller I L 2011 *Corros. Sci.* **53** 1786
[15] Kumar S, Narayanan T S N S, Raman S G S and Seshadri S K 2010 *Corros. Sci.* **52** 711
[16] Nasab M B and Hassan M R 2010 *Trends. Biomater. Artif. Organs.* **24** 69

IOP Conf. Series: Materials Science and Engineering **174** (2017) 012045 doi:10.1088/1757-899X/174/1/012045

# Improving tribocorroson behaviour by electro-codeposition of TiC nano-dispersed particles with nickel as hybrid layers for energy applications

**Lidia Benea[1,*], Eliza Dănăilă[1], Pierre Ponthiaux[2], Jean-Pierre Celis[3]**

[1]Competences Center: Interfaces – Tribocorrosion and Electrochemical Systems, Faculty of Engineering, "Dunarea de Jos" University of Galati, Galati, Romania
[2]Laboratoire Génie des Procédés Matériaux, Ecole Centrale Paris, F–92290 Châtenay-Malabry, Paris, France.
[3]Department of Metallurgy and Materials Engineering, Katholieke Universiteit Leuven, Kasteelpark Arenberg 44, B-3001, Leuven, Belgium.

E-mail: Lidia.Benea@ugal.ro

**Abstract.** In this paper, the results on the surface morphology, nanohardness and tribocorrosion properties of electrodeposited nanostructured Ni/TiC hybrid layers compared with pure Ni layers are presented. The Ni/nano-TiC hybrid layers were obtained by electro-codeposition of TiC nanoparticles (50 nm mean diameter) with nickel from a Watts type bath, on 316L stainless steel support. The combined fretting-corrosion performance was investigated using a reciprocating ball-on-disk tribometer coupled to an electrochemical cell in an electrolyte which simulates the primary water circuit of Pressurized Water Reactors (PWRs). Open circuit potential, as in situ electrochemical technique was performed before, during and after fretting tests in order to obtain information on the changes in surface conditions induced by fretting. The results clearly revealed enhanced nanohardness, improved tribocorrosion properties and reduced coefficients of friction for Ni/nano-TiC hybrid layers as compared to pure Ni layers. The much improved nanohardness and tribocorrosion behaviour can be attributed to the TiC nanoparticles reinforced into Ni matrix.

## 1. Introduction

Wear resistance is an important factor governing the industrial application of surface coatings. The objective of applying coating to a surface for any tribology based application is to impart both smoothness and hardness to the surface so that friction and wear are reduced. The minimization of the friction of the coating is a vital need. The friction characteristics are also dependent on the condition of the tribological testing and the counterface material used [1].

The tribocorrosion process represents a material degradation which involves mechanical wear processes and electrochemical and/or chemical corrosion processes. The tribocorrosion phenomenon exists in different areas of the industry (energy production, transportation, mining and metallurgy, offshore industry, biomedical implants) where it causes severe damage and economic losses.

A variety of particles such as: SiC [2-4], $TiO_2$ [5-7], CNT [8] and WC [9] with mean size diameter from 10 nm to 500 nm, are reported as being incorporated in different metallic matrices such as Ni, Zn and Co to form composite layers and coatings and further to be studied in different friction conditions for diverse applications.

Sun et al [5] investigated the friction and wear properties of the electrodeposited nickel-titania nanocomposite coatings and they report under the same conditions that titania-nickel nanocomposite coatings exhibited smoother surfaces, higher hardness, and lower friction and higher wear resistance as compared with the nanocrystalline nickel coatings electrodeposited at same parameters. Zhou et al. [3] shows in their paper work a lower friction coefficient and better wear resistance of as-codeposited Ni-SiC nanocomposite coatings compared to the as-deposited Ni film, which can be attributed to the incorporation of nanometer SiC particles in the deposit greatly increasing the hardness of the composite coating through grain refinement strengthening and dispersion strengthening. S. K. Ghosh and his collaborators [9] concluded that reciprocating dry sliding tests revealed that a lower coefficient of friction was recorded on CoW-WC coatings than on CoW, with a better wear-resistance compared to CoW alloys.

From special literature the following factors are known to influence the fretting: normal load, amplitude, frequency, number of cycles, relative humidity and temperature, materials involved, roughness [10, 11].

In nuclear power plants, some tribological issues occur due to regular interaction between components (control rod drive mechanisms). Therefore, in order to improve the nanohardness, reduce the friction coefficients and further improve the wear properties in wet conditions of pure Ni layers, Ni/TiC nanostructured hybrid layers were prepared on the surface of stainless steel by electrochemical codeposition process. The present work aims to improve surface mechanical property of pure nickel layers by electro-co-deposition of Ni with dispersed second phase titanium carbide nanoparticles (50 nm mean diameter). It was proved that tribocorrosion resistance of the Ni/nano-TiC hybrid layers were significantly enhanced due to the strong dispersion strengthening effect gained by TiC nanoparticles incorporation into Ni matrix. The results have shown that the TiC nanoparticles reinforcing Ni matrix exhibit a significant higher wear resistance under wet conditions in primary water circuit of Pressurized Water Reactors (PWRs) as compared to pure Ni coatings. The corrosion resistance under fretting conditions was improved and the friction coefficients corresponding to Ni/nano-TiC hybrid layers are reduced compared with the pure Ni electrodeposited layers at all applied forces.

## 2. Materials and methods

### *2.1. Fabrication of Ni/TiC nanocomposite coatings*

A standard Watts' nickel plating bath containing nickel sulphate 200 $gL^{-1}$, nickel chloride 50 $gL^{-1}$ and boric acid 30 $gL^{-1}$ was used as electrolytic solution. TiC particles with mean diameters of 50 nm used in this work were purchased from Hefei Kaier Nanometer Technology & Development Co. Ltd. and the crystalline structure of this nanopowder was cubic with >99.0 % purity. The concentration of TiC nanoparticles in the electrolyte bath was 10 $gL^{-1}$.

A pure Ni sheet was used as the anode, 304 stainless steel sheets with area of 25 $cm^2$ were used as the cathode and a saturated calomel electrode (SCE - $Hg/Hg_2Cl_2$, KCl saturated solution, E=+244mV vs. standard hydrogen electrode) was used as the reference electrode. Before the electrodeposition, cathode was degreased with 5M NaOH and 1:1 HCl, washed in distilled water, and then immersed immediately in the electroplating bath.

The pH of solution was kept at 4.04 ± 0.2. The electrodeposition was carried out at 45 ± 1 °C temperature, by applying a constant current density of 40 $mA/cm^2$ and a time of 15 min. The temperature was kept constant by using a double walls glass cell having 450 mL plating solution. The suspension of TiC nanoparticles was maintained in the bath by continuous agitation using a stirring rate of 450 rpm. Pure Ni layers were also produced under the same experimental conditions for comparison.

The surface morphology and the elemental composition of the pure nickel and nanostructured Ni/TiC hybrid layers obtained electrochemically were characterized by scanning electron microscopy (SEM-Philips XL 30 FEG) with energy dispersive X-ray spectroscopy (EDX).

### *2.2. Nanohardness tests*
The pure Ni layer and nanostructured Ni/TiC hybrid layer were tested by nanoindentation tester (CSM-instruments equipped with a Berkovich diamond indenter) in order to extract elastic modulus (E), indentation hardness (H), Vickers nanohardness (Hv) and maximum depth penetration of the specimen material from load–penetration depth measurements. The applied loading force was 10 mN, with a 5 s pause time after each measurement.

### *2.3. Tribocorrosion test apparatus and solution*
Before the morphological, elemental characterization and also tribocorrosion tests, all the electrodeposited samples were ultrasonically cleaned in ethanol for 10 min and then in distilled water.

Experimental set-up for tribocorrosion experiments (figure 1) consisted in a conventional three-electrode cell composed of: working electrode – pure Ni layer and Ni/TiC nanostructured hybrid layers, reference electrode – Ag/AgCl microelectrode with KCl saturated solution (E = +200 mV vs. standard hydrogen electrode) and auxiliary electrode – Pt foil. The active surface area of the working electrode exposed to the solution was 1 $cm^2$.

The fretting tests were performed in wet conditions, at room temperature (23 °C) on reciprocating fretting motion using a ball-on-plate geometry by applying the normal forces of 1 N and 2 N, a frequency of 1 Hz, under linear displacement amplitude of 200 μm for 10000 numbers of fretting cycles. As a test solution in the tribocorrosion experiments was used an electrolyte which simulates the primary water circuit of Pressurized Water Reactors (PWRs) with the following composition 5.72 $gL^{-1}$ LiOH and 0.4485 $gL^{-1}$ $H_3BO_3$, having a pH of 8.3 [12]. The alumina balls with 10 mm diameter (CERATEC Technical Ceramics BV) were used as a counterbody during the tribocorrosion tests. The friction coefficients versus fretting distance were recorded automatically during the test.

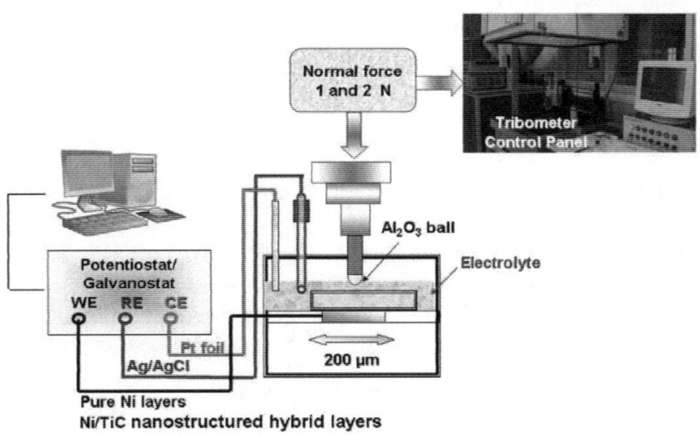


**Figure 1.** Schematic view of the experimental set-up used for tribocorrosion tests.

After the fretting tests, the debris and the wear track morphology was investigated by SEM analysis using a Philips model XL 30 FEG.

## 3. Results and discussion

### *3.1. Morphological and elemental characterization*
The nanocomposite layer with good adherence to stainless steel support, having a thickness of 10 micrometers and about 5.16 wt % of titanium carbide included into nickel matrix was characterized by SEM-EDX, nanoidentation and to tribocorrosion process in bidirectional fretting conditions in water +Li+B solution simulating the cooling water from PWR.

The surface morphology and chemical composition of Ni/nano-TiC nanocomposite layers are presented in figure 2 (b, c) comparatively with pure nickel layers obtained at same electrodeposition parameters, figure 2 (a, b). TiC nanoparticles are homogeneously dispersed in the Ni matrix. This means that nickel ions from the electrolyte are adsorbed on TiC nanoparticles and then are reduced on the TiC surface and their deposition on the surface is uniform. The SEM morphology of nanostructured Ni/TiC hybrid layers also reveals a cauliflower structure, comparatively with pure nickel electrodeposited layers which show defined grain crystallites.

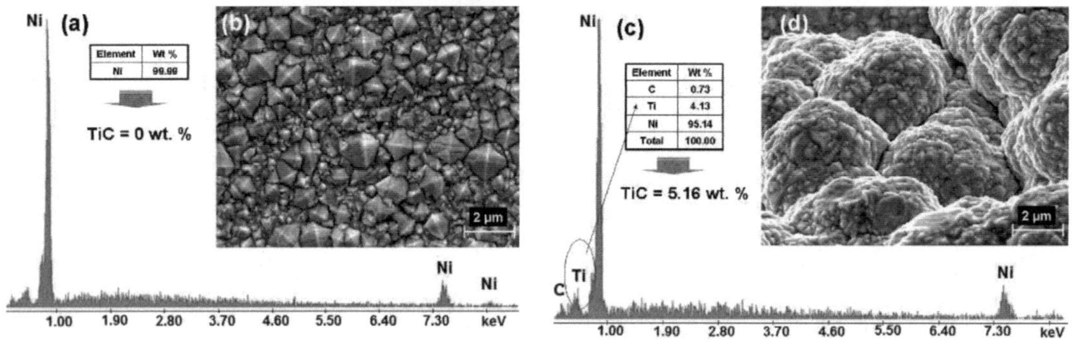


**Figure 2.** EDX spectrum and elemental analysis collected from the entire surface of: (a) pure Ni layer corresponding SEM micrograph (b) and (c) Ni/TiC nanostructured hybrid layers corresponding SEM micrograph (d).

The incorporation percentage of TiC into Ni matrix is determined by transformation of average Ti element wt % in molecular mass of TiC, considering the standard atomic weight of Ti element which is 47.86. The weight percentage for Ti element included into Ni matrix, as shown in figure 2c was determined by EDX analyses being equal to 4.13 wt % (mean value). Therefore the wt. % of TiC reinforcing nanostructured Ni/TiC hybrid layers resulted at an average value of 5.16, percentage that improved the properties of the nanostructured hybrid layer.

*3.2. Nanohardness properties*

The nanohardness properties of the pure Ni layers and nanostructured Ni/TiC hybrid layers were measured by nanoindentation technique using a diamond Berkovich indenter at normal load of 10 mN. The hardness values reported are the statistical average of 3 indentations. From load–penetration depth measurements were extracted and plotted in figure 3 the following parameters: indentation hardness (H), Vickers nanohardness (HV), elastic modulus (E) and maximum depth penetration of the specimen.

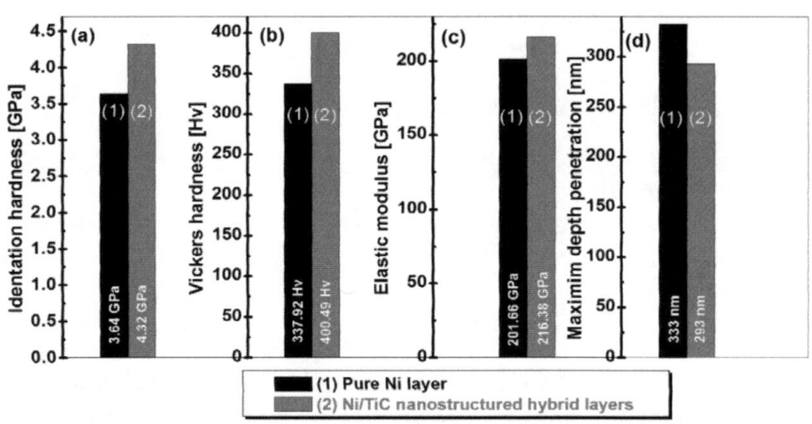


**Figure 3.** Nanoidentation results as (a) indentation hardness, (b) Vickers hardness, (c) Elastic modulus and (d) maximum depth penetration of (1) pure Ni layer and (2) Ni/TiC nanostructured hybrid layer.

IOP Conf. Series: Materials Science and Engineering **174** (2017) 012045 doi:10.1088/1757-899X/174/1/012045

As it is shown in figure 3, the higher nanohardness, Vickers hardness and elastic modulus values correspond to nanostructured Ni/TiC hybrid layers compared with pure Ni layers. Also it can be seen that the maximum depth penetration of the indenter corresponding to nanostructured Ni/TiC hybrid layers reveal a value of 293 nm, while for pure Ni layers show a value of 333 nm.

These results demonstrate the beneficial effect of TiC nanoparticles incorporation into Ni matrix. The inclusion of TiC nanoparticles provokes new nucleation sites on the surface and therefore the grain boundaries expand and obstruct the dislocation.

### *3.3. Tribocorrosion experiments*

#### *3.3.1 Open circuit potential*

The measurement of open circuit potential for materials and coatings gives information on the electrochemical state of a material or coating, for example if the material is in active state and dissolution occurs or if it is in passive state and the dissolution is slowed down by the passive film formed over the surface in the given solution. However, open circuit potential measurements provide limited information on the kinetic and mechanism of surface reactions occurring on the studied material. The open circuit potential recorded during bi-directional friction tests, in which the disc (Working Electrode) from figure 1 is the pure nickel or Ni/TiC nanostructured hybrid layer under investigation, becomes a mixed potential reflecting the combined state of the unaffected material surface and the material from the wear track. A galvanic coupling between rubbed and unrubbed parts on the layer surface may take place. The measured open circuit potential is then an average value depending on current density distribution over the whole layer surface exposed to corrosive environment.

The evolution of the open circuit potential was measured under free surface in the solution without friction and under friction mechanically loaded conditions with two applied normal forces as it is shown in figure 4a for pure nickel layer and figure 4b for Ni/TiC nanostructured hybrid layer.

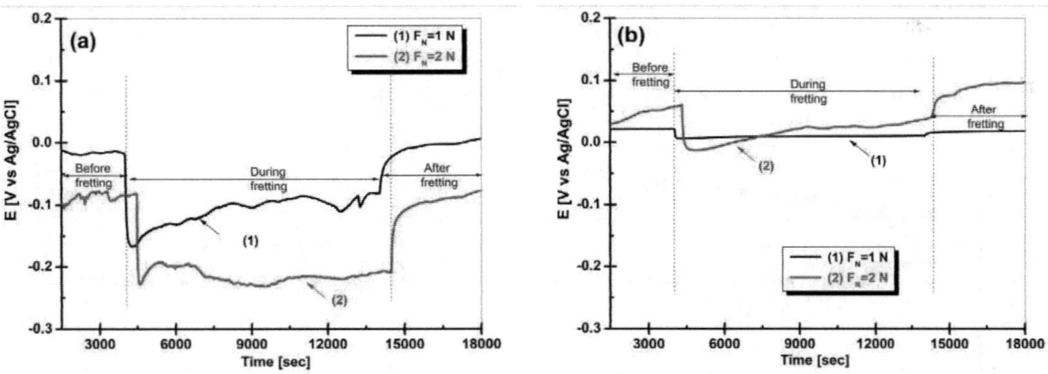


**Figure 4.** Electrochemical measurements of open circuit potential (E) recorded before, during, and after bi-didirectional fretting test in water +B+Li solution with two applied normal forces of 1 N and 2 N, 200 μm dispacement amplitude, 1 Hz fretting frequency and 10000 cycles for: (a) Pure Ni layer, (b) Ni/TiC nanostructured hybrid layer.

From figure 4a it is observed that the open circuit potential of pure nickel layer in fretting contacts is quite sensitive to the loading conditions in aqueous solution. The open circuit potential of pure nickel layer is shifted down during fretting with about 145 mV at a normal force of 1N and 152 mV at applied normal force of 2 N.

The open circuit potential of Ni/TiC nanostructured hybrid layer is not such sensitive to loading conditions and the shutting down of the potential during starting the fretting is only with few mV (0.015 mV) at 1N loading force and increase very slowly at 2 N loading force at about 0.072 mV, as it is shown in figure 4b.

Accordingly, the potential variations in Figs 4a and 4b suggest that the tested Ni/TiC nanostructured hybrid layers fretting against corundum under the considered test conditions mainly undergo a removal of their passive surface film at the start of friction, removal which is much more evident for pure nickel layer. The layers surface remains partly active during the fretting test, and finally progressively reactivates when fretting is stopped (unloading conditions).

This indicates the re-establishment of a passive state on the surface of Ni/TiC nanostructured hybrid layers and pure nickel layers in the worn area. It should be underlined that the rate at which the open circuit potential is shutting down when loading the normal force and increases on unloading conditions is totally different for the two types of layers. The underlying processes are thus different in both cases and have different mechanism and kinetics. Indeed on loading, the mechanical destruction of the passive surface film is taking place, while on unloading the repassivation takes place at a rather limited oxidation rate. After unloading, only a small part of the disc surface in the wear track is active, and a galvanic coupling takes place between this small rubbed active area and the rest of the disc surface, which remain in a passive state. This coupling induces an anodic polarization of the active area, resulting in an increase of the repassivation rate, and a smaller repassivation time constant for Ni/TiC nanostructured hybrid layers.

### *3.3.2. Friction coefficients*

Figures 5 (a and b) provides the coefficients of friction monitored in wet conditions as a function of normal forces of 1 N and 2 N, at the fretting frequency of 1 Hz, for number of cycles equal to 10000 and 200 μm displacement amplitude corresponding to pure Ni layer (5a) compared to Ni/TiC nanostructured hybrid layers (5b).

The figure 5 (a, b) keeps the same general behaviour, namely that the friction coefficients recorded under wet condition of Ni/TiC nanostructured hybrid layers are smaller than that of pure Ni layers for both applied normal forces. For instance, the average value of friction coefficient for Ni/TiC nanostructured hybrid layers at the normal force of 2 N is equal to 0.19 and with 0.28 respectively for pure Ni layer. This could be attributed to the incorporation of the hard TiC nanoparticles into nickel matrix which impedes the damage of the nickel oxides passive film formed on pure Ni layer. Similar observation was made previously by Lekka et al. [13], who compared the performance of pure nickel and Ni–SiC nano-structured composite coatings as far as corrosion, wear and abrasion resistance.

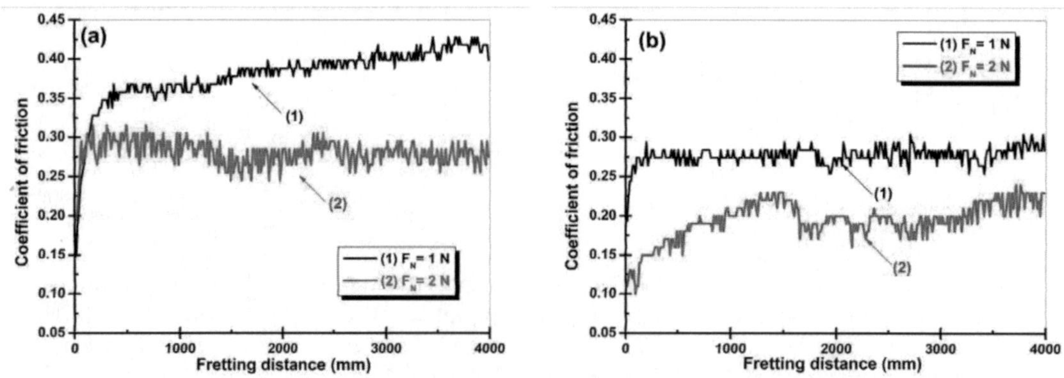


**Figure 5.** Evolution of friction coefficients during bi-bidirectional fretting test in water +B+Li solution with two applied normal forces of 1 N and 2 N, 200 μm dispacement amplitude, 1 Hz fretting frequency and 10000 cycles for: (a) Pure Ni layer, (b) Ni/TiC nanostructured hybrid layer.

### *3.3.3. Wear tracks analysis*

In figure 6 are shown the wear tracks corresponding to pure Ni layer (a, c) and Ni/TiC nanostructured hybrid layer (b, d) at different magnifications view. The wear track of Ni/TiC nanostructured hybrid layer (figure 6b) is significantly smaller compared with that of pure Ni layer (figure 6a). From figure 6d it can be seen that there are no cracks on the wear track of Ni/TiC nanostructured hybrid layer like in the case of pure Ni layer (figure 6c), rather occurred only the flattening of the surface roughness.

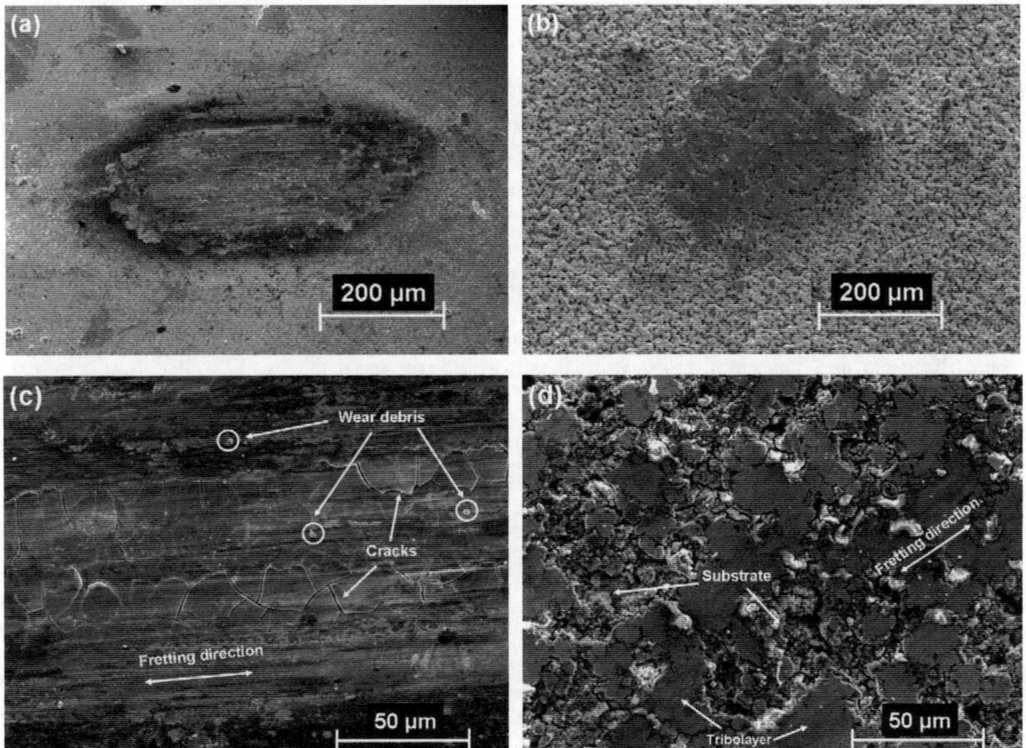


**Figure 6.** SEM worn analysis corresponding to pure Ni layer (a, c) and Ni/TiC nanostructured hybrid layer (b, d) of the entire wear tracks (a, b) and a zoom inside the wear tracks (c, d).

The magnified view of the wear track corresponding to pure Ni layer (figure 6c) reveals the severe deterioration suffered, being obvious small cracks and wear debris caused by the ploughing action of $Al_2O_3$ ball.

From the worn surface analyses it can be concluded that TiC nanoparticles have a good effect on Ni/TiC nanostructured hybrid layer revealing improved resistance to wear in wet conditions as compared with pure nickel layer.

## 4. Conclusions

The TiC nanoparticles could be electro-codeposited with nickel to reinforce the obtained Ni/TiC nanostructured hybrid layer. The inclusion of TiC nanoparticles into nickel matrix during electroplating process was proved by SEM - EDX analysis. The mean wt. % of TiC obtained was 5.16.

By comparatively testing the Ni/TiC nanostructured hybrid layer and pure nickel layers by nanoidentation using 10 mN force the results show higher values for nanohardness, Vickers hardness and elastic modulus induced by TiC nanoparticles reinforcing nickel matrix.

The tribocorrosion behaviour of tested layers in simulating cooling water from nuclear power plant is dependent of surface characteristics.

The open circuit potential is much less influenced by fretting of Ni/TiC nanostructured hybrid layer as compared with pure nickel layer due to higher hardness and formation of a more compact passive protective layer.

The values of friction coefficients recorded in wet conditions also reflect a better behaviour of Ni/TiC nanostructured hybrid layer being smaller at both applied normal forces compared with pure Ni layer.

**Acknowledgements**
UEFISCDI – Ministry of Education and Research is acknowledged for the financial support to Competences Centre Interfaces – Tribocorrosion and Electrochemical Systems (CC–ITES) – Dunarea de Jos University of Galati – Research Project: HyBioElect, contract 10 /30–08–2013 (2013 – 2016) in the frame of National Research Programme Romania – PN II PCE.

**References**

[1] Sahoo P and Das S K 2011 *Mater. Design* **32** 1760
[2] Vaezi M R, Sadrnezhaad S K and Nikzad L 2008 *Colloid. Surface A: Physicochem. Eng. Aspects* **315** (2008) 176-182
[3] Zhou Y, Zhang H and Qian B 2007 *Appl. Surf. Sci.* **253** 8335
[4] Benea L, Wenger F, Ponthiaux P and Celis J P 2009 *Wear* **266** 398
[5] Sun X J and Li J G 2007 *Tribol. Lett.* **28** 223
[6] Baghery P, Farzam M, Mousavi A B and Hosseini M 2010 *Surf. Coat. Tech.* **204** 3804
[7] Li Q, Yang X, Zhang L, Wang J and Chen B 2009 *J. Alloy. Compd.* **482** 339
[8] Carpenter C R, Shipway P H and Zhu Y 2011 *Wear* **271** 2100
[9] Ghosh S K and Celis J P 2013 *Tribol. Int.* **68** 11
[10] Waterhouse R B 1992 *ASM Handbook, Friction, Lubrication, and Wear Technology* vol 18 (ASM International) chapter Fretting Wear pp 449-481
[11] Shaffer S J and Glaeser W A 1996 *ASM Handbook, Fatigue and Fracture* vol 19 (ASM International) chapter Fretting Fatigue pp 801-824
[12] Benea L, Basa S B, Danaila E, Caron N, Raquet O, Ponthiaux P and Celis J P 2015 *Mater. Design* **65** 550
[13] Lekka M, Kouloumbi N, Gajo M and Bonora P L 2005 *Electrochim. Acta* **50** 4551

# Numerical investigation of the effect of friction conditions to increase die life

**M O Mutlu, C G Guleryuz and Z Parlar**

Dept. of Mechanical Engineering, Istanbul Technical University, Gumussuyu, 34437, Istanbul, TURKEY

**Email:** parlarze@itu.edu.tr

**Abstract**. The standard die materials in aluminium extrusion offer good mechanical properties like high tempering resistance, high strength and ductility. On the other hand, they struggle with the problem of sliding wear. As a result, there is a growing interest in using surface treatment techniques to increase the wear resistance of extrusion dies. In this study, it is aimed to observe the effects of the different friction conditions on material flow and contact pressure in extrusion process. These friction conditions can be obtained with the application of a variety of surface treatment. In this way, it is expected to decrease the friction force on the die bearing area and to increase the homogeneity of the material flow which will result in the increase of the quality of the extrudate as well as the improvement of the process economically by extending die life. For this purpose, an extrusion process is simulated with a finite element software. A die made of 1.2344 hot work tool steel-commonly used die material for aluminium extrusion process- has been modelled and Al 1100 alloy used as billet material. Various friction factor values defined on the die surface under the same process parameters and effects of changing frictional conditions on the die and the extrusion process have been discussed.

## 1. Introduction

During the extrusion process high pressures and high relative velocity of extrudates are reached. Wear types or wear mechanisms generally depend on the relative sliding velocity, normal contact pressure, materials of rubbing bodies and the environment [1]. As a result of high stresses of the process, tools and especially die and bearing area of the die are exposed to excessive wear and fatigue. Ma et al. states that, in Al extrusion process, friction inside the bearing channel is highly significant to control the surface quality of the final product. Besides, with a high contact pressure, a fully plastic contact is observed between die bearing area and aluminum workpiece, which cause wear [2]. Thus, increasing the tool service life is important for continuity of high quality production by achieving toughness, hardness, wear resistance at elevated temperatures expected from the tool material.

Beyond the good mechanical properties of die materials, nitriding method has been used commonly to increase wear resistance and also complex coatings have been investigated to enhance performance of dies. However, since the most coatings and nitriding forms brittle compounds, and also open to deformation in the substrate, the efficiency varies according to the coating material, the removal rate of the nitrided die material and so on. Therefore, Bakhtiani, Mounayri and Zhang investigated the wear distribution on an extrusion die for two following cases using numerical simulations: without any surface coating and with a bilayer (TiCN + $Al_2O_3$) CVD coating. Wear simulations show that the coated die had a higher resistance to temperature and wear than an uncoated one [3]. Another study of PVD coatings on extrusion dies by Lukaszkowicz et al. presents the research results of CrAlSiN and CrN coatings on an extrusion die. Both coatings showed high potential to extend die life in several orders of magnitude compared to uncoated steel. The lowest wear was obtained with CrN coated dies [4]. In a similar study with various different coatings of Björk T. et al., bearing surfaces of extrusion dies that will work against aluminum under high temperature conditions were coated with TiN, CrN, (Ti,Al)N or $TiB_2$. The high potential of reducing the extrusion die wear of CrN and especially multilayered-(Ti,Al)N and $TiB_2$ coatings have been shown [5]. On the other hand, Birol also studied to test the performances of CrN, AlCrN and AlTiN coated hot work tool steel samples under actual aluminum extrusion conditions. It has been stated that the adhesive interaction of the coating with aluminum is the primary wear mechanism in CAPVD coated hot work tools. While CrN coating has shown the most extensive wear damage, AlTiN provided the best performance among CrN and AlCrN in sliding contact against aluminum in terms of friction, resistance against aluminum transfer and wear damage [6]. Also another investigation of Björk et al. concentrating on the wear of the surface treated dies with PVD coatings for aluminum extrusion stated that duplex coatings-plasma nitriding then PVD coating with CrN-can increase tool life at least several orders of magnitude compared to nitriding [7].

As the coating methods have been applied to extend the service life of manufacturing tools used under harsh conditions such as high local stresses, high temperature or cycling loading, the effect of surface roughness and geometric features of the surface has been neglected until recent years. Other than nitriding and surface coating methods, surface texturing and patterning methods have been widely studied these days. In this study, the effect of enhanced friction conditions provided by surface texturing on the extrusion process has been investigated.

Features in the applied surface texture on a surface of a substrate can be different in size or shape but the main purpose to apply these features is to create a self-lubricated surface on the die surface by generating micro-traps for wear debris and micro-reservoirs for both solid and liquid lubricants to make them remain within these micro-cavities. In the view of such information, surface texturing methods and various patterns have been investigated to improve the service life of extrusion dies and prevent the possible extrusion defects.

Guleryuz et al. have studied surface texturing of hard coatings for self-lubrication and their applications on machining tools. For this purpose, composite coatings have been developed and surface pattern has been fabricated onto machining tool inserts to sustain lubrication during interaction. In conclusion, it is shown that these surface texturing methods can be applied to manufacturing tools and promising results have been obtained for machining tool inserts [8]. Pettersson et al. developed an original technique to manufacture embossing tool for texturing metallic surfaces. First, silicon etching technology has been used to form surface indentations. A diamond film deposited into these indentations with CVD method and the diamond film was supported by a layer of nickel by electro-deposition. Finally, when the silicon was removed by etching, diamond protrusions were generated. The embossing tool is capable to apply texture on flat and curved steel surfaces [9]. Costa and Hutchings investigated lubrication using surface patterns in drawing process. In this study, circular-shaped pockets and parallel grooves have been used for patterning of the drawing die and the friction forces were compared. It has been observed that the effect on lubrication of the circular pockets is little. However, groove shaped pattern covering a larger area decreased the strip forces and friction relatively, but its performance was dependent on the relative orientation of the grooves, as the grooves parallel to the drawing direction cause to escape of lubricants and result in poorer surface finish on the drawn product [10]. A similar study has been carried out by Suh et al. on the shape of grooves but this time by means of the application of the cross-hatched pattern with different angle orientations and aspect ratios under mixed and elastohydrodynamic lubrication. Their study has showed that decrease of aspect ratio and increase of groove length cause reduction in friction [11].Tang et al. investigated the effect of different area fractions of dimples with liquid lubrication on friction, wear and load capacity. As a result, an optimum area coverage of 5 % of dimples has showed the best performance by providing retention of the lubricant and the generation of hydrodynamic pressure, also by trapping the wear debris [12]. Jiwang et al. studied a method combining micro-indentation and ultra-precision cutting to manufacture micro dimples on the surface [13]. Jianliang et al. studied tribological properties of laser surface texturing and molybdenizing (duplex treating) of stainless steel at elevated temperatures. Surface pattern has been implemented by laser processing and $MoS_2$ was used as a lubricant. Samples with and without molybdenizing have been compared. After molybdenizing, the hardness of stainless steel increased and the friction coefficient of stainless steel covered with $MoS_2$ powder decreased from 0.4 to 0.1 at elevated temperatures [14]. In their study Basnyat et al. have also used $MoS_2$ as a solid lubricant on surface of textured TiAlCN coated metallic surfaces. Dimples on TiAlCN coating have been introduced using reactive ion etching in a mixed $Ar/CF_4$ plasma. An overlayer of $MoS_2$ or $Mo/MoS_2/Ag$ solid lubricants deposited on micro-textured surfaces by magnetron sputtering. The frictional and wear properties are examined for textured and untextured surfaces. As a result, a significant decrease in friction and wear is achieved at both room temperatures and elevated temperatures for textured coatings as the dimples acted as reservoirs which provides a new supply for solid lubricants on the contact surface. [15]. Shipeng Li et al. investigated tribological performance of TiSiN-$WS_2$ composite coatings deposited on WC/TiC/Co cemented carbide. It is shown that the deposition of $WS_2$ layer on top of TiSiN layer significantly improved the tribological performance of the structure providing lower friction coefficient values and smaller wear rates of the steel ball [16].

In this study, it is aimed to investigate the effect of friction conditions on material flow and contact pressure in extrusion process. Here, most of the recently developed or studied surface modification techniques covered above may be exerted to die surfaces. Most of these techniques promise to decrease friction coefficients and/or to improve die life. The finite element simulation was performed for the extrusion process and the numerical analysis around the bearing area have been performed for this purpose.

## 2. 3D modeling

A circular die cavity is relatively easy to create and it is sufficient to observe the effect of the different friction conditions by eliminating other effects that would be interfere with our results when complex part cross-sections used. Due to these reasons, an extrusion die with a circular die opening was selected.

The main objective of this research is to observe the effects of the different friction conditions. These friction conditions can be obtained with the application of a variety of surface modification methods covered above used together with liquid and/or solid lubricants. Most critical section of an extrusion die is the bearing area, therefore we have concentrated our research on this area.

The finite element software, SimuFact forming, developed by MSC Software Company, has been used to numerically model the extrusion process. Extrusion and die cavity are shown in figure 1 given below. Since friction at the bearing area is important for the research, a simple extrusion die tooling has been designed with a 6-mm-length-bearing-area. In the numerical simulation, the pressure is considered constant during the process like in the case of the real process using an hydraulic press. Simulation model consists of a container, a billet and a support to fix the extrusion die as shown in figure 1 (a). Extrusion profile is chosen circular to model under the condition of axisymmetrical modeling. By means of axisymmetrical model, the computation time is decreased significantly.

The billet has been defined as a deformable body whereas other setup parts were defined as rigid objects. Die material defined as 1.2344 hot work tool steel with a hardness of 50 HRC as it is generally used in industry in Al extrusion process. Billet material used was Al coded as 1100 with a diameter of 25 mm and a length of 40 mm. This material was selected due to its sticky behavior during the extrusion process. Table 1. shows the mechanical and physical properties of the tool material (1.2344 hot work tool steel) and billet material (Al 1100) of the simulations.

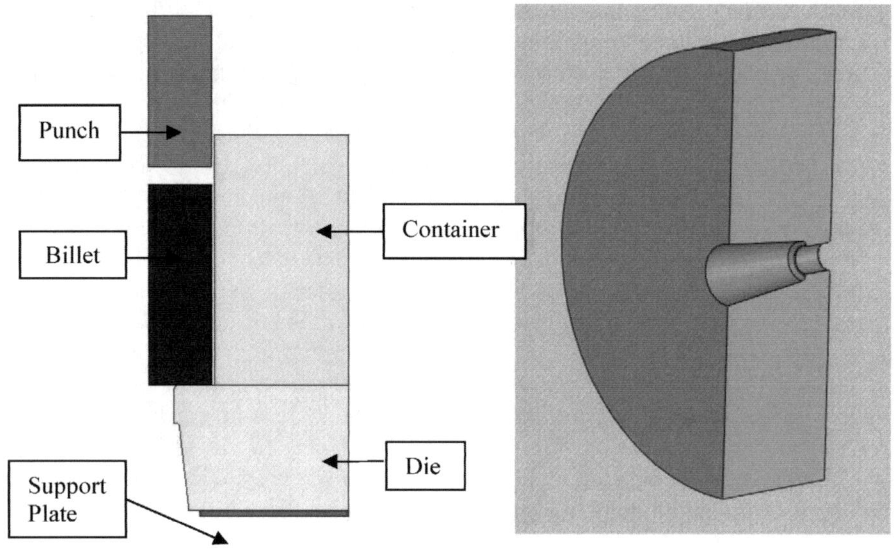


(a) (b)

**Figure 1.** (a) Model used in FE Simulations, (b) 3D Solid Model of extrusion die and die cavity.

**Table 1.** Material data used in simulations.

| Mechanical/physical property | Unit | 1.2344 hot work steel | Al 1100 |
|---|---|---|---|
| Young's modulus | GPa | 215 | 70 |
| Poisson's ratio | | 0,30 | 0,33 |
| Thermal expansion coefficient | 1/ºC | $10.4 \times 10^{-6}$ | $23.6 \times 10^{-6}$ |
| Thermal conductivity | W/(m.K) | 28,6 | 218 |
| Density | $kg/dm^3$ | 7,80 | 2,71 |

To define the die interface correctly, quadtree elements used in the die model. Because of the intense deformation during the extrusion process, finer advancing front quadrilateral elements used in billet model as shown in Figure 2. Friction behavior of billet material, Al 1100, was simulated with friction factors (coefficient of friction values) of 0.05, 0.2, 0.4, 0.7 corresponding various possible friction conditions. In metal forming, well-lubricated processes such as wire-drawing and deep-drawing, a friction coefficient (friction factor) in a range of 0.05 to 0.15 is very common and severe processes such as hot rolling and forging may have friction factor values up to 0.4 [17]. Also, lowest end of this friction factor value range may be represented application of a lamellar solid lubricant such as $MoS_2$ or graphite [18]. Highest end of the friction factor range shows the surface coverage of the liquid and/or solid lubricant has been mostly lost and partial-sticking conditions may be observed. In order to observe effects of different possible friction factors of the die bearing area on extrusion process, billet interface friction was assumed 1 as pure sticking condition (no lubrication assumed) in all scenarios. In the numerical simulation, the pressure is considered constant during the process like in the case of the real process using an hydraulic press and ram speed was defined 2 mm/s with a maximum stroke of 30 mm. Initial temperatures of the container, the billet and the extrusion die were defined as 25 °C.

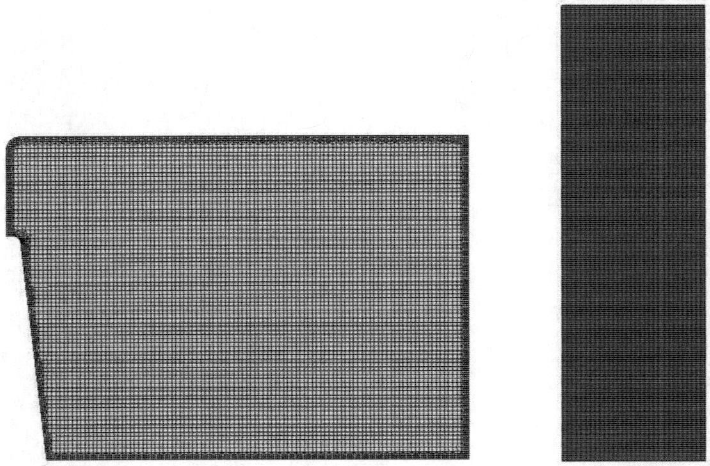

(a) (b)

**Figure 2.** (a) Mesh of the die model and (b) billet model.

## 4. Results and Discussions

Axisymmetric model of deformed billet during extrusion process is shown in figure 3 and figure 4 given below. The thick section is the initial billet that is still preserved in the extrusion container and the thin section of it is the extruded part of the Al 1100 billet.

Contact pressures at the billet-container interface and billet-die interface can be seen in Figure 3. f is the friction factor and it is defined in the range of 0.05-0.7. The highest pressure values have been observed on the container wall, then on the die surface, and on the entering section of the bearing area respectively. As the process quality is especially bound to the die bearing area, the highest pressure values concentrated on the entrance of the die bearing area has been considered. In figure 3 (a), at the entrance of the bearing area, contact pressure starts from a mean value of 730 MPa and it decreases to around 500 MPa further along the bearing area. In figure 3 (b), with a friction factor of 0.4, contact pressure on the die bearing area starts from a mean value of 610 MPa and it decreases to around 280 MPa, for figure 3 (c), with a friction factor of 0.2, it varies from 500 MPa to 280 MPa through the bearing area and pressure drops earlier than figure 3 (b) in start of bearing area, for figure 3 (d) with a friction factor of 0.05, the contact pressure for the entrance is 400 MPa and it quickly decreases to 270 MPa.

Björk et.al. have mentioned in their study on the wear surface treated dies for aluminum extrusion that wear reaches critical rates especially in entrance of the bearing area [19]. This comment is consistent with the simulation results given in figure 3. showing that high pressure values concentrated at this area of the bearing. Also Bombac et al. has stated in their study that wear occurs at the exit of the bearing area due to high contact pressures at this area [20]. As seen in figure 3, contact pressure is concentrated towards to exit edge of the bearing area in (b), (c) and (d) given for a friction factor of 0.7, 0.4, 0.2 and 0.05 respectively. And for these simulations, pressure values at this area decreased by decreasing friction. With decreasing friction factor values, contact pressure decreases, too, and therefore, wear rates of extrusion die are expected to decrease as wear is also related to the load values on surface.

Recent studies show that hard coatings with micro-reservoirs filled with solid or liquid lubricants can provide decreases in friction factor values. The increased effectiveness of the solid lubricants when applied with micro-reservoirs in the hard coatings have been shown in various studies [15,21]. In their work, Zimmerman et.al. have applied carbon coating on the hard coating with micro-reservoirs and achieved friction coefficient values of about 0.1 against aluminum counterface for about 8,000 cycles in pin-on-disk tests. It is observed that no such decrease is possible without surface texture [21].

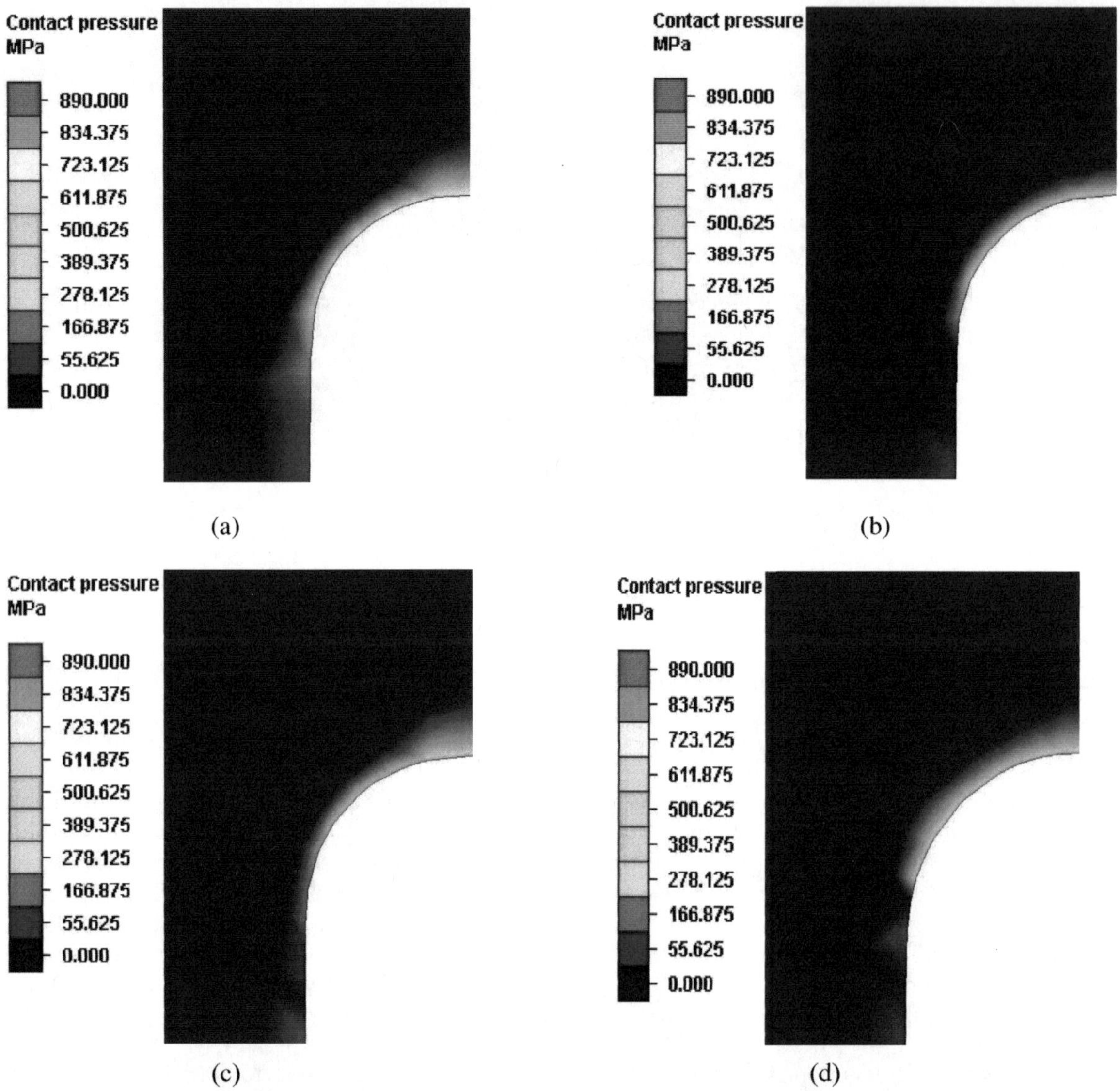


**Figure 3.** Contact pressure distribution at the interface material-die of the FE simulations of directly extruded Al part with friction factors of: (a) f=0.7; (b) f=0.4; (c) f=0.2 and (d) f=0.05.

Figure 4 shows flow rate of the 1100 Al alloy during the extrusion process. Dead metal zone is validated in the simulations as mentioned in Ouwerkerk's study [22]. As seen from the flow rates of material in figure 4, the maximum flow rate of material decreases as the friction decreases. This happens because the friction at the walls that holds the material locally. As a result, the material in the center section of extrudate have higher flow rate. As the friction value on the die section decreases, material flow rate difference between the interface and the center section of the material decreases. So decrease of friction reduce sticking effect of the material on the die bearing surface locally, extrude material flows more homogeneously and this decreases the possible defects in the extruded profile.

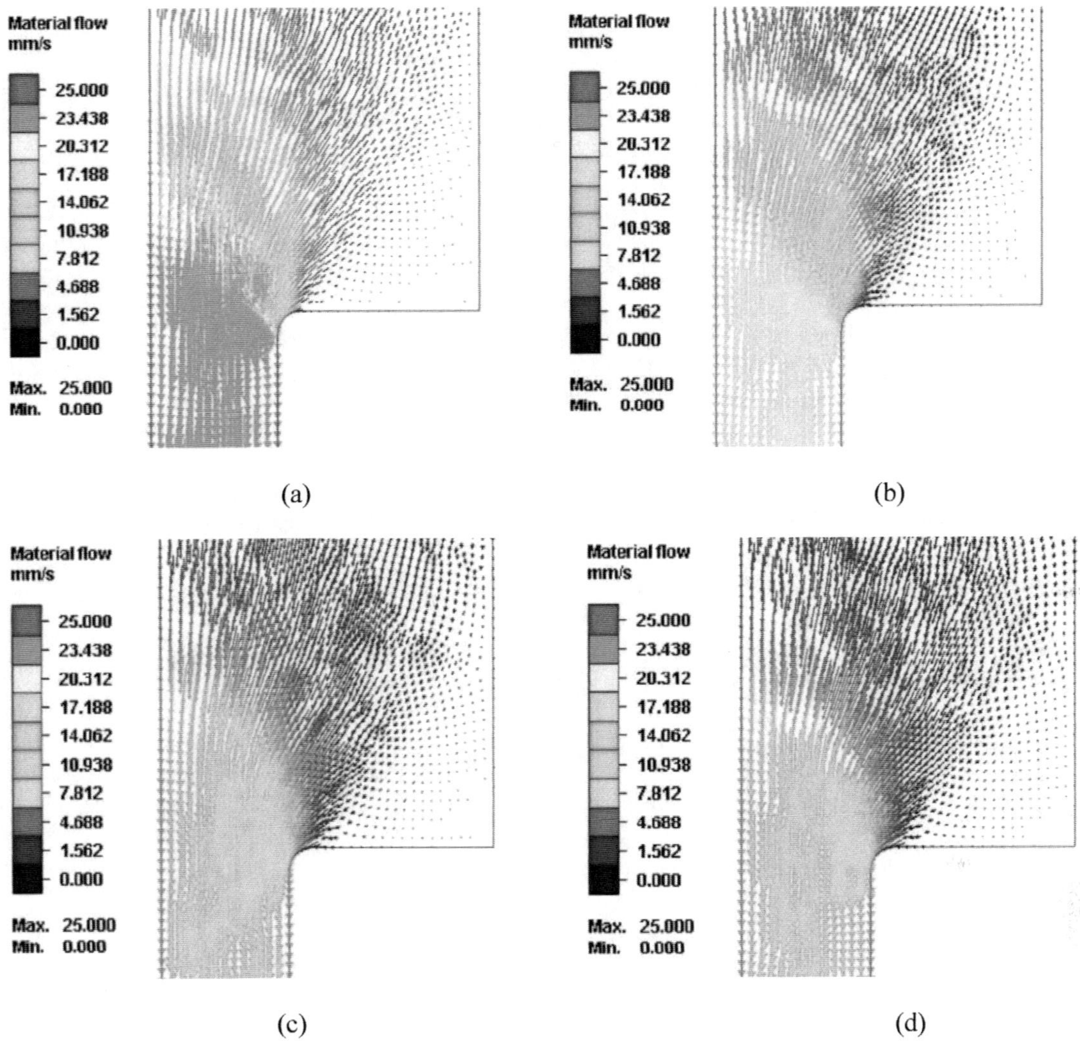


**Figure 4.** Material Flow rate distribution with respect to friction factors: (a) f=0.7; (b) f=0.4; (c) f=0.2 and (d) f=0.05.

## 3. Conclusions

An extrusion process has been modeled for a product with a simple geometry using a die with a circular cross-section. In the model, die-workpiece surface interaction has been taken into account using different friction factor values. Contact pressure especially at the bearing area is important because high contact pressures increase die wear.

In this study, contact pressure change has been investigated for various friction factor values in the range of 0.05-0.7. As the friction decreases, contact pressure values around the die section decreases substantially. Also extrusion process has been investigated by observing the material flow change with respect to the friction values within the same range. With the decreasing friction factor values, material flow rate difference between the surface and the center section decreased. Minimum flow rate difference will decrease possible defects that may occur in the extrudate as the material flow is more homogeneous. According to the results of the finite element simulations performed in the scope of this study, the most appropiate surface condition of the die is the one with a friction factor of 0.05 for 1100 aluminum alloy. It is the optimum value among the other friction values that will prevent inhomogeneous deformation and provide the minimum die wear.

## References

[1] So H, Chen H M and Chen L W 2008 Extrusion wear and transition of wear mechanisms of steel *Wear* **265** 1142-48
[2] Ma X, de Rooij M B and Schipper D J 2012 Friction conditions in the bearing area of an aluminium extrusion process *Wear* **278-279** 1-8
[3] Bakhtiani T, El-Mounayri H and Zhang J 2014 Numerical simulation of aluminum extrusion using coated die *Mater. Today* **1** 94-106
[4] Lukaszkowicz K, Dobrzanski L A, Kokot G and Ostachowski P 2012 Characterization and properties of pvd coatings applied to extrusion dies *Vacuum* **86** 2082-88
[5] Björk T, Berger M, Westguard R, Hogmark S and Bergström J 2001 New physical vapour deposition coatings applied to extrusion dies *Surf. Coat. Tech.* **146-147** 33-41
[6] Birol Y 2013 Sliding wear of CrN, AlCrN and AlTiN coated AISI H13 hot work tool steels in aluminium extrusion *Tribol. Int.* **57** 101-106
[7] Björk T, Westergard R and Hogmark S 2001 Wear of surface treated dies for aluminium extrusion – a case study *Wear* **249** 316-323
[8] Guleryuz C G, Krzanowski J E, Veldhuis S C and Fox-Rabinovich G S 2009 Machining performance of TiN coatings incorporating indium as a solid lubricant *Surf. Coat. Tech.* **203** 3370-76
[9] Pettersson U and Jacobson S 2006 Tribological texturing of steel surfaces with a novel diamond embossing tool technique *Tribol Int* **39** 695-700
[10] Costa H L and Hutchings I M 2009 Effects of die surface patterning on lubrication in strip drawing *J. Mater. Process. Tech.* **209** 1175-80
[11] Suh M, Chae Y, Kim S, Hinoki T and Kohyama A 2010 Effect of geometric parameters in micro-grooved crosshatch pattern under lubricated sliding friction *Tribol. Int.* **43** 1508-17
[12] Tang W Zhou Y Zhu H and Yang H 2013 The effect of surface texturing on reducing the friction and wear of steel under lubricated sliding contact *Appl. Surf. Sci.* **273** 199-204
[13] Yan J, Horikoshi A, Kuriyagawa T and Fukushima Y 2012 Manufacturing structured surface by combining microindentation and ultraprecision cutting *CIRP Journal of Manufacturing Science and Technology* **5** 41-47
[14] Jianliang L, Dangsheng X, Hongyan W, Yongkang Z and Yongkun Q 2013 Tribological properties of laser surface texturing and molybdenizing duplex-treated stainless steel at elevated temperatures *Surf. Coat. Tech.* **228** 219-223
[15] Basnyat P, Luster B, Muratore C, Voevodin A A, Haasch R, Zakeri R, Kohli P and Aouadi S M 2008 Surface texturing for adaptive solid lubrication *Surf. Coat. Tech.* **203** 73-79
[16] Li S, Deng J, Yan G, Zhang K and Zhang G 2014 Microstructure, mechanical properties and tribological performance of TiSiN-$WS_2$ hard-lubricant coatings *App. Surf. Sci.* **309** 209-217
[17] Button S T 2013 Tribology in metal forming processes Ch. 3 in *Tribology in Manufacturing Technology* Ed J P Davim (Berlin Heidelberg: Springer) p 106
[18] Bhushan B 2002 *Introduction to Tribology* (New York: John Wiley and Sons Inc.) ch. 3 p 255
[19] Björk T, Westergard R and Hogmark S 2001 Wear of surface treated dies for aluminium extrusion – a case study *Wear* **249** 316-323
[20] Bombač D, Terčelj M, Peruš I and Fajfar P 2013 The progress of degradation on the bearing surfaces of nitride dies for aluminium hot extrusion with two different relative lengths of bearing surface *Wear* **307** 10-21
[21] Zimmerman J H, Guleryuz C G, Krzanowski J E Fabrication and tribological properties of titanium nitride coatings incorporating solid lubricant microreserviors *Surf. Coat. Tech.* **202** 2023-32
[22] van Ouwerkerk G 2009 *CAD implementation of design rules for aluminum extrusion dies* Doctoral dissertation (PhD thesis) University of Twente

IOP Conf. Series: Materials Science and Engineering **174** (2017) 012047 doi:10.1088/1757-899X/174/1/012047

# Tribo-thermal fatigue of the steel used for the forging die construction

**A C Drumeanu**

Petroleum-Gas University of Ploiesti, Bd. Bucuresti, no. 39, 100690, Ploiesti, Romania

E-mail: drumeanu@upg-ploiesti.ro

**Abstract**. Frequently the durability of the forging dies is firstly determined by the non-isothermal fatigue wear, which causes the cracks appearance on their internal surfaces, much more before their abrasion wear to reach the limit value. In these conditions it is necessary to design the forging dies firstly by the point of view of the non-isothermal fatigue wear. For a correctly choosing and using of metallic material, it is necessary to determine their intrinsic characteristics regarding its cyclic non-isothermal stresses durability. The experimental determination of these characteristics implies a lot of experiments, which are done in specific conditions, different from those used for isothermal mechanical fatigue durability determination. The paper presents the experimental results concerning intrinsic characteristic determination of the forging dies steel. Based on these results there were determined specific equations which characterize this kind of stresses, and the diagrams that represent their graphic image. These data can be used both in designing and exploitation of the forging dies.

## 1. Introduction

Non-isothermal fatigue wear is a kind of degradation, which characterizes the metallic element friction surface of some couples, like forging dies, fire guns, heavy-duty mechanical brakes etc. In case of forging dies the thermal stresses and strains, which appear in the friction surface adjacent zone of these couples, have high values, which are higher than the yield limits of the metallic materials used at their construction [1, 2, 3].

During the drop forging process the die surfaces are heated. Because this heating has a variable or almost cyclic character, the thermal stresses and strains effects are similar to that, which characterize the low cycle fatigue. These effects consist in the cracks appearance on the friction surface. In time, the number and the size of these cracks increase, and the result is the attaining of the fatigue fracture state.

Frequently the durability of the forging dies is firstly determined by the non-isothermal fatigue wear, which causes the cracks appearance on their internal surfaces, much more before their abrasion wear to reach the limit value. In these conditions it is necessary to design the forging dies firstly from the point of view of the non-isothermal fatigue wear.

For a correctly choosing and using of metallic material, it is necessary to determine its intrinsic characteristics regarding its cyclic non-isothermal stresses durability. The experimental determination

IOP Conf. Series: Materials Science and Engineering **174** (2017) 012047 doi:10.1088/1757-899X/174/1/012047

of these characteristics implies a lot of experiments, which are done in specific conditions, different from those used for isothermal mechanical fatigue durability determination [4, 5, 6].

It has to be mentioned, that up to present, most of the researches in the thermal and non-isothermal fatigue field were done for the metallic materials like heat-resisting alloys and steels used for the construction of steam-turbines, turbo-jet engines, nuclear equipment, chemical reactors etc. [5].

Also, there are many structures non-isothermal stressed where there are used for construction middle alloy steels. The non-isothermal fatigue durability of these sorts of steels has been relatively little studied till now [7, 9, 10].

## 2. Experimental conditions

Usually, both isothermal and non-isothermal fatigues of the metallic materials characterize their durability in oligocyclic range (low durability) [4, 8, 11]. The isothermal durability determination in this range presumes the cyclic axial stress in an elastoplastic regime when the maximum and minimum temperature values are maintained constants.

Unlike the isothermal fatigue experiments, which are developed under constant amplitude, the non-isothermal fatigue tests present the strain amplitude variation in time. In the first part of the test, the total elastoplastic strain varies until the hysteresis cycle stabilization, and finally it decreases because of the material hardening state appearance. The cylindrical test pieces are recommended for tests with the strain amplitude up to 2 %. Over this value there are recommended the toroidal test pieces [4, 7, 8].

The experiments were developed on a testing stand, specially built for non-isothermal fatigue of the metallic materials [7].

The sample pieces heating on this stand is carried out through thermal effect of electric current. The testing stand can be used for determining the durability characteristics for the metallic materials tested in the following conditions:

- thermal cycles with the test piece strained with tensile stresses at the maximum temperature of the cycle;
- thermal cycles with the test piece strained with compression stresses at the maximum temperature of the cycle (by its fastening in a frame with determined rigidity).

The concrete conditions of the tests were the following:

- test pieces were tested in non-isothermal regime between a minimum and a maximum temperature values;
- test pieces were fastened by enclosure and they were stressed to compression at the maximum temperature of the cycle;
- elastoplastic strain variation was limited using three rigidity steps of the test piece fastened system (12, 28, 55 MN/m);
- maximum cycle temperature values were 600°, 700°C respectively 800°C in accordance with the maximum temperature values which are reached during the forging process;
- minimum cycle temperature value was 100°C;
- average heating speeds of the test piece were in the range of 35…80 °C/s;
- average cooling speeds of the test piece were in the range of 5.5…7.0 °C/s;
- absence of the keeping time at the maximum temperature of the cycle;
- the using of the full test pieces with 4 mm diameter and 30 mm length of the calibrated part (figure 1);
- the criterion of the testing break was the decrease of the initial strain amplitude value with 10 %.

Experimental determinations were carried out on middle alloyed steel which is frequently used for forging dies construction. Two kinds of sample pieces were used: one type made from normalized steel and the other made from temper hardening steel. Chemical composition and main mechanical characteristics of tested steel are presented in table 1 respectively in table 2.

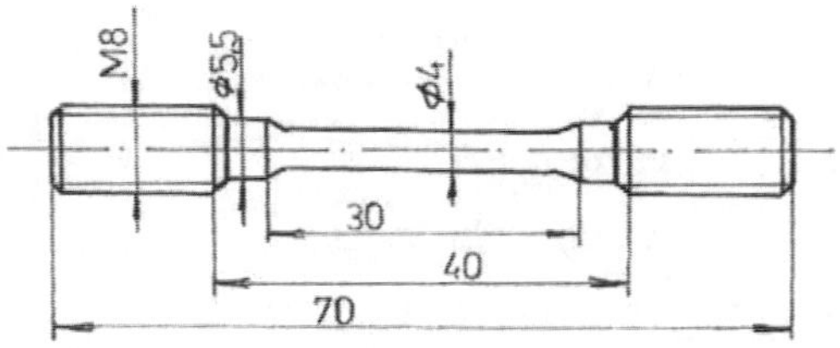


**Figure 1.** Geometry of the sample.

**Table 1.** Chemical composition of tested steel.

| Chemical alloying elements [wt. %] | | | | |
|---|---|---|---|---|
| C | Mn | Cr | Ni | Mo |
| 0.34 | 0.55 | 1.55 | 1.60 | 0.23 |

**Table 2.** Mechanical characteristics of tested steel.

| Mechanical characteristic | Values | |
|---|---|---|
| | Normalized steel | Temper hardening steel |
| Tensile strength [MPa] | 900…1100 | 1200…1400 |
| Yield limit [MPa] | 700 | 1000 |
| Elongation [%] | 12 | 9 |
| Reduction of area [%] | 48 | 40 |
| Hardness, HB | 225 | 248 |

**3. Results and discussions**

A primary evaluation with comparative character of non-isothermal cyclic durability can be done using the following dependence [4]:

$$N = A \cdot t_{\max}{}^{b} \tag{1}$$

where $N$ is the durability, expressed like the number of cycles until testing break; $t_{max}$ - the maximum temperature of the cycle; $A$ and $b$ – material constants.

The dependence expressed by relation (1) is in accordance with experimental results presented in figure 2. The $A$ and $b$ constants were determined as a result of a regression analysis based on experimental data, and their values are shown in table 3.

**Table 3.** The results of the regression analysis (1).

| Regression characteristics | Values | |
|---|---|---|
| | Normalized steel | Temper hardening steel |
| Coefficient $A$ | $17 \cdot 10^3$ | $21 \cdot 10^3$ |
| Coeficient $b$ | -5.311 | -6.611 |
| Determination coefficient $R^2$ | 0.937 | 0.9629 |

IOP Conf. Series: Materials Science and Engineering **174** (2017) 012047 doi:10.1088/1757-899X/174/1/012047

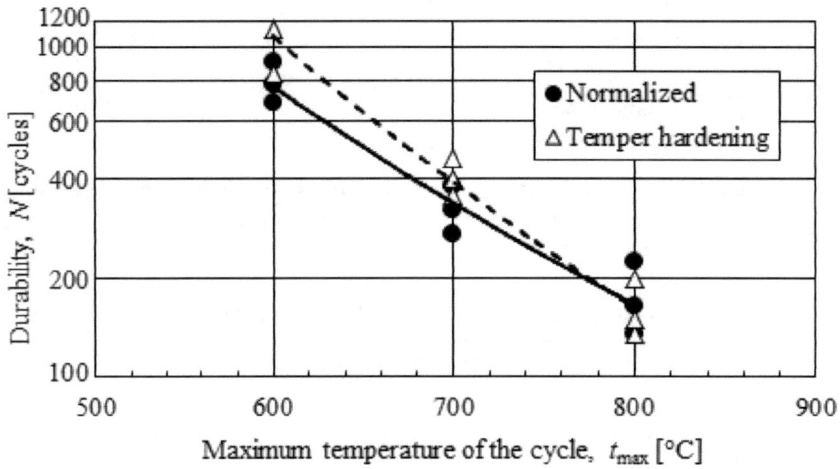


**Figure 2.** Durability vs. maximum temperature of the cycle for tested steel.

It has to be remarked that equation (1) can be used only for comparative calculus of non-isothermal fatigue durability because it does not contain total elastoplastic strain variation.

In figures 3, 4 and figures 5, 6 the influence of the total relative strain variation respectively stress variation on the durability are presented.

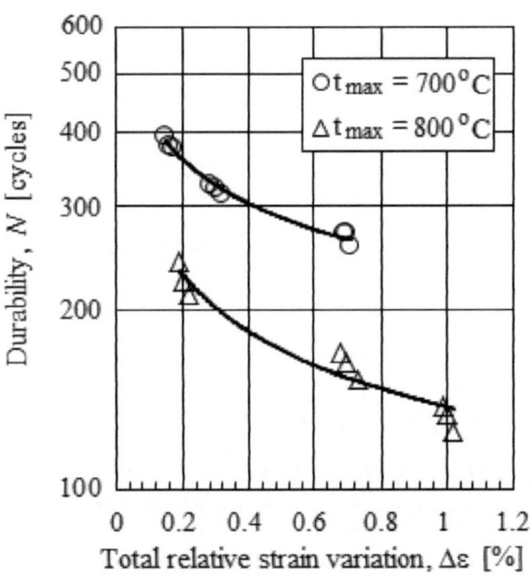


**Figure 3.** The dependence between durability and total relative strain variation for normalized steel and different maximum temperatures of the cycle.

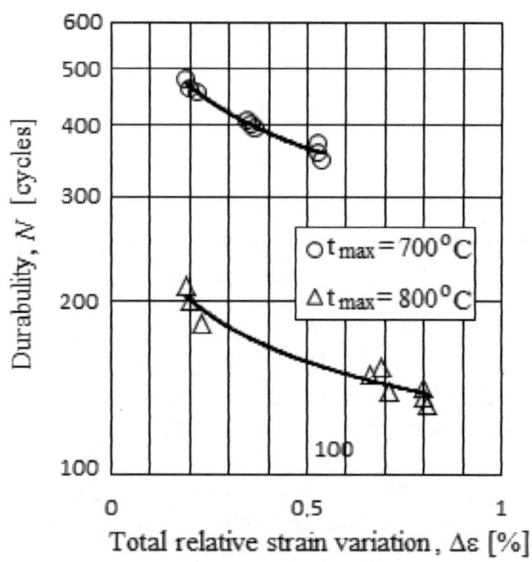


**Figure 4.** The dependence between durability and total relative strain variation for temper hardening steel and different maximum temperatures of the cycle.

The total strain and the stress variation influence on the durability can be evidenced using dependence relations between the number of cycles until the crack appearance and the strain or stress, like:

$$N = B \cdot \Delta\varepsilon^{k_1} \quad (2)$$

$$N = C \cdot \Delta\sigma^{k_2} \quad (3)$$

where: $\Delta\varepsilon$ is the total relative elastoplastic strain variation, %; $N$ – the durability, expressed like the number of cycles until the testing break; $\Delta\sigma$ - the stress variation, MPa; $k_1$, $k_2$ – the durability exponents; $B$, $C$ – the durability factors.

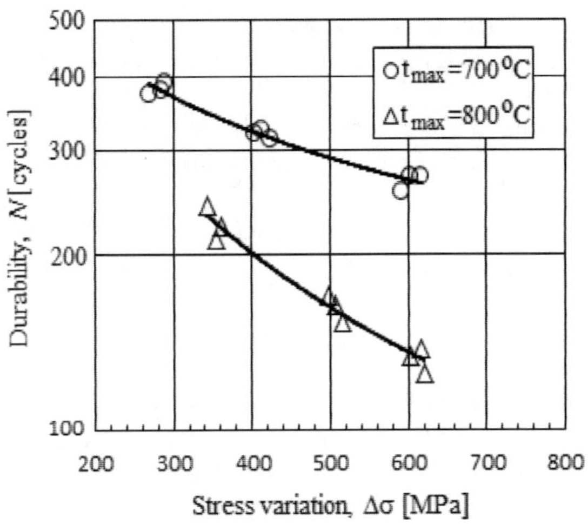


**Figure 5.** The dependence between durability and stress variation for normalized steel and different maximum temperatures of the cycle.

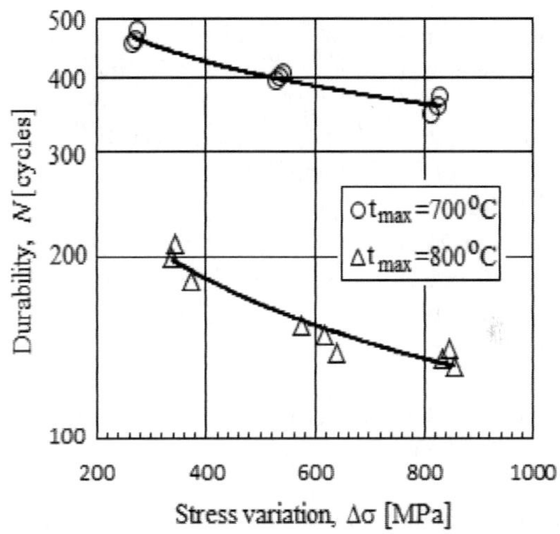


**Figure 6.** The dependence between durability and stress variation for temper hardening steel and different maximum temperatures of the cycle.

The influence of total relative strain and stress variation on non-isothermal fatigue durability is emphasized by regression analysis based on equations (2) and (3), (table 4).

**Table 4.** The results of the regression analyses (2) and (3).

| Regression characteristics | | Normalized steel | | Temper hardening steel | |
|---|---|---|---|---|---|
| Cycle maximum temperature [°C] | | 800 | 700 | 800 | 700 |
| Cycle minimum temperature [°C] | | 100 | | | |
| $\Delta\varepsilon$ variable range [%] | | 0.19 - 1.02 | 0.15 - 0.71 | 0.19 - 0.81 | 0.19 - 0.54 |
| $\Delta\sigma$ variable range [MPa] | | 363.8 – 577 | 289.3 – 590.8 | 363.8 – 853.4 | 277.8 – 814.1 |
| $N$ variable range [cycles] | | 125 - 242 | 256 - 393 | 128 - 218 | 341 - 479 |
| Regression coefficients | $B$ | 137.55 | 241.53 | 130.48 | 301.16 |
| | $k_1$ | - 0.319 | - 0.247 | - 0.267 | - 0.268 |
| | $C$ | 64864 | 5535.8 | 2641.3 | 1716 |
| | $k_2$ | - 0.964 | - 0.474 | - 0.444 | - 0.233 |
| Determination coefficient $R^2$ for equation (2) | | 0.9455 | 0.9858 | 0.9981 | 0.9796 |
| Determination coefficient $R^2$ for equation (3) | | 0.9713 | 0.9664 | 0.9278 | 0.9527 |

IOP Conf. Series: Materials Science and Engineering **174** (2017) 012047 doi:10.1088/1757-899X/174/1/012047

About the above presented experimental results the following observations can be done:

- the sample pieces were tested with a relative elastoplastic strain in the range of 0.19 … 1.02 %, which are in accordance with forging dies thermal strains;
- small durability values ($N \leq 1000$ cycles) were registered; this fact justifies the utilization of equation (2) because the strain deformations character is preponderant plastic;
- for high strain range and for maximum temperature of 700°C the temper hardening steel has the best behaviour, while for maximum temperature of 800°C both types of steel have approximately the same behaviour;
- for $t_{max}$ = 800°C and small total relative strain values, the normalized steel has higher durability than temper hardening steel, while for $t_{max}$ = 700°C an inverse situation is appearing;
- for determination coefficient of equation (2), high values in the range of 0.9455 … 0.9981 were obtained, as a result of the major influence of total strain;
- the stress variation has, also, a major influence on non-isothermal durability; this fact is shown by the determinations coefficient values of the equation (3), which are a bit smaller than those obtained for equation (2);
- the normalized steel, tested at 800°C, confirms its better behavior in plastic range (the stress variation has values in a narrow range of 363 … 577 MPa, and also, for $t_{max}$ = 700°C in the range of 289 … 590 MPa;
- the temper hardening steel tested at $t_{max}$ = 700°C has the worse plastic behavior because of stress variation level that influences its durability in a large measure.

Stress time, which generally is expressed through the cycle period, has a different influence for each material. Thus, for the cyclic non-isothermal stresses with keeping at the maximum cycle temperature, the keeping time has a major influence on the durability, while for the thermal cycles without keeping at the maximum cycle temperature has to be taken into consideration the total cycle period.

For emphasizing the influences of the heating time and total relative elastoplastic deformation on durability, in figures 7 and 8 it was represented the dependence between the durability and heating strain speed.

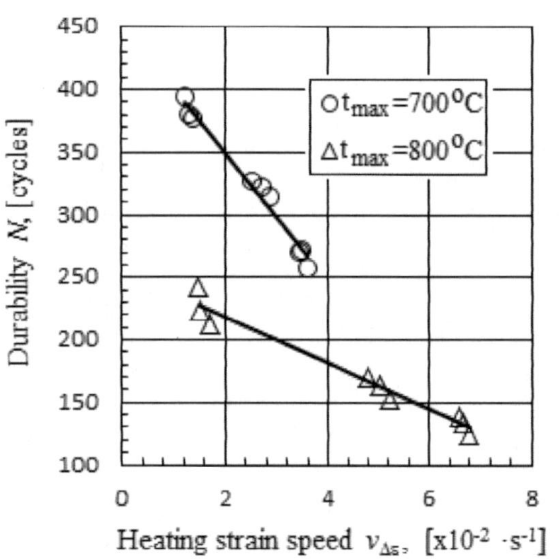


**Figure 7.** Durability vs. heating strain speed for normalized steel and different maximum temperatures of the cycle.

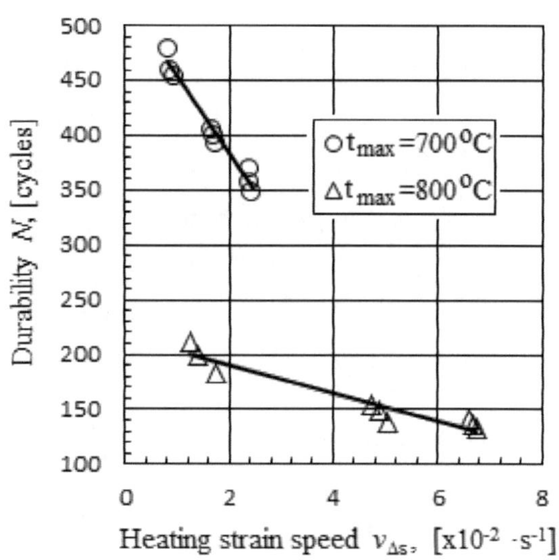


**Figure 8.** Durability vs. heating strain speed for temper hardening steel and different maximum temperatures of the cycle.

For modeling this dependence it was used the relation (4) obtained as a result of a regression analysis:

$$N = D \cdot v_{\Delta\varepsilon} + E \tag{4}$$

where: $v_{\Delta\varepsilon}$ is the heating strain speed, $s^{-1}$; $N$ – durability, cycles; $D$, $E$ – coefficient respectively durability exponent (material constants presented in table 5).

From figures 7 and 8 it can be observed that for maximum cycle temperature of 800°C the heating strain speed does not influence to much the durability of both steel types.

**Table 5.** The results of regression analysis (4).

| Regression characteristics | Normalized steel | | Temper hardening steel | |
|---|---|---|---|---|
| Cycle maximum temperature [°C] | 800 | 700 | 800 | 700 |
| Coefficient $D$ | - 18.451 | - 52.74 | - 12.486 | - 69.68 |
| Coefficient $E$ | 255.17 | 456.14 | 214.82 | 523.92 |
| Determination coefficient $R^2$ | 0.9688 | 0.9807 | 0.9202 | 0.9742 |

For the tests where the cycle maximum temperature was 700°C, the small values of the heating strain speed imply high durability of the temper hardening steel. Also, it can be remarked, from table 5, the high values of the determination coefficient. This fact shows, once again, the strong connection between the considered parameters.

## 4. Conclusions

The starting point for the design and carrying out of the experiments was represented by cyclic thermal stresses of the hot forging dies.

Based on the Manson – Coffin model of metallic materials behavior at thermal fatigue it was used a testing stand specially built for non-isothermal fatigue experiments.

The main conclusions of the paper are the following:

- the determinations with the highest verisimilar degree of non-isothermal fatigue durability were carried out when the sample piece was fastened at both ends and subjected to compression at maximum cycle temperature;
- in the range of high total strain values the temper hardening steel has the best behaviour for cycle maximum temperature values of 700°C, while for the maximum temperature of 800°C both steel types have approximately the same behavior;
- for $t_{max}$ = 800°C and low values of the total relative strain the normalized steel has higher durability than the temper hardening steel, while for the cycle maximum temperature of 700°C the situation is inversely;
- the above determined experimental results represents intrinsic characteristics, by the point of view of the non-isothermal fatigue durability, for the tested steels, and they can be used both in designing and exploitation of hot forging dies.

## References

[1] Persson A, Hogmark S and Bergstrom J 2005 Thermal fatigue cracking of surface engineered hot work tool steels *Surf Coat Tech* **191** p 216

[2] Zych J 2013 Thermal fatigue of the selected tool steel applied for metal moulds elements in die casting of metals *Arch Metall Mater* **58** p 757

[3] Persson A, Hogmark S and Bergstrom J 2004 Simulation and evaluation of thermal fatigue cracking of hot work tool steels *Int J Fatigue* **26** p 1095

[4] Dulnev R A and Kotov P I 1980 *Termiceskaia ustalost metallov* (Moskva: Masinostroenie)

[5] Tuleakov G A 1978 *Termiceskaia ustalost v teploenerghetike* (Moskva: Masinostroenie)

[6] Serensen S V and Kotov P I 1962 *Zavodskaia laboratoryia* **28** p 1233

[7] Drumeanu A C, Antonescu N N and Ripeanu R G 2010 Some Aspects Concerning the Experimental Determinations Methodology for the Non-isothermal Cyclic Durability of the Machine Steels *J Balk Tribol Assoc* **16** p 315

[8] Aleksiuk M M, Borisenko V A and Krascenko V P 1980 *Mehaniceskie ispitanyia materialov pri visokih temperaturah* (Kiev: Naukova Dumka)

[9] Antonescu N N, Drumeanu A C and Nae I 1996 *Proc. of 10th International Colloquium Tribology* (Esslingen: Germany/Wilfried J. Bartz) p. 899

[10] Drumeanu A.C and Antonescu N N 1998 *Proc. of 11th International Colloquium Tribology* (Esslingen: Germany/Wilfried J. Bartz) p 1927

[11] Rusu O, Teodorescu M and Lascu-Simion N 1992 *Oboseala Metalelor* (Bucuresti: Editura Tehnica)

IOP Conf. Series: Materials Science and Engineering **174** (2017) 012048 doi:10.1088/1757-899X/174/1/012048

# Fault tree analysis of most common rolling bearing tribological failures

**Aleksandar Vencl[1], Vlada Gašić[1] and Blaža Stojanović[2]**

[1] University of Belgrade, Faculty of Mechanical Engineering, Kraljice Marije 16, 11120 Belgrade 35, Serbia

[2] University of Kragujevac, Faculty of Engineering, Sestre Janjić 6, 34000 Kragujevac, Serbia

E-mail: avencl@mas.bg.ac.rs

**Abstract.** Wear as a tribological process has a major influence on the reliability and life of rolling bearings. Field examinations of bearing failures due to wear indicate possible causes and point to the necessary measurements for wear reduction or elimination. Wear itself is a very complex process initiated by the action of different mechanisms, and can be manifested by different wear types which are often related. However, the dominant type of wear can be approximately determined. The paper presents the classification of most common bearing damages according to the dominant wear type, i.e. abrasive wear, adhesive wear, surface fatigue wear, erosive wear, fretting wear and corrosive wear. The wear types are correlated with the terms used in ISO 15243 standard. Each wear type is illustrated with an appropriate photograph, and for each wear type, appropriate description of causes and manifestations is presented. Possible causes of rolling bearing failure are used for the fault tree analysis (FTA). It was performed to determine the root causes for bearing failures. The constructed fault tree diagram for rolling bearing failure can be useful tool for maintenance engineers.

## 1. Introduction

Rolling bearings are machine elements found in a wide field of applications, and they are typically very reliable. Nevertheless, any damage or failure of the rolling bearings results in a partial or complete functional failure of the machine and can cause significant economic losses. Although, generally, the price of the rolling bearing is relatively small, a bearing damage that reduces the function of the system or result in its failure can cause a lot of overhead costs. That is way it is very important to be able to identify and predict failure beforehand, if possible, so that preventive measures can be adopted. Under normal operating conditions, i.e. if rolling bearings are used correctly, they usually have a substantial service life. Method of calculating the basic rating life based on 90 % reliability, with commonly used high quality material, good manufacturing quality and with conventional operating conditions is specified in ISO 281 standard [1]. This standard also specifies method of calculating the modified rating life, in which various reliabilities, lubrication condition, contaminated lubricant and fatigue load of the bearing are taken into account. However, it does not take into account particular influence of wear and corrosion on bearing life.

Unfortunately, rolling bearings often fail prematurely due to avoidable mistakes, i.e. the effect of unanticipated wear may be to initiate damage that leads to premature failure. According to some studies [2,3], following common rolling bearing failure causes can be classified, with their percentage shares: (1) insufficient amount of lubricant: 15 %; (2) inadequate selection of lubricant: 20 %; (3) aged lubricant (wear products, additive depletion): 20 %; (4) solid particles contamination of lubricant: 20 %; (5) liquid contamination of lubricant: 5 %; (6) inadequate selection of bearing (design, size, load-carrying capacity): 10 %; (7) handling induced defects: 5 %; (8) mounting faults: 5 %; (9) bearing material and production faults: less than 1 %. As it can be seen, about 35 % of bearing failures are related to irregularities in the lubrication process (causes 1 and 2) and in total, about 80 % of bearing failures are associated with the tribological causes (causes 1 to 5).

According to the ISO 13372 standard failure is defined as "termination of the ability of an item to perform a required function" [4]. In practice, damage of a rolling bearing may often be the result of several mechanisms operating simultaneously. It is the complex combination of many influencing parameters which often causes difficulty in establishing the primary cause of damage. The relevant literature and practical experience show that the rolling bearing damages are mainly caused by wear, and that surface fatigue wear is the most common wear type. Also, very often, in mixed-to-boundary lubrication conditions, adhesive wear can occurs, as well as abrasive wear if lubricant is contaminated with foreign particles. Wear itself is a very complex process initiated by the action of different mechanisms, and can be manifested by different wear types which are often related. The paper presents the classification of most common bearing damages according to the dominant wear type, i.e. abrasive wear, adhesive wear, surface fatigue wear, erosive wear, fretting wear and corrosive wear. In addition, fault tree analysis (FTA) was performed to determine the root causes for bearing failures. Similar analysis was already performed for journal bearing failures [5].

## 2. Appearances and manifestations of the different wear types

### *2.1. Abrasive wear*

Abrasive wear, according to ISO 15243, is classified as wear, i.e. abrasive wear. Terms: particle wear, three body wear, burnishing, and scoring are also used. Abrasive wear in rolling bearings occurs due to the presence of solid contaminants (three-body abrasion) or by rubbing of the bearing elements in direct contact (two-body abrasion) [6]. Rolling bearings are very sensitive to solid contaminants which can happen even during mounting of the bearing. Fine hard particle contamination can remove material from bearing contact surfaces by excessive polishing (burnishing) like a lapping compound, changing internal bearing geometry by abrasive wear [7]. Denting of contact surfaces (figure 1a) can occur when hard and ductile particles are entrained through heavily loaded bearing contacts and plastically deform the surfaces. Dents produced by large ductile particles may lead to pitting and spalling, while dents attributed to small ductile particles can lead to micro-pitting or peeling of bearing raceways [8]. Large hard particles can become embedded into softer cage material and create scratches and groove marks in rolling element body surfaces (figure 1b). Similar appearances show abrasive wear by rubbing of the bearing elements in direct contact. Intensive abrasive wear of rolling bearings will manifests as scoring (severe scratching).

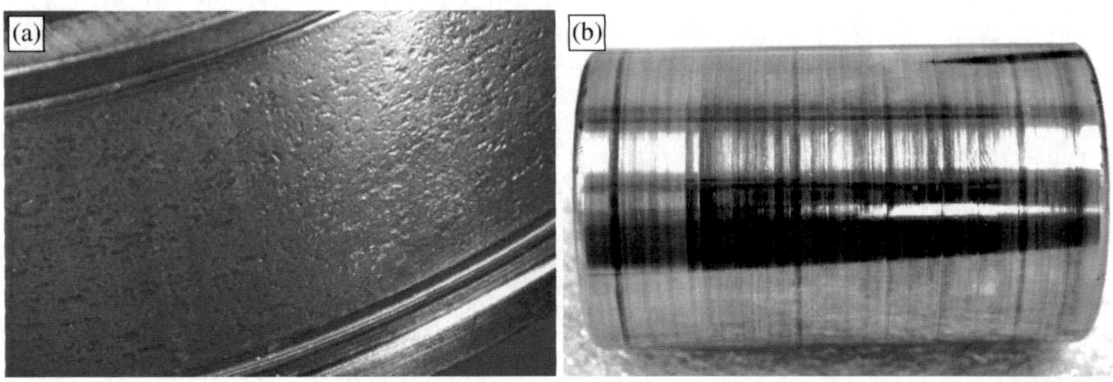


**Figure 1.** Bearing surfaces damaged by abrasive wear: (a) dull surface and visible dents (craters) on inner ring and (b) scratching and pronounced grooves on rollers.

### 2.2. *Adhesive wear*

Adhesive wear, according to ISO 15243, is classified as wear, i.e. adhesive wear. Terms: smearing, skidding, galling, and scuffing are also used. Adhesive wear in rolling bearings occurs when the conditions for formation of a lubricant film to separate bearing contacting surfaces are inadequate (boundary or mixed lubrication). In that conditions, and mainly in sliding contacts, contacting asperities bond in intimate contact (micro welds) and tear. Adhesive wear occurs at locations with high slide-to-roll ratio, i.e. contact of rolling elements with cage, shoulder or other rolling element (figures 2a and b). It also occurs when the rolling elements are subjected to sudden acceleration on their re-entry into the load zone (figure 2c). When the rolling elements are in the unloaded zone, without direct contact with the rings, their rotational speed is lower than when they are in the loaded zone. The rolling elements are therefore subjected to sudden acceleration when they enter the loaded zone, with high slide-to-roll ratio as a result.

Adhesive wear can occurs in the form of mild or "normal" wear or as severe adhesive wear. Mild adhesive wear typically manifests with wiping of the material from the contact surfaces, sometimes accompanied by change of colour of the bearing material (figures 2a and b). It is normal during the running-in of a rolling bearing, which is a natural short process after which the running behaviour stabilizes or even improves [9]. Severe adhesive wear (scuffing or galling) is usually accompanied by high temperatures, and occurs upon loss of lubrication at contacts with a large sliding component. It manifests with change of colour, significant plastic deformation of surfaces due to its softening, and material transfer (figures 2d, e and f). As a result, seizure may occur, when components are not easily separated due to significant welding of the worn surfaces. Smearing (skidding) is somewhat in between mild and severe adhesive wear. It is usually localized on specific areas (figure 2c). In smearing, bearing elements slip also produces plastic deformation, but at the asperity scale, that may be accompanied by material transfer from one surface to the other [8].

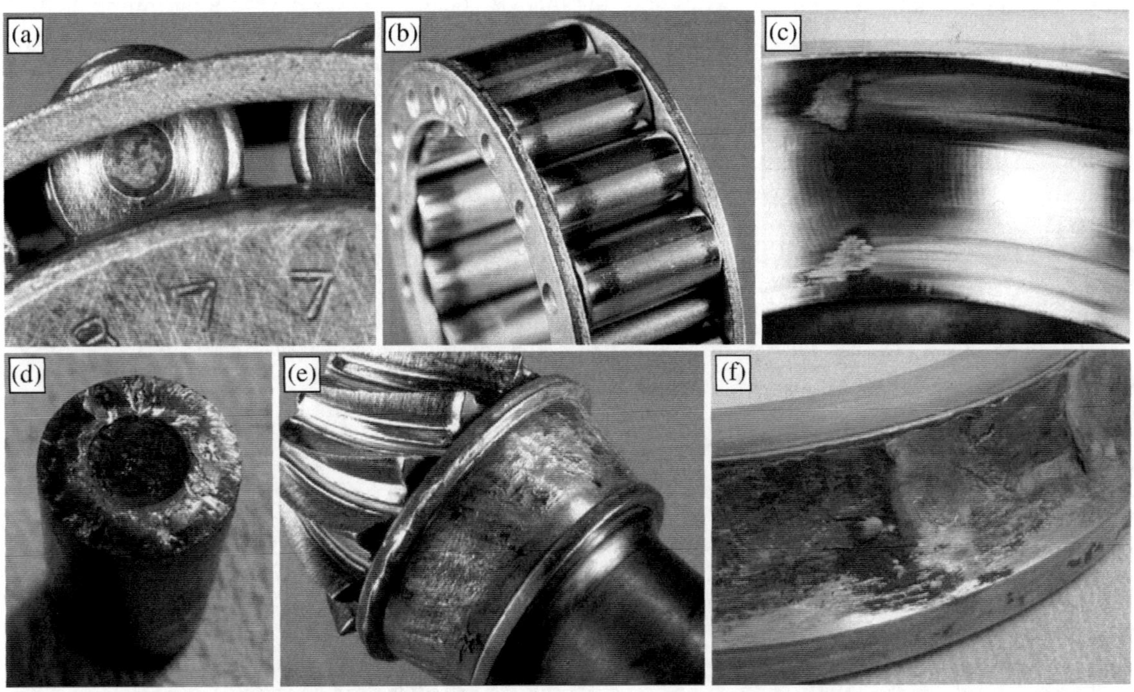


**Figure 2.** Bearing surfaces damaged by adhesive wear: (a) and (b) wiping of rollers (mild adhesive wear), (c) smearing of outer ring and (d), (e) and (f) change of colour and surface softening and flow of roller and inner rings material (severe adhesive wear).

*2.3. Surface fatigue wear*

Surface fatigue wear, according to ISO 15243, is classified as fatigue, i.e. subsurface initiated fatigue, and surface initiated fatigue. Terms: pitting, peeling, spalling, and flaking are also used. Surface fatigue wear in rolling bearings occurs due to the presence of dynamic load conditions. Continued cyclic contact stress may initiate cracks that can then propagate by fracture, leading to the removal of surface material. According to the location of initial cracks formation, two types of fatigue (rolling contact fatigue) can be distinguished, i.e. subsurface and surface initiated fatigue. Location of the initial crack is defined with the position of the highest shear stress, which is not the same for sliding/rolling (figure 3a). In pure rolling initial cracks will be initiated at a certain depth below the surface (figure 3b), usually 0.1 to 0.5 mm [10]. These near-surface cracks will propagate to the surface and ultimately lead to macroscopic surface material removal.

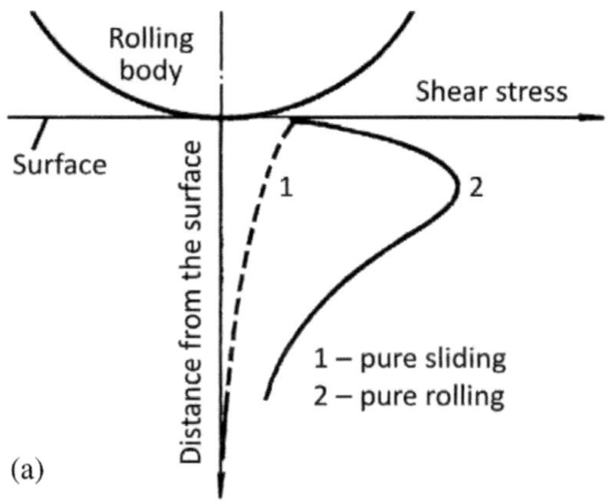


(a)

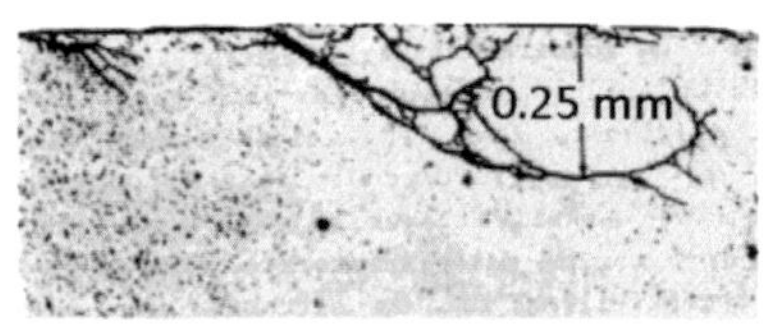


(b)

**Figure 3.** Surface fatigue wear: (a) position of the highest shear stress for pure sliding/rolling and (b) propagation of the initial crack to the surface.

Both, subsurface and surface initiated fatigue manifest like pitting, i.e. formation of the surface craters or voids (pits) by loss of material. If the pitting is caused by overload due to the fault assembly of the bearing (misalignment or small clearance) failures occur at intervals corresponding to rolling element pitch (figure 4a). Overload due to the large clearance during mounting of the bearing will also cause localised failure, but the rolling elements and raceways wear unevenly (figure 4b). On the other hand, inadequate lubrication will cause pitting on the whole contact area (figure 4c). Surface fatigue wear starts as initial pitting (peeling), but over the time pits grow and join together causing progressive pitting (spalling or flaking). Peeling or micro-pitting occurs when thin flakes of material are removed from contact surfaces. The damage area is usually shallow on a microscopic scale, giving the gray stained appearance [10].

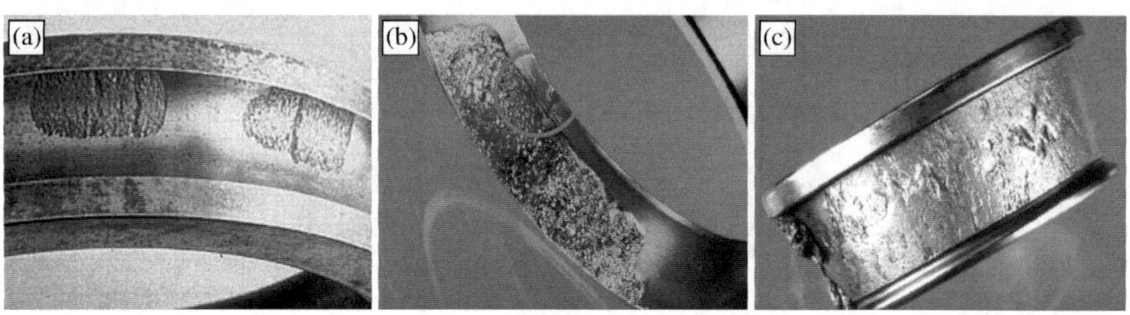


**Figure 4.** Bearing surfaces damaged by surface fatigue wear: (a) and (b) localised pitting of the inner and outer ring due to overload and (c) progressive pitting of the inner ring due to the inadequate lubrication.

### 2.4. Erosive wear

Erosive wear, according to ISO 15243, is classified as electrical erosion, i.e. excessive voltage, and current leakage. Terms: electrical pitting, electrical fluting, and washboarding are also used. Erosive wear (electrical erosion) in rolling bearings occurs due to the passage of an electric current in the area of minimum lubricant film thickness between the rolling elements and raceway. This type of wear most often occurs in bearings of the electric motors. Two electrical erosion damages can be distinguished, depending on the voltage, i.e. high-voltage electrical erosion (excessive voltage) and low-voltage electrical erosion (current leakage). In high-voltage electrical erosion (electrical pitting) the flow lines of the current are condensed, resulting in localized and rapid heating, so that the contact areas melt and weld together. It manifests as a series of small craters (pits) on both contact surfaces (figure 5a). Craters diameters are usually up to 100 μm [9]. In low-voltage electrical erosion (electrical fluting or washboarding), the intensity of the current is relatively low but acts during the long period of time. The damage manifests as corrugation marks on the contact surfaces (figure 5b). Formed groves are shallow, close to each other and equally spaced.

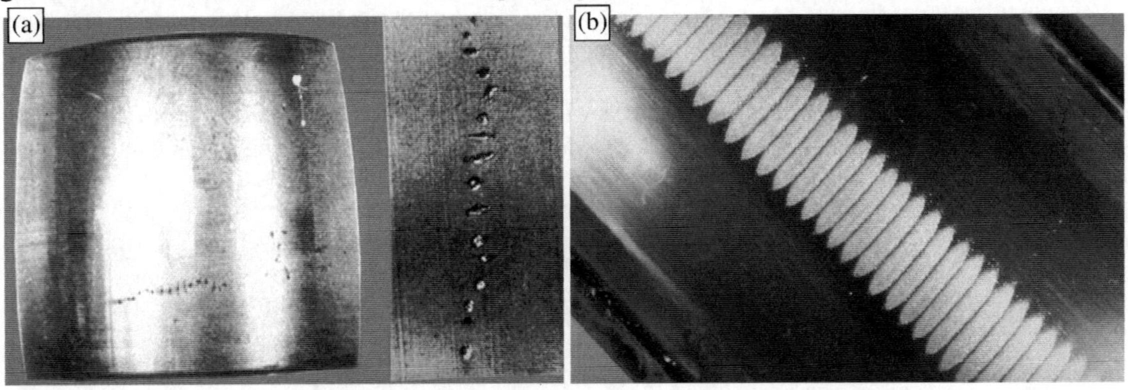


**Figure 5.** Bearing surfaces damaged by erosive wear (electrical erosion): (a) small craters (pits) on roller and inner ring due to the excessive voltage (electrical pitting) and (b) shallow and equally spaced groves (flutes) due to the current leakage (electrical fluting or washboarding).

### 2.5. Fretting wear

Fretting wear, according to ISO 15243, is classified as corrosion (frictional corrosion), i.e. fretting corrosion, and false brinelling. Terms: fretting rust, tribocorrosion, tribo-oxidation, and vibration corrosion are also used. Fretting (vibration) wear in rolling bearings occurs due to the relative micromovements between mating surfaces caused by vibrations. These micromovements can induce the oxidation of material surface. In presence of corrosive environment, corroded layer can be formed as well. Subsequent micromovements remove or disrupt this oxide/corroded surface layer, and the cycle repeats. This type of wear is very common in rolling bearings. According to the location of damage, two types of fretting (vibration) wear can be distinguished, i.e. fretting corrosion/oxidation (fretting rust, tribocorrosion or tribo-oxidation) and false brinelling (vibration corrosion). Fretting corrosion/oxidation occurs between inner ring and shaft or outer ring and housings as a result of vibrations and insufficient press fit (too light interference fit). It manifests with visible wear products in powder form, and very often changes of colour (figure 6a). False brinelling occurs between rolling elements and raceways as a result of vibrations and insufficient clearance. It manifests as shallow depressions in the raceways at rolling element pitch and usually accompanied by loss of surface finish marks in the worn areas (shiny colour) and the presence of reddish or black wear debris (figure 6b).

### 2.6. Corrosive wear

Corrosive wear, according to ISO 15243, is classified as corrosion, i.e. moisture corrosion. Terms: oxidation and rust are also used. Corrosive wear in rolling bearings occurs due to the chemical

**Figure 6.** Bearing surfaces damaged by fretting (vibration) wear: (a) powder wear debris and change of colour on inner ring and shaft, due to the insufficient clearance and (b) equally spaced depressions and shiny worn areas (false brinelling) on inner ring due to the insufficient press fit.

reaction (oxidation/corrosion) of steel surfaces with the environment, and subsequent removing or disrupting of oxide/corroded surface layer, due to the relative movements of the bearing elements. The origin chemically aggressive medias (water or moisture, acids and products from lubricant degradation) is either outside the bearing (contaminants) or they are generated within the bearing itself. This type of wear most often occurs in bearings used in papermaking machines. Corrosion wear manifests with corrosion pits or rust on critical surfaces which change colour and chemical composition (figure 7a). A specific form of corrosion wear in the contact areas between rolling elements and raceways manifests with equally spaced discoloured areas on both contact surfaces (figures 7b and c).

**Figure 7.** Bearing surfaces damaged by corrosive wear: (a) corrosion pits and rust all over the bearing and (b) and (c) discolouration of contact areas on rollers and outer ring.

## 3. Fault tree analysis

Rolling element bearings are typically operated under elastohydrodynamic lubrication (EHL) regime. Elastohydrodynamic lubrication can be defined as a form of hydrodynamic lubrication where the elastic deformation of the contacting bodies and the changes of viscosity with pressure play fundamental roles [11]. A thin layer of EHL lubricant film separates bearing elements, preventing their direct contact (boundary or mixed lubrication), and thus minimizes friction and wear. Lubricants also act as a medium to remove and/or distribute heat, protect the surface from corrosion, and remove wear products from contact zone [2]. There are many parameters that affect lubricant film thickness and obtained lubrication regime, and most important are presented in figure 8.

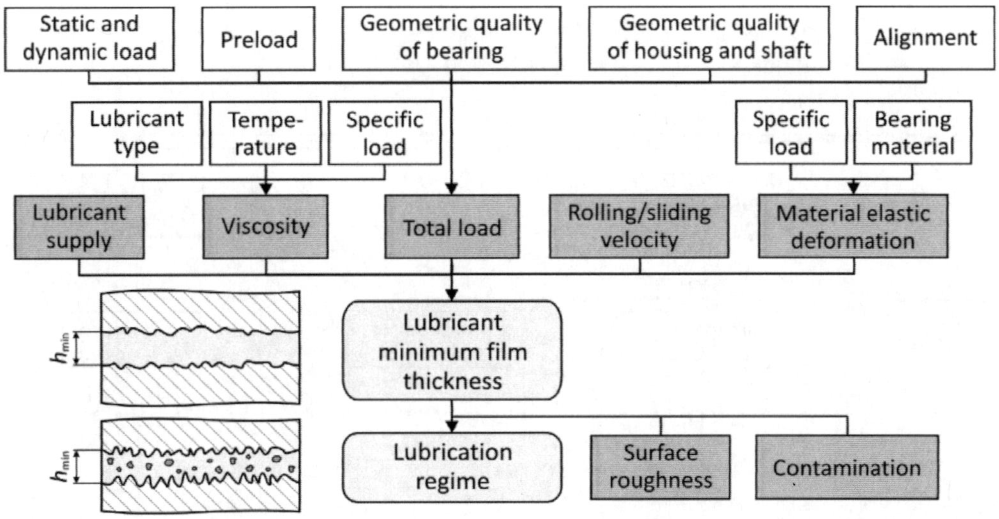


**Figure 8.** Parameters related to the lubricant film thickness and lubrication regime in a rolling bearing.

Possible causes of rolling bearing failure, used for the fault tree analysis (FTA), were taken from the literature data and personal experience. The different types of bearing damages were correlated with the ISO standard 15243 [9], which can be used for the further explanation/description of the constructed fault tree root causes. FTA is a useful tool that can be applied on different mechanical engineering problems [12-14]. A fault tree diagram is a logic diagram which traces the possible events leading to a major event, almost like a family tree. The desired or, in most cases, undesired main event is placed at the top of the tree and called the ''top event'' [15]. In our case, the top event will be entitled ''rolling bearing failure''. In general terms, we can categorize two basic types of failure: (1) tribological and (2) mechanical, but as more dominant, only tribological failures were discussed further. Tribological failures include six broad wear types: (1) abrasive, (2) adhesive, (3) surface fatigue, (4) erosive (electrical erosion), (5) fretting (vibration), and (6) corrosive wear. The discussion of different wear type failures must be more general, since they are intermediate events on fault tree of tribological bearing failure (figure 9).

*Abrasive wear* in rolling bearings is mainly caused by the (1) contamination of lubricant with foreign particles. Lubricant contamination with foreign particles is basically caused by poor sealing of the bearing or poor filtration of the lubricant. Scratches also occur during sliding of the bearing elements in direct contact. Direct contact of the bearing elements can occur due to the (2) insufficient clearance (caused by poor design of the bearing, or its fault assembly, or high surface roughness) or (3) inadequate lubrication (caused by insufficient amount of the lubricant, low viscosity of the lubricant or inadequate quality of the lubricant).

*Adhesive wear* in rolling bearings is mainly caused by the (1) lubricant film breakdown due to excessive load or speed or presence of vibrations (poor balancing) or (2) inadequate lubrication due to the insufficient or excessive amount of the lubricant, low viscosity of the lubricant or inadequate quality of the lubricant. In both cases, lubricant film thickness is insufficient to prevent direct contact between the bearing elements (boundary or mixed lubrication). Adhesive wear can be also caused by (3) improper material selection, i.e. combination of compatible materials (usually connected with seizure) or (4) insufficient clearance due to the poor design of the bearing or its fault assembly).

*Surface fatigue wear* in rolling bearings is mainly caused by the low fatigue strength of the bearing elements material, i.e. when (1) load exceeds fatigue strength. Subsurface initiated fatigue could be also caused by the (2) improper material production, i.e. presence of structural defects and/or inclusions in bearing elements material. Surface initiated fatigue is mostly attributable to surface distress and occurs when a certain percentage of sliding motion is present. Surface distress could be

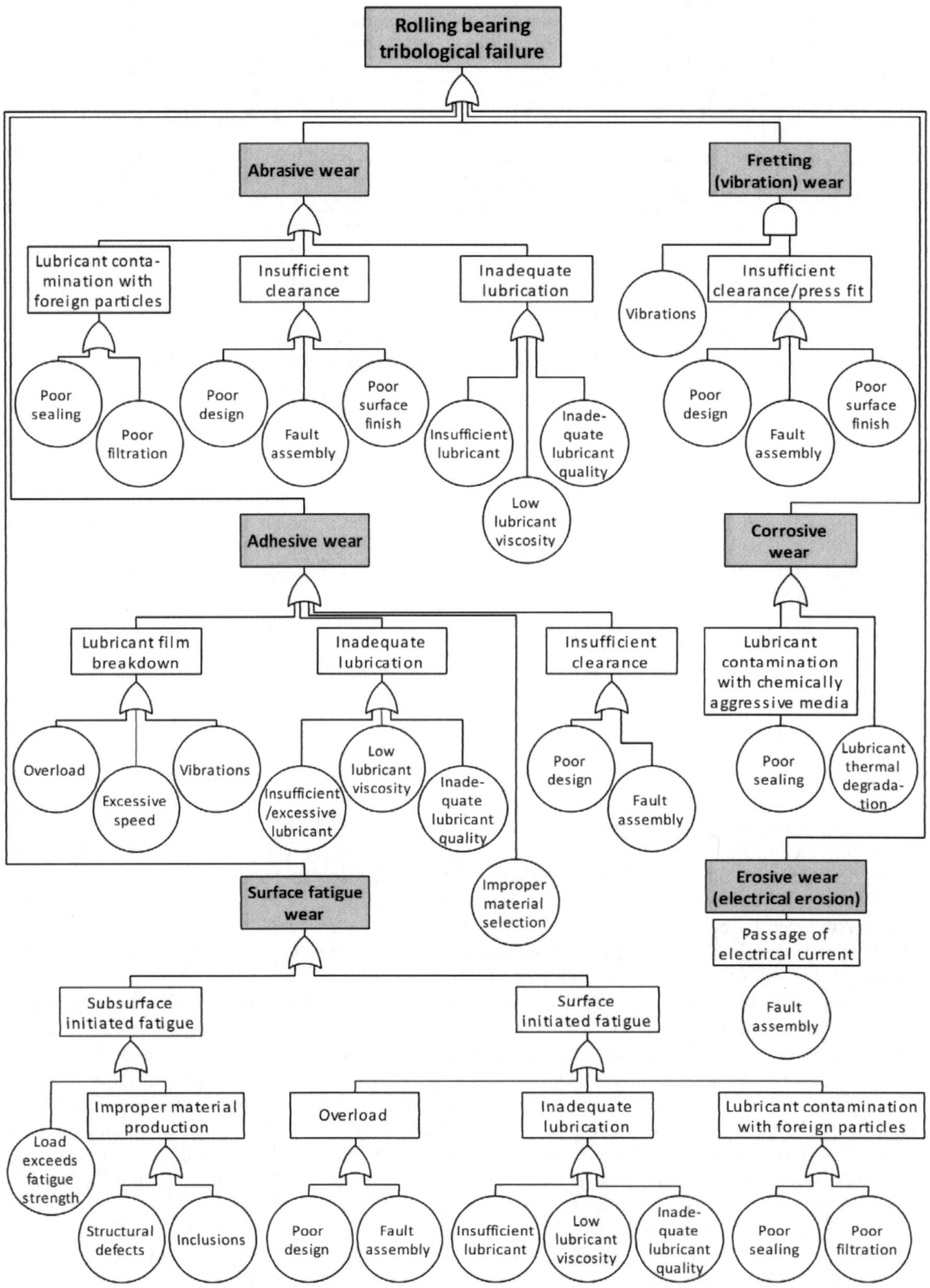


**Figure 9.** Fault tree of tribological rolling bearing failure.

caused by (3) overload due to the poor design of the bearing or its fault assembly, or (4) inadequate lubrication due to the insufficient amount of the lubricant, its low viscosity or inadequate quality of the lubricant, or (5) lubricant contamination with foreign particles due to the poor sealing of the bearing or poor filtration of the lubricant (contamination of the bearing and subsequent indentation of the foreign particles will crate places with high stress concentrations). Surface initiated fatigue could be also caused by the surface defect that might be caused by some other type of wear.

*Erosive wear* (electrical erosion) in rolling bearings is caused by the (1) passage of an electric current in the area of minimum lubricant film thickness between the rolling elements and raceway. Main reason for this is the fault assembly of the bearing (improper grounding of equipment, insufficient or defective insulation of bearing, or static discharge issues).

*Fretting wear* in rolling bearings is caused by the (1) vibrations and (2) insufficient clearance or insufficient press fit due to the poor design of the bearing, or its fault assembly, or high surface roughness.

*Corrosive wear* in rolling bearings is caused by the (1) lubricant contamination with chemically aggressive media due to the poor sealing or (2) thermal degradation of lubricant during service.

## 4. Conclusions

The presented examples of damages represent the main types of wear that occur in rolling bearings. Based on the literature, the most prominent (and thus the most important) types of wear in rolling bearings are surface fatigue wear, adhesive wear and abrasive wear. Other types of wear, i.e. erosive (electrical erosion), fretting (vibration) and corrosive wear are less pronounced, although for certain conditions and applications they can cause significant damage of bearing.

Described visual manifestations of the most common bearing damages assist in determining the dominant wear type in damaged rolling bearing. Establishment of dominant type of wear facilitate the determination of the measures for its reduction or elimination and thus extending the service life and reliability of the rolling bearings. In addition, sometimes confusing terms for rolling bearing damages used in ISO 15243 standard are correlated with the generally accepted and well known terms for wear types.

Further help in determining the root causes for roller bearing failures is provided by the fault tree analysis (FTA) which was performed, and appropriate descriptions of causes are presented. Constructed fault tree diagram for rolling bearing failure should give better perceive to the engineers in practice and help them in maintenance of machines with belonging rolling bearings.

## Acknowledgments

This work has been performed as a part of activities within the projects TR 35021 and TR 35006. These projects are supported by the Republic of Serbia, Ministry of Education, Science and Technological Development, whose financial help is gratefully acknowledged.

## References

[1] ISO 281 2007 Rolling bearings – Dynamic load ratings and rating life
[2] Hannon W M and Ai X 2013 Rolling bearing lubricants, in: Wang Q J and Chung Y - W (Eds.) *Encyclopedia of Tribology* (New York: Springer) pp 2848–56
[3] -- 2015 Selecting the right grease for rolling bearings, *World Pumps* **2015** 31–33
[4] ISO 13372 2004 Condition monitoring and diagnostics of machines – Vocabulary
[5] Vencl A and Rac A 2014 Diesel engine crankshaft journal bearings failures: Case study, *Eng. Fail. Anal.* **44** 217-228
[6] Kandeva M, Karastoyanov D and Vencl A 2016 *Advanced Tribological Coatings for Heavy-Duty Applications: Case Studies* (Sofia: Marin Drinov Academic Publishing House)
[7] Widner R L 1986 Failures of rolling-element bearings, in: *ASM Handbook Volume 11, Failure Analysis and Prevention* (Metals Park: ASM International) pp 490–513
[8] Evans R D 2013 Wear of bearings, in: Wang Q J and Chung Y - W (Eds.) *Encyclopedia of Tribology* (New York: Springer) pp 4061–68
[9] ISO 15243 2004 Rolling bearings – Damage and failures – Terms, characteristics and causes

[10] van Beek A 2009 *Advanced Engineering Design* (Delft: TU Delft)
[11] Stachowiak G W and Batchelor A W 2014 *Engineering Tribology* (Amsterdam: Elsevier)
[12] Moraru R-I and Băbuţ G-B 2013 The use of fault tree in industrial risk analysis: A case study, in: *Proc. 1st Int. Conf. on Industrial and Manufacturing Technologies (Vouliagmeni)* (WSEAS Press) 70–75
[13] Olmi G 2015 Statistical tools applied for the reduction of the defect rate of coffee degassing valves, *Case Stud. Eng. Fail. Anal.* **3** 17–24
[14] Nouri.Gharahasanlou A, Mokhtarei A, Khodayarei A and Ataei M 2014 Fault tree analysis of failure cause of crushing plant and mixing bed hall at Khoy cement factory in Iran, *Case Stud. Eng. Fail. Anal.* **2** 33–38
[15] Strauss B M 1984 Fault tree analysis of bearing failures, *Lubr. Eng.* **40** 674–680

IOP Conf. Series: Materials Science and Engineering **174** (2017) 012049 doi:10.1088/1757-899X/174/1/012049

# Geometric modelling of the contact point between the bushing and sprocket in chain drives

**R Saulescu, R Velicu and M Lates**

Product Design, Mechatronics and Environment Department, Transilvania University of Brasov, 29 Eroilor Blvd, 500036, Brasov, Romania

E-mail: rsaulescu@unitbv.ro

**Abstract.** An important problem of the bush chains dynamics is represented by the calculus of the normal and transversal forces on all the contacts; these forces are producing vibrations in the chain and due to this, the chain is affected by the wear. One aspect of that dynamics is referring directly on the sprockets geometry and on the bushing and sprocket contact. The paper presents a calculus method for the contact angle between the bushing and the sprocket; this angle is a variable one depending on the bushing´s number being in contact (i) and on the specific elongation of the chain (x) due to the functioning of it. Based on the presented calculus model, a comparative analysis is proposed for these factors by using sprockets with different teeth numbers and different specific elongations of the chain. The results of the numerical simulations allow the dissemination of recommendations regarding the contact angle´s evolution, from the beginning to the end of the contact and regarding the influence of the chain´s specific elongations on the out of use of it.

## 1. Introduction

The bush chains are mainly used when the transmission of the high torques with a constant transmission ratio is needed, in the case that due to the big distances between the axes, the use of the gears is not suitable; the main disadvantage in this case is represented by the specific elongation of the chain.

In [1] is concluded that the chain´s elongation is limiting the physical functioning capacity of the chain. In this way, there are needed improvement criteria for the chain drives physical and mechanical characteristics, in particular for the geometry of the friction surfaces by considering the rotations of the bushings in the entering moment of the contact with the sprocket.

The reference [2] is studying the bushings friction in chains and is concluding that the measuring of the wear is hard due to the complex geometry of the surfaces; this wear is influencing the geometry and the configuration of the component elements of the chain.

The paper [3] proposes a new geometry design of a sprocket with the teeth number z=17 instead of the usual case of z=25, by considering some hypotheses as: the calculus are not taking into account the tolerances for the sprockets dimensions; the calculus are no taking into account the limiting misalignments of the chain links lengths; the chain elongation is made by constant pitch. These hypotheses are making possible to set up a new solution for the sprockets profile; this new profile increases the life time and the acoustics due to the vibrations of the transmission. The new profile allows the parametric optimization of the sprocket by investigating the geometry of it.

In [4] are proposed two solutions regarding the modelling of the contact between the bushing and the sprocket. These two solutions are hard to be developed and are referring on a real tooth profile and a circular one. The case study is describing a chain drive transmission used in a diesel engine from a boat. The paper´s conclusion is that the action of the polygonal effect of the transmission in concordance with the impact between the bushing and the sprocket make impossible to avoid the longitudinal and transversal vibrations, so the chain´s flexibility has an important influence. These effects may be decreased by studying the influence of the chain´s elongation on the transmission functioning and on the bushing and sprocket contact.

A graphic analysis of the bushing – sprocket contact is presented in [5]; there are studied two profiles for the sprockets – one from the American standard and one from the European standard.

The sprockets geometry and the elongation of the chains are representing a starting point in order to establish the contact between the bushing and the sprocket which means a right calculus for the contact static forces.

## 2. Problem formulation

A realistic model of the contact forces is achieved by assuming a right geometric model of the contact angles between the sprocket and the bushing.

The paper presents the evolution of the contact angle ($\alpha$) which is influencing the contact forces; this evolution is correlated to the chains pitch deviation (a maximum deviation of 0.2 % from the standard value). This deviation is produced mainly by the chains elongation due to the wear which appears during its running.

The evolution of the contact angle for the gearing between the chain and the bushing is analyzed by considering the sprockets with different teeth numbers (z=16, 24, 96), with the transmission ratio equal with 1; it is recommended that the first position of the chain on the sprocket should be considered the vertical contact between the bushing and the chain (i=0) due to the reason that the contact angle is $\alpha = 0$ and in this point the tangential forces are equal with zero. This evolution is useful in order to identify the maximum contact angle which appears at the last bushing which is coming out from the contact with the chain (approximatively one quarter of sprocket).

In the study presented in the paper, it has been chosen a standardized chain with bushings and small links; for this chain there are known the sprockets teeth numbers (z), the pitch of the chain (p) and the diameter of the bushing ($d_b$).

## 3. Modeling the contact angle

The aim of this modelling is to find out the contact angle between the sprocket and the bushing, according to the sprocket's geometry. In order to achieve this modelling there are considered as known the following parameters: the chain's pitch (p), the sprocket's teeth number (z) and the diameter of the bushing ($d_B$); it is recommended to use an even number of teeth and links for the chain.

According to these parameters the sprocket's geometry can be determined [6]; in order to determine the contact angle it is necessary to know the following geometrical parameters: the radius of the sprocket's pitch circle ($R_A$), the radius of the roller's hole of the sprocket ($r_A$), the radius of the bushing ($r_B$) and the sprocket's angular pitch ($\tau$) [5].

According to figure 1, the eccentricity ($A_0B_0$) between the center of the sprocket's foot circle ($A_0$) and the bushing's center ($B_0$) it may be written as:

$$A_0B_0 = r_A - r_B, \tag{1}$$

where $A_0B_0 = A_1B_1 = A_2B_2 = ... = A_iB_i = A_kB_k$

The longed pitch $B_iB_{i+1}$ is determined by considering that during its running the chain has an elongation with a constant length (x) through its entire length, relative to the chain's pitch ($B_0B_{01}$):

$$B_iB_{i+1} = B_0B_1 = B_0B_{01} + x = p + x \tag{2}$$

where $B_0B_1 = B_1B_2 = B_2B_3 = ... = B_iB_{i+1} = B_{k-1}B_k$

In order to establish the longed pitch it is used the figure 1, which is a deformed one according to the real case due to the reason that it should be seen the difference between the standard pitch and the longed one, for two types of deformations: $x_1$ and $x_2$, where $x_2 > x_1$.

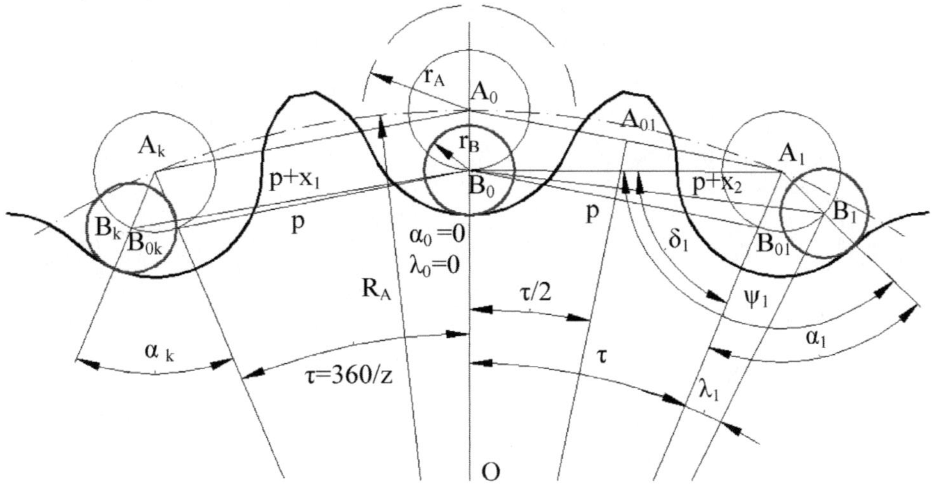


**Figure 1.** The sketch used to find out the main geometrical parameters of the chain drive and specific elongation of the chain.

The geometrical parameters are determined by considering the initial mounting position where the contact angle ($\alpha_0 = 0$) (see figure1) and the first contact between the bushing and the sprocket (i=1) is noted with the index 1.

According to figure 1 the next mathematical model used to determine the contact angle can be proposed. This model is based on the geometrical parameters which have been obtained in the mounting position (the initial position) and by considering, step by step, each contact. The values of the current contact angle are depending on the previous contacts where *i* is considered as the number of the contact between the bushing and the chain; for i=0, in the mounting position, $\alpha_0$=0, $\lambda_0$=0 and $OB_0 = R_A - r_B + r_A$:

$$B_{i-1}A_i = \sqrt{OB_{i-1}^2 + OA_i^2 - 2OB_{i-1}OA_i \cos(\tau - \lambda_{i-1})} \tag{3}$$

$$\delta_i = \arccos\frac{B_{i-1}A_i^2 + OA_i^2 - OB_{i-1}^2}{2B_{i-1}A_i \cdot OA_i} \tag{4}$$

$$\psi_i = \arccos\frac{B_{i-1}A_i^2 + A_iB_i^2 - B_{i-1}B_i^2}{2B_{i-1}A_i \cdot A_iB_i} \tag{5}$$

$$\alpha_i = \psi_i - \delta_i \tag{6}$$

$$OB_i = \sqrt{OA_i^2 + A_iB_i^2 - 2OA_i \cdot A_iB_i \cos\alpha_i} \tag{7}$$

$$\lambda_i = \arcsin\frac{A_iB_i \sin\alpha_i}{OB_i} \tag{8}$$

where $OA_i = R_A$,

IOP Conf. Series: Materials Science and Engineering **174** (2017) 012049 doi:10.1088/1757-899X/174/1/012049

By using the algorithm from above, in the next step, the influence of the specific chain's elongation is determined – figure 2a; this influence is determined by considering a variation of the chain's specific elongation starting from 0.2% from the chain's pitch and ending with a value four times bigger. If the value of the contact angle ($\alpha$) is bigger than the value of the sprocket's roller's hole angle ($\phi$) (figure 2b) the bushing is coming out from the contact with the radius of the roller's hole of the sprocket ($r_A$) and due to that the previous proposed mathematical model is no longer useful.

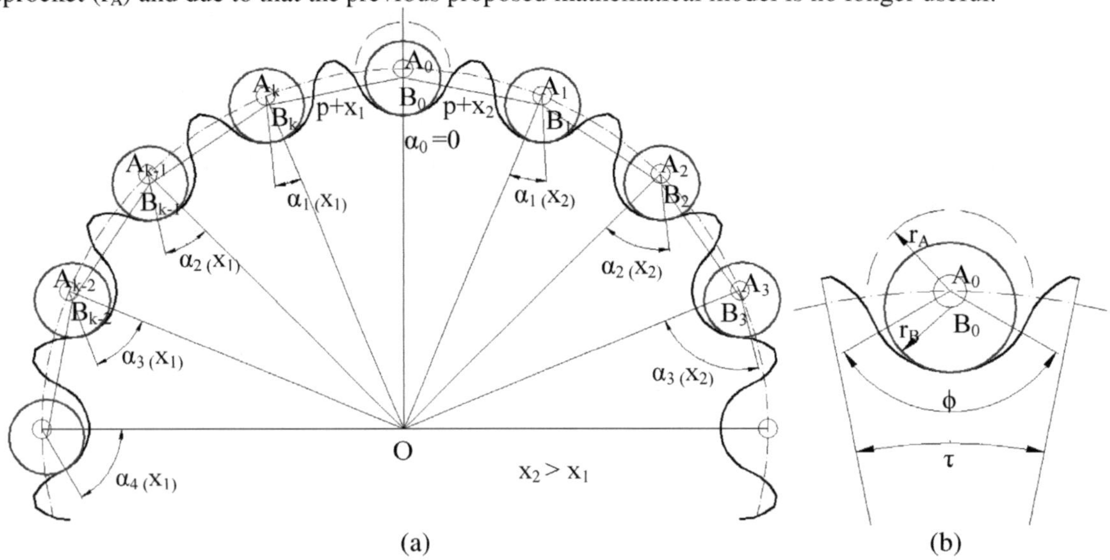


**Figure 2.** Bushing chain transmission: the influence of the chain's specific elongation on the contact angle (a) and a detail of a bushing-sprocket contact by considering the rolling hole's angle (b).

**4. Case study**

The aims of this paper are referring on determining the influence of the specific elongation (x) on the contact angle ($\alpha$); this influence is established by varying the sprocket's teeth number in the hypothesis of the transmission ratio equal with one. In the case study it is proposed a standard chain from STAS 3006-80 which has the pitch p= 9.525 mm and the bushing diameter which is in contact with the chain $d_B$=5.08 mm. By considering z=16 there are established the sprocket's profile geometrical parameters used in the analytical modelling of the contact angle [7, 8]:

- The angular pitch $\tau = \frac{360}{z} = 22.5$ grd
- The radius of the rolling's hole $r_A = 0.505 \cdot d_B + 0.069 \frac{d_B}{3} = 2.68$ mm
- The radius of the pitch circle $R_A = \frac{1}{2}\left[p\left(\sin\frac{\tau}{2}\right)^{-1} - d_B\right] + r_A = 24.55$ mm
- The radius of the bushing $r_B = \frac{d_B}{2} = 2.54$ mm

The three radiuses from above are replaced in the proposed algorithm from before and, according to this, the diagrams with the variation of the contact angle ($\alpha_i$) are obtained; the diagrams show the variation of the contact angle depending on the variation of the specific elongation (x), starting with the allowable value (2 % from the chain's pitch) and ending with the value of 8 % from the pitch. In order to establish the elongation's influence (x) on the chain drive transmission, in the algorithm is modified the value of the elongation:

IOP Conf. Series: Materials Science and Engineering **174** (2017) 012049 doi:10.1088/1757-899X/174/1/012049

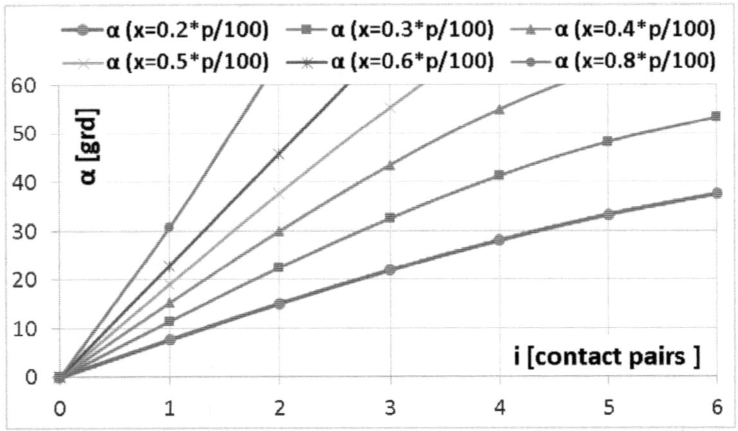


**Figure 3.** The influence of the specific elongation on the contact angle for z=16.

Generally, in the chain drive transmissions the first four contact pairs are influencing the transmission's functionality; due to this there are varied the sprocket's teeth number for the first four contacts between the bushing and the sprocket (see figure 4), for different specific elongations:

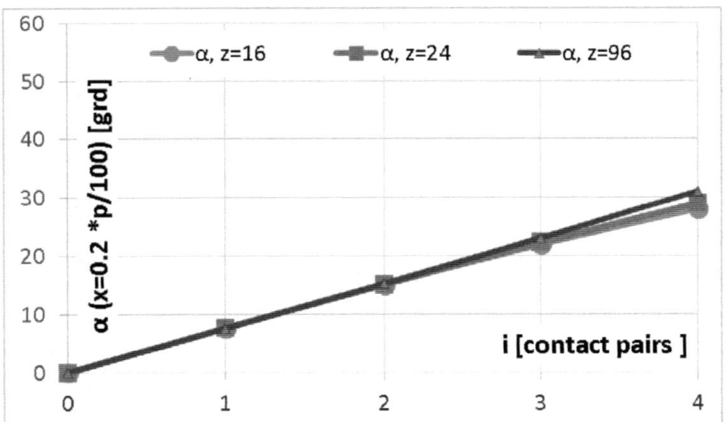


(a)

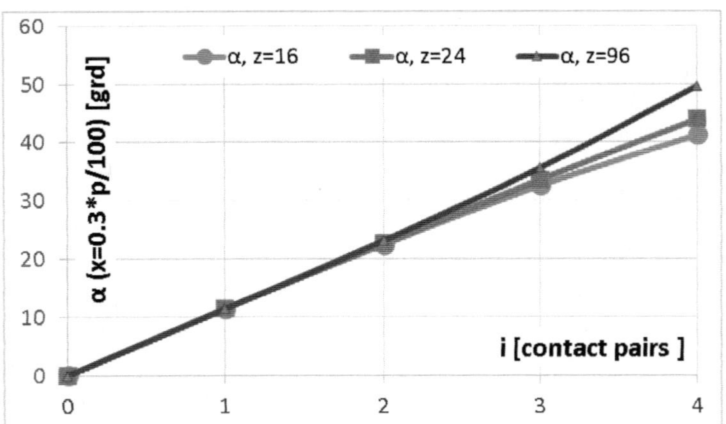


(b)

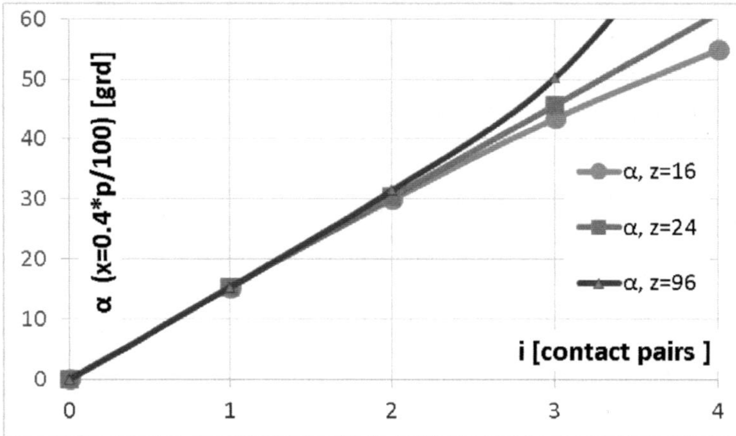
α (x=0.4*p/100) [grd]
i [contact pairs ]
α, z=16
α, z=24
α, z=96
0
10
20
30
40
50
60
0
1
2
3
4

(c)

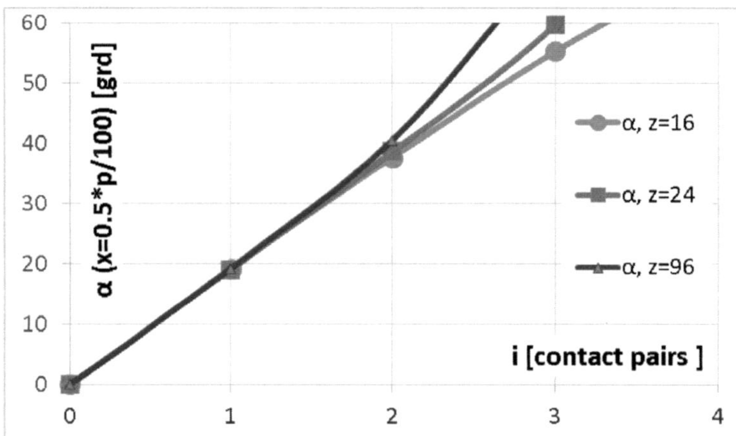
α (x=0.5*p/100) [grd]
i [contact pairs ]
α, z=16
α, z=24
α, z=96
0
10
20
30
40
50
60
0
1
2
3
4

(d)

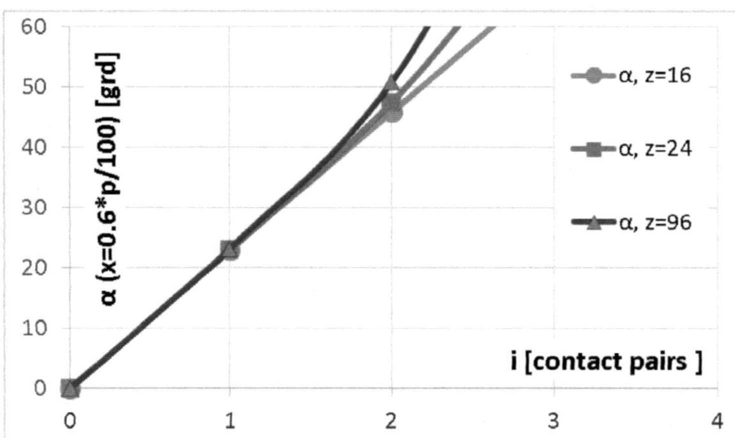
α (x=0.6*p/100) [grd]
i [contact pairs ]
α, z=16
α, z=24
α, z=96
0
10
20
30
40
50
60
0
1
2
3
4

(e)

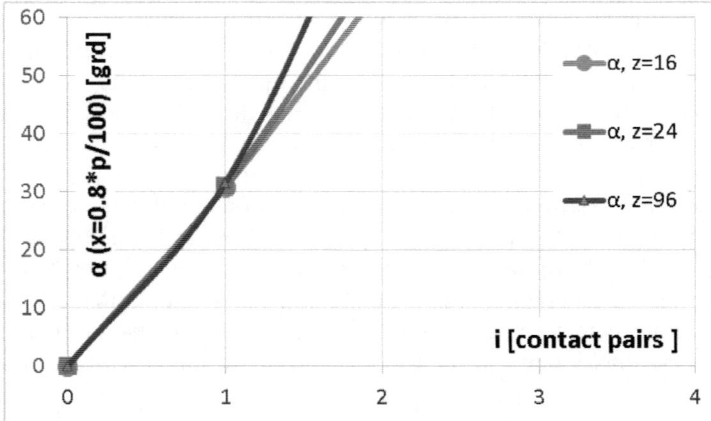


(f)

**Figure 4.** The influence of the specific elongation on the contact angle, for different teeth numbers; elongation (% from the chain's pitch) is: 0.2 (a), 0.3 (b), 0.4 (c), 0.5 (d), 0.6 (e) and 0.8 (f).

## 5. Results and discussions

The main criteria which represents the basics of the chain's functioning safety is the wear behavior because the wear produces the chain's elongation and, as consequence, a bad functioning with supplementary vibrations and noises. Due to this the specific elongation is one of the factors which is influencing the right functioning of the chain drive and is determine the out of use of it. This elongation is allowable up to 0.225 % for the weak resistance chains and up to 0.28 % [9] from the nominal pitch, for the other chains. Due to the specific elongation of the chain, the contact point between the bushing and the sprocket is changing its position so, the contact angle is modifying its value and that is influencing directly the transmission's dynamics, due to the variation of the normal and transversal forces which are inducing vibrations and wear.

In the paper there are analyzed the influence of the chain's specific elongation (x), according to a proposed mathematical model which allows the determining of the contact angle between the sprocket and the bushing ($\alpha_i$), depending on the chain's elongation and the teeth number.

The geometry of the sprocket is based on a standardized bushing chain which has the pitch p= 9.525 [mm] and the diameter of the bushing $d_B$=5.08 [mm]. For these parameters the sprocket's geometry is depending directly on the sprocket's teeth number; these values are presented in the table 1:

**Table 1.** Main geometric parameters.

| z | 16 | 24 | 96 |
|---|---|---|---|
| $R_A$ mm | 24.55 | 36.63 | 145.7 |
| $r_A$ mm | 2.68 | 2.68 | 2.68 |
| $r_B$ mm | 2.54 | 2.54 | 2.54 |

According to figure 3 it can be observed that by increasing the number of the bushings in contact (*i*) and by increasing the specific elongation of the chain (x), the contact angle is increasing, starting with the initial mounting position (see figures 1 and 2). The contacts between the bushings and the sprocket are decreasing with the increasing of the chain's elongation; in this case the contact is achieved between the top of the sprocket and the bushing and due to that, the vibrations and as consequence, the wear are appearing; as example: z=16 for x=0.3% from the nominal pitch there will be in contact 6 bushings (i=6); for x=0.8 % from *p* it could be in contact one tooth (i=1), as exception considering the contact between the bushing and the chain in the initial position i=0.

According to figure 4, the transmission's behavior is influenced mainly by the specific elongation of the chain. The number of bushings being in contact with the sprocket ($i$), by considering the increasing of the teeth number (z), is decreasing with the increasing of the specific elongation (x); as example: for x=0.5% from the nominal pitch and for z=24 there are 3 bushings in contact with the chain; for z=96 there are 2 in contact with the chain (figure 4d).

By concluding, it can be observed that a right functioning of the chain (the chain's life time) is depending on the specific elongation of the chain which is influencing the increasing of the wear. By increasing the specific elongation of the chain (x), the contact angle ($\alpha$) is increasing and the bushing number being in contact ($i$) is decreasing, so the bushings / the chain will coming out from the gearing with the chain. The obtained diagrams represent a starting point to determining the contact angle; in this way, the contact point between the bushing and the sprocket is established and this allows the dynamics study of the transmission and of the friction loses in the chain with small links and bushings.

## 6. Conclusions

As it is known, the bushing chain drives have a longer life time due to the rolling contact between the sprocket and the bushing than the other chain drives; the rolling motion is produced due to the radius of the bushing which is smaller than the radius of sprocket's foot and due to the pitch of the sprocket which is smaller than the chain's nominal pitch. By comparing the results obtained in the paper it can be concluded that with the increasing of the chain's elongation, the contact angle is increasing which has as consequence frictions on bigger surfaces between the bushing and the sprocket; another consequence could be the out coming of the chain from the gearing which produces higher vibrations in the transmission.

The paper is analyzing the influence of the specific elongation of the chain on the number of the bushings being in contact based on a generalized mathematical model used to determine the contact angle.

The wear which appears in a chain drive transmission is produced due to the chain's elongation which is influencing the contact point between the bushing and the chain, so the position of the transversal forces during the functioning of the chain. The determination of these parameters is useful in the optimizing process of the bushing chain drives by considering supplementary criteria with directly connections to minimizing the transversal vibrations and the friction.

## References

[1] Metil'kov S A and Yunin V V 2008 Influence of Wear of a Roller Drive Chain on Transmission Fitness *Russ Eng Res* **28 No. 8** 741-745

[2] Metil'kov S A, Berezhnoi S B and Yunin V V 2008 Wear of Hinges in Roller Drive Chain *Russ Eng Res* **28 No. 9** 839-844

[3] Sergeev S A and Moskalev D V 2009 Parametric Optimization of Chain-Transmission Sprockets *Russ Eng Res* **29 No. 5** 452-455

[4] Pedersen S L 2005 Model of contact between rollers and sprockets in chain-drive systems *Arch Appl Mech* **74** 489-508

[5] Todi-Eftimie A,Velicu R, Saulescu R and Jaliu C 2013 Geometric modeling of power joints from bush chain drives *The 11th IFToMM Intern. Symp. on Science of Mechanisms and Machines- SYROM 2013. Mechanisms and Machine Science* **18** 471-479

[6] Gotovtsev A A and Kotenok I P 1982 *Proektirovanie tsepnykh peredach: Spravochnik (Design of Chain Transmission: A Handbook)* (Moscow: Mashinostroenie)

[7] Papuc R, Velicu R and Gavrilă C C 2014 Geometrical study of guide-chain contact, for general chain transmission *Annals of University of Oradea, Fascicle of Management and Technological Engineering* **XXIII** (XIII) 123-126

[8] Papuc R, Velicu R and Lates M T 2015 Guide-chain contact pressure tribological analysis *Annals of University of Oradea, Fascicle of Management and Technological Engineering,* **XXIV** (XIV) 169-174

[9] Diamond Chain Co. 2016 *Technical guides. Diamond Chain Maintenance Guide* available at http://www.diamondchain.co.uk/tech-guides.php (29.07.2016)

# Penetration tests to study the mechanical tribological properties of chisel type knife

**L Vlăduţoiu[1], G Chişiu[2], T Andrei[2], A Predescu[2], C Muraru[1] and V Vlăduţ[1]**

[1] National Institute of Research - Development for Machines and Installations Designed to Agriculture and Food Industry – INMA, Ion Ionescu de la Brad Blvd. No. 6, Sector 1, 013813, Bucharest, Romania

[2] Dept. of Machine design and tribology, University POLITEHNICA of Bucharest, Splaiul Independentei No. 313, Sector 6, 060042, Romania

E-mail: laurentiuvladutoiu82@gmail.com

**Abstract.** The goal of this study was to analyze the behaviour of chisel knife type penetration in a certain type of sand. A series of penetration tests were carried out with chisel knife type, the answer to penetration depending mainly on nature, shape, size of knife and operating parameters such as speed, depth and working conditions. Tests were conducted in work conditions with wet sand and dry sand and determined force of resistance to penetration of the chisel knife type to a certain depth.

## 1. Introduction

The working components of agricultural machinery are in direct contact with an abrasive mass. The wear of these components depends of the type of soil [1]. Since these working parts are subjected to different high loads the wear intensity is much higher than in other parts of the equipment. These are called large bodies of wear [2],[3],[4]. Construction of a chisel is shown in figure 1a. The active bodies are changeable and of various shapes, most often of the chisel type as in figure 1b.

(a) (b)

**Figure 1.** Agricultural machinery: (a) chisel; (b) chisel knife type.

The chisel type soil breaking machines are designed to perform the soil loosening without overturning the furrow in order to increase the loosen layer thickness, for raising the capacity of water ingress [5].

## 2. Theoretical model

The cutting edge sharpness is particularly important in the chisel geometry. The wear of components of soil-tilling machinery is directly influenced by the mechanical properties of the soil [1]. The intrinsic properties of soil are considered the cohesion ($c_o$) and the internal friction angle ($\phi$) [7],[8],[9]. We consider a model of chisel (figure 2) with the vertical ($\sigma_z$) and horizontal ($\sigma_x$) stresses [7].

$$\begin{cases} \sigma_z = \gamma z \\ \sigma_{xp} = k_p \sigma_z + 2c_0\sqrt{k_p} \\ \sigma_{xa} = k_a \sigma_z - 2c_0\sqrt{k_a} \end{cases} \quad (1)$$

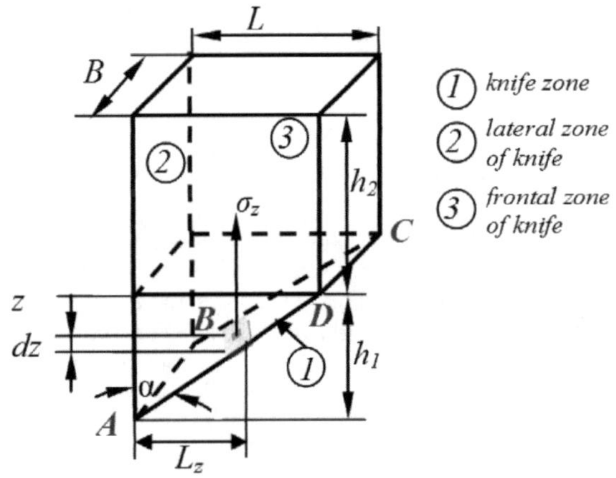


**Figure 2.** Geometry details of the chisel knife.

where $\gamma$ is volumetric weight; $z$- depth below soil surface; $k_a$, $k_p$ coefficients of active or passive earth pressure.

In penetration phenomena of chisel in the soil are distinguish two cases:

- penetration in zone of cutter (0, $h_1$), (figure 3 a);
- penetration in central zone of chisel (0, $h_2$), (figure 3 b).

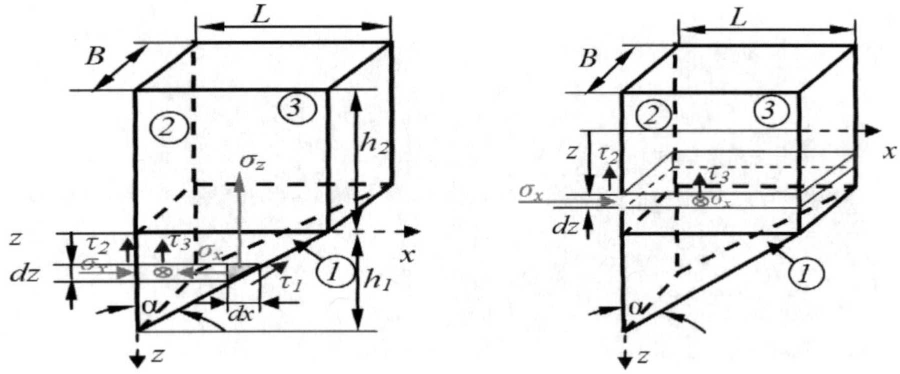


**Figure 3.** Stresses on the wall of knife: (a) knife zone; (b) central zone.

The penetration force is obtained by the mechanical equilibrium. Thus, for first case:
- in *knife zone* ( symbol 1 in figure 1 and figure 3 (a)

$$F_{a1la} = h_a^2 \cdot \left[ \tan(\alpha) + \mu_c \cdot \left( \sin(\alpha) + k_a \cos(\alpha) - \frac{4c_{0a}\sqrt{k_a} \cdot \cos(\alpha)}{h_a} \right) \right] \tag{2a}$$

for the case of active earth pressure (soil is expected to slide in download direction, along the surface of cutter);

$$F_{a1lp} = h_a^2 \cdot \left[ \tan(\alpha) + \mu_c \cdot \left( \sin(\alpha) + k_p \cos(\alpha) + \frac{4c_{0a}\sqrt{k_a} \cdot \cos(\alpha)}{h_a} \right) \right] \tag{2b}$$

for the case of passive earth pressure (soil is expected to slide in upward direction, along the surface of cutter);
- in *lateral zone of knife* (symbol 2 in figure 2 and figure 3 (a))

$$\begin{cases} F_{a12a} = \mu_c \cdot l \cdot h_a^2 \cdot \left( k_a - \dfrac{4c_{0a}\sqrt{k_a}}{h_a} \right) \\ F_{a12p} = \mu_c \cdot l \cdot h_a^2 \left( k_p + \dfrac{4c_{0a}\sqrt{k_p}}{h_a} \right) \end{cases} \tag{3}$$

- in *frontal zone of knife* (symbol 3 figure 2, 3 (a) two surfaces)

$$\begin{cases} F_{a13a} = 2\mu_c \tan(\alpha) l \cdot h_{1a} h_a \cdot \left( 1 - \dfrac{2}{3} k_a h_a^2 - 4c_{0a}\sqrt{k_a} + 2c_{0a}\sqrt{k_a} \cdot h_a \right) \\ F_{a13p} = 2\mu_c \tan(\alpha) l \cdot h_{1a} h_a \left( 1 - \dfrac{2}{3} k_p h_a^2 + 4c_{0a}\sqrt{k_p} - 2c_{0a}\sqrt{k_p} \cdot h_a \right) \end{cases} \tag{4}$$

The total dimensionless force in penetration direction ($z$), for first case (*knife zone*) is:

$$\begin{cases} F_{azla} = F_{a1la} + F_{a12a} + F_{a13a} \\ F_{azlp} = F_{a1lp} + F_{a12p} + F_{a13p} \end{cases} \tag{5}$$

where $\mu_c$ is the friction coefficient between chisel and soil; $\alpha$- the cutter angle (figure 2, figure 3 (a) and 3 (b)), and dimensionless force penetration parameter ($F_a$), geometrical parameters ($h_{1a}$, $h_a$), dimensionless cohesion parameter ($c_{oa}$) respectively:

$$\begin{cases} F_a = \dfrac{F}{\frac{1}{2}\gamma h_1^2 B} \\ h_{1a} = \dfrac{h_1}{B} \\ h_a = \dfrac{h}{h_1} \\ c_{0a} = \dfrac{c_0}{\gamma h_1} \end{cases} \tag{6}$$

If the coefficient of active or passive earth pressure ($k_a$), ($k_b$) respectively [1],[5] is:

$$\begin{cases} k_a = \dfrac{1-\sin(\phi)}{1+\sin(\phi)} \\ k_p = \dfrac{1+\sin(\phi)}{1-\sin(\phi)} \end{cases} \tag{7}$$

it is obtained the penetration force of chisel model for the first case.

Thus, figure 4 shows the dimensionless penetration force as a function to depth for some values of friction internal angle ($\Phi$), friction coefficient ($\mu_c$) and dimensionless cohesion parameter ($c_{oa}$).

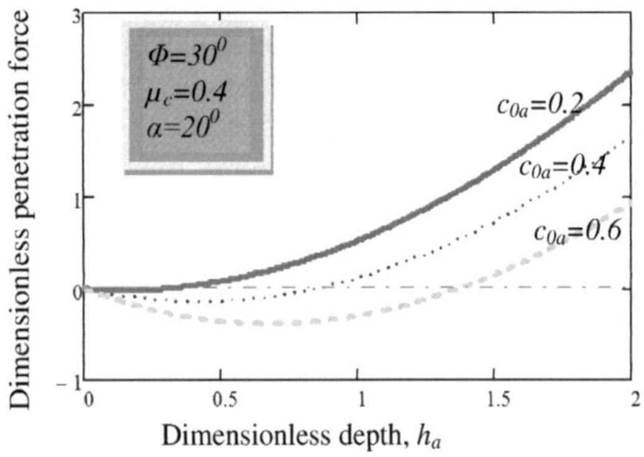


**Figure 4.** The dimensionless penetration force in knife zone for some soil types (active pressure earth).

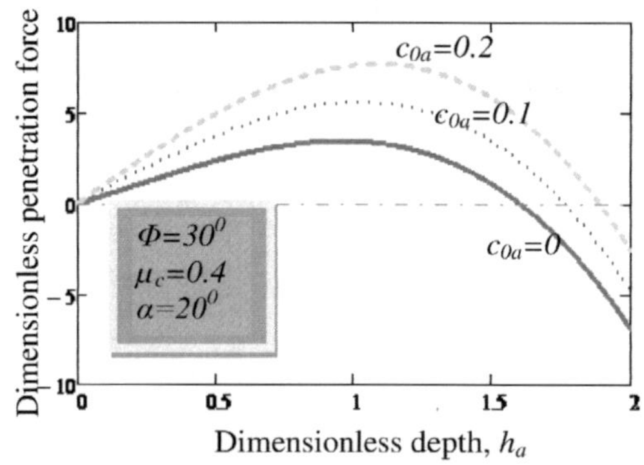


**Figure 5.** The dimensionless penetration force in knife zone for some soil types (passive pressure earth).

The figure 5 shows that fracture of soil appears when the penetration force will be zero. In this case, is observed the positive effect of chisel for the fragmentation of soils with high cohesion.

In the second case ($h_1<h<h_2$), the penetration force is determined similarly:

$$\begin{cases} F_{a21a} = F_{a11a} \\ F_{a22a} = \mu_c \cdot \left(k_a - 4c_{0a}\sqrt{k_a}\right) + 2\mu_c\left(k_a\left(h_a^2-1\right) - 4c_{0a}\sqrt{k_a}\left(h_a-1\right)\right) \\ F_{az2p} = F_{a21p} + F_{a22p} + F_{a23p} \\ F_{a23a} = 2\mu_c\tan(\alpha)\cdot h_{1a}\left(1-\dfrac{2}{3}k_a - 4c_{0a}\sqrt{k_a} + 2c_{0a}\sqrt{k_a} + 2\mu_c\tan(\alpha)\cdot h_{1a}\left(k_a\left(h_a^2-1\right) - 4c_{0a}\sqrt{k_a}\left(h_a-1\right)\right)\right) \end{cases} \tag{8}$$

The chisel penetration force in soil is defined by:

$$F_{aza} = \begin{vmatrix} F_{az1a}, & if \ \ 0 \le h_a \le 1 \\ F_{az2a}, & if \ \ h_a > 1 \end{vmatrix} \tag{9}$$

for active earth pressure and

$$F_{azp} = \begin{vmatrix} F_{az1p}, & if \ \ 0 \le h_a \le 1 \\ F_{az2p}, & if \ \ h_a > 1 \end{vmatrix} \tag{10}$$

for passive earth pressure.

The figures 6, 7 and 8 show the penetration dimensionless forces for some soils in contact with steel material of chisel sample.

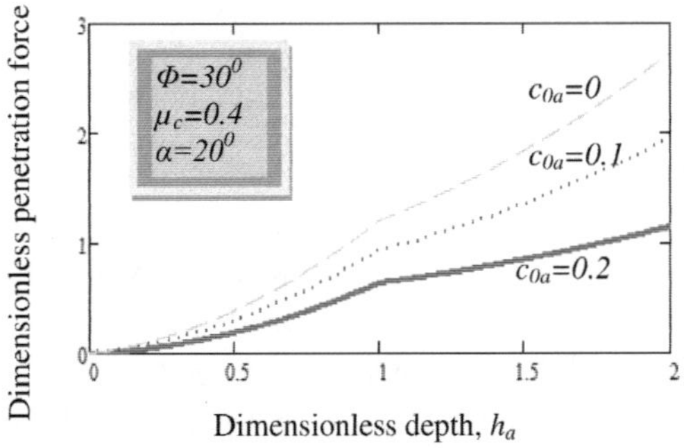


**Figure 6.** The dimensionless penetration force of chisel in sand with some friction angle (active pressure earth).

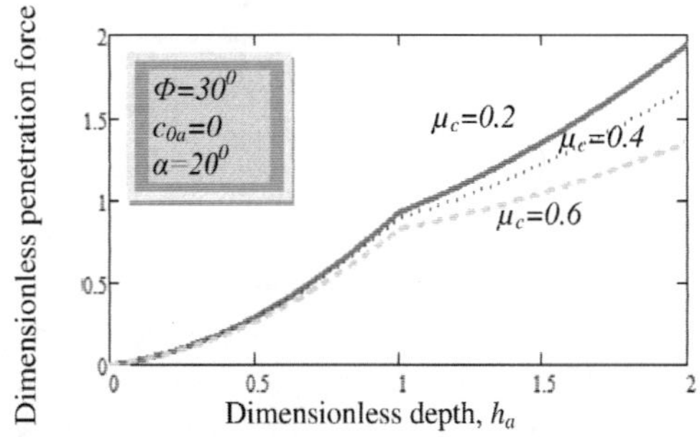


**Figure 7.** The dimensionless penetration force of chisel in sand with some friction coefficient between sand and chisel (active pressure earth).

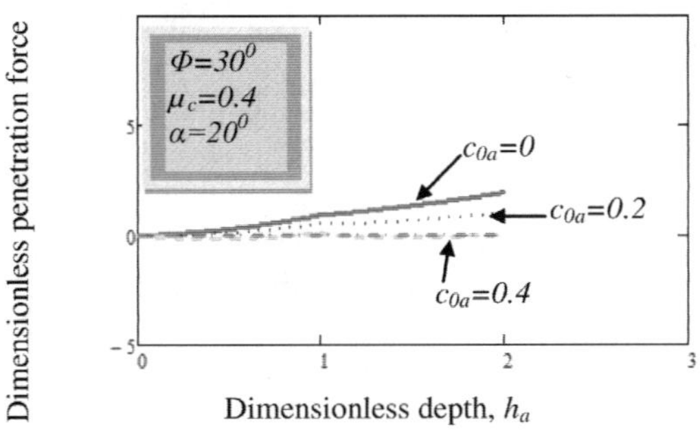


**Figure 8.** The dimensionless penetration force of chisel in soil with some cohesion strength (active pressure earth).

## 3. Experimental procedure

The penetration tests were performed with the UMT 2 tribometer (BRUKER® former CETR, USA) (figure 9). This apparatus can control and measure the applied loads, displacements and velocities of the indentor. Hard indenter used in these tests was made of steel, with prismatic form and an angle of $20^0$.

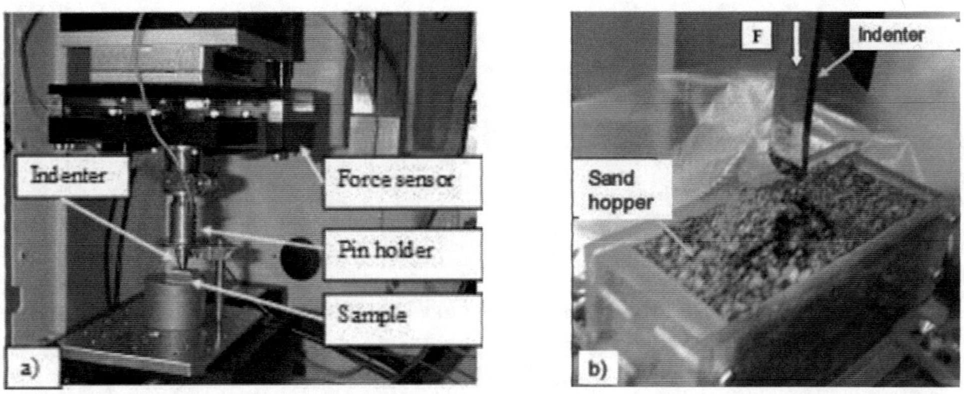


**Figure 9.** Penetration test details: (a) Penetration test set-up of UMT 2; (b) Schematic of penetration test set-up [10][11].

For testing the sand at the penetration resistance we used a plastics box to put the sample of sands. The plastic box was filled with sand with a quantity of 0.168 kg which was levelled to the same height over the entire tank. The surface area of the sand is 36 mm × 85 mm and the thickness is 32 mm.

**Table 1.** Features plastic box with and without sand.

| | Empty box | Sand | | Sand box | |
|---|---|---|---|---|---|
| | | Dry | Wet | Dry | Wet |
| **Weight** (kg) | 0.114 | 0.168 | 0.324 | 0.282 | 0.438 |

For the performed researches a certain type of sand was chosen sand that formed the basis of tool-soil interaction study, this being characterized by the lack of cohesion and high abrasive property.

The maximum load was limited to 19 N assuming that a 20 N (2 kg) force sensor is installed in the tribometer UMT 2. The penetration velocity was 2 mm/s. All tests were performed with penetration depths at 8 mm into the sand. Two test conditions were employed: dry and wet contact. These conditions were used in order to account for the effect of penetration lubrication. Wet contact tests were performed using water, in order to replicate the *in vivo* environment. The plastic box was filled with 0.156 kg of water.

The penetration tests were carried out with chisel knife type in two conditions: same area and different area of the sand.

There have been performed 10 tests for each work condition while eliminating outliers, out of which nine tests remained valid.

## 4. Experimental results

Figure 10 shows the load-displacement curves for the sand material in contact with the hard indenter, under wet conditions. Figure 11 shows the load-displacement curves for the sand material in contact with the hard indenter, under dry conditions.

In figures 10, 11, 12 were used the following notation:

- T1 – the average of penetrations tests in the same area;
- T2 - the average of penetrations tests in different area;
- T3 – the average of penetrations tests in the same area;
- T4 - the average of penetrations tests in different areas.

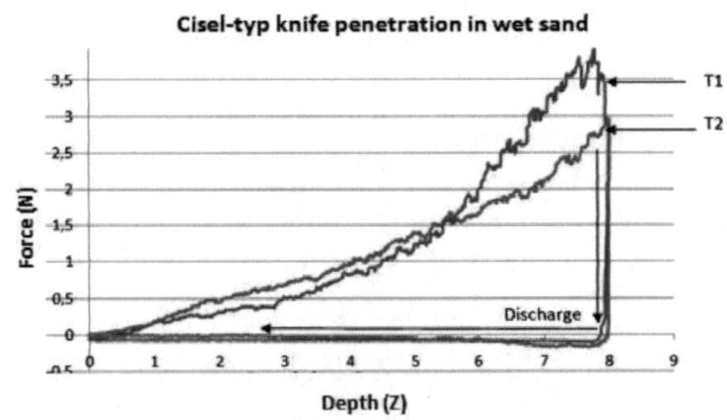


**Figure 10.** The experimental load-displacment curves for average of penetrations tests with chisel type knife under wet conditions.

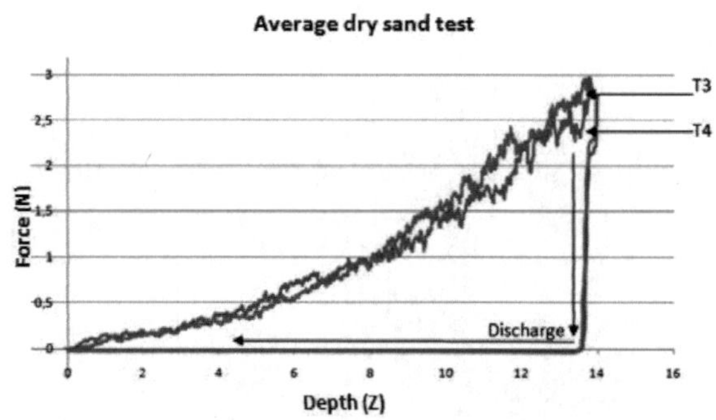


**Figure 11.** The experimental load-displacement curves for average of penetrations tests with chisel type knife under dry conditions.

Figure 12 shows the load-displacement curves for the sand material in contact with the hard indenter under dry and wet conditions.

The response of the surface to penetrations for different lubrication conditions is influenced by the mechanical properties of the material and it is noticeable in the all figures. For example, the peak penetration force of 3.92 N for sand tested in the same area is higher than the peak penetration force of 3.32 N for sand tested in different areas, in dry condition tests. The peak penetration force of 1 N for sand tested in the same area is approximated the same value for the peak penetration force of 0.89 N for sand tested in different areas, in wet conditions.

The penetration force for dry contact is always greater than for wet contact, regardless of material type.

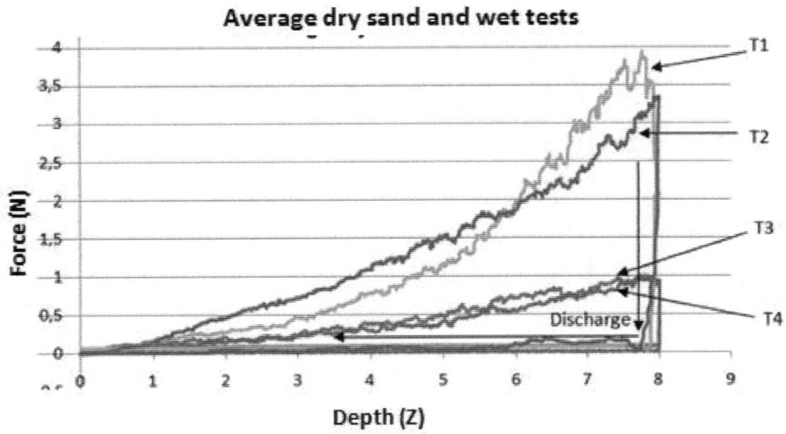


**Figure 12.** The experimental load-displacement curves for average of penetrations tests with chisel type knife under dry and wet conditions.

## 5. Conclusions

The answer to penetration mainly depends on nature, shape, size of knife and operating parameters such as speed, depth and working conditions.

Maximum penetration force in the same area for dry sand is higher than maximum penetration force in different areas for dry sand. Maximum penetration force in the same area for wet sand has approximately the same value as maximum penetration force in different areas for wet sand.

Sand humidity has influenced the resistance force at penetration, therefore the penetration force in case of dry sand is higher than penetration force for wet sand.

## References

[1] Braharu D, Băjenaru S, Vlăduţ V and Matache M 2007 Researches regarding materials selection of the operating parts manufacturing for soil cultivation. Materials and treatments used for theirs design *Annals of University of Craiova – Agriculture Montanology* **XXXVII** (B) 48-55

[2] Matache M, Ganga M, Mihai M, Postelnicu E and Bajenaru S 2008 Researches regarding determination of mechanical and wear characteristics for friction materials *Scientific Papers INMATEH* **28** 120-123

[3] Tanco C, Heraşcu (Roşca) M, Radu (Hanea) C Nicolae I 2011 Research on reconditioning agricultural plough by applying welding hardfacing *The 7th Intern. Conf. on Materials Science and Engineering – BRAMAT* 69-72

[4] Mueller M, Chotěborský R, Valášek P and Hloch S 2013 Unusual possibility of wear resistance increase research in the sphere of soil cultivation *Technical Gazette* **20** **(4)** 641-646

[5] Jankauskas V, Katinas E, Skirkus R and Alekneviciene V 2014 The method of hardening soil rippers by surfacing and technical-economic assessment *J. Frict. Wear* **35** **(4)** 270–277

[6] Verruijt A 2002 *Soil mechanics* ed Barends F B J and Steijger P M P C (Delft: CRC Pres)

[7] Keller T Lamandé M 2010 Challenges in the development of analytical soil compaction models *Soil Till. Res.* **111** 54-64

[8] Cole D M, Mathisen L U, Hopkins M A and Knapp B R 2010 Normal and sliding contact experiments on gneiss *Granul. Matter* **12 (1)** 69–86

[9] Naderi B M, Alimardani R, Hemmat A, Sharifi A, Keyhani A, Tekeste M Z and Keller T 2013 3D finite element simulation of a single-tip horizontal penetrometer–soil interaction. Part I: Development of the model and evaluation of the model parameters *Soil Till. Res.* **134** 153-162

[10] Chişiu G, Tiberiu L Tudor A 2011 Indentation tests to study the mechanical tribological properties of UHMWPE *UPB Scientifical Bulletin* Series D **73 (4)** 209-222

[11] Chişiu G and Tudor A 2014 Wear characteristics of UHMW polyethylene by scratching method *J Balk Tribol Assoc* **20 (1)** 138–150

# Experimental study on guide friction contribution in global power loss of a tooth chain transmission

**R Velicu, R Papuc, C C Gavrila and S Popa**

Product Design Mechatronics and Environment Department, Transylvania University of Brasov, Eroilor 29, 500036 Brasov, Romania

E-mail: rvelicu@unitbv.ro

**Abstract.** The subject of the paper is the friction between a tooth chain and the guide. An experimental study is developed with the aim of determining the contribution of chain-guide friction on the global friction of a basic tooth chain transmission. The measurements have been made on a chain friction rig, testing a basic tooth chain transmission with transmission ratio equal to 1, with a controlled tensioning device. The following parameters can be adjusted and measured: rotational speed, tensioning force in the chain, position of the guide, temperature and pressure of the oil used for lubrication. Friction torque at the input shaft is a sum of friction torques coming from bearings, chain and guide. The paper presents the contribution of the guide in the power loss by friction, as percent of the power loss from friction in chain and guide together. Influences of speed, tensioning force and oil temperature are presented.

## 1. Introduction

One of the first steps in the research for diminishing the friction of automobile engine is the correct evaluation of friction and the influences that really matter. The part of the fuel energy consumed in friction can be divided into groups based on data from [1, 2, 3, 4]: 12–45% is needed to overcome the rolling friction in the tire–road contact, 30–35% are needed to overcome friction in the engine system, 7–18% to overcome friction in the transmission system, and 10–18% to overcome friction in the brakes.

Very few experimental results on chain drive friction have been published. A testing rig, procedure and experimental data on contribution of chain, guide and bearings on global friction losses in the timing chain drive of an engine is presented in [5]. The testing rig is based on a full engine equipped in order to measure separate friction in bearings, chain and the two guides. The results are not showing the influences of tensioning, rotational speed and temperature.

Theoretical approaches on the friction in chain drive system, like [6], needs values of friction coefficients in the links. A correct evaluation of friction depends on the influence of different variables on friction coefficients. Influence of pressure, speed and temperature, with and without lubrication, on the friction between chain plates and guide materials are presented in [7, 8]. The conditions of testing are slightly different than in the real case. This paper presents only experimental results on friction in chain drives, which can be compared with results of theoretical approaches.

This paper is proposing a procedure for determination of the separate friction in the chain and in the guide of a basic chain transmission (transmission ratio equal to 1). It shows the influences of rotational speed, tensioning and lubricating oil temperature on the contribution of guide friction on the global (chain + guide) power losses.

The research developed at Transilvania University of Brasov is using the friction chain rig presented in [9] with the functional diagram from figure 1. It allows to control and measure the rotational speed at input shaft, chain tensioning (F) and temperature of the lubrication oil used for chain and bearing boxes. The measured input torque (T) is a sum of the friction torques from: input bearing box $T_{fbearing1}$, output bearing box $T_{fbearing2}$ and from the chain $T_{fchain}$. Friction in the chain is calculated by subtracting the bearing boxes friction from the measured input torque [9]. The procedure for measuring bearing boxes friction is presented in [10] and the measured bearing boxes friction depending on speed, tensioning and oil temperature is presented in [11].

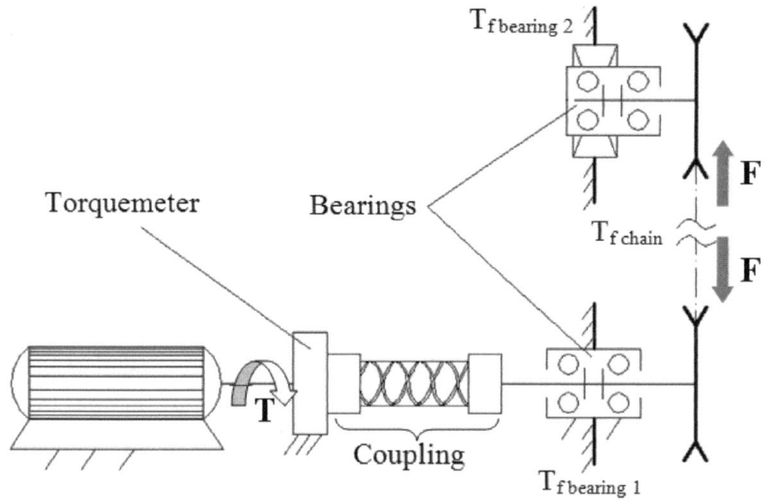


**Figure 1.** Functional diagram of the chain friction rig.

## 2. Equipment and guide friction device

In order to apply a controlled tensioning with a guide, the device presented in figure 2 is built, considering the particularities of the chain rig and the types of chains that will be tested.

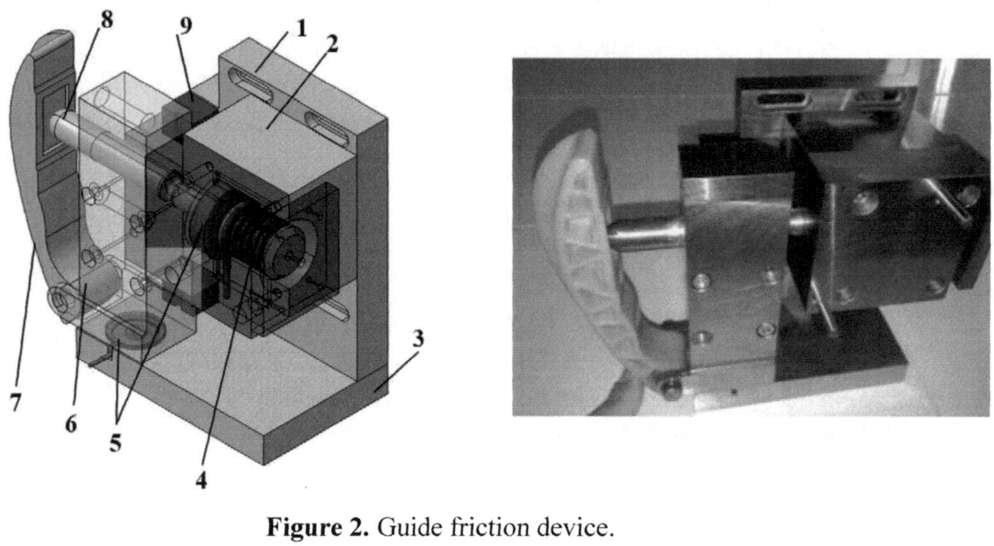


**Figure 2.** Guide friction device.

The guide friction device consists of: 1 – main board for atachment on the rig; 2 – block containing the elements for adjustment and measure the pushing of the guide; 3 – base plate; 4 – tensioning spring; 5 – force sensors; 6 – guide link on the base plate; 7 – articulated guide; 8 – plunger for pushing the guide; 9 – translational bearing.

Figure 3 presents the chain rig equiped with the guide friction device.

**Figure 3.** Chain friction rig with guide friction device.

The components of the chain friction rig equiped for guide friction measurements are: 1 – input bearing box; 2 – oil circuit for bearings lubrication; 3 – oil circuit for chain and guide lubrication; 4 – separate spraying oil circuits for chain and guide; 5 – output chain sprocket; 6 – output bearing box; 7 – slider, support of output bearing box; 8 – output shaft; 9 – chain; 10 – guide friction device; 11 – input shaft; 12 – input chain sprocket.

## 3. Measurement procedure

### *3.1. Testing parts and specifications*

The tested chain is a tooth chain (silent chain) with 100 links and 6.35 mm pitch. The two sprockets are identical (transmission ratio equal to 1) with number of teeth 23. The centre distance of the chain drive is approximately 300 mm.

The guide is made of PA66, circular shape with radius of 200 mm in contact with the chain. The guide is placed so that the chain has a 9.5 mm deviation from the initial line.

### *3.2. Preparation of rig and types of measurements*

Measurements are performed for three separate structures of the rig, looking for separate results.

- Measurements of friction in bearings as input torque depending on speed, tensioning and oil temperature ($T_{fbearings} = T_{fbearing1} + T_{fbearing2}$). The measurements are made for the two bearing boxes mounted together, as presented in [9, 10].
- Measurements of friction in chain and bearings together (see figure 1) as input torque depending on speed, tensioning and oil temperature ($T_{fbearings+chain} = T_{fbearings} + T_{fchain}$), without the guide friction device. Friction in the chain is calculated as $T_{fchain} = T_{fbearings+chain} - T_{fbearings}$.
- Measurements of friction in chain, bearings and guide, together (see figure 1) as input torque depending on speed, tensioning and oil temperature ($T_{fbearings+chain+guide} = T_{fbearings} + T_{fchain} + T_{fguide}$), with the guide friction device. Friction in the guide is calculated as $T_{fguide} = T_{fbearings+chain+guide} - T_{fbearings+chain}$.

### *3.3. Testing program*

The testing program is consisted of steps with constant controlled parameters (rotational speed, tensioning force and oil temperature). The first step is usually longer since it must check and adjust the oil temperature and also stabilize the temperature distribution on all the elements of the rig. The time for each step is minimum 250 seconds. The role of these steps is to stabilize the system and create the steady state conditions. The readings that count in evaluation of bearing friction are only the one of the steady state period [10].

Friction torques ($T_{fbearings}$, $T_{fbearings+chain}$, $T_{fbearings+chain+guide}$) have been measured for: rotational speed, n: 1000, 2200, 3000, 5000 rot/min; tensioning force, F: 0.5, 1, 1.5 kN; oil temperature for chain and guide lubrication, t: 40, 90 °C. All the tests are repeated 3 times and an average of the results is considered.

### *3.4. Oil measurements*

The oil used in bearing lubrication is Castrol Edge 5W30 and the measured viscosity depending on temperature [10] is presented in figure 4, showing important decrease with increase of temperature, especially at lower temperatures.

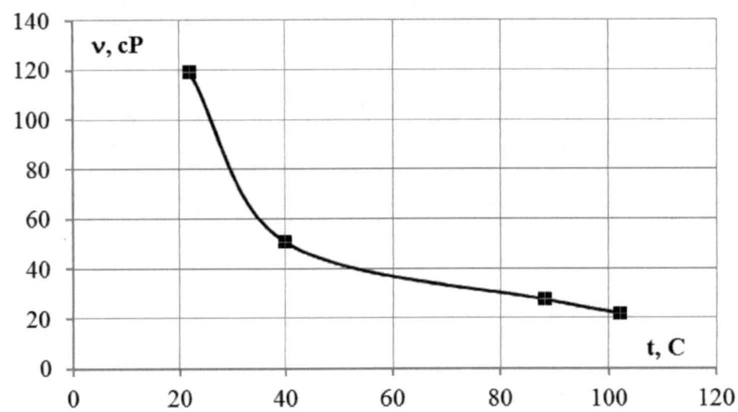


**Figure 4.** Kinematic viscosity versus temperature.

## 4. Results and discussion

By loading the chain with the tensioning force F, the force $F_n$ pushing on the guide changes. Figure 5 presents the variation of the force pushing on the guide depending on the tensioning force of the chain.

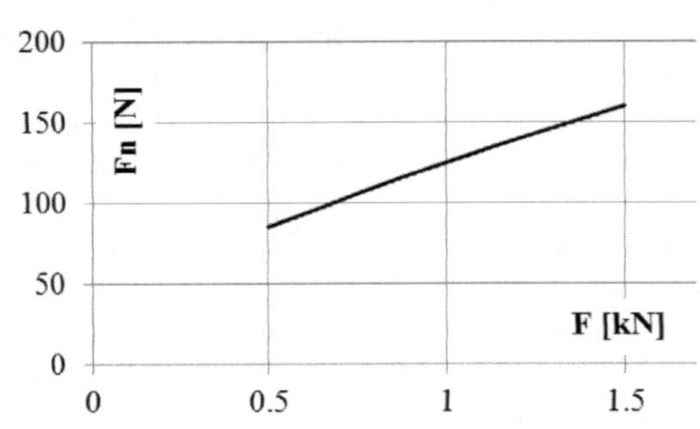


**Figure 5.** Force on guide depending on tensioning force on chain.

Figure 6 presents the steps of determining the friction in guide as friction torque, for constant lubricating oil temperature of 40 °C. The values of friction torques, presented in figure 6, are calculated as percent of the friction torque from bearings and chain at 1000 rot/min rotational speed and F = 0.5 kN tensioning force (see figure 6b). Guide friction (see figure 6c) is obtained by

IOP Conf. Series: Materials Science and Engineering **174** (2017) 012051 doi:10.1088/1757-899X/174/1/012051

subtracting friction in bearings and chain (see figure 6, b) from friction in bearings, chain and guide (see figure 6a). It can be seen that friction in the guide is almost constant with rotational speed, at constant tensioning and it is a very small amount compared with global friction (bearings, chain and guide friction together).

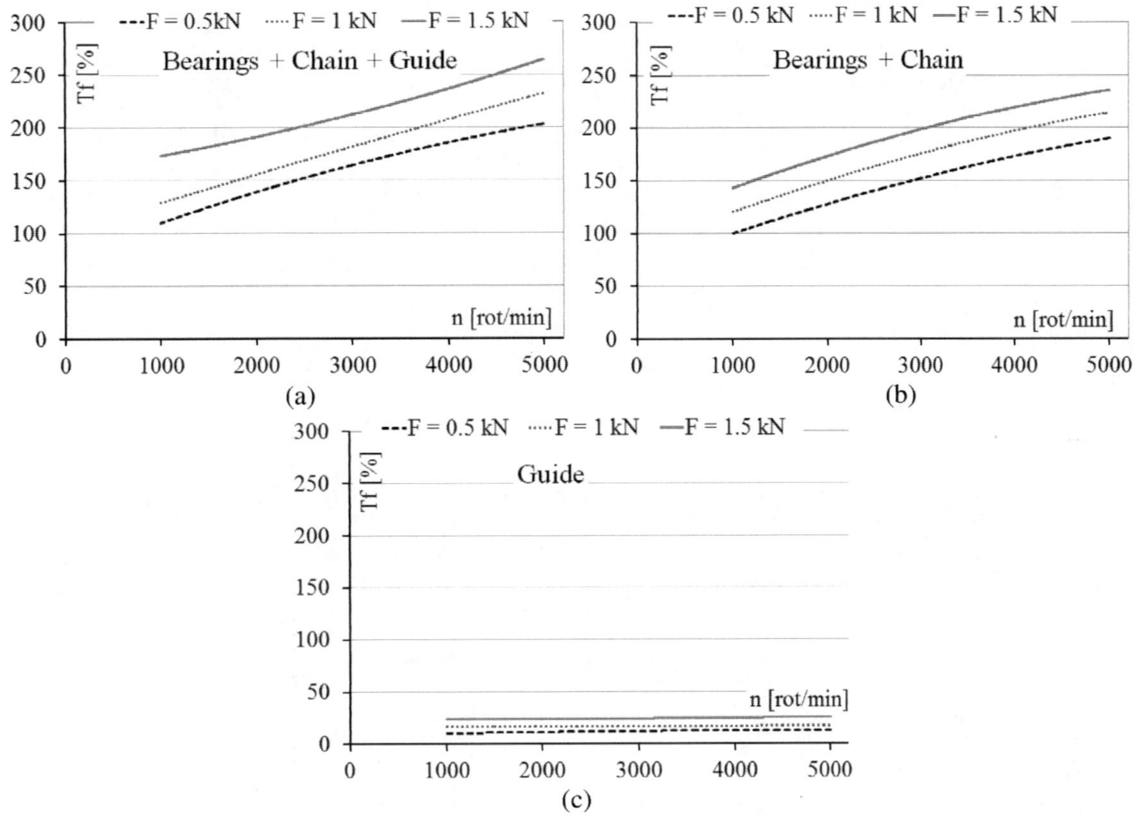


**Figure 6.** Steps for determining friction torque in guide: (a) sum of friction torque in bearings, chain and guide; (b) sum of friction torque in bearings and chain; (c) friction torque in the guide resulted as subtraction of (b) from (a).

In this case, the minimum guide friction contribution on the global (bearings, chain and guide) friction is approximately 6.5%, for maximum rotational speed (n = 5000 rot/min) and minimum tensioning force (F = 0.5 kN).

The maximum guide friction contribution on the global (bearings, chain and guide) friction is approximately 14%, for minimum rotational speed (n = 1000 rot/min) and maximum tensioning force (F = 1.5 kN).

Friction in bearings will not be considered in the further analysis since chain and guide transmissions can work with different bearings depending on applications. Only guide friction and chain friction are compared in the following analyse.

Figure 7 presents friction in guide, measured as torque, in percent of the minimum torque value, obtained at n = 1000 rot/min and F = 0.5 kN. Guide friction is presented depending on rotational speed, for three steps of tensioning and for two values of lubricating oil temperature.

Both diagrams show a slightly increase of guide friction with rotational speed. This can be explained by the influence of centrifugal force which slightly increases the pushing of the guide.

The guide friction increases almost directly with the loading force. For smaller oil temperature (40 °C), guide friction increases with 220 – 230%, for 3 time increase of loading force. For higher oil temperature (90 °C), loading force influence is bigger and guide friction increases with 350 – 380% higher, for 3 time increase of loading force.

Figure 8 presents the contribution of guide and chain friction on a chain transmission with guide, excluding bearing friction, for different rotational speed and two levels of loading force, for 40 °C oil temperature.

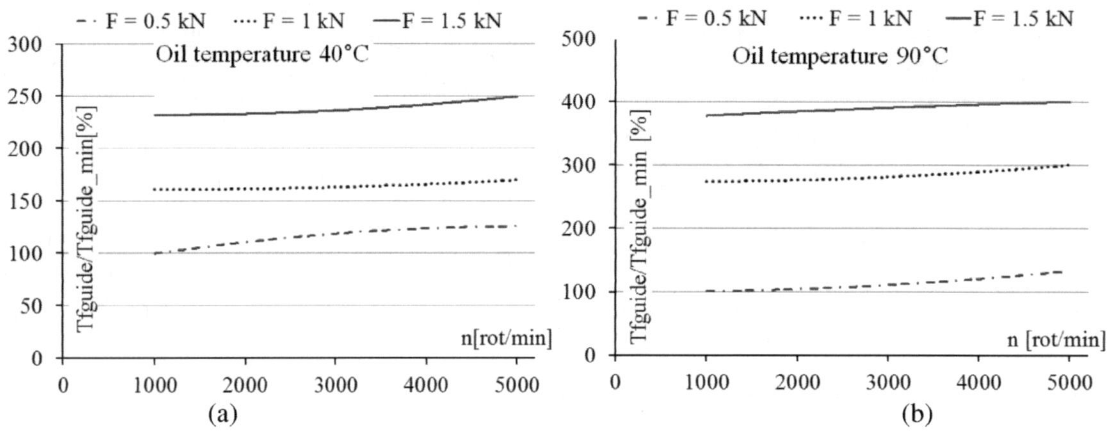


**Figure 7.** Guide friction depending on rotational speed, for constant levels of tensioning force and for 40 °C (a) and 90 °C (b) oil temperature.

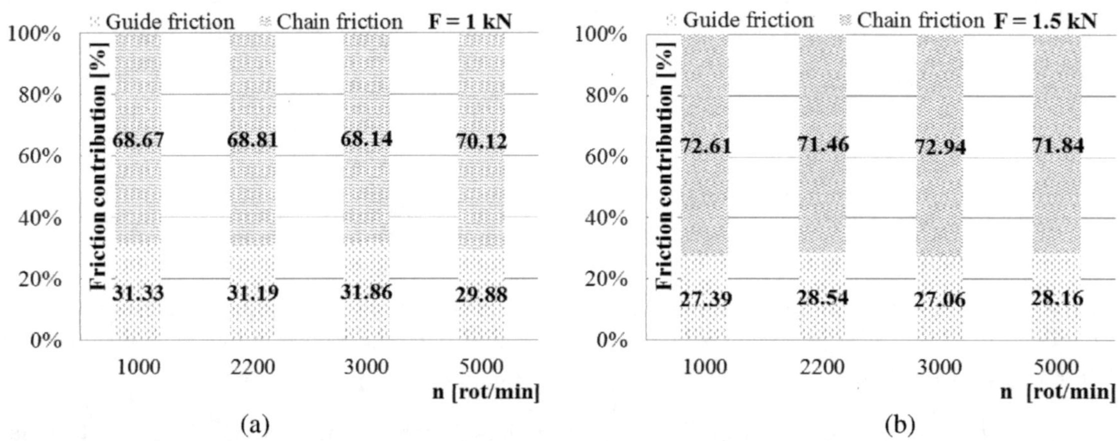


**Figure 8.** Guide friction vs. chain friction contribution depending on rotational speed at 40 °C lubricating oil temperature and for constant levels of tensioning force (a) F = 1kN and (b) F = 1.5 kN.

Figure 9 presents the contribution of guide and chain friction on a chain transmission with guide, excluding bearing friction, for different rotational speed and two levels of loading force, for 90 °C oil temperature.

The influence of rotational speed on guide and chain friction contribution is very small, as seen in all the diagrams from figures 8 and 9.

Guide friction contribution is decreasing with increase of loading force. In the case of smaller temperature (40 °C), an increase of loading force from 1 kN to 1.5 kN determines a decrease of guide friction contribution from 30 – 32% to 27 – 28.5%. In the case of higher temperature (90 °C), for an increase of loading force from 1 kN to 1.5 kN determines a decrease of guide friction contribution from 29 – 30% to 25.5 – 26.5%.

Guide friction contribution is slightly decreasing with increasing temperature.

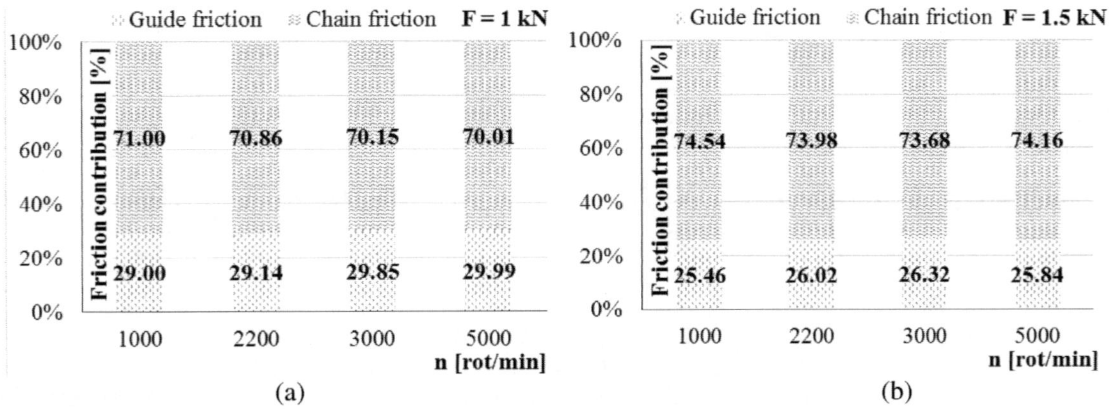


**Figure 9.** Guide friction vs. chain friction contribution depending on rotational speed at 90 °C lubricating oil temperature and for constant levels of tensioning force (a) F = 1kN and (b) F = 1.5 kN.

## 5. Conclusion

The results presented in this paper show that, in the case of chain drives, even if the guide friction is only half of the chain friction, it is still an important aspect to be studied for diminishing power losses and improving mechanical efficiency. Influences from loading and lubrication, presented in this paper, are giving a good start for measures of improvement. Further research should focus on different chains, different geometry of the guide, different materials and different positions of the guide.

## References

[1] Tung S C and McMillan M L 2004 Automotive tribology overview of current advances and challenges for the future *Tribol Int* **37 (7)** 517-536
[2] Holmberg K, Andersson P and Erdemir A 2012 Global energy consumption due to friction in passenger cars *Tribol Int* **47 (14)** 221-234
[3] Schwaderlapp M, Koch F and Dohmen J 2000 Friction reduction - the engine's mechanical contribution to saving fuel *Seoul 2000 FISITA World Automotive Congress* pp 1-8
[4] Taylor C M 1998 Automobile engine tribology—design considerations for efficiency and durability *Wear* **221** 1-8
[5] Hyakutake T, Inagaki M, Matsuda M, Hakamada N and Teramachi Y 2001 Measurement of friction in timing chain *JSAE Review* **22 (3)** 343–347
[6] Lates M T and Gavrila C 2014 Friction phenomenon in polyamide – steel plate front face type contacts *Annals of the Oradea University, Fascicle of Management and Technological Engineering* **XIII (XXIII)** pp75-78
[7] Zielinski F, Abt A, Laer B, Klumpp P and Willand J 2008 Effiziente Methodik zur Berechnung der Reibung von Kettentrieben in Verbrennungsmotoren *MTZ Conf. Virtual Powertrain Creation*
[8] Lates M T and Gavrila C 2016 Study of the friction coefficient in polyamide / steel type contacts in non-lubricated conditions *Appl Mech Mater* **823** 485-488
[9] Todi-Eftimie A, Velicu R, Saulescu R and Jaliu C 2013 Bearing friction vs. chain friction for chain drives *Adv Mat Res* **753-755** 1110-1113
[10] Velicu R and Lates M 2015 On the measurement procedure for testing friction in bearing mountings *Annals of the Oradea University, Fascicle of Management and Technological Engineering* **XIV (XXIV)** 59-64
[11] Velicu R and Popa S 2016 Experimental study of bearing boxes friction depending on load speed and oil temperature *Annals of the Oradea University, Fascicle of Management and Technological Engineering* **XV (XXV)** 5-9

IOP Conf. Series: Materials Science and Engineering **174** (2017) 012052 doi:10.1088/1757-899X/174/1/012052

# Studies on Tribological Behavior of Aluminum Nitride-Coated Steel*

**G C Ionescu[1], I Nae[1], R G Ripeanu[1], A Dinita[1] and G Stan[2]**

[1] Department of Mechanical Engineering, Petroleum-Gas University of Ploiesti, 39 Bucharest Blvd., 100680 Ploiesti, Romania.
[2]National Institute of Materials Physics, 105 bis Atomistilor Street, 077125 Magurele-Ilfov, Romania.

E-mail: ionescu_g_r@yahoo.con, inae59@gmail.com, rrapeanu@upg-ploiesti.ro, alindinita@yahoo.com, george_stan@infim.ro.

**Abstract.** The new opportunities introduced by the large development of the IoT (internet of things) are increasing the demand for sensors to be located as close as possible to the supervised process. The Aluminum Nitride (AIN) is one of the most promising materials for sensors due to its piezoelectric, excellent mechanical properties, chemical inertness and high melting point. Due to these material properties, the AlN sensors are suitable to operate in high temperature and harsh environment conditions and therefore are very promising to be employed in industrial applications. In this article are presented the studies conducted on several Aluminum Nitride-Coated Steel structures with the goal of producing sensors embedded in the ball bearings, bearings and other mobile parts of machine tools. The experiments were conducted on simple coatings structures without lubricating materials and the obtained results are promising, demonstrating that, with some limitations the AIN could be used in such applications.

## 1. Introduction

The Industrial Internet of Things (IIoT), could be considered as an extension of automation and connectivity also known as machine-to-machine (M2M) communication. IIoT is, growing in parallel with Internet of Things (IoT) due to the spectacular development of smart and connected [1].

People working in industrial environment would like to have smart sensors, with low energy consumption - eventually capable of harvesting energy from the machine tool where is installed, capable to withstand harsh environmental condition such as temperature, abrasion etc, wireless connected with the machine automation system and with factory's monitoring system.

Such a multitude of requirements are fulfilled only by some special materials, consequently it is necessary to investigate which are the most suitable and furthermore which could be produced through techniques adequate for the desired applications.

A basic criterion in choosing the material will be its possibility of being used for developing energy harvesting devices, consequently piezoelectric properties of material will be a must. Another criterion will be its possibility to be embedded on the monitored component of the machine tool. Although there it is a large area of piezoelectric materials there are limitations in regard with their capability of working in harsh industrial environment and not the last technical possibility of using them as presented in table 1.

* This paper was accepted for publication in Proceedings after double peer reviewing process but was not presented at the Conference ROTRIB'16

**Table 1.** Comparison of major charactheristics of piezoelectric materials [1].

| material/ properties | $Pb(Zr,Ti)O_3$ | $SiO_2$ | $GaPO_4$ | $La_2Ga_2SiO_{14}$ | AIN | $LiNbO_3$ |
|---|---|---|---|---|---|---|
| usable temperature ($C^0$) | 250 | 350 | 920 | 1000 | 1200 | 1200 |
| piezoelectric constant $d_{33}$ (pC/N) | 6 | 2 | 6 | 6 | 6 | 35 |
| Available technology | Yes | No | No | No | Yes | Yes |

As it could be observed in the above listed table the choice is somehow limited in this area. Although AIN (Aluminum Nitride) is a well known material, extensively studied for sensing applications in the last 20 years [2,3], for industrial applications its applications are not exhausted. AlN piezoelectric layers could be deposited by sputtering directly on the equipment parts needed to be monitored. Therefore it is possible to obtain embedded sensors, avoiding the reliability problems associated with the bonding or sensor attachment. For applications on which we are interested in (sensors applied on the surface of the element to be monitored) radio-frequency magnetron sputtering (RF-MS) is the only technical possibility. The deposition of highly structured AlN is dependent on the underlying films, the topography of the surface, and the deposition process parameters [4,5,6,7].

## 2. Sensor element manufacturing

With the goal of testing mechanical properties of the AIN layer deposited on materials similar to the one used in the machine tools, bearings, and ball bearings two samples with 60X60 mm dimensions were manufactured as presented in figure 1. It was decided to manufacture, one sample on Stainless Steel S3400 substrate and one sample on Inconel substrate. The reason to use two different substrates was also to test the different thermal behavior of AIN layer on these two materials.

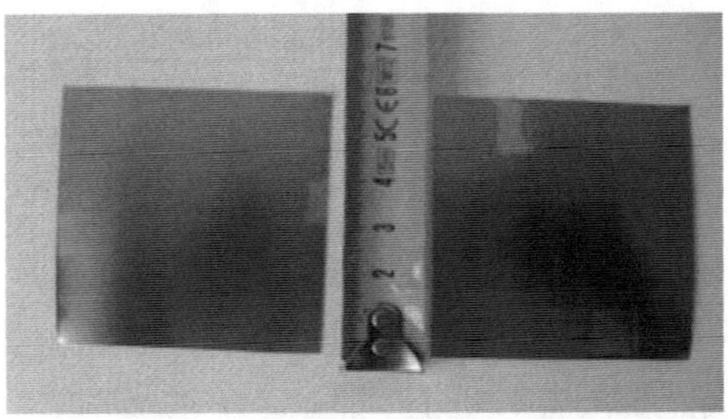

**Figure 1**. The two samples produced (Inconel left, Stainless Steel 3400 right).

Highly oriented (with *c* axis perpendicular on substrate) wurtzitic aluminum nitride (AlN) thin films were deposited by reactive radio-frequency magnetron sputtering (RF-MS) on stainless steel (SS) foil for sensor element fabrication. This type of substrate has few advantages: *(i)* it is flexible, and therefore the obtained sensor could be conformally attached to some mechanical parts that need to be monitored having also non-planar forms; *(ii)* it is resistant to the high temperatures that eventually could be developed in some situations during vibration monitoring; *(iii)* opens the way of realizing in the future the piezoelectric vibration sensor by depositing directly the AlN layer on the steel pieces that need to be monitored.

A high purity aluminum target (99.9999%, Mateck GmbH) with thickness of 2 mm and diameter of 110 mm was used for depositing the AlN films. The target was fixed on a magneton cathod whose plasma ring is about 55 mm. Before their introduction into the deposition chamber, the substrates were successively cleaned ultrasonically in acetone and isopropyl alcohol for 10 min and then dried in argon flow. The target-to-substrate separation distance was set at 35 mm.

Firstly the deposition chamber was evacuated down to a base pressure of ~$2x10^{-4}$ Pa, and then the high purity working gases – argon and nitrogen – were admitted. The selected working pressure was 0.2 Pa and nitrogen partial pressure was 25%. The total gas flow was 40 sccm. The gas flows were controlled with Teledyne Hasting electronic mass flow meters and the gas pressure was monitored by an Alcatel capacitance vacuum gauge.

Prior to the deposition, the target was sputter-cleaned for 20 min in argon and then another 10 min in the working atmosphere (argon nitrogen mix). During target cleaning, the substrates were masked with a stainless steel shield in order to avoid undesired depositions onto them. Further on, the substrates were etched for 15 min in argon plasma produced by a wolfram plasmatron, in order to remove thin native oxide layers or other impurities which might persist after the ultrasonic cleaning.

An AlN layer with thickness of about 1 micron was then deposited on the stainless steel substrate with a deposition rate of about 14 $nm \cdot min^{-1}$ using a radio-frequency (1.78 MHz) generator and maintaining a low RF power (~100 W) in order to avoid overheating of the target surface. The substrates were not heated during deposition and their temperature was only dependent on plasma self-heating (~$50C^0$) [6].

By using the same approach an AlN layer of about 1 micron was deposited on Inconel substrate.

The crystallographic quality of the AlN layers was investigated by X-ray diffraction (XRD) using a Bruker-AXS D8 Advance diffractometer in parallel beam setting, with CuKα radiation. Measurements were performed in the typical symmetric (θ–θ) geometry and by rocking curves (RC) in order to determine the degree of angular dispersion of the c-axis of the AlN crystallites around the surface normal.

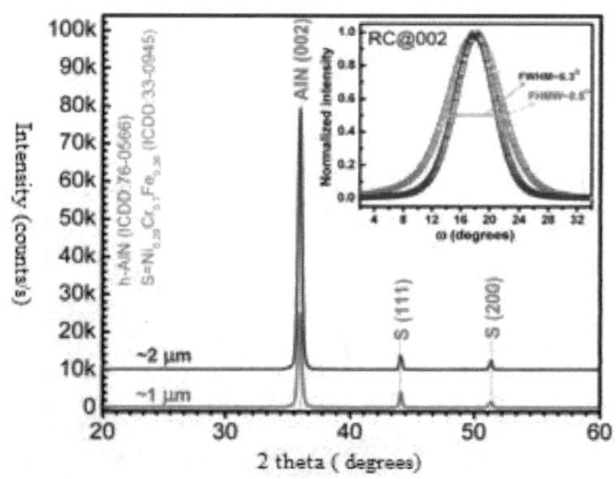


**Figure 2.** Symmetric geometry XRD patterns of AlN/Inconel sputtered films. Insert: rocking curves corresponding to the (002) AlN crystal planes.

The XRD patterns (figure 2), indicated that AlN films deposited on Inconel type substrate elicit only the 002 diffraction line of the hexagonal AlN phase (ICDD: 76-0566). This is indicative of a strongly c-axis textured material. The diffraction maxima of the metallic substrate have also been observed due to the low thickness of the sputtered films.

The full width at half maximum (FWHM) of the RCs, associated to the standard deviation of the crystallites' alignment, decreases from 8.6° to 6.3° (figure 2-insert), when the AlN film thickness is increased from ~1 μm to ~2 μm, suggestive a gradual orientation of crystallites during film growth.

## 3. Tests conducted

The programme of testing the AlN layers was developed based on the challenges predicted to be encountered by the sensing elements to be developed. Thus it was established that only materials which could be coated with AlN and may withstand high temperatures (above $200C^0$) would be further tested for the other requirements.

### *3.1. Temperature testing*

The testing of samples behavior on different temperature were conducted by using a thermal chamber CTS - Thermal Shock Test Chambers, Series TSS capable of delivering temperatures up to 600C$^0$ presented in figure 3.

According to the technical literature the following data for the temperature tests were considered: Thermal expansion coefficient: AlN: 4.15 x10$^{-6}$ K$^{-1}$ [7] Stainless steel S 3400: 17.3 x10$^{-6}$ K$^{-1}$[8], Inconel: 11.5 x10$^{-6}$ K$^{-1}$ [9].

It was interesting to observe that after exposing the sample with base material stainless steel to a temperature around 200C$^0$ on the sample occurred exfoliations of the AlN layer (figure 4). For the sample with base material of Inconel the temperature raised up to 500C$^0$ without occurring any exfoliations. Thus the decision taken was to continue the tests only on the samples which had been manufactured on Inconel base material.

**Figure 3.** CTS Thermal Shock Test Chambers, Series TSS.

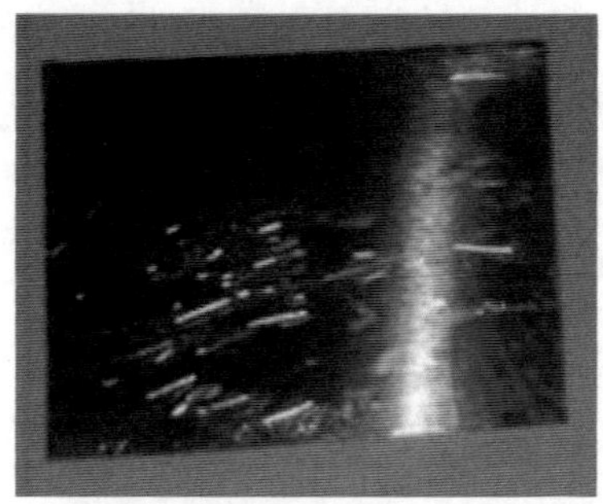

**Figure 4.** Exfoliation occurred on stainless steel sample at 200C$^0$.

### *3.2. Hardness*

The test equipment used for hardness tests was the Dura Scan – Emco TEST Micro hardness tester with extended load range from 10 gf - 10 kgf for Vickers and Knoop tests. This system presents a unique system of combined closed- loop and dead weight technology ensuring accurate and repeatable test loads. In addition the completely automatic test cycles eliminate operator influence. The test machine is presented in figure 5 and the test results in figure 6.

**Figure 5.** Emco TEST Micro hardness tester.

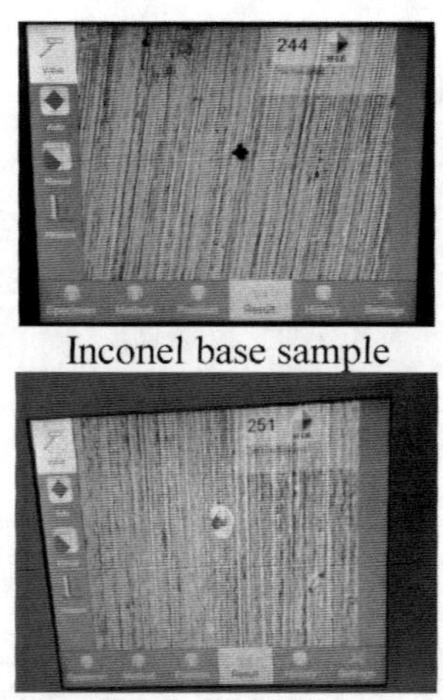


**Figure 6.** The two test results for Inconel and Stainless steel bases.

*3.3. Abrasion testing*

Two Inconel coated with AIN samples were tested, using different testing parameters as presented in figure 7, on CSM Instruments tribometer (figure 8). The block diagram of tribometer active section is presented in figure 11. In addition in order to have a reference two samples of Inconel without coating were tested by using similar parameters.

**AIN sample 2-v0.1**

Tribo parameters

| Acquisition | Sample | Static partner | Environment |
|---|---|---|---|
| Radius : 6.01 [mm]<br>Lin. speed : 0.10 [m/s]<br>Normal load : 1.00 [N]<br>Stop condit. : 50.00 [m]<br>Effective stop : meters<br>Acquisition rate : 2.6 [hz] | -Coating : AIN<br>-Substrate : INCONEL<br>-Cleaning : MEC<br>-Supplier : INOE | -Coating : No<br>-Substrate : 100Cr6<br>-Cleaning : M.E.C.<br>-Supplier : CSM<br>-Dimension : 6.00 [mm]<br>-Geometry : Ball | Temperature : 22.50 [°C]<br>Atmosphere : Air<br>Humidity : 56.60 [%] |

| Sample | Static partner | Calculations |
|---|---|---|
| Worn track section : 403.7 µm2<br>Young's modulus : 310.0 gpa<br>Poisson ratio : 0.290 | Worn cap diameter : 244.0 µm<br>Young's modulus : 200.0 gpa<br>Poisson ratio : 0.300 | Sample wear rate : 0.0003049 mm3/n/m<br>Partner wear rate : 1.16E-006 mm3/n/m<br>Max Herzian stress : 0.7254 gpa |

**AIN sample 2-v0.2**

Tribo parameters

| Acquisition | Sample | Static partner | Environment |
|---|---|---|---|
| Radius : 7.06 [mm]<br>Lin. speed : 0.20 [m/s]<br>Normal load : 1.00 [N]<br>Stop condit. : 50.00 [m]<br>Effective stop : meters<br>Acquisition rate : 4.5 [hz] | -Coating : AIN<br>-Substrate : INCONEL<br>-Cleaning : MEC<br>-Supplier : INOE | -Coating : No<br>-Substrate : 100Cr6<br>-Cleaning : M.E.C.<br>-Supplier : CSM<br>-Dimension : 6.00 [mm]<br>-Geometry : Ball | Temperature : 22.50 [°C]<br>Atmosphere : Air<br>Humidity : 56.60 [%] |

| Sample | Static partner | Calculations |
|---|---|---|
| Worn track section : 173.1 µm2<br>Young's modulus : 310.0 gpa<br>Poisson ratio : 0.290 | Worn cap diameter : 184.0 µm<br>Young's modulus : 200.0 gpa<br>Poisson ratio : 0.300 | Sample wear rate : 0.0001536 mm3/n/m<br>Partner wear rate : 3.751E-007 mm3/n/m<br>Max Herzian stress : 0.7254 gpa |

**INCONEL sample 2-v-01**

Tribo parameters

| Acquisition | Sample | Static partner | Environment |
|---|---|---|---|
| Radius : 6.01 [mm]<br>Lin. speed : 0.10 [m/s]<br>Normal load : 1.00 [N]<br>Stop condit. : 50.00 [m]<br>Effective stop : meters<br>Acquisition rate : 2.6 [hz] | -Coating : No<br>-Substrate : INCONEL<br>-Cleaning : MEC<br>-Supplier : INOE | -Coating : No<br>-Substrate : 100Cr6<br>-Cleaning : M.E.C.<br>-Supplier : CSM<br>-Dimension : 6.00 [mm]<br>-Geometry : Ball | Temperature : 22.50 [°C]<br>Atmosphere : Air<br>Humidity : 56.60 [%] |

| Sample | Static partner | Calculations |
|---|---|---|
| Worn track section : 39.0 µm2<br>Young's modulus : 210.0 gpa<br>Poisson ratio : 0.300 | Worn cap diameter : 152.0 µm<br>Young's modulus : 200.0 gpa<br>Poisson ratio : 0.300 | Sample wear rate : 2.946E-005 mm3/n/m<br>Partner wear rate : 1.746E-007 mm3/n/m<br>Max Herzian stress : 0.6483 gpa |

**INCONEL sample 2-v0.2**

Tribo parameters

| Acquisition | Sample | Static partner | Environment |
|---|---|---|---|
| Radius : 7.03 [mm]<br>Lin. speed : 0.20 [m/s]<br>Normal load : 1.00 [N]<br>Stop condit. : 50.00 [m]<br>Effective stop : meters<br>Acquisition rate : 4.5 [hz] | -Coating : No<br>-Substrate : INCONEL<br>-Cleaning : MEC<br>-Supplier : INOE | -Coating : No<br>-Substrate : 100Cr6<br>-Cleaning : M.E.C.<br>-Supplier : CSM<br>-Dimension : 6.00 [mm]<br>-Geometry : Ball | Temperature : 22.50 [°C]<br>Atmosphere : Air<br>Humidity : 56.60 [%] |

| Sample | Static partner | Calculations |
|---|---|---|
| Worn track section : 83.0 µm2<br>Young's modulus : 210.0 gpa<br>Poisson ratio : 0.300 | Worn cap diameter : 192.0 µm<br>Young's modulus : 200.0 gpa<br>Poisson ratio : 0.300 | Sample wear rate : 7.329E-005 mm3/n/m<br>Partner wear rate : 4.449E-007 mm3/n/m<br>Max Herzian stress : 0.6483 gpa |

**Figure 7.** The testing parameters.

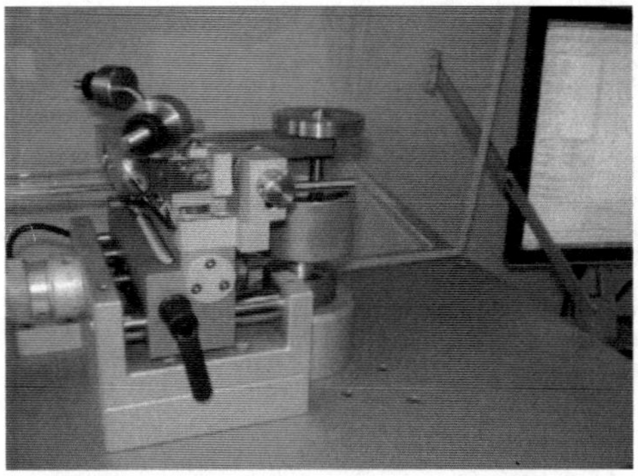

a

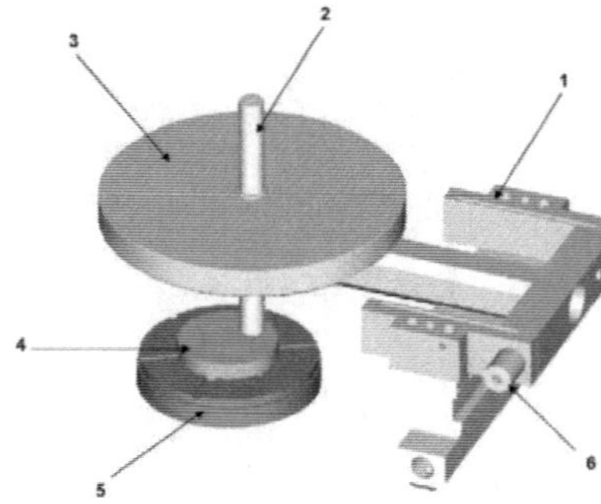


b

**Figure 8.** The CSM Instruments tribometer (a), the block diagram of tribometer active section (b): 1. Elastic shim; 2. Ball fixture; 3.Weights; 4. Fixture for sample; 5. Tangential force. sensor.

The obtained results are presented in figure 9 and 10.

Samples were also observed using the electronic microscope Hitachi and the results are presented in figure 11 and figure 12, where it could be observed that the AIN layer is eroded quite fast.

*(a)*

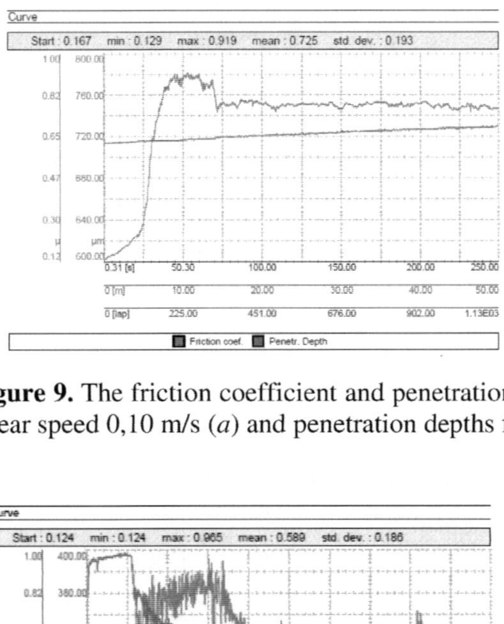


*(b)*

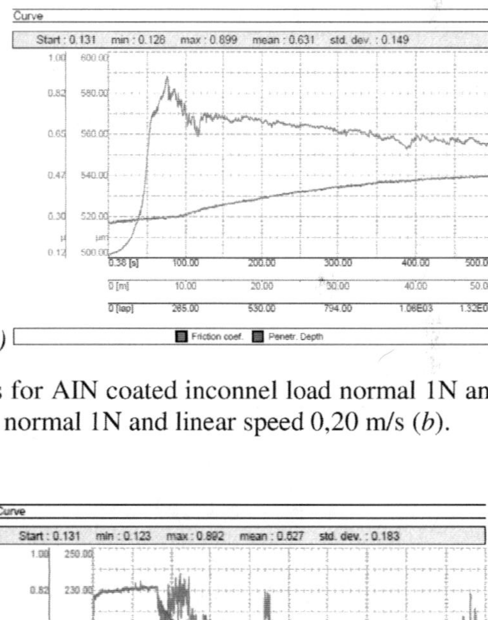


**Figure 9.** The friction coefficient and penetration depths for AIN coated inconnel load normal 1N and linear speed 0,10 m/s (*a*) and penetration depths for load normal 1N and linear speed 0,20 m/s (*b*).

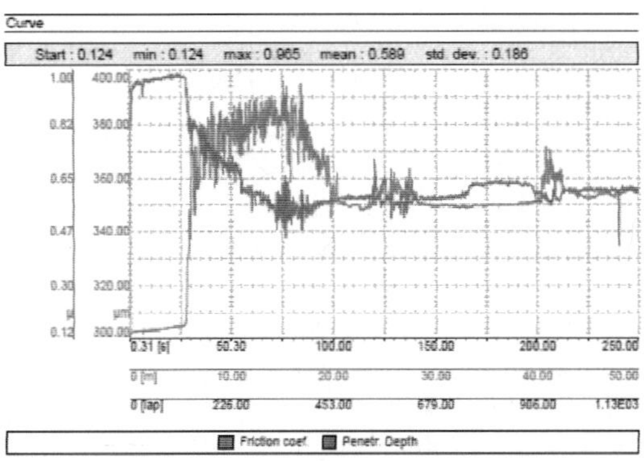


(a)

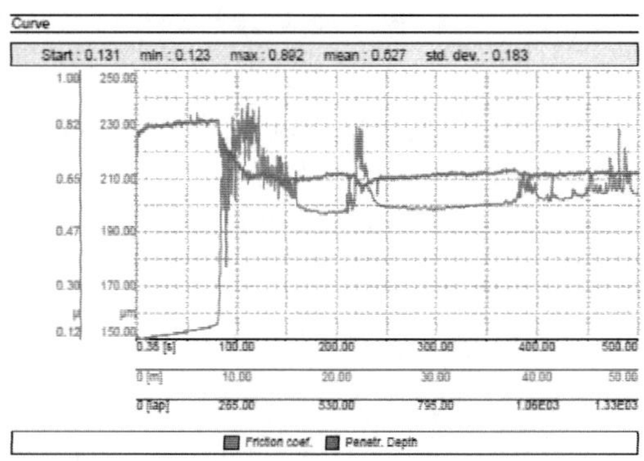


(b)

**Figure 10.** Penetration depths for uncoated Inconnel Load normal 1N and linear speed 0,10 m/s (*a*) and Penetration depths for Load normal 1N and linear speed 0,20 m/s (*b*).

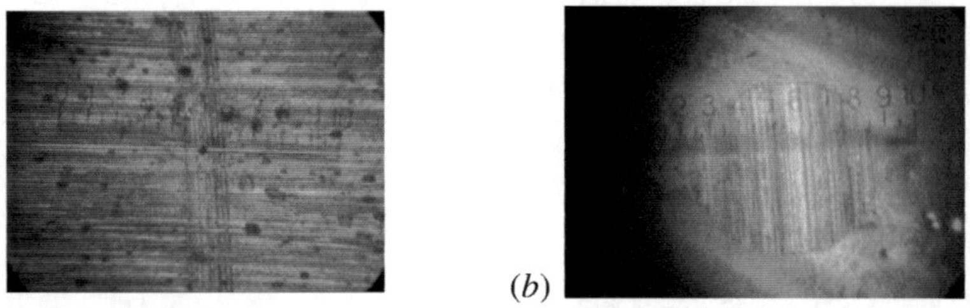

(*a*) (*b*)

**Figure 11.** Images of wearing band on disk and wearing on ball, width of wearing band on disk (*a*); Wearing on ball (*b*).

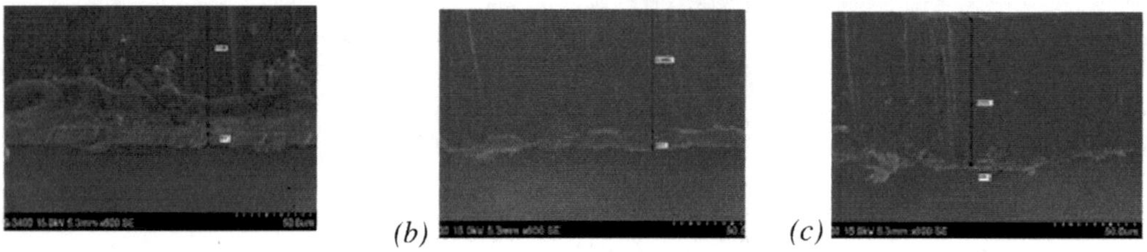

(*a*) (*b*) (*c*)

**Figure 12.** The results tested on CSM Instruments tribometer, initial (*a*); after 50 seconds (*b*); after 100 seconds (*c*).

## 4. Conclusions

The results of the conducted experiments on the manufactured samples are demonstrating that:

- Usable AlN layers of 1 to 2 microns could be successfully deposited by sputtering directly on the equipment parts needed to be monitored (ball bearings, bearings, rotating axes, cutting tools etc).
- There is recommended to avoid exposing stainless steel components coated with AlN layers at temperatures above $200C^0$. If the temperature is expected to increase over $200C^0$ there is recommended to use an interlayer material or to use a Inconel substrate.
- AlN layers do present a constant high friction coefficient and it isn't recommended to be used on contact surfaces.
- AlN layers cannot withstand harsh abrasion condition therefore it is recommended to avoid using them on contact surfaces.
- There are real possibilities of manufacturing sensors embedded on the monitored components of the machine tool; however, more work has to be done in this direction.

## References

[1] Da Xu L, He W, and Li S 2014 Internet of Things in Industries: A Survey, *IEEE Intl. Conf. Ind. I.* **10** (4) 2233-41

[2] Parks D A and Tittmann B R 2011 Piezoelectric Material for Use in Harsh Environments *Review of Progress in Quantitative Nondestructive Evaluation AIP Conf. Proc.* **1335** pp 1671-75

[3] Akiyama M, Tabaru T, and Kishi K 2012 Development of a pressure sensor using a piezoelectric material thin film *Synthesilogy* English edition **5** (3) 162-170

[4] Fraga M A, Furlan H, Pessoa R S, and Massi M 2014 Wide bandgap semiconductor thin films for piezoelectric and piezoresistive MEMS sensors applied at high temperatures: an overview *Microsyst. Technol* **20** (1) 9-21

[5] Stan G E, Pasuk I, Trinca L M, Galca A C, Enculescu M, and Miculescu F 2012 Tilt c axis crystallite growth of aluminium nitride films by reactive rf-magnetron sputtering *Dig. J. Nanomater. Bios.* **7** (1) 41- 50

[6] Ionescu G et al. 2013 Wireless AlN sensor for condition based monitoring of industrial equipment. In *CAS 2013 vol. 1 (International Semiconductor Conference)* pp. 55-58

[7] Dumitru V, Morosanu C, Sandu V, and Stoica A 2000 Optical and structural differences between RF and DC Al x N y magnetron sputtered films. *Thin Solid Films* **359** (1) 17-20
[8] IOFFE Database 2014 AIN – Aluminium nitride. Thermal properties available at: www.ioffe.ru /SVA/NSM/ Semicond/AlN/thermal.html
[9] Agilent Technologies 2012 *Laser and Optics User's Manual* (Ch. 17. Material Expansion Coefficients) available at: psec.uchicago.edu/thermal_coefficients/cte_metals_05517-90143.pdf

IOP Conf. Series: Materials Science and Engineering **174** (2017) 012053 doi:10.1088/1757-899X/174/1/012053

# Theoretical and experimental analysis of the lubricating system of a high speed multiplier

**V G Marian[1], R F Mirică[1] and T Prisecaru[2]**

[1] Department of Machine Elements and Tribology, Faculty of Mechanical Engineering and Mechatronics, University Politehnica of Bucharest, spl. Independentei 313, 060042 Bucharest, Romania

[2] Department of Thermodynamics, Engines, Thermal and Refrigeration Equipments, Faculty of Mechanical Engineering and Mechatronics, University Politehnica of Bucharest, spl. Independentei 313, 060042 Bucharest, Romania

E-mail : victormarian@omtr.pub.ro

**Abstract**. Flywheel-based energy storage systems are used for energy storage in form of kinetic energy using a flywheel rotating at high speed. In order to achieve this high rotating speed a high speed multiplier could be used in order to increase the rotation speed of a conventional motor. This article presents a theoretical and experimental analysis of the lubricating system of a high speed multiplier used in a flywheel-based energy storage system. The necessary oil flow is theoretically computed using analytical formulas. The oil is used for lubricating the gears, the roller bearings and the sliding bearings. An experimental test rig is used to measure the oil flow. Finally the two results are compared.

## 1. Introduction

Flywheel-based energy storage systems are used for energy storage in form of kinetic energy using a flywheel rotating at high speed [1], [2]. This type of mechanical energy storage is used in Formula 1 cars in order to recover the kinetic energy of the cars during the braking process and to use it later in order to overtake the opponent. A small scale prototype of such a system was built in [3]. Different types of hybrid car systems based on flywheel energy storage were developed [4]-[7]. Flywheel energy storage is also used in space applications, as UPS in case of a power failure and as energy regulator in electrical grids [8], [9]. In order to accumulate maximum of energy at high speed, if the flywheel diameter is less than 50 cm it has to operate at speeds above 15000 rpm in a low friction environment like vacuum in order to minimize friction [10]. Composite materials are the first choice in order to withstand the high centrifugal forces [11]-[13]. Hydrodynamic bearings are also used in such systems [14]. As conventional motors turn at a rotation speed of 3000 to 5000 rpm, a high speed multiplier is interposed between the high speed flywheel and the electric motor. Different possibilities for oil jet lubrication for gears are presented in [15]. Also gear lubrication and different constructive factors have an influence on the efficiency of gear systems. The efficiency decreases with increasing speed [16]. The aim of the present paper is to analyze theoretically and experimentally the lubrication system of a high speed multiplier.

## 2. Experimental setup

The experimental test rig is presented in figure 1. The high speed multiplier is not put in rotation by a motor during the experimental tests. The first difference between the functioning system and the stationary one is the oil viscosity. In the stationary system the oil viscosity is higher due to the lower oil temperature. Therefore the oil flow is lower. The second difference is that in order to observe the oil flow there are three roller bearings (8 in figure 3) missing on the top of the multiplier. Therefore there are three roller bearings on the test rig instead of six roller bearings in real functioning.

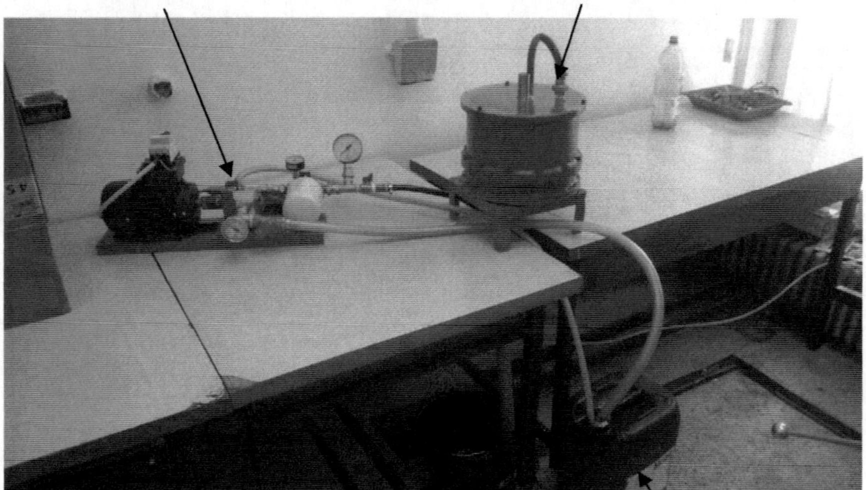


**Figure 1.** Experimental test rig.

The lubricant supply system of the high speed multiplier is presented in figure 2.

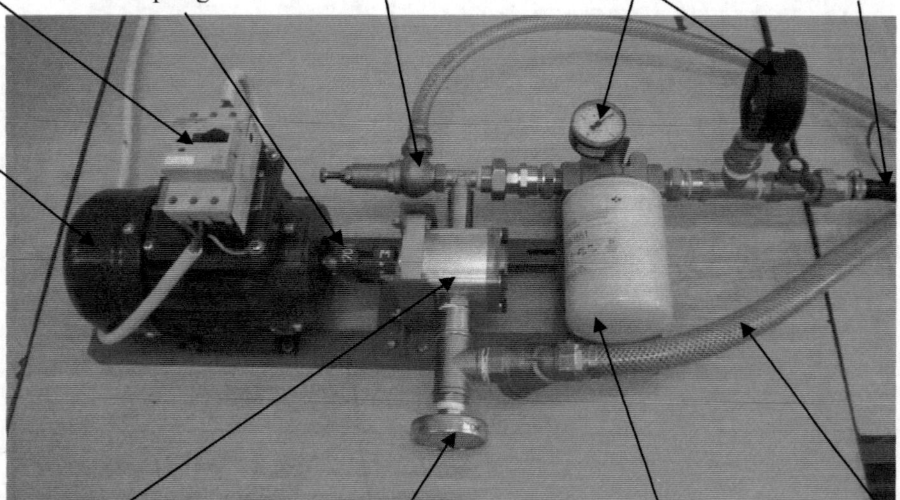


**Figure 2.** Description of the lubricant supply system.

The operating pressure is 10 bars. The main limitation of the operating pressure is due to the hydraulic filter. The electric motor is operating at a rotation speed of 2800 rpm and has a nominal power of 375W. Due to the fact that the necessary computed oil flow is around 8 L/min, a hydraulic pump (7) with a displacement of 4.3 $cm^3$/min and a flow up to 12 L/min is chosen in order to have also an oil flow reserve. The maximal working pressure of the hydraulic pump is 280 bars, far beyond the 10 bars needed working pressure.

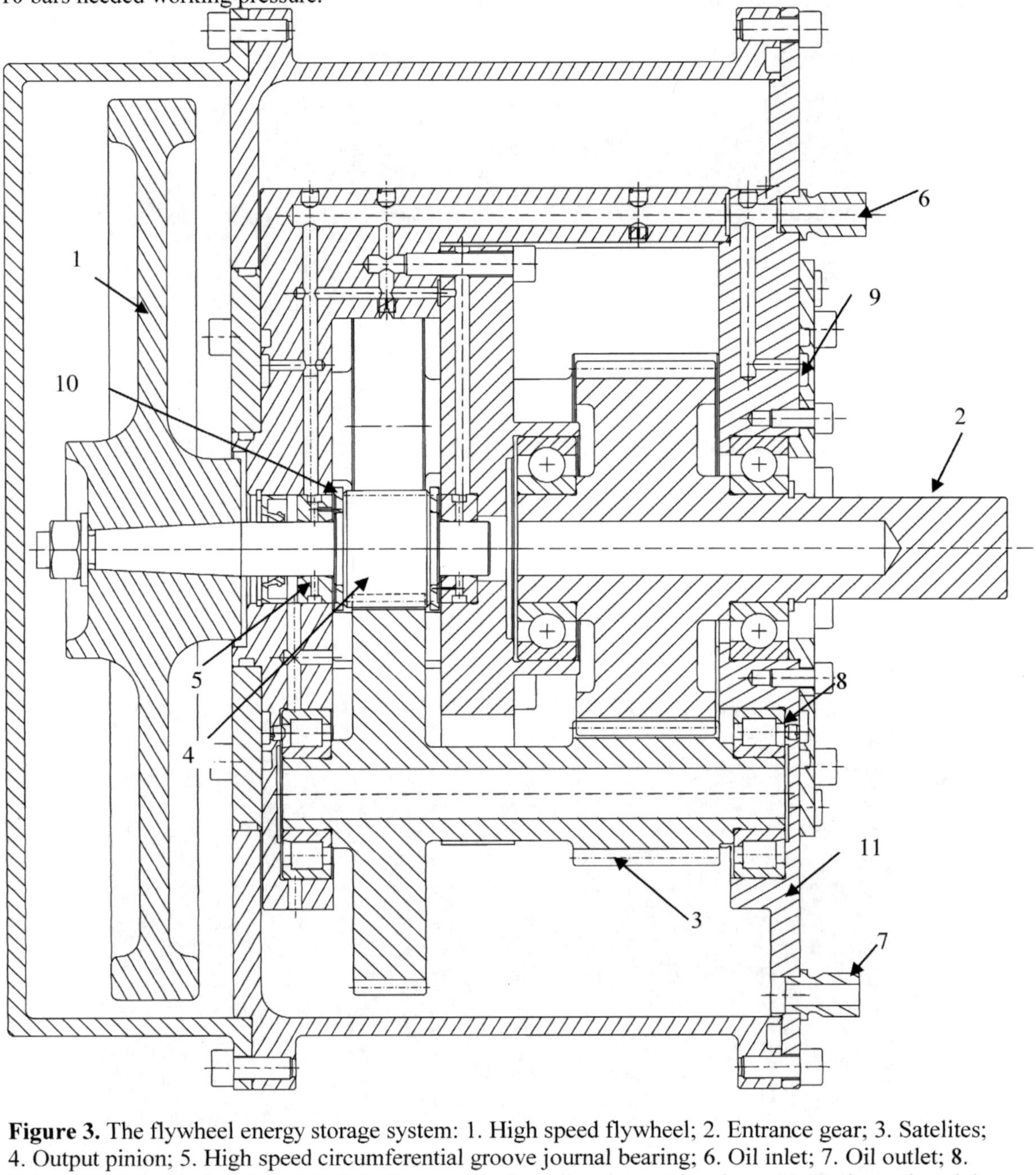


**Figure 3.** The flywheel energy storage system: 1. High speed flywheel; 2. Entrance gear; 3. Satelites; 4. Output pinion; 5. High speed circumferential groove journal bearing; 6. Oil inlet; 7. Oil outlet; **8**. Oil nozzle for roller bearing; 9. Lubrication system for roller element bearings; 10. Hydrostatic axial bearing; 11. Casing cover.

IOP Conf. Series: Materials Science and Engineering **174** (2017) 012053 doi:10.1088/1757-899X/174/1/012053

A jaw coupling (3) is used to compensate the small misalignment of the electric motor with the hydraulic pump. It also protects the hydraulic pump gears from high shocks when the electric motor is started. The supply pressure can be adjusted using the pressure relief valve (4). This pressure will be adjusted at 10 bars. Two manometers (5) are used, one before and one after the oil filter (9). If the pressure difference between the two pressures is more than 1.72 bars, the oil filter must be changed. The filtering element is a cellulose type one, having a maximum filtered particle size of 7 μm. It is important to keep a small particle size in the circuit in order to keep good functioning of the oil nozzles, gears (accuracy grade 4) and bearings. The thermometer (8) indicates the temperature of the oil, temperature which is extremely important in determining the oil viscosity. The on-off switch (2) is a motor protection circuit breaker which is used to manually start and stop the electrical motor, but also to protect the motor in case of overload.

A schematic of the high speed multiplier in the frame of a flywheel energy storage system is presented in figure 3. The high speed multiplier has a multiplication ratio of 11.75 which is realized in two steps, the first gear set having a multiplication ratio of 49/17 and the second a multiplication ratio of 53/13. Due to the high output rotation speed (18000 rpm) hydrodynamic bearings were used for the output gear train. The hydrodynamic bearings have also a good damping capacity. In order to reduce the dimensions of the high speed multiplier the power flow is split in 3 like in the case of a planetary gear set. Helical gears were used in order to have lower noise when functioning.

As lubricating additivated gear oil Mobilgear 600 XP150 was used, having following main properties:

- kinematic viscosity at $40^{\circ}$C: 150 $mm^2/s$
- kinematic viscosity at $100^{\circ}$C: 14.7 $mm^2/s$
- density at 15.6°C : 0.89 kg/L

The oil flow at the entrance gear set in the high speed multiplier is presented in figure 4. Only the first three nozzles from the first gear set are presented. The other three nozzles for the second gear set cannot be observed. Also the oil jets of the three roller bearings which are below could not be observed.

**Figure 4.** Visualization of oil jets for the lubrication of gears (the multiplier is without the casing cover 11 from figure 3).

IOP Conf. Series: Materials Science and Engineering **174** (2017) 012053 doi:10.1088/1757-899X/174/1/012053

The oil flow was measured using the scaled bucket. After three measurements we obtain an oil flow of around 7 L/min (mean value) at a functioning pressure of 10 bar. The measured oil temperature is 32°C.

## 3. Theoretical computations

The total oil flow is computed as following:

$$Q_{total} = Q_{hydrostatic} + Q_{hydrodynamic} + Q_{rollerb} + Q_{gearl} \quad (1)$$

where

$Q_{hydrostatic}$ – oil flow for the hydrostatic bearings
$Q_{hydrodynamic}$ – oil flow for the hydrodynamic bearings
$Q_{rollerb}$ – oil flow for the roller bearings
$Q_{gearl}$ – oil flow for the gear lubrication

### 3.1. *Computation of the oil flow for the hydrostatic bearings*

Due to the fact that the hydrostatic bearings are not externally loaded and the total gap is 0.5 mm, it is considered that the bearings are equally loaded due to the external lubrication pressure and the fluid film thickness is 0.25 mm for each bearing.

Using the conservation of flow (the flow that enters through the capillary restrictor is equal to the flow that leaves through the bearing gap) we obtain [17]:

$$Q_{hs} = \frac{p_r \pi h^3}{6\eta ln\left(\frac{r_a}{r_i}\right)} = \frac{p_s - p_r}{k_c \eta} \quad (2)$$

where

$$k_c = \frac{128\, l}{\pi\, {d_r}^4} \quad (3)$$

In order to minimize the total power of the hydrostatic bearings, following dimensions were computed: $r_i$=15 mm, $r_a$=17 mm, $h$=250 µm, $p_s$=10 bar, $l$=4 mm, $d_r$=0.5 mm. With a restrictor length of 4mm and a diameter of 1mm a value of $k_c$=2.68·10$^{12}$ 1/m$^3$ is obtained.

The oil viscosity needs to be computed at the measured temperature of 32°C. For this it is considered that the dynamic viscosity varies with temperature using the Reynolds equation:

$$\eta = \eta_i \cdot e^{-\beta(T-T_i)} \quad (4)$$

Knowing the dynamic viscosity of the oil Mobilgear 600 XP150 at 40°C $\eta_{40}$=0.133 Pa s and at 100°C $\eta_{100}$=0.013 Pa s , the dynamic viscosity at 32°C is $\eta_{32}$=0.181 Pa s.

From equation 2 a recess pressure $p_r$=5.8·10$^3$ Pa is found. The flow for the hydrostatic bearing is then $Q_{hs}$=0.13 L/min. The total flow for the two hydrostatic bearings is $Q_{hydrostatic}$=2 $Q_{hs}$=0.26 L/min.

### 3.2. *Computation of the oil flow for the hydrodynamic bearings*

The two hydrodynamic bearings are circumferential groove bearings. Due to the fact that there is no load on the bearings during the measurements, the shaft is considered to be concentric with the bushing and only the hydrostatic flow component will be computed [18]:

$$Q_{hydrodymanic} = \frac{\pi p_s D^4 (B_1 + B_2) \Psi^3}{96\eta B_1 B_2} \quad (5)$$

The left bearing has following dimensions: $B_1$=$B_2$=4 mm, $D$=15 mm, $\Psi$=42 µm/15 mm=0.0028, $\eta_{32}$=0.181 Pa s, $p_s$=10 bar. A value of $Q_{left}$=0.006 L/min is obtained.

The right bearing has following dimensions: $B_1$=4 mm, $B_2$=7 mm $D$=15 mm, $\Psi$=72 µm/15mm=0.0048, $\eta_{32}$=0.181 Pa s, $p_s$=10 bar. A value of $Q_{right}$=0.024 L/min is obtained.

The total flow for the two hydrodynamic bearings is $Q_{hydrodynamic}$=$Q_{left}$+$Q_{right}$=0.03 L/min.

### 3.3. Computation of the oil flow for the lubrication of the roller bearings

The roller bearings are lubricated using the lubrication nozzles presented in figure 3 (number 8). The flow through one lubricating nozzle is calculated using the relation:

$$Q_{hs} = \frac{p_s}{k_c \eta} \tag{6}$$

The necessary oil flow for lubricating the roller bearings can be estimated from [19]. It should be between 0.001 and 0.003 L/min. From equation 6 a diameter of 0.18mm for a length of 4mm of the oil nozzle is found. Due to the technical difficulties in executing these diameters, nozzle diameters of 0.5 mm and a length of 2 mm were machined using a lathe machine.

Hence a flow of 0.25 L/min was computed for each roller bearing. Due to the fact that there are only three roller bearings on the test rig instead of six in real functioning, the total roller bearing flow is $Q_{rollerb}$=0.75 L/min.

### 3.4. Computation of the oil flow for the lubrication of the gears

In order to compute the oil flow necessary for the gear lubrication the dissipated power in the gears is necessary [20]:

$$P_{Vzt} = P_{vz} + P_{vz0} \sim P_a \cdot K_a \left[\frac{0,1}{z_1 cos\beta} + \frac{0,03}{v_t + 2}\right] \tag{7}$$

For the first gear train we have $K_A$=1.2, $z_1$=17, $\beta$=10°, $v_t$=8.13 m/s and $P_{Vzt1}$=332.4 W is obtained.

For the second gear train we have $K_A$=1.2, $z_1$=25, $\beta$=10°, $v_t$=25 m/s. We obtain $P_{Vzt2}$=331.9 W.

Due to the fact that the dissipated heat by the outer surface of the multiplier is hard to compute it is estimated (like for journal bearings) that all the heat is absorbed by the cooling oil.

The necessary oil flow for cooling the gear multiplier is [20]:

$$Q_{gearl} = \frac{P_{Vzt}}{c_{oil} \cdot \rho_{oil} \cdot \Delta T_{oil}} \tag{8}$$

By replacing $c_{oil}$=1900 Ws/(kgK), $\rho_{oil}$=900 kg/m$^3$ and $\Delta T_{oil}$=5 K [20] a flow of $Q_{gearl}$=4.65 L/min is obtained.

**Figure 5.** Measurement of pressure behind a gear nozzle.

Using equation 6, an entrance temperature of 40°C (when the multiplier is functioning) and a dynamic viscosity of 0.133 Pas and knowing the fact that there are 6 restrictors the following can be found: $Q_{restrictor}= Q_{gearl}/6$, the restrictors diameter is $d_r$=0.67 mm and their length is $l$=3 mm. The restrictors should be manufactured at $d_r$=0.7 mm. The six restrictors were manufactured with a bore diameter of 0.9 mm in order to compensate the pressure loss in the supply channels of the multiplier. In order to show the pressure difference, a manometer was put behind a nozzle which is different from the entrance nozzle (figure 5). The measured pressure was 5 bar.

Due to the fact that the high speed multiplier is not loaded, the oil temperature is lower (32°C). The oil flow for the gears at this temperature can be computed using equation 6, a dynamic viscosity of $\eta_{32}$=0.181 Pa s and a diameter of 0.7 mm, neglecting the pressure drop in the multiplier oil circuit. The new oil flow is $Q_{gearl}$=3.93 L/min.

### *3.5. Computation of the total flow for the high speed multiplier*

Using equation 1 we find that the total flow necessary to lubricate the high speed multiplier is $Q_{total}=Q_{hidrostatic}+Q_{hydrodynamic}+Q_{rollerb}+Q_{gearl}$=0.26 L/min+0.03 L/min+0.75 L/min+3.93 L/min=4.97 L/min.

## 4. Comparison between theoretical and experimental results

The total flow determined experimentally is $Q_{total\_experimental}$=7 L/min and the theoretical determined oil flow is $Q_{total\_theoretical}$=4.97 L/min. The relative error can be computed using following formula:

$$error = \frac{Q_{total\_experimental} - Q_{total\_theoretical}}{Q_{total\_experimental}} \cdot 100 \tag{9}$$

Hence an error of 29% is obtained.

## 5. Conclusions

After theoretical and experimental determination of the oil flow, following conclusions are obtained:

- the difference between the theoretically determined diameters of the oil nozzles and the manufactured diameters is the main source of the difference between the theoretical and experimentally determined oil flow;
- there is also a pressure drop in the oil channels of the multiplier, which determines a lower entrance pressure in the oil nozzles;
- the hydraulic pump having a maximum flow of 12 L/min was correctly dimensioned, giving some oil flow reserve if necessary;
- the lubrication system, which was designed from scratch, proves to be reliable;

**Acknowledgments**

The authors express their gratitude to the technical staff, Constantin Răducanu and Mihai Rozorea, and to Professor Alexandru Rădulescu for their precious help.

## Appendix

Following notations were used in the article:

$B_1$ – bearing width at the left side of the circumferential groove
$B_2$ – bearing width at the right side of the circumferential groove
$C$ – radial clearance
$c_{oil}$ – oil specific heat
$D$ – diameter of the journal bearing
$d_r$ – diameter of the restrictor
$h$ – fluid film thickness
$K_a$ – application factor
$k_c$ – capillary constant of the hydrostatic bearing restrictor
$l$ – length of the restrictor
$p_r$ – recess pressure of the hydrostatic bearing
$p_s$ – supply pressure
$P_{vz}$ – dissipated power in the gears through friction under load

$P_{vz0}$ – dissipated power in the gears under free motion
$P_{vzt}$ – total dissipated power
$Q_{gearl}$ – oil flow for the gear lubrication
$Q_{hs}$ – oil flow for one hydrostatic bearing
$Q_{hydrodynamic}$ – oil flow for the hydrodynamic bearings
$Q_{hydrostatic}$ – oil flow for the hydrostatic bearings, 2 $Q_{hs}$
$Q_{rollerb}$ – oil flow for the roller bearings
$r_a$ – outer radius of the hydrostatic bearing
$r_i$ – recess radius of the hydrostatic bearing
$v_t$ – circumferential speed
$z_1$ – number of gear teeth in entrance
$\beta$ – helix angle of helical gear
$\Delta T_{oil}$ – oil temperature difference beween inlet and outlet
$\eta$ – dynamic oil viscosity
$\rho_{oil}$ – oil density
$\Psi$–dimensionless clearance, $2C/D$

**References**

[1] Genta G 1985 *Kinetic energy storage* (Cambridge: Butterworth)
[2] Lindley D 2010 Smart grids: The energy storage problem *Nature News* **463** 18–20
[3] Laurian T, Marian V G and Prisecaru T 2013 Experimental analysis of a kinetic energy recovery system intended for small and medium passenger cars *Power Transmissions: Proc. of the 4th Int. Conf. (Sinaia, Romania, 20-23 June 2012), Series Mechanisms and Machine Science* vol 13 ed G Dobre (Dordrecht: Springer Netherlands) pp 347-355
[4] Dhand A and Pullen K 2013 Review of flywheel based internal combustion engine hybrid vehicles *Int. J. Automot. Techn.* **14** 797–804
[5] Hilton J 2012 Schwungrad-Hybridfahrzeuge als Alternative zu Elektrofahrzeugen *ATZ Automobiltech. Z* **114** 11 874–8
[6] Sontheim J 2008 Kinetischer Speicher für Hybridfahrzeuge *ATZ Automobiltech. Z* **110** 226–31
[7] Atkins A and Feulner P 2011 Der mechanische Hybridantrieb von Ricardo *MTZ Motortech. Z* **72** 122–127
[8] Liu H and Jiang J 2007 Flywheel energy storage—An upswing technology for energy sustainability *Energ. Buildings* **39** 599–604
[9] Hebner R, Beno J and Walls A 2002 Flywheel batteries come around again *IEEE Spectr.* **39** 4 46–51
[10] Marian V G, Laurian T, Mirică R F, Pascu A and Seiciu P L 2014 Theoretical and experimental analysis of friction forces acting on a flywheel energy storage system rotating in air *Proc. of the 8th Int. Conf. on Tribology BALKANTRIB'14* Sinaia Romania
[11] Hiroshima N, Hatta H, Koyama M, Goto K and Kogo Y 2015 Optimization of flywheel rotor made of three-dimensional composites *Compos. Struct.* **131** 304–11
[12] Hiroshima N, Hatta H, Koyama M, Yoshimura J, Nagura Y, Goto K and Kogo Y 2016 Spin test of three-dimensional composite rotor for flywheel energy storage system *Compos. Struct.* **136** 626–34
[13] Krack M, Secanell M and Mertiny P 2010 Cost optimization of hybrid composite flywheel rotors for energy storage *Struct. Multidiscip. O.* **41** 779–95
[14] Dai X, Shen Z and Wei H 2001 On the vibration of rotor-bearing system with squeeze film damper in an energy storage flywheel *Int. J. Mech. Sci.* **43** 2525–40
[15] Greiner J 1990 *Untersuchungen zur Schmierung und Kühlung einspritzgeschmierter Stirnradgetriebe* Dissertation Universität Stuttgart
[16] Strasser D 2005 *Einfluss des Zahnflanken- und Zahnkopfspieles auf die Leerlaufverlustleistung von Zahnradgetrieben* Dissertation Ruhr-Universität Bochum
[17] Pascovici M D and Cicone T 2001 *Elemente de Tribologie* (Bucharest: Bren)
[18] Pascovici M D, Carp-Ciocârdia R M and Pârvu C 1987 The automatic computation of hydrodynamic journal bearings with circumferential groove *The 5th Conf. on Friction, Lubrication and Wear TRIBOTEHNICA Bucharest* pp 281-285
[19] Decker K H 1998 *Maschinenelemente* (München Wien: Carl Hanser Verlag)
[20] Niemann G, Winter H 1985 *Maschinenelemente* (Berlin Heidelberg: Springer-Verlag)

IOP Conf. Series: Materials Science and Engineering **174** (2017) 012054 doi:10.1088/1757-899X/174/1/012054

# Correlations of the instantaneous carrying force and pressure distribution in the case of the narrow sliding radial bearing under hard shocks*

**I M Alexandrescu, R Coteţiu and D Darabă**

Technical University of Cluj-Napoca, North University Center of Baia Mare, Faculty of Engineering, Department of Engineering and and Technology Management, 62A Victor Babes St., RO-430083, Baia Mare, Romania

E-mail: ioan.marius.alexandrescu@gmail.com

**Abstract**. We present the determining relationship of carriage in non-dimensional form for narrow radial bearings subjected to shocks and vibrations, as well as the determining relationships of the lubricant minimum thickness in relation to the dynamic loading. Due to the very short time of loading radial bearings subjected to shocks and vibrations, of about 0.5-1 ms, we consider only the approaching motion between spindle/axle and bushing on the direction of the centre line, without the rotation of the spindle/axle (the case of the non-rotating bearing), so that the effect of the lubricant expulsion be prevalent in the achieving of the self-carrying film. The paper is focused on determining relationship of carriage in dimensional form for narrow radial bearings exposed to shocks and vibrations, as well as the determining relationships of the pressure distribution from the film to be lubricated in various places of the bearing's body. It is showed the details during the measuring accomplishments and the experimental results are registered in a record of obtained results.

## 1. Introduction

The study of the behavior of radial bearings with hydrodynamic lubrication, functioning under conditions of shocks and vibrations, is carried out from a tribological point of view, observing the aspect of friction and lubrication, the lubricant film, by which the shock is damped [1]. In the case of bearings subjected to heavy loading (shocks), the difficulty occurring stays in the solution to Reynolds' equation, the equation of energy, the equation of elastic deformations of the axle and bushing surfaces, and the equation of lubricant viscosity and density variation with pressure, and all these together form a non-linear integral and differential system [2].

That is why we consider useful a systemic approach to these problems, with the conviction that the results obtained will contribute to the finding of new solutions, in the qualitative understanding of the phenomena that occur in the functioning of sliding bearings [1].

We consider the closing motion between spindle and bushing on the direction of the center line, without the rotation of the spindle (the case of the non-rotating bearing), so that the lubricant expulsion effect be prevalent in the achieving of the squeeze film [2].

* This paper was accepted for publication in Proceedings after double peer reviewing process but was not presented at the Conference ROTRIB'16

The modeling of the lubricant expulsion effect (squeeze) starts from Reynolds's equation, in which we have to consider the terms that contain the closing speed of the two surfaces ($V = -\partial h / \partial t$ ) [3].

Analytical expression of the Reynolds equation corresponding to this study, with an isothermal approach is [4]:

$$\frac{\partial}{\partial x}\left(h^3 \frac{\partial p}{\partial x}\right) + \frac{\partial}{\partial z}\left(h^3 \frac{\partial p}{\partial z}\right) = 12\eta \frac{\partial h}{\partial t}. \quad (1)$$

where η- viscosity of lubricant (Ns/m$^2$); p-pressure (Pa); h- fluid film thickness (m).

Nomenclature:

*L*- length of bearing (m); *η*- viscosity of lubricant (Ns/m$^2$); *G*- static loading (N); *p*-pressure (Pa); *F*- dynamic load (N); *h*- fluid film thickness (m); *D*- journal diameter (m); *Ai, Bi, Ci* – instantaneous squeeze force in dimensional form (N); *H* - weight launching height (m); *ci* – time of shock (sec.).

The schematic of a narrow hydrodynamic radial bearing with circular bushing subjected to shocks, modeled in 4 areas, is presented in figure 1 [1].

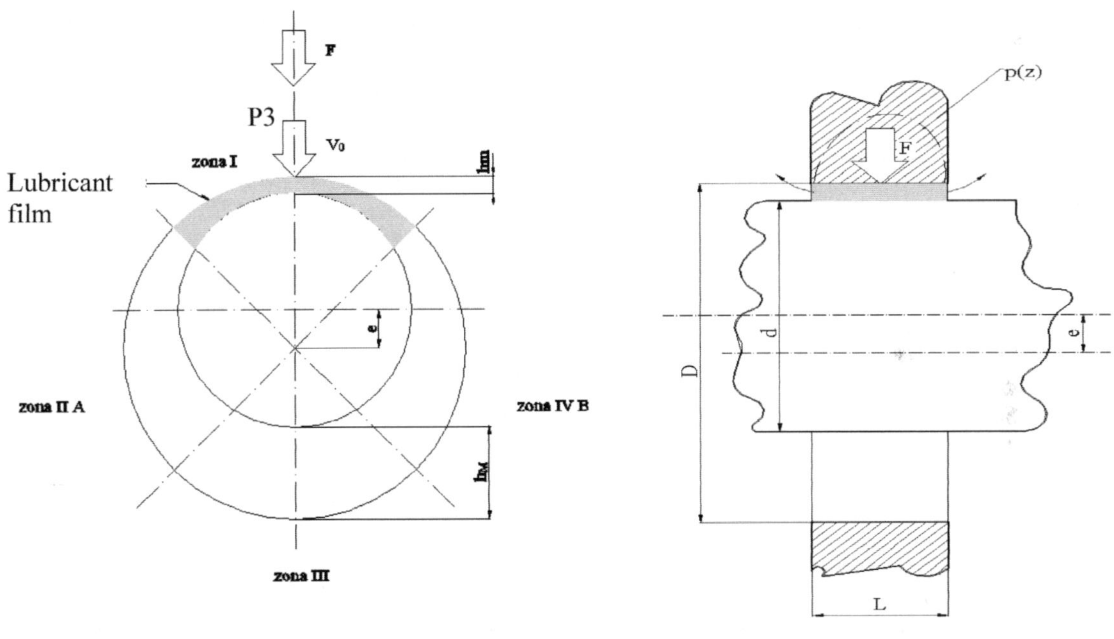


**Figure 1.** The effect of lubricant expulsion under shock for narrow radial bearing.

The simplified modeling of the lubricant film thickness and carriage under the conditions of a closing motion of the spindle/axle and bushing surfaces for the narrow radial bearing exposed to shocks (figure 1) has as starting point the following hypotheses:

- in area III the motion is of separating surfaces, pressure decreases, it can be practically considered constant under the conditions of cavity occurrence;
- in area II A and IV B the section remains "approximately" constant and thus the pressure remains constant;
- area I represents the only area that really opposes the closing motion: the geometry of the lubricant film will be approximated with a constant thickness surface, equal to the minimum thickness of the lubricant film under the condition of static loading, on the basis of the rectangular model of infinite length [5].

The circumferential pressure distribution is [6]:

$$p(\theta)=\frac{12\eta VB^2}{J^3(1-\varepsilon\cos\theta)^3} \tag{2}$$

where θ is the angular coordinate, V is the bushing surfaces velocity immediately before impact and $V_0$ is the velocity immediately after impact:

$$V=-\frac{dh}{dt}=\frac{J}{2}\dot{\varepsilon}=V_0-\frac{\eta\pi DL^3g}{8F}\left(\frac{1}{h_m^2}-\frac{1}{h_{m0}^2}\right). \tag{3}$$

The relative eccentricity given by

$$\varepsilon(t)=1-\frac{2h_m(t)}{J} \tag{4}$$

and

$$h_m=\frac{1}{\sqrt{\frac{1}{h_{m0}^2}+\frac{8F\sqrt{2gH}}{\eta\pi DL^3g}}} \tag{5}$$

where $h_{m0}$ represents the minimum thickness of lubricant under static regime, and $h_m$ represents the minimum lubricant thickness in the dynamic regime [2].

The instantaneous squeeze force has the following expression

$$F^*=2\int_0^{\pi/2}\frac{4\eta VDL^3\cos^2\theta\cdot d\theta}{J^3(1-\varepsilon\cos\theta)^3} \tag{6}$$

We can write

$$\overline{F}_s=\frac{1}{A}\left[\overline{H}_s^{\,3}(1+A)-\overline{H}_s^{\,5}\right] \tag{7}$$

where $A=4\overline{F}\Pi$, $\overline{H}_s=\frac{h_{m0}}{h_m}=H_{s-ad}$ and the parameters of lubricant film expulsion Π have the expression $\Pi=\frac{H}{h_{m0}}$ (H being the height from which the weight dynamically loading the bearing is launched) [6].

## 2. Theoretical results

The variations of the instantaneous carrying force, in relation to the non-dimensional thickness of the lubricant film for the three weight launching heights H are presented in figure 2 and figure 3.

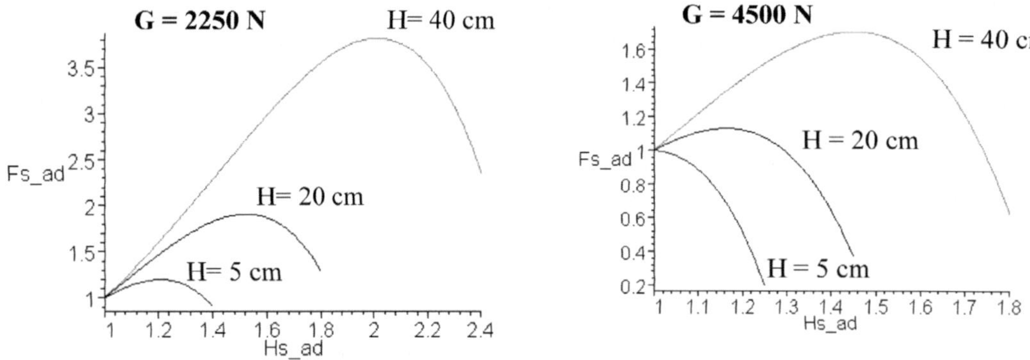


**Figure 2.** The instantaneous carrying force in relation to the non-dimensional thickness of the lubricant film (n=370 rot/min, $p_{in}$=0.5 bar).

IOP Conf. Series: Materials Science and Engineering **174** (2017) 012054 doi:10.1088/1757-899X/174/1/012054

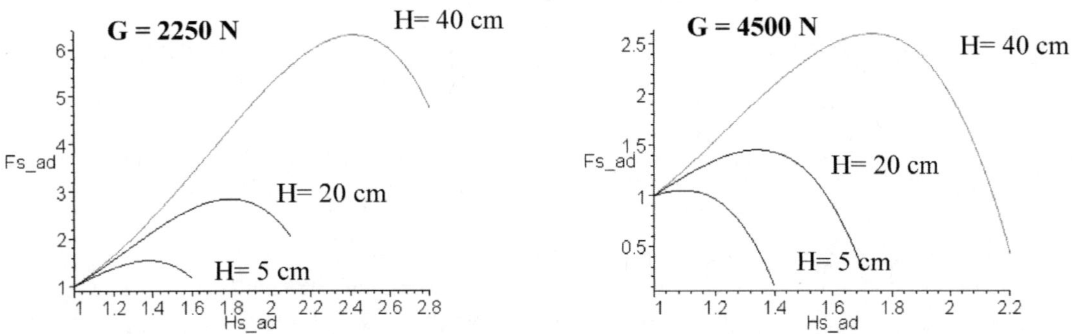


**Figure 3.** The instantaneous carrying force in relation to the non-dimensional thickness of the lubricant film (n=600 rot/min, $p_{in}$=1.5 bar).

The variations of the instantaneous carrying force, in relation to the dimensional thickness of the lubricant film and in relation to the time of shock, for the three weight launching heights H are presented in figure 4 – 7.

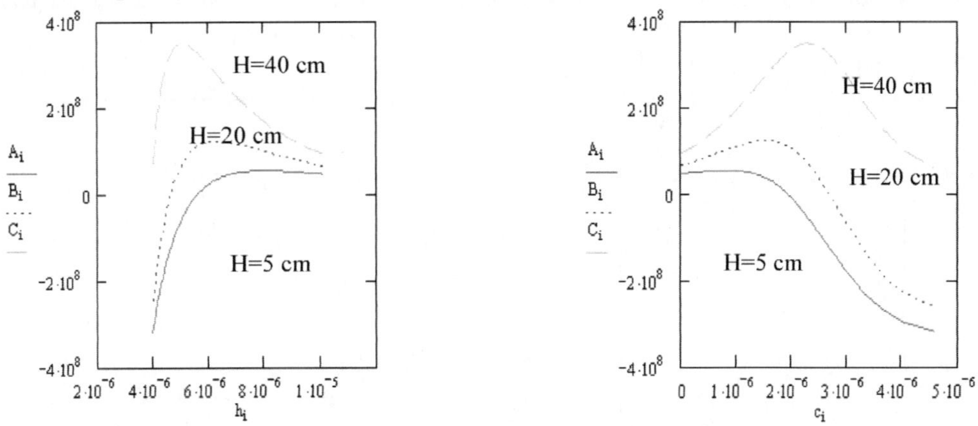


**Figure 4.** The instantaneous carrying force in relation to the dimensional thickness of the lubricant film (n=370 rot/min, $p_{in}$=0.5 bar, G=2250 N, $h_{m0}$=10.175 μm).

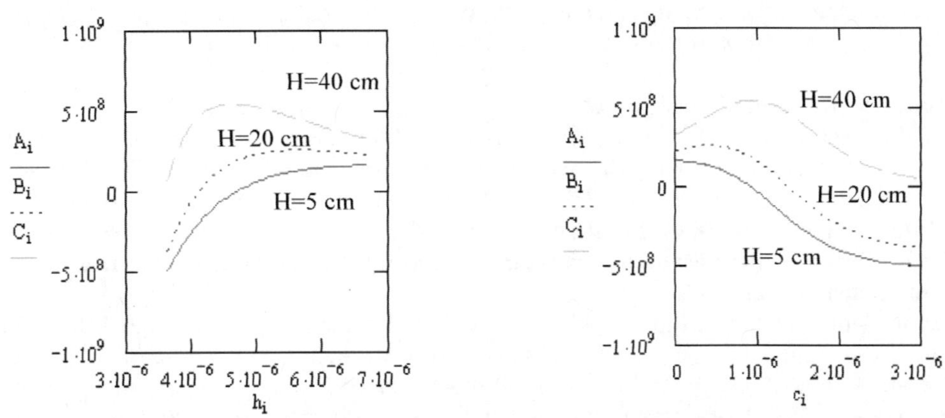


**Figure 5.** The instantaneous carrying force in relation to the dimensional thickness of the lubricant film (n=370 rot/min, $p_{in}$=0.5 bar, G=4500 N, $h_{m0}$=6.723 μm).

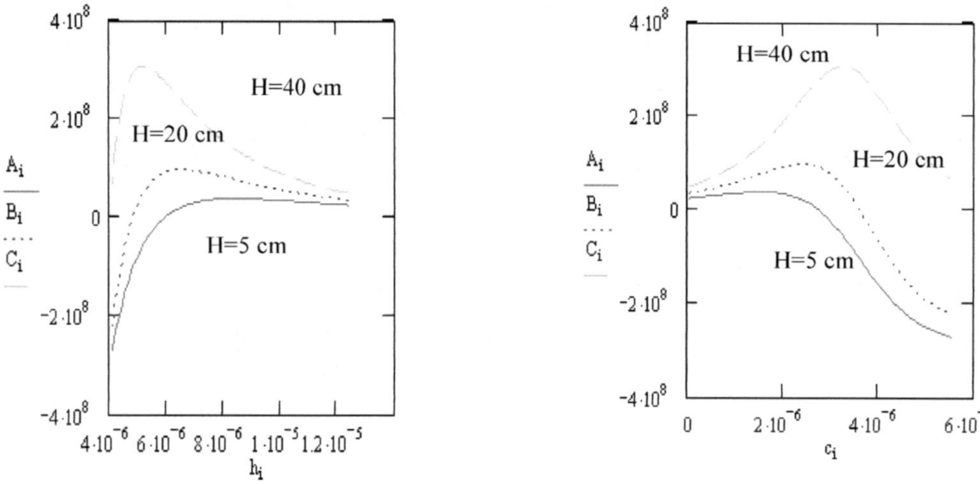


**Figure 6.** The instantaneous carrying force in relation to the dimensional thickness of the lubricant film (n=600 rot/min, $p_{in}$=1.5 bar, G=2250 N, $h_{m0}$=12.554 μm).

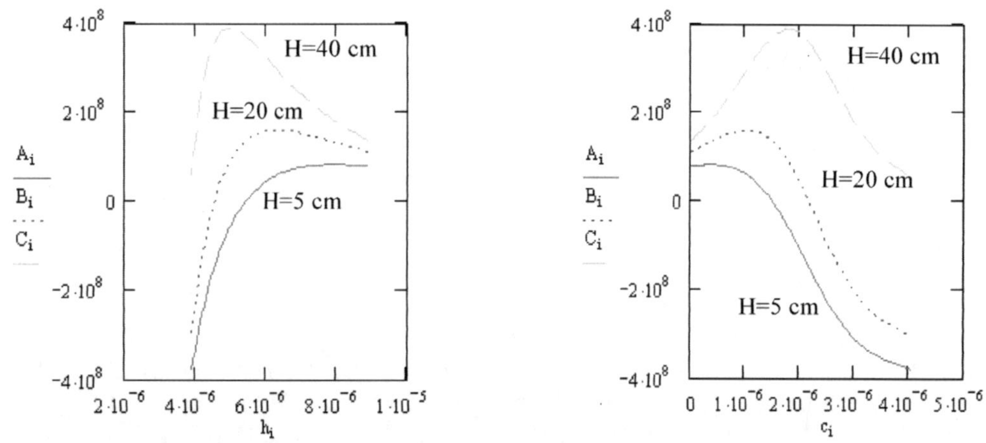


**Figure 7.** The instantaneous carrying force in relation to the dimensional thickness of the lubricant film (n=600 rot/min, $p_{in}$=1.5 bar, G=4500 N, $h_{m0}$=8.493 μm).

## 3. Experimental devices and acquisition chains

The experimental research was carried out using a HD radial bearing with L/D=0.5 and the spindle's diameter $d_e$ =59.86 mm, and the bushing diameter $D_e$ =59.93 mm, spindle's asperity 58-62 HRC, made of 18MoCr10, bronze bushing made of 88% Sn, 8%Sb, 4%Cu. The assessment was made on the experimental stand of the Tribology Engines Lab from the Technical University of Cluj-Napoca, North University Center of Baia Mare, making use of the modern technology concerning the results' processing and acquisition [1].

The dynamic loading of the bearing is achieved through the launching of a weight which hits the bearing at different heights. Assessments for heights between 5 and 40 cm were carried out, using a weight with m=5 kg, so as for H=5 cm we have $F_1$=1665 N, for H=20 cm we have $F_2$=2356 N, and for H=40 cm we have $F_3$=3332 N. The static working conditions is presented for the following value H=0 cm [7].

All the tests were conducted at a 40°C of the lubricant, being constant, pressure distribution $p_{in}$ having the following values, from 0.5 bar to 10 bar.

Using a lubricant oil for bearings of LA 32 STR 5152-89 type, with the viscosity of 31.3 cSt at 40°C, it was focused on the determination pressure distribution from the film to be lubricated in various places of the bearing's body, with the help of pressure measuring dose with tensometric translators put together through an amplifier placed at the acquisition plate ADuC 812 [8].

In dynamic loading conditions, the pressure distribution was determined in the lubricated film in those 5 points on the bearing's body with the help of pressure measuring with tensometric transducers.

Those 4 strain gages are connected in a tensometric bridge and connected to an amplifier at the acquisition plate ADuC 812 [9].

Figure 8 presents the pressure measuring chain in the lubricant film [10].

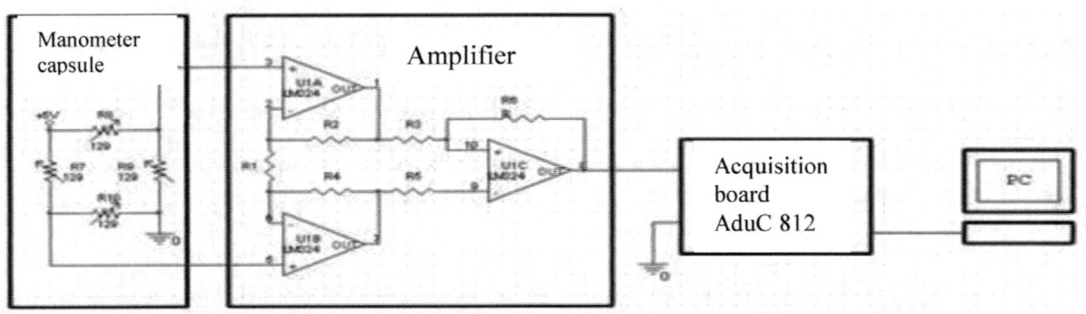


**Figure 8.** The pressure measuring chain in the lubricant film.

The pressure was measured in the case of dynamic charging, on the manometer's exhibition stand, using the above chain, focusing on the variation exit sign and registering amplifier and the acquisition plate ADuC 812. The pressure increase was made bar by bar, the dose distortion being linear with the pressure. It was established the dependency relation between the pressure and the tension in the exit point in mV (2.3 mV = 1bar Δp) [11].

## 4. Experimental results

The pressure distribution at points P3 of the bearing's body, which corresponds to the direction of charging application, depending on the available supply pressure, the static and dynamic charging conditions at different spindle's rotations are presented in figure 9 for n=370 rot/min, $p_{in}$ =0.5 bar, and figure 10, for n=600 rot/min, $p_{in}$ =1.5 bar.

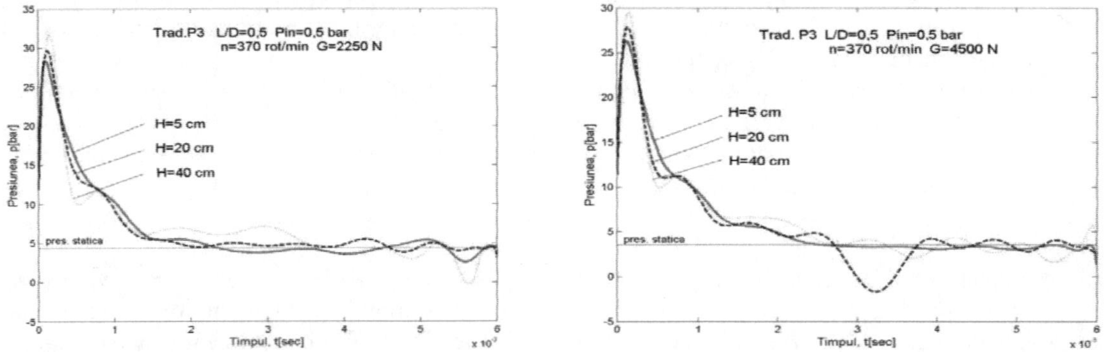


**Figure 9.** The dynamic pressure distribution depending on the static and dynamic charging conditions of the bearing (n=370 rot/min, $p_{in}$ =0.5 bar).

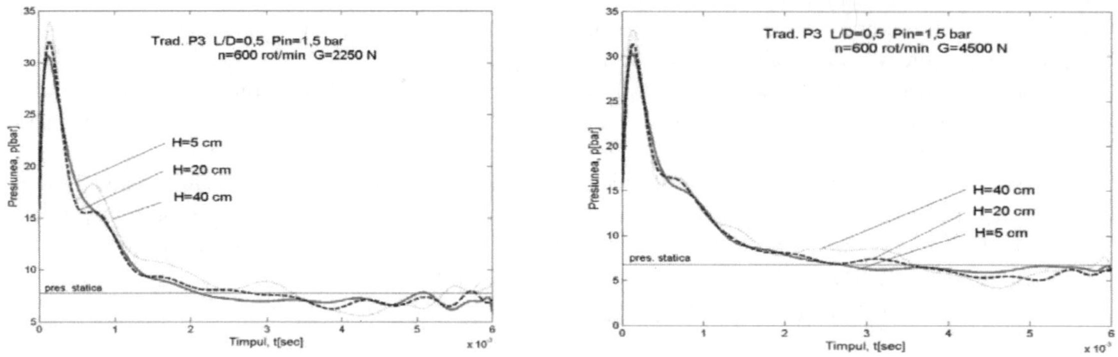


**Figure 10.** The dynamic pressure distribution depending on the static and dynamic charging conditions of the bearing (n=600 rot/min, $p_{in}$ =1.5 bar).

## 5. Conclusions

From the analysis of the theoretical and experimental results, the following observations can be stated:

- the ratio of film thickness Hs_ad sensitively influences carriage: once the area of maximum is outrun, the carriage rapidly decreases;
- the existence of an optimum point from the viewpoint of carriage: any change in the functional parameters of the bearing leads to straying from the optimum value from the viewpoint of carriage;
- in all these situations the following fact is to bare in mind: the short time for pressure variation in dynamic charging (under 0.5 ms);
- the decrease of the lubricant film minimum thickness along with the increase of static loading;
- the existence of an optimum point from the viewpoint of carriage: any change in the functional parameters of the bearing leads to straying from the optimum value from the viewpoint of carriage;
- in all these situations the following fact is to bare in mind: the short time for pressure variation in dynamic charging (under 0.5 ms);
- the dynamic pressure from the moment of shock is increased when increasing the dynamic charging conditions; for the studied position, P3, the dynamic pressure rise at the same time with the rise of dynamic charging, the pressure leap being between 5.95 and 7.45 multiplied with static pressure for the HD bearing with L/D=0.5;
- the static charging conditions of the bearing does not have an important influence regarding the changing in the pressure's values, as the static charging conditions gets bigger, so as the dynamic pressure is bigger;
- the draught's pressure in dynamic conditions has a slightly shifting to the entrance zone of the lubricant when static charging conditions are increasing;
- for the studied P3 position, at the same time with the rise of static charging, the static pressure decrease.

## References

[1] Alexandrescu I M 2005 *Studiul comportării lagărelor radiale cu ungere hidrodinamică în condiţiile funcţionării cu şocuri şi vibraţii.* PhD Thesis, Tehnical University of Cluj-Napoca

[2] Alexandrescu I M, Pay E and Pascovici M D 2005 The effect of lubricant film expulsion under shock in the case of the narrow sliding radial bearing (L/D < 0,6) *Scientific Bulletin, Serie C, Vol. XIX. Fascicle: Mechanics, Tribology, Machine Manufacturing Technology. Part I. North University of Baia Mare* 1-6

IOP Conf. Series: Materials Science and Engineering **174** (2017) 012054 doi:10.1088/1757-899X/174/1/012054

[3] Moore D F 1993 *Viscoelastic Machine Elements. Elastomers and lubricants in machine system.* (Oxford, Butterworth-Heinemann) pp 139-163

[4] Evans R D 2013 Wear of bearings *Encyclopedia of Tribology,* Eds.: Wang Q J and Chung Y (New York: Springer) pp 4061–4068

[5] Khonsari M M and Booser E R 2001 *Applied Tribology. Bearing Design and Lubrication* (New York: John Wiley & Sons)

[6] Alexandrescu I M, Pay E and Ungureanu N 2006 Experimental research concerning the radial bearings with HD lubrication in the case of huge challenging working *Journal of Proc. of the Intern. Conf. on Manufacturing Systems ICMaS*, pp 215-218

[7] Alexandrescu M, Cotetiu R and Ungureanu N 2007 Theoretical correlations regarding the instantaneous carrying force in the case of the narrow sliding radial HD working bearing under hard shocks *Automation in Production Planning and Manufacturing, 8th Intern. Scient. Conf.* Žilina – Valča, pp 9-13

[8] Pascovici M D and Cicone T 2002 *Elemente de tribologie (*Bucureşti: BREN)

[9] Pay E and Alexandrescu I M 2004 The HD lubrication limits of radial bearing *The 8-th Intern. Conf. on Tribology* (Veszprem) pp 247-252

[10] Alexandrescu I M, Pay E and Buchman A 2005 Experimental results regarding hard shocks behaviour to the radial hydrodynamic working bearings *microCAD International Scientific Conference. Machine and Construction Design* (University of Miskolc) pp 1-12

[11] Pay E and Alexandrescu I M 2004 Searches regarding radial journal bearings with HD lubrication *Romanian Journal of Technical Sciences Applied Mechanics* **49** 15-20

IOP Conf. Series: Materials Science and Engineering **174** (2017) 012055 doi:10.1088/1757-899X/174/1/012055

# Wear behavior of the plasma and thermal oxidized Ti-15Mo and Ti-6Al-4V alloys

**I Hacisalioglu[1], F Yildiz[1], A Alsaran[2] and G Purcek[3]**

[1] Department of Mechanical Engineering, Erzurum Technical University, Yakutiye/Erzurum, 25240, Turkey
[2] Department of Mechanical Engineering, Anadolu University, Tepebaşı/Eskisehir, 26555, Turkey
[3] Department of Mechanical Engineering, Karadeniz Technical University, Ortahisar/Trabzon, 61080, Turkey

Email: ihacisalihoglu@gmail.com

**Abstract**. Titanium and its alloys widely used in load bearing applications. Titanium alloys are capable of providing lower elastic modulus and better corrosion resistance with alloying processes. In spite of the modified mechanical properties, the surface degradation is still the main critical defect. Ti-15Mo alloy is one of the alpa+beta titanium alloys with acceptable mechanical and chemical superiority. Recent researches in literature show that the wear performance of base Ti-15Mo is relatively low as compared to Ti-6Al-4V, using in high performance applications. Plasma oxidized surfaces increase the tribological and chemical performance of titanium alloys. In this study the Ti-15Mo alloy and Ti-6Al-4V alloys were compared in terms of wear performance. To obtain alloys with similar microstructure they were solution treated at 800°C 1H and then air-cooled. The plasma and thermal oxidations were applied at 650°C for 1 hour. Wear performance of oxidized surfaces investigated in dry conditions. Oxidized surface characterized with XRD, SEM, 3D profilometer and hardness measurements. Wear volume calculated with 3D profilometer. Results show that oxidizing increased the surface roughness and improved the wear performance of Ti15Mo alloy. The plasma and thermal oxidized Ti-15Mo showed a remarkable increase in wear resistance.

## 1. Introduction

Titanium and its alloys are important structural materials due to their high specific strength and good corrosion resistance. Thus, they have been widely used in many applications ranging from aerospace and defense systems to biomedical industries. However, their wear resistance is very poor due to the unstable oxide film formed on the surface by the effect of tribo-chemical reaction leading to rapid oxidative wear loss. In this regard, various surface modification techniques including thermochemical processes and thin film coatings have been applied commonly to improve wear resistance.

In recent years, on the other hand, some new-generation techniques including anodizing [1,2], ion implantation [3], dc-diode plasma oxidation [4], plasma electrolytic oxidation [5], reactive dc-sputtering [6,7] and thermal oxidation (TO) [8,9] have recently gained interest from researchers. The thermal oxidation and plasma oxidation (PO) techniques are especially promising methods because of their ease-to-use and effectiveness in some specific applications. Both techniques are based on diffusional processes that form a thick oxide layer and a diffusional region between the oxidized layer and the bulk material. However, the plasma oxidation technique, has some advantages over the thermal oxidation such as lower temperatures and shorter times required for processing. In general, these techniques have been used especially on the oxidation of pure titanium and Ti-6Al-4V alloy [10-12]. In recent years, alpha+ beta / beta-type titanium alloys including Zr, Nb, Ta and Mo have been developed and used due to their higher corrosion resistance and also low elastic modulus. However, the wear resistance of such alloys is still a main obstacle to be employed in broader applications where the tribological effects are the major concerns.

Therefore, the main purpose of the current study is to improve the wear resistance of alpha+beta type titanium alloys by applying both plasma and thermal oxidations. In this study, the wear performance of plasma and thermal oxidized Ti-15Mo and Ti-6Al-4V alloys was investigated and compared in terms of their wear properties.

## 2. Experimental Procedure

In this study, Ti-6Al-4V and Ti-15Mo alloys were used for the oxidization processes. Table 1 represet the chemical composition of the alloys.

**Table 1.** Chemical composition (%) of Ti-16-4V and Ti-15Mo.

| Elements | Al | V | Mo | Fe | Ti | Other |
|---|---|---|---|---|---|---|
| Ti-6Al-4V | 6 | 4.1 | - | <0.30 | Balance | 0.45 |
| Ti-15Mo | - | - | 15 | <0.20 | Balance | 0.40 |

Ti-6Al-4V and Ti-15Mo alloys were obtained from their commercial grades. To obtain the same initial conditions, they were solution treated at 800°C for 1 hour and then, air cooled to room temperature. The process was performed under argon atmosphere. Electron Discharge Machine (EDM) was used to cut the disks with thickness of 5 mm and diameter of 10 mm from the ingot. The surfaces were prepared by mechanical grinding using SiC paper up to 1200 mesh followed by polishing using alumina in 1 µm grain size.

The disk samples were treated for 1 hour at 650°C (which is below the Beta transus temperature) in both thermal and plasma oxidation processes. The thermal oxidation treatments were carried out in an atmosphere-controlled furnace with an oxygen flow rate of 10 ppm. Prior to the plasma oxidation treatments, hydrogen sputtering was applied to the surface of all samples for 15 min under a voltage of 500 V and a pressure of $5 \times 10^2$ Pa for eliminating of any contaminants on the surfaces. The plasma treatments were performed in a mixture of 50% Ar – 50% $O_2$ for 1 hour under the same pressure of sputtering.

The wear tests were performed with a reciprocating linear motion of 1 Hz frequency and 4 mm amplitude, a ball-on-plate configuration was used in a tribometer (UMT II, Bruker, Germany). The wear tests were carried out by rubbing a 5 mm alumina ball on the sample surface under 10 N loads for the duration of 1 hour. The oxidation process was performed on 3 samples in one chamber for both thermal and plasma oxidation treatment. Also the wear tests were performed at least 3 times on two different samples to obtain reliability.Before the wear tests, the surface of the plasma oxidized samples was also polished with alumina in 1 µm grain size for removing the any peeling formed.

The phase structure of the oxidized surfaces was determined by using an X-ray diffraction spectrometer (XRD) with Co tube (Explorer, GNR, Italy). A Scanning Electron Microscope (FEI Quanta FEG 250, USA) was also utilized for the evaluation of processed surface and sub-surface morphologies. The wear volume was obtained from 3D profilometer (Bruker Contour GT) and Vision64 software. The surface and cross-sectional hardness of the samples were measured in a microhardness tester (Shimadzu HMV-G, Japan) under a normal load of 25 g for a dwell time of 15 second.

## 3. Results and Discussion

### *3.1. Structure analysis*

Figure 1 and figure 2 show the representative X-ray diffraction patterns for untreated (AR), PO and TO treated Ti–6Al–4V and Ti-15Mo samples. Both AR alloys consist of alpha + Beta microstructures. (Alpha (α) (JCPDS cards #44-1294) and Beta (β) (JCPDS cards #44-1288)). As seen, both anatase (A) (JCPDS cards #21-1272) and rutile (R) (JCPDS cards #21-1276) phases formed on both oxidized alloy surfaces. The intensity of the rutile phases formed by plasma oxidation is higher than that of phased formed by thermal oxidation. This may be due to the relatively high thickness of the layer formed by plasma oxidation and the plasma kinetics (Fig 4). It is known that the rutile phase is more stable compared to the anatase phase [13, 14]. Therefore, more stable structures can be achieved during plasma oxidation at chosen temperatures.

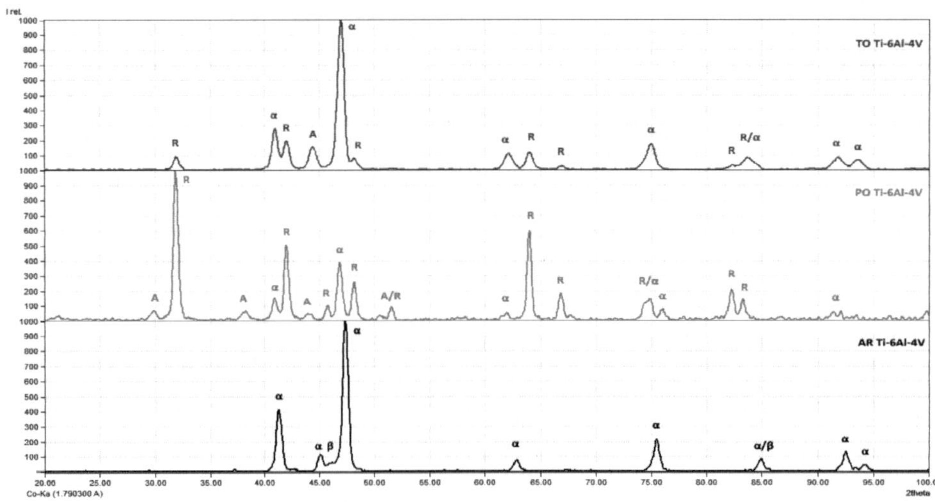


**Figure 1.** XRD patterns of the Ti-6Al-4V alloy samples after plasma and thermal oxidations (α: Alpha, β: Beta, A: Anatase, R: Rutile).

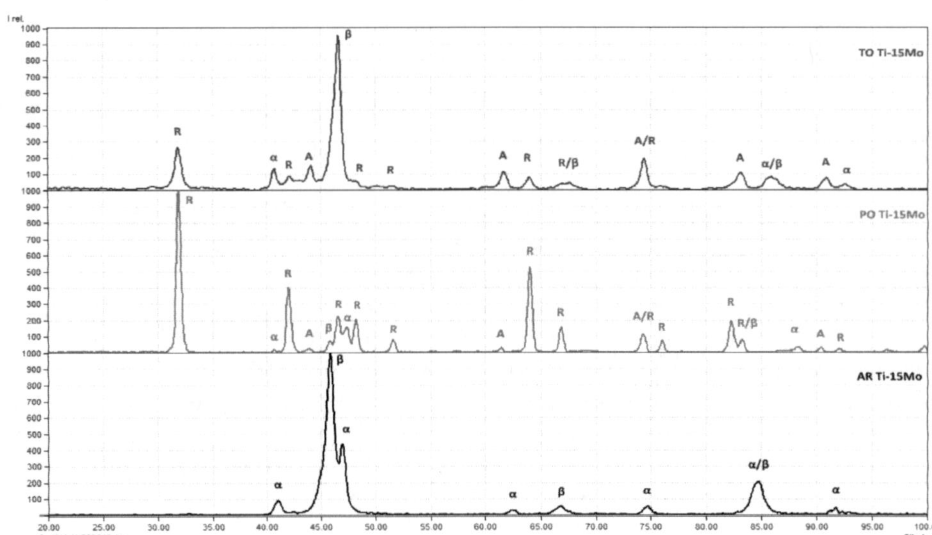


**Figure 2.** XRD patterns of the Ti-15Mo alloy samples after plasma and thermal oxidations (α: Alpha, β: Beta, A: Anatase, R: Rutile).

*3.2. Mechanical Properties*

The surface morphologies of the samples after oxidation processes are shown in figure 3. It is seen from the SEM micrographs that the surface morphology of the samples changed completely after oxidation treatments. While the PO treatment caused a more porous surface morphology, the TO treatments did not change the surface morphology of both alloys. The high magnification micrographs of Ti-15M shows a net-shaped microporous structure, which was formed after the PO.

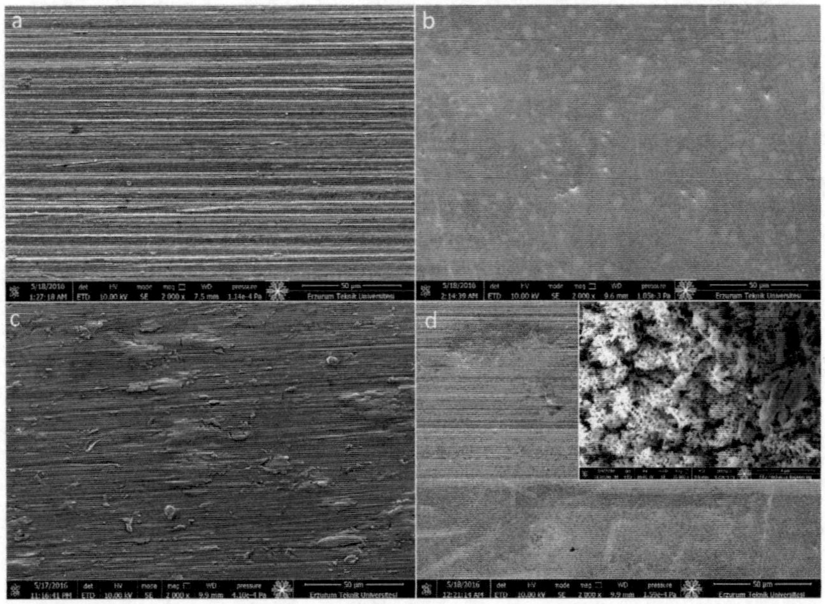


**Figure 3.** SEM images of Plasma and Thermal treated Ti-6Al4V and Ti-15Mo alloys (a: TO Ti-6Al-4V, b: PO Ti-6Al-4V, c: TO Ti-15Mo, d: PO Ti-15Mo and detailed view).

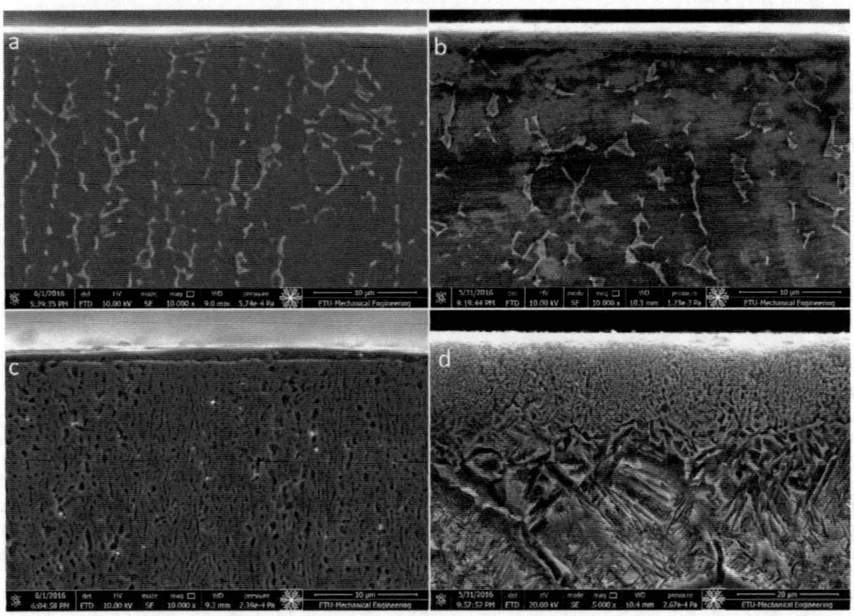


**Figure 4.** Cross-sectional SEM images of Plasma and Thermal treated Ti-6Al4V and Ti-15Mo alloys (a: TO Ti-6Al-4V, b: PO Ti-6Al-4V, c: TO Ti-15Mo, d: PO Ti-15Mo).

The cross-sectional views of the samples, oxidized by PO and TO, are shown in figure 4. It is seen that the sample surfaces are not completely covered by an oxide film during TO. Also the oxidized layer thickness of TO sample is lower than that of PO counterpart. After PO, an oxide layers in thickness of 2-2.5 μm and 20-25 μm were obtained for Ti-6Al-4V and Ti-15Mo, respectively. In addition, beneath the oxide layer, a thick diffusion region formed for both alloys. Furthermore, the oxide layer formed on the Ti-15Mo sample has a different morphology with a grid-shaped appearance connecting with the well-developed grain boundaries inside the diffusional region.

After the TO, a remarkable oxide layer formed on Ti-15Mo surfaces, unlike Ti-6Al-4V. On the other hand, the PO led to a thick oxide layer on the surfaces of both alloys. However the oxide layers obtained from PO were partially peeled off, as reported by other researchers [ 15-16]. Therefore, the diffusion layers were determined by microhardness measurements and the SEM analyses.

The surface hardness of solution treated and air-cooled (AR) Ti-6Al-4V and Ti-15Mo alloys were measured to be 480±22 and 357±17 $HV_{0.025}$, respectively. Hardness distribution through the cross-section of Ti-6Al-4V and Ti-15Mo after TO and PO samples is shown in figure 5. In general, the hardness profile demonstrates that both PO and TO processes significantly increase the hardness of the surface and sub-surface. It is caused by the distortion of the lattice system by the effect of oxygen atoms. The hardness profile is almost the same for both alloy samples under PO process. Both alloy exhibit almost the same profile in terms of hardness and the thickness of diffusional layer. After TO, while the Ti-15Mo is higher surface hardness compared to that of Ti-6Al-4V, it has lower thickness of diffusional layer. These results are consistent with the cross-sectional images (figure 4). It can be said that while the surface hardness of oxidized sample is related to its phase structure, the depth or thickness of the diffusional layer depends on the diffusional capacity of oxygen inside the alloy. As seen from the XRD results (Fig 1 and 2), the intensity of the rutile phase is higher in Ti-15Mo alloy than the TO Ti-6Al-4V in accordance to hardness measurements. However, the thickness of the diffusion layer of Ti-15Mo is relatively low due to the presence of Mo atoms which prevent the easy penetration of oxide atoms through the surface. After the plasma and thermal treatments, the samples were cooled to the ambient temperature in an oven or a plasma chamber. The slow cooling rate leads to decreasing of the base metal hardness. The hardness of base metals reduced to 400±12 and 338±18 $HV_{0.025}$ for the Ti-6Al-4V and Ti-15Mo alloys, respectively.

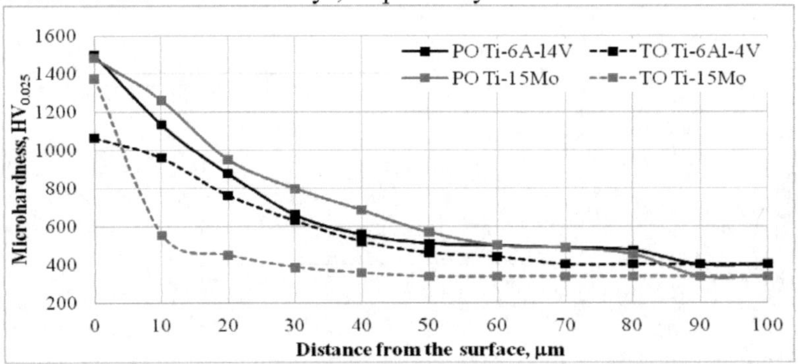


**Figure 5.** The change in microhardness as a function of depth below the surface of PO and TO Ti-15Mo and Ti-6Al-4V.

Friction test results of untreated and oxidized Ti-6Al-4V and Ti-15Mo are given in figure 6 and figure 7, respectively. Both alloys in the as-received states exhibit more stable friction coefficient during sliding time compared to oxidized counterparts. The average friction coefficient of AR Ti-6Al-4V and Ti-15Mo are in the steady state conditions are 0.4 and 0.8, respectively. The friction coefficient of AR Ti-6Al-4V surface gradually increased throughout the rubbing period (Fig 6). On the other hand, the friction coefficient of AR Ti-15Mo was generally stable. Also, a slight decrease was observed at about 1800 sec. and 3200 sec (Fig 7).

At the start of the friction tests, an increase in the friction coefficient of all samples was observed due to mechanical contact. Later, it remained steady-state as the surface became smoother. The PO treatment led to lower friction coefficient value for Ti-6Al-4V in the beginning of rubbing, it reached to higher value after about 400 seconds. Subsequently a frictional fluctuation was observed, due to formation of debris. The more stable friction coefficient of PO may be ascribed to lower debris produced during rubbing of PO (Fig 6) [17]. The difference between the frictional behaviour of Ti-6Al-4V may imply differences in surface morphologies caused by varying processes.

In general, the Ti15Mo had higher friction coefficient than the Ti-6Al-4V and the oxidation processes did not have significant effect on friction coefficient. The PO surface suddenly reached the same value with AR Ti-15Mo after about 150 seconds. The TO gradually attained the same value with others as well (Fig 7).

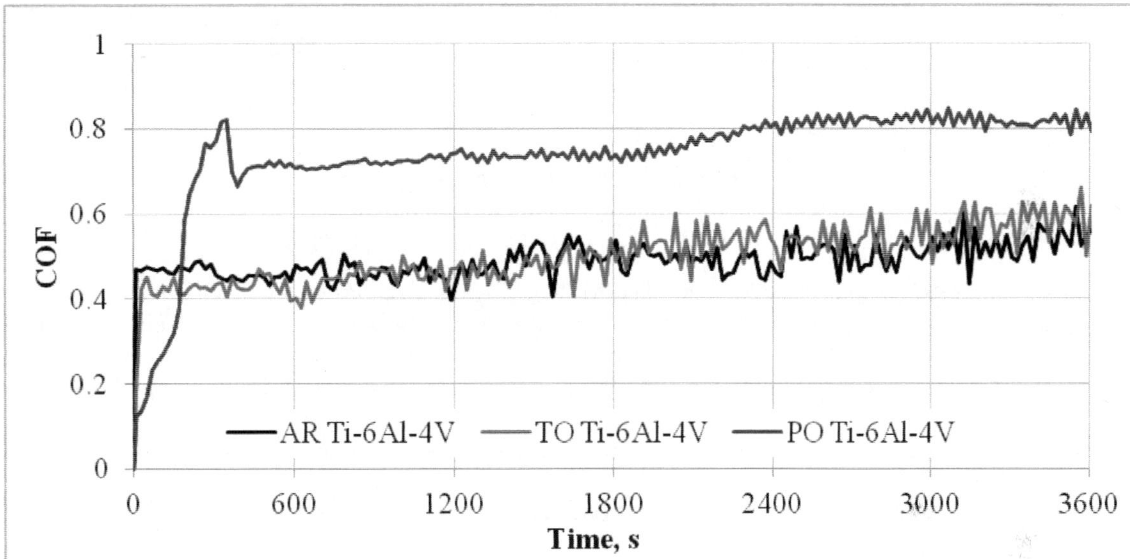


**Figure 6**. The change in friction coefficients of Ti-6Al-4V alloy samples oxidized by TO and PO as a function of sliding time.

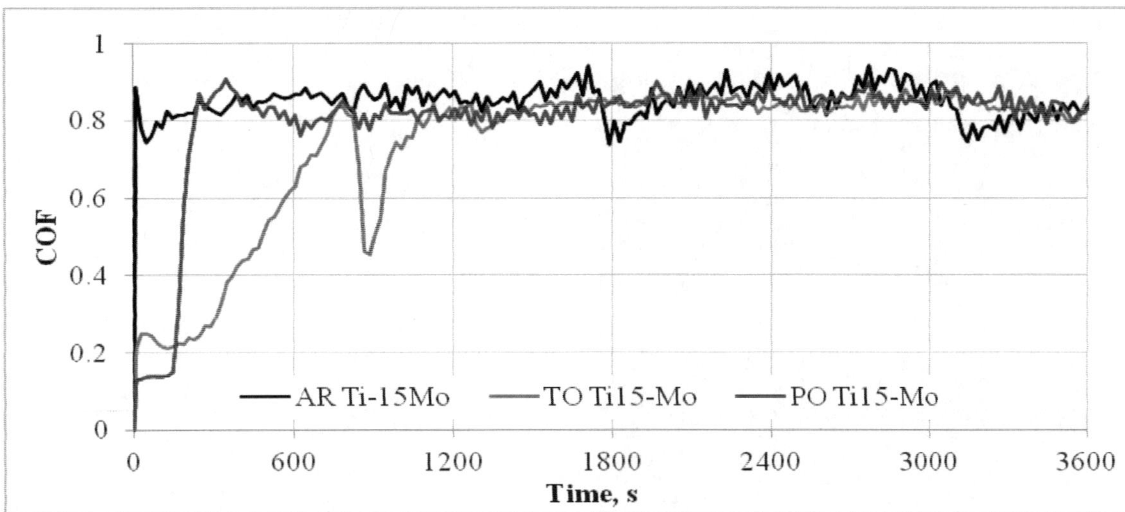


**Figure 7.** The change in friction coefficients of Ti-15Mo alloy samples oxidized by TO and PO as a function of sliding time.

As seen from the results, the PO process has a significantly effect on the wear resistance of Ti-6Al-4V alloy. The highest wear resistance (or the lowest wear rate) was obtained after PO treatments of both alloys. In general, high hardness means high wear resistance as expected. The thermal oxidation led to slightly thin oxide layer on Ti-6Al-4V alloy. The hardness measurements and cross-sectional SEM images claim the diffusion zone increased the surface hardness.

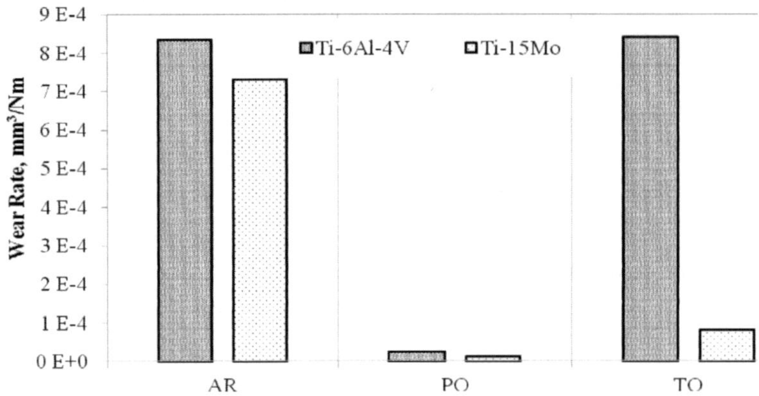


**Figure 8.** Wear rates of Ti-15Mo and Ti-6Al-4V alloys in the AR, PO and TO conditions.

SEM images of both Ti-15Mo and Ti-6Al-4V alloys are shown in figure 9. As considering the width of wear scars, the Ti-15Mo sample exhibited higher wear resistance in all conditions compared to Ti-6Al-4V alloy. Also, it is clearly seen that the PO decreased the wear rates of both alloys as presented in figure 8. The lowest width of the wear track is obtained from the Ti-15Mo sample oxidized by PO. As indicated on the figure 9f, the wear track was wider at the beginning of rubbing and then became narrower. This observation may comply with frictional behaviour after about 130 seconds (Fig 7).

In generally, adhesive, abrasive and oxidative wear mechanisms seem to be operative in all samples.

When SEM images of AR and TO Ti-6Al-4V and their frictional behaviour are considered, it suggests a similar wear mechanism. The rough, wide and deep wear track implies that this can be adhesive mechanism (Fig 9). According to 3D surface profile, the wear track depth of both PO alloys remained in the diffusion layer and an improvement of load-bearing capacity was observed (Fig 10).

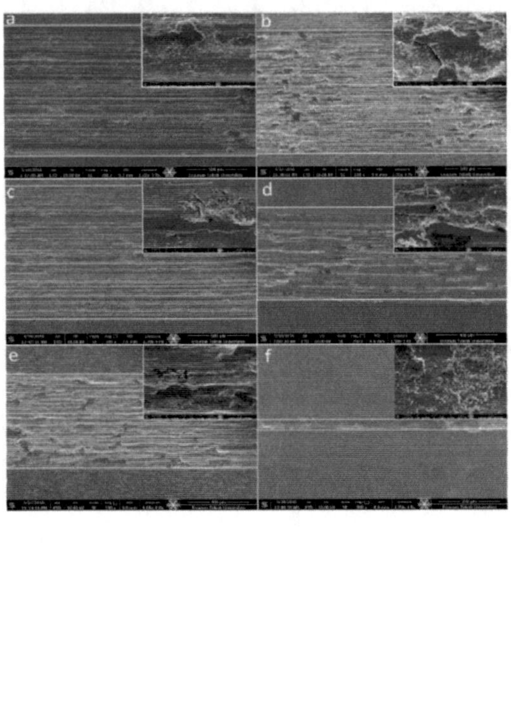


**Figure 9**. SEM images showing the wear tracks and worn surfaces of Ti-15Mo and Ti-6Al-4v alloys: (a: AR Ti-6Al-4V, b: AR Ti-15Mo c: TO Ti-6Al-4V, d: PO Ti-6Al-4V, e: TO Ti-15Mo, f: PO Ti-15Mo).

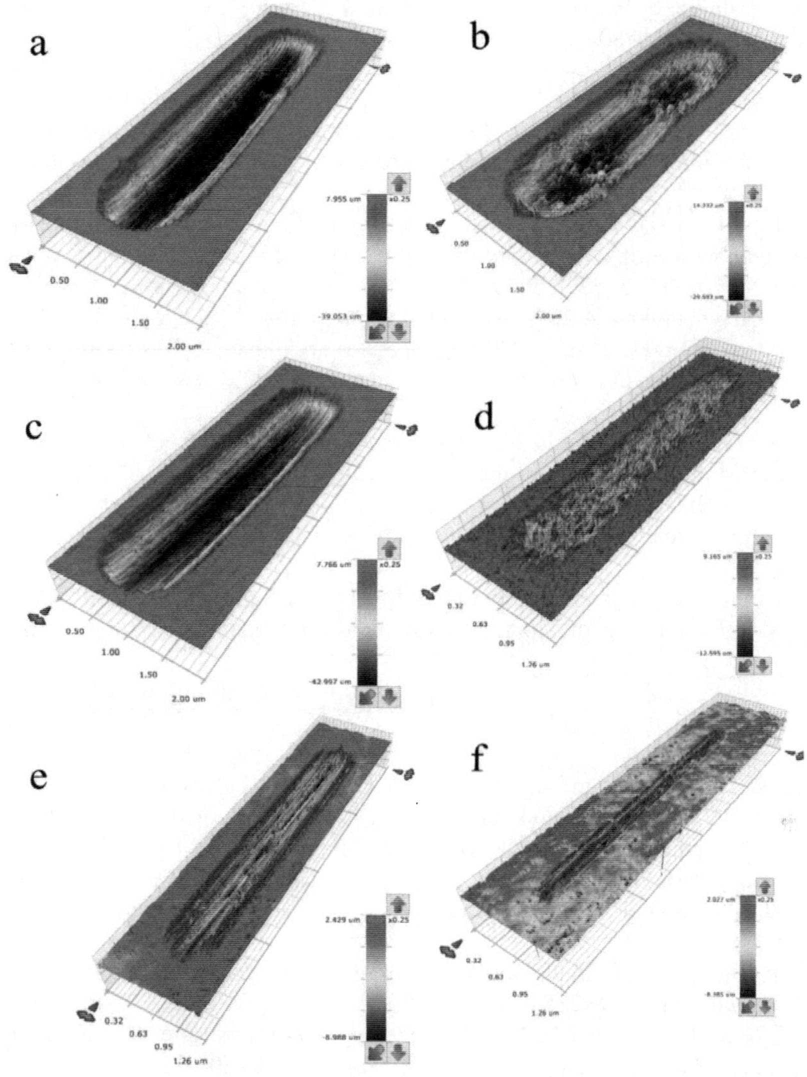


**Figure 10**. 3D profilometer images showing the wear tracks and worn surfaces of Ti-15Mo and Ti-6Al-4v alloys: (a: AR Ti-6Al-4V, b: AR Ti-15Mo c: TO Ti-6Al-4V, d: PO Ti-6Al-4V, e: TO Ti-15Mo, f: PO Ti-15Mo).

## 4. Conclusions

Thermal and plasma oxidation treatments were applied to alpha+beta type Ti-15Mo and Ti-6Al-4V alloys, and their tribological properties as well as structure and hardness were investigated.

- An oxide layer consisting of anatase and rutile phases formed on the surfaces of Ti-15Mo and Ti-6Al-4V after both plasma oxidation and thermal oxidation treatments. After plasma oxidation treatment, the intensity of rutile phase and the thickness of oxidized layer were higher for both Ti-15Mo and Ti-6Al-4V samples than after thermal oxidation.
- The surface hardness and the thickness of diffusion layer of both alloys after plasma oxidation were higher compared to thermal oxidized.

IOP Conf. Series: Materials Science and Engineering **174** (2017) 012055 doi:10.1088/1757-899X/174/1/012055

- The Ti-15Mo alloy exhibited higher wear resistance compared to Ti-6Al-4V alloy under all conditions. Wear resistance of both Ti-15Mo and Ti-6Al-4V alloy oxidized by plasma oxidation was higher than the thermally oxidized.

**References**

[1] Kuromoto N K, Simão R A and Soares G 2007 A Titanium oxide films produced on commercially pure titanium by anodic oxidation with different voltages *Mater. Charact.* **58** 114-121

[2] Babaei M, Dehghanian C and Babaei M 2015 Electrochemical assessment of characteristics and corrosion behavior of Zr-containing coatings formed on titanium by plasma electrolytic oxidation *Surf. Coat. Tech.* **279** 79-91

[3] Stepanov I B, Ryabchikov A I, Ananin PS, Bumagina A I, Shevelev A E, Shulepov I A and Sivin D O 2016 Investigation of filtered vacuum arc plasma application for TiAlN and TiSiB coatings deposition using ion beam and plasma material processing *Surf. Coat. Tech.* **296** 20-25

[4] Yetim A F, Yildiz F, Vangolu Y, Alsaran A and Celik A 2009 Several plasma diffusion processes for improving wear properties of Ti6Al4V alloy *Wear* **267** 2179-85

[5] Yerokhin A L, Nie X, Leyland A and Matthews A 2000 Characterisation of oxide films produced by plasma electrolytic oxidation of a Ti-6Al-4V alloy *Surf. Coat. Tech.* **130** 195-206

[6] Ceschini L, Lanzoni E, Martini C, Prandstraller D and Sambogna G 2008 Comparison of dry sliding friction and wear of Ti6Al4V alloy treated by plasma electrolytic oxidation and PVD coating *Wear* **264** 86-95

[7] Yildiz F, Yetim A F, Alsaran A and Efeoglu I 2009 Wear and corrosion behaviour of various surface treated medical grade titanium alloy in bio-simulated environment *Wear* **267** 695-701

[8] Garcia-Alonso M C, Saldana L, Valles G, Gonzalez-Garrasco J L, Gonzalez-Cabrero J, Martinez M E, Gil-Garay E and Munuera L 2003 In vitro corrosion behaviour and osteoblast response of thermally oxidised Ti6Al4V alloy, *Biomaterials* **24** 19-26

[9] Çelik I, Alsaran A and Purcek G 2014 Effect of different surface oxidation treatments on structural, mechanical and tribological properties of ultrafine-grained titanium *Surf. Coat. Tech.* **25** 842-848

[10] Dong H and Bell T 2000 Enhanced wear resistance of titanium surfaces by a new thermal oxidation treatment *Wear* **238** 131-137

[11] Boettcher C, Bell T and Dong H 2002 Surface engineering of timet 550 with oxygen to form a rutile-based, wear-resistant coating *Metall. Trans. A.* **33** 1201-11

[12] Bailey R and Sun Y 2013, Unlubricated sliding friction and wear characteristics of thermally oxidized commercially pure titanium *Wear* **308** 61

[13] Kumar S, Sankara Narayanan T S N, Ganesh Sundara Raman S and Seshadri S K 2010 Thermal oxidation of Ti6Al4V alloy: Microstructural and electrochemical characterization *Mater. Chem. Phys.* **119** 337-346

[14] Aniołek K, Kupka M and Barylski A 2016 Sliding wear resistance of oxide layers formed on a titanium surface during thermal oxidation *Wear* **356-357** 23-29

[15] Somsanith N, Sankara Narayanan T S N, Kim Y K, Park I S, Bae T S and Lee M H 2015 Surface medication of Ti–15Mo alloy by thermal oxidation: Evaluation of surface characteristics and corrosion resistance in Ringer's solution *Appl. Surf. Sci.* **356** 1117–26

[16] Siva Rama Krishna D, Brama Y L and Sun Y 2007 Thick rutile layer on titanium for tribological applications *Tribol. Int.* **40** 329-334

[17] Cassar G, Avelar-Batista Wilson J C, Banfield S, Housden J, Matthews A and Leyland A 2010 A study of the reciprocating-sliding wear performance of plasma surface treated titanium alloy *Wear* **269** 60–70

IOP Conf. Series: Materials Science and Engineering **174** (2017) 012056 doi:10.1088/1757-899X/174/1/012056

# Determination of residual stresses and natural frequencies of roll-tensioned disc by a dynamic simulation of the rolling process

**G Skordaris, K-D Bouzakis and D Tasoulas**

Laboratory for Machine Tools and Manufacturing Engineering, Mechanical Engineering Department, Aristotle University of Thessaloniki, 54124 Thessaloniki, Greece

E-mail: gskor@eng.auth.gr

**Abstract.** Roll tensioning is a common method for increasing locally the superficial strength of thin circular saws and in this way their dynamic stability. Through roll tensioning, residual stresses are induced into the disc material leading to a significant enhancement of its dynamic stiffness. In this paper, a FEM-methodology is proposed for determining the developed residual stresses in the discs after rolling and for investigating their effects on the circular saw natural frequencies. More specifically, a 3D-FEM model was developed for the dynamic simulation of the rolling process on circular saws, using the LS-DYNA software. This model enables the explicit determination of the developed residual stresses in the roll-tensioned discs. Furthermore, the natural frequencies of the pre-stressed circular saws were calculated by the ANSYS software. In these calculations, the already determined residual stresses were taken into consideration. Different distances of the roll-tensioned zone from the disc centre were taken into account for estimating their effect on the disc's natural frequencies. By the proposed methodology, optimum roll-tensioning conditions can be predicted for improving the dynamic behaviour of thin circular saws during cutting.

## 1. Introduction

Thin circular saws are often applied in wood industry at high cutting speed [1,2]. Due to the small thickness compared to their diameter, such tools are characterized by an insufficient dynamic behaviour. Depending upon the applied rotational speed, the circular saws are subjected to undesirable axial vibrations deteriorating the tribological conditions between the tool and the workpiece material, thus decreasing the tool life and the surface integrity. For enhancing the dynamic behaviour of thin circular saws, roll-tensioning procedures are usually applied on a certain circular area of such tools [3-5]. Through this method, because of the induced local plastic deformation of the tool material, residual stresses are developed in the disc material structure, changing its dynamic characteristics [3-7]. Hereupon, the residual stresses occurred in the circular saw after rolling are located in a limited radial width of a few millimetres [8]. These residual stresses affect the circular saw structure stiffness and thus, its natural frequencies [9].

Various methods have been developed for estimating the induced residual stresses after rolling in the circular saw material and their influence on disc's natural frequencies. In the literature [4,5], three-region theoretical models were proposed for approximating the circumferential and radial stresses in the relaxation stage. In these investigations, it was assumed that two external regions are stressed elastically and the intermediate one is permanently deformed, since the circular saw elastically springs back to a new equilibrium after the load removal [5]. In the literature [7], the induced residual stresses were evaluated with the aid of X-ray measurements inside and outside of rolling paths in both

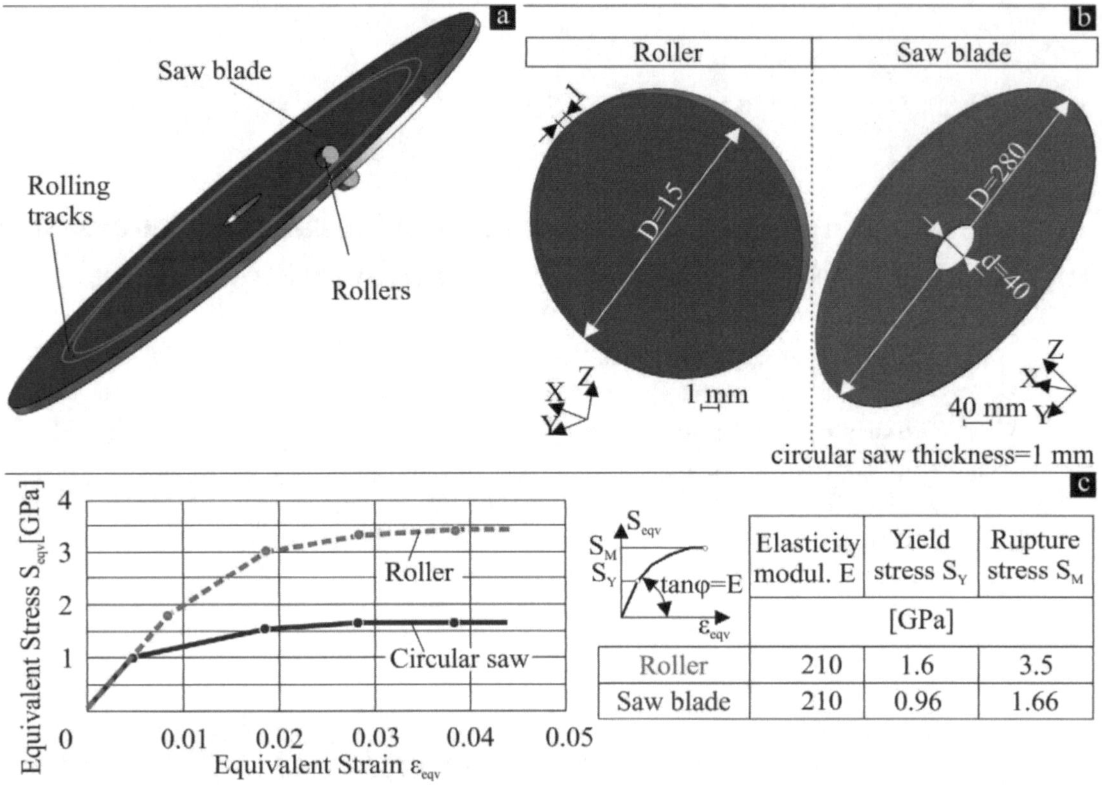


| | Elasticity modul. E | Yield stress $S_Y$ | Rupture stress $S_M$ |
|---|---|---|---|
| | [GPa] | | |
| Roller | 210 | 1.6 | 3.5 |
| Saw blade | 210 | 0.96 | 1.66 |

**Figure 1.** (a) A schematic presentation of the roll-tensioning process.
(b) The geometrical characteristics of the employed circular saw and the rollers.
(c) Mechanical properties of the employed materials.

tangential and radial directions. Moreover, a simulation model was developed for calculating the residual stresses induced by roll-tensioning [8]. The natural frequencies and the mode shapes of a circular saw with or without roll-tensioning were investigated employing modal analysis [5,6] and using appropriate experimental set-ups [4,10,11].

In this work, a methodology is introduced for an efficient assessment of the applied rolling conditions on the dynamic behaviour of circular saws. In this context, a dynamic simulation of the rolling process on a circular saw was developed for the explicit determination of the induced residual stresses using the LS-DYNA software [12]. Considering the determined residual stresses, a further FEM model was created for conducting eigenvalue analysis on the pre-stressed circular saw with the aid of the ANSYS software [13]. The FEM calculations were carried out at different distances of the roll-tensioned zone from the disc centre for revealing their effect on the disc's natural frequencies.

## 2. The dynamic FEM model developed for calculating the induced residual stresses in the circular saw material after rolling

### *2.1. The structure of the developed FEM model*

A schematic presentation of the roll-tensioning process, in which the circular saw is locally deformed between the opposing rollers, is shown in figure 1a. The geometrical characteristics of the employed circular saw and the rollers in the developed FEM model for calculating the induced residual stress after rolling are illustrated in figure 1b. The roller width in the described investigations was intentionally selected to about 1 mm for calculating approximately the specific rolling force. For describing the circular saw and the rollers strength properties in the developed FEM model, materials with piecewise linear plasticity were used [14]. The stress-strain curves employed are illustrated in

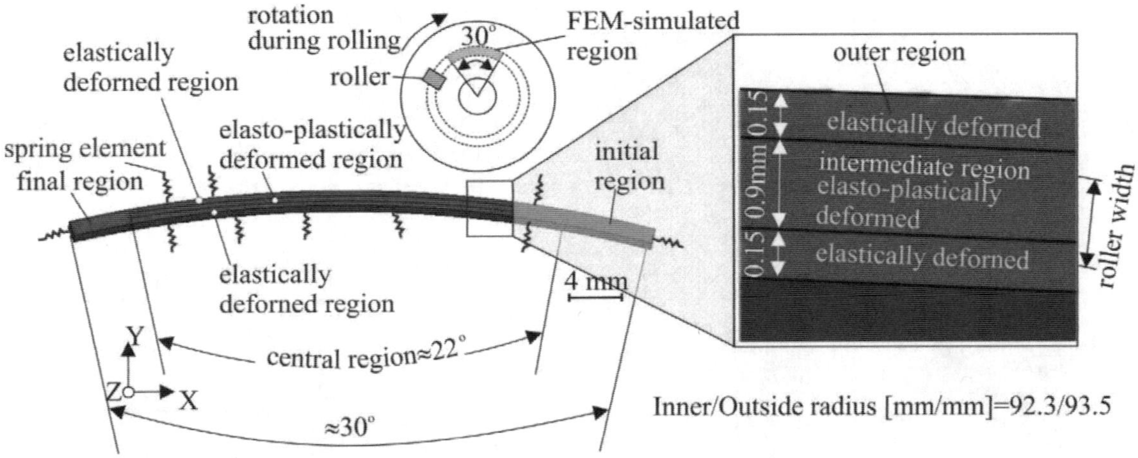


**Figure 2.** The applied geometry of circular saw in the developed FEM-model.

figure 1c. These curves were determined by evaluating nanoindentation results on the rollers and the disc using the procedures documented in [15]. These material properties are characteristic for a high carbon and a high-alloyed steel.

A 3D-FEM model describing the whole disc geometry using a fine discretizartion network for attaining a reasonable accuracy when simulating the rolling process on a circular saw leads to a long calculation time. In order to overcome this problem, only a certain region of the circular saw was employed in the conducted FEM simulation. More specifically, in the circumferential direction, a subtended angle of 30 degrees was used for describing the circular saw geometry, as shown in figure 2. Since only a circular arc was considered in the developed FEM model and not the whole disc, the regions included in the first and the last 4 degrees were not taken into consideration in the FEM calculations. The elements associated with these regions were elastically deformed due to the plastic deformation of neighbouring elements of the central region after rolling. Additionally, elastic spring elements were set as boundary conditions in the circumferential direction, possessing an elastic stiffness corresponding to the steel's elasticity modulus. A three regions model was built in the radial direction, as shown in figure 2. The width of these regions was appropriately selected depending upon the employed rolling conditions. The elements which are located in the intermediate region, are elasto-plastically deformed, whereas the related ones belonging to the outer regions only elastic. The rest circular saw material was substituted by elastic spring elements.

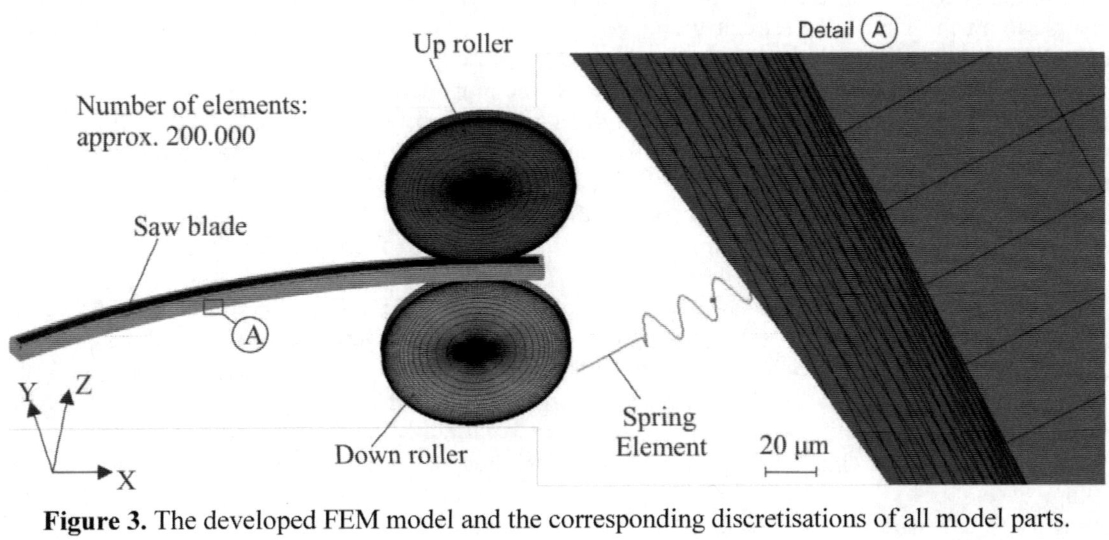


**Figure 3.** The developed FEM model and the corresponding discretisations of all model parts.

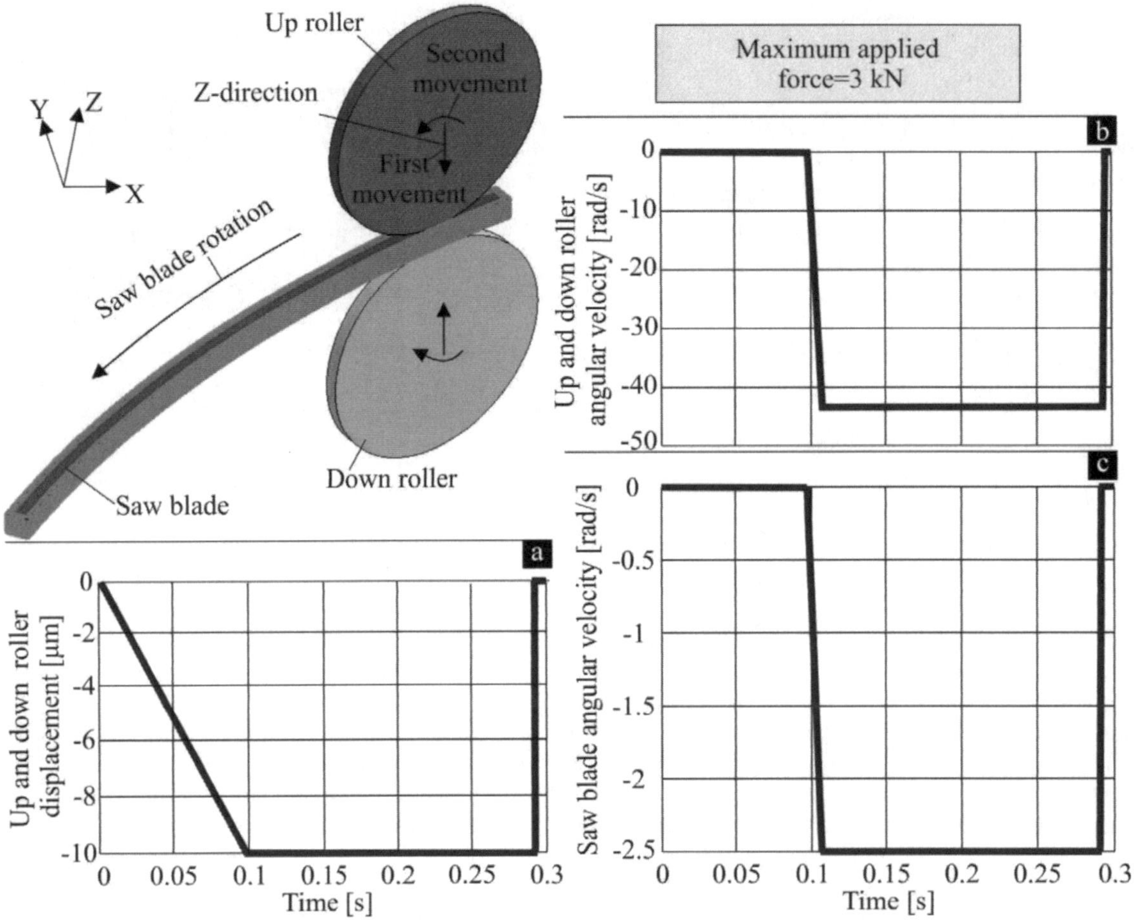


**Figure 4.** The employed kinematics of the circular saw and the rollers in the developed FEM model.

The developed FEM model and the corresponding discretisations of all model parts are shown in figure 3. Convergence studies were conducted to determine the optimum mesh density and obtain a mesh independent grid. Fully integrated solid elements were employed with poor aspect ratio for all parts, which is characterized as an efficient formulation [14]. A surface to surface contact was applied for describing the interface between the circular saw and the rollers in the developed FEM model [12,14]. This is a penalty-based contact with springs placed between all penetrating nodes and the contact surface [12]. A more coarsely meshed part is recommended to be identified as a master part in the LS-DYNA environment. Moreover, the rollers were coarsely meshed compared to the circular saw, since the goal of the conducted investigations was to determine the induced residual stresses in the disc's structure. Thus, the rollers were identified as the master parts, whilst the circular saw as the slave one. The density and the Poisson ratio of the employed parts are equal to $7.85\text{x}10^{-6}$kg/mm and 0.3, respectively. The used static friction coefficient between the rollers and the circular was set equal to 0.15, corresponding to a lubricated rolling case [16,17].

The kinematics of the circular saw and the rollers used in the developed FEM model are presented in figure 4. The motions of the individual parts are introduced as a function of time and are linked to certain curves representing their time dependent values. The first movement is associated with the rollers' displacement up to a desired maximum depth, as shown in figure 4a. When the maximum indentation depth is achieved, then the rollers start rotating with a certain angular velocity (see figure 4b). At the same time, an angular velocity is applied at the circular saw for implementing the rolling process (see figure 4c). By employing the rolling conditions shown in figure 4, the maximum rolling force, calculated by the developed FEM model amounts to 300 daN.

IOP Conf. Series: Materials Science and Engineering **174** (2017) 012056 doi:10.1088/1757-899X/174/1/012056

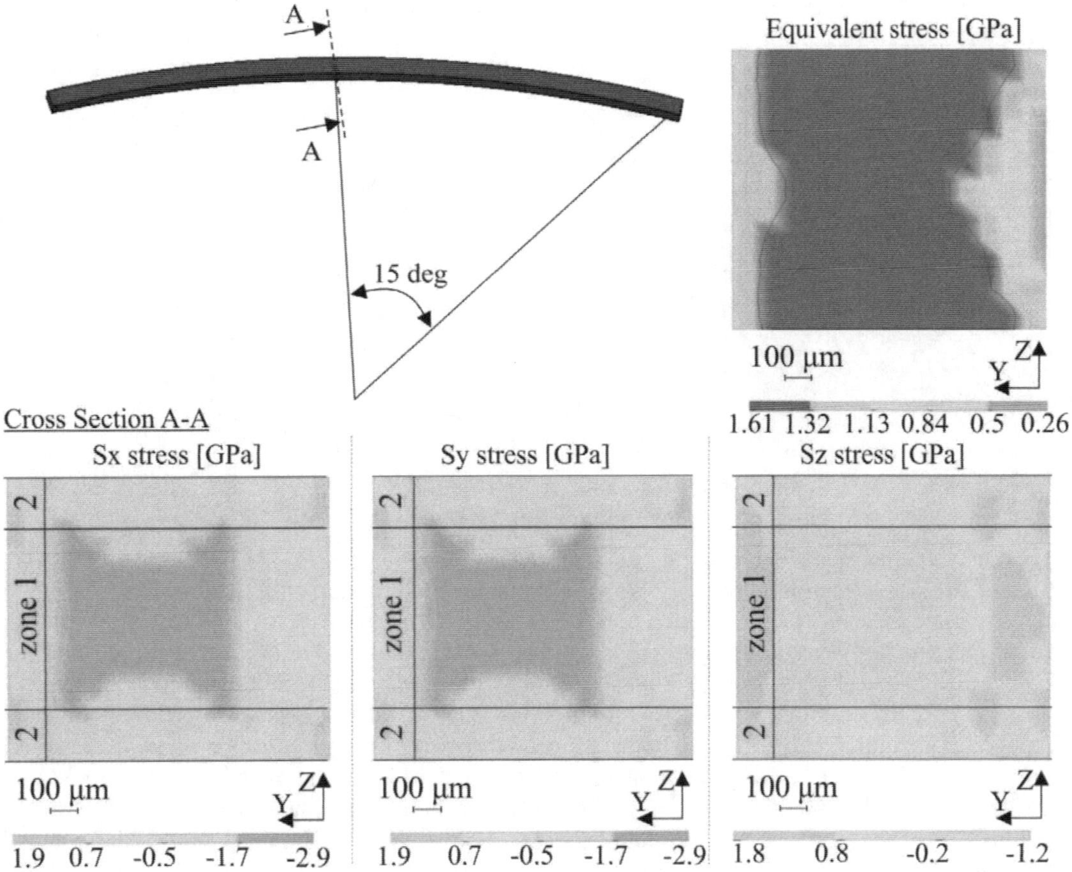


**Figure 5.** Residual stress distributions in the circular saw material in x, y and z directions after roll-tensioning.

*2.2. The determination of the disc's residual stress using the developed FEM model*

Calculations were conducted by the developed FEM model for the explicit determination of the residual stress distributions in the circular saw material in x, y and z directions after rolling. The equilibrium differential equations are integrated for incremental solution time steps of few milliseconds. Each solution step is based on the results of the previous one (explicit method) [12]. Figure 5 shows characteristic stress fields developed at a cross section of the circular saw in the middle of its length. The equivalent stress field is also illustrated. Highly compressive stresses are induced in the central region remaining in all directions after the relaxation. As a consequence, tensile stresses are developed in the outer regions.

This is visible in figure 6, where the stress distributions along the roll-tensioned zone (radial direction) are exhibited. These results correspond to the middle of the disc's thickness which is associated with the most loaded region after rolling and they cover a zone of approximately 60% (zone 1) of the overall circular saw thickness (see figure 5). The related stress distributions along the roll-tensioned zone were also calculated in the zone 2 within the circular saw thickness (see figure 5). The residual compressive stresses are nullified at a distance of approximately 450 μm from the roll-tensioned zone's centre. At a larger distance, tensile stresses are developed in all directions. The width of the roll-tensioned zone in the radial direction was appropriately selected in order to nullify all the residual stresses. The curves describing the course of the residual stresses were approximated by discrete values as shown in figure 6. These values were taken into consideration in a further FEM model for capturing the effect of the residual stresses on the disc's natural frequencies.

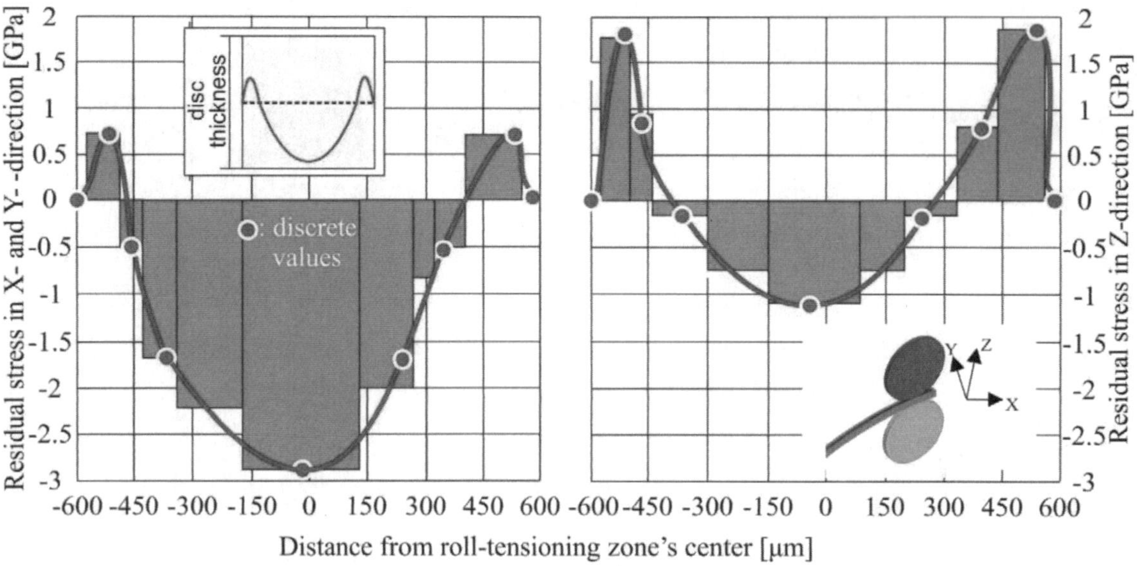


**Figure 6.** Residual stress distributions in the circular saw material along the roll-tensioned zone.

## 3. Effect of the disc's residual stresses after roll-tensioning on its natural frequencies

For investigating the effect of the induced residual stresses in the circular saw structure after rolling on its natural frequencies, eigenvalue analyses were conducted using the ANSYS software [13]. The 3D-FEM model developed for performing a pre-stressed eigenvalue analysis is illustrated in figure 7. The saw teeth do not appear in the developed FEM model, since their effect on the natural frequencies of a circular saw is negligible [5]. The geometrical characteristics of the circular saw were introduced in figure 1. The application of the disc circular fixture is simulated by fixing the nodes of the FEM model located up to a radius of 60 mm from the disc centre corresponding to the circular fixture radius. The distance of the roll-tensioned zone from the disc's centre is a variable and changeable parameter. The procedure for conducting a pre-stressed eigenvalue analysis consists of two steps. Initially, a static solution is performed considering the induced residual stresses due to the rolling process at the individual directions. The developed stresses in the disc material under loading start progressively to increase. If the developed stresses exceed its yield stress, then the remaining plastic deformation during the material relaxation induces residual stresses and strains. Thus, if the disc is reloaded, the plastically deformed material behaves as possessing yield strength greater than the pristine one (see figure 8). In a further step, eigen-frequencies analysis is carried out on the already pre-stressed circular saw. Related calculations were conducted on a stress free circular saw. In the case of a pre-stressed disc, the resulting stresses and strains in the roll-tensioned zone are integrated in the calculations made.

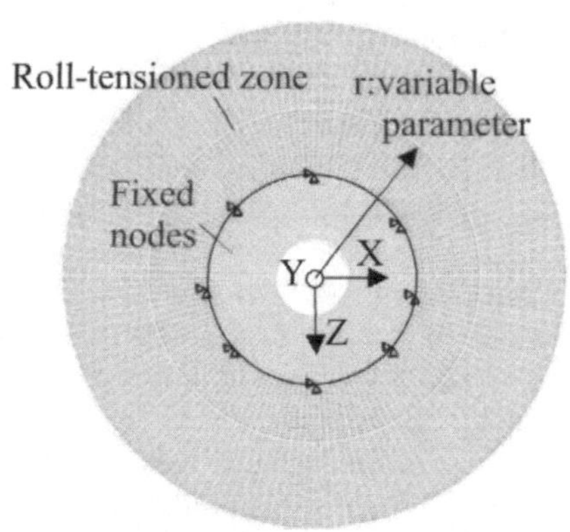


Pre-stressed eigenvalue analysis

Software: ANSYS 14

Number of elements: approx. 500.000

**Figure 7.** The employed FEM-model for conducting eigenvalue analysis.

IOP Conf. Series: Materials Science and Engineering **174** (2017) 012056 doi:10.1088/1757-899X/174/1/012056

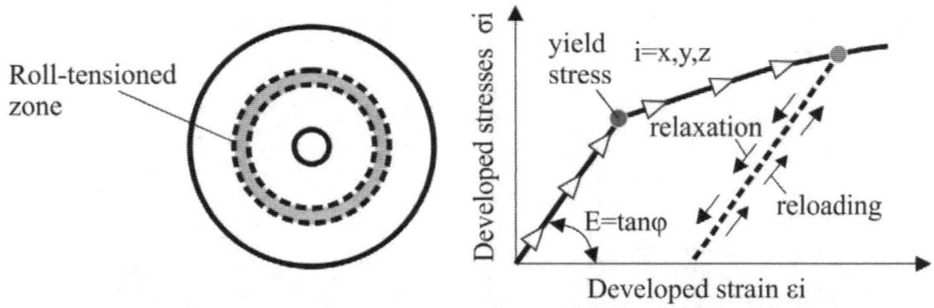


**Figure 8.** Mechanical behaviour of roll-tensioned zone during reloading.

Figure 9 shows characteristic results after obtaining a static solution in the pre-stressed model at certain distance from the disc's centre. The residual stresses induced in the disc's structure after rolling (see figure 5) were considered as input data in the developed FEM model for conducting static analysis. In order to introduce these stresses into the FEM model, the discrete values shown in figure 6 were taken into account along the pre-tensioned zone with respect to the position of each element within the circular saw thickness (see figure 5). These stresses were introduced into the FEM model taking into consideration the local coordinate system of its element within the pre-tensioned zone. Compressive residual stresses are developed x and z directions. As a consequence, tensile stresses are induced in a perpendicular y direction for equilibrium reasons.

The first six natural frequencies of the roll-tensioned circular saw were calculated at various distance of the treated region from the disc's centre through the FEM-supported eigenvalue analysis (see figure 10). The same diagram exhibits the related natural frequencies of an untreated disc. The larger the distance for conducting rolling on a circular saw from the disc centre, the higher the natural frequencies are. The first three natural frequencies decrease roughly to the magnitude of those ones

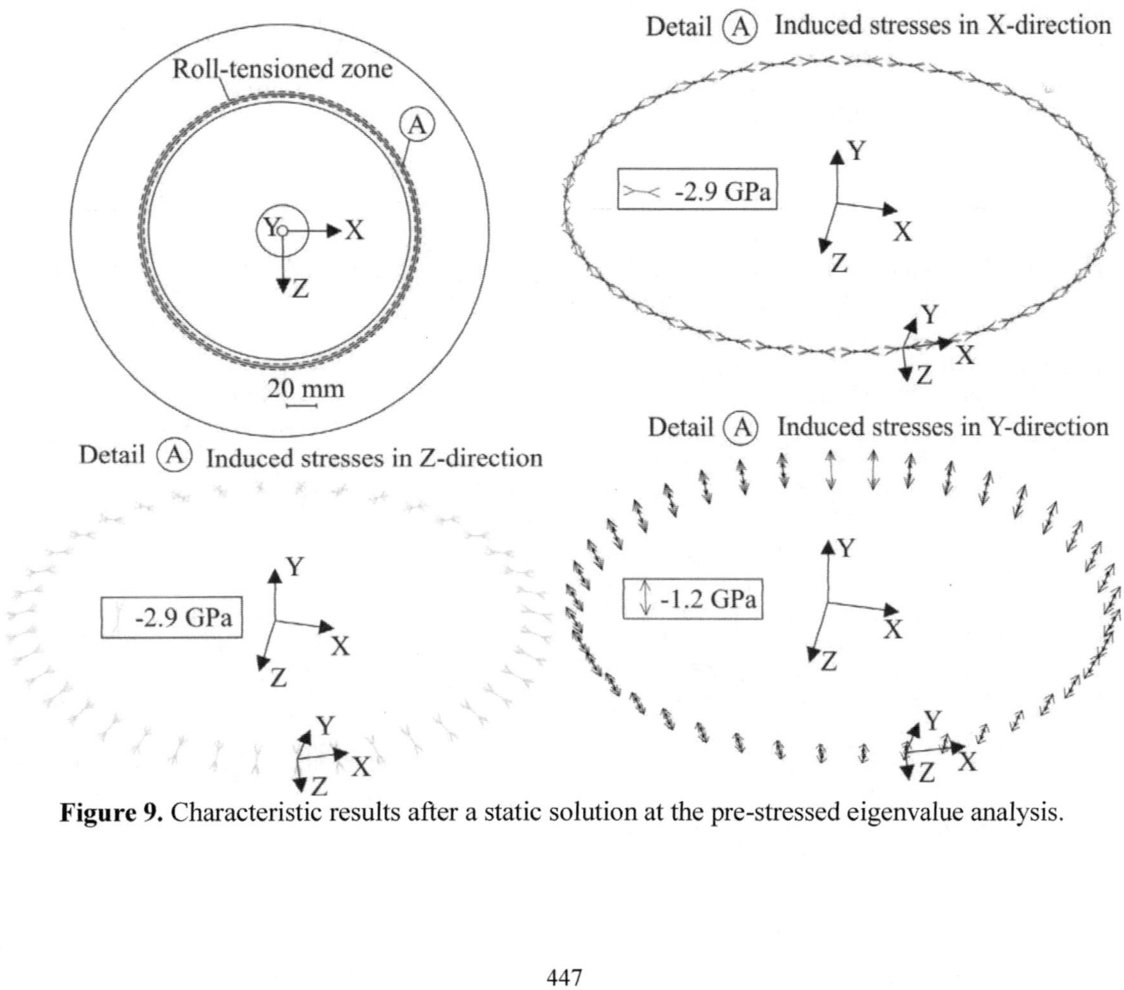


**Figure 9.** Characteristic results after a static solution at the pre-stressed eigenvalue analysis.

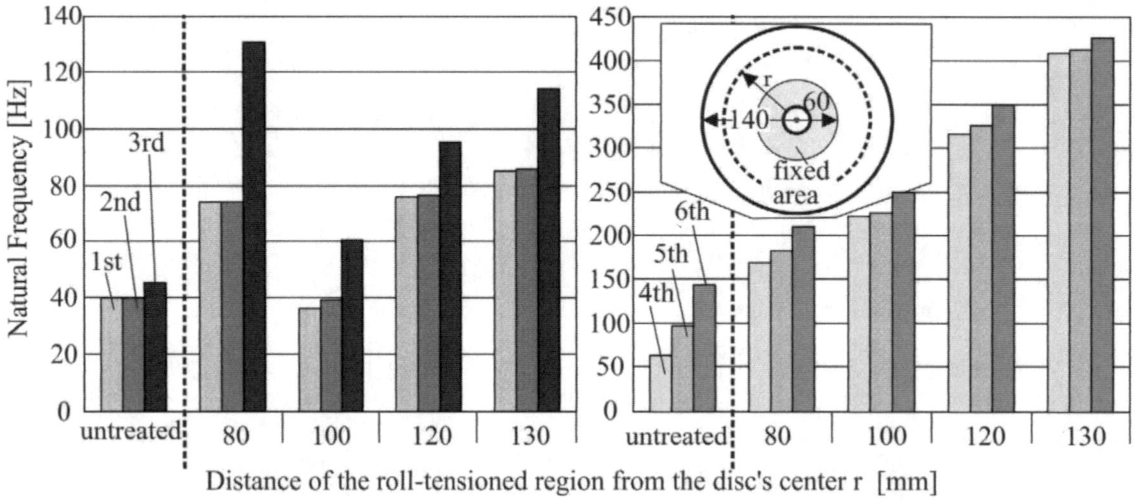


**Figure 10.** Natural frequencies of the circular saw after the conduct of roll-tensioning at various distances from the disc centre.

of the untreated circular saw when the rolling process is conducted on the middle of the rotation disc i.e. up to a distance of approximately 100 mm from the disc centre. This effect has been mentioned in various publications [5,6,9].

## 4. Conclusions

This paper proposes a methodology for predicting appropriate roll-tensioned conditions of circular saws, which can improve their dynamic behaviour. More specifically, a 3D-FEM model was developed for the dynamic simulation of the rolling process on circular saws, enabling the explicit calculation of the induced residual stresses. Furthermore, eigenvalue analysis on the pre-stressed circular saws was conducted by the ANSYS software taking into consideration the residual stresses already determined. Through the methodology developed, optimum rolling conditions can be predicted, thus minimizing the experimentation cost.

## References

[1] Bouzakis K-D and Koutoupas G 2003 *Wood Sci Technol* **37** 141
[2] Bouzakis K.-D, Koutoupas G, Siganos A, Leyendecker T, Erkens G, Papapanagiotou A and Nikolakakis P 2000 *Surf Coat Tech* **133-134** 548
[3] Szymani R and Mote Jr C D 1974 *Wood Sci Technol* **8** 148
[4] Szymani R and Mote Jr C D 1979 *Wood Sci Technol* **13** 211
[5] Schajer G S and Mote Jr C D 1983 *Wood Sci Technol* **17** 287
[6] Nicoletti N, Fendeleur D, Nilly L and Renner M 1996 *Holz als Roh - und Werkstoff* **54** 99
[7] Porankiewicz B, Parantainen J and Ostrowska K 2010 *Engineering* **2** 727
[8] Heisel U, Stehle T and Ghassemi H 2014 *Adv Mat Res* **1018** 57
[9] Kuratani F and Yano S 2000 *Arch Appl Mech* **70** 279
[10] Ukvalbergiene K and Vobolis J 2005 *Wood Res-Slovakia* **50** 47
[11] Georgas P J and Schajer G S 2013 *Exp Mech* **53** 1461
[12] LS-DYNA Keyword Users Manual, Version 971 (Livermore Software Technology Corp.)
[13] ANSYS release 14.0
[14] Skordaris G 2013 *J Mater Eng Perform* **22** 3192
[15] Bouzakis K-D, Michailidis N, Hadjiyiannis S, Skordaris G, and Erkens G 2002 *Mater Charact* **49** 149
[16] *** Coefficient of friction for a range of material combinations - available at http://www.tribology-abc.com/abc/cof.htm#mu_friction (2015)
[17] Han Xinghui and Hua Lin 2014 *Tribol Int* **73** 117

# Evaluating lubricating capacity of vegetal oils using Abbott-Firestone curve

**C Georgescu, G C Cristea, C Dima and L Deleanu**

Department of Mechanical Engineering, "Dunărea de Jos" University of Galaţi, str. Domnească 47, Galaţi, Romania

E-mail: constantin.georgescu@ugal.ro

**Abstract.** The paper presents the change of functional parameters defined on the Abbott-Firestone curve in order to evaluate the surface quality of the balls from the four ball tester, after tests done with several vegetable oils. The tests were done using two grades of rapeseed oil (degummed and refined) and two grades of soybean oil (coarse and degummed) and a common transmission oil (T90). Test parameters were 200 N and 0.576 m/s (1500 rpm) for 60 minutes. For the refined rapeseed oil, the changes in shape of the Abbott-Firestone curves are more dramatic, these being characterized by high values of *Spk* (the average value for the wear scars on the three balls), thus being 40% of the sum *Svk* + *Sk* + *Spk*, percentage also obtained for the soybean oil, but the value *Spk* being lower. For the degummed soybean oil, the profile height of the wear scars are taller than those obtained after testing the coarse soybean oil, meaning that the degumming process has a negative influence on the worn surface quality and the lubricating capacity of this oil. Comparing the surface quality of the wear scars on fixed tested balls is a reliable method to point out the lubricant properties of the vegetable oils, especially if they are compared to a "classical" lubricant as a non-additivated transmission mineral oil T90. The best surface after testing was obtained for the soybean oil, followed by T90 oil and the degummed grades of the soybean oil and rapeseed oil (these three giving very close values for the functional parameters), but the refined rapeseed oil generated the poorest quality of the wear scars on the balls, under the same testing conditions.

## Introduction

Since 1933, when Abbott and Firestone had defined the 'bearing area curve', this has been a useful instrument to evaluate not only the surface texture, but also the changes in surface quality as the shape and parameters related to this curve reflect the modifications in the superficial layer consistency.

The curve has been used to quantify the surface quality in metal cutting and finishing processes [1], [2], [3] (especially for engine cylinders, segments and bearings), in medicine (mainly for femoral stem wear [4] and dentistry [5]) and tribological systems [6]. More recently, it has been used to describe teeth surfaces and worn surfaces (regardless the type of wear, adhesive, abrasive, erosion etc.) [7].

The bearing curve represents the material distribution related to the profile height. Initially, this distribution had a 2D interpretation: the cumulative distribution of the lengths of individual plateaux, normalised by the total assessment length of the investigated profile. Now, due to performant equipment to record the surface texture, the curve is related to a 3D interpretation of the material distribution on the texture height [8], [9].

Even the authors, in previous works [10] have been suggested that the bearing curve would describe better the initial and worn surfaces than Ra, a conclusion also sustained by Torrance [11]. A process (tribological or technological) may remove the peaks, thus, a different texture may be superimposed on the resulting plateaux; the deep valleys may remain or not unaffected, influencing, for instance, the lubricating capability of the surfaces in contact.

The bearing curve could measure the influence of a synergic process, like the tribological ones, when different types of wearing are overlapping. Studying this curve during the triboelements exploitation may offer information on the tendency of the surface changing in the immediate future.

2D studies of the bearing area curve were done by Sosa et al. [12] for studying the texture quality of gear tooth. In another paper, Sosa et al. [13] pointed out the differences in the peak zone (0...30%) of the 2D Abbott-Firestone curves after investigating the running-in process of the tooth, concluding that the valleys seem to remain unchanged while the asperity peaks are worn off.

Affatato et al. [4] characterized the worn surface by comparing Abbott-Firestone curves for unaffected and affected zone of the femoral head made of advanced ceramics.

In 2015, Franco and Sinatora [14] proposed a correlation between *Spk* and the volume *Vmp* (material volume in the hill zone), discussing that it is not correct to designate *Spk* as the average height of the protruding peaks above the core surface, as stated in the standard ISO 25178-2:2012 and *Sk* should be maintained, but that the new volume parameters, through their relationship to the "k series", make it possible to discard *Spk* and *Svk*. *Spk* does not reflect isolated maximum heights of the texture, but this zone of the Abbott-Firestone curve is affecting the tribological behavior.

Stoica et al. [15] presented an interesting study on surfaces of some polyimides, after being rubbed against a cotton velvet and a velvet of cellulose diacetate, using Abbott-Firestone curves and the 3D functional parameters calculated from Abbott–Firestone curves for pointing out the improvements in surface adhesion after rubbing.

Vehmeyer et al. [3] used Abbott-Firestone curves to compare processed surfaces to virtually generated ones and to get optimal feed speed and width of cut of metallic elements.

In a consisting review of the commonly used in vitro methods for measuring surface loss and surface change of teeth, after analysing 69 papers, Field et al. [5] reported that no single technique provides a comprehensive assessment of the remaining tooth surface and each technique has peculiar limitations. Despite a variety of in vitro tests for measuring surface changes, the roughness average (*Ra*) is still the main reported parameter in dentistry studies. The authors of this review proposed to report bearing area parameters as they will allow for a more useful description of the tooth surface quality.

This paper presents the change of functional parameters defined on the Abbot-Firestone curve in order to evaluate the surface quality of the balls from the four ball tester, after tests done with several vegetable oils.

**Testing Methodology**

The tests were done using two grades of rapeseed oil (degummed and refined) and two grades of soybean oil (coarse and degummed) and a common transmission oil (T90). Degumming and refining processes are done at the firm "Prutul" in Galati, Romania. Tests were done under a load of 200 N and a sliding speed of 0.576 m/s (1500 rpm), for 60 minutes.

The balls, as delivered by SKF, are lime polished, made of chrome alloyed steel balls (steel grade EN31), having 12.7 ± 0.0005 mm in diameter, with 64-66 HRC hardness.

The sample oil volume required for each test was 8 ml ± 1 ml.

The test method used to investigate wear preventive characteristic was that presented in EN ISO 20623:2003 Petroleum and related products. Determination of the extreme-pressure and anti-wear properties of fluids - four ball method.

The surface investigations were done with the help of the profilometer Laser NANOFOCUS μSCAN from "Stefan cel Mare" University of Suceava, with an access zone of 150 mm × 200 mm, vertical range of 1.00 μm to 18 mm and a vertical resolution of 25 nm. The measuring step was 5 μm and the distance between lines was also set at 5 μm. The parameters were calculated for raw profiles.

**Results and comments**

Figure 1 presents the significance of functional parameters, discussed for evaluating the surface quality of the wear scars. Functional parameters are calculated according to SPIP roughness parameters [16].

The horizontal lines in figure 1b are drawn for bearing area percentage of 5% and 80%, respectively.

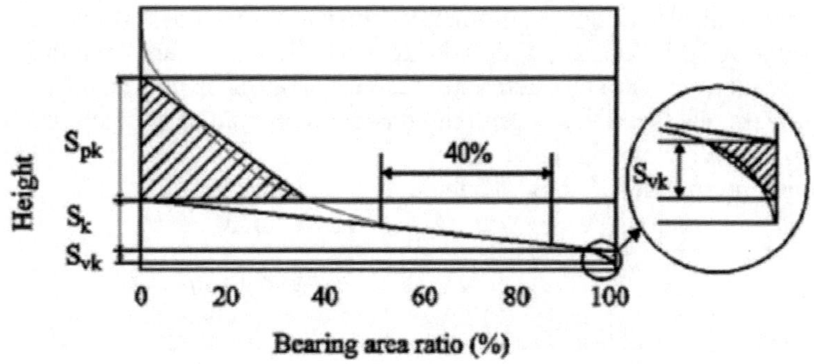


(a) calculation of the reduced summit height (*Spk*), the core roughness depth (*Sk*) and the reduced valley depth (*Svk*)

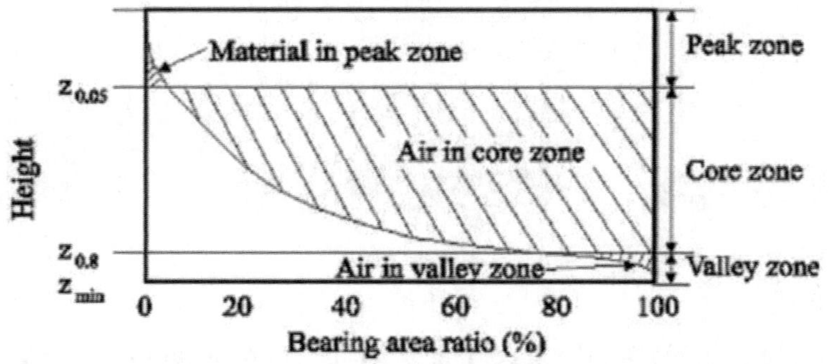


(b) calculation of surface bearing index (*Sbi*), core retention index (*Sci*) and valley fluid retention index (*Svi*)

**Figure 1.** Functional parameters on Abbott-Firestone curve [16].

Figures 2 to 6 present typical Abbott-Firestone curves obtained for the entire area of the wear scars for a tested oil, an original method proposed by the authors, allowing for studying the entire surface of the wear scar. Other studies present only analysis on samples of surface texture, this asking for a statistical approach of the texture quality, including the probability of avoiding high isolated peaks.

Using the bearing area curve, the following parameters may be analyzed. The surface bearing index, *Sbi*, is defined as

$$Sbi = \frac{Sq}{Z_{0.05}} \tag{1}$$

where $Z_{0.05}$ is the height from the top of the investigated surface (the top of Abbott-Firestone curve) to the line marking 5% of the bearing area. For a Gaussian height distribution, *Sbi* approaches 0.60. Large values indicate good bearing properties. Analyzing the values of this parameter from Table 1, one may notice that the values are in the range 0.167...0.310, the greater value being obtained for the degummed coarse soybean oil.

The core retention index, *Sci*, is calculated as:

$$Sci = \frac{V_v(h_{0.05}) - V_v(h_{0.80})}{(M-1)(N-1)\delta x \cdot \delta y} \cdot \frac{1}{Sq} \tag{2}$$

where $V_v(h_{0.05})$ is the area on Abbott-Firestone curve representing the void volume over the bearing area from 0 to $h_{0.05}$ and similarly, $V_v(h_{0.80})$ is the area representing the void volume till $h_{0.80}$. For a Gaussian height distribution *Sci* approaches 1.56. Large values of *Sci* indicate that the void volume in the core zone is large and this is not recommended because sharp asperities will be broken under higher loads. The values of this parameter (1.567...1.75 for the wear scars - see table 1) are around the value for Gaussian surfaces that are characteristic for turned surfaces, but also for surfaces bearing an abrasive wearing process, as it happens when the contact has a boundary or mixt lubrication (abrasive traces are evident on the wear scars in figure 6).

The valley fluid retention index, *Svi*, is defined as:

$$Svi = \frac{V_v(h_{0.80})}{(M-1)(N-1)\delta x \cdot \delta y} \cdot \frac{1}{Sq} \tag{3}$$

For a Gaussian height distribution *Svi* approaches 0.11. Large values of *Svi* indicate large void volumes in the valley zone. In relations (2) and (3), *M* and *N* are the number of investigated lines and the number of points on a line, respectively. As *Svi* for the initial surface is small (0.089), higher values mean that abrasive wear takes places over the entire height of the surface texture.

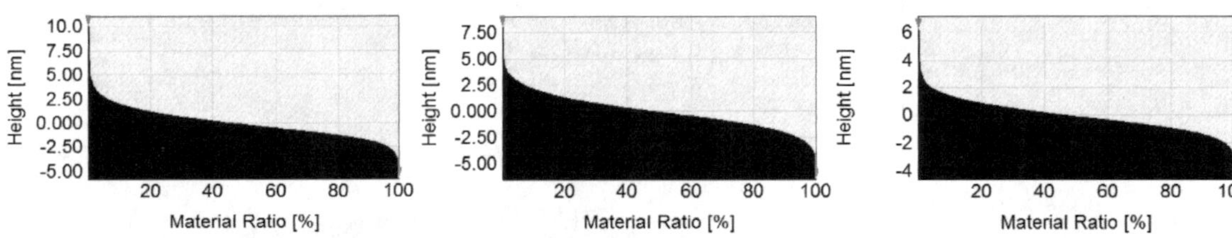


(a) ball 1 (b) ball 2 (c) ball 3

**Figure 2.** Typical Abbott-Firestone curves obtained for the entire area of the wear scars when using coarse degummed rapeseed oil as lubricant.

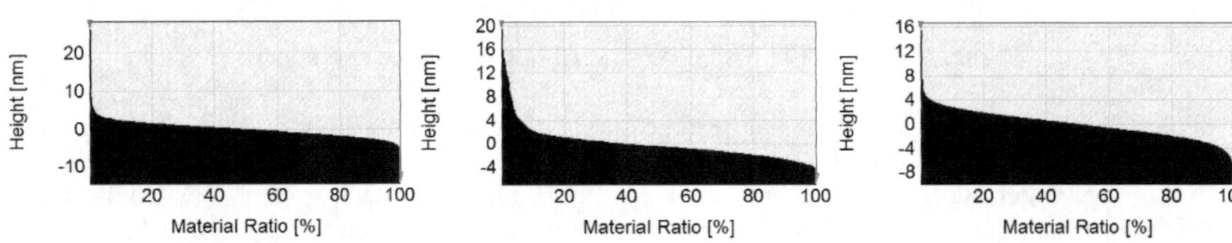


(a) ball 1 (b) ball 2 (c) ball 3

**Figure 3.** Typical Abbott-Firestone curves obtained for the entire area of the wear scars when using refined rapeseed oil as lubricant.

For the refined rapeseed oil (figure 3), the Abbott-Firestone curves reveal a dramatic change in the range of 0...10% of the material ratio, especially for the *Spk* zone, meaning that there are high and rare peaks that increase the risk of direct contact and to scratch the mating surface. In addition, the shapes of the curve are quite different for each ball, meaning that the wear process is not uniform, probably due to the lower lubricating capacity of this oil.

(a) ball 1 (b) ball 2

**Figure 4.** Typical Abbott-Firestone curves obtained for the entire area of the wear scars when using coarse soybean oil as lubricant.

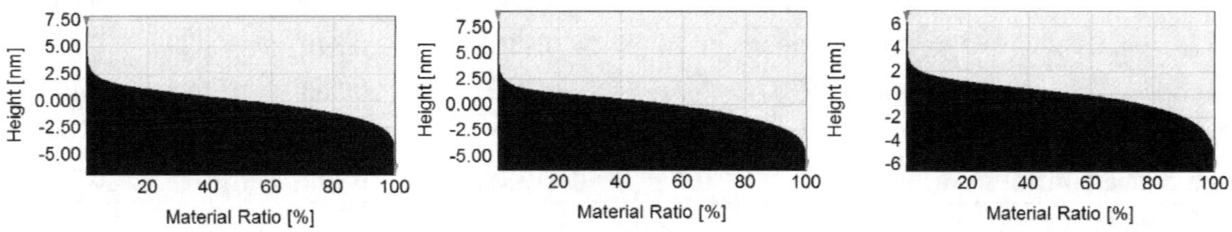


(a) ball 1 (b) ball 2 (c) ball 3

**Figure 5.** Typical Abbott-Firestone curves obtained for the entire area of the wear scars when using degummed soybean oil as lubricant.

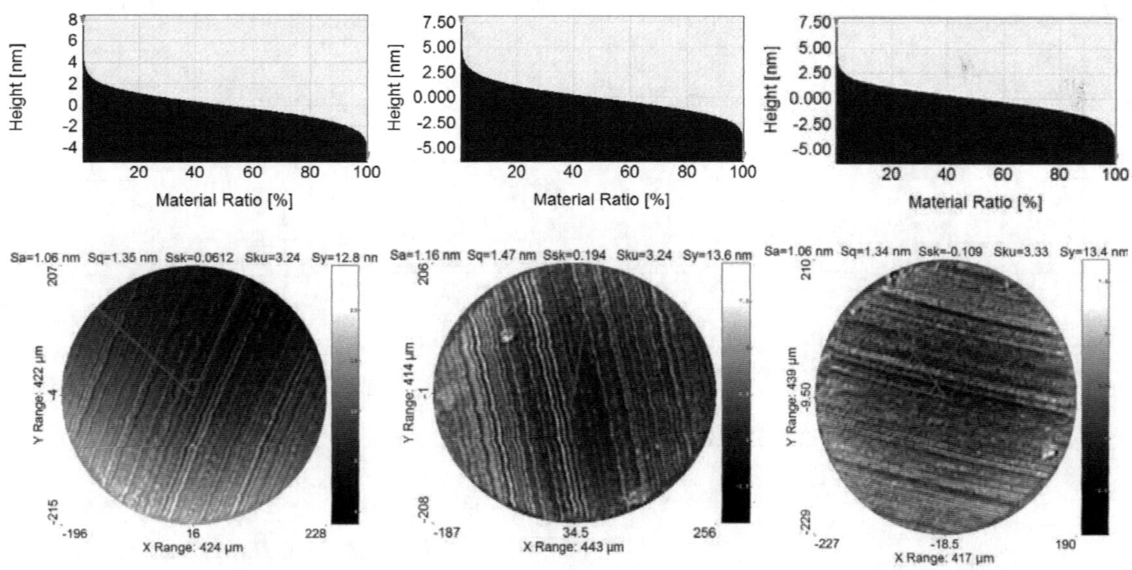


(a) ball 1 (b) ball 2 (c) ball 3

**Figure 6.** Typical Abbott-Firestone curves (up) and virtual images of the worn surfaces (down) obtained for the entire area of the wear scars when using non-additivated transmission oil T90 as lubricant.

The value of the ratio $(Spk+Sk+Svk)/St$ gives a measure of the sharpness of the top of the Abbott-Firestone curve ($St$ is the height difference between the highest and the lowest point of the investigated zone). For higher values of this ratio, the surface has very high rare peaks that give the risk of the direct contact between surfaces. The smallest value for this ratio was obtained for the coarse soybean oil, thus, this vegetable lubricant has a better tribological behavior at least from the point of view of the quality of the worn scars.

**Table 1.** Values as average calculated for the wear scars on the three fixed balls from a test.

| Oil | Code | *Sa* (μm) | $\frac{Spk+Sk+Svk}{St}$ | Hybrid parameters | | |
|---|---|---|---|---|---|---|
| | | | | *Sbi* | *Sci* | *Svi* |
| Non-additivated transmission oil | T90 | 1.09 | 0.4841 | 0.281 | 1.567 | 0.115 |
| Coarse soybean oil | SB | 0.79 | 0.349 | 0.167 | **1.75** | 0.089 |
| Degummed coarse soybean oil | SBD | 1.10 | 0.4851 | **0.310** | 1.429 | **0.135** |
| Degummed coarse rapeseed oil | RBD | 1.21 | 0.483 | 0.263 | 1.673 | 0.104 |
| Refined rapeseed oil | RR | 1.06 | 0.468 | 0.198 | 1.601 | 0.09 |
| Unworn ball (initial surface) | - | 0.43 | 0.398 | 0.171 | 1.915 | 0.089 |

The increase of *Sbi* as compared to the value of the initial surface may suggest that only the top of the very high peaks are cut, thus improving the bearing properties of the superficial layer.

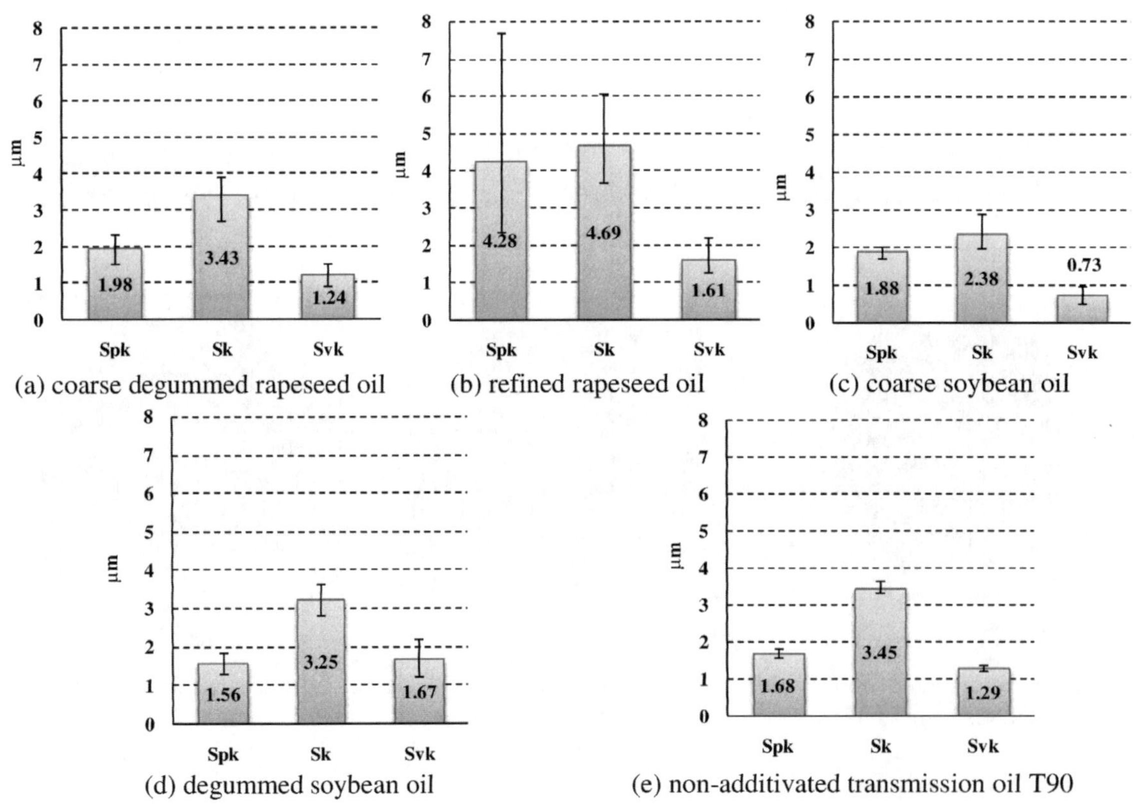


(a) coarse degummed rapeseed oil (b) refined rapeseed oil (c) coarse soybean oil

(d) degummed soybean oil (e) non-additivated transmission oil T90

**Figure 7.** Average values and spread ranges of the functional parameters.

Figure 7 presents not only the average values of the functional parameters *Spk*, *Sk* and *Svk*, but also their spread ranges.

For all tested oils, *Svk* is higher than the value of the unworn surface of the ball (0.558 μm), increasing the capability of having more lubricant on the surface.

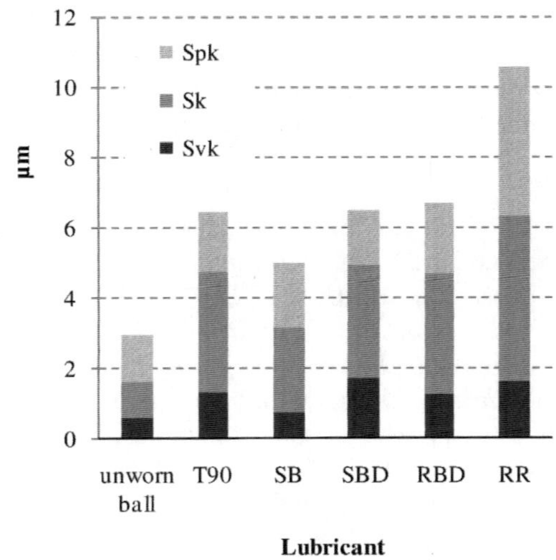


**Figure 8.** The sum of the functional parameters *Spk* + *Sk* + *Svk*.

Figure 8 makes the authors to pay attention that the height of the surface texture of the wear scars increased more than double the initial values of the balls for all the tested oils, the lowest value being obtained for the soybean oil and the highest for the refined rapeseed oil. For the other three tested oils (T90, degummed soybean oil and coarse degummed rapeseed oil), the sum was in a very narrow range. A small value of this sum correlated with small values for *Spk* means a better lubricating capability, but high values of this sum correlated with high value for *Skp* characterize a poorer lubricating capability. The values represented in figure 8 are the average values obtained for the three wear scars on the fixed balls of the tribotester.

## Conclusions

The authors evaluated the lubricating capability of several vegetable oils in comparison to a non-additivated transmission oil T90 by the quality of surfaces of the worn scars generated on the four ball tribotester. As far as the authors search the literature, no study of the surface topography was reported for the entire surface of the wear scars produced on the balls.

Analyzing the Abbott-Firestone curves, the authors formulated the following conclusions.

Even if the *Spk* parameter does not reflect the shape of the Abbott-Firestone curves for the first 5% of the bearing ratio, for the tested oils, its magnitude, but also its percentage from the sum *Spk+Sk+Svk*, help the engineer to rank the oils. Based on this criterion (the lowest sum and lowest value for *Spk*), the best oil among the tested ones would be the coarse soybean oil. The degummed rapeseed oil, the degummed soybean oil and the non-additivated T90 generate a texture quality in narrow ranges for the functional parameters, the worst behavior being attributed to the refined rapeseed oil.

For the refined rapeseed oil, the changes in shape of the bearing curves are more dramatic, these being characterized by high values of *Spk*, its values being 40% of the sum *Svk* + *Sk* + *Spk*, value also obtained for the soybean oil, but the value *Spk* being lower.

IOP Conf. Series: Materials Science and Engineering **174** (2017) 012057 doi:10.1088/1757-899X/174/1/012057

For the degummed soybean oil, the profile heights of the wear scars are taller than those obtained after testing the coarse soybean oil, meaning that the degumming process has a negative influence on the worn surface quality.

Comparing the surface quality of the tested balls is a new reliable method to point out the lubricant properties of the vegetable oils, especially if they are compared to a "classical" lubricant as the non-additivated transmission mineral oil. The best surface after testing was obtained for the course soybean oil, followed by T90 and the degummed grades of the soybean oil and rapeseed oil (these three giving very close values for the functional parameters), but the refined rapeseed oil generated the poorest quality of the balls, under the same testing conditions.

**References**

[1] Lipa Z and Tomanícková D 2011 Utilisation of abbott-firestone curves characteristics for the determination of turned surface properties *Annals of Faculty of Engineering Hunedoara - International Journal of Engineering* **IX (3)** 223-226

[2] Corral I B, Calvet J V and Salcedo C M 2010 Use of roughness probability parameters to quantify the material removed in plateau-honing *Int J Mach Tool Manu* **50** 621–629

[3] Vehmeyer J, Piotrowska-Kurczewski I, Bohmermann F, Riemer O and Maaß P 2015 Least-squares based parameter identification for a function-related surface optimisation in micro ball-end milling *Procedia CIRP* **31** 276-281

[4] Affatato S, Ruggiero A, De Mattia J S and Taddei P 2016 Does metal transfer affect the tribological behaviour of femoral heads? Roughness and phase transformation analyses on retrieved zirconia and Biolox ®Delta composites *Compos Part B-Eng* **92** 290-298

[5] Field J, Waterhouse P and German M 2010 Quantifying and qualifying surface changes on dental hard tissues in vitro *J Dent* **38** 182–190

[6] Moshkovich A, Perfilyev V, Lapsker I and Rapoport L 2014 Friction, wear and plastic deformation of Cu and a/ß brass under lubrication conditions *Wear* **320** 34–40

[7] Las Casas E B, Bastos F S, Godoy G C D and Buono V T L 2008 Enamel wear and surface roughness characterization using 3D profilometry *Tribol Int* **41** 1232–36

[8] Stout K J and Blunt L 1994 *Three-Dimensional Surface Topography* (London: Penton Press)

[9] Dong W P, Sullivan P J and Stout K J 1993 Comprehensive study of parameters for characterising three-dimensional surface topography II: Statistical properties of parameters variation *Wear* **167** 9-21

[10] Tomescu L, Ripa M and Georgescu C 2001 Analyzing abbott curve for composites with polymeric matrix and fibbers *Tribology in Industry* **23** 65-71

[11] Torrance A A 1997 A simple datum for measurement of the Abbott of a profile and its first derivative *Tribol Int* **30** 239-244

[12] Sosa M, Björklund S, Sellgren U and Olofsson U 2015 In situ surface characterization of running-in of involute gears *Wear* **340-341** 41–46

[13] Sosa M, Sellgren U, Björklund S and Olofsson U 2016 In situ running-in analysis of ground gears *Wear* **352-353** 122–129

[14] Franco L A and Sinatora A 2015 3D surface parameters (ISO 25178-2): Actual meaning of Spk and its relationship to Vmp *Precis Eng* **40** 106–111

[15] Stoica I, Barzic A I and Hulubei C 2013 The impact of rubbing fabric type on surface roughness and tribological properties of some semi-alicyclic polyimides evaluated from atomic force measurements *Appl Surf Sci* **268** 442– 449

[16] The Scanning Probe Image Processor SPIP$^{TM}$ Roughness parameters 2016 available at http://www.imagemet.com/WebHelp6/Default.htm#RoughnessParameters/Roughness_Parameters.htm#kanchor778 (6.02.2016)

# Improvement the wear behavior of low carbon steels by friction stir processing

**D M Sekban[1], S M Aktarer[2], H Yanar[3], A Alsaran[4] and G Purcek[3]**

[1]Department of Naval Architecture and Marine Engineering, Karadeniz Technical University, Trabzon, Turkey
[2]Department of Automotive Technology, RecepTayyip Erdogan University, Rize, Turkey
[3]Department of Mechanical Engineering, Karadeniz Technical University, Trabzon, Turkey
[4]Department of Mechanical Engineering, Anadolu University, Eskisehir, Turkey

Email: murat_sekban@hotmail.com

**Abstract.** A low carbon structural steel was surface-hardened by friction stir processing (FSP) through 4 mm thickness from the surface. The hardness of the alloy increased from 140 Hv0.1 to about 240 Hv0.1 after single-pass FSP. This improvement came from the substantial microstructural refinement due to both severe plastic deformation and dynamic recrystallization. Both yield and tensile strength of the alloy increased without a considerable decrease in ductility after FSP. Friction and wear behavior of the alloy before and after FSP was investigated by a pin-on-disk type tribometer according to ASTM-G133. The substantial increase in both hardness and yield strength resulted in a considerable improvement in wear resistance of the alloy depending on applied pressure. In this study, metallurgical and mechanical reasons for such improvement in wear behavior and any change in wear mechanisms after FSP were investigated.

## 1. Introduction

Low carbon steels have been used in a wide range of applications from automotive sector to ship building industry due to their low cost, high workability and good weldability [1]. However, they have relatively low hardness and strength compared to high carbon steels which restricts their broader industrial applications. On the other hand, it is well known that harder materials provide lower wear rate and friction coefficient [2]. There are various ways to enhance hardness and strength of steels [3-4]. Heat treatment may be an effective way to enhance their hardness. Also, various thermomechanical [5-7] and plastic deformation treatments [6-8] are used to enhance the bulk and surface properties of low-carbon steels. Recently, much attention has been paid on surface modification techniques like friction stir processing (FSP). It is a new solid-state surface processing technique that can be used for surface hardening through microstructural modification, which was developed based on the principle of friction stir welding [9-11]. The basic concept of FSP is that a rotating tool including a pin and a shoulder is penetrated into a material surface and traversed along the desired path to cover the region of interest. During FSP, workpiece undergoes intense plastic deformation and high temperature during process, and dynamic recrystallized grains take place occur through that zone. Severe plastic deformation and thermal exposure of the material result in significant changes in the surface microstructure. As a result, three main zones form: the nugget zone (NZ), the thermos-mechanically affected zone (TMAZ) and the heat-affected zone. By the combined effects of severe plastic deformation and dynamic recrystallization bring about a microstructure with refined and equiaxed grains in the size range from 1.0 to 10 μm. The FSP has been used for many studies and thus more detailed information related to its processing principles and advantages can be obtained in earlier literatures [9, 12-16].

There are limited studies on the FSP of steels [17-21]. They mainly focused on the changes in both microstructural and mechanical properties. Also, no studies have been undertaken on tribological properties of FSPed steels.

It is well known that the wear resistance of steels depends mainly on their bulk and surface properties. It is accepted that the wear resistance of a material can be increased by increasing its surface hardness. FSP seems to be a simple way to enhance the mechanical properties of surface material. Thus, the main purpose of the current study is to apply FSP to a low carbon steel for modifying the microstructure and evaluate the changes in wear and friction behavior by the effect of FSP.

## 2. Experimental Procedure

In this study, a hot rolled low carbon steel (ABS-P2-96 Gr A) was used (table 1).

**Table 1**.Chemical composition of the low carbon steel (ABS-P2-96 Gr A) used in this study.

| Elements | C | Mn | P | S | Si | Cu |
|---|---|---|---|---|---|---|
| Chemical Composition (%) | 0.2 | 0.6 | 0.03 | 0.03 | 0.45 | 0.15 |

Samples with the dimensions of 200 mm x 40 mm x 6 mm were cut from the steel plate for FSP. FSP was performed with a processing tool having a convex shoulder with the diameter of 18 mm and a cylindrical pin with the diameter and length of 8 mm and 3 mm, respectively. FSP was conducted with a tool rotation of 635 rpm and a traverse speed of 45 mm/min. The shoulder tilt angle was set at 30, and the tool plunger downforce was kept constant at 11 kN during process.

Optical microscope (OM) was used to observe the microstructure of the samples before and after FSP. The specimens for OM were cross-sectioned on the processed sample perpendicular to the processing direction (figure 1), polished with standard techniques and then etched in %3 Nital (3ml. HNO3 + 97 ml. C2H6O) for 15 s.

Mechanical properties of the samples before and after FSP were determined using tensile test and hardness measurements. Tensile properties of the samples were determined using dog-bone shaped specimens with the dimensions of 2 mm x 3 mm x 26 mm using an electro-discharge machining (EDM). The tensile specimens were cross-sectioned parallel to the process direction (figure 1). The tests were performed using an Instron-3382 electro-mechanical load frame with a video type extensometer at a strain rate of $5.4x10^{-4}$ $s^{-1}$ to be able to stay into the quasi-static deformation range. Hardness measurements were performed using a Vickers micro-hardness tester with a load of 300 g and a dwell time of 10 s. Hardness values were scanned along the cross-section of the sample as shown figure 1.

The dry sliding tests were carried out in a ball-on-disk type wear testing device (UTS Tribometer T10-20) at room temperature (20°C) and a relative humidity of about 75%. The samples (15 mm x 15 mm x 5 mm) were cut from processed an unprocessed material using an EDM. The wear experiments were conducted using an Al2O3 ball (6 mm in diameter). The wear test was carried out at sliding speeds of 0.125 m/s under normal loads of 5, 10 and 15N for a sliding distance of 250 m corresponding to a period of 30min. Prior to wear testing, all the specimen surfaces were ground with 1200 grit emery paper, cleaned with alcohol and then dried. The wear resistance was determined primarily by measuring the weight loss using a microbalance with an accuracy of ±0.1 mg. For the measurements, the samples were cleaned before and after each test with acetone in an ultrasonic bath for 5 min and subsequently dried with hot air. Also, 2D and 3D scanning of the worn tracts were also undertaken by a Nanofocus µscan in order to clarify the morphological features and determine the volume of the worn tracts (grooves).

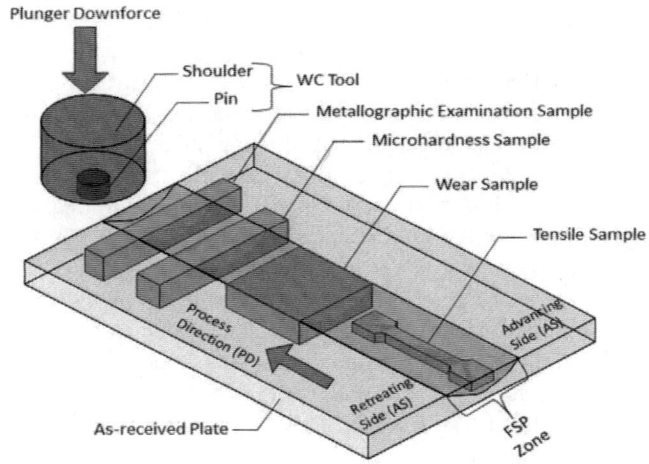


**Figure 1.** Schematic illustration of the FSP of the sample and the position and shape of specimens inside the processed regions.

## 3. Results and Discussion

### *3.1. Microstructure*

Optical micrographs showing the microstructures of base and FSPed samples of the steel are given in figure 2. The initial microstructure of base steel plate is consisted of coarse ferrite+perlite grains with an average grain size of 25 µm (figure 2(a)). FSP resulted in a considerable refinement in the microstructure especially inside the nugget zone (NZ) (figure 2(b-c)). After FSP, the average grain size decreased from 25 µm down to about 3.0 µm in the NZ. The coarse ferrite and perlite grains were fragmented and refined by the effect of both severe plastic deformation and dynamic recrystallization during FSP [22]. TEM micrograph given in figure 2(c) showed that dislocation density increased during FSP into the stir zone. Also, most of the dislocations are accumulated at grain boundary regions, and the amount of dislocation decreases through the grain interiors.

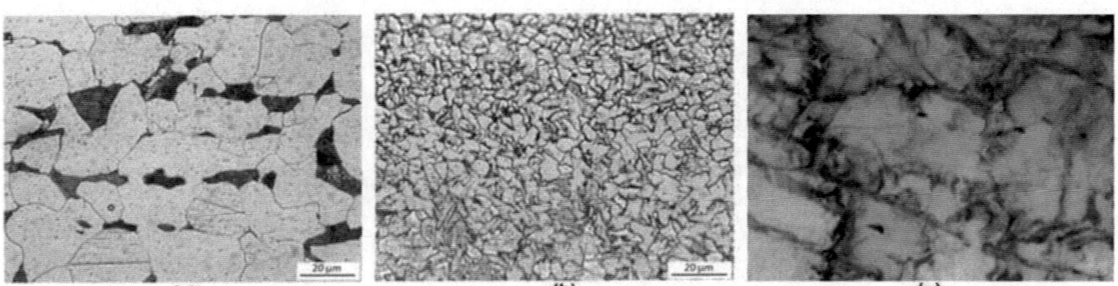


**Figure 2.** Micrographs showing the microstructures of: (a) based material and (b-c) FSPed region.

*3.2. Mechanical Properties*

The hardness profile taken from the vertical cross-section of the FSPed samples is shown in figure 3(a)-(b)). As shown, the hardness of steel increased from 140 Hv0.3 to about 200 Hv0.3 in the NZ after FSP due to the substantial grain refinement and increase in dislocation density [23]. However, the top most surface layer just beneath the shoulder has relatively low hardness. This may be due to the early recrystallization and grain coarsening effect due to the high pressing effect of shoulder [24]. Ultrafine grained microstructure formed around the pin increased the hardness to about 245 Hv in that zone (figure 3). This probably originates from the effectiveness of the pin end which not only rotate but also behaves like a forging affect and causes more refined microstructure in that zone compared to other region of NZ. After peak value around the pin end, the hardness starts decreasing through the rest of FSPed plate. However, the hardness value is still higher than that of the base material in the TMAZ (figure 3(a)-(b)). Because, only temperature-induced structural alteration without intense deformation is valid in that zone. After then, the hardness continuous to decrease and reaches a stable level near the other surface of the plate with the hardness values higher than that of the base material. As considering the sheet thickness, this result is normal. As seen from the cross-sectional micrographs (Fig. 3(a)) of the plate-type sample, all cross section of that sample was thermo-mechanically affected. So the effect of plastic deformation and also temperature generated during deformation caused grain refinement even up to other side of the steel plate. Thus, this microstructural alteration brings about higher hardness even opposite surface of the sample compared to the un-processed sample.

The strength and ductility values taken from the stress-strain curves of steel are given in table 1. As clearly seen, the steel before FSP exhibited low yield strength, good ductility due to the CG microstructure. FSP of the steel increased considerably its strength values without considerable decrease in ductility into the NZ. Both yield and tensile strength values increased from 256 MPa and 435 MPa to about 334 MPa and 525 MPa, respectively (table 1). FSP decreased slightly the elongation to failure from 44% to about 32%. The increase in strength values after FSP can be attributed to the substantial grain refinement (Hall-Petch effect) and increase in dislocation density (strain hardening effect) (figure 3(c)) [25]. It is well known that an increase in dislocation density occurs during FSP, and this formation make a further effect on strengthening of the FSPed microstructure [23]. Such microstructural changes also brought about a decrease in the ductility of the materials due to decreasing in their strain hardening effect [26]. However, it should be noted that ductility of the steel sheet was not radically decreased although the strength values increased considerably.

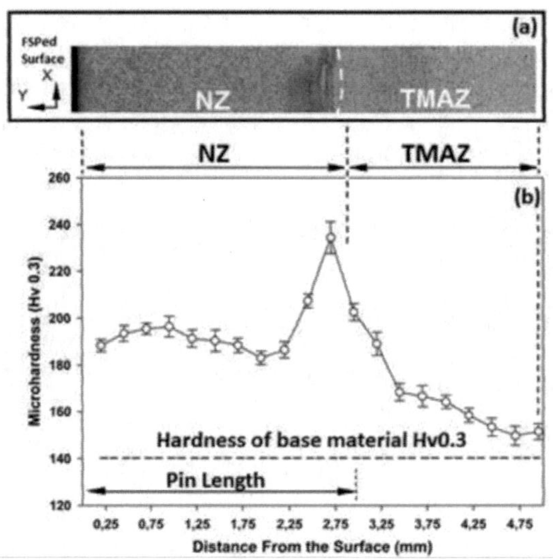


**Figure 3.** (a) Optical micrograph of vertical cross-section of FSPed sample and (b) hardness profile through the thickness of the plate.

**Table 1.** Mechanical properties the low carbon steel before and FSP.

| Condition | Yield Strength (MPa) | Tensile Strength (MPa) | Uniform Elongation (%) | Failure Elongation (%) |
|---|---|---|---|---|
| Base | 256±07 | 435±06 | 18.4±0.3 | 44.2±2.4 |
| FSPed | 334±06 | 525±13 | 13.9±0.6 | 32.4±1.7 |

### *3.3. Friction and Wear Behavior*

Figure 4 shows the changes in friction coefficient of low carbon steel as a function of sliding distance before and after FSP. The curves indicate a sharp increase in friction coefficient during the initial sliding period with steady-state levels reached after about 50 m for both conditions. As carefully investigate, the FSPed sample shows slightly lower friction coefficient values inside the state-state period. Also, the scatter or variation in the friction coefficient values is relatively higher in the sample without FSP. After FSP, the sample exhibits more steady variations. The average coefficient of friction values in the state-steady-condition are 0.46 for the un-processed sample and 0.42 for FSPed sample.

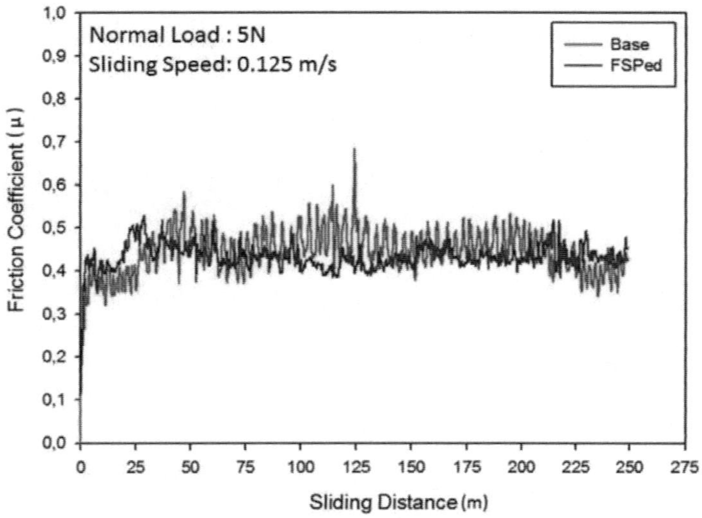


**Figure 4.** The coefficient of friction vs. sliding distance curves for the low carbon steel before and after FSP at room temperature and under 5N load.

Figure 5 shows a typical relationship between the weight loss and applied normal load for the base and FSPed samples of low carbon steel. Generally, it can be seen that amount of wear in both conditions increased with increasing normal load. The FSP increased the wear resistance and decrease the weight loss of steel under all loads. The difference in weight losses between base and FSPed samples is more pronounced under low loads, and it decreased with increasing normal load. The 2D and 3D profilometric views of wear tracts of base and FSPed samples are also shown in figures. 6(a)-(d). These profilometric results also confirm the results given in figure 6. Both diagrams demonstrate the same variation under all loads. As shown the wear depth increases with increasing applied load for all conditions. Also the depth of wear in the base material is higher than that of FSPed sample. This result is expected as considering the hardness and also yield strength of both conditions. As explained above, the hardness and strength increased substantially after FSP especially inside the NZ. This increase brings about an increase in resistance against to plastic deformation during rubbing, and thus the wear resistance of material increases [27]. Also, this improvement in wear resistance is higher under low normal loads. As the loads increases, the heat generated by the rubbing action increases which may cause recrystallization of processed zone. This microstructural alteration decreases the hardness of the rubbing surfaces with deterioration of the wear behaviour

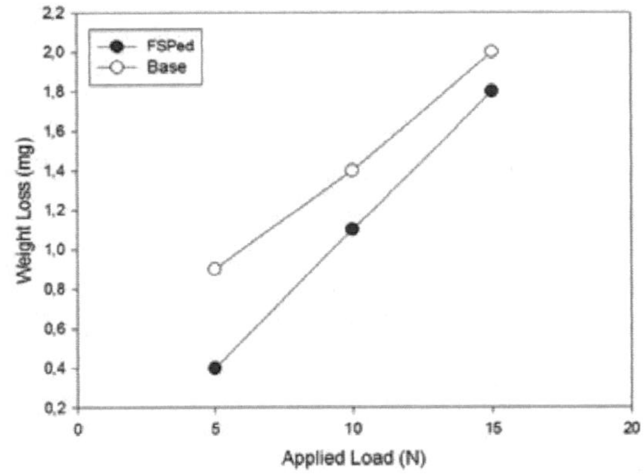


**Figure 5.** Weight loss of low carbon steel before and after FSP as a function of applied loads.

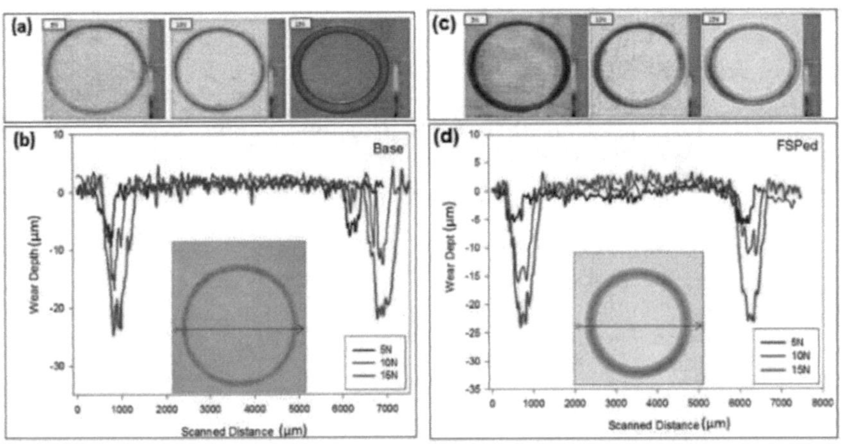


**Figure 6.** 3D profilometric view of wear tracts of steel under different loads and 2D view of worn tracts took place under different loads: (a)-(b) Base sample before FSP and (c)-(d) FSPed sample.

## 4. Conclusions

In the present study, the effect of friction stir processing (FSP) on the microstructure, mechanical properties and friction and wear behavior of a low carbon steel was investigated. The main results and conclusions of this study can be summarized as follows:

- A fine grained microstructure having mean grain size of 5 μm were obtained from the CG (grain size: 25 μm) steel with FSP. Such grain refinement leads to a significantly enhancement in strength and hardness without notably sacrificing the elongation to failure.
- FSPed sample has slightly lower coefficient of friction than un-processed sample. The average coefficient of friction values in the steady-steady-condition are about 0.46 for un-processed sample and 0.42 for FSPed sample.
- FSP increases the wear resistance of steel. This is associated with a considerable increase in its strength and hardness values after one-pass FSP.

**References**

[1] Lampman H F, Crankovic, G M, Lampman S R, Zorc T B 1990 *ASM Handbook, vol. 1: Properties and Selection: Irons, Steels and High-Performance Alloy* (Ohio: ASM International)
[2] Chichinadze A V, Berliner E M, Brown E D, Bushe N A and Buyanovskii I A 2003 *Friction, Wear and Lubrication - Tribology and Triboengineering* (Moscow: Mashinostroenie)
[3] Shabashov V A, Korshunov L G, Mukoseev A G, Sagaredze V V, Makarov A V, Pilyugin V P, Novikov S I and Vildanova N P 2003 *Mat. Sci. Eng. A-Struct.* **346** 196–197
[4] Garnham J E and Beynon J H 1992 *Wear* **157 (1)** 81–109
[5] Fox-Rabinovich G S, Weatherly G C, Kovalev A J, Shuster L S 2003 *Proc. of the Intern. Conf. on Metallurgical Coatings and Thin Films* (San Diego) p 52
[6] Luzhnov Yu M 2002 *Proc. of the 6th Intern. Symp. "Insycont-02"* (Krakow) pp129–135
[7] Johnson K L 1987 *Contact Mechanics* (Cambridge: Cambridge University Press) p 452
[8] Odintsov L G 1987 *Strengthening and Finishing of Parts via Surface Plastic Deformation: Reference Book* (Moscow: Mashinostroenie)
[9] Mishra R S, Mahoney M W, McFadden S X, Mara N A and Mukherjee A K 1999 *Scripta Mater.* **42** 163-168
[10] Mishra R S and Mahoney M.W 2001 Mater. Sci. Forum **357-3** 507
[11] Thomas W M, Nicholas E D, Needham J C, Murch M G, Templesmith P and Dawes C J 1991 *GB Patent Application No. 9125978.8*
[12] Mishra R S and Ma Z Y 2005 *Mat. Sci. Eng. R* **50** 1-78
[13] Su J Q, Nelson T W and Sterling C J 2005 *Scripta Mater.* **52** 135-140
[14] Chang C I, Du X H and Huang J C 2007 *Scripta Mater.* **57** 209-212
[15] Xue P, Xiao B L and Ma Z Y 2012 *Mat. Sci. Eng. A-Struct.* **532** 106-110
[16] Xue P, Xiao B L and Ma Z Y 2013 *Scripta Mater.* **68** 751-754
[17] Lienert T J, Stellwag W L, Grimmett B B and Warke R W 2003 *Weld. J.* **82** 1-9
[18] Ueji R, Fujii H, Cui L, Nishioka A, Kunishige K and Nogi K 2006 *Mat. Sci. Eng. A-Struct.* **423** 324- 330
[19] Cui L, Fuji H, Tsuji N, Nakata K, Nogi K, Ikeda R and Matsushita M 2007 *ISIJ Int.* **47(2)** 299–306
[20] Choi D H, Lee C Y, Ahn B W, Choi J H, Yeon Y M, Song K, Park H S, Kim Y J, Yoo C D and Jung S B 2009 *Int. J. Refract.Met. H.* **27** 931–936
[21] Nandan R, Roy G G, Lienert T J and Debroy T 2007 *Acta Mater.* **55** 883–895
[22] Xue P, Xiao B L, Wang W G, Zhang Q, Wang D, Wang Q Z and Ma Z Y 2013 *Mat. Sci. Eng. A-Struct.* **575** 30-34
[23] Su J Q, Tracy W N and Colin J S 2005 *Mat. Sci. Eng. A-Struct.* **405** 277-286
[24] Prangnell P B and Heason C P 2005 *Acta Mater.* **53** 3179-3192
[25] Hajian M, Abdollah-Zadeh A, Rezaei-Nejad S S, Assadi H, Hadavi S M M, Chung K and Shokouhimehr M 2015 *Mater. Design* **67** 82-94
[26] Chabok A and Dehghani K 2013 *J. Mater. Eng. Perform.* **22** 1324-1330
[27] Huang S J, Semenov V I, Shuster L S and Lin P C 2011 *Wear* **271** 705-711

# Coating multilayer material with improved tribological properties obtained by magnetron sputtering

**A O Mateescu[1], G Mateescu[1], M Balasoiu[1,2], G O Pompilian[3] And M Lungu[3,4]**

[1]Horia Hulubei National Institute for Research and Development in Physics and Nuclear Engineering, 30 Reactorului Street, 077125 Magurele, Ilfov, Romania
[2]Joint Institute for Nuclear Research, Frank Laboratory of Neutron Physics, 6 Joliot-Curie, 141980 Dubna, Moscow Region, Russia
[3]National Institute of Physics, Plasma and Radiation Physics, 409 Atomistilor Stret, 077125 Magurele, Ilfov, Romania
[4]University of Bucharest, Faculty of Physics, 405 Atomistilor Stret, 0777125 Magurele, Ilfov, Romania

e-mail: mateescu@nipne.ro

**Abstract**. This work is based on the Patent no. RO 128094 B1, granted by the Romanian State Office for Inventions and Trademarks. The goal of the work is to obtain for investigations tribological coatings with multilayer structure with improved tribological properties, deposited by magnetron sputtering process from three materials *(sputtering targets)*. Starting from compound chemical materials *(TiC, $TiB_2$ and WC)*, as sputtering targets, by deposition in argon atmosphere on polished stainless steel, we have obtained, based on the claims of the above patent, thin films of multilayer design with promising results regarding their hardness, elastic modulus, adherence, coefficient of friction and wear resistance. The sputtering process took place in a special sequence in order to ensure better tribological properties to the coating, comparing to those of the individual component materials. The tribological properties, such as the coefficient of friction, are evaluated using the tribometer test.

## 1. Introduction

It is well known that, for the terrestrial mechanisms, friction has a strong contribution to: decreased performance; increased energy consumption; added maintenance; wear damage; shortened lives or catastrophic failure of the mechanical systems. Lubrication is indispensable for achieving reduced friction between two moving surfaces and to fulfill these requests only fluid lubricants were used in the first period. The increasing demand for low friction and wear in aircraft and other tribological applications that, involve severe sliding conditions and the limitations of fluid lubricants at extreme environmental conditions, involve the necessity for using only of the reliable lubricants that were for a long time tested and verified in real working conditions.

For a given tribological application with given working and environment conditions *(temperature, pressure, contact load, sliding or rotating speed, presence of corrosive gases, etc.)* it must be chosen the correct coating type in terms of deposition methods and types of materials used *(that have a great influence on friction couple, for: A. working life time, respectively wear resistance and B. financial cost),* because the response of a coated system *(friction couple)* depends on the following factors [1]:

1. Substrate properties *(roughness, hardness, toughness, cleaning state before deposition)*;
2. Counterpart properties *(roughness, hardness, toughness, coefficient of friction);*
3. Coating properties *(hardness, toughness, porosity, thermal oxidation/ chemical corrosion - resistance)* that mainly depends on the used materials and deposition method;
4. Properties of substrate-coating interface *(adhesion)*.

According to the claims of above Patent the materials for the tribological coatings must be selected by taking into account the fact that tribological coatings must fulfill *(in accordance with the working conditions)* the following maximal properties: 1. Low coefficient of friction; 2. High hardness; 3. High toughness; 4. Good adhesion to the substrate; 5. High thermal conductivity; 6. High corrosion resistance and high thermal oxidation resistance.

WC is a material with: very good thermal *conductivity* ($\lambda$ = *80-100 W/m.K)*; good coefficient of friction *(μ/ COF = 0,4-0,6*), high hardness *(*$HV_{30}$*=2100)*, good elasticity *(E=550GPa*). Coatings with carbon compounds generate tribofilms, reducing the friction coefficient. Variations in the number of layers and thicknesses, as well as the distribution within the coating are determinant to obtain the adequate mechanical properties in order to reduce the wear rate [2].

. $TiB_2$ is well known as a ceramic compound with relatively high melting point, hardness, strength to density ratio, abrasion, oxidation resistance and wear resistance. Although $TiB_2$ coatings have been widely studied by many researchers, their real applications have been very limited. The fact is that the adhesion of the $TiB_2$ coatings is poor for the coating-substrate system. The main reasons for this are that the $TiB_2$ coating deposited is very brittle and that it accommodates very high compressive residual stresses. These limit both the practical adhesion and the thickness of the $TiB_2$ coatings [3]. In order to enhance the $TiB_2$ coating adhesion $TiB_2$-based nanostructured engineering coatings should be developed. A possible way to obtain lower stress levels could be to utilize multilayer coatings that is, to combine the stressed material with a more ductile material in a layered structure [4].

TiC is a material that has a relatively good adhesion to the metallic substrates and allows obtaining by reactive magnetron sputtering in DC *(with* $N_2$ *as reactive gas)* of the TiCN coatings that have been used to improve wear resistance of metallic materials in a wide range of engineering applications, such as cutting and forming tools.

According to the patent claims, using TiC *(material with good adhesion)*, $TiB_2$ *(ceramic material with high hardness, high oxidation and wear resistance)* and WC *(material with low coefficient of friction, high thermal conductivity)* deposited successively, in the above order and simultaneously at different concentrations, we obtained thin films of multilayer design with promising results regarding their hardness, elastic modulus, adherence, coefficient of friction and wear.

## 2. Experimental procedures

According to the patent claims, for achievement of the tribological coatings with multilayer structure, we have used standard magnetron sputtering deposition process in DC. We have used 3 sputtering targets: Titanium carbide *(TiC),* Titanium diboride *(*$TiB_2$*)* and Tungsten carbide *(WC)* with the following properties:

- TiC, used for the $1^{st}$ layer, with acceptable COF and adhesion to the metallic substrate.
- $TiB_2$, used for the $2^{nd}$ layer, with high hardness and also with a high wear resistance.
- WC, used for the $3^{rd}$ layer, with very good thermal conductivity and low COF.

The experiments took place in a vacuum deposition system equipped with four DC magnetrons of 600 W, having 2' targets that can be adjusted for co-focal deposition and with a controller with quartz crystals sensors for measure film thickness and deposition rates. The distance between the targets and substrates was around 15 cm. The polished stainless steel substrates were ultrasonically cleaned before introducing them in the deposition process chamber. Then they were heated up to 550°C for a better adhesion of the coated film to the metallic substrates and they were in a 20 m/s rotation motion in order to obtain coatings with high uniformity for their entire surface. We reached a pressure less than

$10^{-5}$ mbar in the technological chamber in the first step and around $3x10^{-3}$ mbar when we introduced 100 sccm of Ar. A deposition control ensured process controlling and in – situ measurement of deposition rate and coating thickness.

In the above configuration of the deposition process we made coatings using 550°C for the heating temperature, in a multilayer design consisting in: 10 and 15 packages of three constituent layers 10x($TiC/TiB_2/WC$) – Sample 1 and 15x($Ti/TiB_2/WC$) – Sample 2 (Figure 1-a) and a multilayer structure of 3 constituent compound layers of 3 materials with different concentrations – Sample 3 (Figure 1-b).

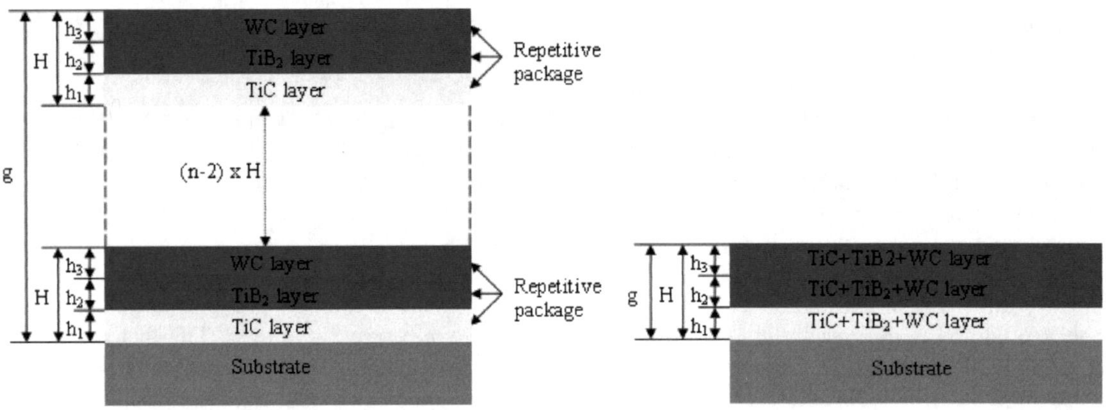


a) b)

**Figure 1.** Structure of the multilayer tribological coatings
a) multilayer structure (n = number of the repetitive packages)
b) multilayer structure made of three constituent compound layers with different concentrations

The power injected in magnetrons was the same for all the targets for Sample 1 and 2 and for Sample 3 it was double in each layer for one material than it was for the other two used (Table 1). So, in this way we made tribological coatings with multi layer structure and improved tribological properties, with quaternary composition (Ti-B-W-C), respectively with pseudo-ternary composition ($TiC-TiB_2-WC$), starting from the 3 sputtering targets: TiC, $TiB_2$ and WC, by standard DC magnetron sputtering process.

The working process parameters for achievement of the 3 samples are presented in Table 1.

**Table 1.** Working process parameters for deposition of the samples

| Sam-ple No. | Materials | | Power injected in the gun's plasma [%x$P_{max}$]* | Deposition time [min] | Magnetron parameters | |
|---|---|---|---|---|---|---|
| | Target | Deposited layer | | | Voltage [V] | Current [mA] |
| | TiC | 10x(TiC/ | 30 | | 338- 339 | 84 -83 |
| 1 | $TiB_2$ | $TiB_2$/ | 30 | 150 | 466 -560 | 175-174 |
| | WC | WC) | 30 | | 557- 559 | 177- 176 |
| | TiC | 15x[Ti/ | 30 | | 338 - 303 | 84-83 |
| 2 | $TiB_2$ | TiB2/ | 30 | 150 | 580 - 454 | 175-174 |
| | WC | WC] | 30 | | 578 - 520 | 177 -175 |
| | TiC+$TiB_2$+WC | TiC+$TiB_2$+WC/ | 60+30+30 | | 286 -280 | 25 - 24 |
| 6 | TiC+$TiB_2$+WC | TiC+$TiB_2$+WC/ | 30+60+30 | 150 | 432 - 440 | 55 - 54 |
| | TiC+TiB2+WC | TiC+$TIB_2$+WC | 30+30+60 | | 462 - 470 | 57 - 56 |

* $P_{max}$ = 600W

## 3. Evaluation methods

The relative concentration of W and Ti of the samples 1 and 2 was determined using the micro-beam X-ray fluorescence (μ-XRF) method, developed at NILPRP [5-7]. The samples were analyzed at low value excitation energies (<50 KeV). This method is optimal for thinner layers (< 10 μm) due to the fluorescence auto saturation effect. As we already know the integrated area peak for the experimentally investigated bulk Ti (152970±10%) and also for the Ti contained in the ceramic compound (6737±10%), we made conversion to layer thickness. Experimental conditions for μ-XRF and source parameters were kept fixed, integration time of 30 seconds, Mo target X-ray source with current of 700 μA and a 45 kV voltage. Using bulk materials, energetic calibration of the spectrum was carried out in order to conduct area integration upon characteristic energetic line especially for W (L-line 9.67 keV) and Ti (K-line 4.5 keV).

The tribological behavior of the prepared coatings was tested using a Ball-on disc CSM, Switzerland tribometer. The following parameters were set: sliding speed: 2cm/s, bearing stainless steel 6 mm diameter, sliding radii: 2.5 mm, 3 mm and 3.5 mm respectively, loading force: 0.5 N, 1 N and 2 N respectively.

Vickers hardness was measured using a „Microhardness tester FM-700" with an indenter with rhomboidal prophile for Sample 1. The loading forces used were: 25, 10 and 50 gf. and the Dwell time was 10 sec.

## 4. Results and Discussions

Based on the calibration curves, we conducted also the conversion from the integrated peak area (Gaussian fit) to layer thickness of W and Ti on the investigated sample and the results are presented in Table 2. We mention that by XRF method, B concentration could be not measured due to the low atomic mass number.

**Table 2**. Relative concentration of W and Ti determined by μ-XRF method

| Investigated sample | Integrated peak area Ti Kα | Gaussian fit W Lβ | Calibrated thickness (nm) ± 10% relative error | Relative integral concentration (%) |
|---|---|---|---|---|
| Sample 1 | 6737 (Ti 33%) | 31095 | ~700 (W), ~780 (Ti) | 47 (W), 53 (Ti) |
| Sample 2 | 6890 (Ti 33%) | 31003 | ~720(W), ~780 (Ti) | 48 (W), 52 (Ti) |

For sample 1 the results for the coefficient of friction are presented in Figure 2.

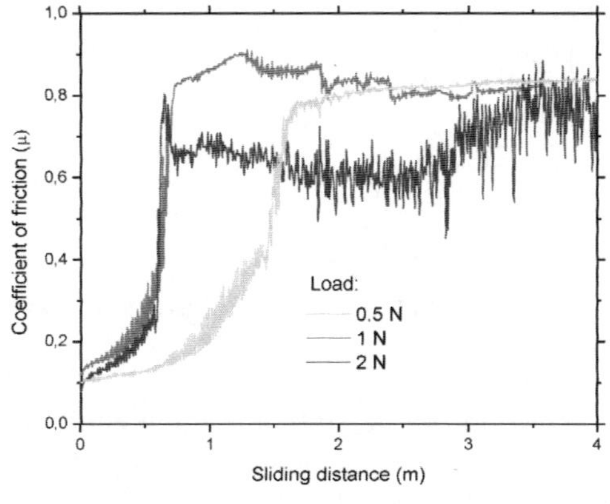


**Figure 2.** Coefficient of friction for Sample 1

IOP Conf. Series: Materials Science and Engineering **174** (2017) 012059 doi:10.1088/1757-899X/174/1/012059

One could observe a reduction of the coefficient of friction of about 8 times ($\mu \sim 0.1$) compared with the coefficient of friction of the substrate ($\mu \sim 0.8$) in the case when the loading force was 0.5 N and a little bit higher ($\mu \sim 0.15$) when the load was 1 N or 2 N. We observe also that the deposited layer resisted without breaking 0.75 m in the case of the 0.5 N load and about 0.2 m in the case of the 1 N or 2 N loads.

After about 1.5 m sliding in dry conditions, the coefficient of friction reached the value of 0.8, the substrate characteristic. The same situation is observed for the case of 1 N loading force while in the case of the loading force of 2 N, between 0.6 m and about 3 m sliding the coefficient of friction was lower than of the substrate ($\mu \sim 0.6$ instead of $\mu \sim 0.8$) due to the rolling friction mechanism (the film debris acted as some ball bearings, and the coefficient of friction was reduced).

In the Hardness Test, a variation of the hardness with the loading force as it is presented in Figure 3 was observed. (the hardness increases with the load decreasing). The $HV_{0.01}$ was arround 300 that means arround 2,94 GPa and this value was higher with 20% than the substrate hardness.

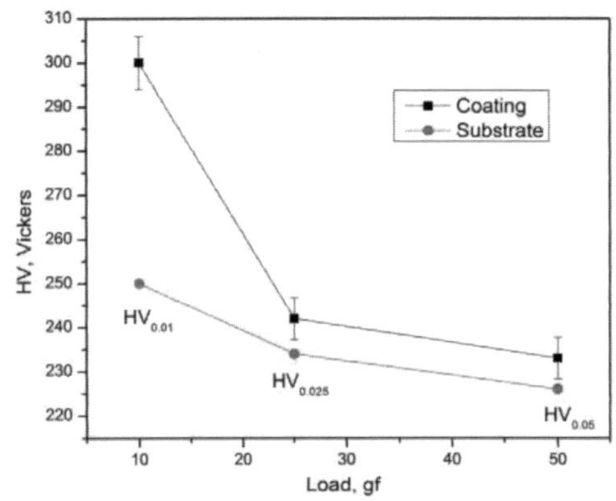


**Figure 3.** Hardness of the Sample 1

The coefficient of friction obtained by Ball-on-Disc-Tribometer Test for the Samples 2 and 3 is presented in Figure 4 and 5 and has the same evolution as for the Sample1.

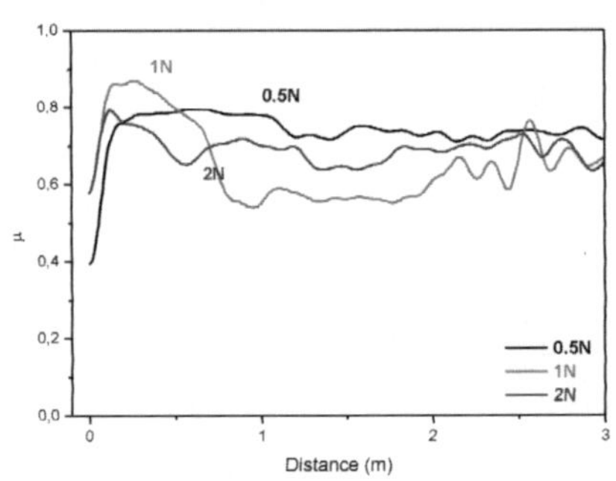


**Figure 4.** Coefficiecient of friction for Sample 2

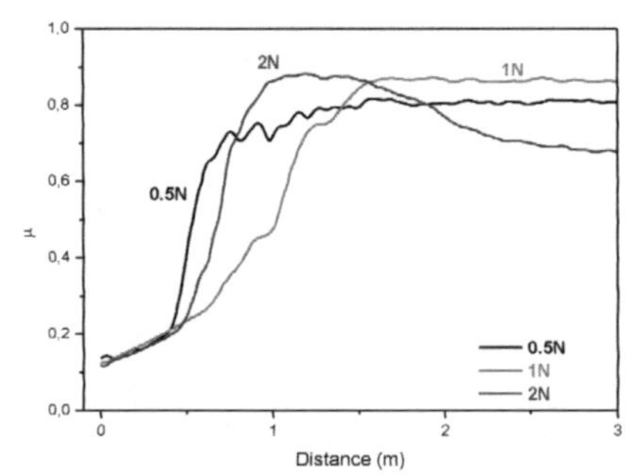


**Figure5.** Coefficient of friction for Sample 3

## 5. Conclusions

For the multilayer design, the deposition drastically reduces the coefficient of friction of the moving components coated with this film, as one used on sample 1. Because of the reduced thickness of the coating, the coefficient of friction for sample 1 is under 0.2 only for a sliding distance of 1 m for Ball-on Disk Tribometer Test, with loading force of 0.5 N and sliding speed of 2 cm/s.

The reduced sliding distance with low coefficient of friction *(when the deposited layer resists without breaking)* represents a consequence of the reduced adhesion of deposited layers to the metallic substrate and of the reduced cohesion forces between the constituent layers of the multilayer.

Multilayer from $TiC/TiB_2/WC$ start from a lower value of COF (under 0.1) than the multilayer made fom $TiN/TiB_2/WC$ (that start at a COF arround 0.3) when more deposited packages led to lower coefficients of friction [8].

Also, the wear resistance and the adherence to the substrate for $TiC/TiB_2/WC$ multilayer structures is very low and this is proven by a short sliding disctance (maximum 0.6 m at a loading force of 0.5N) with low coefficient of fiction (under 0.2).

**Acknowledgement**
The authors acknowledge for the financial support from Nucleu Programme PN 16 42 01 04 and Collaboration Protocol between JINR Dubna and IFIN-HH, No. 4406-4-2015/2017.

**References**

[1] Mateescu A O and Mateescu G, 2014 Multilayer coating material with tribological properties and procedure for its achievement *Patent no.* RO 128094 B1

[2] A Rivera-Tello C D, Broitman E, Flores-Ruir F J Jiménes 2016 Mechanical properties and tribological behavior at micro and macro-scale of WC/WCN/W hierarchical multilayer coatings *Tribol. Int.* **101** 194–203

[3] Panich N, Wangyao P, Vattanaprateep N and Yong S 2006 Techniques to Improve Coating Adhesion of Superhard Coatings *Journal of Metals, Materials and Minerals,* **16** (2) 19-23

[4] Berger M and Larsson M 2000 Mechanical properties of multilayer pvd $Ti/TiB_2$ coatings *Surf. Eng.* **16** (2) 122-126

[5] Tiseanu I, Craciunescu T and Möslang A 2009 Assessment of X-ray tomography for irradiated IFMIF/HFTM RIG *Fusion Eng. Des.* **84** 1847-51

[6] Tiseanu I Craciunescu T, Pegourier B, Maier H, Ruset C, Mayer M, Dobrea C and Sima A 2011 Advanced x-ray imaging of metal-coated/impregnated plasma-facing composite materials *Phys. Scr.* (T145) 014073

[7] Tiseanu I, Mayer M, Craciunescu T, Hakola A, Koivuranta S, Likonen J, Ruset C and Dobrea C 2011 X-ray microbeam transmission/fluorescence method for non-destructive characterization of tungsten coated carbon materials. *Surf. Coat. Technol.* **205** S192-S197

[8] Jinga V, Cristea D. Mateescu A O, Samoila C, Ursutiu D, Mateescu G. and Munteanu D, 2015 Dry lubricant materials deposted by magnetron sputtering and friection coefficients evaluation *Key Eng Mater.* **660** 75-80

IOP Conf. Series: Materials Science and Engineering **174** (2017) 012060 doi:10.1088/1757-899X/174/1/012060

# Wear resistance of WC/Co HVOF-coatings and galvanic Cr coatings modified by diamond nanoparticles

**M Kandeva[1], T Grozdanova[2], D Karastoyanov[3] and E Assenova[4]**

[1]Faculty of Industrial Technology, Technical University of Sofia, Bulgaria
[2]Space Research and Technologies Institute, Bulgarian Academy of Sciences, Bulgaria
[3]Institute for Information and Communication Technologies, Bulgarian Academy of Sciences, Bulgaria
[4]Society of Bulgarian Tribologists, Sofia, Bulgaria

E-mail: kandevam@gmail.com

**Abstract**. The efforts in the recent 20 years are related to search of ecological solutions in the tribotechnologies for the replacement of galvanic Cr coatings in the contact systems operating under extreme conditions: abrasion, erosion, cavitation, corrosion, shock and vibration loads. One of the solutions is in the composite coatings deposited by high velocity gas-flame process (HVOF). The present paper presents comparative study results for mechanical and tribological characteristics of galvanic Cr coatings without nanoparticles, galvanic Cr coatings modified by diamond nanoparticles NDDS of various concentration 0.6; 10; 15 и 20% obtained under three technological regimes, and composite WC-12Co coating. Comparative results about hardness, wear, wear resistance and friction coefficient are obtained for galvanic Cr-NDDS and WC-12Co coatings operating at equal friction conditions of dry friction on abrasive surface. The WC-12Co coating shows 5.4 to 7 times higher wear resistance compared to the galvanic Cr-NDDS coatings.

## 1. Introduction

For more than 70 years galvanic Cr coatings played the role of „gold standard" for high wear resistance of details operating under extreme conditions: abrasion, erosion, cavitation, corrosion, shock and vibration loads.

Modifying galvanic Cr coatings by nanosize particles improves their mechanical and tribological characteristics. This is the reason for their multiple studies in the field of nanotechnologies and nanotribology [1,2,3,4].

The specific production conditions of galvanic Cr coatings containing hexavalent chromium require higher measures of labor safety and environment protection from the toxic wastes of Cr electrolyte [5,6].

Basic disadvantages of galvanic Cr coatings are:

- Irregular distribution of coating thickness on the coated surface;
- High porosity and internal stress, which worsen the resistance against fatigue wear and corrosion;
- High sensitivity of structure, mechanical and tribological characteristics to electrolyte composition, parameters and the kinetics of the electrolysis;
- Impossibility to regenerate highly worn details because of the brittleness of thicker Cr coatings;

- Impossibility for deposition of galvanic Cr coatings on special steels with higher W and Co content, as well as on cast irons with higher hardness.

Promising replacements of galvanic Cr coatings are WC/Co composite coatings deposited by high velocity oxy-flame process (HVOF). HVOF-technology is a modern high technology with higher productivity and better ecological and economic characteristics compared to galvanic chrome plating.

WC/Co coatings deposited by HVOF-technology are new generation powder composite coatings with essential application in aircraft, space and power industry, petrol and chemical industry, etc. Table 1 shows characteristics of WC/Co composite coating and galvanic Cr coating from literature [5].

**Table 1.** Comparative values of some of the parameters of galvanic Cr coating and HVOF WC/Co coating.

| Parameter | Galvanic Cr coating | HVOF WC/Co coating |
|---|---|---|
| Hardness, HRc | 60-70 | >70 |
| Microhardness, DPH 300 | 750-850 | >1050 |
| Adhesion strength, MPa(psi) | 41 (60 00) | > 80 (10 000) |
| Porosity, % | Sub-microcracks | < 1 |
| Thickness, mm | < 0.13 | > 1 |
| Oxides, % | > 1 | <1 |

The present paper aims a comparative study of abrasive wear and friction coefficient of galvanic Cr coatings without nanoparticles, galvanic Cr coatings modified by diamond nanoparticles of different concentration and WC/Co (88/12) coating operating at equal friction conditions [6].

## 2. Experimental procedures

Galvanic Cr-diamond nanoparticles composite coatings are deposited on cylindrical steel specimens of 30 mm diameter and 10 mm height. Table 2 shows the steel composition.

**Table 2.** Composition (wt. %) of the steel specimens, basis for the galvanic Cr-diamond nanoparticles composite coatings.

| Element | C | S | Mn | P | Si | Cr | Ni | Fe |
|---|---|---|---|---|---|---|---|---|
| % | 0.4 | 0.045 | 0.55 | 0.45 | 0.20 | 0.30 | 0.30 | Balance |

Cr electrolyte is composed by: $CrO_3$ – 220 g/l, $H_2SO_4$ – 2.2 g/l. Cr-diamond nanoparticles coatings are obtained under three technological regimes by variation of current density: 45 and 60 A/dm$^2$ and deposition duration: 35 and 45 minutes (table 3).

**Table 3.** Technological regimes of deposition of galvanic Cr coatings with different diamond nanoparticles concentration.

| Regimes | R1 | R2 | R3 |
|---|---|---|---|
| Current density, A/dm$^2$ | 45 | 45 | 60 |
| Duration, min | 35 | 45 | 35 |

Diamond nanoparticles of equal size 25 nm with variation of nanoparticles concentration: 0.6 g/l, 10 g/l, 15 g/l and 20 g/l are used for the three regimes R1, R2 and R3. The nanoparticles are added to the electrolyte in the form of water suspension. The process of coatings deposition runs along with intensive electrolyte stirring. The temperature in the electrolyte tray is kept stable in the interval 52-55°C.

IOP Conf. Series: Materials Science and Engineering **174** (2017) 012060 doi:10.1088/1757-899X/174/1/012060

**Table 4.** Technological regime parameters for HVOF coating deposition.

| № | Parameter | Technological regime |
|---|---|---|
| 1. | Propylene/oxygen ratio, % | 55/100 |
| 2. | Jet velocity, m/s | 1000 |
| 3. | Distance „nozzle-coating" L, mm | 120 |
| 4. | Angle between orifice axis and surface, $\alpha$ ° | 90 |
| 5. | Air pressure from compressor, bar | 5 |
| 6. | $N_2$ pressure in the proportioning device, bar | 4 |
| 7. | Velocity of powder material feeding, tr/min | 1.5 |
| 8. | Mass flow rate of the powder material, g/min | 22 |

The galvanic chromium-nanodiamond coatings and the WC/Co coating were not subjected to additional heat treating. Coatings thickness is been measured by means of the Pocket LEPTOSKOP 2021Fe device using eight point of the contact surface, and the coatings hardness – by Equotip Bambino 2 measuring device. Abrasive wear of galvanic chromium coating, chromium-nanodiamond coating and WC/Co (88/12) HVOF-coating are studied under conditions of dry friction on the surface with fixed abrasive particles using the „pin-on-disk" tribotester shown schematically in figure 1.

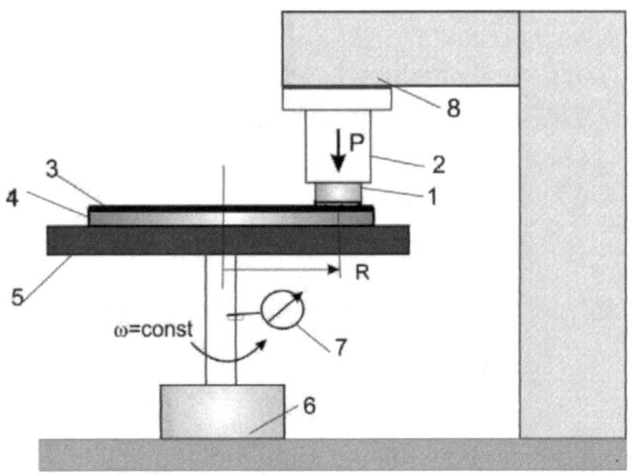


**Figure 1**. Schematic drawing of the „pin-on-disk" tribotester applied for the study of abrasive wear of coating during friction on the surface with fixed abrasive particles.

The studied cylindrical specimen 1 with the coating is placed in a holder 2 of the loading box 8, so that the coating stays in contact with the abrasive surface 3, fixed on horizontal disk 4 and mounted on the support 5. The disk is driven by the motor 6 and is rotating around its vertical central axis with rotational speed ω = const.

The sliding distance is measured by number of revolutions read on the revolution-counter 7. The device allows sliding speed variation by changing the disk rotational speed through the control unit and/or by changing the distance R between disk axis and specimen axis.

All tests are carried out at equal contact interaction conditions: load, sliding speed, sliding way, type and parameters of the abrasive (see table 5).

The test procedure embraces the measurement of mass wear $m$ for each specimen after a given sliding way L followed by wear resistance $I$ calculation using the formula:

$$I = \frac{\rho A_a L}{m} \tag{1}$$

where $\rho$ is coating matrix material density, in our case the density of chromium $\rho = 6.92.10^3$ kg/m$^3$ and tungsten $\rho = 19.25.10^3$ kg/m$^3$.

**Table 5.** Abrasive wear test parameters.

| Normal load | P = 10.3 N |
|---|---|
| Apparent (nominal) contact area | $A_a = 706.5.10^{-6}\ m^2$ |
| Apparent (nominal) contact pressure | $P_a = 1.46\ N/cm^2$ |
| Average sliding speed | V = 0.82 m/s |
| Sliding way | L = 630 m |
| Abrasive surface | Corundum P 320 |

The coefficient of friction is calculated for each test according to Amontons' law:

$$\mu = \frac{T}{P} \tag{2}$$

where T is the friction force, and P the normal load in specimen's center.

The coatings are compared by means of two relative criteria: $K_{ND}$ and $K_{WC}$. The first one: $K_{ND}$ characterizes the influence of diamond nanoparticles concentration upon the wear resistance of chromium-nanodiamond coatings:

$$K_{ND} = \frac{I(Cr - ND)}{I(Cr)} \tag{3}$$

It is the dimensionless ratio between the wear resistance of Cr coating containing diamond nanoparticles $I(Cr - ND)$ and the wear resistance of Cr coating without diamond nanoparticles $I(Cr)$ under equal friction regimes.

The second criterion $K_{WC/Co}$ indicates how many times the wear resistance of WC/Co (88/12) coating deposited by HVOF-technology is higher than the wear resistance of the chromium-nanodiamond coating. The criterion is given by the ratio of $I(WC / Co)$: the wear resistance WC/Co (88/12) coating and $I(Cr - ND)$: the wear resistance of electrolyte Cr coating with nanodiamond particles, i.e.

$$K_{WC/Co} = \frac{I(WC / Co)}{I(Cr - ND)} \tag{4}$$

## 3. Experimental results

Figure 2 shows the microstructure of galvanic Cr coating deposited under technological regime R1: without diamond nanoparticles (a) and with diamond nanoparticles with 10 g/l concentration.(b).

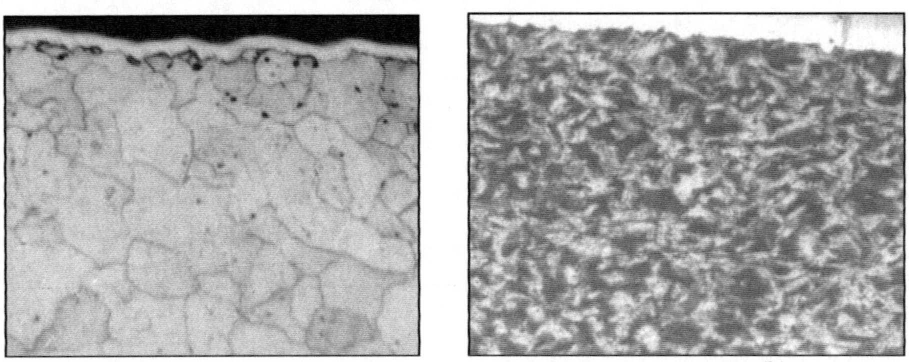

(a) (b)

**Figure 2.** Microstructures of galvanic chromium coating: (a) without diamond nanoparticles and (b) with diamond nanoparticles of size 25 nm and concentration 10 g/l, current density – 45 A/dm$^2$, time of the process – 35 min.

IOP Conf. Series: Materials Science and Engineering **174** (2017) 012060 doi:10.1088/1757-899X/174/1/012060

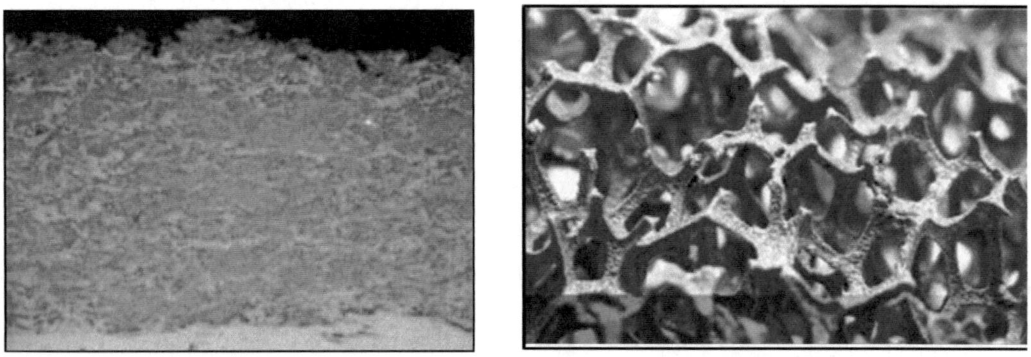

(a) (b)

**Figure 3.** Morphology (a) and microstructure, x200; (b) of WC/Co (88/12) HVOF coating on steel.

Based on the described procedures and devices, experimental results are obtained for the thickness, hardness, mass wear, wear resistance and friction coefficient of all coatings (see table 6).

**Table 6.** Experimental results for thickness, hardness, mass wear, wear resistance and friction coefficient of the studied galvanic coatings and WC/Co (88/12) coating.

| № | Speci-men | Coating | NDDS g/l | Thickness µm | Micro-hardness MPa | Wear g | Wear-resistance m/kg | COF |
|---|---|---|---|---|---|---|---|---|
| 1 | 0-R1 | Cr | 0 | 20÷35 | 5906 | $42.2x10^{-3}$ | $1.5x10^{7}$ | 0.35 |
| 2 | 1-R1 | Cr-(0.6)ND | 0.6 | 12÷16 | 8170 | $89x10^{-3}$ | $0.7x10^{7}$ | 0.46 |
| 3 | 2-R1 | Cr-(10)ND | 10 | 6÷23 | 7907 | $32.4x10^{-3}$ | $2x10^{7}$ | 0.24 |
| 4 | 3-R1 | Cr-(15)ND | 15 | 10÷20 | 8260 | $34.3x10^{-3}$ | $1.8x10^{7}$ | 0.26 |
| 5 | 4-R1 | Cr-(20)ND | 20 | 13÷17 | 8401 | $40x10^{-3}$ | $1.6x10^{7}$ | 0.32 |
| 6 | 0-R2 | Cr | 0 | 7÷11 | 6010 | $40.4x10^{-3}$ | $1.6x10^{7}$ | 0.31 |
| 7 | 1-R2 | Cr-(0.6)ND | 0.6 | 33÷40 | 8172 | $72x10^{-3}$ | $0.9x10^{7}$ | 0.42 |
| 8 | 2-R2 | Cr-(10)ND | 10 | 23÷38 | 8196 | $68.4x10^{-3}$ | $0.9x10^{7}$ | 0.40 |
| 9 | 3-R2 | Cr-(15)ND | 15 | 22÷25 | 8455 | $53.5x10^{-3}$ | $1.2x10^{7}$ | 0.38 |
| 10 | 4-R2 | Cr-(20)ND | 20 | 22÷26 | 8613 | $44.4x10^{-3}$ | $1.4x10^{7}$ | 0.36 |
| 11 | 0-R3 | Cr | 0 | 18÷32 | 6890 | $80.2x10^{-3}$ | $0.8x10^{7}$ | 0.42 |
| 12 | 1-R3 | Cr-(0.6)ND | 0.6 | 29÷34 | 8075 | $49.1x10^{-3}$ | $1.3x10^{7}$ | 0.37 |
| 13 | 2-R3 | Cr-(10)ND | 10 | 26÷29 | 7618 | $51.2x10^{-3}$ | $1.2x10^{7}$ | 0.38 |
| 14 | 3-R3 | Cr-(15)ND | 15 | 20÷25 | 8520 | $66.4x10^{-3}$ | $0.9x10^{7}$ | 0.40 |
| 15 | 4-R3 | Cr-(20)ND | 20 | 18÷24 | 9820 | $96.8x10^{-3}$ | $0.6x10^{7}$ | 0.58 |
| 16 | WC/Co | WC/Co (88/12) | | 220 | 72 HRC | $5.8x10^{-3}$ | $10.8x10^{7}$ | 0.14 |

Figures 4 and 5 give graphically the variation of thickness and micro hardness of galvanic Cr-nanodiamond coatings with the concentration of the diamond nanoparticles in the electrolyte under the three deposition regimes R1, R2 and R3.

IOP Conf. Series: Materials Science and Engineering **174** (2017) 012060 doi:10.1088/1757-899X/174/1/012060

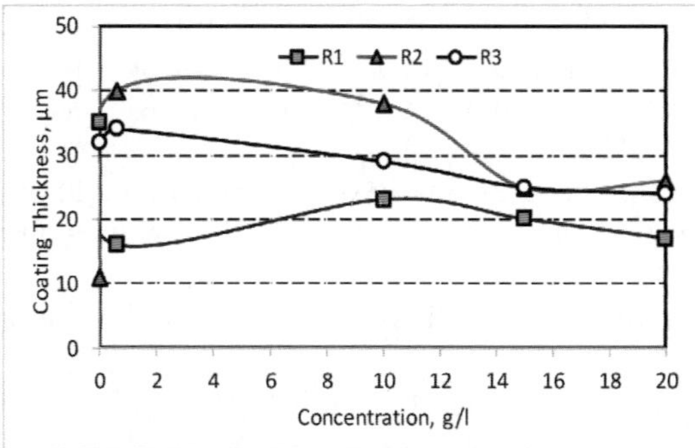


**Figure 4.** Influence of NDDS concentration on coating thickness.

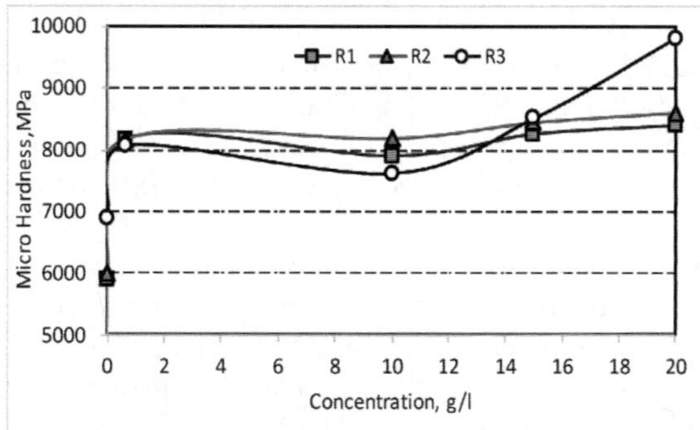


**Figure 5.** Influence of NDDS concentration on coating microhardness.

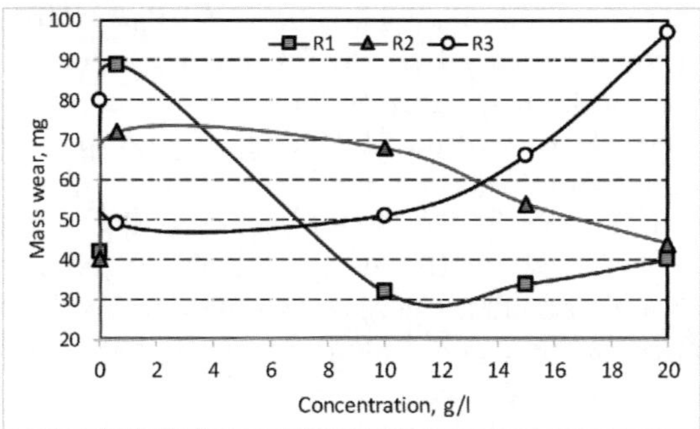


**Figure 6.** Influence of NDDS concentration on mass wear.

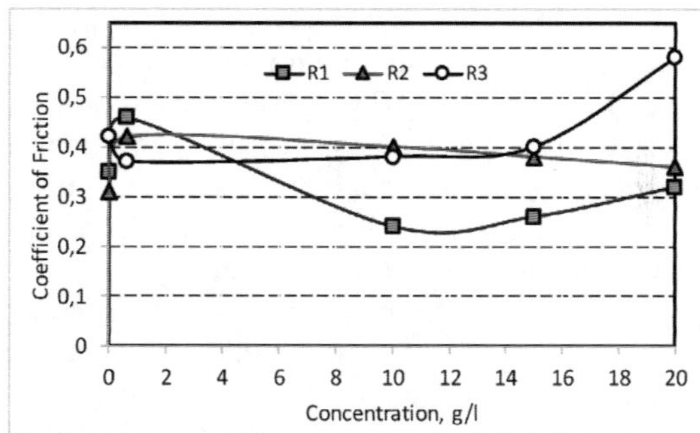


**Figure 7.** Influence of NDDS concentration on friction coefficient.

Figures 6 and 7 give graphically the variation of mass wear and friction coefficient of galvanic Cr-nanodiamond coatings with the concentration of the diamond nanoparticles in the electrolyte under the three deposition regimes R1, R2 and R3.

Figures 8 and 9 show diagrams of the relative criteria for comparison, $K_{ND}$ and $K_{WC}$, at different nanodiamond particles concentrations in the Cr coatings for regime R1 and in WC/Co (88/12) coating.

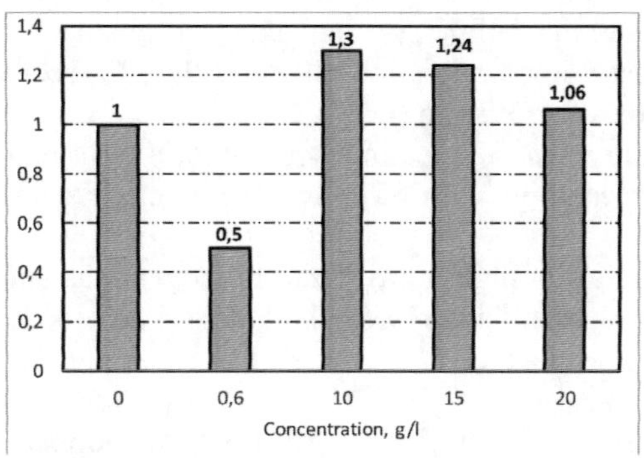


**Figure 8.** Diagram of criterion $K_{ND}$.

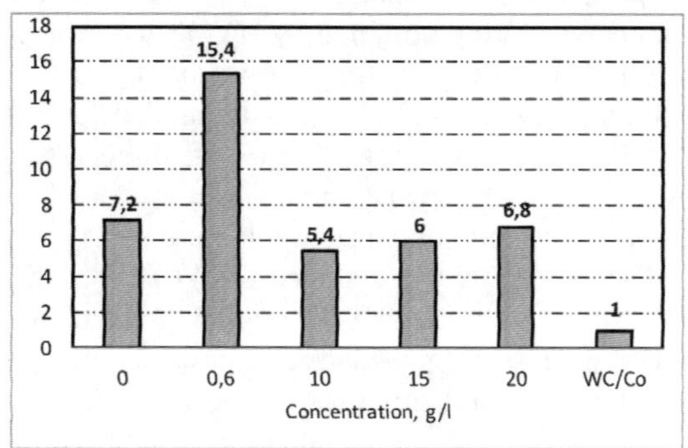


**Figure 9.** Diagram of criterion $K_{Wc/Co}$.

## 4. Discussion and conclusions

The comparative study for galvanic Cr and Cr-nanodiamod coatings provides the following interpretations and conclusions:

- Adding nanodiamond particles (NDDS) in electrolytic Cr coating affects the thickness, hardness, wear and friction coefficient. The nature and value of this influence depend on the technological regime of coatings deposition: current density and time of deposition, as well as on nanodiamond particles concentration;
- For small NDDS concentration – 0.6 g/l, wear increases compared to wear of Cr coating without NDDS for the three regimes R1, R2, R3 (figure 6). The wear resistance decreases of 50% compared to wear resistance of Cr coating without NDDS;
- The dependence of mass wear on NDDS concentration is nonlinear, the type of the curve depending on deposition regime – figure 6;
- Lowest wear shows Cr-nanodiamod coating deposited under regime R1 for 10-15 g/l NDDS concentration (figure 6). The wear resistance of these coatings increases by 30% compared to wear resistance of Cr coating without NDDS (figure 8);
- The variation of friction coefficient with NDDS concentration (figure 7) for the same regime R1 has similar character as the variation of wear (figure 6). Increasing NDDS concentration leads to increase of COF from 0.35 to 0.46; then it decreases, passes the minimum value 0.24 at 10 g/l NDDS concentration and increases gradually to 0.32 for concentration 20 g/l (figure 7). By technological regimes R2 and R3 COF keeps its constant value, except for the following: at concentration 20 g/l in regime R3 the COF increases abruptly to maximal value 0.58 for all studied coatings (figure 7). The last corresponds to the maximal wear of the same coating – figure 6, curve R3;
- Electrolyte Cr and Cr-nanodiamond coatings exhibit irregularly distributed thickness. The value depends on the technological regime and NDDS concentration (figure 4). Increasing NDDS concentration leads to thickness increase up to a given value, and then decreases to minimum values for concentration 20 g/l. Minimal thickness show Cr-nanodiamond coatings deposited at technological regime R1, which show the least wear among all studied galvanic coatings (figure 6);
- The presence of NDDS leads to increment in Cr coatings hardness (figure 5). At small concentration 0.6 g/l the hardness of all coatings increases by 42% compared to Cr coatings without NDDS. Further increment of NDDS concentration does not influence the hardness under different regimes. An exception is the coating obtained in regime R3 and concentration 20 g/l, where the hardness increases sharply with 20% compared to the hardness of all other coatings (figure 5). The higher hardness does not however result in increase of the wear resistance of the coating. Its wear resistance $I = 0.6.10^7$ is lower than the wear resistance of coatings obtained at regimes R1 and R2 (table 6). This result confirms again the fact that there is no proportionality between wear resistance and hardness of the surface layers.

Comparing the results for galvanic chromium, Cr-nanodiamond coatings and WC/Co (88/12) composite coating obtained by HVOF-technology points to the following conclusions:

WC/Co (88/12) coating shows the highest wear resistance compared to all electrolyte Cr coatings without and with NDDS, deposited under the three regimes R1, R2 and R3.

WC/Co coating shows 7.2 times higher wear resistance than Cr coating without NDDS in regime R1 and 5.4 to 15.4 times higher wear resistance than Cr coatings with NDDS in the whole interval of variation of their concentration (figure 9).

The higher wear resistance of WC/Co coating is due to its high density and homogeneous small-grain structure made of WC grains linked in an internal contact network of cobalt (Co). The properties and size of this network are of great significance for the contact interaction of the coating during friction: small tangential resistance and corresponding small wear.

The results of the comparative study allow the following basic conclusion to be done: the composite WC/Co (88/12) coating deposited by HVOF-technology can replace the non-ecological chromium and Cr-nanodiamond coatings in tribological systems.

**Acknowledgments**
This study is connected with the completion of the Project **DUNK-01/3** "University R&D Complex for innovation and transfer of knowledge in micro/nano-technologies and materials, energy efficiency and

virtual engineering" funded by the Bulgarian Ministry of Education and Science, and with CEEPUS III Network: CIIIBG0703 "Modern Trends in Education and Research on Mechanical Systems – Bridging Reliability, Quality and Tribology"

## References

[1] Bhushan B (ed.) 1999 *Handbook of Micro/Nanotribology*, $2^{nd}$ ed. (CRC Press Series Mechanics and Material Science) (London, New York, Washington D.C.: CRC Press LLC)
[2] Bhushan B 2011 Nanotribology, nanomechanics and materials characterization studies and applications to bio/nanotechnology and biomimetics in *Proc. Int. Conf. on Tribology SERBIATRIB`11* (Kragujevac, Serbia) pp 3-4
[3] Nguen Vict-Hue, Hoang Thi-Nam, Nguyen Ngoe-Thong, Kwon Sik-Choa, Man-Kim and Joo-Yui Lee 2009 Cr/Nanodiamond Composite Plating with Cobalt Cation additive *Trans. Nonferrous Met. Soc. China* **19** 975-978
[4] Kandeva M, Gidikova N, Valov R, Petkov V, Kalchevska K and Torbova M 2011 in *Proc. of* $9^{th}$ *International Conference The "A"Coatings* (Tessaloniki, Greece) pp 371-380
[5] Nestler M C, Prenzel G and Seitz T. 1998 in *Proc. of* $15^{th}$ *ITSC Vol 2* (Nice, France) pp 1073-1078
[6] Kandeva M, Karastoyanov D and Vencl A 2016 *Advanced Tribological Coatings for Heavy-Duty Applications: Case Studies* (Sofia: Marin Drinov Academic Publishing House)

IOP Conf. Series: Materials Science and Engineering **174** (2017) 012061 doi:10.1088/1757-899X/174/1/012061

# Effect of friction stir processing on tribological properties of Al-Si alloys

**S M Aktarer[1], D M Sekban[2], H Yanar[3] and G Purçek[3]**

[1] Department of Automotive Technology, RecepTayyip Erdogan University, Rize,Turkey
[2] Department of Naval Architecture and Marine Engineering, Karadeniz Technical University, Trabzon,Turkey
[3]Department of Mechanical Engineering, Karadeniz Technical University, Trabzon, Turkey

Email: purcek@ktu.edu.tr

**Abstract**. As-cast Al-12Si alloy was processed by single-pass friction stir processing (FSP), and its effect on mainly friction and wear properties of processed alloy was studied in detail. The needle-shaped eutectic silicon particles were fragmented by intense plastic deformation and dynamic recrystallization during FSP. The fragmented and homogenously distributed Si particles throughout the improve the mechanical properties and wear behavior of Al-12Si alloy. The wear mechanisms for this improvement were examined and the possible reasons were discussed.

## 1. Introduction

Al-Si alloys has been extensively used in the automotive, aerospace and military applications duo to their superior wear resistance, low coefficient of thermal expansion, high corrosion resistance, high strength-to-weight ratio and excellent castability [1, 2]. However, they generally have insufficient strength, low fracture toughness and poor ductility which limit their broader applications in the as-cast state [3]. Some traditional processes like alloying, variation in solidification rates, heat treatments and stirring during solidification have been applied to solve or decrease their aforementioned problems [1-4]. Although some improvements in their properties have been achieved with such traditional processes, further improvements are needed in their strength, ductility, impact toughness and wear resistance to be able to enhance their applications. Also, the casting defects like porosity, compositional inhomogeneity, coarse and inhomogeneous Si distribution could not be eliminated completely by traditional processes [3,5,6]. Thus, some novel processes including severe plastic deformation (SPD) have been applied to such kind of alloys so far. Among them, equal-channel angular extrusion/pressing (ECAE/P) [1,2,7], high pressure torsion (HPT) [8,9], accumulative roll-bonding (ARB) [10] and recently friction stir processing (FSP) [3-5,11-25] seem to be the viable methods to improve the mechanical properties of Al-Si alloys.

Considering the surface properties and microstructural modification in the large-scale plate- or sheet-type materials, the FSP can be said the best one among others. FSP is a method of changing the properties of a metal through severe, localized plastic deformation which is produced by forcibly inserting a non-consumable tool into the work piece, and revolving the tool in a stirring motion as it is pushed laterally through the work piece [5,26]. This process mixes a material without changing the phase and creates a microstructure with fine, equiaxed and uniformly distributed grains and/or particles. After this process, the processed zone is constituted generally by recrystallized fine grains, fragmented primary particles and uniformly distributed second phase particles. FSP method has been applied to Al-Si alloys in the hypoeutectic [18,19,23], near eutectic [13-15] and hypereutectic compositions [3, 20]. These studies have generally focused on the microstructural evolution and mechanical characteristics of plastic deformation during FSP. In addition, the fatigue and crack growth behaviors [6,9,19,26], microstructural modification [5, 22], mechanical properties [11,22,24,26], superplasticity [13] and wear behavior [3,15,17] of Al-Si alloys after FSP have been investigated.

In view of above, there are many reports on Al-Si alloys including different additional alloying elements. However, few of them were on the as-cast binary Al-Si alloys. Also, the effect of FSP on the microstructural evolution including the combination of strength, ductility and also wear properties of the as-cast Al–Si alloys has not been investigated systematically [3]. Furthermore, more studies are needed for getting further improvement in mechanical properties of Al-Si alloys by applying FSP with optimum parameters. Therefore, the main purpose of this study is to modify the surface structure of eutectic as-cast Al-12Si alloy by FSP, and investigate its microstructural, mechanical and tribological properties.

## 2. Experimental Procedure

Eutectic Al-12wt.%Si alloy was produced by ingot metallurgy with a chemical composition of 12.2%Si, 0.6%Fe, 0.1%Cu, 0.4%Mn, 0.1%Mg, 0.1%Zn, 0.1%Ni, 0.15%Ti, 0.1%Pb, 0.05%Sn and balance Al. The ingots were re-melted in a graphite crucible for getting uniform as-cast microstructure and poured at 650 $^0$C into a mild steel permanent mold stayed at room temperature. For FSP, the plates having dimensions of 200mm x 40mm x 5mm were produced from the re-casted ingots. Two-pass FSP was performed using a tool steel with a shoulder diameter of 16 mm, a threaded pin diameter of 5 mm and length of 2.5 mm. During the experiments, the tool rotation and traverse speeds were fixed at 1250 rpm and 65 mm.min-1, respectively. The tool was tilted 3°, and downforce of the tool was kept constant at 3kN.

Samples for metallographic investigation were sectioned perpendicular to the process direction from the as-cast ingot by wire-EDM technique (figure 1). After then, they were prepared using standard polishing techniques and then etched in a 10%NaOH solution. The microstructural characteristics of the alloy before and after FSP were investigated using an optical microscope. Tensile properties of the samples were determined using dog-bone shaped specimens with the dimensions of 2 mm x 3 mm x 26 mm using the same wire-EDM technique. The tensile specimens were cross-sectioned parallel to the process direction (figure 1). The tests were performed using an Instron-3382 electro-mechanical load frame at a strain rate of 5.4 x 10-4 s-1. This strain rate was especially chosen to be able to stay into the into the quasi-static deformation range. Hardness specimens were cross-sectioned parallel to the process direction (figure 1) with 2 mm depth. Hardness measurements were performed using a Vickers micro-hardness tester operated at HV0.5. The measurements were carried out on both longitudinal and transverse sections of the FSPed plates.

Dry sliding wear tests of the as-cast and friction stir processed alloy samples were conducted under dry sliding conditions using a pin-on-disc type tribometer (UTS Tribometer T10/20) at room temperature (20°C) and a relative humidity of about 75%. The samples (15 mm x 15 mm x 5 mm) were cut from processed and unprocessed material using the wire-EDM technique. The wear experiments were conducted using a 6 mm diameter Al2O3 ball. In this study, working with a ball shaped counter-face (it is suitable especially for such kind of comparative studies according to ASTM G-99) were preferred. The wear tests were performed for a total sliding distance of 100 m corresponding to a period of 16 min under four different constant loads (2.5N, 5N, 7.5N and 10N) using a sliding speed of 0.1 $ms^{-1}$. Wear rate was defined by measuring mass loss of alloy at the end of every sliding distance with using a microbalance with an accuracy of ±0.1 mg. Prior to the wear tests, all sample surfaces were ground using emery paper with a grit size of 1200, which yielded an approx. 0.1 µm roughness (Ra) on the sample surfaces for all conditions before the wear tests.

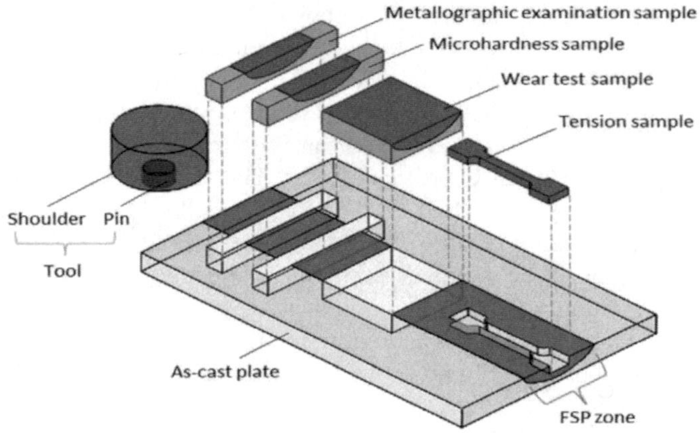


**Figure 1.**Schematic illustrations of the FSPed plate and the sample geometries machined from that plate.

## 3. Results and Discussion

### *3.1. Microstructure*

Figure 2 (a) shows optical micrograph of as-cast Al-12Si alloy. It consists of eutectic Si particles (dark), α-Al phase (white), and few coarse primary silicon particles (dark). Eutectic Si particles exhibited a fibrous morphology and it is in the form of acicular or needle-shaped plates dispersed throughout the α-Al matrix. The presence of large primary Si particles was also observed in the microstructure due to the non-equilibrium cooling and slight variations in the eutectic composition. This is a typical structure of as-cast eutectic Al-Si alloy [1,4,5]. The effect of FSP on the microstructure of as-cast Al-12Si alloy is shown in Figure 2(b). The effect of FSP on the shape, size and size distribution of the eutectic Si particles dispersed in the matrixes are more pronounced as compared to other constituents. The Si particles after two-pass FSP are nearly equiaxed in shape and uniform in size and distribution than the particles in the as-cast condition. Finally, the microstructure of the FSPed alloy in the nugget zone (NZ) consists of fine eutectic silicon particles with uniformly distribution throughout the NZ [23]. The fracture in coarse primary Si particle seems to be inconsiderable. The structure in the Figure 2 (b) appears to be partially recrystallized where some fine Si particles started to form within the elongated Si particles. Such structural formation has also been reported in many Al alloys [3,23].

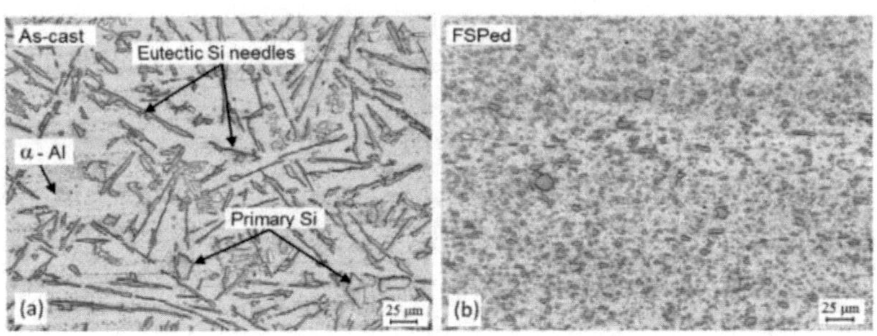


**Figure 2.** Microstructures: (a) Microstructure of as-cast Al-12Si alloy and (b) microstructure of friction stir processed Al-12Si alloy.

### *3.2. Mechanical Properties*

The engineering stress-engineering strain curves of the Al-12Si alloy before and after FSP are shown in figure 3, and the values of strength and ductility that obtained from these curves are given in table 1. As-cast alloy exhibited a typical stress–strain curve for a brittle material with poor uniform and total elongations. After FSP, this curve changed to a typical ductile material's curve with a large strain hardening region. Therefore, the strength and ductility of the alloy increased simultaneously by the effect of FSP. The alloy sample showed an increase in the ultimate tensile strength ($\sigma_{UTS}$) of about 48%, from 148 MPa for the as-cast sample to about 220 MPa for the FSPed sample. Similarly, the yield strength increased about 32% after FSP, from 100 MPa to 132 MPa. In addition, while the $\sigma_{UTS}/\sigma_Y$ ratio is 1.48 in the as-cast alloy, this ratio increased after FSP to about 1.66, which meant that the strain hardening region increased with FSP. The ductility of the alloy also increased substantially after FSP. The FSPed sample exhibited considerably higher elongation to fracture ($\varepsilon_f$) of about 23% in comparison with no processed sample having only 3.7%, which was almost 6.2 times higher than that of the as-cast sample. More importantly, uniform elongation ($\varepsilon_u$) increased considerably after FSP in contrast to other UFG materials. It increased from 2.9% in the as-cast state to about 14% after FSP. The simultaneous increase in strength and ductility of the alloy after FSP was attributed to the changes in microstructure during FSP. It is well known that the mechanical properties of Al-Si alloys strongly depend on the shape, size and distribution of silicon particles, porosities and also the size of α-Al matrix phase [1, 5, 15, 24]. Needle shaped silicon particles dispersed in the matrix acts as an internal stress raiser, which raises the cracking tendency of the alloy. Therefore, the as-cast alloy exhibited low strength and very poor ductility at room temperature. During FSP, the as-cast morphology is completely broken up especially in the NZ (figure 2). The extensive friction stir effect during processing results in significant changes in Si particle morphology by breaking them into smaller particles. Finally, disintegration/fragmentation and better distribution of eutectic silicon particles form in the matrix after FSP, which leads to extraordinary improvement in strength and ductility of the alloy.

The as-cast Al-12Si alloy exhibited hardness values varying between 52Hv0.5 and 66Hv0.5 which indicates highly scattered hardness distribution inside the as-cast structure. The average value is 58 Hv0.5 for the as-cast structure. This is a normal as considering the inhomogeneous distribution of coarse eutectic Si particles as well as casting defects inside the as-cast microstructure (figure 2(a)). Because, there is a more chance that the hardness indenter may press on the places with or without Si particles in coarse-grained as-cast structure. Two-pass FSP increased the hardness of the alloy inside the NZ. The sample shows an increase in the hardness of about 15%, from 58 Hv0.5 for the un-processed as-cast zone to about 67 Hv0.5 for the NZ. This is primarily due to the increase in dispersion hardening effect of refined and more uniformly distributed eutectic silicon particles in the α-Al matrix by the effect of FSP [11,15]. Also, the α-Al matrix phase refines due to the dynamic recrystallization during FSP, which provides additional improvement in hardness of this zone [25, 26]. The NZ zone, on the other hand, have more uniform hardness distribution with fewer scatter as reported in some previous studies on Al-Si alloys [3, 5,11,15,18, 20, 24]. The hardness profile is relatively smooth (ranges from 63 HV to 67 HV), and less scatters form inside the NZ. This can be attributed to the fine and more uniformly distributed eutectic Si particles in the NZ of FSPed sample.

**Table 1.** Tensile properties and hardness of the as-cast and FSPed Al–12Si alloy (σUTS: Ultimate Tensile Strength, σY: Yield Strength, εf: Elongation to Failure, εu: Uniform Elongation, HV0.5: Hardness Vickers).

| | $\sigma_{UTS}$ (MPa) | $\sigma_Y$ MPa) | $\varepsilon_f$ (%) | $\varepsilon_u$ (%) | $\sigma_{UTS}/\sigma_Y$ | HV0.5 |
|---|---|---|---|---|---|---|
| **As cast** | 148 ± 6 | 100 ± 5 | 3.7 ± 0.3 | 2.9 ± 0.5 | 1.48 | 58 |
| **FSPed** | 220 ± 3 | 132 ± 4 | 23.0 ± 1.1 | 14.0 ± 0.6 | 1.66 | 67 |

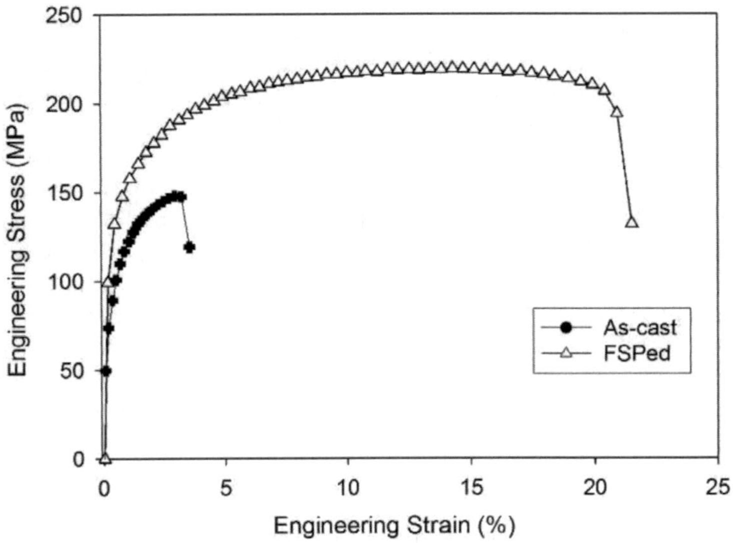


**Figure 3.** Engineering stress–strain curves of the as-cast and FSPed Al–12Si alloy.

### *3.3. Friction and Wear Behavior*

Figure 4 shows the coefficient of friction versus sliding distance curves for the alloy before and after FSP. The figure shows a high initial value, about 1.09 in friction coefficient during the initial transient-state sliding period with steady-state levels reached after about 10 m for both conditions, which demonstrates that no considerable changes occur after microstructural alteration by FSP. The average coefficient of friction in the steady-state condition for both conditions is about 0.5.

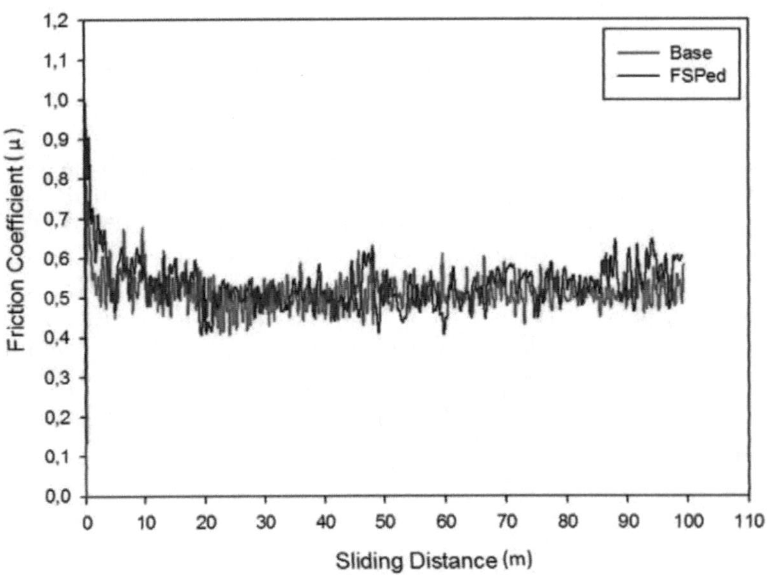


**Figure 4.** The coefficient of friction vs. sliding distance curves for Al-12Si alloy before and after FSP.

Figure 5 shows a typical relationship between the weight loss and applied normal load of the as-cast and FSPed Al–12Si alloy. It reveals that the weight loss increases almost linearly with increasing applied load for the Al–12Si alloy in both conditions. The results show that the FSPed samples exhibited significantly lower weight loss for all normal loads compared to the as-cast Al–12Si alloy. But, the difference is more distinct at higher loads. The comparison of the wear behavior of the as-cast and FSPed alloy show that the FSP increased the wear resistance of the alloy considerably by decreasing the weight loss due to the increase in strength and hardness after FSPed for especially high applied load. These results are also confirmed by the 2D and 3D profilometric maps. The effect of applied load on the wear depth and 3D profilometric view of wear tracts of as-cast Al-12Si and FSPed Al-12Si are shown in figure 6. It is clearly seen that both diagrams demonstrate almost the same variation in each case. The wear depth increases with increasing applied load for all conditions, and the increment is more pronounced at higher loads. It is observed that the groove shapes of the base sample and FSPed sample are more-or-less different under different loads, and grove shapes of the samples widens with increasing applied load. The base sample has deeper grooves and wider ridges, which causes more weight/volume loss than FSPed sample under all loads. Also, the 2D profiles of FSPed are smoother than those of base alloy (figure 6). As shown, the base sample has about 70 μm of wear depth while the FSPed sample has about 50 μm of wear depth for 10N applied load.

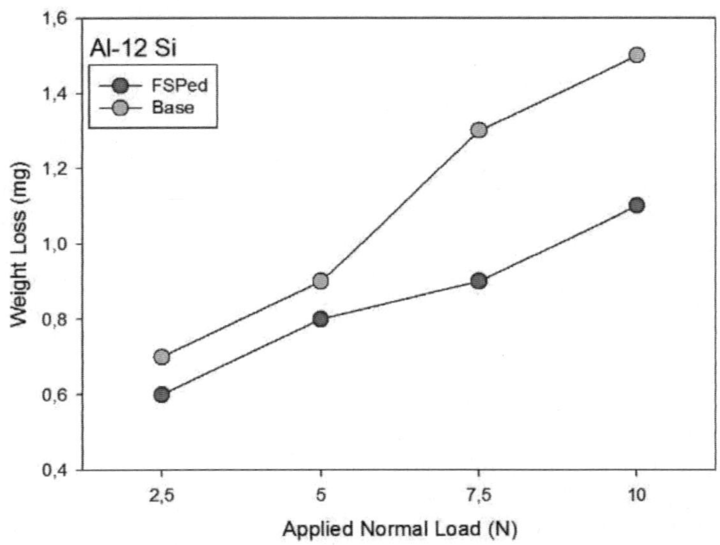


**Figure 5.** The change in weight loss as a function of applied load for as-cast and FSPed Al–12Si alloy.

The advancement of the wear resistance after FSP might be connected to the radical microstructural alterations. It is well known that the mechanical properties and wear resistance of Al-Si alloys strongly depend on the shape, size and distribution of silicon particles, porosities and also the size of α-Al matrix phase [1, 3-5, 24]. Needle shaped silicon particles dispersed in the matrix acts as an internal stress raiser, which raises the cracking tendency of the alloy. High stress concentrations along the Si/matrix interfaces, which are responsible for the nucleation and propagation of cracks, may be caused by Si particles with high aspect ratio. When the cracks take place, they may lead to the fragmentation of the large Si particles into smaller and homogenous ones. After a wearing process, those particles may be disengaged from the surface and work as an additional abrasive, thus increase the wear rates [17]. As a result of the Al soft matrix's direct relation to the counterface, the alloy's ability to support the load is reduced by the void left by the disengaged Si particles. The extensive friction stir effect during FSP results in significant changes in Si particle morphology by breaking them into smaller particles as shown in figure 2. Finally, disintegration/fragmentation and better distribution of eutectic silicon particles form in the matrix after FSP, which leads to improvement in wear resistance of the alloy. The elimination of the porosity, as well as the Si particles refining, contributes notably to the improvement of the wear resistance of Al-12Si alloy. In addition, the fact that stress concentrations may occur around the pores results in an increase in the wear resistance. Also, the hardness of the alloy increases by the effect of FSP, and this bring about a further improvement in wear resistance of FSPed alloy [1,3,4,22]

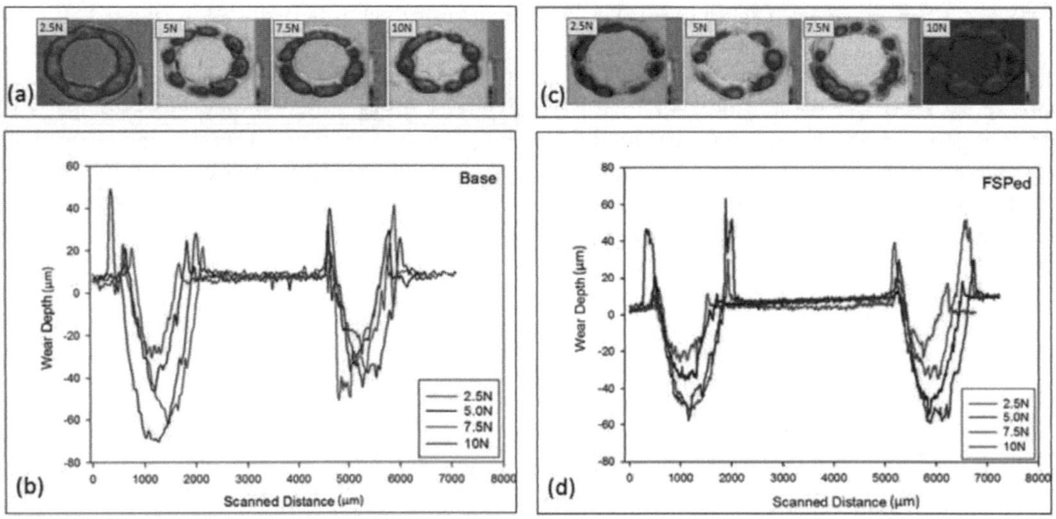


**Figure 6.** 3D profilometric view of wear tracts of Al-12Si alloy under different loads and 2D view of worn tracts took place under different loads: (a)-(b) As-cast Al-12Si alloy before FSP and (c)-(d) FSPed alloy.

## 4. Conclusions

Friction stir process technique was applied to the as-cast eutectic Al–12Si alloy to improve their friction and wear properties. The main findings and conclusions of this study can be summarized as follows:

1. Both strength and ductility of the as-cast eutectic Al–12Si alloy increased considerably as well as hardness after a two-pass process. This simultaneous increase in strength and ductility was mainly attributed to the grain refinement and morphological changes in eutectic coarse acicular Si particles as well as coarse primary Si particles. The alloy after FSP exhibited 220 MPa tensile strength and 132 MPa yield strength, which were about 48% and 32% higher than those of the as-cast alloy, respectively.

2. After FSP, there is no significant change in the average coefficient of friction. The FSPed samples and as-cast samples exhibited average coefficient of friction of about 0.5.

3. The weight loss (also volume loss) increased almost linearly with increasing applied load for the as-cast and FSPed Al-12Si alloy. The FSPed samples exhibited significantly lower weight loss especially under high loads compared to the as-cast Al–12Si alloy. The FSPed sample exhibited better wear resistance than the as-cast Al-12Si alloy for especially high applied loads.

4. It can be concluded that the use of microstructural modification by FSP is a simple and effective procedure for providing remarkable improvement in strength, ductility, hardness and wear resistance of as-cast Al-12Si alloys.

## References

[1] Purcek G, Saray O and Kul O 2010 *Met. Mater. Int.* **16** 145-154
[2] Kucukomeroglu T 2010 *Mater. Design* **31** 782-789
[3] Mahmoud T S 2013 *Surf. Coat. Tech.* **228** 209-220
[4] Aktarer S M, Sekban D M, Saray O, Kucukomeroglu T, Ma Z Y and G Purcek 2015 *Mat. Sci. Eng. A-Struct.* **636** 311-319
[5] Ma Z Y, Sharma S R and Mishra R S 2006 *Metall. Mater. Trans. A* **37a** 3323-3336
[6] Mishra R S and Ma Z Y 2005 *Mat. Sci. Eng. R* **50** 1-78
[7] Zhang Z G, Hosoda S, Kim I S and Watanabe Y 2006 *Mat. Sci. Eng. A-Struct.* **425** 55-63
[8] Gode C, Yilmazer H, Ozdemir I and Todaka Y 2014 *Mat. Sci. Eng. A-Struct.* **618** 377-384
[9] Abd El Aal M I and Kim H S 2014 *Mater. Design* **53** 373-382
[10] Kitazono K, Nishizawa S, Sato E and Motegi T 2004 *Mater. Trans.* **45** 2389-2394
[11] Choi D H, Kim Y H , Ahn B W, Kim Y I and Jung S B 2013 *T. Nonferr. Metal. Soc.* **23** 335-340
[12] Yang C W, Chang Y H, Lui T S and Chen L H 2012 *Mater. Design* **40** 163-170
[13] Tutunchilar S, Besharati M K, Haghpanahi M and Asadi P 2012 *Mat. Sci. Eng. A-Struct.* **534** 557-567
[14] Tsai F Y and Kao P W 2012 *Mater. Lett.* **80** 40-42
[15] Mahmoud T S and Mohamed S S 2012 *Mat. Sci. Eng. A-Struct.* **558** 502-509
[16] El-Rayes M M and El-Danaf E A 2012 *J. Mater. Process. Tech.* **212** 1157-1168
[17] Alidokht S A, Abdollah-zadeh A, Soleymani S, Saeid T and Assadi H 2012 *Mater. Charact.* **63** 90-97
[18] Karthikeyan L, Senthilkumar V S and Padmanabhan K A 2010 *Mater. Design* **31** 761-771
[19] Jana S, Mishra R S, Baumann J B and Grant G 2010 *Acta Mater.* **58** 989-1003
[20] Rao A. G, Rao B R K, Deshmukh V P, Shah A K and Kashyap B P 2009 *Mater. Lett.* **63** 2628-2630
[21] Jana S, Mishra R S, Baumann J B and Grant G 2009 *Scripta Mater.* **61** 992-995
[22] Ma Z Y, Sharma S R and Mishra R S 2006 *Scripta Mater.* **54** 1623-1626
[23] Ma Z. Y, Sharma S R and Mishra R S 2006 *Mat. Sci. Eng. A-Struct.* **433** 269-278
[24] Santella M L, Engstrom T, Storjohann D and Pan T Y 2005 *Scripta Mater.* **53** 201-206
[25] Sharma S 2004 *Scripta Mater.* **51** 237-241
[26] Ma Z Y, Pilchak A L, Juhas M C and Williams J C 2008 *Scripta Mater.* **58** 361-366

# HVOF sprayed WC-CoCr coatings on aluminum: tensile and tribological properties

**A Koutsomichalis[1], M Vardavoulias[2] and N Vaxevanidis[3]**

[1] Hellenic Air-Force Academy, Department of Aeronautical Studies, Dekelia Air Force Base, TGA 1010, Attica, Greece
[2] Pyrogenesis S.A, Technological Park of Lavrio, GR 195 00, Lavrio, Greece
[3] School of Pedagogical & Technological Education, Dept. of Mechanical Engineering Educators, GR 141 21 Heraklion, Athens, Greece

E-mail: vaxev@aspete.gr

**Abstract.** This paper provides a comprehensive assessment of the tensile and sliding wear behaviour of WC-10Co4Cr agglomerated and sintered powder deposited on aluminum by Hyper Velocity Oxy-Fuel (HVOF) process. Microstructural analysis (SEM) identified grains of tungsten carbide (WC) in the metal matrix of the cobalt-chromium (Co-Cr). A transformation of the WC in the $W_2C$ phase was observed and decomposition of WC in the metal matrix. The HVOF WC-Co-Cr coating was found to decrease tensile strength of the aluminum substrate. Transverse cracks were observed to initiate on the coating surface, increasing rapidly with the increase in tensile strain and stopped on the coating-substrate interface causing decohesion. Tribological properties were examined using the pin-on-disk method under various loads. The friction coefficient rose abruptly at the start-up phase and stabilized at almost the same sliding distance independently of the applied load. Both the friction coefficient and the wear volume were found to increase with increasing applied load. Study of the wear mechanisms revealed surface micro-cracking and fragmentation of flattened coating layers with subsequent gradual pull out of the carbide particles.

## 1. Introduction

Tungsten carbide coatings find a large variety of industrial applications for the protection of mechanical components against sliding and abrasive wear at various temperatures [1-3]. Tungsten carbide either in coated or sintered form is used for structural applications like cutting tools, plungers, bearings, gears, steam turbine, hydro turbines etc. [4]. The sprayed WC-Co coatings exhibit multi-phase microstructures, which are formed by the decomposition of WC-Co powders [5]. The degree of decomposition of WC-Co powders during spraying is related to carbide dissolution in the molten matrix with subsequent decarburization; it is also dependent on the temperature, the carbide grain size and the time (velocity) and is reported to influence wear characteristics [6]. The properties of WC-CoCr coatings depend on the powder size and volume fraction of WC powder feedstock and matrix composition [7]. The wear resistance properties of WC-Co-Cr coatings is dependent on the presence of hard, wear resistant WC grains in a CoCr based metallic binder phase, with the latter providing some toughness in the coating in comparison with pure ceramic coatings. Recent studies are focused on the tribological behaviour of composite cermet/metal coatings, in which the metallic phase participates not as binder metal, but as a distinct matrix of the product in percentages higher than 20% [8]. HVOF is widely used for the deposition of WC-based cermet coatings, since it retains a larger fraction of WC in the coating because of the short powder-flame interactions due to its hypersonic velocity that limits WC decomposition [9]. The WC decomposition to $W_2C$ and metallic W is reported to decrease coating hardness and deteriorate wear resistance [10,11-16]. Recent studies are focused on the optimization of the deposition process parameters (e.g. spray distance, oxygen and flue flow rates) towards enhanced coating hardness [15,16].

The objective of this work was to investigate the friction and wear behaviour of HVOF WC-CoCr coating at different loads and identify the effect of load on the wear mechanism of the carbide coating. This paper also attempts to provide an in situ microscopic observation on the crack initiation, propagation and fracture behaviour of the coated system under uni-axial tensile loading.

## 2. Experimental researches

A commercially available WC-Co-Cr agglomerated and sintered powder (WOKA 3653) was deposited on the surface of Al-4%Cu alloy. The powder characteristics are given in table 1. SEM micrograph of agglomerated micro-structured WC-10Co-4Cr powder with EDX analysis is shown in figure 1.

**Table 1.** Coating powder characteristics.

| | |
|---|---|
| Composition | WC: 86%, Co:10%, Cr: 4%, Fe<0.3% |
| particle size | -45+11 μm |
| shape | spherical |
| density | 14.7 g/cm$^3$ |

Two different types of specimens were prepared by the aluninum substrate: tensile specimens with dimensions 130x90x2.1 mm$^3$ and flat cylindrical test pieces for wear tests, 38.1 mm in diameter and 10mm thick. All specimens were grit blasted, prior to spraying, with aluminum oxide mesh 90.

The cermet powder was deposited by HVOF equipment. This step was accomplished by means of a Diamond Jet 2700 Sulzer Metco gun, employing a mixture of propane as fuel and oxygen. The spraying conditions are listed in table 2.

WC-CoCr coating was deposited on both sides of the tensile dog bone shaped specimens in equal thicknesses in order to enable balanced loading during tensile testing. Tensile testing was performed in accordance with ASTM E8 standard on an Instron 4482 apparatus using three specimens for each set.

Cylindrical wear test pieces were sprayed to an average coating thickness of 400 μm. The microstructure of the coated samples was observed under a metallographic microscope (Leica DMR) and SEM (Jeol); see figure 1 with image analysis software (Image Pro).

For the evaluation of friction and wear characteristics sliding friction tests were performed (according to ASTM G 99-05 standard) on a state-of-the-art pin-on-disc apparatus using, as counterbody, a cutting insert with a diamond-coated tip. Four normal loads were selected, namely 1N, 2N, 5N and 10N. All tests were conducted at 16 mm track radius and with constant linear speed of 0.6 m/s.

**Table 2.** Spraying parameters.

| | |
|---|---|
| Pressure | Oxygen: 220 psi, Propane: 30 psi; Air 320 psi, Argon (carrier): 140 psi |
| Flow rate | Oxygen: 240 lt/min, Propane: 74 lt/min; Air: 800 lt/min, Argon (carrier): 30 lt/min; Coolant: 24 lt/min |
| Powder feed | 2,5 rpm |
| Spraying distance | 200 mm |
| Revolutions | 53 rpm |
| Gun speed | 0,044 m/s |
| Flame temperature | 2500 K |

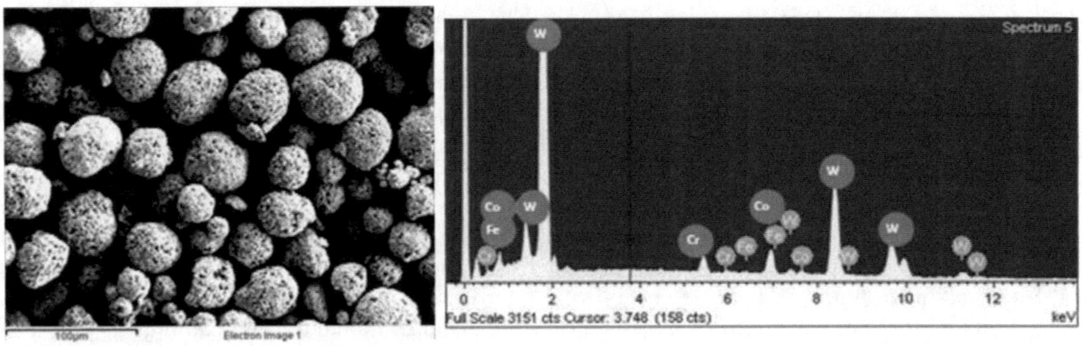


**Figure 1.** SEM micrograph of the WoKA 3653 micron-sized feedstock powder showing spherical geometry.

## 3. Results and discussion

Figure 2 shows the cross section of the coating which is exhibiting dense structure, low porosity and is macro-crack free. Coating-substrate interface shows no gaps or cracks indicating good adhesion between the coatings and the substrate. The coating is built up by a deposit of lenticular splats, one over the other, in a uniform manner, throughout the coating. Pores appear black in the micrographs. The coating microstructure is observed to be composed of WC particles, which are uniformly distributed within the coating.

(a) (b)

**Figure 2.** (a) Optical and (b) SEM migrographs of the WC-Co-Cr coating cross section.

The coating microstructure typically consists of adjacent layers of dark and light areas (figure 3). The microstructure of the dark areas was revealed to be mostly Co particles with aggregate of WC, while the microstructure of the light areas was mostly consisted of sintered aggregate WC particles. Close examination of light areas of coatings showed very small WC particles forming larger particles. Tungsten carbide grains with sharp edges are attributed to their incomplete melting, while rounded carbide grains result from the dissolution of WC, due to local overheating, in the Co matrix.

It has been reported [17-21], that during spraying of WC-Co powder, particles are subjected to decarburization leading to a decrease of the total carbon content. The extent of the WC transformation has been associated with the starting powder type (size, morphology, and carbide size), the type of spray process, the amount of oxygen in the environment, and the spray parameters [22,23].

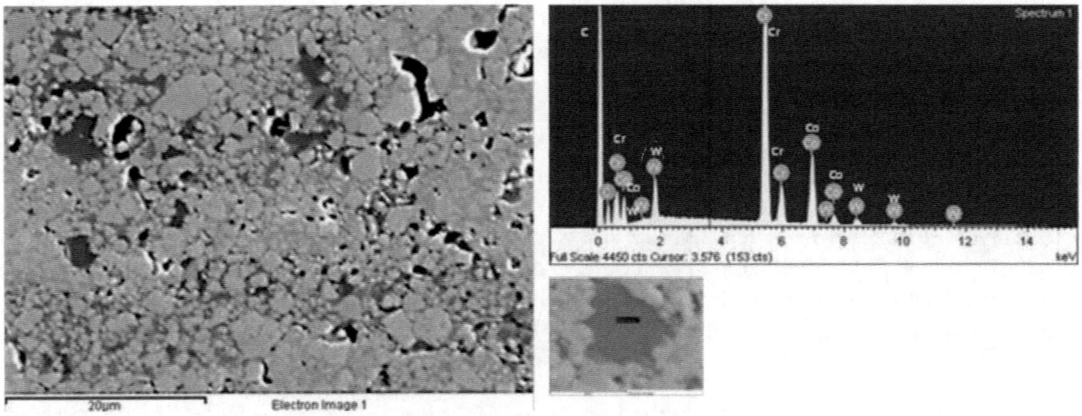


**Figure 3.** SEM/EDS analysis of HVOF coatings.

During decarburization tungsten carbide (WC) is transformed to tungsten semicarbide ($W_2C$) according to the following reactions [17]:

$$2WC \rightarrow W_2C + C$$
$$W_2C \rightarrow 2W + C$$

The XRD pattern of the coating, shown in figure 4, indicates the presence of both WC and the decarburized brittle $W_2C$ phase formed during deposition. These brittle phases have been correlated with reduced toughness and overall wear resistance [24] and are normally distributed in the peripheries of splats, leading to a weak interface between the splats [25]. The formation of hard crystalline phases and tough crystalline Co, without the consumption of a great deal of the WC phase, results in higher wear resistance coatings [26].

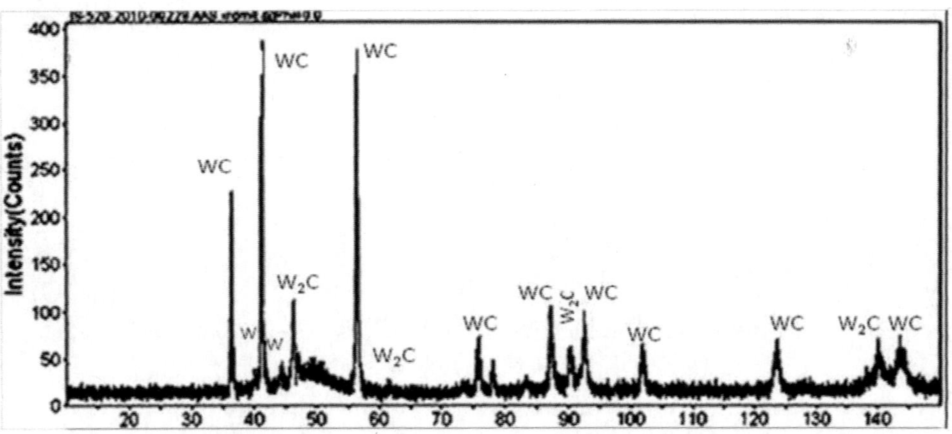


**Figure 4.** XRD diffractogram of the WC-Co-Cr coating.

Figure 5 shows the engineering stress – strain curves for the WC-Co-Cr coated Al substrate with various coating thickness. The presence of the WC-Co-Cr coating results in a lower tensile strength of the coated aluminum in comparison with the tensile strength of the uncoated aluminum. The reduction of the tensile properties may be attributed to the brittle characteristics of the coating, while fracture analysis revealed that brittle cracks, formed in the coating, are the primary cause of the tensile failure of the coated specimens. Tensile tests were conducted for a number of different coating thicknesses. The coated steel substrate with equal WC-Co-Cr coating thickness on both sides can be treated as a laminated composite stressed parallel to the lamellas with the coating and the substrate being the constituents, which determine the elastic and elastic-plastic behaviour.

Since the applied strain in the loading direction is same for all the components of the specimen, an iso-strain condition in the substrate and coating layers is assumed. According to the mixture law for composite materials, following relationship can be given for the present coating system:

$$E_{cs}(h_{Al} + h_{WC}) = E_{WC}h_{WC} + E_{Al}h_{Al} \quad (1)$$

where $E_{cs}$, $E_{Al}$ and $E_{WC}$ are the Young's modulus of the coated system, the substrate (Al) and coating layer (WC-Co-Cr), respectively, and $h_{Al}$, $h_{WC}$ are the thicknesses of the substrate and the coating layers. From the measured uniaxial stress–strain curves, the average Young's modulus of three coated specimens, $E_{cs}$, is 372 GPa. The Young modulus of the substrate is 70 GPa, as measured, and the reported Young modulus of the coating [27] is 310 GPa, the estimated Young's modulus of the coated system, according to equation (1), is 400 GPa, which is in good approximation with the experimental values.

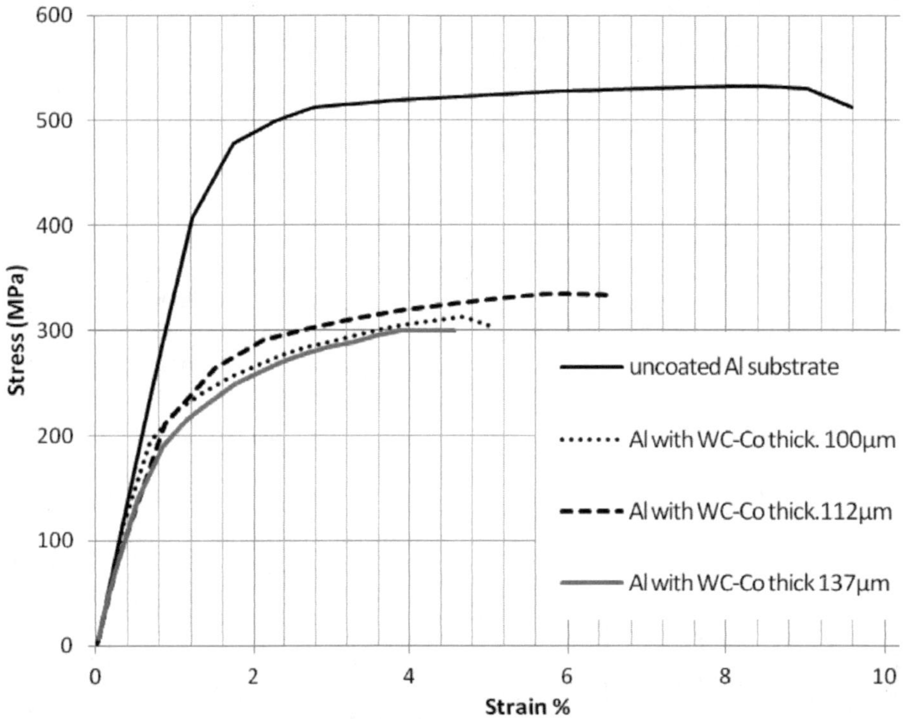


**Figure 5.** Tensile stress–strain curves of HVOF WC-Co-Cr coated Al substrate with various coating thicknesses.

During tensile testing the coating exhibited cracking in the transverse direction of the tensile load. After the saturation of the multiple cracks, the cracks stopped on the interface between the top coat and substrate. Cracks in the coating surface were observed prior to the transition in the plastic region and are attributed to its brittle nature. While tensile testing was deployed, cracks multiplied (figure 6a). Multiplication of transverse cracks in the coating proceeds until the applied strain reaches ~1.6% and then the crack-to-crack spacing rapidly decreases. As tensile strain increased, the initiated cracks propagate transversely towards the coating-substrate in tree like manner. When the initially nucleated cracks approach the interface, they stop and then new cracks form and propagate (figure 6b). The cracks upon reaching the interface deflect into parallel microcracks and propagate causing delamination of the coating especially if the coating thickness is increased. After the cracks propagate and pass through the coating-substrate interface, decohesion occurs just after the transverse crack tips arrive at this interface (figure 6c). The interfacial delamination cracks have been studied in coated systems and are attributed to the linkage of previously formed cracks, are relative to the residual stresses and are associated with the fracture toughness of the coating [28,29]. The coating-substrate adhesion influences the tensile performance of the HVOF WC-Co-Cr coatings and is related to the substrate surface roughness (accomplished by grit blasting) which enables mechanical interlocking of the WC-Co-Cr splats with the substrate surface. Embedment of grit into the substrate or the presence of grit remnants may act as stress raisers and influence interfacial crack propagation and delamination of the coating upon stress induction.

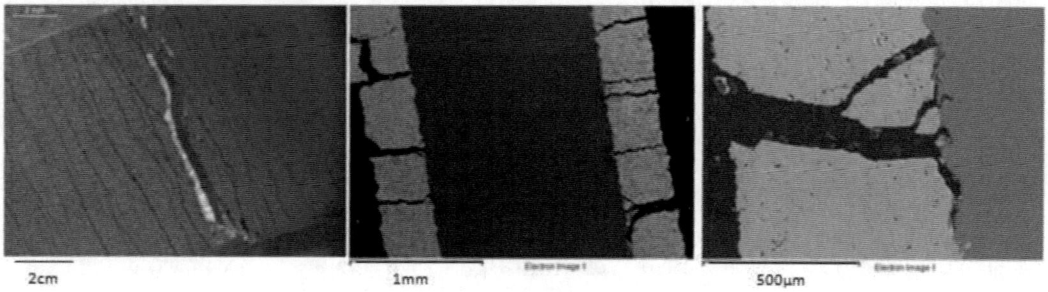


(a) ..... (b) (c)

**Figure 6.** (a) Plan view of the fractured coated system (b) transverse cracks in the cermet coating passing through the interface (c) enlargement of crack tip region in (b).

Figure 7 shows the friction coefficient of WC-Co-Cr coating for the four different loads 1 N, 2 N, 5 N and 10 N as a function of the sliding distance (or equivalently, time). Initially and for the first 250 m approximately the friction coefficient rapidly increases. After this initial running-in period [30], the friction coefficient tends to a constant steady-state value, for all the four loads applied. This constant value was found to be increasing with increasing normal load. More precisely, the steady-state friction coefficient value increases from 0.36 to 0.44 and to 0.55, as the normal load increases from 2 to 5 and to 10 N, respectively. During the application of the 10N load an increase of temperature was observed affecting the ductility of the coating thus leading in higher friction coefficient since as it has been elsewhere reported [31] even a relatively modest temperature increment can have a large effect on the tribological properties. The friction coefficient is in the same range with most literature values [32-34].

The wear volume loss after 10000 m sliding versus the applied load is shown in figure 8. It is evident that wear volume (equivalently wear mass losses) increases with increasing applied normal load following the Lancaster model [35]:

$$V = k\,s\,N \tag{2}$$

where V is the volume wear, s is the total sliding distance, N is the normal load and k is the wear coefficient or specific wear rate.

For relatively low normal load applied (2 N), the pressure at the contact area does not exceed the fracture stress of the ceramic coating; during sliding, friction obeys to the classical mechanical mechanisms leading to higher friction coefficient values due to the high roughness of these coatings. At the level of the contact area, sliding follows the topography of the coating's protrusions and wear takes place through micro-cutting. For high normal load applied (10 N), the pressure at the contact area exceeds the fracture stress of the ceramic leading to lower friction coefficient values, since extensive micro-fragmentation at the level of the coating's protrusions results in an intermediate layer of debris that facilitates sliding. For intermediate normal load applied (5 N), the pressure at the contact area exceeds only locally the fracture stress of the ceramic, e.g. at the level the higher protrusions leading to a mixed friction wear mechanism.

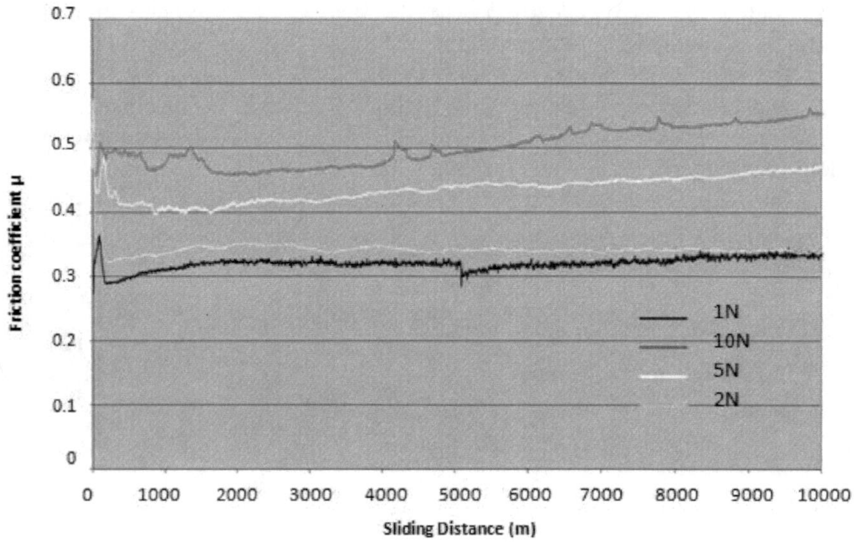


**Figure 7.** Variation of the friction coefficient with the sliding distance for various loads.

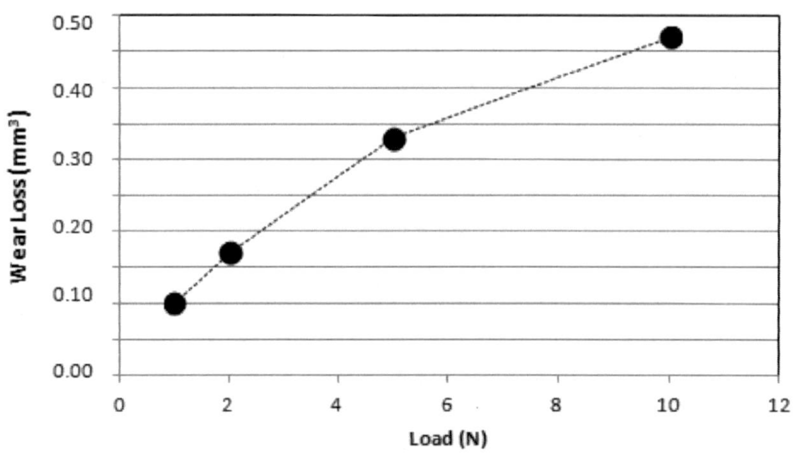


**Figure 8.** Wear volume as a function of applied load.

Stereoscopic analysis showed that the wear tracks become wider with increasing normal load and during sliding material is removed and deposited on the groove edges (figure 9). These are indicative of typical abrasion wear, encountered when a ceramic counterbody (diamond tip) is sliding against an elastoplastic metallic surface (WC-Co-Cr coating) [36].

The WC-Co-Cr coating consists of hard particles into a tough matrix and consequently the wear behaviour is related to the initial removal of the binder phase and the subsequent uprooting and exclusion of the hard WC particles and thus the wear loss can be explained by the delamination from the interface [37]. As a consequence of that the wear debris are trapped between the coating surface and the counterbody. While the wear debris is small, damage along the wear groove is low. Polishing lines along the sliding direction, together with micro-ploughing traces are clearly seen in figure 10a.

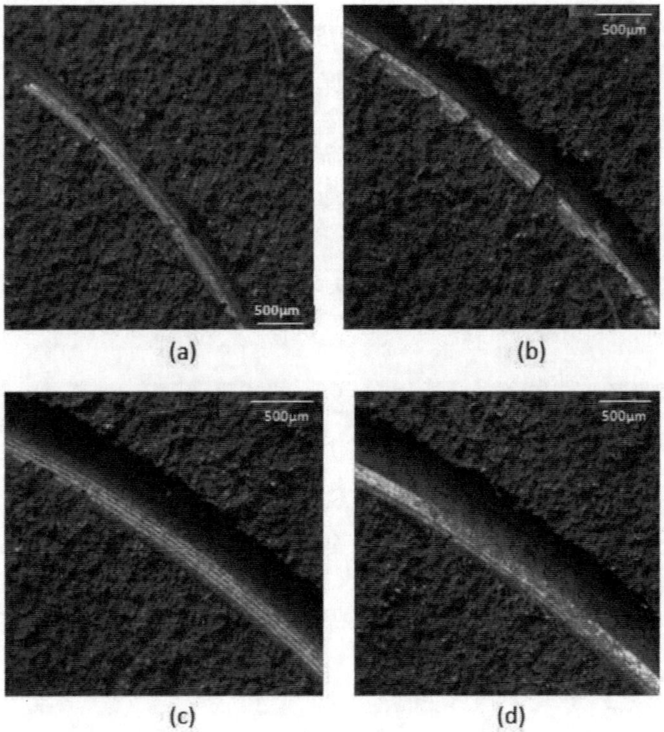


**Figure 9.** Plan view of the wear tracks (a) 1 N, (b) 2 N, (c) 5 N, (d) 10 N.

The decohesion of the metallic matrix led to pull-out of the tungsten carbides, as also observed previously [6], that remained on the worn surface during testing (figure 10b). Further observation of the wear tracks revealed that the coating was worn out through the normally expected material removal via debris creation.

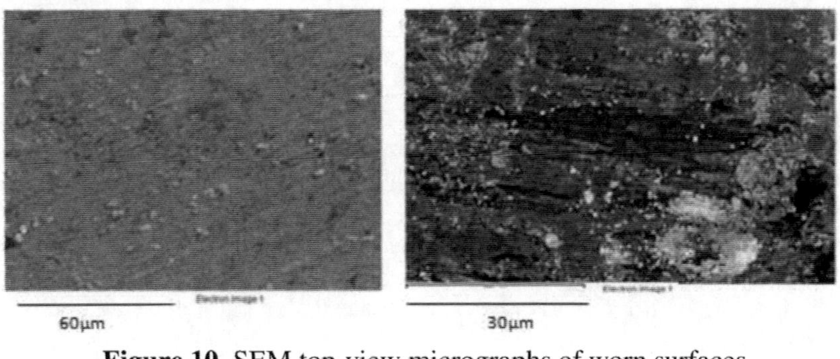


**Figure 10.** SEM top-view micrographs of worn surfaces.

SEM observations on composite coatings cross sections revealed subsurface coating damage as well. Cracks were developed at the wear track boundaries on the contact surface and propagated parallel to the coating surface (figure 11a and b), practically splitting them and resulting in partial decohesion. Such cracks have been reported elsewhere [34, 38] but are found to propagate perpendicular leading to eventual splat exfoliation and consequently to relatively increased wear rates. In this case cracks are localized to surface and were not found to reach the coating-substrate interface. EDX microanalysis showed that the debris layer consists of a mixture of metallic and carbide components occurring from the simultaneous wear of both (figure 11c, d). The carbides particles in the debris act as a third abrasive body, destroying locally the tribo-layer that acts as an in-situ lubricant.

IOP Conf. Series: Materials Science and Engineering **174** (2017) 012062 doi:10.1088/1757-899X/174/1/012062

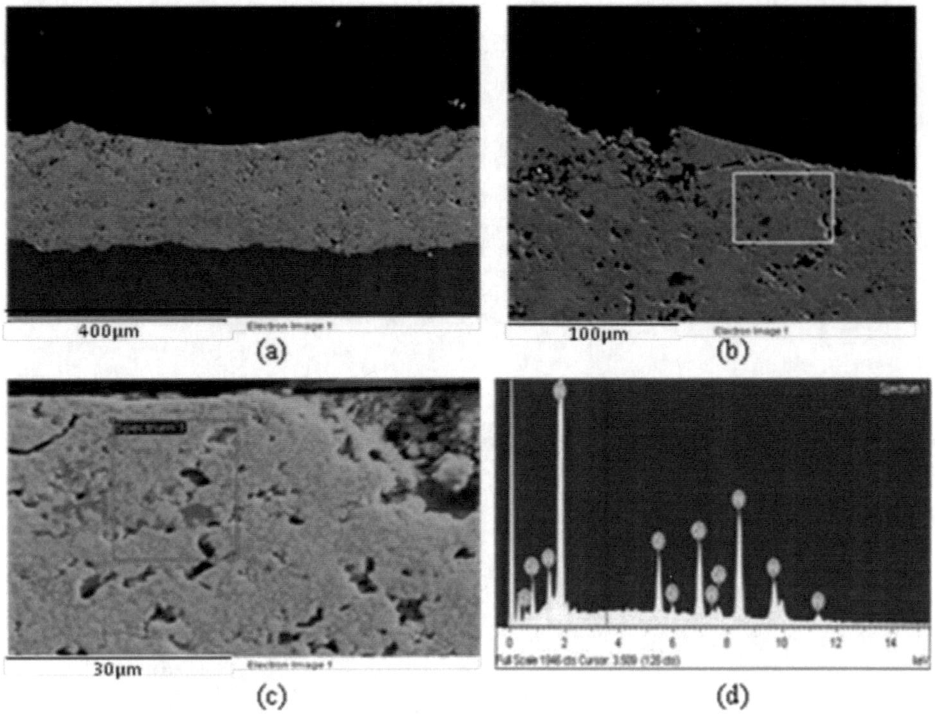


**Figure 11.** SEM cross section of micrographs of the WC-Co/Cr composite coating: (a) Typical view of the entire wear track width; (b) Cracks at the boundaries of the wear track (c) Magnification of relevant marked area in (b) and (d) EDX analysis of marked area in (c).

## 4. Conclusions

1. The tensile strength of the WC-Co-Cr coated aluminum was lower than that of the uncoated one indicating that the presence of the WC-Co-Cr coating reduces the tensile strength. During tensile loading, cracks first initiated in the outer layer of the coating and transversely propagated towards the substrate, and then cracks multiplied and saturated in the coating; finally interfacial decohesion occurred at the coating - substrate interface.

2. The variation of the friction coefficient with the sliding distance followed similar patterns for four loads applied (1, 2, 5, 10N). After an initial running-in period, the friction coefficient tends to a constant steady-state value, for all the cases examined. This steady-state value was found to be increasing with the applied normal load.

3. Wear volume increases with increasing applied normal load. The wear mechanism was micro-cracking of the metal binder leading to tungsten carbide particles pull-out. Cracks were developed at the wear track boundaries on the contact surface and propagated parallel to the coating surface causing partial decohesion.

## Acknowledgments

The authors wish to express their thanks to the Laboratory of ICE and Tribology, Department of Mechanical Engineering, Piraeus University of Applied Sciences (Director Assoc. Prof. P. Psyllaki) for providing the tribology testing facilities.

## References

[1] Davis J R 2004 *Handbook of Thermal Spray Technology* (Ohio: ASM International)

[2] Legg K, Graham M, Chang P, Rastagar F, Gonzales A and Sartwell B 1996 *Surf. Coat. Technol.* **81** 99-105

[3] Bolelli G et al 2015 *Surf. Coat. Technol.* **265** 125-144
[4] Stewart D, Shipway P and McCartney D 2000 *Acta Mater.* **48** 1596-04
[5] Qiao Y, Liu Y and Fischer T 2001 *J. Therm. Spray Technol.* **10** 118-125
[6] Kumari K, Anad K, Bellacci M and Giannozzi M 2010 *Wear* **268** 1309-19
[7] Strachowiak B and Stachowiak W 2010 *Int. J. Refract. Met. Hard Mater.* **28** 95-105
[8] Kekes D, Psyllaki P, Vardavoulias M and Vekinis G 2014 *Tribology in Industry* **36** 375-383
[9] Verdon C, Karimi A and Martin J 1998 *Mater. Sci. Eng., A* **246** 11-24
[10] Venter A, Oladijo O, Luzin V, Cornish L and Sacks N 2013 *Thin Solid Films* **549** 330-339
[11] Morks M, Gao Y, Fahim N, Yingqing F and Shoeib M 2005 *Surf. Coat. Technol.* **199** 66-71
[12] Morks M, Gao Y, Fahim N and Yingqing F 2006 *Mater. Lett.* **60** 1049-53
[13] Sahraoui T, Fenineche N, Montavon G and Coddet C 2003 *Mater. Des.* **24** 309-313
[14] Xie M, Zhang S and Li M 2013 *Appl. Surf. Sci.* **273** 799-805
[15] Fang W, Cho T, Yoon J, Song K, Hur S, Youn S and Chun H 2009 *J. Mater. Process. Technol.* **209** 3561-67
[16[ Hong S, Wu Y, Wang B, Zheng Y, Gao W and Li G 2014 *Mater. Des.* **55** 286-291
[17] Li H, Khor K, Yu L and Cheang P 2005 *Surf. Coat. Technol.* **194** 96-102
[18] Bolelli G, Cannill V, Lusvarghi L, Rosa R, Valarezo A, Choi W, Dey R, Weyant C and Sampath S 2012 *Surf. Coat. Technol.* **206** 2585-01
[19] Al-Mutairi S, Hashmi M, Yilbas B and Stokes J 2015 *Surf. Coat. Technol.* **264** 175-186
[20] Venter A, Oladijo O, Luzin V, Cornish L and Sacks N 2013 WC-Co *Thin Solid Films* **549** 330-339
[21] Xie M, Zhang S and Li M 2013 *Appl. Surf. Sci.* **273** 799-805
[22] Li C, Ohmori A and Harada Y 1996 *J. Mater. Sci.* **31** 785-794
[23] Guilemany J, Dosta S and Miguel J 2006 *Surf. Coat. Technol.* **201** 1180-96
[24] Adamson A W 1982 *Physical Chemistry of Surfaces* (New York: John Wiley) p 404
[25] Wang H, Song X, Liu X, Wei C, Gao Y and Fu J 2012 *Surf. Coat. Technol.* **207** 117-122
[26] Wang Q, Li L, Yang G, Zhao X and Ding Z 2012 *Surf. Coat. Technol.* **206** 4000-10
[27] Watanabe M, Owada A, Kuroda S and Gotoh Y 2006 *Surf. Coat. Technol.* **201** 619-627
[28] Rabiei A and Evans A 2000 *Acta Mater.* **48** 3963-76
[29] Qian L, Zhu S, Kagawa Y and Kubo T 2003 *Surf. Coat. Technol.* **173** 178-184
[30] Zhu S, Guan X, Shibata K and Iwasaki K 2002 *Metall. Mater. Trans. A* **33** 1292-95
[31] Veverkova J and Hainsworth S 2008 *Wear* **264** 518-525
[32] Zois D, Lekatou A and Vardavoulias M 2009 *Surf. Coat. Technol.* **204** 15–27
[33] Cho J, Kim T, Jung Y and Lee K 2009 *J. Ceram. Proc. Res.* **10** 344-350
[34] Kekes D, Psyllaki P and Vardavoulias M 2014 *Tribology in Industry* **36** 361-374
[35] Lancaster J 1967 *Wear* **10** 103-107
[36] Mourlas A, Psyllaki P and Vardavoulias M 2014 *Proc. 8th Intern. Conf. on Tribology*, Sinaia pp 506-512
[37] Lee C, Han J, Yoon J, Shin M and Kwun S 2010 *Surf. Coat. Technol.* **204** 2223-29
[38] Sharma P and Majumdar J 2013 *Metall. Mater. Trans. A* **44** 372-380

IOP Conf. Series: Materials Science and Engineering **174** (2017) 012063 doi:10.1088/1757-899X/174/1/012063

# Study on surface finish of AISI 2080 steel based on the Taguchi method

**S Yalcinkaya**[1] **and Y Şahin**[2]

[1] Department of Mechanical Engineering, Faculty of Technology, Marmara University, Kadikoy, 34722 Istanbul, Turkey
[2] Department of Manufacturing Engineering, Faculty of Technology, Gazi University, Besevler-06500, Ankara, Turkey

E-mail: ysahin@gazi.edu.tr

**Abstract.** Surface finish and dimensional accuracy play a vital role in manufacturing engineering applications. Grinding is one of the most important methods for producing a better surface quality. This paper describes a study of the influences of cutting parameters such as table speed, depth of cut and feed rate on surface finish of AISI 2080 steels, based on the Taguchi (L27) method. The experimental results showed that the table speed was the machining parameter, which had a greater effect on the surface finish, followed by depth of cut, whereas feed rate showed no significant effect. Analysis of variance indicated that a better surface finish was obtained at 190 m/min speed, 0.003 mm depth of cut and 0.08 mm/rev feed rate.

## 1. Introduction

In the manufacturing industry, surface quality is one of the most important parametres in order to evaluate the product quality. A better surface finish is desired in many engineering applications because it minimize friction and wear, improved the fatigue strength, corrosion resistance, precision fits, thereby enhancing the operating hours of machined parts. Obtaining the better surface finish in any machining processes such as lathe, milling, drilling and grinding depends on various parameters like machining parameters, tool and workpiece material properties and cutting conditions. In grinding operation, the surface roughness also depends on cutting speed, feed rate, depth of cut, lubrication, machine vibrations etc. However, an appropriate selection of the grinding parametres is more important [1, 2, 3].

Taguchi designs are used to analyze and optimize the machining parametres because a desirable product quality can be achieved with system and parametric design [4, 5]. Several researchers have conducted on development of matheamatical models of surface roughnesses for different steels such as carbon steel [6], En8 steel [7], En18 steel [8], AISI 1040 [9], AISI 4140 [10, 11] and use of oil effect [12] using a Taguchi technique. There are numbers of studies carried out on different steels by using the statistical method [5-11]. However, no study is performed on grinding of AISI 2080 steel with statistically and experimentally. Therefore, the aim of the present work was to investigate the surface finish of under different girinding conditions. The Taguchi L27 ($3^{13}$) method and ANOVA were adopted to identify the effect of process parameters on the surface finish of tested materials.

## 2. Experimental

### *2.1. Materials*

Chemical composion of AISI 2080 steel is shown in table 1. Its size is about 40x80x120 mm. Due to 11.5%Cr and 2%C content, it has a high thermal stability of tool steel. It is known as higher wear resistance material. Thus, they are prefered to use cutting tool, burr cutting, cutting tool for plastic moulding and paper, deep pressing tool and pressing brace.

**Table 1.** Chemical composions of AISI 2080

| C | Si | Mn | Cr | Mo | Ni | V | W |
|---|---|---|---|---|---|---|---|
| 2.00 | 0.20 | 0.30 | 11.5 | - | - | - | - |

Table 2 indicates standard use of grinding wheel surface and their some properties. A series of parametres were devised to quantify the information from stylus measurement data such as $R_a$ (an average surface roughess/center line average), $R_z$ (an average of maximum peaks to minimum valleys) and $R_q$ (root mean square roughness). Among these parameters, it proves too general to describe the surface's functional nature. Figure 1 shows a Mohr Pertometer types of stylus equipment used for measuring the surface roughness of the steel.

**Table 2.** Standard use of grinding wheel surface.

| Type of abrasives material | SiC-EKR | Silicon carbide |
|---|---|---|
| Grain size | 60 | Medium size |
| Hardness of grinding Wheel | K | Medium hardness |
| Degree of wheel's structure | 6 | Microstructure of wheels |
| Combining materials | V | Ceramic |
| Dimension of grinding wheel | 350*40*127 | Diemeter of wheel/thickness of wheel/diameter of flange |

Taguchi experimental plan with notation of L27 ($3^{13}$) was chosen. This experiment specifies three principal wear testing conditions, including the table speed (*A*), depth of cut (*B*), feed rate (*C*) of tested materials as process parameters. The codes and levels of control parameters are shown in Table 3. Each combination of experiments was repeated three times to acquire a more accurate result. Orthogonal array and analysis of variance (ANOVA) were employed to investigate the influence of process parameters on the grinding of the steel.

**Table 3**. Grinding parametres and their levels

| Symbol | Parameters | Level 1 | Level 2 | Level 3 |
|---|---|---|---|---|
| A | Table speed (m/min) | 190 | 290 | 390 |
| B | Depth of cut (mm) | 0.003 | 0.007 | 0.010 |
| C | Feed rate (mm/rev) | 0.04 | 0.08 | 0.12 |

## 3. Results and Discussion

### *3.1. Analysis of wear results*

The experimental layout and the results of surface roghness of the steels considered are shown in Table 4 and figure 2. The tests relevant to this table were carried out at a fixed material, but with the parameters indicated. It is evident from the figure that the table speed showed a higher effect than the other ones. On the other hand, the surface roughness was maximum for trial 27 due to using the third-level of speed (390 m/min), the third level of depth of cut (0.010 mm) and the third-level of feed rate (0.12 mm/rev).

**Figure 1.** Mohr Pertometer types of stylus equipment used for measuring the surface roughness of the steel.

Figure 2 shows the mean surface roughness as a function of speed, depth of cut, and feed rate. The surface roughness increased more or less linearly with table speed on the tested samples, indicating an improved surface finish at lower speed. This figure also illustrates the surafce finish as a function of depth of cut, whose values were in the range of 0.80 μm to 1.22 μm). As seen, the roughness decreased with lowering the depth of cut. The decrease in the surface finish was more affected by the speed than by the depth of cut. It was observed that, feed rate had a very slight effect on the surface finish.

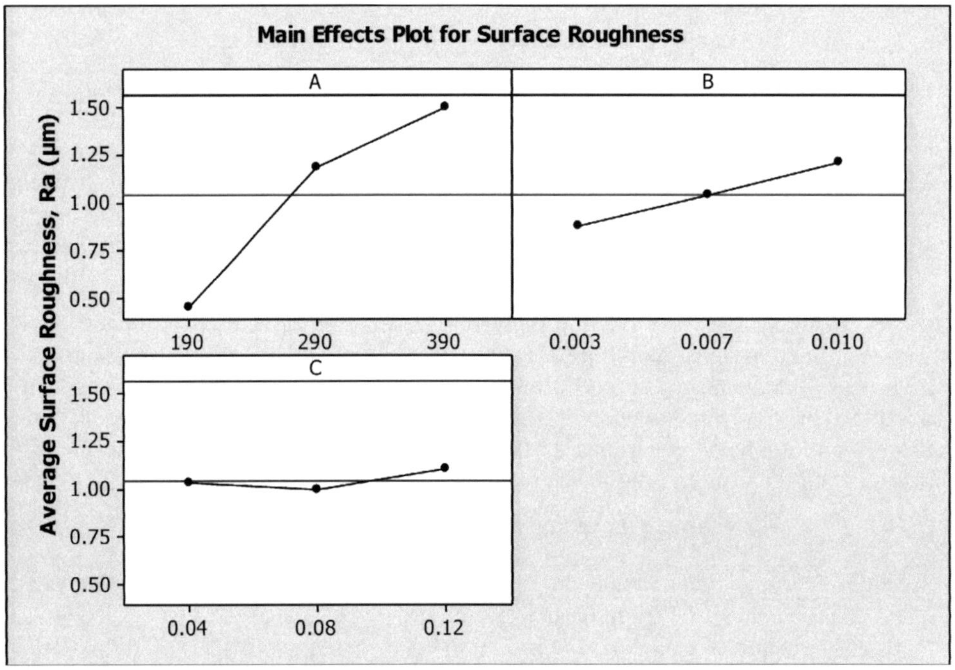


**Figure 2.** Mean surface roughness of steel as functions of factors of table speed (A), depth of cut (B), and feed rate (C).

The relation between the normal probability *Pr* and the residual *R* for the steels tested is shown in figure 3. The distributions of errors was assumed to be normal because it lays down very closely around the line. In general, the residual *R* was lower than ±0.5. However, sometimes the variance of observation grew as the magnitude of observation increased, because the error or noise in the experiment is a constant percentage of the size of observation. Figure 4 indicates a residual vs. fitted value for surface finish of the steel. The residual is mostly cumulated around zero except a few points over that. This also confirms the previous figure.

**Table 4.** Experimental design and results for average surface roughness of steel at grinding process

| Exper. No. | Table speed (m/min) | Depth of cut (mm) | Feed rate (mm/rev) | Mean surface roughness ($R_a$), μm |
|---|---|---|---|---|
| 1 | 190 | 0.003 | 0.04 | 0.379 |
| 2 | 190 | 0.003 | 0.08 | 0.426 |
| 3 | 190 | 0.003 | 0.12 | 0.525 |
| 4 | 190 | 0.007 | 0.04 | 0.283 |
| 5 | 190 | 0.007 | 0.08 | 0.314 |
| 6 | 190 | 0.007 | 0.12 | 0.409 |
| 7 | 190 | 0.010 | 0.04 | 0.436 |
| 8 | 190 | 0.010 | 0.08 | 0.624 |
| 9 | 190 | 0.010 | 0.12 | 0.709 |
| 10 | 290 | 0.003 | 0.04 | 0.966 |
| 11 | 290 | 0.003 | 0.08 | 0.881 |
| 12 | 290 | 0.003 | 0.12 | 1.175 |
| 13 | 290 | 0.007 | 0.04 | 1.398 |
| 14 | 290 | 0.007 | 0.08 | 1.158 |
| 15 | 290 | 0.007 | 0.12 | 1.252 |
| 16 | 290 | 0.010 | 0.04 | 1.454 |
| 17 | 290 | 0.010 | 0.08 | 1.192 |
| 18 | 290 | 0.010 | 0.12 | 1.184 |
| 19 | 390 | 0.003 | 0.04 | 1.155 |
| 20 | 390 | 0.003 | 0.08 | 1.184 |
| 21 | 390 | 0.003 | 0.12 | 1.233 |
| 22 | 390 | 0.007 | 0.04 | 1.621 |
| 23 | 390 | 0.007 | 0.08 | 1.409 |
| 24 | 390 | 0.007 | 0.12 | 1.513 |
| 25 | 390 | 0.010 | 0.04 | 1.634 |
| 26 | 390 | 0.010 | 0.08 | 1.787 |
| 27 | 390 | 0.010 | 0.12 | 1.921 |

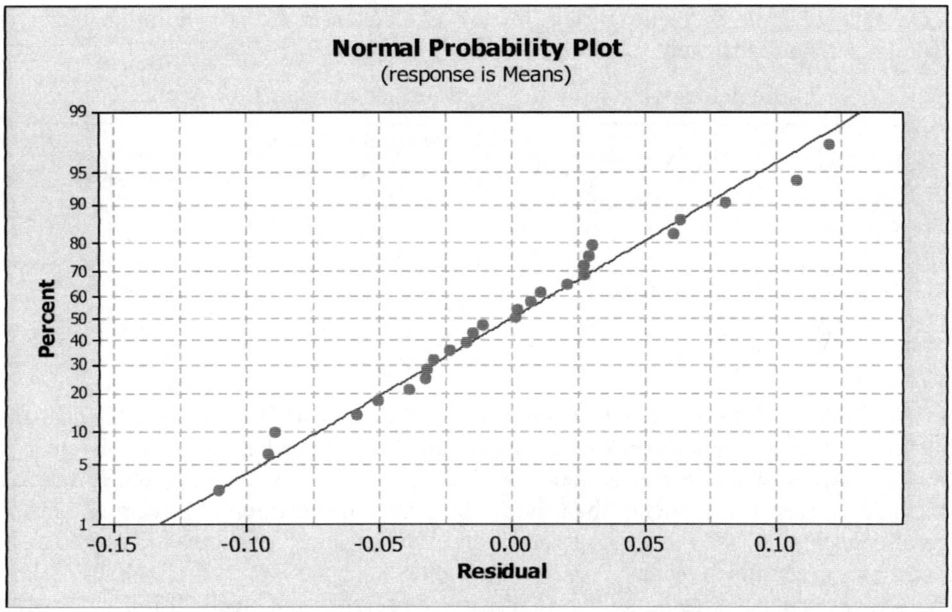


**Figure 3.** The normal probability *Pr* vs. residual *R* of steels.

IOP Conf. Series: Materials Science and Engineering **174** (2017) 012063 doi:10.1088/1757-899X/174/1/012063

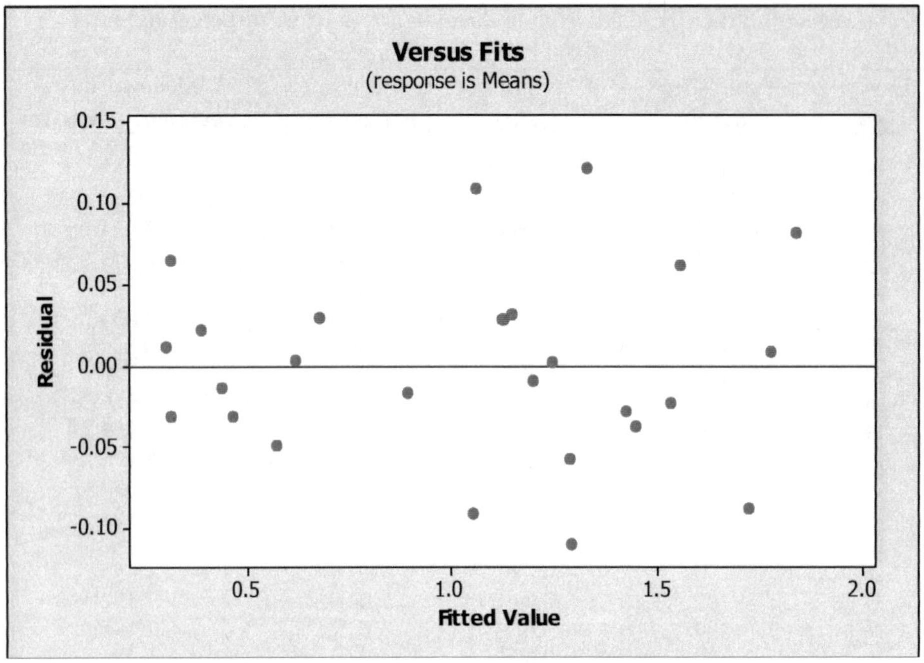


**Figure 4.** Residual vs.fitted value.

Table 5 shows the response for the means of surface roughness of tested steel specimens. The response table can enable one to decide the significance of the parameters of performance characteristics. In the response table, surface roughness are shown for different levels of parameters. According to the calculated values of delta, the parameters were ranked. The control factors were sorted in relation to the values of differences. The strongest influence was found to exerted by the factor *A* (1.039 µm), followed by the factor *B* (0.335 µm). The factor C showed the least effect, since the difference was about 0.105 µm. The minimum surface roughness gave the combination of level 3 of the factor *A*, level 3 of the factor *B*, and level 2 of the factor *C*. The optimum mean values of response for the surface finish were found to be A1, B1, and C2.

**Table 5.** Response table for mean surface roughness of steels.

| Levels | Main machining factors | | |
|---|---|---|---|
| | A (Table speed) | B (Depth of cut) | C (Feed rate) |
| Level 1 | 0.4565 | 0.8809 | 1.0367 |
| Level 2 | 1.1849 | 1.0401 | 0.9976 |
| Level 3 | 1.4957 | 1.2162 | 1.1029 |
| Delta | 1.0391 | 0.3353 | 0.1053 |
| Rank | 1 | 2 | 3 |

### *3.2. ANOVA analysis*

The ANOVA results for the surface roughness of steels are listed in table 6. This analysis was performed for the 5% significance level, that is, for the 95% confidence level. It is clear from the Table 5 that the factors *A* and *B* had a statistical and physical significance for the grinding process, because the value of test-F was higher than the *F*5% value and the value of error was lower than the percentage of contribution (*P*%) of each factor. In the ANOVA table, *P* shows the degree of influence of each factor on surface finish results. As seen, the factors *A* ($P$ = 83.4%), *B* ($P$ = 8.24%), and *C* ($P$ = 0.83%) showed an acceptable level on the rate of surface roughness of steels. It follows from the table that the interaction factor *A*C*, and *B*C* affected the surface roughness of tested specimens negligibly because they are lower than residual values.

**Table 6**. Analysis of variance for the mean surface roughness of the steel.

| Source | DF | Adj SS | Adj MS | Test-F | Table-F (95%) | P | P(%) |
|---|---|---|---|---|---|---|---|
| A | 2 | 5.1200 | 2.5600 | 241.93 | 5.310 | 0.000 | 83.40 |
| B | 2 | 0.5060 | 0.2530 | 23.92 | 5.310 | 0.000 | 8.24 |
| C | 2 | 0.0510 | 0.0255 | 2.41 | 5.310 | 0.152 | 0.83 |
| A*B | 4 | 0.2560 | 0.0642 | 6.07 | 7.708 | 0.015 | 4.17 |
| A*C | 4 | 0.0744 | 0.0186 | 1.76 | 7.708 | 0.230 | 1.20 |
| B*C | 4 | 0.0379 | 0.0094 | 0.90 | 7.708 | 0.509 | 0.62 |
| Residual | 8 | 0.0846 | 0.0105 | | | | 1.54 |
| Total | 26 | 6.1390 | | | | | 100 |

In the present study, the surface finish of steel when applied the grinding process was investigated using the Taguchi approach with L27 orthogonal array. It was revealed that the average surface roughness of the materials tested depended not only on the materials type, hardness, speed and lubrication, but also on the machining parameteres and the type of steel specimens. The optimum surface finish was obtained for the AISI 2080 steels under 190 m/min speed, 0.03 mm depth of cut and 0.08 mm/rev feed rate conditions applied.

## 4. Conclusions

The surface roughness of AISI 2080 steels were studied using a surface grinding machine. The Taguchi robust design method was employed to investigate the different grinding parametres on steel. The following conclusions can be drawn:

1. The average surface roughness decreased with decreasing speed and depth of cut, but no significant changes occurred with feed rate when machined with EKR grinding disc.
2. Among the tested parametres, the best surface finish was obtained when the test was carried out at a speed of 190 m/min, depth of cut of 0.03 mm and feed rate of 0.08 mm/rev conditions for plane grinding the steels.
3. ANOVA analysis results indicated that the table speed was the most significant factor ($P$ = 83.4%) affecting the surface finish of steels, which was followed by depth of cut ($P$ = 8.24%), whereas feed rate (P = 0.83%) had the least effect.

## References

[1] Rudrapati R *et al* 2013 *Advanced Materials Manufacturing & Characterization* **3** 335
[2] Kahraman F 2009 *Materials and Technology* **43** 267
[3] Boothroyd G and Knight WA 2005 *Fundamentals of Machining and Machine Tools* (Boca Raton: CRC Press, Taylor and Francis group)
[4] Roy R K 1990 *A Primer on the Taguchi method* (Dearborn: Society of Manufacturing Engineers)
[5] Rao P V and Agarwal S A 2005 *Int. J. Mach. Tool. Manu.* **45** 609
[6] Malkin S 1976 *Journal of Engineering for Industry* **98** 56
[7] Thakor P S and Patel D M 2014 *International journal of Engineering Development and Research* **2** 2486
[8] Pal K and Shivhare V 2014 *International Journal of Basic and Applied Sciences* **1** 11
[9] Singh K *et al* 2014 *American Journal of Mechanical Engineering* **2** 58
[10] Singh B and Singh B 2014 *IOSR Journal of Mechanical and Civil Engineering*. **10** 61
[11] Bianchi E C *et al* 2004 *Materials Research-Ibero-american Journal of Materials* **7** 385
[12] Şahin Y and Yalcinkaya S 2016 *Key Engineering Materials* **689** 7

IOP Conf. Series: Materials Science and Engineering **174** (2017) 012064 doi:10.1088/1757-899X/174/1/012064

# Increasing in the wear resistance of injection molds made of 1.2343 steel using Ti/TiN/TiCN/nc-TiCN:a-C/nc-TiC:a-C/a-C nanocomposite coating

**Velko Rupetsov[1], Georgi Mishev[2], Stefan Dishliev[3], Viktor Kopanov[4], Vassiliy Chitanov[4], Lilyana Kolaklieva[4] and Chavdar Pashinski[2,4]**

[1] Technical College of Smolyan, PU "Paisii Hilendarski", 28 Dicho Petrov St, 4700 Smolyan, Bulgaria
[2] Technical University of Sofia, branch Plovdiv, 25 Tsanko Diustabanov St, 4000 Plovdiv, Bulgaria
[3] University of Food Technologies, 26 Maritsa Blvd, 4000 Plovdiv, Bulgaria
[4] Central Laboratory of Applied Physics, Bulgarian Academy of Sciences, 61 Sankt Petersburg Blvd, 4000 Plovdiv, Bulgaria

E-mail: stefan_dishliev@hotmail.com

**Abstract**. Injection molds used in production of plastic components are subject of heavy abrasion wear. The increase of their wear resistance significantly reduces the production cost. In the current work are presented research results of the wear resistance of injection molds made of steel 1.2343, coated with Ti/TiN/TiCN/nc-TiCN: a-C/nc-TiC:a -C/a-C. The study of the wear rate was done using the volumetric method and the influence of the trace length was investigated. The coating thickness, nanohardness, elastic modulus and adhesion were also tested. The coating was applied on unhardened ground specimens, hardened ground specimens and hardened polished specimens.

## 1. Introduction

The wear resistance of the tools in machine industry is a basic factor influencing on the productivity and production cost. Also, the wear of the tools results in a loss of accuracy of manufactured goods.

The injection molds work under very heavy conditions: heavy loads; high temperatures; significant wear. The tool's lifespan is directly linked to the wear of the working surfaces. Before tool's elaboration it is important to be known in advance the main factors influencing on the wear and working performance: the annealing depth (for annealed steels), the hardness of the tool's surface and the abrasive characteristics of the molded material (polymer) [1, 2, 3].

In order to increase the resource (life) of the injection molds, which going to lead to significant economic effect, it is necessary to take measures for increasing the hardness of the working surfaces of the tool and improving the wear resistance thereof. About the enhancement of these two properties, the methods for deposition of hard wear-resistant coatings are the most effective. Recently, the coatings obtained by the process of Physical Vapor Deposition (PVD) possess the most widely application [4, 5, 6, 7, 8, 9].

## 2. Aim of the work

This work aims to study the tribological characteristics of one PVD coating: Ti/TiN/TiCN/nc-TiCN:a-C/nc-TiC:a-C/a-C applied on samples of 1.2343 steel (usual material for molds making). To achieve this goal it was necessary to solve the following tasks:

- One manufacturing of test samples of 1.2343 steel and covering them with multilayer nanocomposite coating Ti/TiN/TiCN/nc-TiCN:a-C/nc-TiC:a-C/a-C by PVD method;
- A choice of methods for experimental investigation of tribological characteristics of the test samples with the above mentioned coating;
- A conducting of experimental studies;
- One analysis of results and conclusions.

## 3. Test samples preparation

To carry out the experimental studies, test samples of W. Nr. 1.2343 (DIN X38CrMoV5-1) steel which is used for the manufacture of injection molds were made in the form of a rectangular parallelepiped with dimensions shown in figure. 1.

The samples were prepared as unhardened ground (denoted as 2343 A), hardened ground (denoted as 2343 B) and hardened polished (denoted as 2343 C). After their preparation they were covered with a PVD nanocomposite coating Ti/TiN/TiCN/nc-TiCN:a-C/nc-TiC:a-C/a-C. This coating was deposited by Closed Field Unbalanced Magnetron Sputtering (CFUBMS) industrial unit HVP100RHD in Nanotech Group Ltd., Plovdiv. Coating details could be found in [10]. Data about the test samples are listed in table 1.

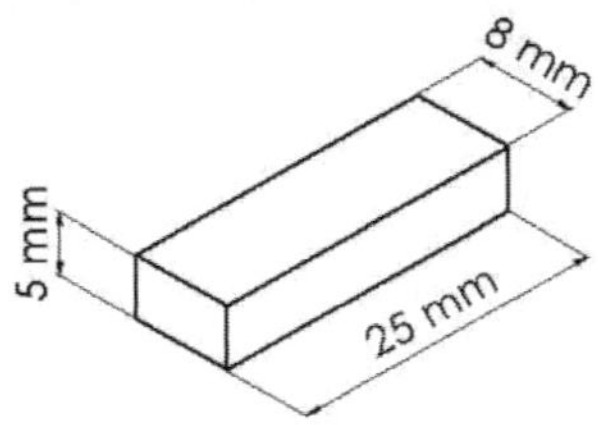


**Figure** 1. Shape and dimensions of the test samples.

**Table 1**. Test samples data.

| Sample | | Coating type | Hardness before deposition | Roughness of bare surface ($Ra$) | Roughness of coated surface ($Ra$) |
|---|---|---|---|---|---|
| **11_2343A** | unhardened ground | carbon-based | 145 HB | 0.191 μm | 0.199 μm |
| **11_2343B** | hardened ground | nanocomposite *Ti/TiN/TiCN/* | 52 HRC | 0.172 μm | 0.180 μm |
| **11_2343C** | hardened polished | *nc-TiCN:a-C/ nc-TiC:a-C/a-C* | 52 HRC | 0.039 μm | 0.040 μm |
| **E23_2343A** | unhardened ground | without coating | 145 HB | 0.191 μm | - |
| **E23_2343B** | hardened ground | | 52 HRC | 0.172 μm | - |
| **E23_2343C** | hardened polished | | 52 HRC | 0.039 μm | - |

## 4. Methods for experimental investigation of tribological characteristics of the test samples

### *4.1. Method for experimental investigation of wear resistance of thin hard coatings*

The experimental studies were conducted by "Ball on Flat Sliding Wear Test" friction scheme in a horizontal orientation of the test surface [11, 12]. One cemented-oxide ball ($Al_2O_3$) with diameter of 3 mm, fixed in holder, was used as a counter-part. It rubs a linear reciprocating sample without lubricant, working in air at room temperature [13, 14]. The width of the arise grooves was measured by a microscope: PC based contactless measurement system TESA VISIO-300 at 100x magnification (resolution: 0.001 mm).

The average width value $b_{cp}$ was calculated:

$$b_{cp} = \frac{1}{n}\sum_{i}^{n} b_i \text{ , mm} \tag{1}$$

where:
$n$ – number of the grooves (traces);
$b_i$ – width of every particular groove.
The wear rate $I_w$ was worked out by the equation:

$$I_w = \frac{V}{F.L}\text{, mm}^3\text{/N.m} \tag{2}$$

where:
$V$ – the wear volume (volume of the trace), mm$^3$;
$F$ – the normal load upon the ball, N;
$L$ - the sliding distance between the ball and sample, m.
The volume of the track was determined by the methodology described in [11].

### 4.2. *Methods for investigation of thickness, nanohardness, elastic modulus and adhesion of the coating*

The thickness of the coating is a key complex indicator and was measured by one calotester elaborated in CLAP-Plovdiv. It was implemented a ball with a diameter of 30 mm and the coating thickness $h$ was calculated by the well-known formula [15, 16]:

$$h = \frac{D^2 - d^2}{8R}.10^{-3}\left[\mu m\right] \tag{3}$$

where:
$D$ - the outer section diameter, mm;
$d$ - the inner section diameter, mm;
$R$ - the ball radius, mm (here: 15 mm).
Nanohardness and elastic modulus of the coating were examined using Compact Platform CPX (MHT/NHT) CSM Instruments in CLAP-Plovdiv. A diamond indenter (Berkovich type) was used and the results were interpreted by the Oliver and Pharr method [17,18].

The adhesion was tested by Micro Scratch Tester (MST) module included in the same apparatus. A diamond indenter (Rockwell type) with rounded apex of 200 μm was used [19].

## 5. Experimental researches and results

### 5.1. *Investigation of influence of the sliding distance on the wear rate*

Table 2 contains summary data about the volume of wear traces over test samples as a function of sliding distance $V=f(L)$ in the respective regimes.

**Table 2**. Volume of traces as a function of sliding distance $V=f(L)$.

| Sliding distance $L$, m | **11_2343A** $10^{-6}$, mm$^3$ | **11_2343B** $10^{-6}$, mm$^3$ | **11_2343C** $10^{-6}$, mm$^3$ | **E23_2343A** $10^{-6}$, mm$^3$ | **E23_2343B** $10^{-6}$, mm$^3$ | **E23_2343C** $10^{-6}$, mm$^3$ |
|---|---|---|---|---|---|---|
| **50** | 901.729 | 418.526 | 29.532 | 7247.592 | 4713.143 | 4528.021 |
| **75** | 940.461 | 868.726 | 72.978 | 9673.319 | 8790.414 | 8232.896 |
| **100** | 1583.903 | 1000.620 | 77.524 | 9885.18 | 9580.145 | 9603.382 |

Diagrams of the volume of wear traces of coated and reference (uncoated) samples in the respective regimes are shown in figures 2÷5.

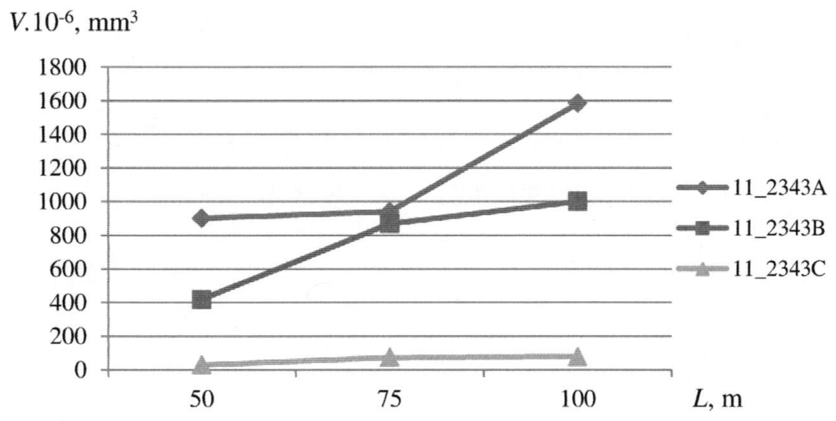


**Figure 2**. Graphical dependence of the volumes of wear traces of coated samples as a function of sliding distance $V=f(L)$.

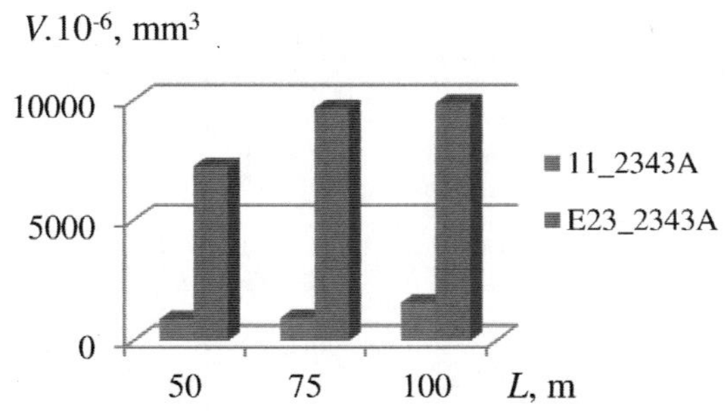


**Figure 3**. Graphical dependence of the volumes of wear traces of coated and reference (uncoated) samples made of unhardened ground 1.2343 steel as a function of sliding distance $V=f(L)$.

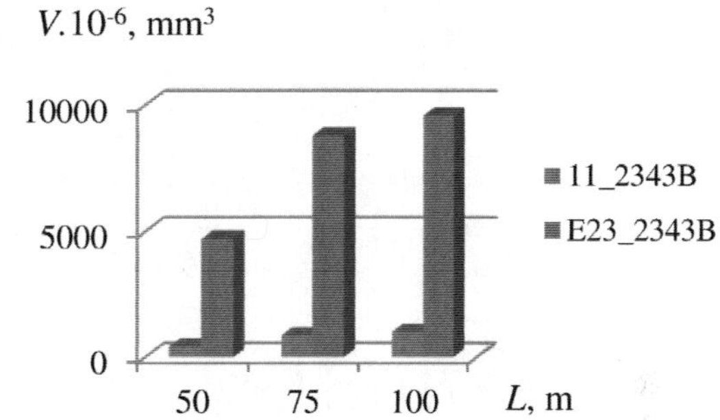


**Figure 4**. Graphical dependence of the volumes of wear traces of coated and reference (uncoated) samples made of hardened ground 1.2343 steel as a function of sliding distance $V=f(L)$.

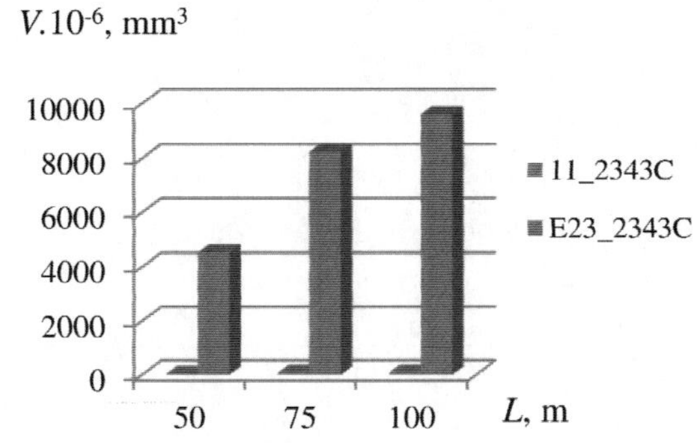


**Figure 5**. Graphical dependence of the volumes of wear traces of coated and reference (uncoated) samples made of hardened polished 1.2343 steel as a function of sliding distance $V=f(L)$.

Table 3 contains summary data about the wear rate as a function of sliding distance $I_w=f(L)$ in the respective regimes.

**Table 3**. Wear rate as a function of sliding distance $I_w=f(L)$.

| Sliding distance *L*, m | **11_2343A** $10^{-6}$, $mm^3/N$ m | **11_2343B** $10^{-6}$, $mm^3/N$ m | **11_2343C** $10^{-6}$, $mm^3/N$ m | **E23_2343A** $10^{-6}$, $mm^3/N$ m | **E23_2343B** $10^{-6}$, $mm^3/N$ m | **E23_2343C** $10^{-6}$, $mm^3/N$ m |
|---|---|---|---|---|---|---|
| **50** | 18.035 | 8.370 | 0.591 | 144.952 | 94.263 | 90.560 |
| **75** | 12.539 | 11.583 | 0.973 | 128.978 | 117.206 | 109.772 |
| **100** | 15.839 | 10.006 | 0.775 | 98.852 | 95.801 | 96.034 |

Diagrams of the wear rate as a function of sliding distance $I_w=f(L)$ are shown in figures 6÷9.

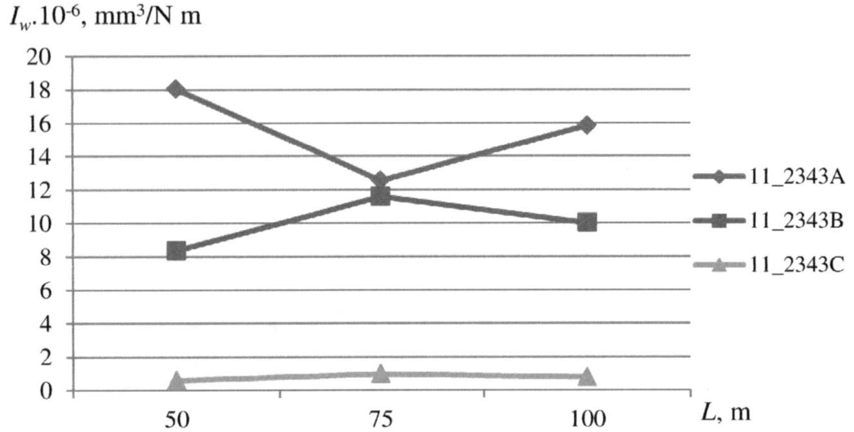


**Figure 6**. Graphical dependence of the wear rate of coated samples as a function of the sliding distance.

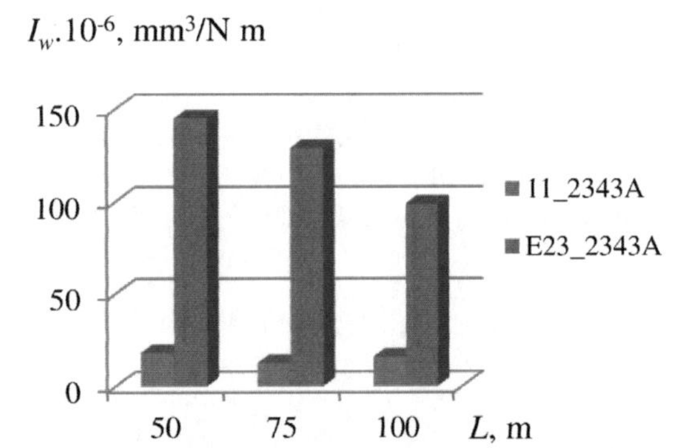


**Figure 7**. Graphical dependence of the wear rate of coated and reference (uncoated) samples made of unhardened ground 1.2343 steel as a function of sliding distance $I_w=f(L)$.

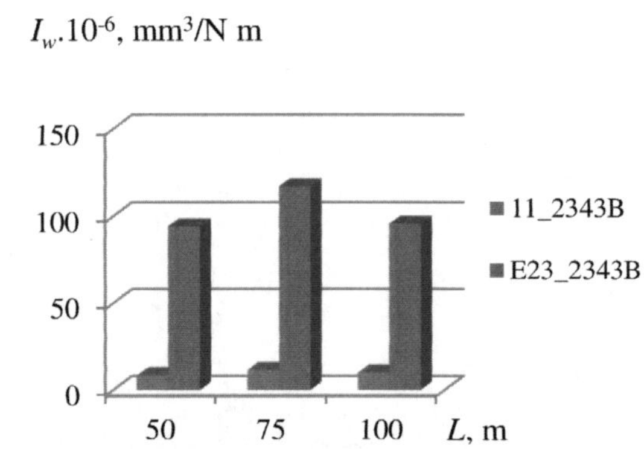


**Figure 8**. Graphical dependence of the wear rate of coated and reference (uncoated) samples made of hardened ground 1.2343 steel as a function of sliding distance $I_w=f(L)$.

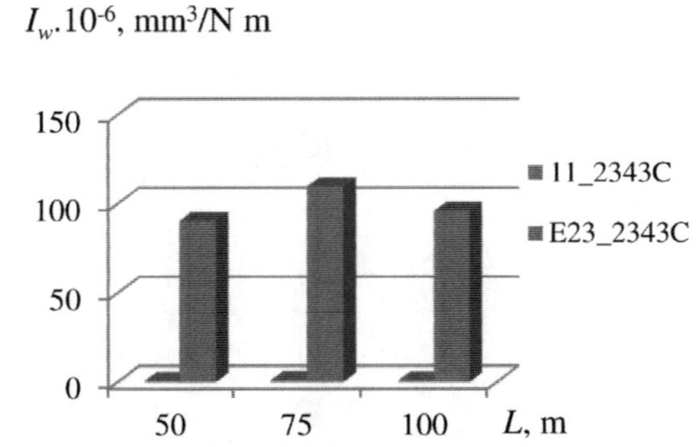


**Figure 9**. Graphical dependence of the wear rate of coated and reference (uncoated) samples made of hardened polished 1.2343 steel as a function of sliding distance $I_w$=*f(L)*.

Experimental results indicate that the sliding distance influences on the wear of the carbon-based nanocomposite Ti/TiN/TiCN/nc-TiCN:a-C/nc-TiC:a-C/a-C coating, thereby increasing the sliding distance causes a wear increase also. The wear rate of this coating applied on a hardened polished surface is average 12.8 times less than that on a hardened ground surface and average 19.8 times less than that on an unhardened ground surface.

*5.2. Investigation of thickness, nanohardness, elastic modulus and adhesion of the coating*

On the base of the calotester measurement, the calculated thickness is 2.32 µm. A nanoindentation curve of the test sample 11_2343C is shown in figure 10. These parameters were specified: nanohardness $H$ = 18 GPa; elastic modulus $E$ = 320 GPa; maximal penetration depth $h_m$ = 309 nm (at maximal indentation load of 30 mN). The maximal penetration depth is less than 15% of the entire coating thickness (2.32 µm), which guarantees that the measured nanohardness is related mostly to coating without a substrate impact.

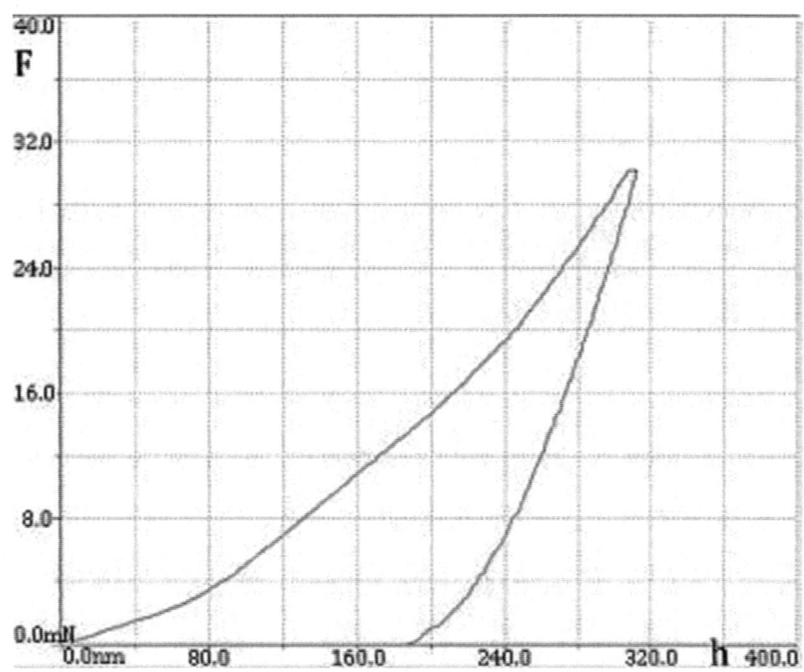


**Figure 10**. Nanoindentation load/unload curve of sample 11_2343C: Load $F$ versus indenter displacement $h$.

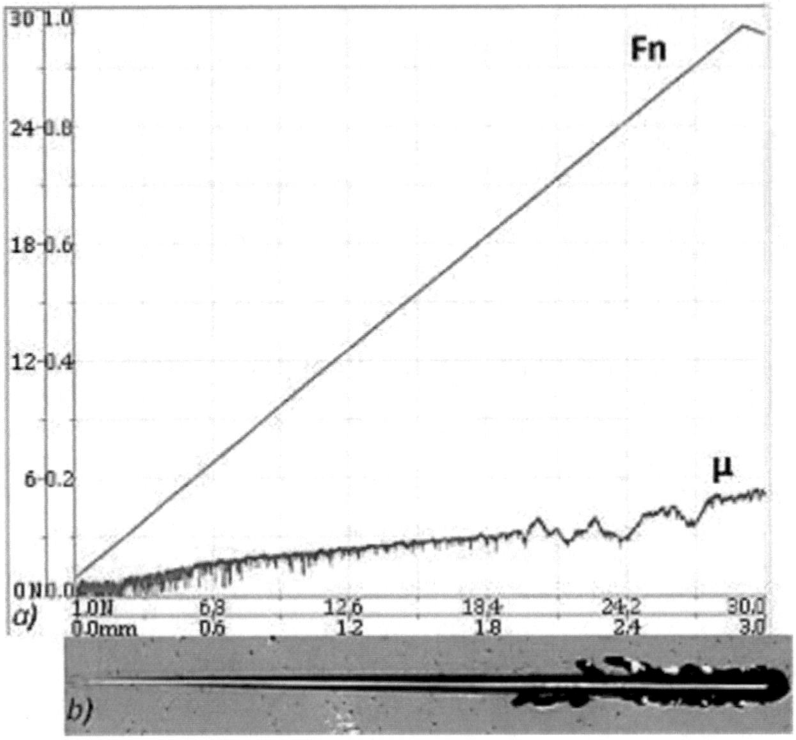


**Figure 11**. Scratch test of sample 11_2343C: a) curves of the applied force *Fn* and friction coefficient $\mu$ as a function of scratch distance *s*; b) view of the trace.

Figure 11a shows the results obtained by the adhesion test (scratch distance of 3 mm). The force *Fn* which presses the indenter was changed linearly to a maximum value of 30 N (the maximum load allowed by this equipment). The first obvious damages in the coating were originated by applied force of ca. 19 N (figure 11b and curve of $\mu$ in figure 11a). These damages are internal (cohesive) and their appearance marks the first critical load: Fc1. Since a delamination of the coating (an exposure of the substrate) in the channel is not noticed, it is proven that the second critical force Fc2 is not reached up to maximal applied force (30 N).

## 6. Conclusions

- The discussed carbon-based nanocomposite Ti/TiN/TiCN/nc-TiCN:a-C/nc-TiC:a-C/a-C coating does not change the primary surface roughness. This is mainly due to a lack of droplets during the coating deposition when using the methods of Magnetron Sputtering (MS). The appearance of droplets is a major shortcoming of other PVD methods: Vacuum (Cathodic) Arc Deposition (VAD, CAD), Pulsed Laser Deposition (PLD), etc. which increases the roughness;
- The quality of the surfaces on which a carbon-based nanocomposite Ti/TiN/TiCN/nc-TiCN:a-C/nc-TiC:a-C/a-C coating is deposited, has a significant impact on the wear of the coating. When surface roughness decreases and/or hardness increases, the wear rate of the coating is reduced;
- The calculated wear rate of the considered carbon-based nanocomposite Ti/TiN/TiCN/nc-TiCN:a-C/nc-TiC:a-C/a-C coating deposited on a hardened polished surface is average 7 ÷ 11 times less than that on a hardened ground surface and average 16 ÷ 20 times less than that on an unhardened ground surface;

- The measured hardness and adhesion show a coating with a relative lower hardness (on the edge of conventional limit for hard coatings: 20 GPa) and a passable adhesion. This low hardness shows the apparent presence of graphite phase in the coating, which phase, however, favors a low coefficient of friction. The relatively weak adhesion is probably due to the significant stress in the coating. It is mainly due to the differences in the nature and parameters of the bonds Ti-N and Ti-C and such stress is often observed in coatings based on TiCN.

**References**

[1] Rees H 2002 *Mold Engeneering* (Munich: Hanser Publishers)

[2] Osswald T A, Turng L-S, and Gramann P J 2008 *Injection Molding Handbook* (Munich: Hanser Publishers)

[3] Gastrow H, and Unger P 2006 *Gastrow injection molds – 130 Proven Designs* (Munich: Hanser Publishers)

[4] Pashinski Ch, Angelov M, Rupetsov V, Petrov D, Shindov P and Dishliev S 2014 Arc bond sputtering equipment for deposition of innovative industrial coatings *Proc. of the IV Int. Conf. Industrial Engineering and Environmental Protection 2014 (IIZS 2014)* (Zrenjanin, Serbia) 356-361

[5] Jones M.I. McColl I.R. Grant D.M. 2000 Effect of substrepresentation preparation and deposition conditions on the preferred orientation of TiN coatings deposited by RF reactive sputtering *Surf. Coat. Tech.* 143-151

[6] Jung M J, Nam K H, Shaginyan L R and Han J G 2003 Deposition of Ti thin films using the magnetron sputtering method *Thin Solid Films* **435** (1) 145-149

[7] Kakaš D, Škorić B, Miletić A, Terek P, Vilotić M, and Kovačević L 2011 Influence of dry sliding conditions and coating roughness on the wear of titanium nitride coatings deposited at low temperature *Proc. of the 7th Intern. Conf. on Tribology "BALKANTRIB'11", 3-5 October, Thessaloniki – Greece*

[8] Veprek S, Veprek-Heijman M G, Karvankova P and Prochazka J 2005 Different approaches to superhard coatings and nanocomposites *Thin Solid Films* **476** (1) 1–29

[9] Zhang S, Sun D, Fu Y and Du H 2003 Recent advances of superhard nanocomposite coatings: a review*Surf. Coat. Tech.* **167** 113-119

[10] www.nanotechgroup.dir.bg

[11] Dishliev S. 2013 Metodika za eksperimentalno izsledvane iznosoustoichivostta na tanki tvardi iznosoustoichivi pokritiq *Dvadeset i vtora mejdunarodna nauchno – tehnicheska konferencia "ADP" Sozopol*

[12] ASTM G133 – 05 (2010) *Standard Test Method for Linearly Reciprocating Ball-on-Flat Sliding Wear* available at http://www.astm.org/Standards/G133.htm

[13] Levashov E. A. 2010 Obespechenie edinstva izmerenii mehanicheskih I tribologicheskih svoistva nanostrukturirovannaeh poverhnostei *MIS iS – ISMAN*, Moskva

[14 www.csm-instruments.com

[15] www.nanovea.com

[16] www.platit.com

[17] Oliver W.C. Pharr G.M. 1992 An improved technique for determining hardness and elastic modulus using load and displacement sensing indentation experiments *J Mater. Res.* *7* (06), 1564-1583

[18] Sveshtarov P. et al. 2007 Microhardness Measurement by Indentation of TiN, Ti(CN), ZrN, (ZrTi)N Hard Coatings *Intern. Workshop NANOHARD* 49-54

[19] www.nanoscan.info

# A method for determining the thickness of tribological performing thin layers formed by selective transfer

**Filip Ilie[1], Georgiana Chisiu[1] and George Ipate[2]**

[1] Department of Machine Elements and Tribology, POLITEHNICA University of Bucharest, Spl. Independentei 313, 060042 Bucharest, Romania
[2] Department of Biotechnical Systems, POLITEHNICA University of Bucharest, Spl. Independentei 313, 060042 Bucharest, Romania

E-mail: iliefilip8@gmail.com

**Abstract.** A new stage in the research of the unconventional friction couples (alloys or pseudo-alloys in thin layers) to implement them in the designing and execution of machines is represented by the modern friction couples which are based on selective transfer (transfer of a material from one element of the friction couple to the other in the presence of a lubricant forming a superficial layer, antifriction, very thin, the order of several microns, which behaves very well to friction and wear). A selective transfer can be achieved with certainty in a friction couple, lubricated with glycerine or with a special lubricant, if in the friction area there is a material from alloys on based copper. The thin superficial layer formed through selective transfer in the friction process of a friction couple is made of the elements of the alloy based on copper, where the copper is predominant. Hence results the practical necessity to determine the thickness of superficial thin layers (0.1 - 4 μm) obtained in the friction couples, by selective transfer (mass selective transfer through diffusion from one element of the friction couple to another, in conditions of local energies favourable to the transfer process and in the presence of relative motion). The aim of this paper is presenting and explaining a methodology for determining the thickness of layers formed by selective transfer, in the friction process, on the surfaces of elements friction couples.

## 1. Introduction

By bombarding a thin object with a focused beam of electrons, a part of the electrons passes through the object or disperses without loss of energy. The other part of the electrons lose partially or wholly the energy and afterwards are either leaving from the object, or are absorbed by the material of the object.

Electrons beam energy is consumed at the ionisation or the excitation of the atoms and molecules of the object material.

Since the early 20th century, the braking law of the electrons was known, and described through a simple exponent known under the name Lenard's law [1].

For the analysis of the thin layers, it is required to take into consideration this law, in order to present as accurately as possible the processing mechanism of the electrons interaction with the thin layers. Viatskin showed in [1] through an experiment as the absorption curve of electrons with an energy of 0.4 to 500 keV has a character, not of simple exponent, but much more complex, mathematically expressed by the following relationship:

$$I = I_0 e^{\mu x q(z)}, \quad (1)$$

where: $I$ - is the intensity of X-rays generated by the substrate, in the pulses per sec; $I_0$ - intensity of X-rays generated by the base material, in the pulses per sec; $\mu$ - linear absorption coefficient in $m^{-1}$; $x$ - thickness of the object after the direction of dispersion radiation, in m; $q(z)$ - coefficient that depends on the order number of the elements composing the object, with values between $5 < q < 30$ (for beryllium $q = 5$, and $q = 27$ for gold).

The relation (1) allows to specify the average route of electrons in the substance of the object and therefore, we can get closer to the excitation area of X-rays in the object [1-5] with greater accuracy. Therefore this paper presents a practical method for determining the thickness of superficial thin layers (0.1-4 μm) obtained in the friction couples and which is functioning with selective transfer [6-12].

A practical possibility for measuring the thickness of thin layers consists in the focusing of the incident radiation of electrons beam in the thickest area, respectively thinnest area of the selective layer, its penetration and reading the number of pulses per second, i.e. $I$ for each case, and $I_0$ for basic material - steel [13-15]. The penetration depth, $x_m$ of the electrons beam is the dependence upon the atomic mass, $A$ of the element, the order number $z$ of the elements composing the object, the density, $\rho$ of the sample, the tension of acceleration $U$, according to of the relationship [6, 8]:

$$x_m = 1.1 \cdot 10^{-13} \frac{A \cdot U^2}{z \cdot \rho} \text{ [mm]}, \quad (2)$$

and $x$ - the thickness of the object after the direction of dispersed radiation, results from relation (1).

The moment in which the electrons had penetrated in the layer is determined by the apparition of the dispersed radiation, as the answer of the substrate.

The principal inconveniences of this method are:

- there is the possibility that the thickness of the layer could be so small, that the tension for which the penetration occurs may be of the critical potential for the excitation of the substrate;
- there is the possibility that the absorption in the layer may be very high, for the wavelength of the X-radiation, in order for the substrate to answer, and the radiation can't penetrate the layer;
- the precision of the measurement of the acceleration tension is pretty diminished.

## 2. Description of the method

An alternative method to determine the thickness of thin superficial layers was using an own method for assessing the thickness, by quantitative investigations. By this method, the investigations have been performed using an electron microprobe (type JXA – 5A JEOL), a device capable of such investigations. Electronic microprobe it can also use for analysis the composition of the superficial thin layers which are obtained in conditions of a selective transfer and can detect elements that are in concentrations of $10^{-3}$ % [8]. With the help of electron microprobe, steel samples (OL37 which is similar to the european/international steel, S235JR) on which there is transferred a thin superficial layer in the conditions of selective transfer, in the areas of contact with the samples of brass (based on aluminium), were bombarded with an electrons beam.

The layer thickness has been measured with a proper precision (± 10 Å) using the special device of the installation.

The microprobe is able to read the number of pulses per sec through the electronic display. Thus, it recorded the number of pulses for the $K_\alpha$ radiation of iron emitted by a clean area (where there is no transferred layer), respectively the number of pulses in the same period of time, for the same radiation, emitted by an area on which there is transferred layer [8, 9]. Using, the absorption law of radiations $K_\alpha$ of iron, for a layer with complex composition (layer deposited by selective transfer), given by relation (1), we managed to assess the thickness of these layers [8, 9].

The incident radiation of the electron beam was focused in the thickest area, respectively the thinnest area of the selective layer, by reading the number of pulses/sec, i.e. $I$ for each case, and $I_0$ for basic material from steel (figure 1). For each area there are three records in order to establish $I$ and $I_0$, as an average value. Must be mentioned that the areas thickest and most thin of the transferred film are chosen by many measurements of the thickness in the same area, marked on the surface of the sample.

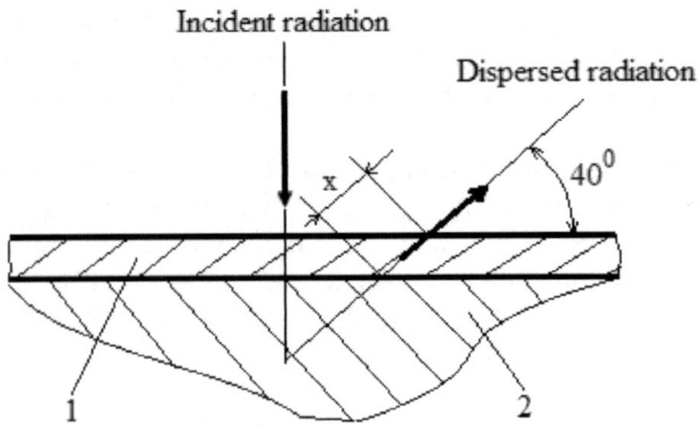


**Figure 1**. Schematic diagram for determining layer thickness formed by selective transfer with the help of microprobe electron: 1 - layer formed by selective transfer; 2 - base material.

## 3. Experimental results

For most samples (from aluminium bronze/S235JR; chemical composition of aluminium bronze is: about 83.5% Cu, 10% Al, 35% Zn, 2% Fe, 2% Mn, 0.3% Pb, 0.1% Sn, 0.05% As) tested on Timken tribometer to determine the thickness of selective layer after 2 hours of operation by friction at speeds and different loads, lubricated with glycerine and water, was followed and the time evolution of the selective layer thickness for different loads and the same speed, at different speeds and for the same load, is shown in figure 2 a, b.

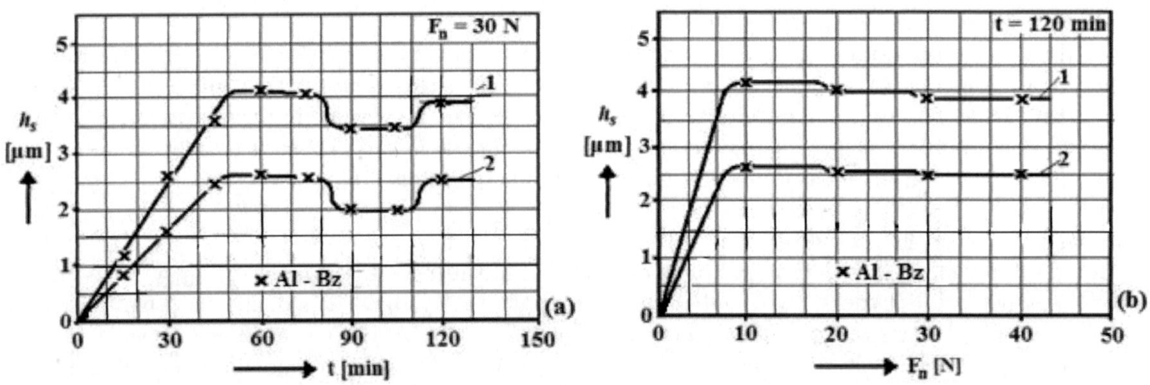


**Figure 2.** Variation of selective layer's thickness $h_s$ function of the friction duration **t** at constant load (a); function of load $F_n$ and constant friction time (b) of aluminium bronze at different sliding speeds (curves 1 to 0.93 m/s, curves 2 to 1.86 m/s).

It must be mentioned that the measurement was made after tribological tests and tribological conditions were used only to obtain the transfer films.

Images of the secondary electrons in figure 3 shows the topography of the surface of steel samples, on which was transferred a thin superficial layer of copper in the aluminium bronze in the conditions of selective transfer, based on which one can appreciate the uniformity of the layer. In figure 3 are showed the profiles of the variation of the concentration of copper (in the bronze aluminium composition) and iron (in the steel composition) on the surface areas that the film was transferred in the conditions of selective transfer after the direction marked with a white line.

The bottom half of the profiles shown in figure 3 represents the variation in the concentration of Fe in the base material (steel), in areas where the film was deposited by selective transfer in the friction process, and the upper part (of the profiles) represents the concentration variation Cu, in the film obtained by selective transfer.

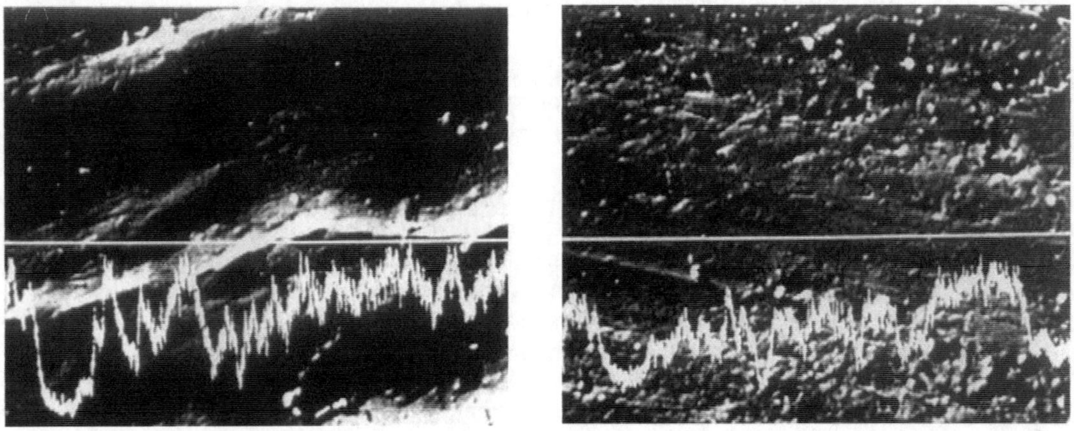

(a) (b)

**Figure 3**. Images of the secondary electrons x 300 at the couple aluminium bronze/steel OL37 lubricated with glycerine: for v = 0.93 m/s (a), to v = 1.86 m/s (b).

If $I$ and $I_0$ are read to determine the thickness transferred layer $x$ (figure 3), we must know the value $\mu$ of the absorption coefficient, which is determined as follows: in the literature [16-22] is given the ratio $\mu/\rho$ [$m^2/kg$], called mass absorption coefficient (table 1).

**Table 1**. Parameter values needed to calculate the mass absorption coefficient.

| Alloy | Element | $z$ | $c_i$ | $\mu_i/\rho_i$ | $\rho_i$ | $\mu/\rho$ | $\rho$ | $\mu$ | $q(z)$ |
|---|---|---|---|---|---|---|---|---|---|
| | Cu | 29 | 85.00 | 7.98 | 8930 | | | | |
| | Al | 13 | 9.00 | 7.34 | 2710 | | | | |
| | Zn | 30 | 2.00 | 8.85 | 7140 | | | | |
| Aluminium bronze | Fe | 26 | 2.00 | 5.95 | 7870 | 8.562 | 8360 | $7.16 \cdot 10^4$ | 9.9 |
| | Mn | 25 | 1.60 | 43.1 | 8140 | | | | |
| | Pb | 82 | 0.25 | 35.4 | 11340 | | | | |
| | Sn | 50 | 0.10 | 38.2 | 7310 | | | | |
| | As | 33 | 0.05 | 12.1 | 5700 | | | | |

In the case alloys, $\mu/\rho$ is calculated with the relation:

$$\frac{\mu}{\rho} = \sum_{i=1}^{n} c_i \frac{\mu_i}{\rho_i}, \tag{3}$$

in which: $c_i$ - is the concentration of the alloying element $i$, in %; $\mu_i$ - the coefficient of absorption of the element $i$, in $m^{-1}$; $\rho_i$ - the density of the alloying element $i$, in $kg/m^3$.

From experimental results and from the picture above (figure 3) it can be observed the uniformity of the format layer through a selective transfer. Figure 3, respectively, represents the image of the

IOP Conf. Series: Materials Science and Engineering **174** (2017) 012065 doi:10.1088/1757-899X/174/1/012065

secondary electrons increase by 300, where it can be observed the profiles of variation of the Cu and Fe concentration from the surface of the zone where it had been realized the transfer of a thin superficial layer in condition of a selective transfer after a direction marked by the white line.

The image of secondary electrons presents the relief of steel sample on which it had been transferred a thin superficial layer of Cu from the bronze which had functioned in the condition of the selective transfer and it can appreciate the uniformity of the layer. The inferior part below the profiles reproduced in figure 3 represents the variation of the Fe concentration from the base material (steel) in the zone where it has been deposed the superficial layer through the selective transfer in the friction process, and the superior part of the profiles represents the variation of the Cu concentration from the obtained layer through the selective transfer.

The more luminous zones are details of the surface bulged out, and the darkness zones are depressions in the surface of the sample. The results obtained by this method (figure 2) are in agreement with those determined in the previous method.

A quantitative assessment of selective layer thickness shows the increase in time until it reaches a maximum (optimum), after that the transfer occurs in the reverse, remaining in function of the friction duration in the range of 0.1- 4 μm, but varies much more.

Layer thickness can be reduced 10-fold and to increase again. These changes in thickness correspond to observations at samples from steel OL37, which has worked together with the alloys on base copper at stress through friction, on which it is formed a layer of copper, which is related to the transfer of copper on the steel surface.

Figure 4 shows the transferred layer, in depth, with a porous structure, clearly delimiting the layer 1, interface 2 between layer 1 and the structure of the base material 3 (steel).

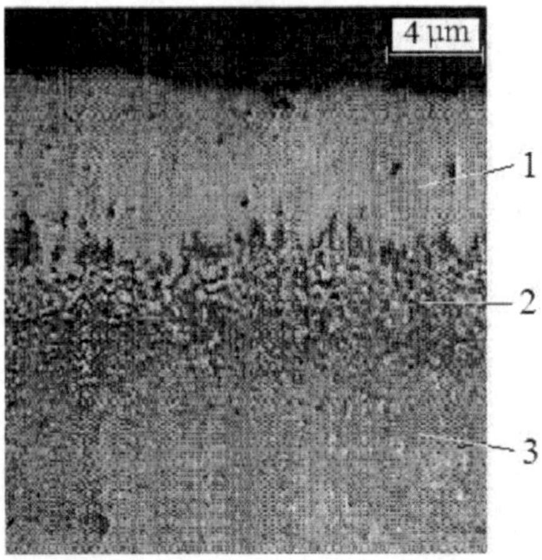


**Figure 4**. Selective layer structure in depth, (increased 12500:1): 1 - transferred layer, 2 - interface, 3 - base material.

## 4. Conclusions

The practical necessity to measure the thickness of the thin layers with a thickness of 0.1 - 4 μm obtained through different procedures, including the friction in the condition of a selective transfer, had imposed this method using the electronic microprobe.

Practically, this method can't be replaced in this particular case if it is necessary to determine the thickness of the thin layer of deficient and expensive materials like brass, bronze, copper (Cu), tin (Sn), zinc (Zn) and lead (Pb).

The reason that this method was selected from other methods of measurement, of the thickness of the thin layers, available (measurement on an optical microscope) is that the others do not enable the measurement of the layer with such a thickness.

The impossibility to measure layers so thin appears first of all because of the limited magnification of the optical microscope, and secondary because of the problems that occur at the samples preparations.

IOP Conf. Series: Materials Science and Engineering **174** (2017) 012065 doi:10.1088/1757-899X/174/1/012065

The results obtained by this method are in agreement with those determined in the previous method.

A quantitative assessment of selective layer thickness shows the increase in time until it reaches a maximum (optimum), after that the transfer occurs in the reverse, remaining in function of the friction duration in the range of 0.1- 4 μm, but varies much more.

Layer thickness can be reduced 10-fold and to increase again. These changes in thickness correspond to observations at samples from steel OL37, which has worked together with the alloys on base copper at stress through friction, on which it is formed a layer of copper, which is related to the transfer of copper on the steel surface.

The irregularity of the layer may be caused by the irregularity of the asperity of the steel probe's surface on which it had been transferred a thin superficial layer.

## References


[1] Vasicev B N 1977 *Elektronnozondovii micronaliz tronkih plenok* (Moskva: Metallurghia)

[2] Procop M, Radtke M, Krumrey M, and Hasche K 2002 Electron probe microanalysis (EPMA) measurement of thin-film thickness in the nanometer range *Anal. Bioanal. Chem.* **237** 631-634

[3] Libo Z, Shengxiang B, Rong W, Shilan L, Lili M, and Dechun L 2009 Thin film thickness measurement using electron probe microanalyzer *Proc. IEEE Int. Conf. on Applied Superconductivity and Electromagnetic Devices* (Chengdu: China, September 25-27)

[4] Pouchou J L 2002 X-ray microanalysis of thin surface films and coatings *Mikrochim. Acta* **138** 133-152

[5] Conroy M 2009 Advances in thick and thin film analysis using interferometry *Wear* **266(5-6)**, 502-506

[6] Ilie F 2012 A method of determining the thickness of the thin layers performing tribologically *J. Balk. Tribol. Assoc.* **18(3)** 357-364

[7] Ilie F and Tita C 2009 Study of copper CMP mechanisms with atomic force microscopy *J. Balk. Tribol. Assoc.* **15(2)** 163-169

[8] Ilie F 2002 *Tribologia straturilor subtiri formate prin transfer selectiv* (Tribology of thin superficial layers formed in the friction couples through selective transfer) (Bucureşti: Ed. Tehnică)

[9] Garkunov D N 1981 Erhonung der Verschleissfestigkeit auf der selektiven Ubertragung (Berlin: VEB Verlag Technik)

[10] Ilie F 2006 Comparative analysis of tribological performances of materials that can achieve thin layers from selective transfer *J. Balk. Tribol. Assoc.* **12(4)** 404-411

[11] Rajora O S and Curzon A E 1985 A simple method for the determination of film thickness from electron image contrast in a scanning electron microscope *Thin Solid Films* **123** 235-238

[12] Hore D and Jena K 2011 A simple transmission-based approach for determining the thickness of transparent films *Am. J. Phys.* **79** 256-260

[13] Morosanu E C 1986 *Depunerea chimica din vapori a straturilor subtiri* (Chemical deposition of thin films by vapour) (Bucureşti: Ed. Tehnică)

[14] Schwartz B and Schwartz N 1968 *Measurements techniques for thin films* (New York: Johnson Reprint Corporation)

[15] Murt M and Gulodner W G 1969 *Physical measurement and analysis of thin films* (New York: Plenum Press)

[16] Phillip S R 2012 *Measuring the thickness of thin metal films* (Idaho: A senior thesis submitted to the faculty of Brigham Young University)

[17] Piegari A and Masetti E 1985 Thin film thickness measurement: A comparison of various techniques *Thin Solid Films* **124** (3–4) 249–257

[18] Habiger W and Stein C 1992 Thickness measurements of thin film: Comparison of techniques using characteristic X-ray ratio techniques *Thin Solid Films* **215** 108-114

[19] Siderov V, Mladenova D, Yordanov R, Milenkov V, Ohlidal M Salyk O, Zhivkov V and Weiter M 2013 Film thickness measurement by optical profilometer MicroProf® FRT *Bulg. Chem. Commun.* Special Issue B **45** 194-197

[20] Ng F L, Wei J, Lai F K and Goh K L 2006 Metallic thin film depth measurements by X-ray microanalysis *Appl. Surf. Sci.* **252** 3972-3976

[21] Pascual R, Curz L R and Ferreira C L 1990 Thin film thickness measurement using the energy-dispersive spectroscopy technique in a scanning electron microscope *Thin Solid Films* **185** 279-288

[22] Kossivas F, Doumanidis Ch and Kyprianou A 2012 *Thickness measurement of photoresist thin films using interferometry*, Interferometry – Research and Applications in Science and Technology, ed. dr Ivan Padron, (New Jersey: InTech USA)

IOP Conf. Series: Materials Science and Engineering **174** (2017) 012066 doi:10.1088/1757-899X/174/1/012066

# Some considerations on noise monitoring for air handling equipments

**C Bujoreanu and M Benchea**

Mechanical Engineering, Mechatronics and Robotics Department, "Gheorghe Asachi" Technical University of Iasi, Bd. Dimitrie Mangeron 43, Iasi, Romania

E-mail: carmen.bujoreanu@gmail.com

**Abstract.** The HVAC (Heating, Ventilating and Air Conditioning) beneficiaries are in particular annoyed by the noise generated from the radiant unit and the air circulating ducts, since they are located inside the rooms and buildings. The comparatively experimental results highlight the relations between the air flow, pressure, power-charging and the sound level. The measurements are carried out at different fan's speeds, ranging the power-charge from 30-100% while the duct air flow is slowly adjusted from full open to full closed, between 0-500 Pa. Third-octave band analysis of random noise of the handling units is realized in an anechoic room, using the measurement procedures that agrees the requirements of the ISO 3744:2011 and ISO 5136:2010 standards. For an accurate design of the HVAC system, the designer needs to know not only the sound power of the radiant unit, but also from all of the air paths, since the sound travels along with the conditioned air. The experimental methodology used in the paper is of real interest for the HVAC manufacturers, in order to rate the sound level of their products and to improve the noise attenuation.

## 1. Introduction

The design for buildings, offices and rooms requires a clear specification of acoustical systems and products location, needing some quantitative information because the noise criteria regulations become more and more strictly. HVAC (Heating, Ventilating and Air Conditioning) systems from the modern buildings are indispensable for the human comfort and their performances must also agree the imposed noise regulations.

The air handling equipment (AHU) involves any type of HVAC system that contains a fan and is used to condition and move air through a duct system [1]. The air handling unit is a good example to prove the difficulty in accurate sound data evaluation since there are many possible paths between the noise source and the receiver.

Each fan, as component of an air handling product is operating at various speeds and within a range of flow and static pressure conditions. However, the fan inside the air handler changes the sound performances due to its housing that modifies the airflow pattern at the inlet and outlet [2]. Moreover, the sound may have different ways to leave the air handler, as it is reported in this paper. There are supply, extract, exhaust and outdoor air ducts along which the sound also travels. Also, the air velocity, the material roughness and the fittings of the ducts significantly increase the friction in the air handler system and obviously the noise generation [2]. Depending on the geometrical inlet/outlet openings configuration of the air handler, the sound passes through the ducts in the direction of airflow or against it. In this case, it travels back out with the return air inlet and the regenerated noise modifies the sound level from the source. Therefore, these paths together with the air handler radiant unit including fan and dampers provide a complex acoustic situation, difficult to be predicted by theoretical algorithms. The computer codes predicting the air handler noise may have potential errors as they calculate the sound level contributions of each element of a system. Nevertheless, numerical techniques for acoustic propagation in dissipative ducts and radiant unit, such as two-port transfer matrix method [3, 4], finite and boundary elements methods [5, 6] have been developed. There are also algorithms and procedures based on sound power computation [7].

In this paper there are experimentally investigated two types of commercial air handler units (AHU) with the same ducts diameters, but different inlet-outlet configuration. The comparatively results highlight the relations between the air flow, pressure, power-charging and the sound level. Third-octave band analysis of random noise of the handling units is realized, using measurement procedures that agree the requirements of the ISO 3744:2011 and ISO 5136:2010 standards [8, 9]. For an accurate design of the HVAC system, the designer needs to know not only the sound power of the radiant unit, but also from all of the air paths, since the sound travels along with the conditioned air.

## 2. Experimental methodology

The HVAC manufacturers are gathering information about the fan power, duct pressure vs. air flow variation, temperature, sound data and these performance parameters are included in their catalogues [10, 11, 12]. Sound power levels are usually determined in acoustics laboratory, by measuring sound pressure levels in test facilities using some specific procedures [8, 9] in order to realize data compatibility between different manufacturers.

HVAC system designers evaluate the sounds in the frequencies between 45 and 11200 Hz. Using the octave bands, this frequency range is separated into eight octave bands with center frequencies of 63, 125, 250, 500, 1000, 2000, 4000 and 8000 Hz. Also, the human ear is most sensitive to sound in the frequency range 1000 Hz to 4000 Hz than to sound at very low or high frequencies. This means that the noise at high or low frequencies will not be as annoying as it would be when its energy is concentrated in the middle frequencies. A higher sound pressure is therefore acceptable at lower and higher frequencies. This knowledge is important in acoustic design and sound measurement [13].

The sound power level *SWL*, according to ISO 3744: 2011, is the sound energy constantly transferred per second from the sound source. A sound source has a constant sound power that does not change if it is placed in a different room environment. Sound power is a theoretical value that is not measurable.

$$SWL = 10\log_{10}\left(\frac{p}{p_0}\right). \tag{1}$$

with: measured sound pressure $p$ and the reference sound power level $p_0 = 10^{-12}$ *W*.

The sound pressure level, *SPL* is a measurable sound level that depends upon environment. It is expressed in decibels at a specified distance and position [14]. Usually, both sound parameters are calculated and expressed in decibels although they are completely different and should not be confused.

$$SPL = SWL + 10\log\left[\frac{Q}{4\pi r^2} + \frac{4}{R_C}\right]. \tag{2}$$

where *SWL* (dB) is the sound power re $10^{-12}$ W; *SPL* (dB) is the sound pressure re 2 x $10^{-5}$ Pa; $r$ (m) is the distance from the source; $Q$ is the directivity factor of the source in the direction $r$; $R_C$ ($m^2$) is the room constant of the following form: $R_C = \dfrac{S \cdot \alpha_W}{1-\alpha_W}$ with $S$ ($m^2$) total room surface and $\alpha_W$ the average absorption coefficient in the room [13]. The two procedures of determining the sound power of air handlers equipment are the reverberant room and free field methods.

The reverberant room method has the aim to create a diffuse sound field through multiple reflections of sound waves, realizing the same sound pressure inside the room. The free field method needs an homogenous medium without boundaries or reflecting surfaces. In this case, the sound pressure values are recorded on a hemispherical surface surrounding the equipment and then, the sound power values are calculated.

For our measurement, we have used the free field method because we dispose of an anechoic room, in our department, with dimensions of 10m x 10m x 8m and a large access door (2m x 1.5m). In the anechoic room the walls, roof and floor are covered in a highly sound absorptive material to eliminate any reflections. The average absorption coefficient is 99% for 125Hz ÷ 20kHz frequency band. Thus, the sound pressure level in any direction may be measured and the directivity factor $Q$ is 1.

The test facility realized in accord with ISO 3744:2011 and ISO 5136:2010 standards is presented in figure 1.

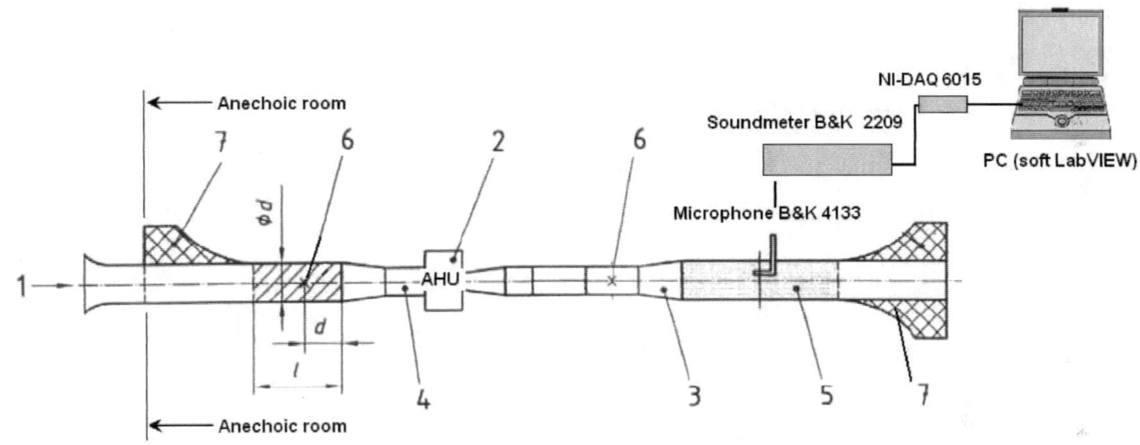


**Figure 1**. Test arrangement for in-duct noise measurements, according to ISO 5136: 2010.

| | |
|---|---|
| 1 flow direction | 5 test duct |
| 2 air handler unit AHU | 6 pressure taps |
| 3 transition duct | 7 anechoic termination |
| 4 intermediate duct | with $l \geq 4d$ or $l \geq 1m$ |

The measurement chain includes: device for measuring the pressure drop (Fluke 922 pressure differential meter), microphones Bruel&Kjaer type 4133, soundmeter Bruel&Kjaer type 2209, connected with the microphone and NIDAQ board, multifunctional external data acquisition board type-NIDAQPad-6015 on USB, laptop with LabVIEW soft compatible with National Instruments DAQPad for data processing.

For various fan's speeds, expressed in power-charging variation between 30 and 100%, the sound level data are acquired. The ducts air flow is slowly adjusted from full closed to full open for each range of charging.

The next step is to analyze the sound level dB(A), A weighted, both on radiant unit and the air ducts, using the third-octave analysis with LabVIEW soft, evaluating the pressure levels (dBA) for nominal frequencies of 63, 125, 250, 500, 1000, 2000, 4000, 8000 Hz. Also, the analysis evidenced the equivalent continuous sound level that is an averaged noise in a specified time interval.

Figure 2 describes typical air paths of handling unit, namely the ducts, in order to a better understanding of the air travel way and the sound that accompanies it.

The sound level tests where performed on two air handling commercial types denoted with AHU 1 and AHU 2. The main geometrical data are depicted in the figures 3 and 4, respectively. Also, we mention: the air flow capacity (350 $m^3/h$ for AHU 1 and 300 $m^3/h$ for AHU 2), the same weight (50 kg) and the air ducts diameters of Φ160 mm.

The important difference between the two air handlers consists in the air flow direction through the ducts system, which can modify the sound level due to the regenerated noise occurrence in a reverse flow. The sound may leave the air handler in multiple ways because the supply, exhaust, extract and outdoor air ducts can have different ways to cross the radiant unit.

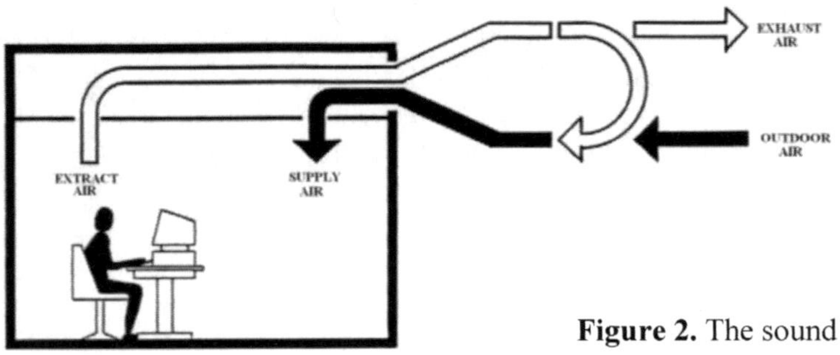


**Figure 2.** The sound paths in the air ducts.

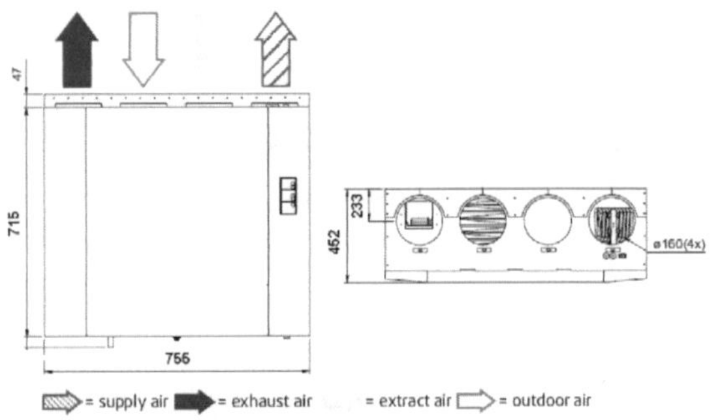


**Figure 3.** AHU 1 geometrical data.

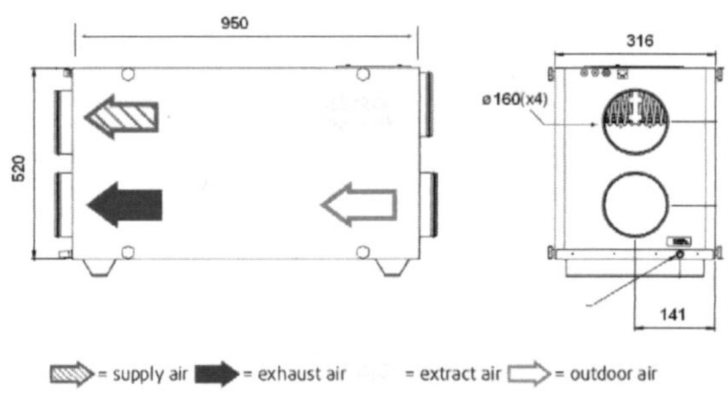


**Figure 4.** AHU 2 geometrical data.

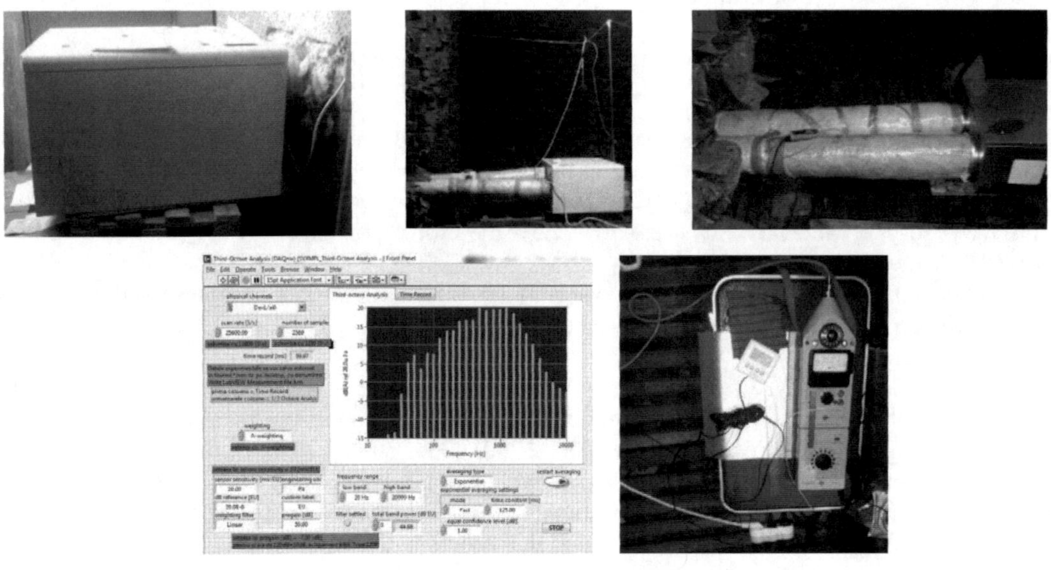

**Figure 5.** Experimental setup - outside and inside the anechoic room.

Figure 5 depicts the experimental setup realized in accord to ISO 3744:2011 and ISO 5136:2010 standards, using the anechoic room facilities from our department.

## 3. Results and discussions

Figure 6 represents a comparative evaluation of the total sound level for the tested radiant units running at different air flow pressure from 0 Pa to 500 Pa (the worst case in exploitation) taking into account of the fan power-charging. AHU 1 radiant unit has the lowest noise level and uniform noise behaviour with charging increase. A very quiet functioning at low charging and P0 air flow pressure can be also noted. AHU 1 sound level values range between 35÷45 dB and it can be observed that it is a little less noisy than the AHU 2 radiant unit (40÷50 dB).

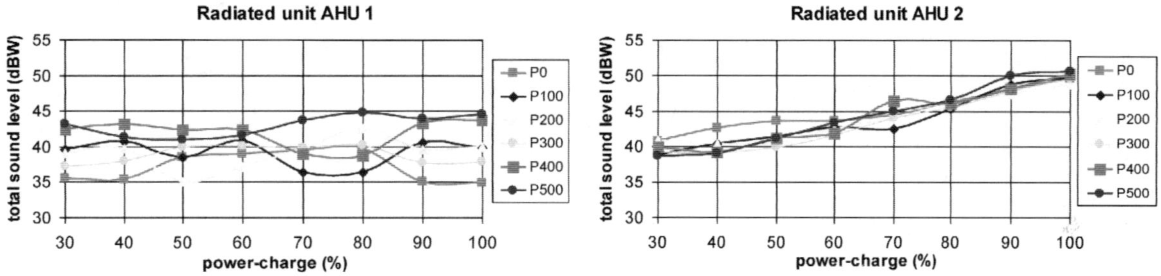


**Figure 6.** Sound level for the radiant units.

The HVAC manufacturers are interested in ducts noise in order to optimize their design and efficiency from this point of view. This optimization means the in-duct noise control, generally solved through the silencers use, thus adding adequate attenuation to the system. Therefore, the silencer selection is imposed by the duct noise level.

In this regard, the total sound level variation related to the fan power-charging is expressed in figure 7 for the supply air ducts of the tested AHU 1 and AHU 2 units, ranging the air flow pressure from 0 Pa to 500 Pa. The total sound level values are much higher (65÷105 dB) than those for the radiant units and the sound increasing tendency with fan power-charging preserved. Also, the supply air duct of AHU 2 has the best performance in sound attenuation and similar to the supply air duct of AHU 1 at 500 Pa air flow pressure. The supply air duct of AHU 1 it is noisier than the supply air duct of AHU 2 at high fan power-charging and low air flow pressure.

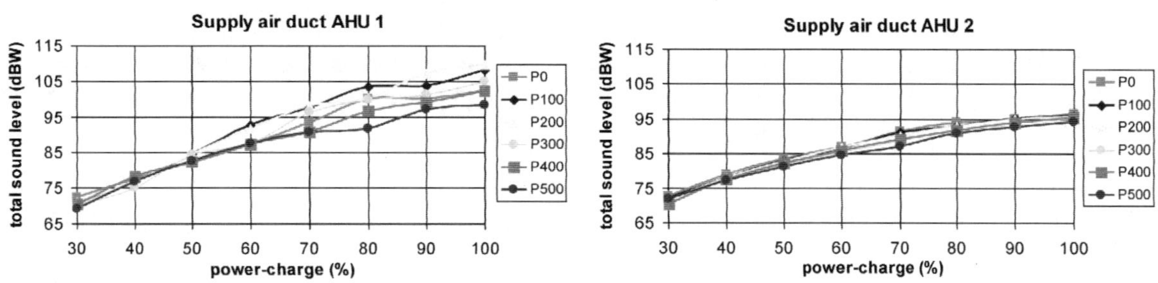


**Figure 7.** Sound level for supply air ducts.

The same tendency is recorded for the others air ducts (figures 8, 9 and 10). The total sound level is higher in the air ducts of AHU 1 unit at high fan power-charging and low air flow pressure, and it is decreasing at high air flow pressures, similar to the air ducts of AHU 2 unit, where the total sound level doesn't varies very much with the air flow pressure.

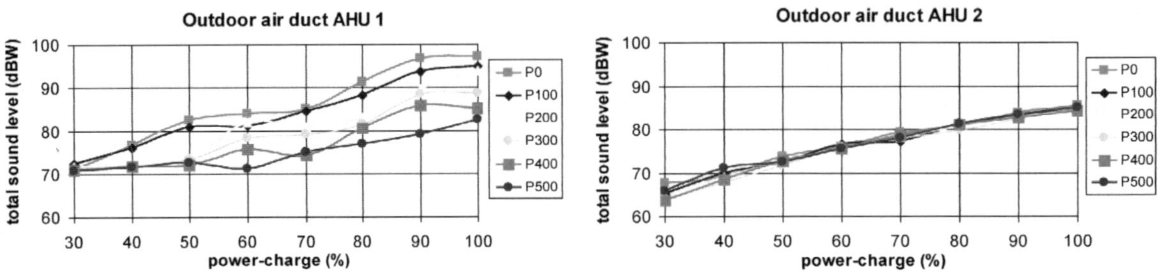


**Figure 8.** Sound level for the outdoor air ducts.

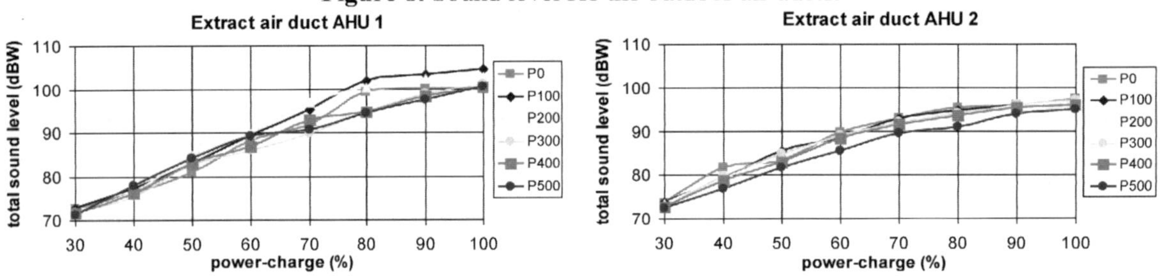


**Figure 9.** Sound level for extract air ducts.

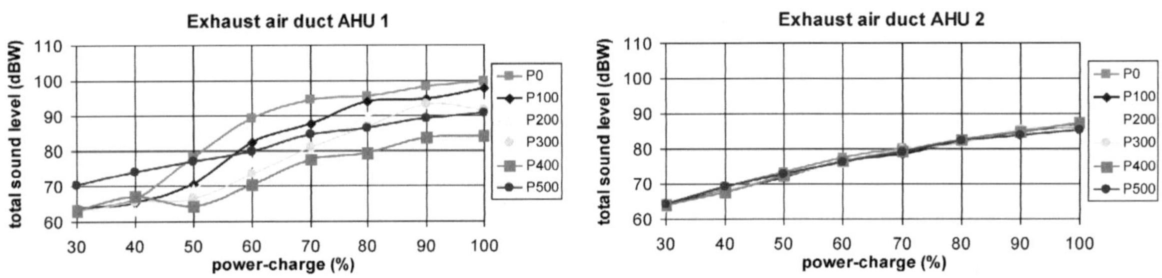


**Figure 10.** Sound level for exhaust air ducts.

A possible explanation of the fact that AHU 1 noise level in the air ducts is greater than the sound level of AHU 2 ducts consists in the air paths configuration. Figures 3 and 4 show the difference in the air flow directions through the ducts system, which modify the sound level. The regenerated noise in the AHU 1 air ducts is more intense since there is reverse air flow due to the disposition of the ducts (AHU 1 has the ducts inlet-outlet on the same side). On the other hand, the AHU 1 radiant unit is a little less noisy than the AHU 2 radiant unit, thus the inside attenuation elements do their job more efficient compared to the AHU 2 unit.

## 4. Conclusions

The HVAC beneficiaries are in particular annoyed by the noise generated from the radiant unit and the air circulating ducts, since they are located inside the rooms and buildings. The paper focuses on the analysis of the equivalent continuous sound level in two commercial air handlers units. The noise level of the radiant unit complies with the standards requirement (about 50 dB) and the ducts sound level tendency emphasizes the importance of the ducts inlet-outlet mounting. They are influencing the air flow direction and, consequently, the sound level. The fan power-charging, the air flow pressure, the noise generated in the system constitute acoustic performances of the HVAC. In order to optimize their design and efficiency from this point of view it is necessary the in-duct noise control, generally solved through the silencers use, thus adding adequate attenuation to the system. A suitable silencer selection is imposed by the duct noise level. Moreover, identifying the noise sources due to frictional causes inside the air handler equipment is a necessary step in the HVAC acoustic study. A future work will be focused on the friction rate. Therefore, the experimental methodology used in the paper is of real interest for the HVAC manufacturers, in order to rate the sound level of their products and to improve the noise attenuation according to the specified standards.

**References**

[1] ASHRAE 2011 *Handbook-Sound and Vibration Control* chapter 48 (USA) pp 279
[2] Air Conditioning Clinic 2012 User's Manual of Trane TRG-TRC007-EN *Fundamentals of HVAC Acoustics* pp 84
[3] Bhatia A B E 2012 Overview of Noise Control and HVAC Acoustics in Buildings, *PDHOnline*, pp 1-72
[4] Borelli D and Schenone C 2014 Application of a simplified "source-path-receiver" model for HVAC noise to the preliminary design of a ship: A Case Study *21st International Congress on Sound and Vibration (ICSV21)* (Beijing) pp 1-8
[5] Kühnelt H, Bäuml T. and Haumer A 2009 Sound Duct Flow: A Modelica Library for Modeling Acoustics and Flow in Duct Networks *Proc. of the 7th Modelica Conference (Como, Italy)* pp 1-16
[6] Borelli D and Schenone C 2012 A finite element model to predict sound attenuation in lined and parallel baffle rectangular ducts *HVAC&R Research* **18** (3) 390-405
[7] Ryherd S R and Wang L M 2006 Measuring Sound Power in Ducted Heating, Ventilating, and Air-conditioning (HVAC) Systems for Use in Verifying Acoustical Prediction Methods *Architectural Engineering -- Faculty Publications.* paper 15 pp 1-6
[8] ISO 3744: 2011 Determination of sound power levels and sound energy levels of noise sources using sound pressure. Engineering methods for an essentially free field over a reflecting plane.
[9] ISO 5136: 2010 Determination of sound power radiated into a duct by fans and other air-moving devices. In-duct method.
[10] Systemair Catalogue 2012 *Air Handling Units* pp 1-138
[11] Lindab Catalogue 2008 *Lindab Ventilation Guide* **(2)** pp 1-22
[12] Air Change Catalogue 2014 *Next-AHU$^{TM}$ Technical Data* pp 1-4
[13] Desmond S D 2000 *Fans in air handling unit* (Woods Air Movement Limited UK) pp 23
[14] Cory W T W 2005 Fans & Ventilation: A Practical Guide, *Elsevier* pp 424

IOP Conf. Series: Materials Science and Engineering **174** (2017) 012067 doi:10.1088/1757-899X/174/1/012067

# Optimization of continuous triboelectrification process for polymeric materials in dry contact

**Y E Prawatya**[1, 2], **M B Neagoe**[1], **T Zeghloul**[1] **and L Dascalescu**[1]

[1]Institut PPRIME, UPR 3346 CNRS - Université de Poitiers - ENSMA, IUT, 4 avenue de Varsovie, 16021 Angoulême, France

[2]Department of Industrial Engineering, Tanjungpura University, Jl. Ahmad Yani, 78124 Pontianak, Indonesia

E-mail: yopa.eka.prawatya@univ-poitiers.fr

**Abstract.** Triboelectrification (i.e., generation of electric charge by friction between two materials) is a complex process. Besides the nature and condition of the surfaces in contact, several factors can have an influence on charge generation: pressure load and relative velocity between the two bodies, number of friction cycles, ambient temperature and humidity, condition and type of material surface. This paper aims at demonstrating that associating the experimental response surface methodology and genetic algorithms is an effective technique for the optimisation of triboelectrification process. The quadratic model derived from the experiments is used in a genetic algorithm program to find the optimal combination of factor values (10 sliding cycles; normal force: 10 N; sliding speed: 55 mm/s) that maximize the average potential at the surface of the tribocharged materials: -1633 V. A final experiment confirmed the prediction of the genetic algorithm. The conclusions of this experimental study can be applied to the optimisation of industrial triboelectrification processes, and contribute to the reduction of the related maintenance, energy and raw-material costs.

## 1. Introduction

Electric charge generation by frictional contact between two bodies is a complex process, the outcome of which is affected by many factors, namely: normal force, number of friction cycles, friction velocity, ambient temperature and humidity, surface condition and nature of the materials in contact.

The purpose of the present work is to determine the optimal values of these factors, so that to obtain maximum electrical charge, evenly distributed on the contact surfaces [1-5]. From a mechanical point of view, frictional contact is highly correlated with wear, load, number of sliding cycles and temperature. Changes in one or more of these factors will alter the condition of the surface of the material and indirectly affect the outcome of the electrification process [6-11].

In the present paper, the response surface methodology (RSM) is used for experimental modelling of the process. A quadratic polynomial model is obtained by means of a central composite face-centred (CCF) experimental design [12-14], performed on a laboratory bench designed for testing translational tribocharging [15-17]. The distribution of charge is evaluated by measuring the electric potential at the surface of the samples, with the auto-compensated induction probe of an electrostatic voltmeter, in a room with controllable temperature and humidity.

## 2. Materials and methods

Acrylonitrile butadiene styrene (ABS) as top sample with the dimensions of 5 mm x 15 mm x 100 mm and Polyvinyl chloride (PVC) as bottom sample, 5 mm x 50 mm x 180 mm, are employed.

The residual charge at the surface of these samples is neutralized using a commercial ionizing system (electrode ECA 88 BS and high-voltage supply SC 04 B, 5 kV, 7 mA, manufacturer: ELCOWA, Mulhouse, France).

Triboelectrification is done using a custom-designed experimental bench [15] to generate the electric charge by friction between a pair of samples (1) and (2), in conformal contact (figure 1). The back-and-forth sliding motion of the bottom sample with respect to the top sample is controlled by a reversible variable speed electric motor drive. The relative velocity between the bodies in conformal contact can be varied between 55 and 275 mm/s, with the amplitude of the strokes ranging from 36 to 55 mm. The top sample holder (4) is fixed in a vertical guiding system, connected with an adjustable force arm. The normal contact force can be adjusted between 1 and 10 N. This triboelectric laboratory bench enables the control of several factors that influence the electrostatic charge generation: polymer rubbing speed, contact force, and number of friction cycles [15-17].

Mapping of the electric potential at the surface of the tribocharged sample, disposed on grounded plate, is made with an auto-compensated induction probe (model 3450, TREK Inc.), connected to an electrostatic voltmeter (model 341B, TREK Inc.). A computer-controlled *x-y* positioning system moves the sample with respect to the probe. Thus, the measurements are carried out in 95 points, which reflect the electric potential on the sample entire surface (figure 2). The distance between two points is 10 mm. The probe is at rest with respect to the sample, at the moment when the result of the measurement is recorded (figure 3). Surface temperature is observed with a thermal camera FLIR E60 to ensure that the requirement of uniform rubbing contact pressure has been respected (figure 4).

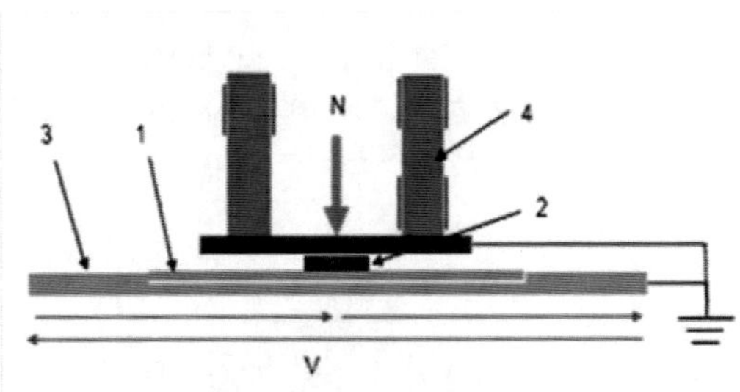


**Figure 1.** Schematic representation of the rubbing process: (1) bottom sample; (2) top sample; (3) bottom sample holder; (4) top sample holder.

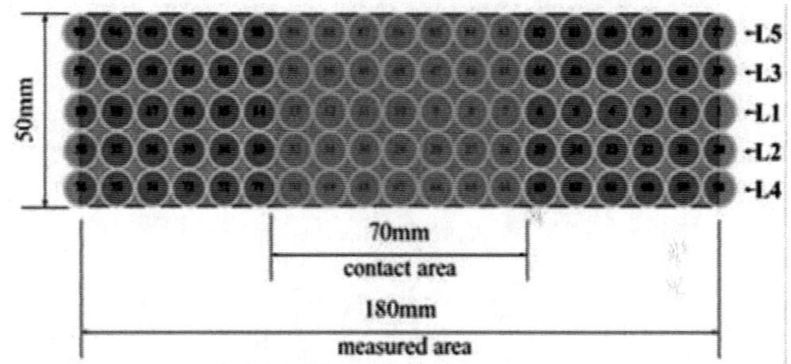


**Figure 2.** Positions of the points in which the electric potential is measured at the surface of a bottom sample.

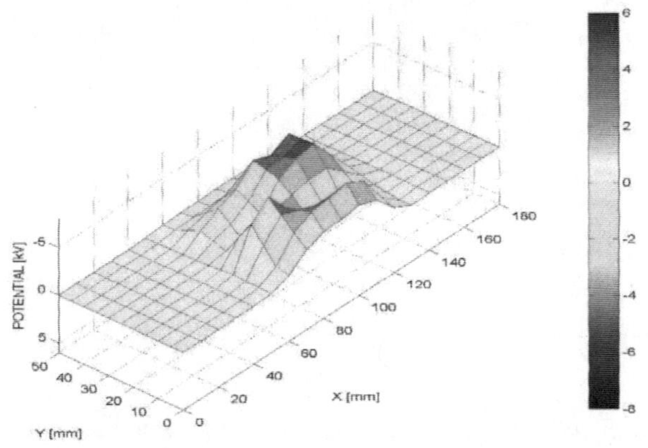


**Figure 3.** Surface potential cartography of polymer sample type B (bottom).

**Figure 4.** Surface temperature cartography of sample type B (bottom).

The factors under study in this experiment are the following: number of tribocharging cycles (10, 55, and 100), normal force applied on type B sample (2 N, 6 N, and 10 N) and frequency of back- and-forth sliding motion (0.5 Hz, which corresponds to a sliding speed of 55 mm/s), 1.5 Hz or 165 mm/s, and 2.5 Hz or 275 mm/s). The responses are: average surface electric potential ($V_{AVG}$), standard deviation of surface potential ($\sigma_V$) and temperature increase during tribocharging ($\Delta temp$). Response Surface Methodology (RSM) recommends the use of a Central Composite Face-Centered (CCF) experimental design as plotted in figure 5. The quadratic models that can be obtained with this experimental design are flexible and can mimic many different types of response function.

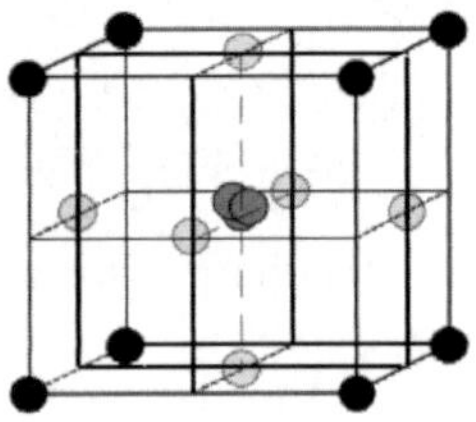

**Figure 5.** Schematic representation of the CCF experimental design for three factors.

The objective is to simultaneously maximize $V_{AVG}$, minimize $\sigma_V$ and minimize $\Delta temp$. Genetic Algorithm is chosen as optimization method. The association of this method with the Response Surface Methodology is expected to be an efficient way to find the optimal values of the three control factors of the triboelectrification process [18-21].

## 3. Results and Discussion

### 3.1. Surface potential mapping

The electric potential is mapped for each of the 17 experiments of the CCF design recommended by the response surface methodology. The experimental data plotted in figure 6, for instance, show that using maximum sliding speed leads to higher uniformity of electric surface potential.

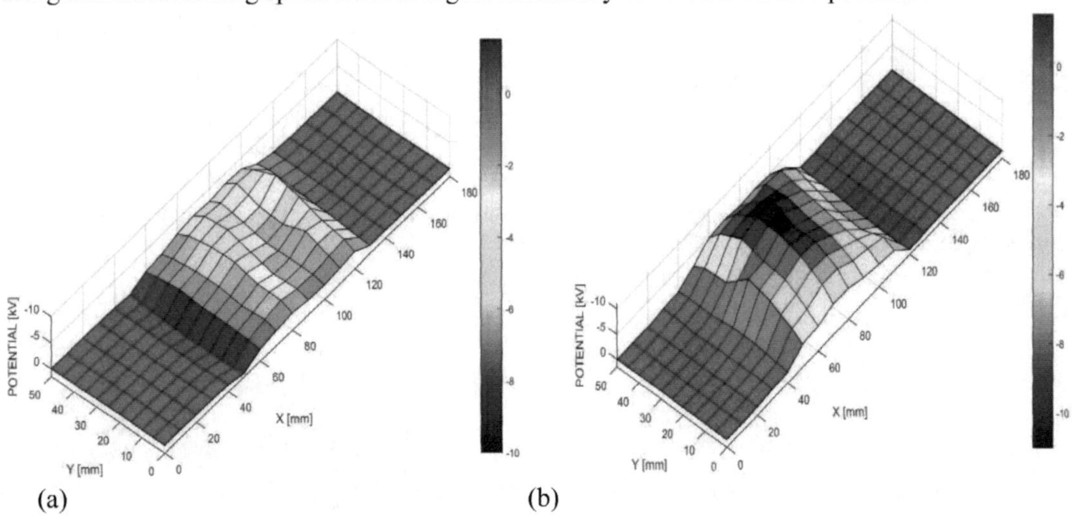


**Figure 6.** Surface potential mapping: (a) 10 Cycle, 10 N, 2.5 Hz; (b) 10 Cycle, 10 N, 0.5 Hz.

### 3.2. Regression model

The commercial software MODDE 5.0 (Umetrics, Sweden) is used to obtain the experimental model of the process. The "summary of fit plot" (figure 7 a) gives the values of the statistical indexes $R^2$ ("goodness of fit") and $Q^2$ ("goodness of prediction"). A good model is characterized by both these indexes close to 1.

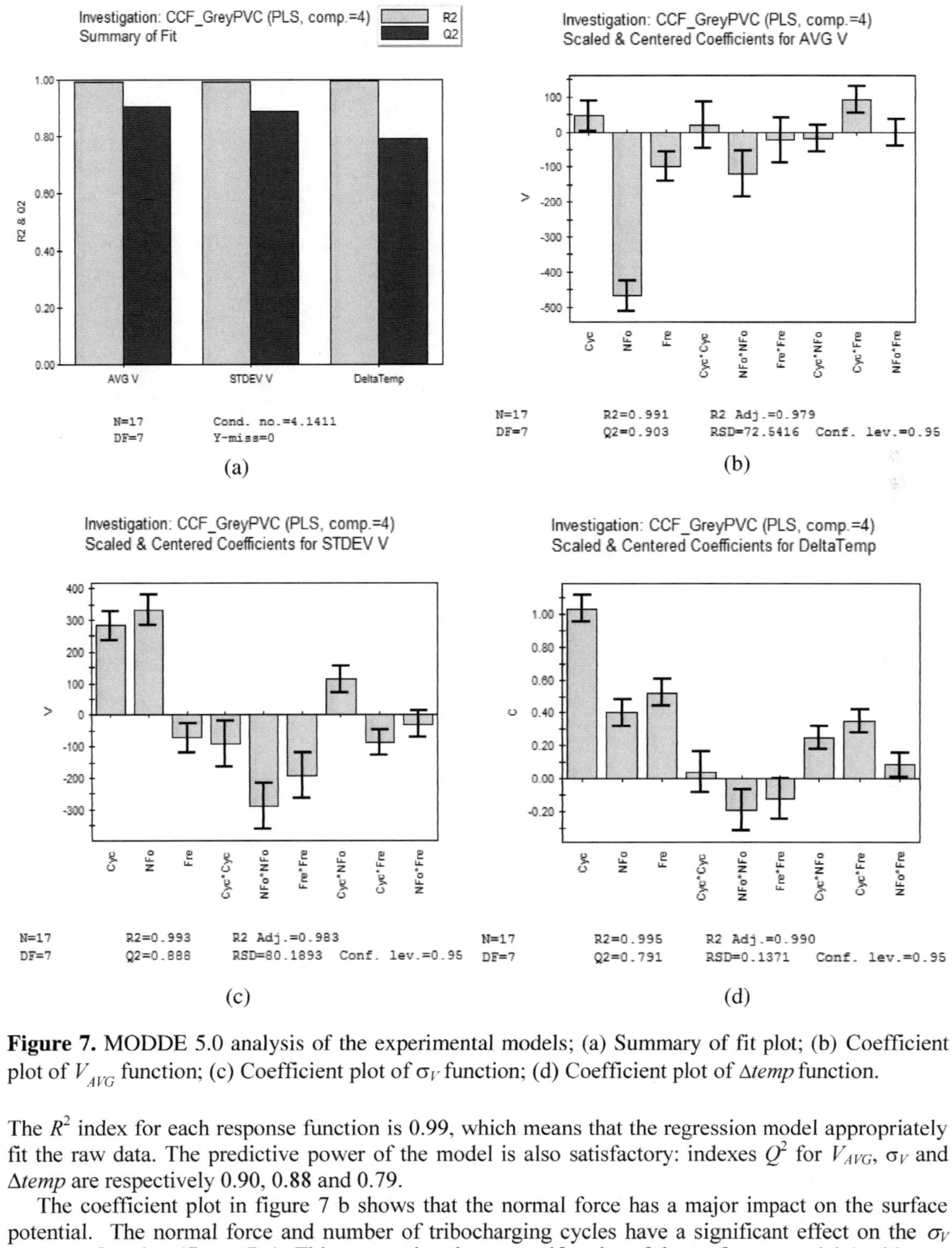


**Figure 7.** MODDE 5.0 analysis of the experimental models; (a) Summary of fit plot; (b) Coefficient plot of $V_{AVG}$ function; (c) Coefficient plot of $\sigma_V$ function; (d) Coefficient plot of $\Delta temp$ function.

The $R^2$ index for each response function is 0.99, which means that the regression model appropriately fit the raw data. The predictive power of the model is also satisfactory: indexes $Q^2$ for $V_{AVG}$, $\sigma_V$ and $\Delta temp$ are respectively 0.90, 0.88 and 0.79.

The coefficient plot in figure 7 b shows that the normal force has a major impact on the surface potential. The normal force and number of tribocharging cycles have a significant effect on the $\sigma_V$ response function (figure 7 c). This means that the non-uniformity of the surface potential (and hence of the respective electric charge) increases with the normal force and the number of tribocharging cycles. The latter factor is the most influential on surface temperature (figure 7 d).The response surface models fit to $V_{AVG,}$ $\sigma_V$ and $\Delta temp$ are:

$$V_{AVG} = -616 + 46\times Cyc - 468\times NFo - 99\times Fre - 119\times NFo\times NFo + 92\times Cyc\times Fre$$

$$\sigma_V = 1740 + 282\times Cyc + 332\times NFo - 72\times Fre - 92\times Cyc\times Cyc - 288\times NFo\times NFo - 191\times Fre\times Fre + 113\times Cyc\times Fre - 30\times NFo\times Fre$$

$$\Delta temp = 1.9 + 1\times Cyc + 0.4\times NFo + 0.5\times Fre - 0.2\times NFo\times NFo + 0.2\times Cyc\times NFo + 0.3\times Cyc\times Fre + 0.1\times NFo\times Fre$$

### *3.3. Interactions between process variables*

The response contour plots in figure **8** illustrate the effects of the normal force *NFo* and of the number of cycles *Cyc*, at high frequency *Fre* = 2.5 Hz. Graphical representations can help to explain the situation around the optimal point.

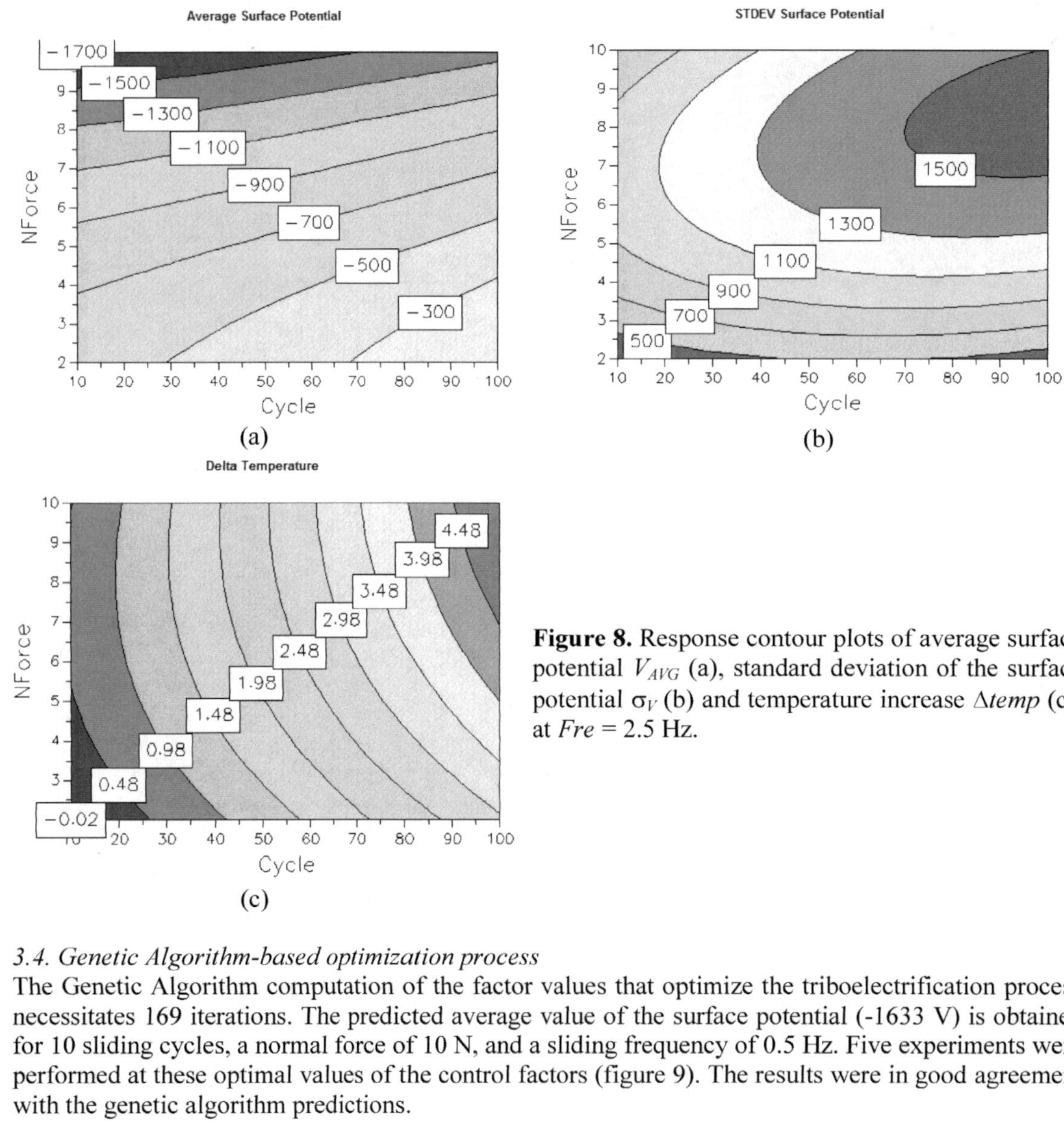


**Figure 8.** Response contour plots of average surface potential $V_{AVG}$ (a), standard deviation of the surface potential $\sigma_V$ (b) and temperature increase $\Delta temp$ (c), at *Fre* = 2.5 Hz.

### *3.4. Genetic Algorithm-based optimization process*

The Genetic Algorithm computation of the factor values that optimize the triboelectrification process necessitates 169 iterations. The predicted average value of the surface potential (-1633 V) is obtained for 10 sliding cycles, a normal force of 10 N, and a sliding frequency of 0.5 Hz. Five experiments were performed at these optimal values of the control factors (figure 9). The results were in good agreement with the genetic algorithm predictions.

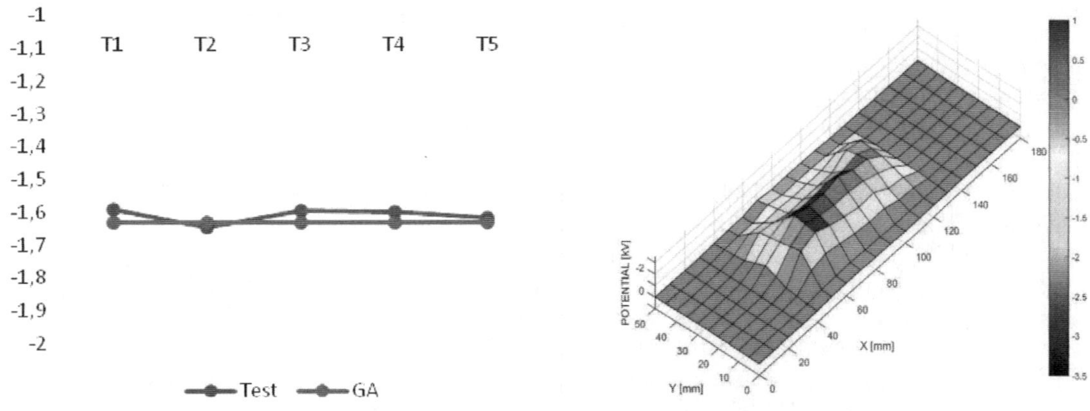


(a) (b)

**Figure 9.** Experimental result based on genetic algorithm optimal value; (a) Comparison of tribocharge results and GA prediction; (b) Surface potential mapping after tribocharging.

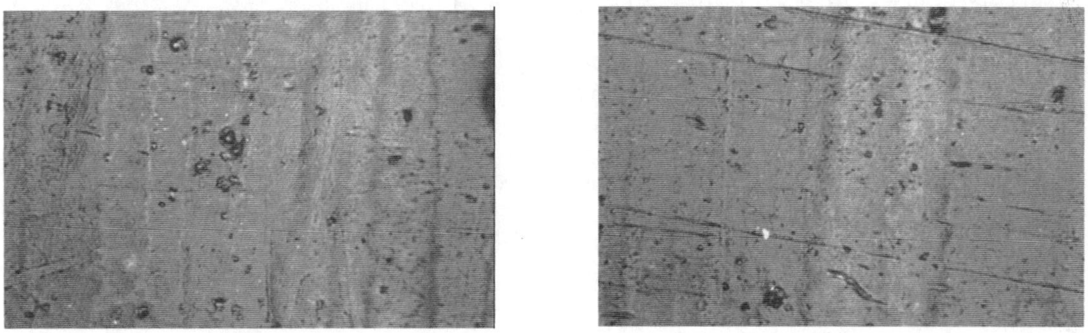

(a) (b)

**Figure 10.** Material microscopic investigation pre (a) and post (b) tribocharging process.

These experimental conditions do not significantly change the aspect of the material surface. However, slight scratches appear on the surface (figure 10).

## 4. Conclusions

1) The number of friction cycles, the normal force at the surface of contact, and the relative sliding speed between the two bodies, are three of the major factors that affect the outcome of the triboelectrification process of polymers in dry conformal contact.
2) The association of response surface modelling with genetic algorithm techniques is an efficient way to optimize a triboelectrification process. The experimental results are in good agreement with the theoretical predictions.
3) Further experiments, with different polymer pairs of samples, will enable a better understanding of the physical mechanisms of triboelectrification.

## Acknowledgments

This work was partially funded by the French Government program "Investissements d'Avenir" (LABEX INTERACTIFS, reference ANR-11-LABX-0017-01) and the Directorate General of Resources for Science, Technology and Higher Education Ministry of Research, Technology and Higher Education of Indonesia.

## References

[1] Zeghloul T, Dascalescu L, Roaugdia K, Fatihou A, Renoux P and Souchet D 2016 *IEEE Trans. Ind. Appl.* **52** 1808-13
[2] Alahmadi A 2014, *Int. J. Sci. & Eng. Res.* **5** 22-29
[3] Harper W R 1998 *Contact and Friction Electrification* (Morgan Hill, CA: Laplacian Press) pp. 64-65
[4] Maegawa S, Itoigawa F and Nakamura T 2015 *Tribol. Inter.* **92** 335-343
[5] Tamai Y, Inazumi T and Manabe K 2016 *J. Mater. Proc. Technol.* **227** 161-168
[6] Okonkwo P C, Kelly G, Rolfe B F and Pereira M P 2012 *Wear* **282–283** 22–30
[7] Okonkwo P C, Kelly G, Rolfe B F and Pereira M 2016 *Tribol. Inter.* **97** 218–227
[8] Nip K H, Gardner L, Davies C M and Elghazouli A Y 2010 *J. Constr. Steel Res.* **66** 96–110
[9] Kennedy F E, Lu Y and Baker I 2015 *Tribol. Inter.* **82** 534–42
[10] Ruiz-Andres M, Conde A, de Damborenea J and Garcia I 2015 *Tribol. Inter.* **90** 32–42
[11] Bowden F P and Tabor D 1964 *The friction and lubrication of solids* (London: Oxford Univ. Press).
[12] Su L, Zhang J, Wang C, Zhang Y, Li Z, Song Y, Jin T and Ma Z 2016 *Appl. Energy* **163** 201-10
[13] Miloudi M, Medles K, Tilmatine A, Brahami M and Dascalescu L 2011 *J. Phys.: Conf. Series* **301** doi:10.1088/1742-6596/301/1/012067
[14] Hammadi N, Ouiddir R, Tilmatine A, Remaoun S M and Medles K 2012 *Scientia Iranica, Trans. D: Computer Sci. & Electr. Eng.* **19** 836-40
[15] Neagoe B, Prawatya Y, Zeghloul T, Souchet D and Dascalescu L 2015 *J. Phys.: Conf. Series* **646** doi: 10.1088/1742-6596/646/1/012058
[16] Liu L, Oxenham W and Seyam A M 2013 *Indian J. Fibre & Textile Res.* **38** 265-269
[17] Liu L, Oxenham W and Seyam A M 2013 *J. Engineered fibers and fabrics* **8** 126-136
[18] Álvarez M J, Ilzarbe L, Viles E and Tanco M 2009 *Quality Technol. & Quantitative Management* **6** 295–307
[19] Kucukkoc I, Karaoglan A D and Yaman R 2013 *Int. J.Prod. Res.* **51** 5039–54
[20] Karimi M, Keyhani A, Akram A, Rahman M, Jenkins B and Stroeve P 2013 *Environ. Technol.* **34** 2201–11
[21] Dahou O, Medles K, Touhami S, Boukhoulda M F, Tilmatine A and Dascalescu L 2010 *Industry Applications Society Annual Meeting (IAS), IEEE* doi:10.1109/ias.2010.5615645

IOP Conf. Series: Materials Science and Engineering **174** (2017) 012068 doi:10.1088/1757-899X/174/1/012068

# Functioning condition monitoring of industrial equipment

**N S Ungureanu, A Petrovan, M Ungureanu and M Alexandrescu**

Technical University of Cluj Napoca, North University Center of Baia Mare, Faculty of Engineering, Department of Engineering and Management of Technology, Victor Babes St., 62A, Baia Mare, RO-430083, Romania

E-mail: nicolae.ungureanu@cunbm.utcluj.ro

**Abstract**. The paper analyses the theoretical aspects related to monitoring industrial equipment. Are treated issues that concern the choosing of industrial equipment to be monitored, the parameters to be monitored, monitoring mode (local or remote) and the mode of collection and transmission of data.

## 1. Introduction

The current national economic situation is characterized by fierce competition between different manufacturers. In this competition the continuous and proper functioning of industrial equipment is an extremely important objective. This can be achieved by purchasing some equipment with a high reliability level and by implementing a performing maintenance system.

The two directions previously mentioned have one major restriction: the purchase price or implementation price. Variation of acquisition costs of equipment and maintenance costs depending on the level of reliability is known and is mentioned in specialty literature [1][2][3].

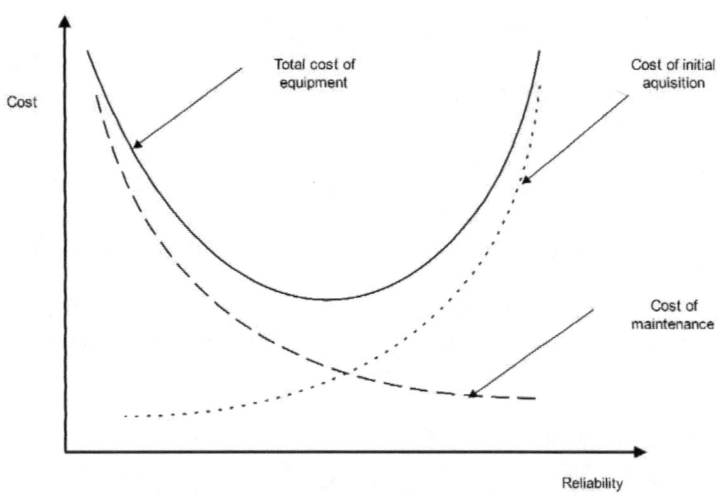


**Figure 1.** Variation of acquisition costs of equipment and maintenance costs depending on the level of reliability [1].

IOP Conf. Series: Materials Science and Engineering **174** (2017) 012068 doi:10.1088/1757-899X/174/1/012068

In theory the maintenance is defined as the actions necessary for retaining or restoring a piece of equipment, machine, or system to the specified operable condition to achieve its maximum useful life, when the maintenance actions are performed in logistic conditions defined. For maximum efficiency the maintenance actions and the related logistics support are structured in maintenance systems.

The main scopes of the maintenance systems is to keeping the machine functionality at the required level, maximizing the machine capacity and enhancing the machine live-time at the lowest cost. The maintenance systems are based on the logical organization of the works according to several criteria:

- criterion of the resources used;
- criterion of the use of the means for automatic data processing;
- criterion of the cost reduction for maintenance, repairs and storage;
- criterion of the necessity of eliminating the accidental stoppage by preventing some damages, mistuning, permanent wear, etc.

Each of the maintenance systems have advantages or disadvantages that make them preferred for each industrial enterprise. It is also possible the use of different maintenance systems, simultaneous, in a single enterprise, in order to get the full benefits of different maintenance systems. At this time the Romanian industry use some maintenance systems that can be illustrated as in figure 2.

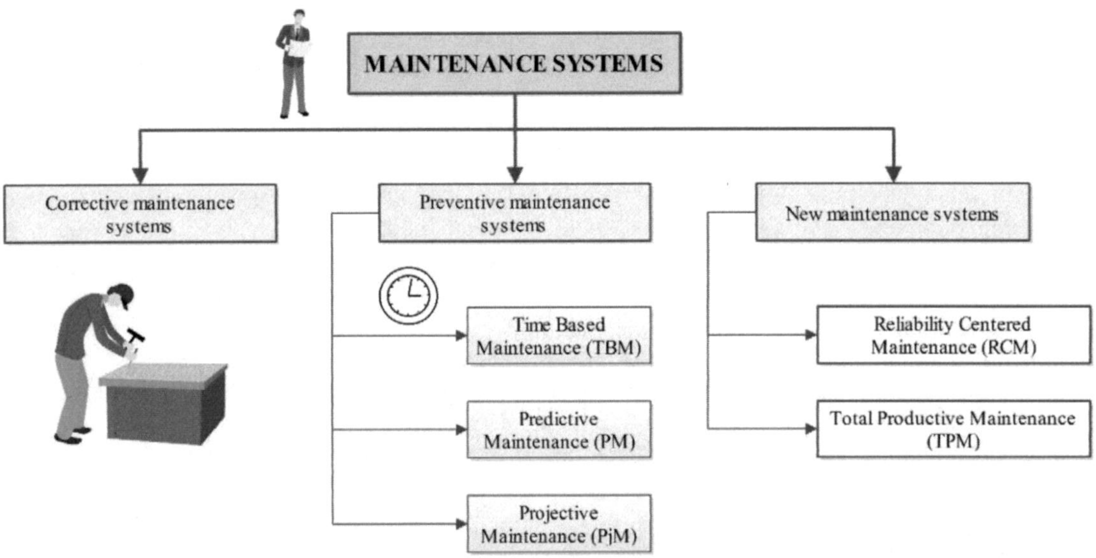


**Figure 2.** Maintenance systems [4].

## 2. Condition monitoring

The functioning condition monitoring of the equipment's can be defined in many ways. The definition of condition monitoring is according to how it is perceived: as part of a maintenance system or as a standalone system.

In the authors' view the functioning condition monitoring is a part of a maintenance system through which is made a collection of data regarding the operation of equipment, transmission or storage of this data, data processing, preparation of reports followed by maintenance action at time recommended.

This opinion is based on the previous experience and on analysis of maintenance systems. In predictive maintenance system and in total productive maintenance (TPM) the functioning condition monitoring is found as a distinct component part, but also in other maintenance systems are found parts of condition monitoring. An example can be found in all maintenance systems: the measuring operating parameters and comparing them with standard values. Measurement is made either before entering the repair (when the equipment is functional) or after completing repairs.

Data that is collected must be representative for the state of operation of the equipment and are selected after an analysis of equipment. In figure 3 are presents some of the most monitored parameters of the equipment. The figure is not exhaustive and can be completed depending on the type of equipment.

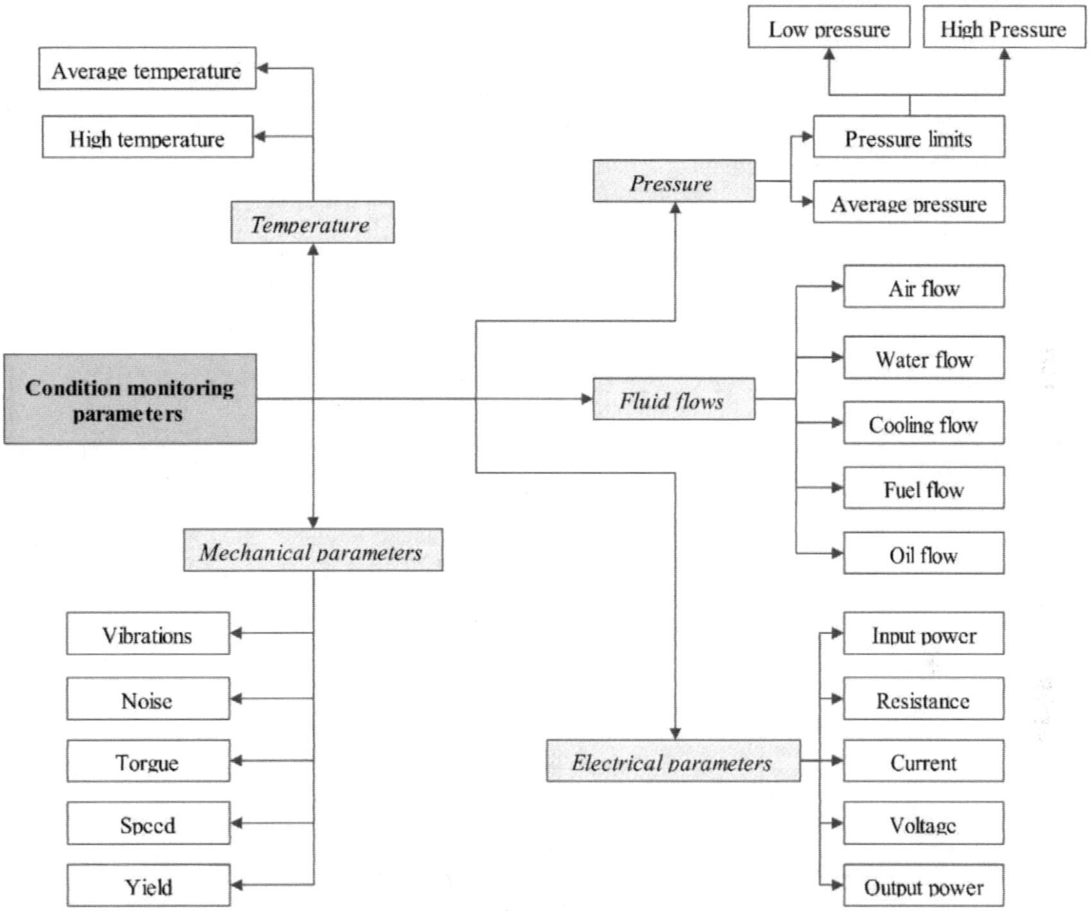


**Figure 3.** Condition monitoring parameters.

An issue of particular importance is the choice of equipment selected for condition monitoring. Selection must be based on a set of objective criteria. Of these [5]:

- cost of machine down-time
- lost production costs
- cost of replacement of machine
- cost of maintenance and cost of spare
- consequential damages
- cost of functioning condition monitoring system
- cost of safety of work
- cost of environmental impact

Each of these factors has a certain degree of influence in determining prioritization regarding the implementation of functioning condition monitoring system. The degree of influence is specific to each situation.

Even if every situation of implementation of the functioning condition monitoring system of equipment should be treated individually can be developed a general logical diagram.

The logical diagram shown in figure 4 has few distinct parts:

- analysis of equipment
- choice of parameters how will be monitored
- choice of monitoring method
- analysis of data
- drafting of reports
- maintenance actions

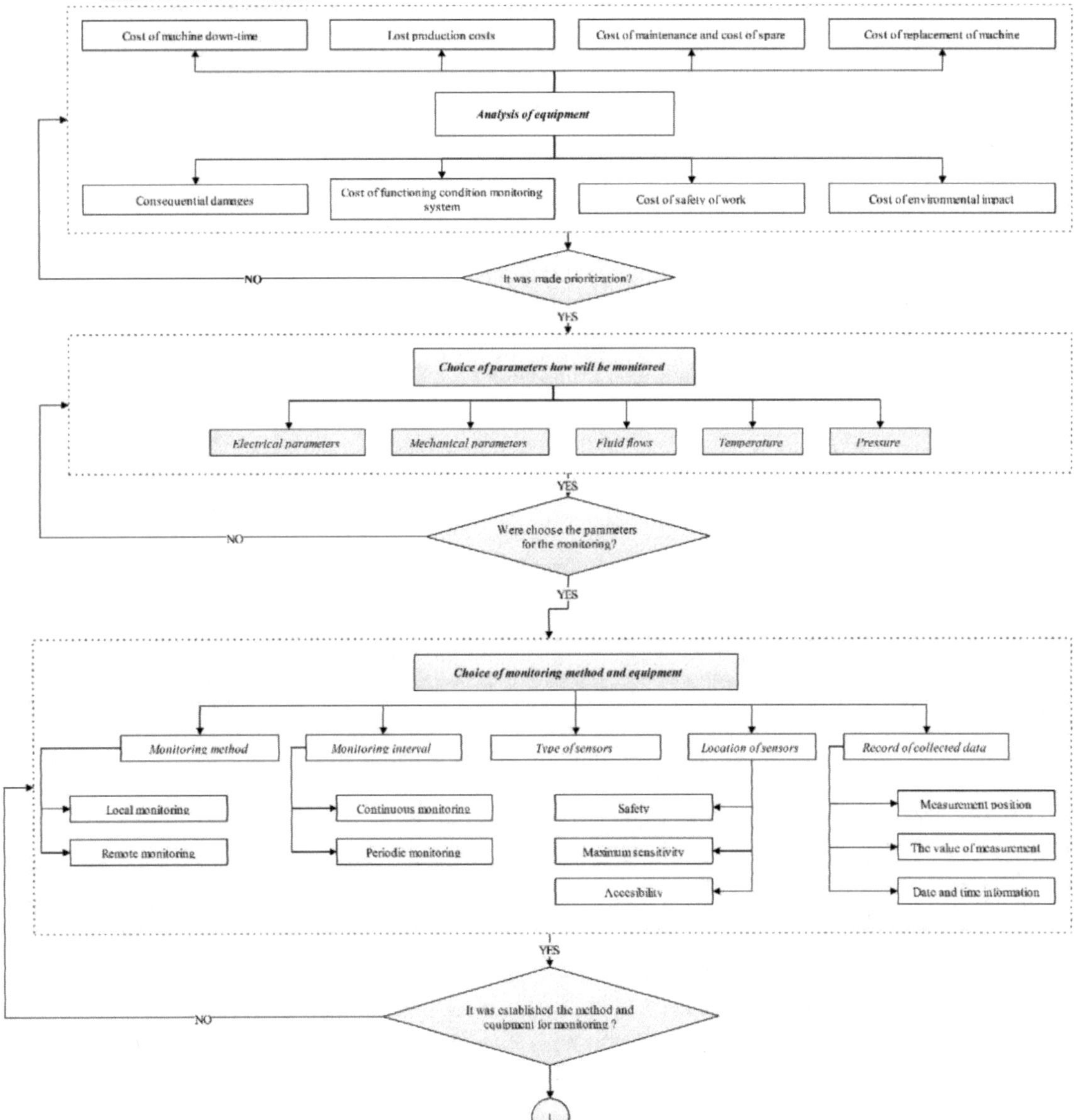


**Figure 4.** Logical diagram for implementation of the functioning condition monitoring system [4].

IOP Conf. Series: Materials Science and Engineering **174** (2017) 012068 doi:10.1088/1757-899X/174/1/012068

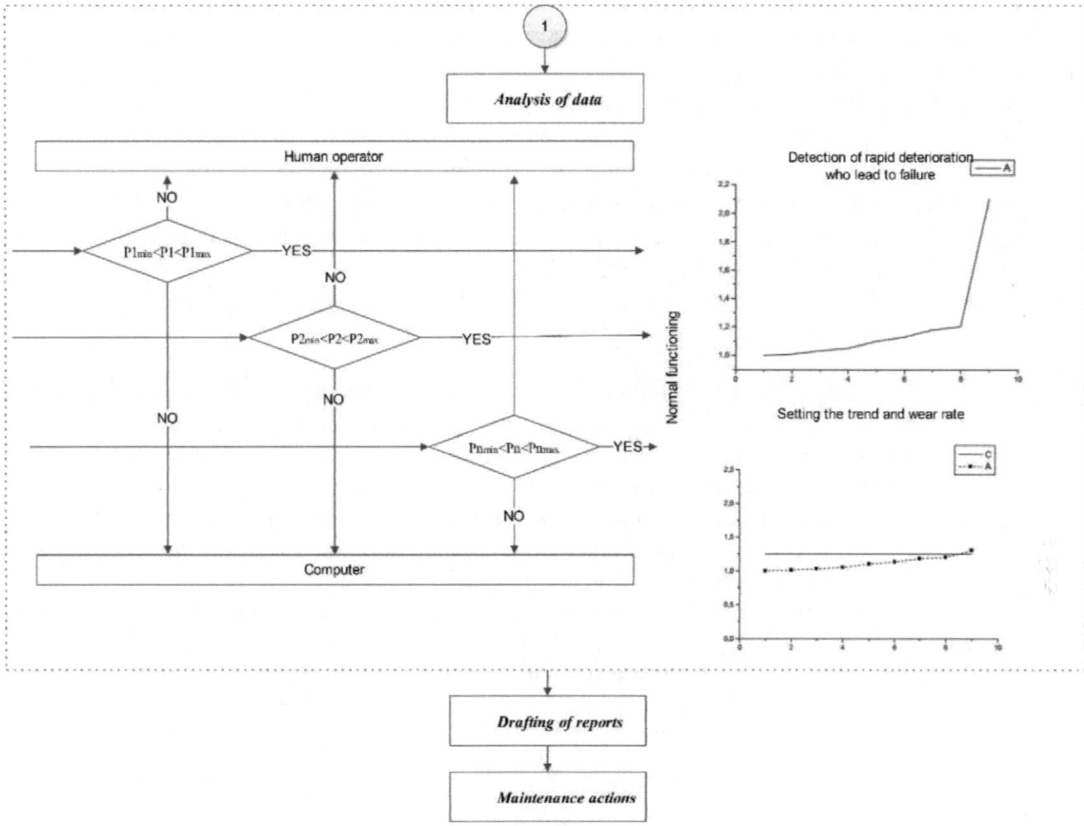


**Figure 4.** Logical diagram for implementation of the functioning condition monitoring system [4].

## 3. Conclusion

The main idea to start the implementation of functioning condition monitoring is that the maintenance remains one of the very few areas where significant increases to company profits can be archived. The benefits of functioning condition monitoring are:

- possibility to early detection of damages;
- moving of maintenance actions before the moment of failure;
- Improves equipment reliability;
- minimizes time spent on maintenance and optimized maintenance intervals (more optimal than manufacturer recommendations) ;
- minimizes overtime costs by scheduling the activities;
- better prediction of maintenance requirements;
- minimizing possible and often even the elimination of accidental malfunctions and stops of production;
- increasing of production time and thus increase the profits;
- less maintenance workload;
- sometimes increase of the working regime for equipment;
- save the costs of spares and minimizes requirement for emergency spare parts,
- improves worker safety;
- improve quality of work;
- improve the factory design;
- increase the efficiency of company.

It could be also mentioned the main drawbacks of functioning condition monitoring:

- high implementation costs;
- the need for a superior training for maintenance staff;
- sometimes, especially for the newest equipment the databases for analyze are expensive;
- uniform wear or fatigue failures is not easily detected with measurements CBM;
- the sensors cannot survive sometimes in the operating environment;
- asset modifications might be required to retrofit the system with sensors;
- the maintenance periods can be unpredictable.

## References

[1] Antonescu V and Stichiţoiu D 1988 *Elemente de teorie şi culegere de probleme de fiabilitate, mentenabilitate, disponibilitate (Elements of theory and collection of problems of reliability, maintainability, availability)* (Bucureşti: Institutul central pentru industria electrotehnică Oficiul de informare documentară)
[2] Kokcharov I and Burov A 2013 *Structural Integrity Analysis* (Kindle Edition, Amazon)
[3] Bengsston M 2007 *On Condition Based Maintenance ant Its Implimentation in Industrial Settings* Ph.D. Thesis Mälardalen University Sweden
[4] Ungureanu N S 2003 *Fiabilitatea şi diagnoză (Reliability and diagnosys)* (Cluj Napoca: Risoprint)
[5] *** ISO 17359:2011 Condition monitoring and diagnostics of machines - General guidelines

# Conformity assessment of the measurement accuracy in testing laboratories using a software application*

**A Diniţă**

Department of Mechanical Engineering, Petroleum-Gas University of Ploiesti, 39 Bucharest Blvd., 100680 Ploiesti, Romania.

E-mail: adinita@upg-ploiesti.ro

**Abstract**. This article presents a method for assessing the accuracy of the measurements obtained at different tests conducted in laboratories by implementing the interlaboratory comparison method (organization, performance and evaluation of measurements of tests on the same or similar items by two or more laboratories under predetermined conditions). The program (independent software application), realised by the author and described in this paper, analyses the measurement accuracy and performance of testing laboratory by comparing the results obtained from different tests, using the modify Youden diagram, helping identify different types of errors that can occur in measurement, according to ISO 13528:2015, Statistical methods for use in proficiency testing by interlaboratory comparison. A case study is presented in the article by determining the chemical composition of identical samples from five different laboratories. The Youden diagram obtained from this study case was used to identify errors in the laboratory testing equipment.

## 1. Introduction

In recent years, interlaboratory comparisons (***ILC***) have gained a special importance both as a tool for verification of traceability in the process of quality assurance, as well as validation tool for working procedures of related testing laboratories accredited or pending accreditation activities.

The need for reliable continuous performance of a laboratory is not only essential for laboratories and their customers, but also for other entities such as regulators, laboratory accreditation institute and other organizations that specify requirements for laboratories. The correct operation of testing laboratories requires an internal or external quality control. Internal quality control is provided by calibration set by different standards and assessment of uncertainty, while external quality control can be established through a modern method – ***ILC*** [1,2].

Interlaboratory comparisons are reliable and efficient mechanisms to certify the competence of laboratories because they allow ensuring quality of results delivered by the laboratory and helps demonstrate its competence as a mechanism for progress. An interlaboratory comparison is designed to answer one of the following objectives: 1. to assess the skills laboratories, 2. to evaluate the characteristics of a material, 3. to evaluate the characteristics of a method of analysis, testing or calibration. Participation in proficiency testing is an appropriate way to train the laboratory and risk management. For technical and metrological level alignment of national and references standards with the performance of European metrology institutes, is needed extension and integration of research

* This paper was accepted for publication in Proceedings after double peer reviewing process but was not presented at the Conference ROTRIB'16

activities carried out within an institutional framework at national and European level through involvement in inter-laboratory comparisons [7,8].

## 2. Organizing interlaboratory comparison tests

In ISO/IEC 17025, participation in interlaboratory comparison tests ***ILC***, also known as proficiency testing is explicitly required for the accreditation of testing and calibration laboratories. ISO/IEC 17043 standard provides general requirements for conducting the proficiency tests. The results of tests obtained based on ISO/IEC 17043 protocols are usually analysed by statistical methods laid down in ISO 13528. In many cases Youden diagram is used to evaluate the test results provided by ***ILC***. Scheme comparison test specifies that two samples (with slightly different properties) are distributed to a participant in a scheme of interlaboratory comparison [5].

Accuracy (truth value and precision) of measurement should be determined in a series of tests or measurement results reported by laboratories participating in the study organized for this purpose. To avoid any discrepancy and to reduce dispersion, all participants must have the same detailed measurement or test method which describes how the measurement should be performed and how the specimens should be obtained or prepared. Measurement method under investigation should be an optimized method in terms of procedure used, such a method should be robust (for example: small variations in procedure should not produce major changes in the final results).

The document specifies the method of measurement should be clearly and completely. All essential operations related to working environment, equipment and metrological traceability of test and specimen preparation method must be included in the ***ILC*** procedure [3,4].

## 3. Youden diagram - model for comparing the test results of *ILC*

Youden diagram is a graphical method to analyse the experimental data provides by different participants in ***ILC***. Chart view variability in laboratory testing and interlaboratory variability. In this graphical method for the analysis of inter-laboratory experimental data, all laboratories analysed two samples. Youden diagram with two variables is a useful method for certain types of interlaboratory comparisons and test results. For Youden diagram, the two samples may be similar and have measurement values close to the average of measurements taken, but may have differences from the average, in this case is used the standard deviation for framing values on the two axes. Axes in this chart are drawn to the same scale: a unit on the x axis has the same length as a unit on the y axis. Each point on the chart corresponds to the results of one laboratory tests and is defined by a variable on the horizontal axis and a second response variable on the vertical axis [6].

The diagram can be drawn using three types of models that take into account the laws of normal distribution, whereby a standard deviation, or sigma ($1\sigma$), define a region that evaluates 68% of all data points (measured values), two sigma ($2\sigma$) will comprise 95% percent of data and three sigma ($3\sigma$) will comprise 97.7%. The most common types of Youden diagram are:

- diagram using framing squares, as shown in figure 1;
- diagram using framing circles, as shown in figure 2;
- diagram using framing ellipse, as shown in figure 3.

Youden diagram is applied to identify laboratories with large systematic errors and high random errors. Interpretation of data obtained from experimental tests from the perspective of integrating them in a Youden diagram, can be made as follows [9,10]:

a. experimental values pairs obtained from experimental tests, conducted in ***ILC,*** focus on arithmetic - *small random and systematic error*;

b. the pairs of values determined by experimental tests are scattered along the line of 45° in the upper right quadrant and the lower left quadrant – *small random error and large systematic error*;

c. the pairs of values determined by experimental tests are located away from the line of $45^0$ in the upper left quadrant and lower right quadrant – *large random error and small systematic error*;

d. the pairs of values determined by experimental tests are situated outside diagram (rectangle, circle or ellipse) - *large systematic and random errors*.

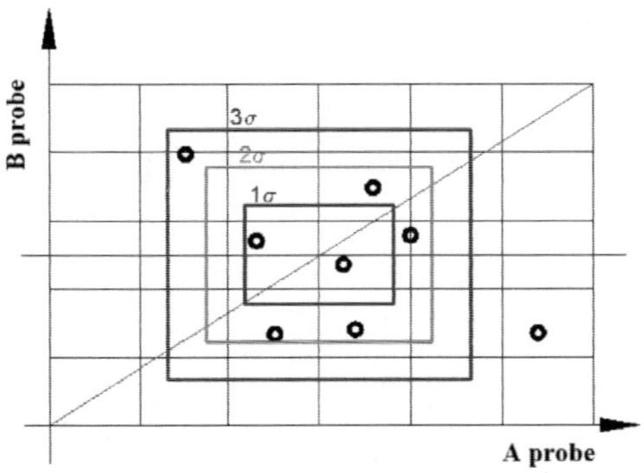


**Figure 1.** Youden diagram using the model with squares.

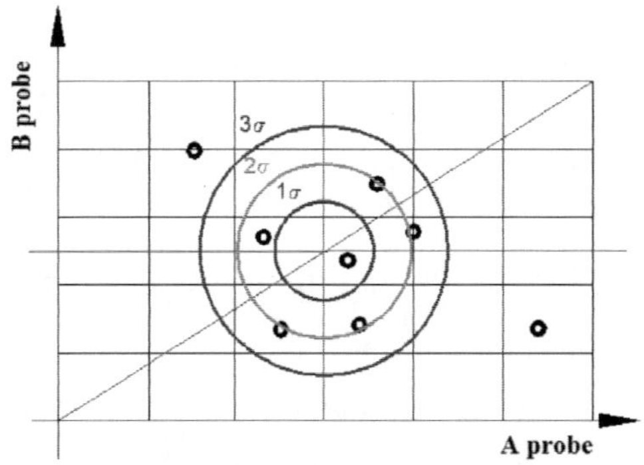


**Figure 2.** Youden diagram using the model with circles.

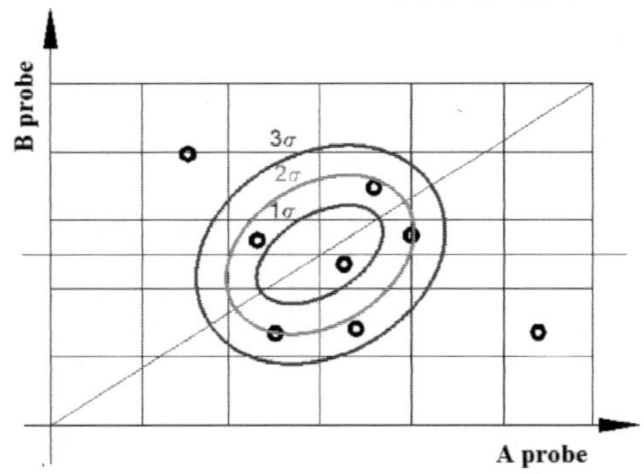


**Figure 3.** Youden diagram using the model with ellipse.

For the analysis of experimental data, in this research, will be used Youden diagram using ellipses framing type ($1\sigma$, $2\sigma$, $3\sigma$). To determine parametric equations of the ellipse (*ellipse is the locus of points in the Euclidean plane whose sum of distances from two fixed points is constant*) is used the following parametric equations, considered model is shown in figure 4 [4,5]:

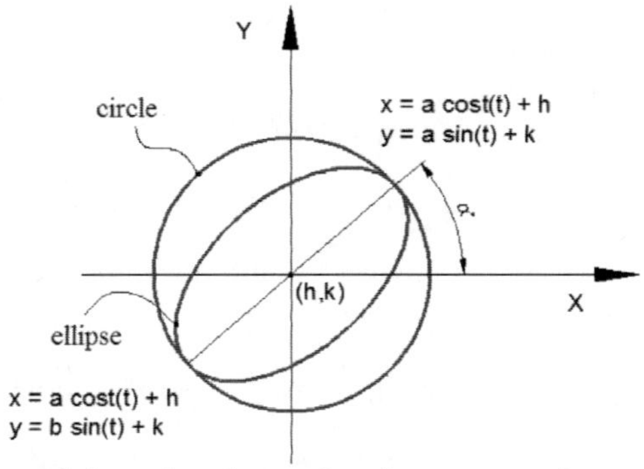


**Figure 4.** Analysis model used to determine the parametric equations of an ellipse.

- parametric equation of a circle:

$$\begin{cases} x = a \cdot \cos(t) + h \\ y = a \cdot \sin(t) + k \end{cases} \quad (1)$$

Coordinates point (*h, k*) is the centre of the circle, *a* is the circle radius, and $t = 0 ... 2\pi(360^0)$ is a parametric variable, being geometrically interpreted as the angle formed by the circle radius with x axis.

- parametric equation of an ellipse rotated at an angle *α*, having semi-major axis equal to *a* and semi-minor axis equal to *b*:

$$[X \;\; Z] = \begin{bmatrix} \cos(\alpha) & \sin(\alpha) \\ -\sin(\alpha) & \cos(\alpha) \end{bmatrix} [X' \;\; Y'] \tag{2}$$

In which $\begin{bmatrix} \cos(\alpha) & \sin(\alpha) \\ -\sin(\alpha) & \cos(\alpha) \end{bmatrix}$ it is rotation matrix.

- parametric equations of an ellipse centred on the point (*h, k*) and rotated at an angle *α* are obtained from the developing equation 2:

$$\begin{cases} x = h + a \cdot \cos(t) \cdot \cos(\alpha) - b\sin(t)\sin(\alpha) \\ y = k + a \cdot \cos(t) \cdot \sin(\alpha) + b\sin(t)\cos(\alpha) \end{cases} \tag{3}$$

Parametric equations of an ellipse rotated at an specific angle used to the analyses results of experimental tests carried out during ILC involves using the arithmetic average of the two series of experimentally determined values as the origin of the ellipse and the standard deviation for **A probe** is considered semi-major axis of the ellipse and standard deviation calculated for the **B probe** is semi-minor axis of the ellipse.

Custom equations used as a tool in evaluating ***ICL*** by the implementation of Youden diagram are:

- *Case I.* The standard deviation (*1σ*) is used to define a region which includes 68% of all the values determined by experimental tests

$$\begin{cases} x = \bar{x} + 1\sigma_{A\,probe} \cdot \cos(t) \cdot \cos(\alpha) - 1\sigma_{B\,probe} \sin(t)\sin(\alpha) \\ y = \bar{y} + 1\sigma_{A\,probe} \cdot \cos(t) \cdot \sin(\alpha) + 1\sigma_{B\,probe} \sin(t)\cos(\alpha) \end{cases} \tag{4}$$

- *Case II.* The standard deviation (*2σ*) is used to define a region which includes 95% of all the values determined by experimental tests

$$\begin{cases} x = \bar{x} + 2\sigma_{A\,probe} \cdot \cos(t) \cdot \cos(\alpha) - 2\sigma_{B\,probe} \sin(t)\sin(\alpha) \\ y = \bar{y} + 2\sigma_{A\,probe} \cdot \cos(t) \cdot \sin(\alpha) + 2\sigma_{B\,probe} \sin(t)\cos(\alpha) \end{cases} \tag{5}$$

- *Case III.* The standard deviation (*2σ*) is used to define a region which includes 97.7% of all the values determined by experimental tests

$$\begin{cases} x = \bar{x} + 3\sigma_{A\,probe} \cdot \cos(t) \cdot \cos(\alpha) - 3\sigma_{B\,probe} \sin(t)\sin(\alpha) \\ y = \bar{y} + 3\sigma_{A\,probe} \cdot \cos(t) \cdot \sin(\alpha) + 3\sigma_{B\,probe} \sin(t)\cos(\alpha) \end{cases} \tag{6}$$

$\bar{x}$, $\bar{y}$ - average values for **A probe** and respectively for **B probe**.

- the rotation angle of the ellipse corresponding to Youden diagram

$$\alpha = atan\frac{\sum_{i=1}^{n}\left(x_i-\bar{x}\right)^2-\sum_{i=1}^{n}\left(y_i-\bar{y}\right)^2+\sqrt{\left(\sum_{i=1}^{n}\left(x_i-\bar{x}\right)^2-\sum_{i=1}^{n}\left(y_i-\bar{y}\right)^2\right)^2+4\cdot\left(\sum_{i=1}^{n}\left(x_i-\bar{x}\right)\cdot\left(y_i-\bar{y}\right)\right)^2}}{2\cdot\sum_{i=1}^{n}\left(x_i-\bar{x}\right)\cdot\left(y_i-\bar{y}\right)^2} \quad (7)$$

## 4. Case study on interlaboratory comparison

Verification of research developed in this paper was realized for a proficiency test - ***chemical composition analysis / identification code CC*** - using optical emission spectrometer to determine quantitative analysis of chemical elements in solid samples of specify alloy, taking into considering only masic concentration of carbon and silicon. Analysis of ***CC*** proficiency test was started by submitting an ***ILC*** invitation to 8 companies (laboratories) which are equipped with optical emission spectrometer, due to copyright the company names are replaced with the following codes: ***ILC_1…8_CC***. After receipt confirmation of acceptance for participating in ***ILC***, each laboratory received two samples marked as **A probe** and respectively **B probe**. Mass concentrations of $\%C_m$ and $\%Si_m$ provided by participating laboratories to ***ILC*** are summarized in table 1.

**Table 1**. The values obtained by laboratory test.

| Laboratory code | A probe | | B probe | |
|---|---|---|---|---|
| | $\%C_m$ | $\%Si_m$ | $\%C_m$ | $\%Si_m$ |
| ***ILC_1_CC*** | 0.150 | 0.243 | 0.105 | 0.197 |
| ***ILC_2_CC*** | 0.212 | 0.229 | 0.138 | 0.172 |
| ***ILC_3_CC*** | 0.210 | 0.242 | 0.130 | 0.214 |
| ***ILC_4_CC*** | 0.220 | 0.270 | 0.140 | 0.220 |
| ***ILC_5_CC*** | 0.178 | 0.242 | 0.120 | 0.199 |
| ***ILC_6_CC*** | 0.205 | 0.241 | 0.125 | 0.195 |
| ***ILC_7_CC*** | 0.200 | 0.250 | 0.140 | 0.185 |
| ***ILC_8_CC*** | 0.215 | 0.245 | 0.135 | 0.198 |

The calculation results of parameters used to analyse the data received from participating laboratories are presented in figure 5. The final report was submitted to the 8 participating laboratories in interlaboratory comparison to take in consideration some measures for corrections of systematic errors as well as random errors.

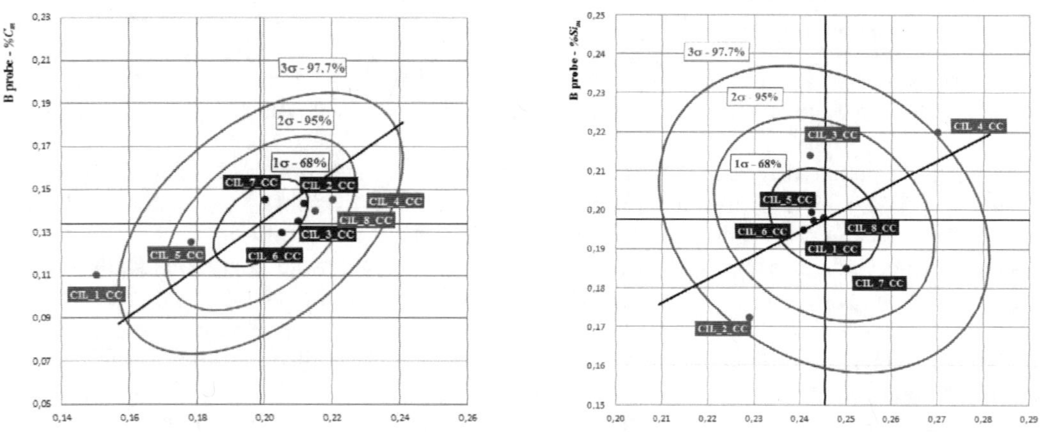


(a) (b)

**Figure 5.** Youden diagram applied for a ***CC*** proficiency test considered, **A probe** (*a*); **B probe** (*b*).

The results obtained and the need to use a program that is not limited by the installation of specialized software required the development of a executable software (*Interlaboratory Comparison software was developed using Matlab*) on any computer with minimal technical specifications, the working interface of the program created by the author is shown in figure 6. The software developed by the author was created using MATLAB (Matrix Laboratory), which is a development environment for numerical analysis and statistical analysis, the programming language that contains the same name, created by MathWorks. MATLAB allows the manipulation of matrices, visualization functions, implementation of algorithms, and creation of interfaces and can interact with other applications. The procedure for the realization of interlaboratory comparisons, using the software, includes the following steps: *a. Organizing interlaboratory comparison*: Establish frequency and data of ***ILC***; selection of participants; protocol development and documentation; sending documents to participants; *b. Preparation of samples used in **ILC***: manufacture or acquisition of samples; calibration / testing and assigning the reference value of sample; handling and storage of samples; *c. Running round **ILC***: preparation and distribution / sending of samples; monitoring activities of participants; data collection and processing; *d. Analysis and Decision:* Comparison of the results obtained by well-established and documented methods; performance evaluation of participants; issue views and interpretations; development and approval of competence test reports; publication of reports.

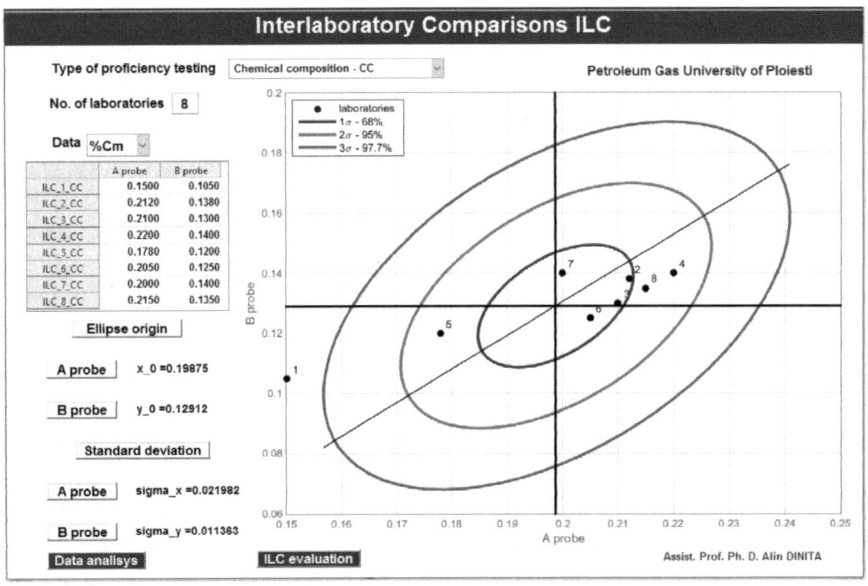


**Figure 6.** Specialized software for ***ILC*** data evaluation.

Analysing the Youden diagram, obtained by processing the data obtained from 8 participating laboratories in ***ILC,*** provides useful information to eliminate possible errors appeared in the experimental tests. Using the application, the processed data from laboratory no. 1 is situated outside from diagram, which indicates a *large systematic and random errors*, which may drastically influence the measurement process. After delivering results to all laboratories involved in the study, the laboratory no. 1 made a recalibration of equipment used to determine the chemical composition and provided second set of values that were within the allowable limits, confirming the usefulness of this software application created.

The main aim of the research was conducted to develop a standalone application that can be used in the performance evaluation laboratories. The application was verified by conducting a comparison scheme involving 8 laboratories in Romania, using as the reference test, determining the chemical composition for two samples provided by the author researches. The study is valuable for laboratories doing quantitative studies.

## 5. Conclusions

Proficiency testing ***ILC*** is used to determine the performance of individual laboratories for some tests or measurements, to monitor the continuing performance and quality of laboratories and the confidence provided by a laboratory evaluation. The software developed in this paper can be used as a simple numerical and graphical criterion to identify warning signals provided by systematic and random errors.

Proficiency testing is gaining increasing importance in many laboratories from different countries and becomes extremely important for quality assurance when used in relation with certified reference materials, validated methods.

ILC schemes and programs are organized primarily for the benefit of laboratories, the results obtained in these schemes are used by third-party organizations, including national accreditation organizations, national and international regulators and customer's laboratories. This is reflected in the quality of services provided by ILC schemes and confirmation of fulfilment of technical requirements and quality assurance specified in the applicable referential is objectively necessary

Software created can easily be used for many types of proficiency tests. It is a necessary tool for laboratories to check the compliance of the test results. The software application can be used to improve the analytical performance of the interlaboratory comparisons and it is a useful tool for getting quick results achieved by different laboratories in laboratories schemes. The results of the intercomparison conducted in this paper improve confidence in the equipment and procedures used to perform various tests and measurements.

The main objective of a software developed is to help the participating laboratory to assess the accuracy of its test results. Intercomparison of measurements obtained from different tests are one of the main method of proving realistic estimates of measurement uncertainty.

## 6. References

[1] ISO / IEC 17025:2005 *General requirements for the competence of testing and calibration laboratories*
[2] ISO / IEC 17043: 2010 *Conformity assessment – General requirements for proficiency testing*
[3] ISO/IEC Standard ISO13528: 2015, *Statistical Methods for use in Proficiency Testing by Interlaboratory Comparisons*
[4] ISO 5725:1994: *Accuracy (trueness and precision) of measurement methods and results - Part 1: General principles and definitions. Part 2: Basic method for the determination of repeatability and reproducibility of a standard measurement method*
[5] ISO/DTS 21748:2002 *Guide for the use of repeatability, reproducibility, and trueness estimates in measurement uncertainty estimation*
[6] Katsuhiro S *et al.* 2013 A study on the utilization of the Youden plot to evaluate proficiency test results *Accred. Qual. Assur.* **18** (3)161–174
[7] Becker R. *et al.* 2006 Mineral oil content in sediments and soils: Comparability, traceability and a certified reference material for quality assurance *Anal Bioanal Chem* **385** (3) 645-651
[8] Thompson M and Ellison S LR 2006 *Fitness for purpose – the integrating theme of the revised Harmonised Protocol for Proficiency Testing in Analytical Chemistry Laboratories*. Accred. Qual. Assur. 11 373-378
[9] Rosario P, Martínez J L, and Silván J M 2008 Comparison of different statistical methods for evaluation of proficiency test data *Accred. Qual. Assur.* **13** (9) 493-499
[10] Greg Miller W 2009 The role of proficiency testing in achieving standardization and harmonization between laboratories *Clin. Biochem.* **42** (4) 232–235

**Institute of Physics**
Dirac House, Temple Back
Bristol BS1 6BE UK

ISSN: 1757-8981
ISBN 978-1-5108-3714-0